Signals and Stochastic Processes

Signals and Stochastic Processes

Y Mallikarjuna Reddy
Giri Babu Kande

Department of Electronics and Communication Engineering
Vasireddy Venkatadri Institute of Technology, Guntur, A.P.

SIGNALS AND STOCHASTIC PROCESSES

UNIVERSITIES PRESS (INDIA) PRIVATE LIMITED

Registered Office
3-6-747/1/A & 3-6-754/1, Himayatnagar, Hyderabad 500 029, Telangana, India
email: info@universitiespress.com, www.universitiespress.com

Distributed by
Orient Blackswan Private Limited

Registered Office
3-6-752 Himayatnagar, Hyderabad 500 029, Telangana, India

Other Offices
Bengaluru, Bhopal, Chennai, Guwahati, Hyderabad, Jaipur, Kolkata,
Lucknow, Mumbai, New Delhi, Noida, Patna, Vijayawada

ISBN 978 93 86235 31 2

501347

23062017

Typeset in Times 11/13 by
Techastra Solutions Private Limited, Hyderabad 500 029

Printed in India by
Graphica Printers & Binders, Hyderabad 500 013

Published by
Universities Press (India) Private Limited
3-6-747/1/A & 3-6-754/1, Himayatnagar, Hyderabad 500 029, Telangana, India

Contents

7. The Laplace Transform

8. The Z-Transform

9. Random Processes — Temporal Characteristics

10. Random Processes — Spectral Characteristics

Preface

This book is prepared for the use of undergraduate engineering students specialising in electronics and communication engineering and meets the syllabus requirements of most technical universities in India. Students taking the AMIE and IETE courses as well as those preparing for GATE, UPSC and other similar competitive examinations will also find the book very useful.

The study of signals and stochastic processes is considered a prerequisite for many fields of science and engineering such as circuit design, power systems, automatic control, communications, and signal processing. The theory of the subject is supported by an elegant collection of mathematical results based on the works of Fourier, Laplace and many other famous mathematicians and engineers. That the Fourier transform, for instance, can be generalised to Laplace transform in the continuous time and to Z-transform in discrete time, and that these transforms exist even for signals that tend to infinity with time reveals the versatility of these tools and mathematical results.

Nowadays, predicting a system's behaviour is usually accomplished through computer simulation. A simulation typically involves the recursive computation of the output signal of a discretised version of the continuous-time system model. MATLAB is a very popular software for simulation in science and engineering, and we have used it to compute and display the results of some of the examples in the book to help students to visualize difficult concepts such as the complex signals, convolution, Fourier series and transforms.

For undergraduate students with a good foundation in mathematics, the theory of signals and stochastic processes should be an easy and interesting subject. A sound understanding of the basic theories of calculus, trigonometry, complex numbers, probability and random variables is indeed a necessary prerequisite for the subject, and students are therefore encouraged to review these topics before reading this book.

The material covered here is divided into four main parts. The first part, comprising Chapters 1 and 2, focuses on the mathematical representation of signals and systems; this knowledge is a requirement for all courses related to electrical engineering. The second part, comprising Chapters 3 to 5, introduces the concepts of Fourier series, Fourier transform and LTI systems. The third part—Chapters 6 to 8—discusses sampling, Laplace and Z-transforms. The fourth part comprising Chapters 9 and 10 provide a detailed analysis of stochastic processes essential for dealing with random signals encountered in signal processing systems in communications.

The book includes many solved examples. Several unsolved problems are also included as exercises in all chapters—the emphasis is on problems of the kind that generally feature in university examinations. A large number of problems solved using MATLAB (the latest version) are provided in Appendix A. Also included are a brief review of probability and random variables that will come handy in the analysis of stochastic processes (Appendix B) and some useful mathematical expressions (Appendix C).

Our wish is that the reader enjoys learning the theory of signals and stochastic processes using this book. Our lecture notes on the subject and the feedback/suggestions received from students who have taken this course have helped us greatly in structuring the contents in a direct and straightforward manner. Additional illustrations and solutions to problems were added with the help of my colleagues and students. Great care has been taken to minimise errors and inconsistencies in the book, and we are indebted to the faculty members who helped us in this task with their critical reviews.

We sincerely thank Sri V Vidya Sagar, Chairman, Sri Badari Prasad, Secretary, Sri M Srikrishna, Joint Secretary, and the management and staff of Vasireddy Venkatadri Institute of Technology, Nambur, Guntur, for their encouragement and support of our work. We particularly wish to thank friends and colleagues, Dr I A Pasha, Dr P Nageswara Rao, Dr E Srinivasa Reddy, P Ammi Reddy, M R N Tagore, Dr M V Raghuram, M Pardha Saradhi, N Sivarama Krishna and K Suresh Babu for their invaluable contribution in the form of reviews. Our sincere thanks are also due to the publishers for their interest in the book. Finally, we thank our family for their patience and affection.

It is our sincere hope that the expectation of a textbook catering to academic requirements of engineering students will be well met with this endeavour of ours. We also welcome feedback for improving the book in its next edition.

Outline of the book

Chapter 1 elucidates various standard continuous-time and discrete-time signals, the classification of signals and the basic operations on signals. The classification and properties of systems are also presented with suitable examples.

In Chapter 2, the representation of signals using vectors is analysed. Also, standard signals and impulse functions are presented.

Chapter 3 presents the Fourier analysis of continuous-time periodic signals, which forms the basis for frequency domain analysis. Fourier series expansion of periodic signals in trigonometric form, exponential form, and cosine form and the conversion from one form to another are provided along with the properties of Fourier series and Fourier spectrum.

Chapter 4 explains the development of the Fourier transform from Fourier series. The Fourier transforms of various types of signals, the properties of Fourier transform and frequency spectrum via Fourier transform are also dealt with.

Chapter 5 describes the basic concepts of linear time-invariant (LTI) systems and their analysis using Fourier transforms. Distortionless transmission through a system, ideal filter characteristics of linear systems and the Paley–Wiener criterion for physical realisation are also discussed. The concept of convolution in both time domain and frequency domain are also presented.

Chapter 6 describes the basic concepts of the sampling process. Different types of sampling and the reconstruction of signals from their samples and the effects of undersampling are also discussed.

Chapter 7 is on the analysis of continuous-time signals using Laplace transforms. It discusses properties of Laplace transforms, region of convergence (ROC) of Laplace transform and the properties of ROC. Inverse Laplace transform, Laplace transform of certain signals using waveform synthesis and system analysis using Laplace transform are also covered.

Chapter 8 deals with the analysis of discrete-time signals using Z-transforms. It covers properties of Z-transforms, the region of convergence of Z-transform and its properties, the inverse Z-transform and system analysis using Z-transforms.

Chapter 9 deals with the concept, classification, distribution and density functions of random processes; the concepts of stationarity, ergodicity, correlation functions, mean and mean-squared value of system response, and the correlation function of the system response are also covered in detail.

Chapter 10 discusses the spectral characteristics of the system response. Power spectrum, cross-power density spectrum, power density spectrum of the response and cross-power density spectrums of the input and the output are illustrated.

The three appendixes deal with the solving of problems with computer simulation using MATLAB, probability and random variables concepts, and handy mathematical expressions.

Dr Y Mallikarjuna Reddy and
Dr Giri Babu Kande

Road Map to JNTUH Syllabus

This book has been prepared in accordance with the syllabus requirements of Signals and Stochastic Processes, a theory paper in the second year of the BTech programme in Electronics and Communication Engineering, JNTU Hyderabad. The road map to using the book is as follows:

UNIT – I

Signal Analysis - Refer to Chapters 1 and 2

Signal Transmission through Linear Systems - Refer to Chapter 5

UNIT – II

Fourier Series - Refer to Chapter 3

Fourier Transforms - Refer to Chapter 4

Sampling - Refer to Chapter 6

UNIT – III

Laplace Transforms - Refer to Chapter 7

Z–Transforms - Refer to Chapter 8

UNIT – IV

Random Processes — Temporal Characteristics - Refer to Chapter 9

UNIT – V

Random Processes — Spectral Characteristics - Refer to Chapter 10

1

Introduction to Signals and Systems

Learning Objectives

The main objective of this chapter is to understand thoroughly the basic concepts of signals and systems. We will learn how signals are represented and classified and how systems are integral to signal processing. By reading this chapter students will be able to get a firm grasp of the following topics:

— Basic concepts of signals and systems
— Continuous and discrete time signals
— Classification of signals
— Operations on signals
— Classification of systems
— Mathematical and graphical representations

1.1 Introduction

A signal, in the broadest sense, is defined as a single-valued function of one or more independent variables such as time, frequency or space variables, which may be single or multi-dimensional, that carry information about a physical quantity of interest. Any physical quantity that show *variation*, be it with respect to time (such as speech, music) or with respect to space (such as a video, ECG image), is a signal. A signal carries energy and/or power to initiate some action; it may exist in varied forms such as physical, mechanical or electrical. For example, the variation of temperature with respect to time in a day represents signal carrying information about heat radiation; the traffic light signal tells us when to stop and when to proceed; two-dimensional image on a screen is a signal with respect to spatial coordinates and provides information about a picture/video. Hand signals, smoke signals, light, temperature variations, force, velocity, sound, etc., are examples of physical signals, whereas voltage, current, power variations, etc., are electrical signals. Since electrical quantities are easy to process because of the advances in technology, information in the physical form is converted to electrical and vice versa. Sensors, displays, etc., are used for signal conversion. For example, a speech signal which represents variation in acoustic pressure as a function of time in the physical form can be converted to voltage/current variations using microphones and speakers.

A system is defined as a group of objects or a set of interconnected components that takes a signal as input and produces an output signal for achieving some desired response. The system could be made up of physical components of, say, an electrical, mechanical or hydraulic system, or it could be an algorithm that computes an output from a given input data. Many real-time systems can be established by interconnecting several sub-systems. The characteristics of the input and output signals of a system depend on the application of the system. For example, in an automatic speaker recognition system the input signal is a speech signal, the system is a processor and the output signal is the identity of the speaker. Likewise, in a communication system the input signal could be a speech/video signal or computer data, the system itself being made up of a combination of a transmitter, channel and receiver, and the output signal is an estimate of the information contained in the original input signal.

Two kinds of problems are generally encountered in signal–system interaction, namely, (i) given an input signal and a system, what will be the output signal (called the analysis problem) and (ii) given that the input and desired output signals are known, what should be the system like (called the synthesis/design problem). In this course we will be dealing with the analysis part, largely, looking at the tools of analysis. The concepts of signals and systems form the basis for several other fields of study such as signal and image processing, communication systems, control systems, medical systems and robotics, to name a few.

Some instances of signals and systems and their interaction are given below.

- Extracting information from a signal – Extracting useful information from a signal for further processing. For example, estimating heart condition from a system that accepts ECG signal as input and processes it using programs/algorithm for giving further useful information.
- Communication system – For transmitting signals over long distances using a transmitter, channel and receiver systems; some of the processes carried out by the system are modulation, demodulation, coding and decoding.
- Image restoration and enhancement – Processing of degraded images, say, those obtained from remote sensing space probes or earth-observing satellites that are degraded due to limitations of the equipment, atmospheric effect, etc., by a system using techniques for extraction of features like lines and regional boundaries to compensate for some of the degradation.
- Modify and control the characteristics of a system –Using sensor signals to regulate the functioning of, say, a plant that has a variety of sensors to measure parameters like temperature, humidity and chemical composition in order to control the system.

In this book we will restrict ourselves to the mathematical description and analysis of electrical signals and systems, the signal being a single-valued function of time or frequency used for representing either a voltage, current or power waveforms, as necessitated by a specific situation. If a signal is a function of time, it is called a time domain signal and is graphically represented as a waveform. On the other hand, if a signal is a function of frequency, it is called a frequency domain signal and is graphically represented as a frequency spectrum.

1.2 Classification of Signals

Signals can be classified variously depending upon their characteristics under consideration. To begin with, we will restrict ourselves to classifying one-dimensional electrical signals that are single-valued functions of the independent variable time. The following dual pair categorizations will be explored.

- Continuous-time and discrete-time signals
- Analog and digital signals
- Periodic and aperiodic signals
- Real and complex signals
- Deterministic and random signals
- Even and odd signals
- Energy and power signals
- Causal and non-causal signals

1.2.1 Continuous-time and Discrete-time Signals

A continuous-time signal $f(t)$ is a function of a continuous time variable t. The magnitude of the signal is defined for all values of continuous time t. It is also called analog signal. The magnitude of a continuous time signal is defined continuously for all values of time t. For example, a noise or a sound signal in the atmosphere as a function of time or a signal defined by, say $f(t) = t^2 + e^{-2t}$, $-\infty < t < \infty$, are continuous time signals. On the other hand, if a signal is represented only at discrete values of the independent time variable $t (t = nT)$, it is called a discrete-time signal or a discrete time sequence. Between the discrete time instants, the value of the signal may be undefined, zero, or of no interest.

Conventionally, to distinguish between continuous-time and discrete-time signals we use the symbol t to denote the continuous time variable and n to denote the discrete-time variable. Further, for continuous-time signals the independent variable is enclosed in parentheses $(\cdot)$, while for discrete-time signals the independent variable is enclosed in bracket $[\cdot]$. Thus a discrete-time signal may be represented by$\{ f[nT], n = 0, \pm 1, \pm 2, ...\}$, or a sequence of numbers like

$$f[n] = \{.....f[-2], f[-1], \underset{\uparrow}{f[0]}, f[1], f[2], ...\}. \tag{1.1}$$

The arrow indicates the $n = 0$ term.

The same may be represented in tabular form as:

n	-2	-1	0	1	2
$f[n]$	$f[-2]$	$f[-1]$	$f[0]$	$f[1]$	$f[2]$

In the $f\ [nT]$ or $f[n]$ representation, n is an integer number representing discrete time with value of $T = 1$. For example, the signal $f[n] = 2^n$, $n \geq 0$ is a discrete time signal. The weekly stock market index over a period of 12 months and the daily high temperature over a period of one week are some examples of discrete time signals. For such signals, the magnitude is defined only at specific instants of time intervals and is undefined at other time instants.

Figure 1.1(a) shows a continuous-time signal, while Fig. 1.1(b) shows a discrete-time signal. A discrete-time signal is often derived from a continuous-time signal by sampling it at a uniform rate.

Note 1: A continuous-time signal is a function of continuous-time variable but may not be continuous in magnitude.

Note 2: Piecewise continuous signals possess different expressions over different time intervals.

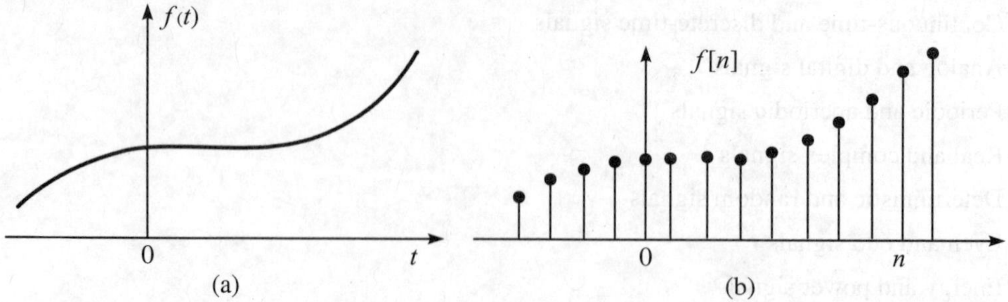

Fig. 1.1 *(a) Continuous-time signal and (b) discrete-time signal*

1.2.2 Analog and Digital Signals

An analog signal $f(t)$ is a continuous function of time in the interval (a, b), where the values a, and b may be from $-\infty$ to $+\infty$. That is, $f(t)$ can be uniquely defined for all t. If a discrete-time signal $f[n]$ is quantized into a finite number of distinct values and encoded into a digital number (word) format, then it is called a digital signal. That is, a digital signal takes discrete values for both time and magnitude. The signals may be of finite or infinite duration. For example, the music you hear from your player due to changes in the air pressure at the speaker's diaphragm is an analog signal because the pressure variation is a continuous function of time. However, the same information stored in a computer is in digital form.

Figure 1.2(a) shows an example of an analog signal and Fig. 1.2(b) that of a digital signal.

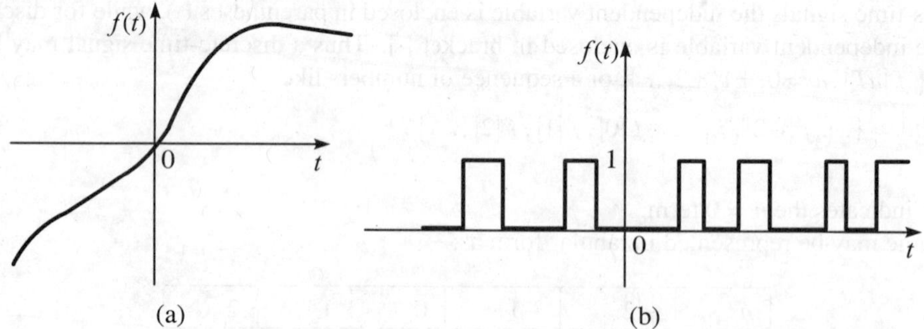

Fig. 1.2 *(a) Analog signal and (b) digital signal*

1.2.3 Periodic and Aperiodic Signals

A signal is said to be periodic if and only if a specific shape of the signal with time T_0 repeats regularly after a specific amount of time T, where $T_0 < T$. A periodic signal $f(t)$ is a function of time t with period T and satisfies the condition

$$f(t) = f(t + nT), \text{ for all } t \text{ and integer values of } n, \tag{1.2}$$

where the period T in seconds is called the fundamental time period of $f(t)$.

The reciprocal of T is $f = \dfrac{1}{T}$ Hz and is called the fundamental frequency, and $\omega = \dfrac{2\pi}{T}$ rad/seconds is the angular frequency.

Any signal $f(t)$ not satisfying the above condition is called aperiodic signal or non-periodic signal. That is, for non periodic signals there is no period for repetition. For example, a sine wave is a periodic signal whereas noise is a non-periodic signal. Figure 1.3 shows an example of a periodic signal and a non-periodic signal.

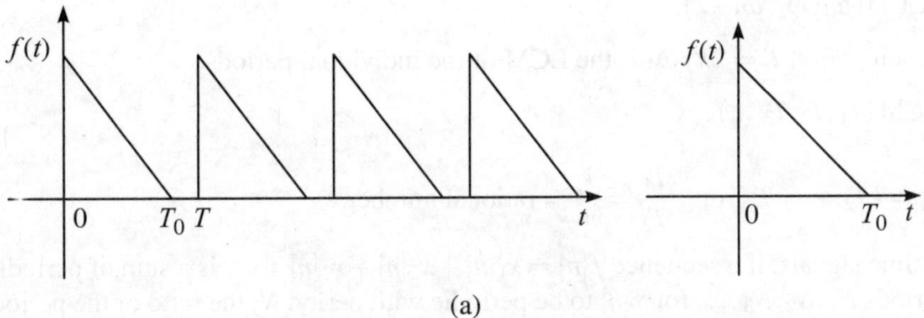

Fig. 1.3 *(a) Continuous time periodic signal and non-periodic signal*

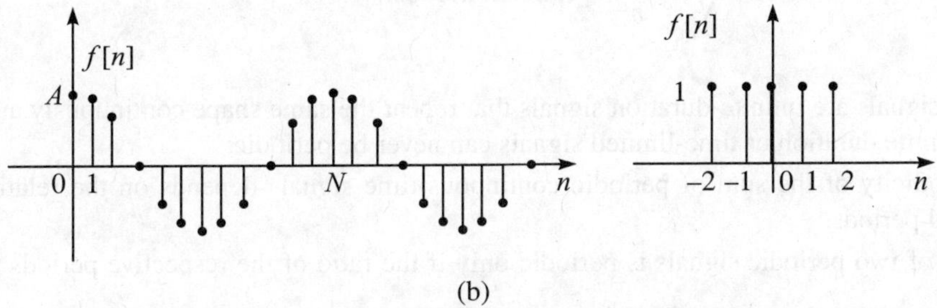

Fig. 1.3 *(b) Discrete time periodic and non-periodic signals*

For discrete-time signals, $f[n]$ is said to be periodic signal if it satisfies the condition

$$f[n] = f[n \pm N], \tag{1.3}$$

where N is a positive integer.

This condition is true if $N = \dfrac{2\pi k}{\omega}$ is integer for any integer k or

$$\frac{N}{k} = \frac{2\pi}{\omega} = \text{rational number.} \tag{1.4}$$

The smallest integer N is called fundamental period. $\omega = \dfrac{2\pi}{N}$ rad/s is the fundamental angular frequency. Figure 1.3(b) shows an example of a discrete-time periodic signal and a non-periodic signal.

Sum of Periodic Signals

The sum of two or more periodic signals may or may not be periodic. The common time period T of the combination of these periodic signals is equal to the smallest duration over which each signal completes an integer number of cycles and is given by the LCM (least common multiple) of the individual time periods. The fundamental frequency ω_0 equals the GCD (greatest common divisor) of the individual frequencies. We can find a common period or fundamental frequency only for a commensurate combination in which the ratio of any two periods (or frequencies) is a rational fraction (ratio of integers with common factors cancelled out).

If a function $f(t) = x_1(t) + x_2(t) + x_2(t) + ...$ is a sum of harmonic signals with time periods $T_1, T_2, T_3 ....$, and frequencies $\omega_1, \omega_2, \omega_3, ...$, for $f(t)$ to be periodic, the ratio of each pair of individual frequencies (or periods) must be a rational number. The fundamental frequency ω_0 is the GCD of the individual frequencies

$$\omega_0 = \mathrm{GCD}\,(\omega_1, \omega_2, \omega_3, ...) \tag{1.5}$$

and the common period $T = 2\pi/\omega_0$ is the LCM of the individual periods.

$$T = \mathrm{LCM}\,(T_1, T_2, T_3 ....)\,. \tag{1.6}$$

$$\text{That is, } mT_1 = kT_2 .... = T \quad \text{or} \quad \frac{T_1}{T_2} = \frac{k}{m} = \text{rational number.} \tag{1.7}$$

For discrete-time signals, if a sequence $f[n] = x_1[n] + x_2[n] + x_2[n] + \ldots$ is a sum of periodic sequences with time periods $N_1, N_2, N_3 \ldots$, for $f[n]$ to be periodic with period N, the ratio of the periods should be a rational number. That is,

$$mN_1 = kN_2 \cdots = N \text{ or } \quad \frac{N_1}{N_2} = \frac{k}{m} = \text{ rational number.} \tag{1.8}$$

Note:

1. Periodic signals are infinite-duration signals that repeat the same shape continuously at every finite period. Finite duration or time-limited signals can never be periodic.
2. The periodicity of the sum of periodic continuous-time signals depends on the relation between individual periods.
3. The sum of two periodic signals is periodic only if the ratio of the respective periods is a rational number.
4. The sum of continuous time periodic signals may not be periodic, whereas sum of discrete time periodic signals is always periodic.

Example 1.1

Determine whether or not each of the following signals is periodic.

(i) $x(t) = \cos t + \sin\sqrt{2}t$; (ii) $x(t) = \sin^2 t$;

(iii) $x(t) = \cos\frac{\pi}{3}t + \sin\frac{\pi}{4}t$; (iv) $x(t) = \cos 10\pi t + \sin 30t$.

Solution

(i) Given $x(t) = \cos t + \sin\sqrt{2}t$.

The signal has two frequencies $\omega_1 = 1$ rad/s and $\omega_2 = \sqrt{2}$ rad/s

The periods are $T_1 = \frac{2\pi}{\omega_1} = \frac{2\pi}{1} = 2\pi$ seconds and $T_2 = \frac{2\pi}{\omega_2} = \frac{2\pi}{\sqrt{2}} = \sqrt{2}\pi$ seconds.

Here, the ratio of the two time periods $\frac{T_1}{T_2} = \frac{2\pi}{\sqrt{2}\pi} = \sqrt{2}$ is not rational number. Hence the signal is non periodic.

(ii) Given $x(t) = \sin^2(t) = \frac{1}{2} - \frac{1}{2}\cos 2t$.

Here the signal has one period $T = \frac{2\pi}{\omega} = \frac{2\pi}{2} = \pi$ seconds. The signal is periodic signal.

(iii) Given $x(t) = \cos\frac{\pi}{3}t + \sin\frac{\pi}{4}t$

The signal has two frequencies $\omega_1 = \pi/3$ rad/s and $\omega_2 = \pi/4$ rad/s.

The signal has two periods: $T_1 = \frac{2\pi}{\omega_1} = \frac{2\pi}{\pi/3} = 6$ s and $T_2 = \frac{2\pi}{\omega_2} = \frac{2\pi}{\pi/4} = 8$ sec.

The ratio $\frac{T_1}{T_2} = \frac{6}{8} = \frac{3}{4}$ is a rational number.

The time period of the signal is

$$T = 4T_1 = 3T_2$$

or $T = \text{LCM of } (8,6) = 2\times 3\times 4 = 24$ sec.

Thus, the signal is periodic with a time period of 24 sec.

(iv) Given $x(t) = \cos 10\pi t + \sin 30t$.

The signal has two frequencies, $\omega_1 = 10\pi$ rad/s and $\omega_2 = 30$ rad/s.

The signal has two periods, $T_1 = \frac{2\pi}{\omega_1} = \frac{2\pi}{10\pi} = \frac{1}{5}$ rad/s $T_2 = \frac{2\pi}{\omega_2} = \frac{2\pi}{30\pi} = \frac{\pi}{15}$ sec.

The ratio $\frac{T_1}{T_2} = \frac{1/5}{\pi/15} = \frac{3}{\pi}$ is not a rational number and hence the signal is aperiodic.

1.2.4 Deterministic and Random Signals

A deterministic signal is a signal whose values are completely specified for any given time and thus it is a known function of the independent variable time; there is no uncertainty about the signal values at any time. For example, any signal represented by a functions such as sine, square, ramp or exponential in a given time interval is a deterministic signal. $f(t) = e^{-2t}$, $-T < t < T$, is a deterministic signal as its value at any t is determinable.

In contrast, a random signal or a stochastic signal is a signal whose values take a random value at any given time, and there is an uncertainty about the signal values at any given time. For example, the noise generated in a system is a random signal. Likewise, all information-bearing signals are random signals. For such signals, their values can be characterized by probability of occurrence, and hence random signals are represented with their probability distribution functions. Shown in Fig. 1.4 are sample sketches of a deterministic signal and a random signal.

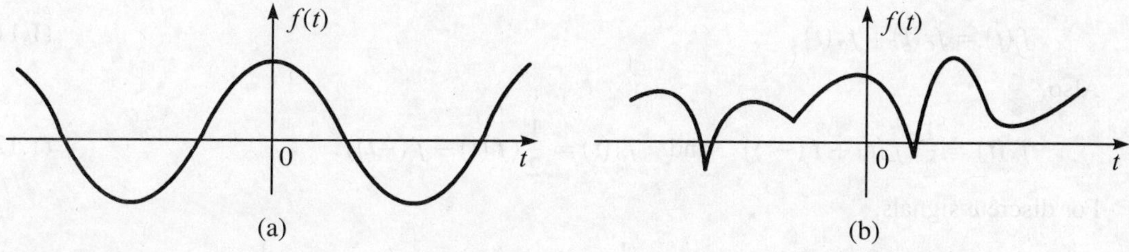

Fig. 1.4 *(a) Deterministic signal and (b) random signal*

1.2.5 Real and Complex Signals

Any signal $f(t)$ is said to be a real signal if its value is a real number, and a *complex* signal if its value is a complex number. A general complex signal $f(t)$ is a function of real and imaginary parts expressed as $f(t) = f_R(t) + jf_I(t)$, where the subscripts R and I indicate the real and imaginary components respectively. An example of a complex expression of a signal is $f(t) = e^{-(\alpha+j\omega)t}$, where $f_R(t) = e^{-\alpha t}$ is the real part of the signal and $f_I(t) = e^{-j\omega t}$ is the imaginary part.

Everything in the real world can be represented as real numbers. However, since signal operations in the complex domain is more suitable in many situations, particularly for the description of propagation of signals through dynamic systems, it is useful to express real functions by complex exponentials or complex transforms, for example using the Euler relations.

1.2.6 Even and Odd Signals

If a signal $f(t)$ is satisfies the condition

$$f(t) = f(-t) \text{ for all values of } t, \tag{1.9}$$

i.e., if the signal is identical to its own reflection or folded version, the signal is called an even signal or symmetric signal. We denote such signals by $f_e(t)$.

If on the other hand, a signal $f(t)$ satisfies the condition

$$f(t) = -f(-t) \quad \text{for all values of } t, \tag{1.10}$$

it is called odd signal or antisymmetric signal. We can denote it by $f_o(t)$.

As an example, $f(t) = A\cos t$ is an even signal while $f(t) = A\sin t$ is an odd signal (Fig. 1.5).

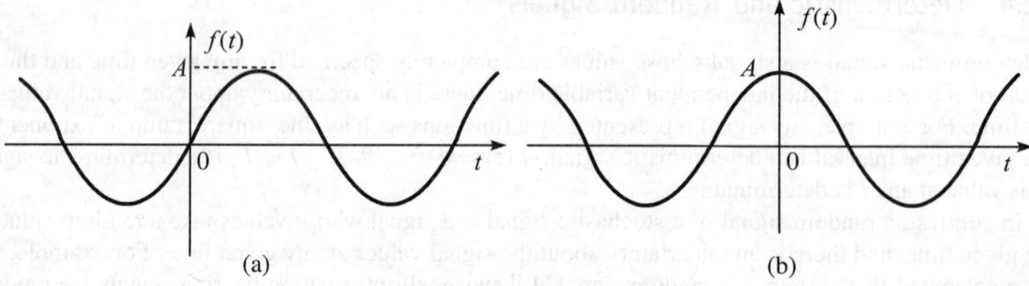

Fig. 1.5 *(a) Odd signal and (b) even signal*

Properties

(1) Any signal $f(t)$ can be expressed as a sum of even and odd signals. That is,

$$f(t) = f_e(t) + f_o(t). \tag{1.11}$$

Also,

$$f_e(t) = \frac{1}{2}(f(t) + f(-t)) \quad \text{and} \quad f_o(t) = \frac{1}{2}(f(t) - f(-t)). \tag{1.12}$$

For discrete signals,

$$f[n] = f_e[n] + f_o[n] \text{ also, } f_e[n] = \frac{1}{2}(f[n] + f[-n]) \text{ and } f_o[n] = \frac{1}{2}(f[n] - f[-n]). \tag{1.13}$$

(2) If a signal is identical to its folded version, with $f(t) = f(-t)$ it is called even symmetric. We see mirror symmetry about the vertical axis passing through the origin t = 0.

(3) If a signal and its folded version differ only in sign, with $f(t) = -f(-t)$, it is called odd symmetric. In either case, the signal extends over symmetric limits about the origin.

(4) For discrete time signals, the conditions for:

$$\text{even signal is} \quad f[-n] = f[n].$$

$$\text{odd signal is} \quad f[-n] = -f[n]. \tag{1.14}$$

(5) For an even symmetric signal as shown in Fig. 1.6(a), the signal values at $t = \alpha$ and $t = -\alpha$ are equal. The area of an even symmetric signal is twice the area on either side of the origin.

(6) For an odd symmetric signal, as shown in Fig. 1.6(b), the signal values at $t = \alpha$ and $t = -\alpha$ are equal but opposite in sign and the signal value at the origin equals zero. The area of an odd symmetric signal over symmetric limits $(-\alpha, \alpha)$ is always zero.

(7) Sum of two even signals $x_e(t) + y_e(t)$ is always of even symmetry; the sum of two odd signals $x_o(t) + y_o(t)$ has odd symmetry; and sum of odd and even signals $x_e(t) + y_o(t)$ has no symmetry.

(8) For the multiplication of two symmetric signals, the following holds:

$x_e(t).y_e(t)$ even symmetry

$x_o(t).y_o(t)$ even symmetry

$x_e(t).y_o(t)$ odd symmetry (1.15)

(9) For even symmetry, $$\int_{-T}^{T} f_e(t)dt = 2\int_{0}^{T} f_e(t)dt \tag{1.16}$$

and for odd symmetry, $$\int_{-T}^{T} f_o(t)dt = 0. \tag{1.17}$$

Figure 1.6 shows an example of even and odd symmetry.

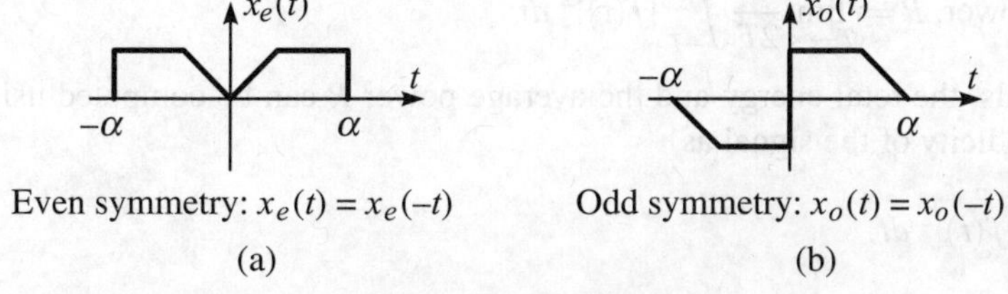

Even symmetry: $x_e(t) = x_e(-t)$
(a)

Odd symmetry: $x_o(t) = x_o(-t)$
(b)

Fig. 1.6 *(a) Even symmetry and (b) odd symmetry*

Half-wave symmetry: Half-wave symmetry is defined only for periodic signals. For a periodic signal $f(t)$ with a period t, the condition for half-wave symmetry is

$$f(t) = -f\left(t + \frac{T}{2}\right) = -f\left(t + nT + \frac{T}{2}\right) \text{ for all } t \text{ and integer values of } n. \tag{1.18}$$

That is, half a period away, $f(t)$ differs only in sign. A half-wave symmetric signal, as shown in Fig. 1.7, say, always show two half-cycles over one period with each half-cycle an inverted replica of the other and the area of one period equaling zero.

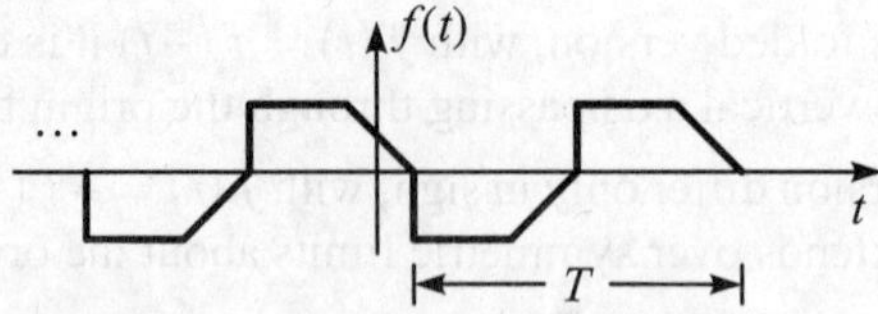

Fig. 1.7 *Half-wave symmetric signal*

Example 1.2

Find out even and odd parts of the signal $x(t) = (1 + \sin t)^2$.

Solution Given the signal $x(t) = (1 + \sin t)^2$,

$$x(t) = (1 + \sin t)^2 = 1 + \sin^2 t + 2\sin t.$$

$$\text{Even part is } x_e(t) = \frac{1}{2}[x(t) + x(-t)] = \frac{1}{2}[1 + \sin^2 t + 2\sin t + 1 + \sin^2(-t) + 2\sin(-t)]$$

$$= \frac{1}{2}[1 + \sin^2 t + 2\sin t + 1 + \sin^2(t) - 2\sin t] = 1 + \sin^2 t.$$

And the odd part is $x_0(t) = x(t) - x_e(t) = 2\sin t$.

1.2.7 Energy and Power Signals

The total energy of a continuous time signal $f(t)$ across a resistor of unit resistance can be expressed as

$$\text{Total energy, } E = \lim_{T\to\infty} \int_{-T}^{T} |f(t)|^2 dt \quad \text{or} \quad E = \int_{-\infty}^{\infty} |f(t)|^2\, dt, \tag{1.19}$$

$$\text{and the average power, } P = \lim_{T\to\infty} \frac{1}{2T} \int_{-T}^{T} |f(t)|^2\, dt. \tag{1.20}$$

For periodic signals, the total energy and the average power P can be computed using a simpler form based on the periodicity of the signal as

$$E = \int_{t_0}^{T+t_0} |f(t)|^2\, dt,$$

$$\text{and} \quad P = \frac{1}{T} \int_{t_0}^{T+t_0} |f(t)|^2 dt, \tag{1.21}$$

where T is the period of the signal and t_0 an arbitrary time instant chosen to simplify the integration (to reduce the functions to be integrated over one period).

Similarly, for a discrete-time signal $f[n]$, the total energy or the normalised energy content E of $f[n]$ is

$$E = \sum_{n=-\infty}^{\infty} |f[n]|^2. \tag{1.22}$$

The normalised average power P is

$$P = \lim_{n\to\infty} \frac{1}{2N+1} \sum_{n=-N}^{N} |f[n]|^2. \tag{1.23}$$

If $f[n]$ is periodic with period N, then

$$E = \sum_{n=0}^{N-1} |f[n]|^2 \quad \text{and} \tag{1.24}$$

$$P = \frac{1}{N} \sum_{n=0}^{N-1} |f[n]|^2. \tag{1.25}$$

The root mean square (RMS) value of the signal is

$$v_m = \sqrt{P} \text{ volts.} \tag{1.26}$$

Most signals can be classified either as energy signals or power signals according to the following criteria:

Energy signal: A signal is said to be an energy signal if and only if it has finite energy and zero average power ($0 < E < \infty$, $P = 0$). That is, energy signals have zero signal power since the finite energy is averaged over all (infinite) time. All time-limited signals of finite amplitude are energy signals. Deterministic and aperiodic signals such as damped exponentials and damped sinusoids are some examples of energy signals.

Power signal: A signal is said to be power signal if and only if it has infinite energy but finite average power ($0 < P < \infty$, $E \to \infty$). Power signals possess finite (nonzero) average power and infinite energy. The commonest examples of power signals are the periodic signals and their combinations.

Note:

1. If a signal value approaches zero as t approaches ∞, then the signal is an energy signal. This is true in most cases, but not always.
2. All periodic signals with infinite duration are power signals, but not all aperiodic signals are energy signals.
3. Any signal $f(t)$ that has a finite amplitude ($|f(t)| < \infty$) within a finite duration and zero elsewhere is an energy signal.
4. Some signals may neither be power signals nor energy signals. They may have either infinite energy and infinite power or infinite energy and zero power. Examples are exponential or parabolic growing signals.
5. Any signal may be either energy signal or power signal. Energy and power signals are mutually exclusive, that is, no signal can be both energy signal and power signal.

Example 1.3

Find the average power and total energy for the signal

$$x(t) = 3\sin(2\pi t), \quad -\infty < t < \infty.$$

Solution Given $x(t) = 3\sin(2\pi t), \quad -\infty < t < \infty$

Since $x(t)$ is periodic with period $T = 2\pi/2\pi = 1$ second, the average power of the signal is

$$P = \frac{1}{T}\int_0^T |x(t)|^2 dt$$

$$P = \frac{1}{1}\int_0^1 |x(t)|^2 dt = \int_0^1 |3\sin(2\pi t)|^2\, dt = 9\int_0^1 \frac{1}{2}[1-\cos(4\pi t)]\, dt$$

$$= 9\int_0^1 \frac{1}{2} dt - 9\int_0^1 \cos(4\pi t) dt \;= \frac{9}{2} - \left[\frac{9}{4\pi}\sin(4\pi t)\right]_0^1 = \frac{9}{2} \text{ watts.}$$

The total energy of the signal is

$$E = \int_{-\infty}^{\infty} |x(t)|^2 dt = \int_{-\infty}^{\infty} |3\sin(2\pi t)|^2 dt$$

$$= 9\int_{-\infty}^{\infty} \frac{1}{2}[1-\cos(4\pi t)]\, dt$$

$$= 9\int_{-\infty}^{\infty} \frac{1}{2} dt - 9\int_{-\infty}^{\infty} \cos(4\pi t) dt = \infty.$$

So, the energy of the signal is infinite and its average power is finite (9/2). Therefore the signal is a power signal.

Example 1.4

Find the average power and total energy for the signal

$$x(t) = 5e^{-2|t|}, \qquad -\infty < t < \infty$$

Solution

Given $x(t) = 5e^{-2|t|}, \qquad -\infty < t < \infty$

Since $x(t)$ is an aperiodic signal, the average power of the signal is

$$P = \lim_{T\to\infty} \frac{1}{T}\int_{-T/2}^{T/2} |x(t)|^2 dt = \lim_{T\to\infty} \frac{1}{T}\int_{-T/2}^{T/2} |5e^{-2|t|}|^2 dt$$

$$= 25 \lim_{T\to\infty} \frac{1}{T}\int_{-T/2}^{0} e^{4t} dt + 25 \lim_{T\to\infty} \frac{1}{T}\int_{0}^{T/2} e^{4t} dt$$

$$= \frac{25}{4} \lim_{T\to\infty} \frac{1}{T}\left[e^{4t}\right]_{-T/2}^{0} + \frac{25}{4} \lim_{T\to\infty} \frac{1}{T}\left[e^{4t}\right]_{0}^{T/2}$$

$$= \frac{25}{4} \lim_{T\to\infty} \frac{1}{T}\left[1-e^{-2T}\right] + \frac{25}{4} \lim_{T\to\infty} \frac{1}{T}\left[e^{2T}-1\right] = 0$$

and the total energy of the signal is

$$E = \int_{-\infty}^{\infty} |x(t)|^2 dt = \int_{-\infty}^{\infty} \left|5e^{-2|t|}\right|^2 dt$$

$$= 25\int_{-\infty}^{0} e^{4t}dt + \int_{0}^{\infty} e^{-4t}dt = \frac{25}{4}\left[e^{4t}\right]_{-\infty}^{0} + \frac{25}{4}\left[e^{-4t}\right]_{0}^{\infty}$$

$$= \frac{25}{4} + \frac{25}{4} = \frac{50}{4} \text{ joules.}$$

So, the energy of the signal is finite and its average power is zero. Therefore the signal is an energy signal.

Example 1.5

Find the average power and total energy for the signal

$$x(t) = \begin{cases} \frac{1}{\sqrt{t}}, & t > 1 \\ 0, & t \leq 1 \end{cases}$$

Solution

Since $x(t)$ is aperiodic signal, the average power of the signal is

$$P = \lim_{T\to\infty} \frac{1}{T}\int_{-T/2}^{T/2} |x(t)|^2 dt$$

$$= \lim_{T\to\infty} \frac{1}{T}\int_{1}^{T/2} \frac{1}{t}dt = \lim_{T\to\infty}\left(\frac{1}{T}\ln[t]_1^{T/2}\right)$$

$$= \lim_{T\to\infty}\left(\frac{1}{T}\ln\left[\frac{T}{2}\right] - \frac{1}{T}\ln[1]\right) = \lim_{T\to\infty}\left(\frac{1}{T}\ln\left[\frac{T}{2}\right]\right).$$

Using L'Hospital's rule, differentiating with respect to T

$$P = \lim_{T\to\infty}\left(\frac{\ln\left[\frac{T}{2}\right]}{T}\right) = \lim_{T\to\infty}\left(\frac{\frac{2}{T}}{1}\right) = 0.$$

And the total energy of the signal is

$$E = \int_{-\infty}^{\infty} |x(t)|^2 \, dt = \int_{1}^{\infty} \frac{1}{t}dt = \ln[t]_1^{\infty} = \infty.$$

So, the energy of the signal is infinite and its average power is zero. Therefore the signal is neither an energy signal nor a power signal.

1.2.8 Causal and Non-causal Signals

Any continuous time signal $f(t)$ is said to be causal if and only if

$$f(t) = 0, \quad \text{for} \quad t < 0 \tag{1.27}$$

otherwise the signal is called non-causal signal. That is for causal, the signal exists only for positive time values. The condition for anti-causal is

$$f(t) = 0, \quad \text{for} \quad t > 0. \tag{1.28}$$

Similarly a discrete signal $f[n]$ is said to be causal if and only if

$$f[n] = 0 \quad \text{for} \quad n < 0. \tag{1.29}$$

Otherwise, the signal is non-causal or anti signal for which the condition is

$$f[n] = 0 \quad \text{for} \quad n > 0. \tag{1.30}$$

Note that if a signal has positive time values, it is causal and if a signal has negative time values, it is non-causal. $x(t) = 5e^{-2t}$, $t > 0$, for example, is a causal signal, whereas $x(t) = 5e^{-2|t|}$ is non causal.

Example 1.6

Which of the following signals are causal or non-causal?
(a) $x(t) = 3\cos(4t)$; (b) $x(t) = e^{2(t-1)}u(t-1)$; (c) $x(t) = \cos^2(2t)u(t)$;
(d) $x(t) = u(-t-1) + u(t-1)$; (e) $x[n] = e^{-\alpha n}u[-n]$.

Solution

(a) Given $x(t) = 3\cos(4t)$. The signal is non-causal since it extends from $-\infty < t < \infty$ and $x(t) \neq 0$ for $t < 0$.
(b) Given $x(t) = e^{2(t-1)}u(t-1)$. The signal is causal since $x(t) = 0$ for $t < 1$.
(c) Given $x(t) = \cos^2(2t)u(t)$. The signal is causal since $x(t) = 0$ for $t < 0$.
(d) $x(t) = u(-t-1) + u(t-1)$. The signal is non-causal since it extends from $-\infty < t < \infty$ and $x(t) \neq 0$ for $t < 0$.
(e) $x[n] = e^{\alpha n}u[-n]$. The sequence is non-causal since it extends along the negative axis and $x[n] \neq 0$ for $n < 0$.

1.3 Standred Types of Signals

Several basic types of signals such as impulse, step, ramp and sinusoidal signals that are commonly encountered in signal processing will be introduced in this section. These signals occur frequently in nature and moreover, form the building blocks from which other signals are constructed.

Unit Impulse Function
The continuous unit impulse function is defined as

$$\delta(t) = \begin{cases} 1 & t = 0 \\ 0 & \text{otherwise.} \end{cases} \tag{1.31}$$

For discrete signals the unit impulse function is defined as

$$\delta[n] = \begin{cases} 1, & n = 0 \\ 0, & n \neq 0. \end{cases} \tag{1.32}$$

That is, the impulse function is zero everywhere except at $t = 0$. Also,

$$\int_{-\infty}^{\infty} \delta(t)\, dt = 1, \tag{1.33}$$

that is, the area under the curve is equal to one. Graphically, the unit impulse functions are shown in Fig. 1.8.

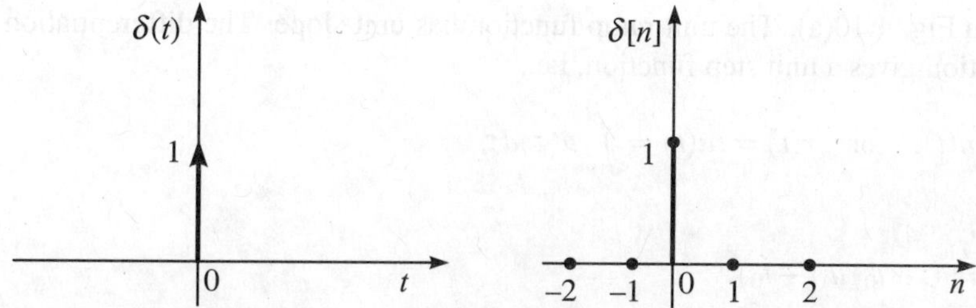

Fig. 1.8 *(a) Continuous unit impulse function, (b) discrete unit impulse function*

Note: The properties of impulse function will be discussed in Chapter 2.

Unit Step Function

The unit step function $u(t)$ is defined as

$$u(t) = \begin{cases} 1 & t \geq 0 \\ 0 & t < 0. \end{cases} \tag{1.34}$$

which is shown in Fig. 1.9(a). The function is discontinuous at $t = 0$ and the value at $t \geq 0$ is unity. The unit step function is very useful in many instances, such as at the input of a system as it can represent the closing of a switch at $t = 0$. The dc power supply given to a system, for example, by closing the switch at $t = 0$ represents a unit step function. The unit step function is also used to construct discontinuous functions.

The unit step function for discrete time signals is

$$u[n] = \begin{cases} 1 & n \geq 0 \\ 0 & n < 0. \end{cases} \tag{1.35}$$

It is also called unit step sequence as shown in Fig. 1.9(b).

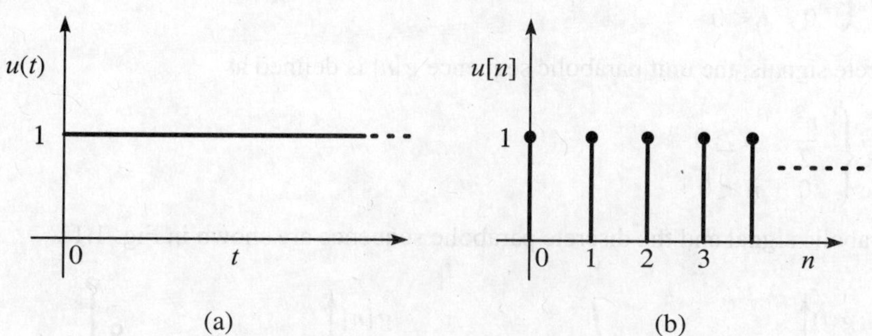

Fig. 1.9 *Unit step function: (a) continuous signal, (b) discrete sequence*

Unit ramp function

The continuous unit ramp function $r(t)$ is defined as

$$r(t) = \begin{cases} t & t \geq 0 \\ 0 & t < 0 \end{cases}, \tag{1.36}$$

and is shown in Fig. 1.10(a). The unit ramp function has unit slope. The differentiation of continuous unit ramp function gives a unit step function, i.e.,

$$\frac{dr(t)}{dt} = u(t), \quad \text{or} \quad r(t) = tu(t) = \int_0^t u(\tau)d\tau \tag{1.37}$$

Also,

$$r(t-t_0) = (t-t_0)\,u(t-t_0). \tag{1.38}$$

Similarly for discrete signals the unit ramp sequence is defined as

$$r[n] = \begin{cases} n, & n \geq 0 \\ 0, & n < 0. \end{cases} \tag{1.39}$$

Figure 1.10 (b) shows the discrete ramp sequence $r[n]$.

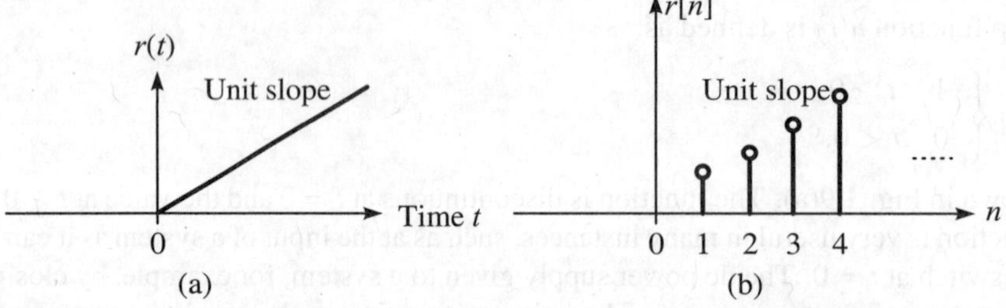

Fig. 1.10 *(a) Unit ramp signal, (b) discrete ramp sequence*

Unit Parabolic Function

The continuous unit parabolic function $g(t)$ is defined as

$$g(t) = \begin{cases} \dfrac{t^2}{2} & t \geq 0 \\ 0 & t < 0 \end{cases}; \tag{1.40}$$

and for discrete signals, the unit parabolic sequence $g[n]$ is defined as

$$g[n] = \begin{cases} \dfrac{n^2}{2} & n \geq 0 \\ 0 & n < 0. \end{cases} \tag{1.41}$$

The unit parabolic signal and the discrete parabolic sequence are shown in Fig. 1.11.

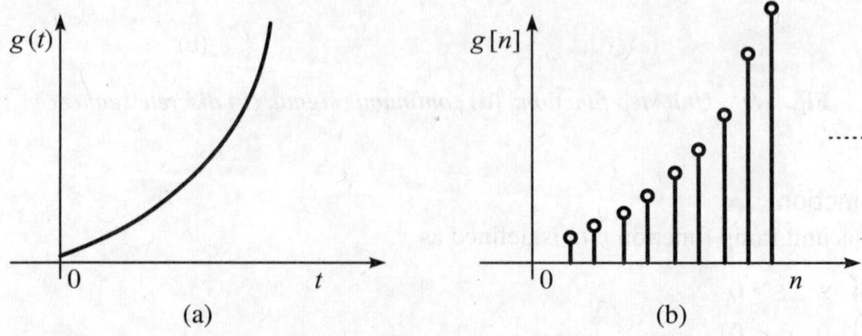

Fig. 1.11 *(a) The unit parabolic signal and (b) discrete parabolic sequence*

The differentiation of unit parabolic function gives unit ramp signal, i.e.,

$$\frac{dg(t)}{dt} = r(t). \tag{1.42}$$

Real Exponential Function

The continuous time real exponential function is defined as

$$f(t) = ce^{at}. \tag{1.43}$$

where c and a, are real numbers. As illustrated in Fig. 1.12, if a is positive, then as t increases, $f(t)$ is a growing exponential, and if a is negative, then as t increases, $f(t)$ is a decaying exponential. Note that for $a = 0$, $f(t) = c$ is constant.

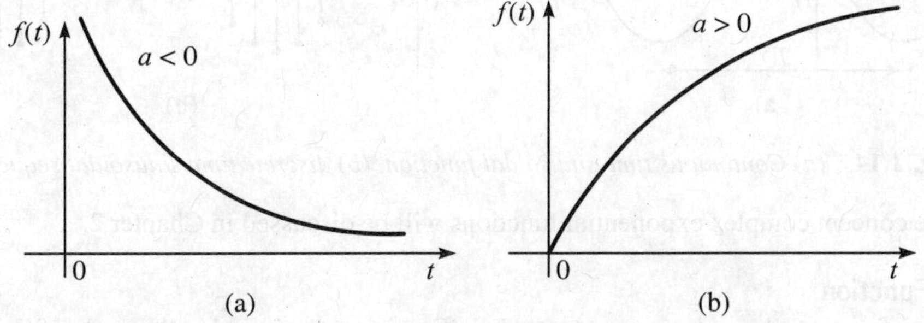

Fig. 1.12 *Real exponential function: (a) decaying exponential, (b) growing exponential*

For discrete signals, the real exponential sequence is given by

$$f[n] = ce^{an}. \tag{1.44}$$

For real values of c and a the waveforms are shown in Fig. 1.13.

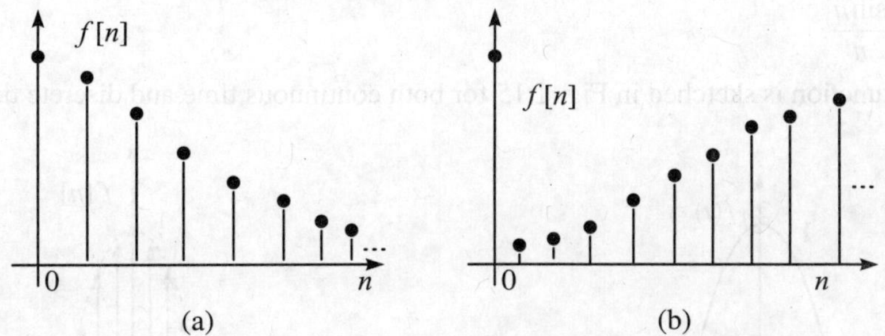

Fig. 1.13 *Real exponential sequence for (a) $a < 0$ and (b) $a > 0$*

Sinusoidal Function

A continuous sinusoidal function is expressed as

$$f(t) = A\cos(\omega_0 t + \phi), \tag{1.45}$$

where A is the amplitude, ω_0 is the angular frequency in radians per second, ϕ is the phase shift in radians and $\omega_0 = 2\pi f_0 = \dfrac{2\pi}{T}$.

For discrete signals, the sinusoidal sequence with period N is expressed as

$$f[n] = A\cos(\omega_0 n + \phi), \tag{1.46}$$

where $\omega_0 = 2\pi f_0 = \dfrac{2\pi}{N}$, N being an integer.

Figure 1.14 shows a sketch of the continuous time sinusoidal signal and the discrete time sinusoidal sequence.

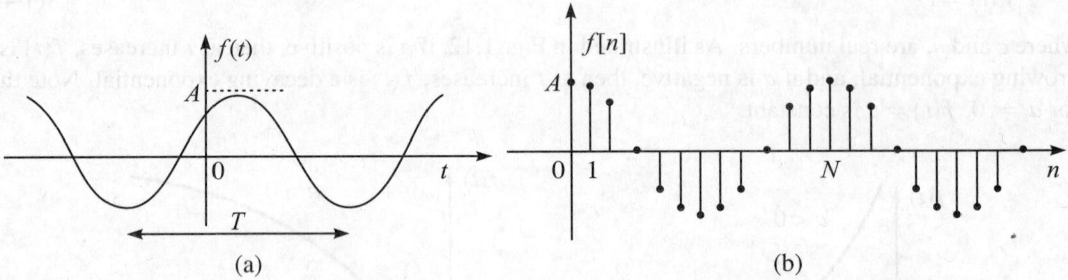

Fig. 1.14 *(a) Continuous time sinusoidal function, (b) discrete time sinusoidal sequence*

Note: The concept complex exponential functions will be discussed in Chapter 2.

Sampling Function

The sampling function play a very important role in Fourier analysis and in the study of linear systems. The sampling function or signal is defined as

$$\text{Sa}(t) = \frac{\sin t}{t}. \tag{1.47}$$

The sampling function has even symmetry with t. The function is zero at $t = \pm\pi, \pm 2\pi, \pm 3\pi, \ldots$ and has a maximum value of unity at $t = 0$. The amplitude of the oscillations decreases as $|t| \to \infty$.

For discrete time signals, the sampling sequence is given by

$$\text{Sa}[n] = \frac{\sin n}{n}. \tag{1.48}$$

The sampling function is sketched in Fig. 1.15 for both continuous time and discrete time signals.

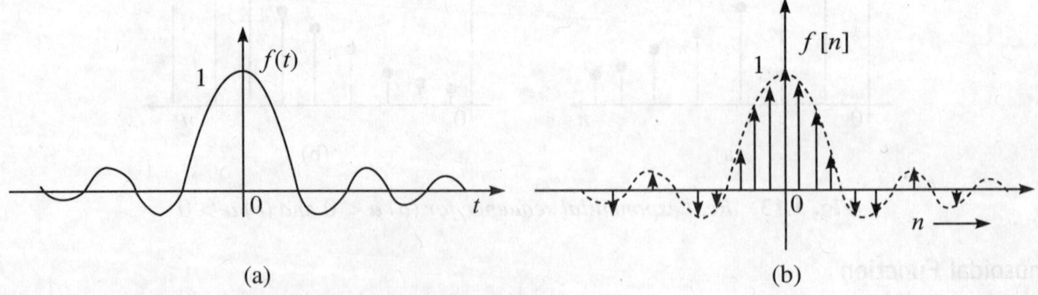

Fig. 1.15 *Sampling function: (a) continuous time signal, (b) discrete time sequence*

The integration of $\text{Sa}(t)$ yields

$$\int_{-\infty}^{\infty} \text{Sa}(t)\, dt = \pi$$

$$\text{or} \quad \int_{-\infty}^{0} \text{Sa}(t)dt = \int_{0}^{\infty} \text{Sa}(t)dt = \frac{\pi}{2}. \tag{1.49}$$

Another definition of the sampling signal yields the *sinc* function. It may be mentioned that sinc and sampling function differ in that the sinc function is a normalized function while sampling is not. Thus x axis representing independent variable t will be integer values for sinc function; for sampling it will be multiples of π. Sinc function is defined as

$$\text{sinc}\,(t) = \frac{\sin \pi t}{\pi t}. \tag{1.50}$$

For discrete signals, the sinc sequence is given by

$$\text{sinc}\,[n] = \frac{\sin \pi n}{\pi n}. \tag{1.51}$$

Rectangular Pulse Function

The rectangular pulse function is defined as

$$\prod\left(\frac{t}{\tau}\right) = \text{rect}\left(\frac{t}{\tau}\right) = \begin{cases} 1 & |t| < \tau/2 \\ 0 & \text{elsewhere.} \end{cases} \tag{1.52}$$

The rectangular pulse function is also called gate function $G_\tau(t)$.

The signal is even symmetric of width τ and unit height as shown in Fig. 1.16(a). For discrete signals, the rectangular pulse sequence is defined as

$$\prod\left[\frac{n}{N}\right] = \text{rect}\left[\frac{n}{N}\right] = \begin{cases} 1 & |n| \leq N \\ 0 & \text{elsewhere.} \end{cases} \tag{1.53}$$

There are $2N+1$ impulses in the sequence shown in Fig. 1.16(b).

Note: The gate sequence $G_N[n]$ is a rectangular pulse sequence.

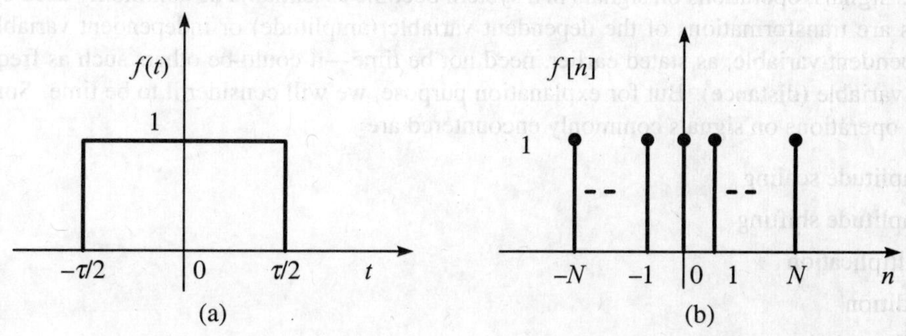

Fig. 1.16 *(a) Rectangular pulse function, (b) rectangular pulse sequence*

The signal $f(t) = \text{rect}\left(\dfrac{t-a}{b}\right)$ describes a rectangular pulse of width b centered at $t = a$.

Triangular Pulse Function

A continuous time triangular pulse function is defined as

$$\Delta\left(\frac{t}{\tau}\right) = \text{tri}\left(\frac{t}{\tau}\right) = \begin{cases} 1 - |t|/\tau & |t| \leq \tau \\ 0 & \text{elsewhere.} \end{cases} \tag{1.54}$$

An even symmetric triangular pulse of width = 2τ and unit height is shown in Fig. 1.17(a).

For discrete signals, the triangular pulse sequence is defined as

$$\Delta[n] = \text{tri}\,[n] = \begin{cases} N - |n| & |n| \leq N \\ 0 & \text{elsewhere.} \end{cases} \tag{1.55}$$

A triangular pulse sequence has $2N+1$ impulses as shown Fig. 1.17(b).

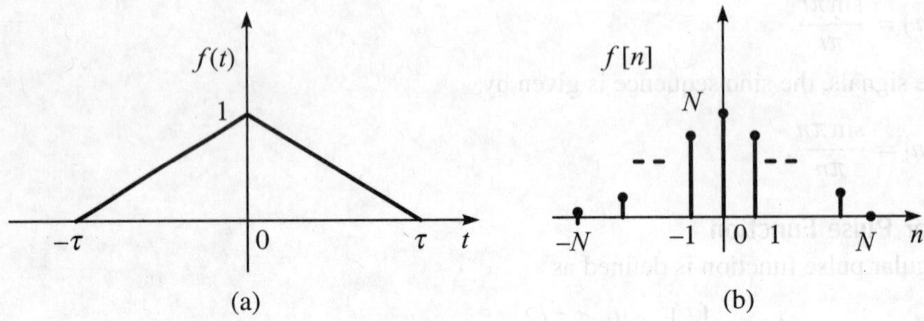

Fig. 1.17 *(a) triangular pulse signal (b) triangular pulse sequence*

The signal $f(t) = \text{tri}\left(\dfrac{t-a}{b}\right)$ describes a triangular pulse of width = $2b$ centered at t = a.

Note: The pulse and triangle signals can be used as windows to truncating signals.

1.4 Operations on Signals

To process signals, operations on signals in a system become essential. The commonly used operations on signals are transformations of the dependent variable (amplitude) or independent variable (time). The independent variable, as stated earlier, need not be time—it could be others such as frequency or the space variable (distance). But for explanation purpose, we will consider it to be time. Some of the important operations on signals commonly encountered are

- Amplitude scaling
- Amplitude shifting
- Multiplication
- Addition
- Differentiation
- Integration
- Time scaling
- Time shifting
- Time reversal or reflection

Amplitude scaling

For a given continuous time signal x(t), amplitude scaling is defined as

$$y(t) = cx(t), \tag{1.56}$$

where c is a constant, called the scaling factor. As a result of amplitude scaling, the signal $x(t)$ may be amplified ($c > 0$) or attenuated ($c < 0$) without the signal shape being altered.

For discrete time signals, amplitude scaling is given by

$$y[n] = cx[n]. \tag{1.57}$$

For example if the signal $x(t) = \cos t$ is amplified by 2, then $y(t) = 2x(t)$ is an amplified signal. The concept of amplitude scaling is shown in Fig. 1.18.

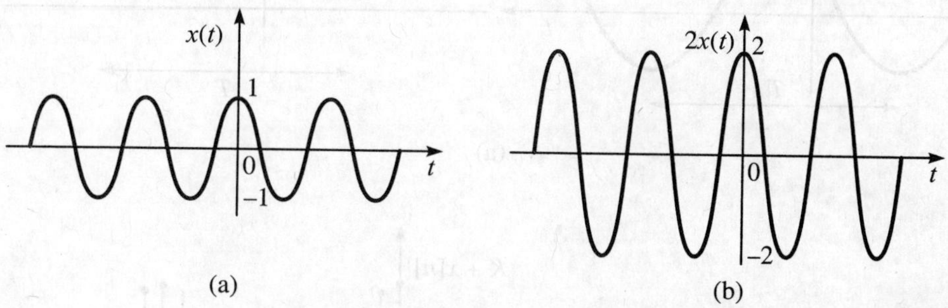

Fig. 1.18 *Amplitude scaling for continuous signal*

Similarly, if a discrete time sequence $x[n] = 2\delta[n+1] + \delta[n] + 2\delta[n-1]$ is scaled by 2, it yields $y[n] = 2x[n] = 4\delta[n+1] + 2\delta[n] + 4\delta[n-1]$. The sequences are shown in Fig. 1.19.

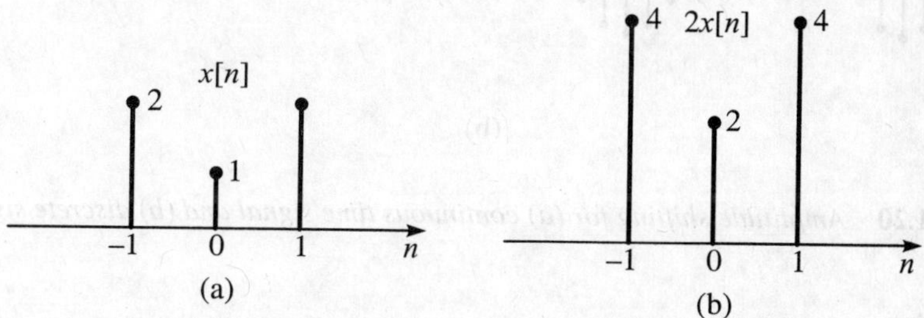

Fig. 1.19 *Amplitude scaling for discrete signal*

Amplitude shifting

The amplitude shift of a continuous time signal $x(t)$ adds a constant K to $x(t)$. The shifted signal is

$$y(t) = K + x(t). \tag{1.58}$$

The operation is in essence equivalent to adding a dc signal component to the given signal. For discrete time signals, amplitude shifting is given by

$$y[n] = K + x[n]. \tag{1.59}$$

The amplitude shifting operation for continuous and discrete time signals are shown in Fig. 1.20.

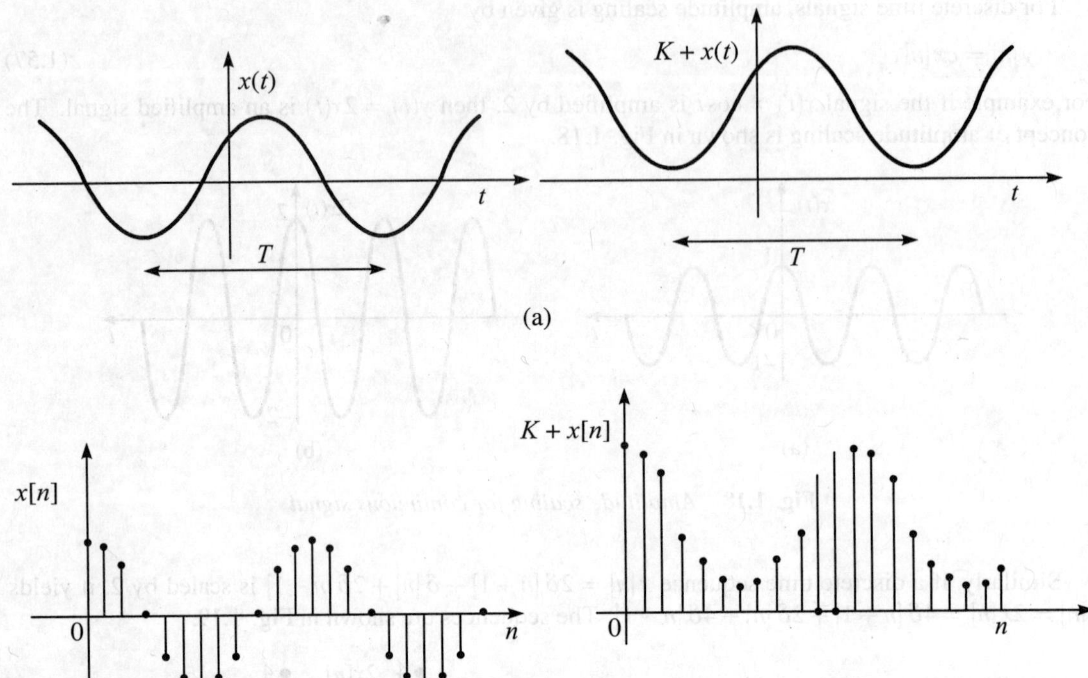

Fig. 1.20 *Amplitude shifting for (a) continuous time signal and (b) discrete signals*

Addition

The addition of two signals $x_1(t)$ and $x_2(t)$ is defined as

$$y(t) = x_1(t) + x_2(t). \tag{1.60}$$

For discrete time signals,

$$y[n] = x_1[n] + x_2[n]. \tag{1.61}$$

For example, if $x_1(t) = u(t)$ and $x_2(t) = u(t-2)$, then $y(t) = u(t) + u(t-2)$.

Also, if $x_1[n] = [1,1,1,1,1]$ for $n = -2,-1,0,1,2$ and $x_2[n] = [2,1,2,1,2,2]$ for $n = -2,-1,0,1,2,3$, then

$y[n] = x_1[n] + x_2[n] = [3,2,3,2,3,2]$ for $n = -2,-1,0,1,2,3$.

The addition of two signals for continuous and discrete signals is shown in Fig. 1.21.

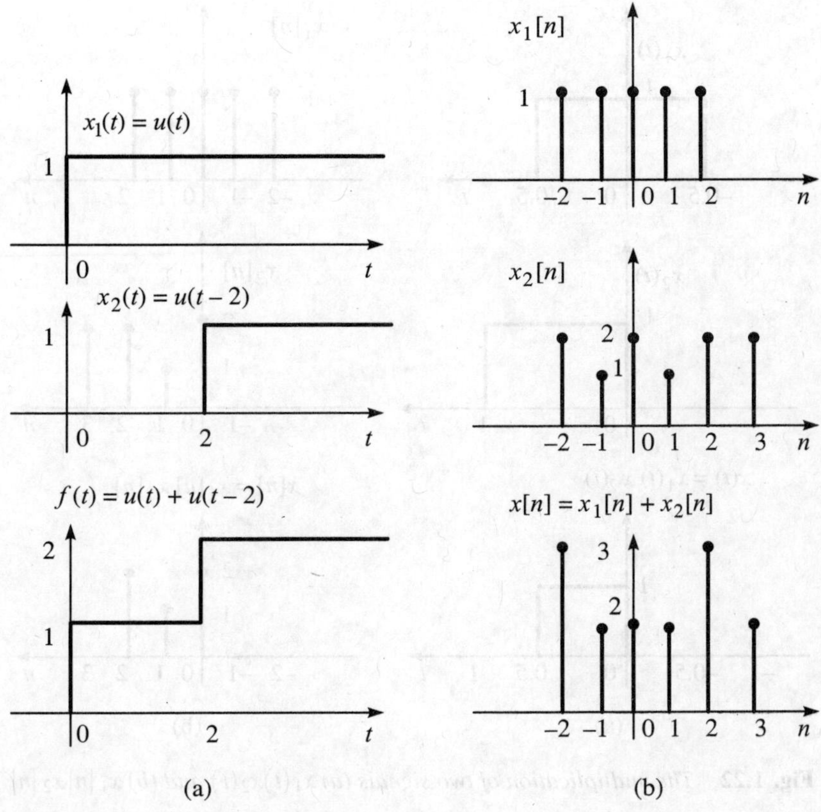

Fig. 1.21 *The addition of two signals (a) $x_1(t) + x_2(t)$ and (b) $x_1[n] + x_2[n]$*

Multiplication

The multiplication of two continuous time signals $x_1(t)$ and $x_2(t)$ is defined as

$$y(t) = x_1(t) \cdot x_2(t). \tag{1.62}$$

When the signals are discrete-time signals, the operation is given by

$$y[n] = x_1[n] \cdot x_2[n]. \tag{1.63}$$

For example, if $x_1(t) = \begin{cases} 1 & |t| \leq 0.5 \\ 0 & \text{otherwise} \end{cases}$ and $x_2(t) = \begin{cases} 1 & 0 \leq t \leq 1 \\ 0 & \text{otherwise,} \end{cases}$ then

$y(t) = x_1(t)x_2(t) = \begin{cases} 1 & 0 \leq t \leq 0.5 \\ 0 & \text{otherwise} \end{cases}$ is the common area of the two signals.

If $x_1[n] = [1,1,1,1,1]$ for $n = -2,-1,0,1,2$ and $x_2[n] = [2,1,2,2]$ for $n = 0,1,2,3$, then

$$y[n] = x_1[n]x_2[n] = [2,1,2] \quad \text{for } n = 0,1,2.$$

The multiplication process for these two signals is shown in Fig. 1.22.

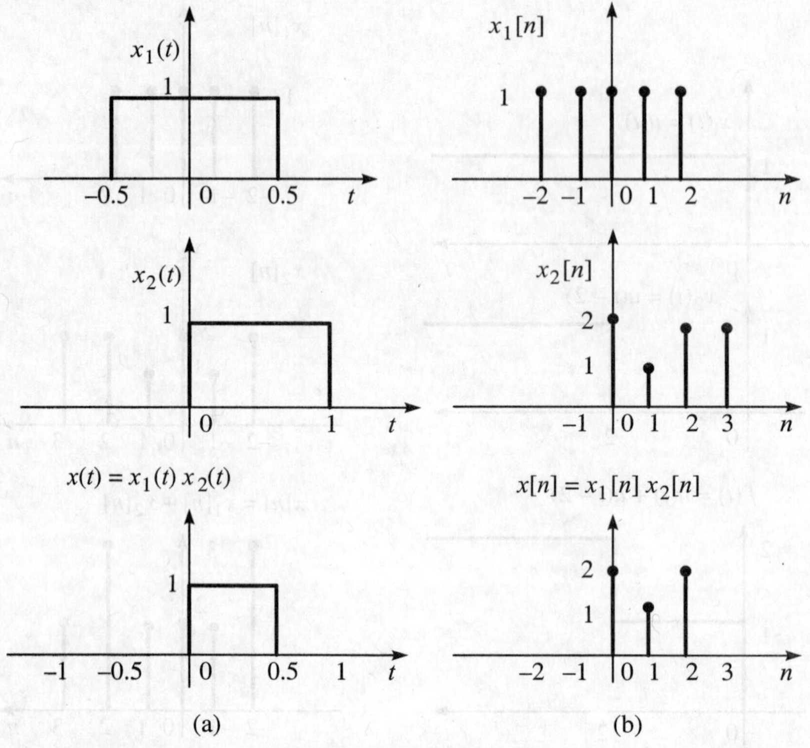

Fig. 1.22 *The multiplication of two signals (a) $x_1(t)x_2(t)$ and (b) $x_1[n]x_2[n]$*

Differentiation

For a given continuous time signal $x(t)$, differentiation with respect to time t is defined as

$$y(t) = \frac{dx(t)}{dt}. \tag{1.64}$$

We can similarly define higher order differentiations such as

$$y(t) = \frac{d^2x(t)}{dt^2}. \tag{1.65}$$

For example, the differentiation of unit ramp function gives the unit step function, i.e.,

$$\frac{dr(t)}{dt} = u(t) \tag{1.66}$$

and the differentiation of the unit step function gives the unit impulse function:

$$\frac{du(t)}{dt} = \delta(t). \tag{1.67}$$

Integration

For a given continuous time signal $x(t)$, the integration with respect to time t is defined as

$$y(t) = \int_{-\infty}^{t} x(\tau)d\tau, \tag{1.68}$$

where τ is the time variable. Higher order integration can be similarly expressed.

For an odd symmetrical signal,

$$\int_{-T}^{T} x(t)dt = 0 \tag{1.69}$$

and for even symmetrical signal,

$$\int_{-T}^{T} x(t)dt = 2\int_{0}^{T} x(t)dt. \tag{1.70}$$

For example the integration of the unit step function gives the unit ramp function and the integration of the unit impulse function gives the unit step function.

Note: Differentiations and integrations are defined only for continuous time variables.

Time Scaling

For a given signal $x(t)$ the time scaling is defined as

$$y(t) = x(at). \tag{1.71}$$

where a is a positive constant called the scaling factor. The signal $x(t)$ through the system may be stretched $a < 1$ or compressed $a > 1$ without changing the signal shape.

Time scaling speeds up or slows down time and results in signal compression or stretching. For example, the signal $y(t) = x(t/2)$ describes a two-fold expansion of $x(t)$ since the time, t is slowed down to $t/2$. Similarly, the signal $y(t) = x(2t)$ describes a two fold compression of $x(t)$ since the time t is speeded up to $2t$ as shown in Fig. 1.23.

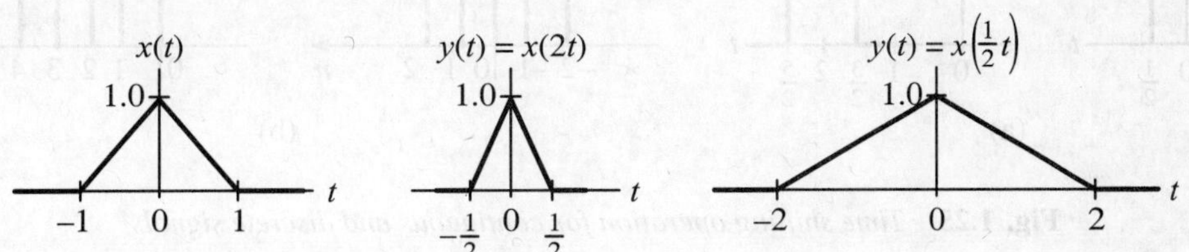

Fig. 1.23 *Time scaling operation for continuous signal*

For discrete time signals, a similar operation is defined by

$$y[n] = x[an] \tag{1.72}$$

and is shown in Fig. 1.24.

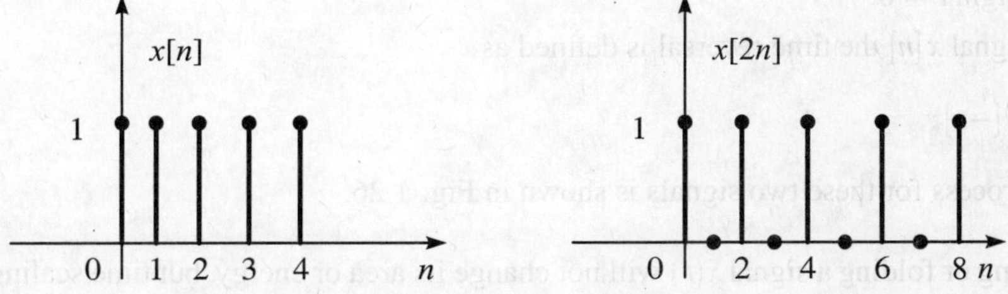

Fig. 1.24 *Time scaling operation for discrete signal*

Time Shifting

A given continuous time signal $x(t)$ is said to be time shifted when

$$y(t) = x(t - t_0), \tag{1.73}$$

where t_0 is a time delay or shift in time. A time shift displaces the signal $x(t)$ in time without changing its shape, i.e., the value of $y(t)$ at $t - t_0$ corresponds to the value of $x(t)$ at $t = t_0$. The signal $y(t)$ is a delayed version of $x(t)$ shifted right by time t_0. Similarly, the signal $y(t) = x(t + t_0)$ is an advanced version of $x(t)$ shifted left by time t_0. For example if $x(t)$ is a square pulse, then with a time delay of 2 units, the new signal $y(t) = x(t - 2)$ will be as shown in Fig. 1.25(a).

For discrete time signals,

$$y[n] = x[n - n_0] \tag{1.74}$$

is a delayed version of $x[n]$. Essentially, the signal is shifted left by time n_0; similarly, $y[n] = x[n + n_0]$ is an advanced version of $x[n]$ or the signal shifted right by time n_0.

For example if $x[n]$ is a rectangular pulse sequence, then a time delay of 2 units yields $y[n] = x[n - 2]$ as shown in Fig. 1.25(b).

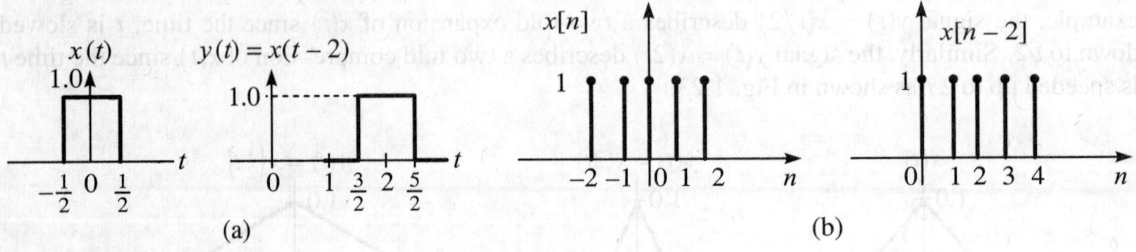

Fig. 1.25 *Time shifting operation for continuous and discrete signals*

Time Reversal or Reflection or Folding

For a given signal $x(t)$, the time reversal is defined as

$$y(t) = x(-t). \tag{1.75}$$

Time reversal creates the folded signal $x(-t)$ as a mirror image of $x(t)$ about the vertical axis passing through the origin $t = 0$.

For discrete signal $x[n]$ the time reversal is defined as

$$y[n] = x[-n]. \tag{1.76}$$

The folding process for these two signals is shown in Fig. 1.26.

Note: Shifting or folding a signal $x(t)$ will not change its area or energy, but time scaling $x(t)$ to $x(at)$ will reduce both its area and energy by a.

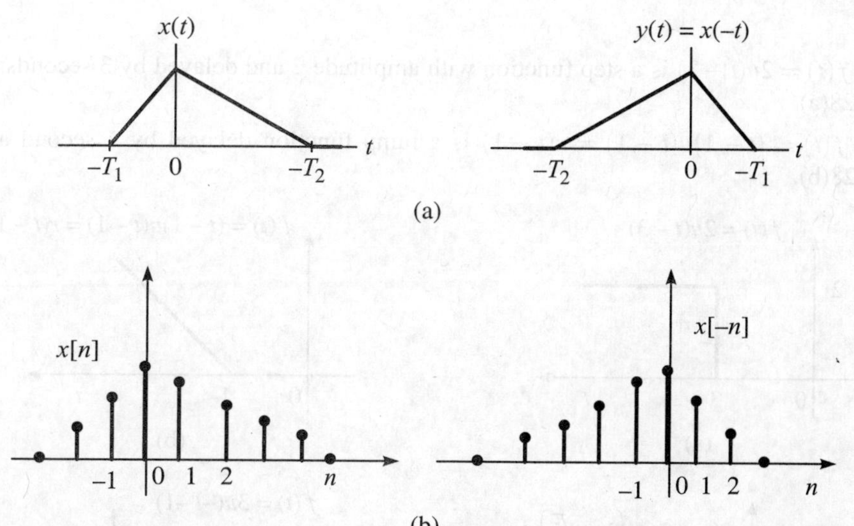

Fig. 1.26 *Signal folding operation for (a) continuous and (b) discrete time signals*

Operations in Combination

It is possible to use a combination of basic operations to arrive at a given signal shape. Let $y(t) = x(at - b)$. To correctly obtain $y(t)$ from $x(t)$, time-shifting and time-scaling operations must be performed in the correct order. The time-shifting operation is performed first on $x(t)$, resulting in an intermediate signal $v(t) = x(t - b)$. Then the time-scaling operation is performed on $v(t)$ giving $y(t) = v(at) = x(at - b)$. An example of such operation is shown in Fig. 1.27.

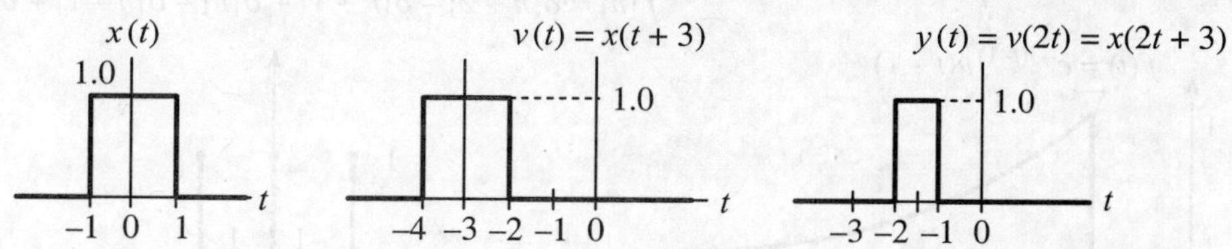

Fig. 1.27 *Time-shifting and the time-scaling operations performed in combination*

Example 1.7

Sketch the following signals:

(i) $f(t) = 2u(t-3)$;

(ii) $f(t) = (t-1)\,u(t-1)$;

(iii) $f(t) = 2\sin\left(3t + \dfrac{\pi}{3}\right)$;

(iv) $f(t) = 3u(-t-1)$;

(v) $f(t) = e^{-3(t-1)}u(t-1)$;

(vi) $f[n] = \delta[n+2] - \delta[n+1] + \delta[n] - \delta[n-1] + \delta[n-2]$ and

(vii) $f[n] = \delta[n-1] + 2\delta[n-2] + 3\delta[n-3] + 2\delta[n-4] + \delta[n-5]$.

Solution

(i) Signal $f(t) = 2u(t-3)$ is a step function with amplitude 2 and delayed by 3 seconds as shown if Fig. 1.28(a).

(ii) Signal $f(t) = (t-1)u(t-1) = r(t-1)$ is a ramp function delayed by 1 second as shown in Fig. 1.28(b).

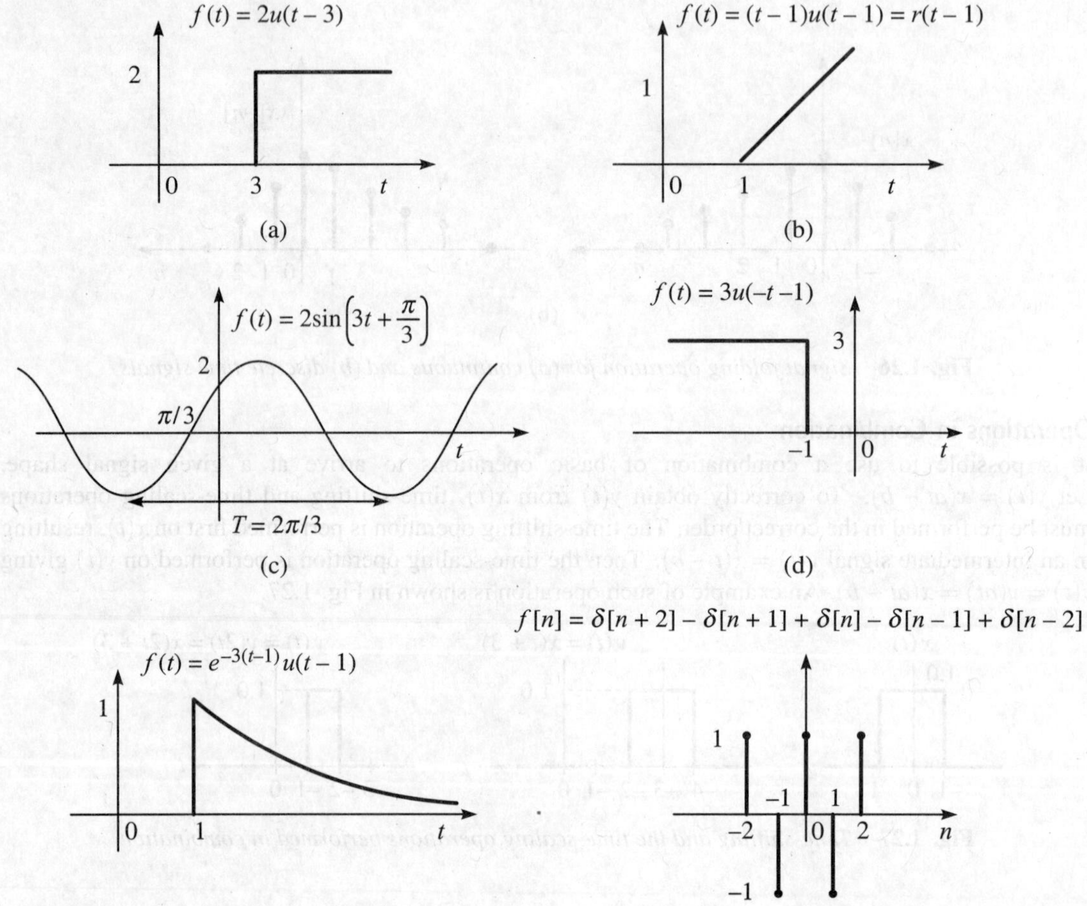

$f(t) = 2u(t-3)$

(a)

$f(t) = (t-1)u(t-1) = r(t-1)$

(b)

$f(t) = 2\sin\left(3t + \frac{\pi}{3}\right)$

$T = 2\pi/3$

(c)

$f(t) = 3u(-t-1)$

(d)

$f(t) = e^{-3(t-1)}u(t-1)$

(e)

$f[n] = \delta[n+2] - \delta[n+1] + \delta[n] - \delta[n-1] + \delta[n-2]$

(f)

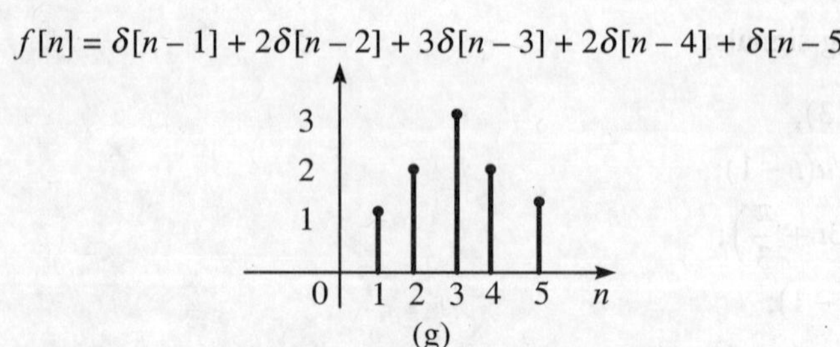

(g)

Fig. 1.28 *Various signals of Example 1.7*

(iii) Signal $f(t) = 2\sin\left(3t + \frac{\pi}{3}\right)$ is a sinusoidal signal with amplitude 2, frequency 3 rad/s and phase shifted by $\frac{\pi}{3}$ degrees as shown in Fig. 1.28(c).

(iv) Signal $f(t) = 3u(-t-1)$ is a step function with amplitude 2 in the interval $-1 < t < -\infty$ as shown in Fig. 1.28(d).

(v) Signal $f(t) = e^{-3(t-1)}u(t-1)$ is an exponential decaying function delayed by 1 second as shown in Fig. 1.28(e).

(vi) $f[n] = \delta[n+2] - \delta[n+1] + \delta[n] - \delta[n-1] + \delta[n-2]$ is a discrete sequence; $f[n] = [1, -1, 1, -1, 1]$ for $n = -2, -1, 0, 1, 2$ as shown in Fig. 1.28(f).

(vii) $f[n] = \delta[n-1] + 2\delta[n-2] + 3\delta[n-3] + 2\delta[n-4] + \delta[n-5]$ is a discrete sequence; $f[n] = [1, 2, 3, 2, 1]$ for $n = 1, 2, 3, 4, 5$ as shown in Fig. 1.28(g).

Example 1.8

Given two signals $x_1(t) = u(t-2)$ and $x_2(t) = u(t-4)$, draw the waveforms for the following operations:

(i) $f(t) = x_1(t) + x_2(t)$,

(ii) $f(t) = 3x_1(t) - 2x_2(t)$,

(iii) $f(t) = x_1(t)x_2(t)$,

(iv) $f(t) = x_1(t+5) + 2x_2(t-2)$.

Solution

Given two signals $x_1(t) = u(t-2)$ and $x_2(t) = u(t-4)$ shown in Fig. 1.29(a) and 1.29(b). The operations on the signals are,

(i) $f(t) = x_1(t) + x_2(t) = u(t-2) + u(t-4)$ shown in Fig. 1.29(c)

(ii) $f(t) = 3x_1(t) - 2x_2(t) = 3u(t-2) - 2u(t-4)$ shown in Fig. 1.29(d)

(iii) $f(t) = x_1(t)x_2(t) = u(t-2)u(t-4)$ shown in Fig. 1.29(e)

(iv) $f(t) = x_1(t+5) + 2x_2(t-2) = u(t+3) + 2u(t-6)$ shown in Fig. 1.29(f)

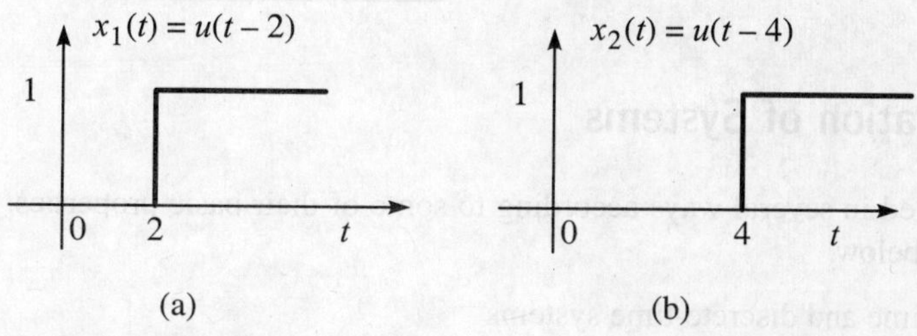

Fig. 1.29 *(a) Signals $x_1(t) = u(t-2)$ and $x_2(t) = u(t-4)$*

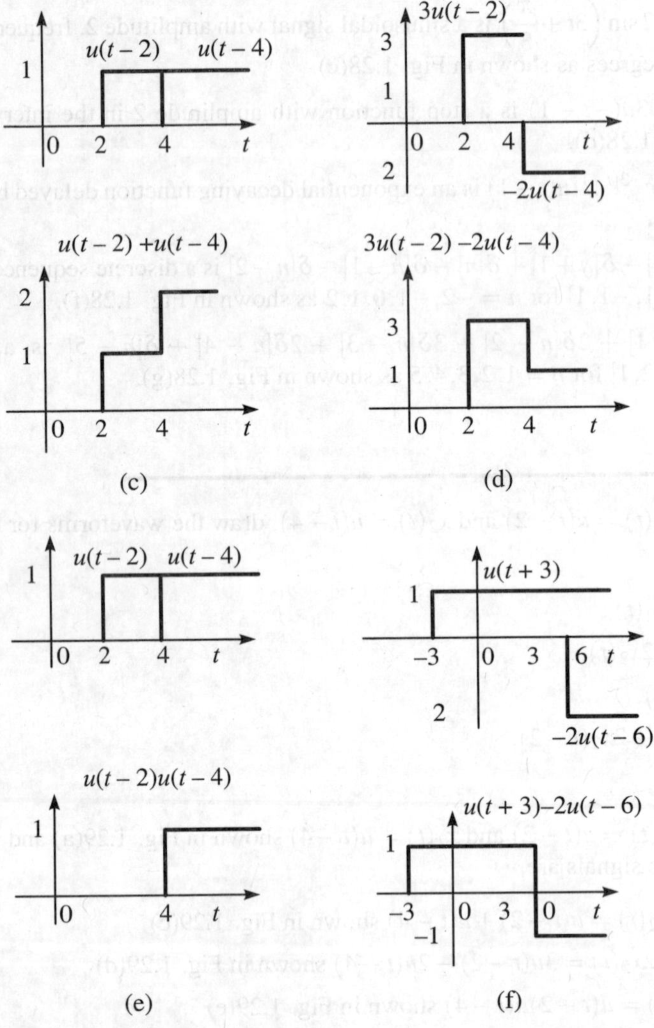

Fig. 1.29 (Cont.) *The given signals and the various operations on them*

1.5 Classification of Systems

Systems are classified in several ways according to some of their basic properties. Those most useful to us are discussed below.

- Continuous-time and discrete-time systems
- Linear and nonlinear systems
- Time-invariant and time-varying systems
- Causal and non-causal systems
- Static and dynamic systems

- Invertible and noninvertible systems
- Stable and unstable systems
- Lumped and distributed systems

Continuous-time and Discrete-time Systems

A continuous-time system is one in which continuous-time input signals are transformed into continuous-time output signals. Similarly, discrete-time system is one that transforms a discrete-time input into a discrete-time output. In other words, a system where the input and the output signals are continuous (analog signals) is a continuous system, while a system where the input and output signals are discrete is a discrete system. In general, a continuous-time system is modelled as a linear differential equation, whereas a discrete-time system is modelled as a linear difference equation. These models describe the relationship between the input and output signals. A pictorial representation of the same is shown in Fig. 1.30.

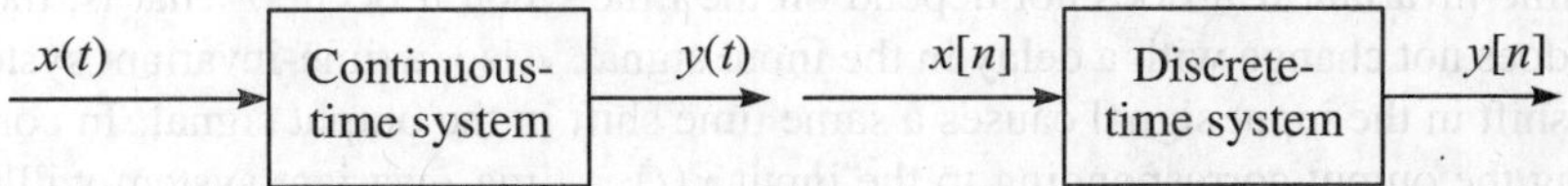

Fig. 1.30 *(a) A continuous-time system and (b) a discrete-time system*

Linear and Nonlinear Systems

A system is said to be linear if it satisfies the scaling property (homogeneity) and principle of superposition (additivity). If a system violates any one of these properties, it is called a nonlinear system.

Additivity: Given any two signals, $y_1(t) = T\{x_1(t)\}$ and $y_2(t) = T\{x_2(t)\}$, the principle of superposition gives

$$y_1(t) + y_2(t) = T\{x_1(t) + x_2(t)\}, \tag{1.77}$$

where the symbol $T\{.\}$ stands for the transformation by the system to which the signals are applied.

Homogeneity: For any signal $x(t)$, the homogeneity property is given by

$$ay(t) = T[ax(t)]. \tag{1.78}$$

Combining the above two properties, the condition for a system to be linear is

$$a_1y_1(t) + a_2y_2(t) = T\{a_1x_1(t) + a_2x_2(t)\}. \tag{1.79}$$

In general, the response of a linear system to a weighted sum of input signals is equal to the same weighted sum of output signals, each output signal being associated with particular input signal acting on the system independently of all other input signals.

If a system is linear, then for the weighted sum of n input signals $x(t) = \sum_{i=1}^{n} a_ix_i(t)$, the principle of superposition and homogeneity gives the system output as $y(t) = \sum_{i=1}^{n} a_iy_i(t)$, where, $y_i(t) = T\{x_i(t)\}$ is the output of the system in response to the input $x_i(t)$ acting alone and $a_i, i = 1, 2, 3, \ldots$ are arbitrary constants. That is,

$$\sum_{i=1}^{n} a_iy_i(t) = T\left\{\sum_{i=1}^{n} a_ix_i(t)\right\}. \tag{1.80}$$

Similarly, for discrete time systems, if the system is linear, then for the weighted sum of n input signals $x[n] = \sum_{i=1}^{n} a_i x_i[n]$, the principle of superposition and homogeneity gives the system output as $y[n] = \sum_{i=1}^{n} a_i y_i[n]$, where $y_i = T\{x_i[n]\}$ is the output of the system in response to the input $x_i[n]$ acting alone. That is,

$$\sum_{i=1}^{n} a_i y_i[n] = T\left\{\sum_{i=1}^{n} a_i x_i[n]\right\}. \tag{1.81}$$

For example, $y = ax + b$ is a linear system, whereas $y = ax^2 + b$ is a nonlinear system; $y(t) = ax(t) + b$ is not linear because a zero input does not lead to a zero output unless $b = 0$.

Time-invariant and Time-varying Systems

A system is time-invariant if it does not depend on the time when it occurs. That is, the shape of the output signal does not change with a delay in the input signal. Also, a time-invariant system is one that where a time shift in the input signal causes a same time shift in the output signal. In continuous time, with $y(t)$ being the output corresponding to the input $x(t)$, a time-invariant system will have $y(t - t_0)$ as the output when $x(t - t_0)$ is the input.

In other words, a continuous-time system described by $y(t) = T\{x(t)\}$ is time invariant if for all time the following holds true.

$$y(t - t_0) = T\{x(t - t_0)\}. \tag{1.82}$$

A discrete system described by $y[n] = T\{x[n]\}$ is time invariant if and only if the following holds true.

$$y[n - n_0] = T\{x[n - n_0]\} \text{ for all time } n_0. \tag{1.83}$$

A system which does not satisfy the above conditions for a continuous-time system or a discrete-time system is called a time-varying system.

For example, a system described by $y(t) = x(t) + 2x(t-1)$ is a time-invariant system since if the input is shifted by time t_0, the output also shifts by the same time as $y(t - t_0) = x(t - t_0) + 2x(t - t_0 - 1)$, whereas the discrete time system $y[n] = nx[n]$ is a time-variant system since if the input $x[n]$ is shifted by n_0, then $y_1[n] = nx[n - n_0]$.

However if the output $y[n]$ is shifted by the same time n_0, then

$$y_2[n - n_0] = (n - n_0)x_1[n - n_0] \neq y_1[n].$$

Thus the system is a time-varying system as the output gain varies with time shift n_0.

Causal and Non-causal Systems

A system is said to be *causal* if and only if its output $y(t)$ at an arbitrary time $t = t_0$ depends on only the input $x(t)$ for $t \leq t_0$. That is, the output of a causal system at the present time depends on only the present and/or past values of the input, not on its future values. Thus, in a causal system, it is not possible to obtain an output before an input is applied to the system. For *non-causal* system the output depends one or more future values of the input signal. Similarly, discrete system is said to be *causal* if and only if its its output $y[n]$ at an arbitrary time $n = n_0$ depends on only the input $x[n]$ for $n \leq n_0$. A causal system is a physically realisable system that cannot have an output response before an input signal is applied. For example, the systems defined by

$$y[n] = x[n] + x[n-1] + x[n-2] \text{ and } y(t) = x(t) + x(t-1)$$

are causal systems whereas

$$y[n] = x[n] + x[n+1] \text{ and } y(t) = x(t+1)$$

are non-causal systems.

Static and Dynamic Systems

A system is said to be static or instantaneous or memoryless system if and only if the output at any time $t = t_0$ depends on only the input at that same time $t = t_0$. Otherwise if the output of the system depends on past or future values of the input signal, the system is called dynamic or said to possess memory.

For example the systems defined by

$$y[n] = x[n] + x^2[n] \quad \text{and} \quad y(t) = x(t) + 2x^2(t)$$

are static systems whereas

$$y[n] = x[n-1] + x[n] + x[n+1] \text{ and } y(t) = x(t+1)$$

are dynamic systems or systems with memory.

Note:

1. All static or memoryless systems are causal.
2. All systems described by either differential or difference equations are dynamic systems.

Stable and Unstable Systems

A system is said to be bounded-input, bounded-output (BIBO) stable if and only if for any bounded input the output is bounded. For stability, both input and output signals should not diverge, i.e., a system must produce a finite output when subjected to a finite input.

The condition for BIBO stability for any bounded input signal $x(t)$ is,

$$|x(t)| \le K_1 < \infty, \quad -\infty < t < \infty; \tag{1.84}$$

the corresponding output $y(t)$ is also bounded, i.e.,

$$|y(t)| \le K_2 < \infty, \quad -\infty < t < \infty. \tag{1.85}$$

Similarly, for a discrete system, for any bounded input signal $x[n]$,

$$|x[n]| \le K_1 < \infty, \quad -\infty < t < \infty, \tag{1.86}$$

the corresponding output $y[n]$ is also bounded, i.e.,

$$|y[n]| \le K_2 < \infty, \quad -\infty < t < \infty. \tag{1.87}$$

For example, a system described by $y(t) = T\{x(t)\} = x(t)e^{at} \sin \omega_c t$ is BIBO stable if $a < 0$ and is unstable if $a > 0$.

Note: The conditions for stability and the methods to find stable systems will be discussed in further chapters.

Invertible and Noninvertible Systems

A system is said to be invertible if the input signal applied the system T can be uniquely recovered from the output signal. The inverse system T^{-1} is such that the cascade interconnection shown in Fig. 1.31 is equivalent to the identity system.

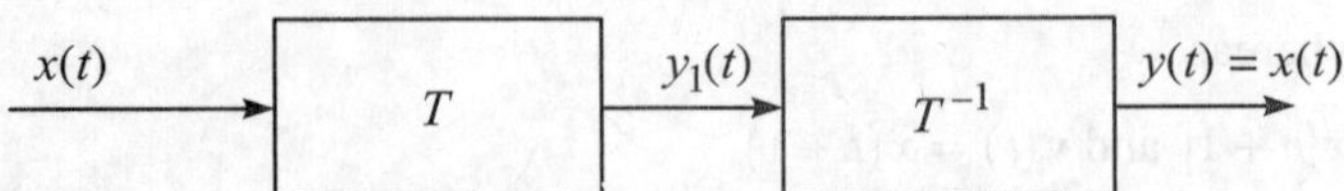

Fig. 1.31 *Cascade for identity system*

The output of the first system is $y_1(t) = T\{x(t)\}$.

The output of the second system is

$$y(t) = T^{-1}\{y_1(t)\} = T^{-1}T\{x(t)\} = x(t), \tag{1.88}$$

that is, $y(t) = x(t)$ is the recovered input signal. Here, $T^{-1}T$ is an identity operator.

Similarly for a discrete system,

$$y[n] = T^{-1}\{y_1[n]\} = T^{-1}T\{x[n]\} = x[n]. \tag{1.89}$$

A non-invertible system cannot determine the input signal from the knowledge of the output signal. For example, for a linear amplifier with gain A, the system $y_1(t) = Ax(t)$ is an invertible system since the inverse system is $y(t) = \frac{1}{A}y_1(t)$, an attenuator. A filter, on the other hand, is a non-invertible system since there is no inverse system.

Lumped and Distributed Systems

A system is said to be lumped system if the components within the system are placed at one place. The total length covered by the components should be very small compared to the wavelength of the applied signal. The signals are functions of only time/frequency. Lumped systems obey Ohm's and Kirchhoff's laws. They can be expressed with ordinary differential equations. Devices like phones, TVs, computers, motors, or any packed systems are lumped systems.

Distributed systems are the systems in which the components are distributed over long distances and the total length covered by the components is very large compared to the wavelength of the applied signal. Distributed systems do not obey Ohm's and Kirchhoff's laws. The signals are functions of both time/frequency and space and can be expressed with partial differential equations. Systems like telephone networks, computer networks, transmission lines, optical fiber systems, waveguides, etc., are examples of distributed systems.

Example 1.9

Consider the system output $y(t) = T\{x(t)\} = x(t)\cos\omega_c t$. Determine whether it is memoryless, causal, linear, time-invariant or stable.

Solution Given the system output $y(t) = T\{x(t)\} = x(t)\cos\omega_c t$.

(a) Since the output depends only on present value of input, the system is memoryless.

(b) Since the output does not depends on the future value of input, the system is causal and static system.

(c) To check for linearity, let, $x(t) = a_1x_1(t) + a_2x_2(t)$. The corresponding outputs are $y_1(t) = x_1(t)\cos\omega_c t$, $y_2(t) = x_2(t)\cos\omega_c t$. Then,

$$\begin{aligned} y(t) &= T\{x(t)\} = [a_1x_1(t) + a_2x_2(t)]\cos\omega_c t \\ &= a_1x_1(t)\cos\omega_c t + a_2x_2(t)\cos\omega_c t \\ &= a_1y_1(t) + a_2y_2(t). \end{aligned}$$

As the linearity theorem holds, the system is linear.

(d) To check for time invariance, let, $y_1(t) = T\{x_1(t)\} = T\{x(t-t_0)\}$.
Then, $y_1(t) = x(t-t_0)\cos\omega_c t$.
But, $y(t-t_0) = x(t-t_0)\cos\omega_c(t-t_0) \neq y_1(t)$. Hence the system is time-variant.

(e) To check for stability,
since $|\cos\omega_c t| \leq 1, |y(t)| = |x(t)\cos\omega_c t| \leq |x(t)|\,|\cos\omega_c t| \leq |x(t)|$.
Thus, if the input is bounded, the output is bounded and the system is BIBO stable.

Example 1.10

Determine whether the following signals are causal or non-causal.

(a) $x(t) = 5\cos(4t) + 4\sin(2t)u(t)$
(b) $x(t) = \cos(2t)\,u(t)$
(c) $x(t) = u(-t+1) + u(t-1)$
(d) $x[n] = u[-n] + u[n-1]$

Solution

(a) Given $x(t) = 5\cos(4t) + 4\sin(2t)u(t)$. The signal is non causal since it extends from $-\infty < t < \infty$ and
$x(t) \neq 0$ for $t < 0$.
(b) Given $x(t) = \cos(2t)\,u(t)$. The signal is causal since $x(t) = 0$ for $t < 0$.
(c) $x(t) = u(-t+1) + u(t-1)$. The signal is non causal since it extends from $-\infty < t < \infty$ and $x(t) \neq 0$ for $t < 0$.
(d) $x[n] = u[-n] + u[n-1]$. The sequence is non causal since it extended in the negative axis and $x[n] \neq 0$ for $n < 0$.

Additional Problems

Example 1.11

Find out the even and odd parts of the following signals:

(i) $x(t) = \cos^2 t + \sin^3 t + \cos t \sin t$
(ii) $x[n] = [1,2,3,4,5]$ for $n = -2,-1,0,1,2$

Solution

(i) Given the signal $x(t) = \cos^2 t + \sin^3 t + \cos t \sin t$.

We know that, the even part is $x_e(t) = \frac{1}{2}[x(t) + x(-t)]$

$$= \frac{1}{2}\left[\cos^2 t + \sin^3 t + \cos t \sin t + \cos^2(-t) + \sin^3(-t) + \cos(-t)\sin(-t)\right]$$

$$= \frac{1}{2}\left[\cos^2 t + \sin^3 t + \cos t \sin t + \cos^2 t - \sin^3 t - \cos t \sin t\right] = \cos^2 t$$

and the odd part is $x_0(t) = x(t) - x_e(t) = \sin^3 t + \cos t \sin t$.

(ii) Given the signal $x[n] = [1,2,3,4,5]$ for $n = -2,-1,0,1,2$.

We know that $x_e[n] = \frac{1}{2}[x[n]+x[-n]]$.

$$x_e[0] = \frac{1}{2}[x[0]+x[-0]] = \frac{1}{2}[3+3] = 3; \quad x_e[1] = \frac{1}{2}[x[1]+x[-1]] = \frac{1}{2}[4+2] = 3;$$

$$x_e[2] = \frac{1}{2}[x[2]+x[-2]] = \frac{1}{2}[5+1] = 3.$$

And also, $x_e[-n] = x_e[n]$. Then

$$x_e[n] = [3,3,3,3,3] \text{ for } n = -2,-1,0,1,2.$$

The odd part is $x_o[n] = \frac{1}{2}[x[n]-x[-n]]$.

$$x_o[0] = \frac{1}{2}[x[0]-x[-0]] = \frac{1}{2}[3+3] = 0; \quad x_o[1] = \frac{1}{2}[x[1]-x[-1]] = \frac{1}{2}[4-2] = 1;$$

$$x_e[2] = \frac{1}{2}[x[2]-x[-2]] = \frac{1}{2}[5+1] = 2.$$

And also, $x_o[-n] = -x_o[n]$. Then

$$x_o[n] = [-2,-1,0,1,2] \text{ for } n = -2,-1,0,1,2.$$

Example 1.12

Determine whether the following signals are energy signals, power signals or neither of these.

(i) $x(t) = e^{-at}u(t),\ a > 0$;

(ii) $x(t) = A\cos(\omega_0 t + \theta)$;

(iii) $x(t) = tu(t)$.

Solution

(i) Given $x(t) = e^{-at}u(t),\ a > 0$. This is a non-periodic signal since $x(t)$ decreases to zero as time t tends to infinity.

The signal energy is $E = \int_{-\infty}^{\infty} |x(t)|^2\, dt = \int_0^{\infty} e^{-2at}\, dt = \frac{-1}{2a}\left[e^{-2at}\right]_0^{\infty} = \frac{1}{2a} < \infty.$

Thus $x(t)$ is an energy signal.

(ii) Given $x(t) = A\cos(\omega_0 t + \theta)$, a periodic signal with period $T_0 = \frac{2\pi}{\omega_o}$ with infinite duration. So, the signal energy is infinity.

The signal power is $P = \frac{1}{T_0}\int_0^{\infty} |x(t)|^2\, dt = \frac{\omega_0}{2\pi}\int_0^{T_0} A^2\cos^2(\omega_0 t + \theta)dt$

$$= \frac{\omega_0}{2\pi}\int_0^{T_0} \frac{A^2}{2}[1+\cos(2\omega_0 t + 2\theta)]dt = \frac{\omega_0}{2\pi}\frac{A^2}{2}\frac{2\pi}{\omega_0} = \frac{A^2}{2}.$$

If A is finite, then $P < \infty$. Thus $x(t)$ is a power signal.

(iii) Given the signal $x(t) = tu(t)$, an aperiodic signal. As t tends to infinity,

$$E = \lim_{T\to\infty}\int_{-T/2}^{T/2}|x(t)|^2dt = \lim_{T\to\infty}\int_{0}^{T/2}t^2dt = \lim_{T\to\infty}\frac{(T/2)^3}{3} = \infty.$$

$$P = \lim_{T\to\infty}\frac{1}{T}\int_{-T/2}^{T/2}|x(t)|^2dt = \lim_{T\to\infty}\frac{1}{T}\int_{0}^{T/2}t^2dt = \lim_{T\to\infty}\frac{1}{T}\frac{(T/2)^3}{3} = \infty.$$

Since both energy and power are infinite, x(t) is neither power signal nor energy signal.

Example 1.13

Consider a signal $x(t) = 2\sin\left(\frac{2}{3}t\right) + 4\cos\left(\frac{1}{2}t\right) + 4\cos\left(\frac{1}{3}t - \frac{\pi}{5}\right)$. If it is periodic, find out the fundamental frequency and signal power.

Solution The individual frequency components of the signal $x(t) = 2\sin\left(\frac{2}{3}t\right) + 4\cos\left(\frac{1}{2}t\right) + 4\cos\left(\frac{1}{3}t - \frac{\pi}{5}\right)$ are $\omega_1 = 2/3$ rad/s, $\omega_2 = 1/2$ rad/s, and $\omega_3 = 1/3$ rad/s. The time periods (in seconds) are $T_1 = \frac{2\pi}{\omega_1} = \frac{2\pi}{2/3} = 3\pi$; $T_2 = \frac{2\pi}{\omega_2} = \frac{2\pi}{1/2} = 4\pi$ and $T_3 = \frac{2\pi}{\omega_3} = \frac{2\pi}{1/3} = 6\pi$.

The ratio of time periods is a rational number. The signal is periodic.

The fundamental frequency ω_0 is GCD $\left(\frac{2}{3}, \frac{1}{2}, \frac{1}{3}\right) = \frac{1}{6}$ rad/seconds.

Signal power $= \frac{1}{2}(2^2 + 4^2 + 4^2) = 36$ watts.

Example 1.14

If the signal is periodic, determine its fundamental period.

(a) $x(t) = 3\cos\left(4t + \frac{\pi}{3}\right)$

(b) $x(t) = e^{j(\pi t - 1)}$

(c) $x(t) = \left(\cos\left(2t - \frac{\pi}{3}\right)\right)^2$

(d) $x(t) = \sum_{n=-\infty}^{\infty} e^{(2t-n)}$

(e) $x(t) = e^{-\alpha|t|}$

Solution

(a) Given $x(t) = 3\cos\left(4t + \frac{\pi}{3}\right)$, frequency is $\omega = 4$ rad/s. The signal is periodic.

Period is $= 2\pi/4 = \pi/2$ seconds.

(b) Given $x(t) = e^{-j}e^{j\pi t}$, frequency is $\omega = \pi$ rad/s. The signal is periodic.
Period = $2\pi/\pi$ = 2 seconds.

(c) Given $x(t) = \cos^2\left(2t - \frac{\pi}{3}\right) = \left[\frac{1+\cos\left(4t - \frac{2\pi}{3}\right)}{2}\right]$, frequency is $\omega = 4$ rad/s. The signal is periodic with period $2\pi/4 = \pi/2$ seconds.

(d) Given $x(t) = \sum_{n=-\infty}^{\infty} e^{(2t-n)}$. Since it is a hyperbolic function, it is not periodic.

(e) Given $x(t) = e^{-\alpha|t|}$. The signal is an exponential signal and reaches zero as time approaches infinity. Therefore it is a non-periodic signal.

Example 1.15

Determine whether or not each of the following discrete-time signals are periodic. If the signal is periodic, determine its fundamental period.

(a) $x[n] = \sin\left(\frac{2\pi}{7}n + 1\right)$

(b) $x[n] = \cos\left(\frac{\pi}{8}n^2\right)$

(c) $x[n] = \cos\left(\frac{\pi}{2}n\right)\cos\left(\frac{\pi}{4}n\right)$

(d) $x[n] = \cos\left(\frac{\pi}{2}n\right) + \sin\left(\frac{\pi}{8}n\right) - 2\cos\left(\frac{\pi}{2}n + \frac{\pi}{6}\right)$

Solution

(a) Given $x[n] = \sin\left(\frac{2\pi}{7}n + 1\right)$. Fundamental frequency is $\omega_0 = 2\pi/7$ rad/s. The period is $N = 2\pi k/\omega_0 = \frac{2\pi k}{2\pi/7} = 7k$. For $k = 1$, the minimum period is $N = 7$. The signal is periodic.

(b) Given $x[n] = \cos\left(\frac{\pi}{8}n^2\right)$.

The fundamental frequency is $\omega_0 = \pi/8$ rad/s; fundamental period is $N = 2\pi k/\omega_0 = \frac{2\pi k}{\pi/8} = 16k$. For $k = 1$, the minimum period is $N = 16$. The signal is periodic.

(c) Given $x[n] = \cos\left(\frac{\pi}{2}n\right)\cos\left(\frac{\pi}{4}n\right) = \frac{1}{2}\left[\cos\left(\frac{3\pi}{4}n\right) + \cos\left(\frac{\pi}{4}n\right)\right]$.

The signal frequencies are: $\omega_1 = 3\pi/4$ and; $\omega_2 = \pi/4$; period $N_1 = 2\pi k/\omega_1 = \frac{2\pi k}{3\pi/4} = \frac{8k}{3}$. For $k = 3$, the minimum period is $N_1 = 8$.

Period $N_2 = 2\pi k/\omega_2 = \frac{2\pi k}{\pi/4} = 8$. For $k = 1$, the minimum period is $N_2 = 8$.

The ratio $\frac{N_1}{N_2} = \frac{8}{8} = 1$ is a rational number.

Thus, the fundamental period $N = 8$, and the signal is periodic.

(d) Given $x[n] = \cos\left(\frac{\pi}{2}n\right) + \sin\left(\frac{\pi}{8}n\right) - 2\cos\left(\frac{\pi}{2}n + \frac{\pi}{6}\right)$.

The signal frequencies are $\omega_1 = \pi/2$ and $\omega_2 = \pi/8$; period $N_1 = 2\pi k/\omega_1 = \frac{2\pi k}{\pi/2} = 4k$. For $k = 1$, the minimum period is $N_1 = 4$.

Period $N_2 = 2\pi k/\omega_2 = \frac{2\pi k}{\pi/8} = 16k$. For $k = 1$, minimum period is $N_2 = 16$.

The ratio $\frac{N_1}{N_2} = \frac{4}{16} = \frac{1}{4}$ is a rational number.

The time period of the signal is $N = 4N_1 = N_2$ or the fundamental period $N = \text{LCM}[N_1, N_2] = 16$. The signal is periodic.

Example 1.16

Determine whether or not each of the following discrete-time signals is periodic. If the signal is periodic, determine its fundamental period.

(a) $x[n] = \cos\left(\frac{n}{6}\right)\cos\left(\frac{n\pi}{6}\right)$;

(b) $x[n] = \cos(0.5n + 30)$;

(c) $x[n] = e^{jn\pi/2}$;

(d) $x[n] = 1 + e^{j2n\pi/3} - e^{j4n\pi/5}$.

Solution

(a) Given $x[n] = \cos\left(\frac{n}{6}\right)\cos\left(\frac{n\pi}{6}\right) = \frac{1}{2}\left[\cos\left(\frac{\pi+1}{6}n\right) + \cos\left(\frac{\pi-1}{6}n\right)\right]$.

Signal frequencies are $\omega_1 = (\pi+1)/6$ rad/s and $\omega_2 = (\pi-1)/6$ rad/s.

The period $N_1 = 2\pi k/\omega_1 = \frac{2\pi k}{(\pi+1)/6} = \frac{8\pi k}{(\pi+1)}$ is not a rational number for all k.

The period $N_2 = 2\pi k/\omega_2 = \frac{2\pi k}{(\pi-1)/6} = \frac{8\pi k}{(\pi-1)}$ is not a rational number for all k.

Therefore, the signal is non periodic.

(b) Given $x[n] = \cos(0.5n + 30)$.

The fundamental frequency is $\omega_0 = 0.5$ rad/s; the period $N = 2\pi k/\omega_0 = \frac{2\pi k}{0.5} = 4\pi k$ is not a rational number for all k. Therefore the signal is non periodic.

(c) Given $x[n] = e^{jn\pi/2}$

The fundamental frequency is $\omega_0 = \pi/2$ rad/s; the period is $N = 2\pi k/\omega_0 = \frac{2\pi k}{\pi/2} = 4k$. For $k = 1$, the minimum period is $N = 4$. The signal is periodic.

(d) Given $x[n] = 1 + e^{j2n\pi/3} - e^{j4n\pi/5}$

The signal frequencies are $\omega_1 = 2\pi/3$ rad/s and $\omega_2 = 4\pi/5$ rad/s; the period $N_1 = 2\pi k/\omega_1 = \frac{2\pi k}{2\pi/3} = 3k$. For $k = 1$, the minimum period is $N_1 = 3$.

The period $N_2 = 2\pi k/\omega_2 = \dfrac{2\pi k}{4\pi/5} = \dfrac{5k}{2}$. For $k = 2$, minimum period is $N_2 = 5$.

The fundamental period $N = \text{LCM}[N_1, N_2] = 15$. The signal is periodic.

Example 1.17

Find the energies of the following signals. Comment on the effect on energy due to sign change, time shifting or doubling of the signal. What is the effect on the energy if the signal is multiplied by k?

(a) $f(t) = \sin t \quad 0 \le t \le 2\pi$

(b) $f(t) = -\sin t \quad 0 \le t \le 2\pi$

(c) $f(t) = \sin t \quad 2\pi \le t \le 4\pi$

(d) $f(t) = 2\sin t \quad 0 \le t \le 2\pi$

Solution

(a) Given $f(t) = \sin t, \quad 0 \le t \le 2\pi$, the energy is

$$E = \int_0^{2\pi} |f(t)|^2\, dt = \int_0^{2\pi} |\sin t|^2\, dt$$

$$= \int_0^{2\pi} \sin^2 t dt = \frac{1}{2}\int_0^{2\pi} dt - \frac{1}{2}\int_0^{2\pi} \cos 2t dt = \pi \text{ joules.}$$

(b) Given $f(t) = -\sin t, \quad 0 \le t \le 2\pi$, the energy is

$$E = \int_0^{2\pi} |f(t)|^2\, dt = \int_0^{2\pi} |-\sin t|^2\, dt$$

$$= \int_0^{2\pi} \sin^2 t dt = \frac{1}{2}\int_0^{2\pi} dt - \frac{1}{2}\int_0^{2\pi} \cos 2t dt = \pi \text{ joules.}$$

Comment: If the sign changes in the signal, there is no change in energy.

(c) Given $f(t) = \sin t, \quad 2\pi \le t \le 4\pi$, the energy is

$$E = \int_{2\pi}^{4\pi} |f(t)|^2\, dt = \int_{2\pi}^{4\pi} |\sin t|^2\, dt = \frac{1}{2}\int_{2\pi}^{4\pi} dt - \frac{1}{2}\int_{2\pi}^{4\pi} \cos 2t dt = \pi \text{ joules.}$$

Comment: If the signal shifts in time, there is no change in the energy.

(d) Given $f(t) = 2\sin t, \quad 0 \le t \le 2\pi$, the energy is

$$E = \int_0^{2\pi} |f(t)|^2\, dt = \int_0^{2\pi} |2\sin t|^2\, dt$$

$$= 4\left[\frac{1}{2}\int_0^{2\pi} dt - \frac{1}{2}\int_0^{2\pi} \cos 2t dt\right] = 4[\pi + 0] = 4\pi \text{ joules.}$$

Comment: If the signal is multiplied K times, the resultant energy of the signal will be increased K^2 times.

Example 1.18

Determine the power and RMS value for each of the following signals:

(a) $10\cos\left(100t+\dfrac{\pi}{3}\right)$

(b) $10\cos\left(100t+\dfrac{\pi}{3}\right)+16\sin\left(150t+\dfrac{\pi}{5}\right)$

(c) $(10+2\sin 3t)\cos 10t$

Solution

(a) Given $10\cos\left(100t+\dfrac{\pi}{3}\right)$, we know that the power of a sinusoidal signal having amplitude A is $P=A^2/2$ watts.

Therefore, the power $P=\dfrac{A^2}{2}=\dfrac{10^2}{2}$ = 50 watts, and the rms value is $v_{rms}=\sqrt{P}=7.07$ volts.

(b) Given $10\cos\left(100t+\dfrac{\pi}{3}\right)+16\sin\left(150t+\dfrac{\pi}{5}\right)$.

Since, power of a sum of sinusoids is equal to the sum of the powers of the sinusoid, in this case,

$$P=\frac{10^2}{2}+\frac{16^2}{2}=178 \text{ watts.}$$

The rms value is $v_{\text{rms}}=\sqrt{P}=13.34$ volts.

(c) Given $(10+2\sin 3t)\cos 10t$

$$\begin{aligned}(10+2\sin 3t)\cos 10t &= 10\cos 10t+2\sin 3t\cos 10t\\ &= 10\cos 10t+\sin 13t-\sin 7t\end{aligned}$$

Therefore, in this case $P=\dfrac{10^2}{2}+\dfrac{1^2}{2}+\dfrac{1^2}{2}=51$ watts, and the rms value is

$v_{\text{rms}}=\sqrt{P}$= 7.1414 volts.

Example 1.19

Determine the power for each of the following signals:

(a) $10\cos 5t\cos 10t$;

(b) $10\sin 5t\sin 10t$;

(c) $e^{j\alpha t}\cos\omega_o t$.

Solution

(a) Given $10\cos 5t\cos 10t$= $5(\cos 5t+\cos 15t)$.

$$P=\frac{5^2}{2}+\frac{5^2}{2}=25 \text{ watts.}$$

(b) Given $10\sin 5t\sin 10t$= $5(\cos 5t-\cos 15t)$.

$$P=\frac{5^2}{2}+\frac{5^2}{2}=25 \text{ watts.}$$

(c) $e^{j\alpha t}\cos\omega_0 t = \frac{1}{2}\left[e^{j(\alpha+\omega_0)t} + e^{j(\alpha-\omega_0)t}\right]$

$$P = \frac{1}{4} + \frac{1}{4} = \frac{1}{2} \text{ watts.}$$

Example 1.19

Sketch the following signals:

(a) $f(t) = r(t) - r(t-1)$,
(b) $f(t) = r(t-1) - r(t-2) - u(t-2)$.

Solution

(a) The given signal $f(t) = r(t) - r(t-1)$ is a truncated ramp signal. It is a subtraction of two ramp signals at different time instants as shown in Fig. 1.32(a).

(b) $f(t) = r(t-1) - r(t-2) - u(t-2)$ is a triangular pulse. It is a subtraction of a truncated ramp signal and a unit step function as shown in Fig. 1.32(b).

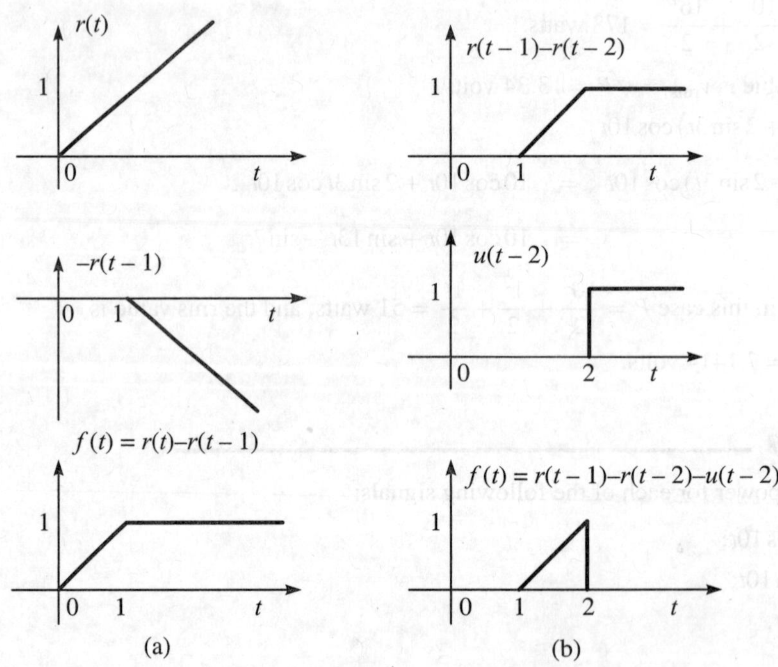

Fig. 1.32 *Sketches of the signals*

Example 1.20

Sketch the following rectangular signals:

$$x_1(t) = \begin{cases} 1 & |t| \le 0.5 \\ 0 & \text{otherwise} \end{cases} \quad \text{and} \quad x_2(t) = \begin{cases} 1 & |t| \le 1 \\ 0 & \text{otherwise} \end{cases}$$

and the sequences:

$$x_1[n] = \begin{cases} 1 & |n| \le 2 \\ 0 & \text{otherwise} \end{cases} \quad \text{and} \quad x_2[n] = \begin{cases} 1 & |n| \le 4 \\ 0 & \text{otherwise.} \end{cases}$$

Also, draw the waveform for

$$x(t) = x_1(t) + x_2(t) \quad \text{and} \quad x[n] = x_1 + x_2[n].$$

Solution

Given two signals $x_1(t) = \begin{cases} 1 & |t| \le 0.5 \\ 0 & \text{otherwise,} \end{cases}$ $x_2(t) = \begin{cases} 1 & |t| \le 1 \\ 0 & \text{otherwise,} \end{cases}$ and the sequences $x_1[n] = \begin{cases} 1 & |n| \le 2 \\ 0 & \text{otherwise,} \end{cases}$ $x_2[n] = \begin{cases} 1 & |n| \le 4 \\ 0 & \text{otherwise} \end{cases}$

The waveforms for $x(t) = x_1(t) + x_2(t)$ and $x[x] = x_1[n] + x_2[n]$ are shown in Fig. 1.33.

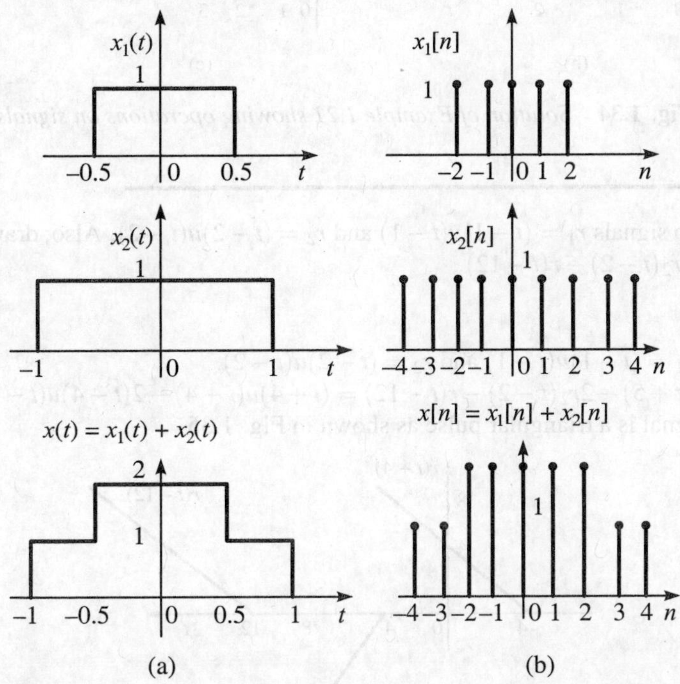

Fig. 1.33 *Addition process for (a) two rectangular signals and (b) two discrete sequences*

Example 1.21

Sketch the two ramp signals $r_1(t) = (t-1)\,u(t-1)$ and $r_2(t) = (t-2)\,u(t-2)$. Also, draw the waveforms for the following operations: (i) $f(t) = r_1(t) + r_2(t)$, (ii) $f(t) = 3r_1(t) - 2r_2(t)$.

Solution

Given two ramp signals $r_1(t) = (t-1)\,u(t-1)$ and $r_2(t) = (t-2)\,u(t-2)$ shown in Fig. 1.34(a). The signals corresponding to operations (i) and (ii) are shown in Fig. 1.34(b) and (c).

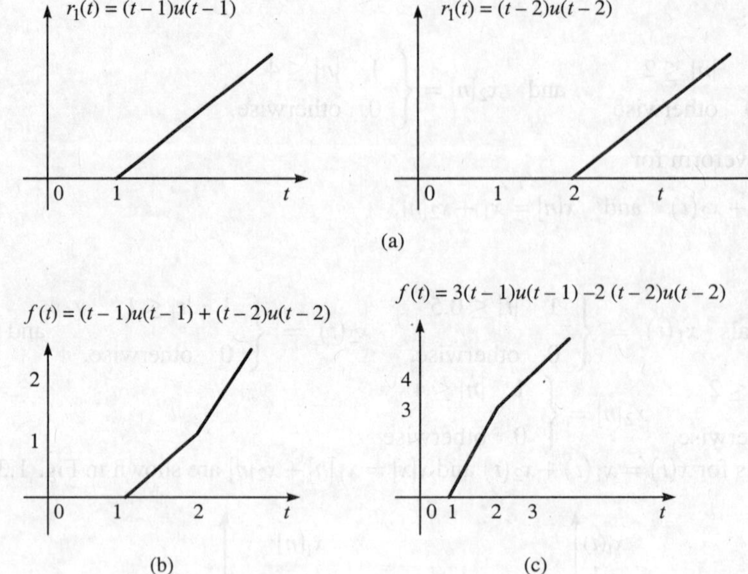

Fig. 1.34 *Solution of Example 1.21 showing operations on signals*

Example 1.22

Sketch the two ramp signals $r_1 = (t-1)u(t-1)$ and $r_2 = (t-2)u(t-2)$. Also, draw the waveform for $f(t) = r_1(t+5) - 2r_2(t-2) - r(t-12)$.

Solution

Given two signals $r_1 = (t-1)u(t-1)$ and $r_2 = (t-2)u(t-2)$.

Then $f(t) = r_1(t+5) - 2r_2(t-2) - r(t-12) = (t+4)u(t+4) - 2(t-4)u(t-4) + r(t-12)$.

The resultant signal is a triangular pulse as shown in Fig. 1.35

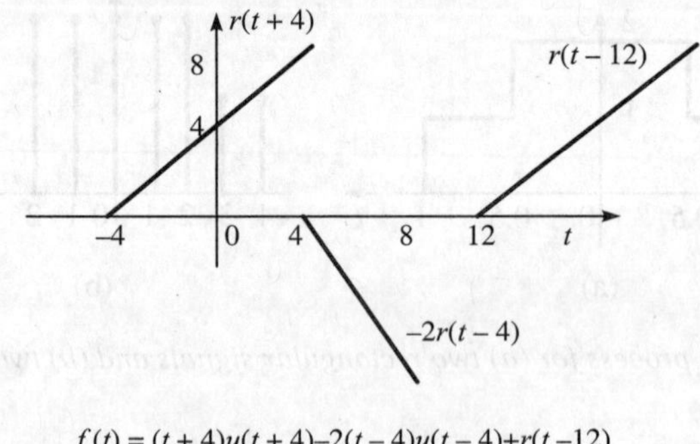

$f(t) = (t+4)u(t+4) - 2(t-4)u(t-4) + r(t-12)$

8

4

–4 0 4 8 12 t

Fig. 1.35 *The formation of the triangular pulse of Example 1.22*

Example 1.23

Sketch the following signals: (a) $f(t) = r(t+2) - r(t+1) - r(t-1) + r(t-2)$; (b) $f(t) = r(t) - r(t-1) - u(t-1)$; (c) $f(t) = r(-t)u(t+1)$; (d) $f(t) = u(t+1)u(-t+1)$.

Solution:

(a) The given signal $f(t) = r(t+2) - r(t+1) - r(t-1) + r(t-2)$ results in a trapezoidal pulse as shown in Fig. 1.36(a).

(b) $f(t) = r(t) - r(t-1) - u(t-1)$ results in a triangular pulse as shown in Fig. 1.36(b).

(c) $f(t) = r(-t)u(t+1)$ results in a triangular pulse as shown in Fig. 1.36(c).

(d) $f(t) = u(t+1)u(-t+1)$ results in a rectangular pulse as shown in Fig. 1.36(d).

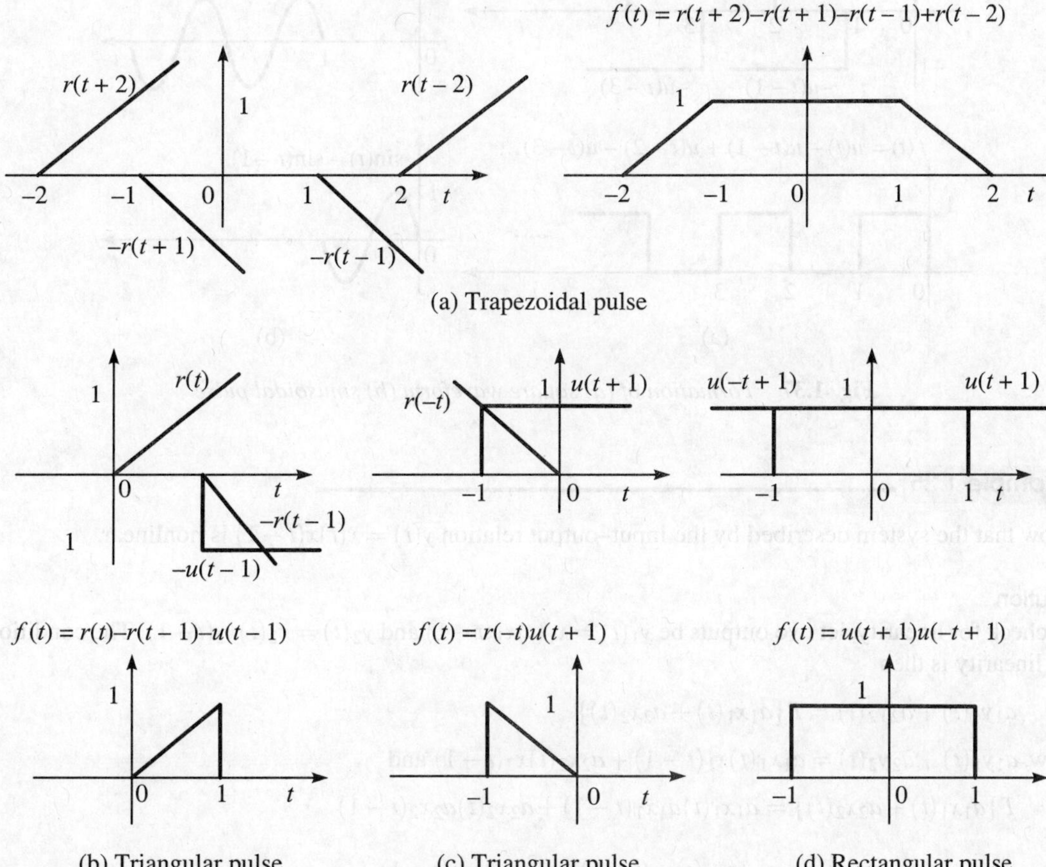

Fig. 1.36 *The formation of various pulses of Example 1.23*

Example 1.24

Sketch the following signals:

(a) $f(t) = u(t) - 2u(t-1) + 2u(t-2) - 2u(t-3) + \ldots$, and (b) $f(t) = \sin(t) - \sin(t-1)$

Solution

(a) The given signal $f(t) = u(t) - u(t-1) + u(t-2) - u(t-3) + \dots$ results a square waveform as shown in Fig. 1.37(a).

(b) $f(t) = \sin(t) - \sin(t-1)$ results a sinusoidal pulse as shown in Fig. 1.37(b).

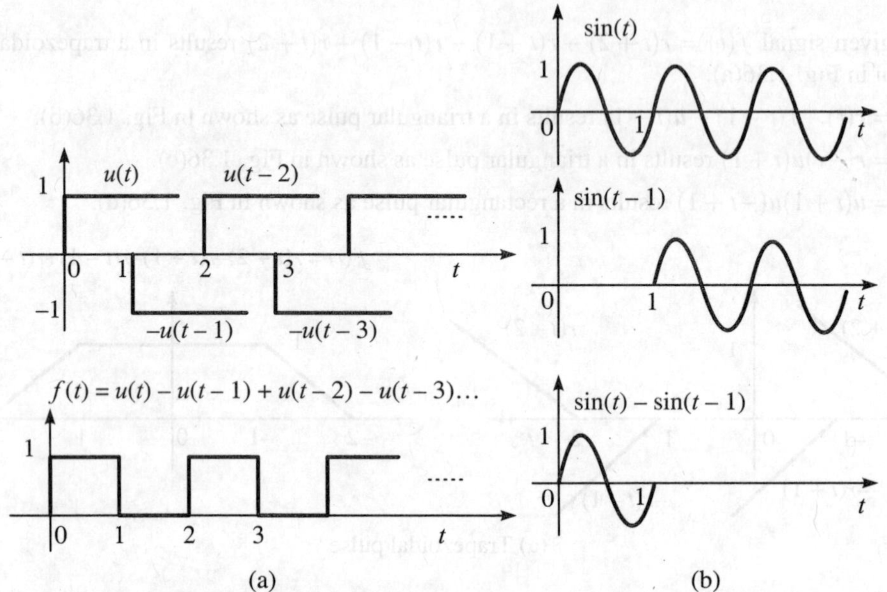

Fig. 1.37 *Formation of (a) square waveform (b) sinusoidal pulse*

Example 1.25

Show that the system described by the input–output relation $y(t) = x(t)x(t-1)$ is nonlinear.

Solution

To check for linearity, let two outputs be $y_1(t) = x_1(t)x_1(t-1)$ and $y_2(t) = x_2(t)x_2(t-1)$. The condition for linearity is then

$$a_1y_1(t) + a_2y_2(t) = T\{a_1x_1(t) + a_2x_2(t)\}.$$

Now, $a_1y_1(t) + a_2y_2(t) = a_1x_1(t)x_1(t-1) + a_2x_2(t)x_2(t-1)$ and

$$T\{a_1x_1(t) + a_2x_2(t)\} = a_1x_1(t)a_1x_1(t-1) + a_2x_2(t)a_2x_2(t-1)$$

$$= a_1^2x_1(t)x_1(t-1) + a_1^2x_2(t)x_2(t-1).$$

Therefore, $T\{a_1x_1(t) + a_2x_2(t)\} \neq a_1y_1(t) + a_2y_2(t)$.

The system does not satisfy linearity property. Hence the system is nonlinear.

Example 1.26

A system is characterised by the equation $y(t) = x(t-2) + x(2-t)$. Determine whether the system is memoryless, time-invariant, linear, causal and stable or not stable.

Solution Given the system $y(t) = x(t-2) + x(2-t)$.

(a) To check for linearity, let $x(t) = a_1x_1(t) + a_2x_2(t)$.

Then,

$$\begin{aligned} y(t) = T\{x(t)\} &= a_1x_1(t-2) + a_1x_1(2-t) + a_2x_2(t-2) + a_2x_2(2-t) \\ &= a_1[x_1(t-2) + x_1(2-t)] + a_2[x_2(t-2) + x_2(2-t)] \\ &= a_1y_1(t) + a_2y_2(t). \end{aligned}$$

As the superposition theorem holds, the system is linear.

(b) Since the output depends on the previous value of the input, the system has memory.

(c) To check for the time invariance, the condition is $y(t-t_0) = T\{x(t-t_0)\}$.

Now, $y(t-t_0) = x(t-2-t_0) + x(2-t-t_0)$ and $T\{x(t-t_0)\} = x(t-t_0-2) + x(2-t+t_0)$.

Then, $y(t-t_0) \neq T\{x(t-t_0)\}$.

Hence, the system is time-invariant.

(d) Since $|y(t)| \leq |x(t-2)| + |x(2-t)|$, if the input is bounded, the output is bounded and the system is BIBO stable.

(e) Since the output depends on the future value of input, the system is non-causal.

Example 1.27

Determine whether the following systems are linear, memoryless, time-invariant, causal and stable or not.

(i) $y[n] = 4x[n] + 5$

(ii) $y(t) = x^2(t-2) + x(t+2)$

Solution

(i) Given the system $y[n] = 4x[n] + 5$.

(a) To check for linearity, let the corresponding outputs be $y_1[n] = 4x_1[n] + 5, y_2[n] = 4x_2[n] + 5$. The condition for linearity is $a_1y_1[n] + a_2y_2[n] = T\{a_1x_1[n] + a_2x_2[n]\}$. Now

$$a_1y_1[n] + a_2y_2[n] = a_14x_1[n] + a_15 + a_24x_1[n] + a_25 \quad \text{and}$$

$$\begin{aligned} T\{a_1x_1[n] + a_2x_2[n]\} &= 4a_1x_1[n] + 5 + 4a_2x_1[n] + 5 \\ &= 4(a_1x_1[n] + a_2x_1[n]) + 10. \end{aligned}$$

Therefore $T\{a_1x_1(t) + a_2x_2(t)\} \neq a_1y_1(t) + a_2y_2(t)$.

The system does not satisfy linearity property. Hence the system is nonlinear.

(b) Since the output depends only on the present value of input, the system is memoryless.

(c) To check for the time invariance, the condition is $y[n-n_0] = T\{x[n-n_0]\}$.

Now, $y[n-n_0] = 4x[n-n_0] + 5$

and $T\{x[n-n_0]\} = 4x[n-n_0] + 5$.

Then $y[n-n_0] = T\{x[n-n_0]\}$. Hence the system is time invariant.

(d) Since the output does not depends on the future value of input, the system is causal.

(e) Since, $|y[n]| \leq 4|x[n]| + 5$, if the input is bounded, the output is bounded and the system is BIBO stable.

(ii) Given the system $y(t) = x^2(t-2) + x(t+2)$.

(a) To check for linearity, let the corresponding outputs be $y_1(t) = x_1^2(t-2) + x_1(t+2)$, $y_2(t) = x_2^2(t-2) + x_2(t+2)$.
The condition for linearity is $a_1y_1(t) + a_2y_2(t) = T\{a_1x_1(t) + a_2x_2(t)\}$. Now,

$$a_1y_1(t) + a_2y_2(t) = a_1x_1^2(t-2) + a_1x_1(t+2) + a_2x_2^2(t-2) + a_2x_2(t+2) \quad \text{and}$$

$$T\{a_1x_1(t) + a_2x_2(t)\} = a_1^2x_1^2(t-2) + a_1x_1(t+2) + a_2^2x_2^2(t-2) + a_2x_2(t+2)$$

$$= a_1^2x_1(t)x_1(t-1) + a_1^2x_2(t)x_2(t-1).$$

Therefore $T\{a_1x_1(t) + a_2x_2(t)\} \neq a_1y_1(t) + a_2y_2(t)$.
The system does not satisfy linearity property. Hence the system is nonlinear.

(b) Since the output depends on previous and future values of input, the system has memory.

(c) To check for the time invariance, the condition is $y(t-t_0) = T\{x(t-t_0)\}$,
Now $y(t-t_0) = x^2(t-2-t_0) + x(t+2-t_0)$
and $T\{x(t-t_0)\} = x^2(t-t_0-2) + x(t-t_0+2)$.
Then, $y(t-t_0) = T\{x(t-t_0)\}$. Hence the system is time invariant.

(d) Since the output depends on the future value of input $x(t+2)$ and $y(t) \neq 0$ for $t < 0$, the system is non-causal.

(e) Since $|y(t)| \leq |x^2(t-2)| + |x(t+2)|$, if the input is bounded, the output is bounded and the system is BIBO stable.

Example 1.28

A system is characterised by the equation, $y(t) = x(t/3)$. Determine whether the system is time-invariant, memoryless, causal and stable or not.

Solution Given the system $y(t) = x(t/3)$.

(a) To check for the time invariance, the condition is $y(t-t_0) = T\{x(t-t_0)\}$.

Now, $y(t-t_0) = x\left(\frac{t}{3} - t_0\right)$ and $T\{x(t-t_0)\} = x\left(\frac{t-t_0}{3}\right) = x\left(\frac{t}{3} - \frac{t_0}{3}\right)$.

Therefore, $y(t-t_0) \neq T\{x(t-t_0)\}$ and the system is time variant.

(b) Since the output depends only on the present value of input, the system is memoryless.

(c) Since the output does not depend on the future value of input, the system is causal.

(d) Since, $|y(t)| \leq |x(t/3)|$, if the input is bounded, the output is bounded and the system is BIBO stable.

Example 1.29

Show that the system described by the differential equation $\frac{dy(t)}{dt} + ty(t) = x(t)$ is linear.

Solution Given the system described by $\frac{dy(t)}{dt} + ty(t) = x(t)$.

(a) Let $x(t) = a_1x_1(t) + a_2x_2(t)$.

Then, $\frac{dy_1(t)}{dt} + ty_1(t) = x_1(t)$ and $\frac{dy_2(t)}{dt} + ty_2(t) = x_2(t)$.

$$T\{x(t)\} = a_1\left(\frac{dy_1(t)}{dt} + ty_1(t)\right) + a_2\left(\frac{dy_2(t)}{dt} + ty_2(t)\right)$$

$$= \frac{da_1y_1(t)}{dt} + ta_1y_1(t) + \frac{da_2y_2(t)}{dt} + ta_2y_2(t)$$

$$= T\{a_1x_1(t)\} + T\{a_2x_2(t)\}.$$

As the superposition theorem holds, the system is linear.

Example 1.30

Determine if the system $y(t) = \int_{-\infty}^{t} e^{-(t-\tau)}x(\tau)d\tau$ is invertible.

Solution

Differentiating, $\frac{dy(t)}{dt} = \frac{d}{dt}\left[e^{-t}\int_{-\infty}^{t} e^{\tau}x(\tau)d\tau\right] = x(t) - e^{-t}\int_{-\infty}^{t} e^{\tau}x(\tau)d\tau.$

or, $\frac{dy(t)}{dt} + y(t) = x(t)$. This is the system equation.

The system is invertible and the inverse system would be $\frac{dx(t)}{dt} + x(t) = y(t)$.

Summary

1. A signal is defined as a single-valued function of one or more independent variables that contains information about a physical quantity.
2. A system is defined as a group of objects that can carry one or more signals and combine them in a manner intended to achieve some desired response.
3. Signals are classified as: continuous-time and discrete-time signals, analog and digital signals, periodic and aperiodic signals, real and complex signals, deterministic and random signals, even and odd signals, energy and power signals, and causal and non-causal signals.
4. A continuous-time signal $f(t)$ is a function of a continuous time variable t. It is also called analog signal.

5. A discrete-time signal is one that can be represented by a sequence of numbers. It is represented as

$$f[n] = \{.....f[-2], f[-1], f[0], f[1], f[2], ...\}$$

6. A digital signal takes discrete values for both time and magnitude.
7. A signal is said to be periodic if and only if $f(t) = f(t+nT)$; the condition for discrete signals is $f[n] = f[n+N]$.
8. If a function $f(t) = x_1(t) + x_2(t)$ is a sum of harmonic signals with time periods T_1, T_2, $f(t)$ is periodic with period T only when $\frac{T_1}{T_2} = \frac{k}{m}$= rational number.
9. If a sequence$f[n] = x_1[n] + x_2[n]$ is a sum of sequences with time periods N_1, N_2 and if $f[n]$ is periodic with period N, then $\frac{N_1}{N_2} = \frac{k}{m}$= rational number. Here N_1, N_2 and N are integers.
10. Sum of two continuous time periodic signals may not be periodic but sum of two periodic discrete time signals is always periodic.
11. If a signal $f(t)$ is even signal or symmetric signal $f_e(t)$, then $f(t) = f(-t)$ and If a signal $f(t)$is odd signal or anti-symmetric signal $f_o(t)$ then $f(t) = -f(-t)$.
12. For discrete time signals, the condition for even signal is $f[-n] = f[n]$ and for odd signal is $f[-n] = -f[n]$.
13. Any signal $f(t)$ can be expressed as a sum of even and odd signals. That is

$$f(t) = f_e(t) + f_o(t);\ \text{also, } f_e(t) = \frac{1}{2}(f(t) + f(-t)) \text{ and } f_o(t) = \frac{1}{2}(f(t) - f(-t)).$$

For discrete signals,

$$f[n] = f_e[n] + f_o[n];\ \text{also, } f_e[n] = \frac{1}{2}(f[n] + f[n]) \text{ and } f_o[n] = \frac{1}{2}(f[n] - f[-n]).$$

14. The sum of two even signals, $x_e(t) + y_e(t)$, is always even symmetry; the sum of two odd signals, $x_O(t) + y_O(t)$, is odd symmetry; and the sum of odd and even signals, $x_e(t) + y_O(t)$, has no symmetry.
15. For the multiplication of two symmetrical signals, the following holds: $x_e(t).y_e(t)$ even symmetry; $x_O(t).y_O(t)$ even symmetry; and $x_e(t).y_O(t)$ odd symmetry.
16. For even symmetry, $\int_{-T}^{T} f_e(t)dt = 2\int_{0}^{T} f_e(t)dt$.
17. For odd symmetry, $\int_{-T}^{T} f_O(t)dt = 0$.
18. The condition for half-wave symmetry is $f(t) = -f\left(t + \frac{T}{2}\right) = -f\left(t + nT + \frac{T}{2}\right)$, for all t and integer values of n.
19. The complex signal $f(t)$ is a function of real and imaginary signals: $f(t) = f_R(t) + jf_1(t)$.
20. For a continuous-time signal $f(t)$, the total energy, $E = \lim_{T\to\infty} \int_{-T}^{T} |f(t)|^2 dt$ or $E = \int_{-\infty}^{\infty} |f(t)|^2 dt$ and the average power, $P = \lim_{T\to\infty} \frac{1}{2T} \int_{-T}^{T} |f(t)|^2 dt$.

21. For a discrete-time signal $f[n]$, the normalised energy is $E = \sum_{n=-\infty}^{\infty} |f[n]|^2$ and the normalised average power P is $P = \lim_{n\to\infty} \frac{1}{2N+1} \sum_{n=-N}^{N} |f[n]|^2$.

22. A signal is said to be an energy signal if and only if it has finite energy and zero average power ($0 < E < \infty$ and $P = 0$). All time-limited signals of finite amplitude are energy signals.

23. A signal is said to be a power signal if and only if it has infinite energy but finite average power ($0 < P < \infty$ and $E \to \infty$).

24. Some signals may be neither power signals nor energy signals. They may have either infinite energy and infinite power or infinite energy and zero power. Examples are exponential or parabolic growing signals.

25. Any continuous signal $f(t)$ is said to be causal if and only if $f(t) = 0$, for $t < 0$; otherwise the signal is called non-causal or anti-causal signal and the condition is $f(t) = 0$, for $t > 0$. Similarly, a discrete time signal $f[n]$ is said to be causal if and only if $f[n] = 0$, for $n < 0$; for non-causal signal the condition is $f(n) = 0$, for $n > 0$.

26. Some of the standard type of signals that commonly encountered in signal analysis are the described by the impulse, step, ramp, parabolic, exponential, sinusoidal, sampling, pulse and triangular functions. These functions for continuous time and discrete time are given below.

The unit impulse function: $\delta(t) = \begin{cases} & t = 0 \\ 0 & \text{otherwise} \end{cases}$ $\quad \delta[n] = \begin{cases} 1 & n = 0 \\ 0 & \text{otherwise} \end{cases}$

The unit step function: $u(t) = \begin{cases} 1 & t \geq 0 \\ 0 & t < 0 \end{cases}$ $\quad u[n] = \begin{cases} 1 & n \geq 0 \\ 0 & n < 0 \end{cases}$

The unit ramp function: $r(t) = \begin{cases} t & t \geq 0 \\ 0 & t < 0 \end{cases}$ $\quad r[n] = \begin{cases} n, & n \geq 0 \\ 0, & n < 0 \end{cases}$

The unit parabolic function: $g(t) = \begin{cases} \frac{t^2}{2} & t \geq 0 \\ 0 & t < 0 \end{cases}$ $\quad g[n] = \begin{cases} \frac{n^2}{2} & n \geq 0 \\ 0 & n < 0 \end{cases}$

The exponential signal: $f(t) = ce^{at}$ and $f[n] = ce^{an}$

The sinusoidal function: $f(t) = A\cos(\omega_0 t + \phi)$ and $f[n] = A\cos(\omega_0 n + \phi)$

The sampling signal: $\mathrm{Sa}(t) = \frac{\sin t}{t}$ and $\mathrm{Sa}[n] = \frac{\sin n}{n}$

The sinc function: $\mathrm{sinc}(t) = \frac{\sin \pi t}{\pi t}$ and $\mathrm{sinc}[n] = \frac{\sin \pi n}{\pi n}$

The rectangular function:

$$\prod(t/\tau) = G_\tau(t) = \mathrm{rect}(t) = \begin{cases} 1 & |t| < \tau/2 \\ 0 & \text{elsewhere} \end{cases} \quad \text{and}$$

$$\prod[n/N] = G_N[n] = \mathrm{rect}[n/N] = \begin{cases} 1 & |n| < N \\ 0 & \text{elsewhere} \end{cases}$$

The triangular function:

$$\Delta(t) = \text{tri}(t) = \begin{cases} 1-|t| & |t| \le 1 \\ 0 & \text{elsewhere} \end{cases} \quad \text{and} \quad \Delta[n] = \text{tri}[n] = \begin{cases} N-|n| & |n| \le N \\ 0 & \text{elsewhere} \end{cases}$$

27. Some important operations on signals are: amplitude scaling, amplitude shifting, multiplication, addition, differentiation, integration, time scaling, time shifting, and time reversal or reflection.
28. Systems are classified variously as: continuous-time and discrete-time systems; linear and non-linear systems; time-invariant and time-varying system; causal and non-causal systems; static and dynamic systems; invertible and noninvertible systems; stable and unstable systems; and lumped and distributed systems.
29. A system is said to be linear if it satisfies the scaling property (homogeneity) and principle of superposition (additivity). If a system violates any one of these properties, it is called a nonlinear system.
30. For linear continuous system, the condition is $\sum_{i=1}^{n} a_i y_i(t) = T\left\{\sum_{i=1}^{n} a_i x_i(t)\right\}$, where, $y_i(t) = T\{x_i(t)\}$ is the output of the system in response to the input $x_i(t)$ acting alone and for discrete system $\sum_{i=1}^{n} a_i y_i[n] = T\left\{\sum_{i=1}^{n} a_i x_i[n]\right\}$.
31. For a continuous time invariant system, if $y(t) = T\{x(t)\}$, then $y(t-t_0) = T\{x(t-t_0)\}$ and for discrete system, $y[n-n_0] = T\{x[n-n_0]\}$.
32. A system is said to be *causal* if its output $y(t)$ at an arbitrary time $t = t_0$ depends on only the input $x(t)$ for $t \le t_0$. Similarly, for a discrete system, a system is said to be *causal* if its output $y[n]$ at an arbitrary time $n = n_0$ depends on only the input $x[n]$ for $n \le n_0$.
33. A system is said to be static or memoryless system if the output at any time $t = t_0$ or $n = n_0$ depends on only the input at that same time $t = t_0$ or $n = n_0$.
34. The condition for BIBO stability is: for any bounded input signal $x(t)$, $|x(t)| \le K_1 < \infty$, $-\infty < t < \infty$, the corresponding output $y(t)$ is also bounded, i.e., $|y(t)| \le K_2 < \infty$, $-\infty < t < \infty$. Similarly, for a discrete system, for any bounded input signal $x[n]$, $|x[n]| \le K_1 < \infty$, $-\infty < t < \infty$, the corresponding output $y[n]$ is also bounded i.e., $|y[n]| \le K_2 < \infty$, $-\infty < t < \infty$.
35. A system is said to be invertible if the input signal applied the system can be uniquely recovered from the output signal, that is, $y(t) = T^{-1}\{y_1(t)\} = T^{-1}T\{x(t)\} = x(t)$. Similarly, for a discrete system, $y[n] = T^{-1}\{y_1[n]\} = T^{-1}T\{x[n]\} = x[n]$.
36. A system is said to be a lumped system if the components within the system are placed at one place, while in distributed system the components are distributed at long distances.

Review Questions

1. Define signal. Classify signals with examples.
2. Distinguish between continuous-time and discrete-time signals.
3. Explain the methods to represent a discrete time signal.
4. What is a digital signal? Give an example.
5. What are the basic operations on signal? Explain with some examples.

6. State and explain any three properties of signals.
7. Define causal and non-causal signal. Give some examples.
8. Define periodic signal. Give the condition for periodicity of a discrete-time periodic signal.
9. Explain the condition for the sum of periodic signals to be periodic.
10. Derive the condition for the sum of two periodic signals to be periodic
11. State and explain symmetry properties of a signal.
12. State and explain periodic properties of a signal.
13. State and explain the following properties of a signal: (i) time shifting, (ii) time scaling.
14. Give examples for energy and power signals.
15. Define rms value of a signal.
16. State and explain even and odd signals.
17. Define a system. Classify systems with examples.
18. State and explain any three properties of systems.
19. Define the following in relation to a continuous-time system:
    (i) causal and non-causal system,
    (ii) time variant and time invariant systems,
    (iii) linear and nonlinear systems,
    (iv) stable and unstable systems.
20. Define and explain time variant and time invariant systems.
21. What is discrete-time system?
22. Define lumped and distributed systems. Give examples.
23. Explain how a system is represented with differential equations.
24. Define and explain static and dynamic systems.

Problems

1. Determine whether or not each of the following signals is periodic.

   (a) $x(t) = \cos 6\pi t + \sin 3t$; (b) $x(t) = 1 + \sin^2 t$; (c) $x(t) = \cos 2t + \sin 7t$.

2. Determine whether or not each of the following signals is periodic

   (a) $x(t) = \cos(6t - 1) + \sin(3t + 1)$; (b) $x(t) = 2u(t) + \sin t$;

   (c) $x(t) = u(t) - 1$; (d) $x(t) = \cos(5t) + \sin\left(\sqrt{5}t\right)$.

3. Find the fundamental period of the following:

   (a) $x(t) = \cos(5t) + \sin(8t)$; (b) $x(t) = e^{j6(t-1)}$; (c) $x(t) = \cos\left(5\pi t + \dfrac{\pi}{3}\right)$.

4. Determine the fundamental period of the signal $x(t) = 2\cos(10t+1) - \sin(4t-1)$.
5. Determine the fundamental period of the signal $x[n] = 1 + e^{j4\pi n/7} - e^{j2\pi n/5}$.
6. Determine whether or not each of the following continuous-time signals is periodic. If the signal is periodic, determine its fundamental period.
   (a) $x(t) = 3\cos\left(4t + \frac{\pi}{3}\right)$
   (b) $x(t) = \exp j(\pi t - 1)$
   (c) $x(t) = \left[\cos\left(2t - \frac{\pi}{3}\right)\right]^2$
   (d) $x(t) = \cos(4\pi t)u(t)$
7. Which of the following signals or functions are periodic and if yes, what is its fundamental period? (May 2011)
   (i) $g(t) = e^{-j60\pi t}$ (ii) $g(t) = 10\sin(12t) + 4\cos(18t)$
8. Determine whether each of the following sequences are periodic or not. If periodic, determine the fundamental period. (Nov 2013)
   (i) $x_1(n) = \sin\left(\frac{6\pi n}{7}\right)$; (ii) $x_2(n) = \sin\left(\frac{n}{8}\right)$
9. Which of the following signals or functions are periodic and if yes, what is the fundamental period? (Nov 2015)
   (i) $g(t) = e^{-j60\pi t}$ (ii) $g(t) = 10\sin(12\pi t) + 4\cos(18t)$
10. An even function $g(t)$ is described by (Nov 2010)
$$g(t) = \begin{cases} 2t & 0 \le t \le 3 \\ 15 - 3t & 3 \le t \le 7 \\ -2 & 7 \le t \le 10. \end{cases}$$
   (i) What is the value of $g(t)$ at time $t = 5$?
   (ii) What is the value of the first derivative of $g(t)$ at time $t = 6$?
11. Determine whether the following functions are periodic or non-periodic: (Nov 2010)
   (i) $a\sin 5t + b\cos 8t$
   (ii) $a\sin\left(\frac{3t}{2}\right) + b\cos\left(\frac{16t}{15}\right) + c\sin\left(\frac{t}{29}\right)$
   (iii) $a\cos t + b\sin\sqrt{2}t$, where a, b and c are real integers.
12. Determine the energy and power of the following signals:
   (a) $x(t) = 5\cos(6t) + \sin(3t)$; (b) $x(t) = 2u(t) + 4\sin t$; (c) $x(t) = 20\cos(100\pi t - 30)$
13. Find the signal energy for the following signals: (Nov 2014)
   (i) $x(t) = 2\text{rect}(-t)$; (ii) $x(t) = 2\sin(300\pi t)$
14. Find the signal energy for the following signals: (May 2014)
   (i) $x(t) = 2\sin(200\pi t)$; (ii) $x(t) = \text{rect}\left(\frac{t}{4}\right)$
15. Find the power and rms value of signal $x(t) = 20\cos 2\pi t$. (Dec 2015)
16. Test whether the signal $x[n] = (1/2)^n u[n]$ is an energy or a power signal. (Dec 2015)

17. Test whether the signal $x(t) = e^{-2t}u(t)$ is an energy or a power signal. (Dec 2015)
18. Determine the power and rms value of the following signals:
    (i) $y(t) = 5\cos\left(50t + \frac{\pi}{3}\right)$; (ii) $y(t) = 10\cos(5t)\cos(10t)$
19. Determine the power and rms value of the signal $x(t) = u(t)$.
20. Determine the power and rms value of the signal $x(t) = e^{j4t}\sin 7t$.
21. Determine the power and rms value of the signal $x(t) = u(t)$.
22. Determine the power and rms value of the signals (a) $x(t) = \cos 6\pi t + \sin 3t$ and (b) $x(t) = 1 + \sin^2 t$.
23. Find the energy of the signal $x(t) = e^{-at}u(t)$.
24. Find whether the following signals are energy signals or power signals:
    (a) $x[n] = 0.3u[n]$; (b) $x[n] = \cos(5\pi n) + \sin(8\pi n)$; (c) $x[n] = u[n] - u[n-2]$
25. Find whether the following signals are even or odd:
    (a) $x(t) = 0.3e^{2t}u(t-2)$
    (b) $x[n] = 0.3u[n-2] + 0.6u[n-5]$
    (c) $x(t) = 0.3e^{2t}u(t-2)$
26. Find odd and even components of the following signals:
    (a) $x(t) = \cos t + \sin 3t$
    (b) $x(t) = 1 + \sin t + 2\sin^2 t \cos t$
    (c) $x(t) = \cos t + \cos 2t \sin t$
27. Find whether the following signals are even or odd:
    (a) $x[n] = \sin(2\pi n)$ (b) $x[n] = \cos(5n) + \sin(8n)$
28. Sketch the two ramp signals $x_1(t) = u(t-1)$ and $x_2(t) = u(t-2)$. Also, draw the waveforms for the following operations:
    (a) $f(t) = x_1(t) + x_2(t)$; (b) $f(t) = 3x_1(t) - 2x_2(t)$; (c) $f(t) = x_1(t+5) + 2x_2(t-2)$.
29. Sketch the two signals, $r_1(t) = (t-1)u(t-1)$ and $r_2(t) = u(t-3)$. Also draw the waveforms for the following operations:
    (a) $f(t) = x_1(t) + x_2(t)$; (b) $f(t) = 3x_1(t) - 2x_2(t)$; (c) $f(t) = x_1(t+5) + 2x_2(t-2)$.
30. Let two functions be defined by: (May 2011)
    $$x_1(t) = \begin{cases} 1, & \sin(20\pi t) \geq 0 \\ -1, & \sin(20\pi t) < 0 \end{cases} \qquad x_2(t) = \begin{cases} t, & \sin(20\pi t) \geq 0 \\ -t, & \sin(20\pi t) < 0. \end{cases}$$
    Sketch the product of these two functions vs time over the time interval $-2 < t < 2$.
31. Let two signals be defined by (May 2011)
    $$x_1(t) = \begin{cases} 1, & \cos(2\pi t) \geq 1 \\ 0, & \cos(2\pi t) < 1 \end{cases} \qquad x_2(t) = \sin\left(\frac{2\pi t}{10}\right)$$
    Sketch these products over the time interval $-5 < t < 5$.

32. Examine the following systems given by

(a) $y(t) = \cos(x(t))$; (b) $y(t) = \log|x(t)|$

with respect to the properties:

(i) causal or non causal, (ii) time variant or time invariant,
(iii) linear or nonlinear, (iv) stable or unstable.

33. A system is characterized by the equation $y(t) = tx(t) + x(t-1) + x(2-t)$. Determine whether the system is memoryless, time-invariant, linear, causal and stable or not.

34. Sketch the following signal $x(t) = A[u(t+a) - u(t-a)]$ for $a > 0$. Also, determine whether the given signal is a power signal or an energy signal or neither. (Nov-2013)

35. Check whether the following systems are invertible or not.

(i) $f(t) = \sin(x(t))$; (ii) $f(t) = \dfrac{dx(t)}{dt}$; (iii) $f(t) = x(t-3) + 4$ (iv) $f[n] = m[n] + 1$

Answers

1. (a) Aperiodic (b) Periodic (c) periodic
2. (a) Periodic (b) Aperiodic (c) Aperiodic (d) Aperiodic
3. (a) 2π (b) $\dfrac{\pi}{3}$ (c) $\dfrac{2}{5}$
4. Fundamental period $= \pi$ sec.
5. Fundamental period $= 35$ sec.
6. (a) Fundamental period $= \dfrac{\pi}{2}$ sec. (b) Fundamental period $= 2$ sec.
   (c) Fundamental period $= \dfrac{\pi}{2}$ sec. (d) Aperiodic
7. (i) Periodic with period of 1/30 sec. (ii) Periodic with period of $\pi/3$ sec.
8. (i) Periodic, with fundamental period $N = 7$ (ii) Non-periodic
9. (i) Periodic, with fundamental period $T = \dfrac{1}{30}$ sec. (ii) Non-periodic
10. (i) $g(t) = 0$ at time $t = 5$ sec. (ii) $\dfrac{d}{dt}[g(t)] = -3$ at time $t = 6$ sec.
11. (i) Periodic (ii) Periodic (iii) Non-periodic
12. (a) $P = 13$ watts ; $E = \infty$ (b) $P = \infty; E = \infty$ (c) $P = 200$ W; $E = \infty$
13. (i) 4 joules (ii) ∞
14. (i) ∞ (ii) 16 joules
15. Power = 200 watts, rms value = 14.14 volts
16. Energy signal with an energy of $\dfrac{4}{3}$ watts
17. Energy signal with an energy of $\dfrac{1}{4}$ watts

18. (i) Power = 12.5 watts, rms value = $\frac{5}{\sqrt{2}}$ volts

(ii) Power = 25 watts, rms value = 5 volts

19. Power = 0.5 watts, rms value = $\frac{1}{\sqrt{2}}$ volts

20. $P = \frac{1}{2}$ watts, rms value = $\frac{1}{\sqrt{2}}$ volts

21. $P = \frac{1}{2}$ watts, rms value = $\frac{1}{\sqrt{2}}$ volts

22. (a) $P = \frac{1}{2}$ watts, rms value = $\frac{1}{\sqrt{2}}$ volts

(b) $P = 1$ watts, rms value = $\frac{11}{8}$ volts

23. $E = \frac{1}{2a}$ joules

24. (a) Power signal (b) Power signal (c) Energy signal

25. (a) Contains both even and odd parts

(b) Contains both even and odd parts

(c) Contains both even and odd parts

26. (a) $x_e(t) = 5\cos 6t$; $x_o(t) = \sin 3t$

(a) $x_e(t) = 1 + 2\sin^2 t \cos t$; $x_o(t) = \sin t$

(b) $x_e(t) = \cos t$; $x_o(t) = \cos 2t \sin t$

27. (a) Odd; (b) The signal contains both even and odd parts $x_e[n] = \cos(5n)$; $x_o[n] = \sin(8n)$

32. (a) Causal, time invariant, nonlinear and stable

(b) Causal, time invariant, linear and stable

33. Memoryless, time-invariant, nonlinear, non-causal and stable

34. Energy signal with an energy of $2aA^2$

35. (i) Non-invertible (ii) Non-invertible (iii) invertible (iv) invertible

Multiple-choice Questions

1. A signal in time domain can be represented as
   (a) frequency spectrum (b) frequency
   (c) waveform (d) none of these
2. Graphical representation of a signal in frequency domain is called
   (a) frequency spectrum (b) frequency
   (c) waveform (d) none of these
3. Which of the following signals is non-periodic?
   (a) $s(t) = \cos 2t + \cos 3t + \cos 5t$ (b) $s(t) = \exp(j\,8\pi t)$
   (c) $s(t) = \exp(-7t) \sin 10\pi t$ (d) $s(t) = \cos 2t.\cos 4t$

4. The sum of two periodic signals is periodic only if the ratio of their respective periods $\frac{T_1}{T_2}$ is

(a) a rational number (b) a complex number

(c) an irrational number (d) a real number

5. The signal $f(t) = e^{-2t}u(t)$ is

(a) power signal (b) energy signal

(c) neither energy nor power signal (d) none of these

6. For energy signals, the average power is

(a) zero (b) 1

(c) infinity (d) none of these

7. For power signals, the total energy is

(a) zero (b) 1

(c) infinity (d) none of these

8. The signal $f(t) = u(t) - u(t-2)$ is

(a) power signal (b) energy signal

(c) neither energy nor power signal (d) none of these

9. The power of the signal $f(t) = A\sin\left(10t + \frac{\pi}{2}\right)$ is

(a) $\frac{A^2}{2}$ (b) $\frac{A}{\sqrt{2}}$

(c) A (d) None of these

10. The rms value of the signal $f(t) = A\sin\left(10t + \frac{\pi}{2}\right)$ is

(a) $\frac{A^2}{2}$ (b) $\frac{A}{\sqrt{2}}$

(c) A (d) none of these

11. Periodic signals are

(a) $x(t+T) = x(t)$ (b) $x(t-T) = x(t)$

(c) $x[n+mN] = x[n]$ (d) all the above

12. Any signal $x(t)$ can be represented as

(a) $x_e(t) + x_o(t)$ (b) $x_e(t) - x_o(t)$

(c) $\frac{x_e(t)}{x_o(t)}$ (d) $x_e(t) * x_o(t)$

13. Even signal satisfies

(a) $x(-t) = x(t)$ (b) $x[-n] = -x[n]$

(c) $x[n+1] = ax[n] + b$ (d) $\frac{dx(t)}{dt} = c$

14. Odd signal satisfies
(a) $x(-t) = -x(t)$
(b) $x[-n] = x[n]$
(c) $x[n+1] = ax[n] + b$
(d) $\frac{dx(t)}{dt} = c$

15. Which system is nonlinear in nature?
(a) $y(t) = cx(t)$
(b) $y(t) = x^2(t) + x(t) + c$
(c) $y(t) = x(t) + x(t-1) + x(t-2)$
(d) $y(t) = x(t) - x(t-1) + x(t-2)$

16. A system whose output depends on future inputs is a
(a) dynamic system
(b) static system
(c) non-causal system
(d) both (a) and (c)

17. A system with an input $x(t)$ and an output $y(t)$ is described by the relation: $y(t) = tx(t)$. This system is
(a) linear and time-invariant
(b) linear and time-varying
(c) nonlinear and time-invariant
(d) nonlinear and time-varying

18. The input and output of a continuous time system are respectively denoted by $x(t)$ and $y(t)$. Which of the following descriptions corresponds to a causal system?
(a) $y(t) = x(t-2) + x(t+4)$
(b) $y(t) = (t-4)x(t+1)$
(c) $y(t) = (t+4)x(t-1)$
(d) $y(t) = (t+5)x(t+5)$

19. An excitation is applied to a system at $t = T$ and its response is zero for $-\infty < t < T$. Such a system is a
(a) non-causal system
(b) stable system
(c) causal system
(d) unstable system

20. The type of the system described by $y^2(t) + 2y(t) = x^2(t) + x(t) + C$ is
(a) linear and dynamic
(b) linear and static
(c) nonlinear and dynamic
(d) nonlinear and static

21. A dynamic system is characterised by
(a) linear equation
(b) quadratic equation
(c) differential equation
(d) statistical parameters

22. A distributed parameter system is described by
(a) linear differential equations
(b) partial differential equations
(c) both (a) and (b)
(d) none of the above

23. A lumped parameter system is described by
(a) linear differential equations
(b) partial differential equations
(c) both (a) and (b)
(d) none of the above

24. Invertible systems are those systems whose
(a) output can be uniquely obtained from the knowledge of the input signal
(b) input signal can be uniquely determined by observing the output signal

(c) (a) and (b) both
(d) system output is always constant.

25. The invertible system in the following is:

(a) $y(t) = x(t) + 5$ (b) $y(t) = x^2(t)$

(c) $y(t) = \int_{-\infty}^{t} x(\tau)d\tau$ (d) (a) and (c)

26. Let $x(t)$ be the input and $y(t)$ be the output of a continuous time system. Match the system properties P1, P2 and P3 with system relations Rl, R2, R3, R4.

Properties	*Relations*
PI: Linear but NOT time-invariant	R1: $y(t) = t^2 x(t)$
P2: Time-invariant but NOT linear	R2: $y(t) = t\lvert x(t)\rvert$
P3: Linear and time-invariant	R3: $y(t) = \lvert x(t)\rvert$
	R4: $y(t) = x(t-5)$

(a) (PI, Rl), (P2, R3), (P3, R4) (b) (PI,R2), (P2, R3), (P3,R4)

(c) (PI, R3), (P2, Rl), (P3, R2) (d) (P1,R1), (P2, R2), (P3.R3)

27. $y(t) = x(t^2)$ is *a*

(a) non-causal, dynamic system
(b) non-causal, time variant system
(c) non-causal, time variant, nonlinear system
(d) non-causal, time invariant, linear system

28. $y[n] = x[n]u[n]$ is *a*

(a) static, linear system (b) causal, time variant system

(c) both (a) and (b) (d) none of these

29. $y[n] = x[n] + nx[n-3]$ is *a*

(a) dynamic system (b) causal system

(c) linear system (d) all the above

30. $\frac{dy(t)}{dt} + 8ty(t) = t^3 x(t)$ is *a*

(a) linear, time invariant system (b) nonlinear, dynamic system

(c) time variant, static, linear system (d) time variant, static system

Answers

1. (c)	2. (a)	3. (c)	4. (a)	5. (b)	6. (a)	7. (c)	8. (b)	9. (a)	10. (b)
11. (d)	12. (a)	13. (b)	14. (a)	15. (b)	16. (d)	17. (b)	18. (c)	19. (c)	20. (d)
21. (c)	22. (b)	23. (a)	24. (c)	25. (d)	26. (a)	27. (b)	28. (a)	29. (d)	30. (d)

2

Signal Analysis

Learning Objectives

The main objective of this chapter is to understand thoroughly the basic concepts of vectors and signals. We will learn how the signals are represented by using vectors. By reading this chapter, students will be able to relate to the following topics.

- Analogy between vectors and signals
- Orthogonal signal space and orthogonal functions
- Signal approximation using orthogonal functions
- Mean square error
- Complete set of orthogonal functions
- Orthogonality in complex functions
- Exponential and sinusoidal signals
- Concepts of impulse function
- Unit step function and signum function

2.1 Introduction

In previous chapter we looked at the basic types of signals and systems. Signals existing in nature can be expressed in terms of many basic types of functions. Here we explore the analogy between vectors and signals, the orthogonal signal space, orthogonal functions, the complete set of orthogonal functions and orthogonality in complex functions. Since signals are vectors, the basic mathematical formulations of signals can be obtained from vector algebra.

2.2 Analogy between Vectors and Signals

A vector is a quantity that has both magnitude and direction and is represented either using the convention of a letter with a superscribed arrow such as $\vec{A}$, $\vec{B}$, etc, or with boldface type **A, B,** etc. The magnitude of a vector is represented as $\left|\vec{A}\right|$ or simply A.

A signal is broadly defined as a single-valued quantity that is a function of independent variables as explained in Section 1.2. This quantity has both magnitude and direction, and therefore, all signals are vectors. Just as a vector can be expressed as a sum of its components resolved along the direction of the coordinates of the system, signals too can be represented as a sum of its components oriented along the direction of the defined coordinate system.

From the basic concepts of vector algebra we know that in a two-dimensional vector space the dot product (inner product) of two vectors $\vec{A}$ and $\vec{B}$ is defined as the product of the magnitudes of $\vec{A}$ and $\vec{B}$ and the cosine of the angle between them. The dot product yields a scalar quantity.

$$\text{Thus,}\quad \vec{A}\cdot\vec{B} = \left|\vec{A}\right|\left|\vec{B}\right|\cos\theta = AB\cos\theta, \tag{2.1}$$

θ being the smaller of the angles between $\vec{A}$ and $\vec{B}$ and less than 180^0. One of the important uses of dot products is in projections.The physical significance of the dot product can be seen in the projection or the component of one vector along the other vector as shown in Fig. 2.1.

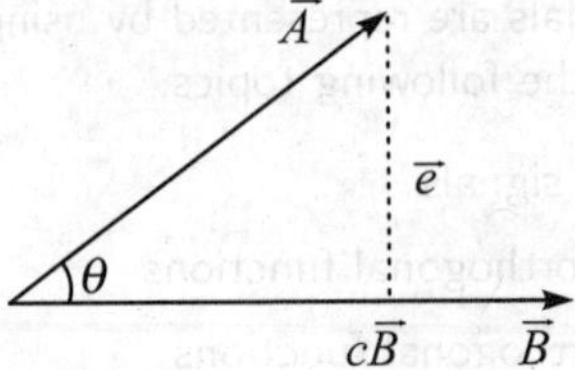

Fig. 2.1 *Projection of $\vec{A}$ along $\vec{B}$*

If $c\vec{B}$ is a vector component of $\vec{A}$ along the direction of $\vec{B}$, where $c < 1$ is a scalar constant, and $\vec{e}$ is another vector called the error vector, then vector $\vec{A}$ can be expressed in terms of vector $\vec{B}$ as

$$\vec{A} = c\vec{B} + \vec{e}$$

$$\text{or}\quad \vec{e} = \vec{A} - c\vec{B}. \tag{2.2}$$

It may be observed that the length of the vector $\vec{e}$ will be minimum if vector $\vec{e}$ is normal to vector $\vec{B}$. In the two-dimensional space the two orthogonal vectors $\vec{B}$ and $\vec{e}$ are also called basis vectors.

Hence, any vector $\vec{A}$ in the two-dimensional space can be expressed in terms of two orthogonal vectors, $\vec{B}$ and $\vec{e}$, subject to the condition that the length of the vector $\vec{e}$ is a minimum. That is, we have to minimise

$$e = |\vec{e}| = \left|\vec{A} - c\vec{B}\right|. \tag{2.3}$$

If $c = 0$, there is no component of $\vec{A}$ along $\vec{B}$ and the vectors $\vec{A}$ and $\vec{B}$ are said to be orthogonal to each other. To evaluate the value of the constant c at the minimum of $\vec{e}$, the scalar component of $\vec{A}$ along the direction of $\vec{B}$ is taken. From Fig. 2.1,

$$cB = \left|c\vec{B}\right| = A\cos\theta$$

$$cB^2 = AB\cos\theta = \vec{A}\cdot\vec{B}$$

$$c\vec{B}\cdot\vec{B} = \vec{A}\cdot\vec{B}$$

$$\therefore \quad c = \frac{\vec{A}\cdot\vec{B}}{\vec{B}\cdot\vec{B}} = \frac{1}{B^2}\left(\vec{A}\cdot\vec{B}\right). \tag{2.4}$$

Substituting the value of c from Eq. (2.4) in Eq. (2.2), the error vector is

$$\vec{e} = \vec{A} - c\vec{B} = \vec{A} - \frac{1}{B^2}\left(\vec{A}\cdot\vec{B}\right)\vec{B}.$$

$$\begin{aligned}\vec{e}\cdot\vec{B} &= \vec{A}\cdot\vec{B} - c\vec{B}\cdot\vec{B} = \vec{A}\cdot\vec{B} - \frac{1}{B^2}\left(\vec{A}\cdot\vec{B}\right)B^2 \\ &= \vec{A}\cdot\vec{B} - \vec{A}\cdot\vec{B} = 0.\end{aligned} \tag{2.5}$$

Thus, vector $\vec{e}$ is minimum if it is normal to vector $\vec{B}$.

Note: If the two vectors $\vec{A}$ and $\vec{B}$ are orthogonal to each other, then $\vec{A}\cdot\vec{B} = 0$ and from Eq. (2.4), $c = 0$. That is, there is no component of $\vec{A}$ on $\vec{B}$.

The concept of vector orthogonality can be extended in a similar manner to signals. Consider any two independent real signals $f(t)$ and $x(t)$ which are functions of time in the interval $t_1 \le t \le t_2$. The function $f(t)$ can be approximated in terms of the function $x(t)$over the interval $t_1 \le t \le t_2$ as

$$f(t) \approx cx(t); \qquad t_1 \le t \le t_2 \tag{2.6}$$

The difference between $f(t)$ and $cx(t)$ is called the error signal $f_e(t)$, and it is given by

$$f_e(t) = f(t) - cx(t). \tag{2.7}$$

For the best approximation, the error should be a minimum. Since error may be a positive or a negative variable with respect to time, to avoid confusion the mean square value of the error signal is taken as the quantity to be minimised.

That is, minimise:

$$e = \frac{1}{(t_2 - t_1)}\int_{t_1}^{t_2} f_e^2(t)dt = \int_{t_1}^{t_2} |f(t) - cx(t)|^2\, dt \quad t_1 \le t \le t_2, \tag{2.8}$$

where e is called the mean square error (MSE) or power of the error signal.

We know that the necessary condition to minimise the mean square error with respect to the constant c is $\frac{de}{dc} = 0$.

$$\frac{de}{dc} = \frac{d}{dc}\left[\frac{1}{(t_2 - t_1)}\left(\int_{t_1}^{t_2} e^2(t)dt\right)\right] = \frac{d}{dc}\left[\frac{1}{(t_2 - t_1)}\int_{t_1}^{t_2} |f(t) - cx(t)|^2\, dt\right] = 0$$

$$\frac{d}{dc}\left(\int_{t_1}^{t_2} f^2(t)dt\right) + \frac{d}{dc}\int_{t_1}^{t_2} c^2x^2(t)dt - \frac{d}{dc}\int_{t_1}^{t_2} 2cx(t)f(t)dt = 0$$

$$0 + 2c\int_{t_1}^{t_2} x^2(t)dt - 2\int_{t_1}^{t_2} x(t)f(t)dt = 0.$$

$$c = \frac{\int_{t_1}^{t_2} x(t)f(t)dt}{\int_{t_1}^{t_2} x^2(t)dt}. \tag{2.9}$$

It is to be noted that the c obtained for signals in Eq.(2.9) is similar to the c obtained for vectors in Eq.(2.4). Substituting Eq.(2.9) in Eq.(2.7), the condition for the minimum of the error signal is

$$f_e(t) = f(t) - cx(t)$$

$$= f(t) - \frac{\int_{t_1}^{t_2} x(t)f(t)dt}{\int_{t_1}^{t_2} x^2(t)dt} x(t).$$

$$\int_{t_1}^{t_2} x(t)f_e(t)dt = \int_{t_1}^{t_2} x(t)f(t)dt - \frac{\int_{t_1}^{t_2} x(t)f(t)dt}{\int_{t_1}^{t_2} x^2(t)dt} \int_{t_1}^{t_2} x^2(t)dt$$

$$= 0. \tag{2.10}$$

That is, the condition for the minimum of the error signal demands that the signals $f_e(t)$ and $x(t)$ be orthogonal to each other over the interval $t_1 \leq t \leq t_2$. These signals are called the basis signals in two-dimensional space.

From Eq. (2.5) and Eq. (2.10) it is observed that signals have properties similar to vectors. The dot or inner product of the vectors gives the area common to the product of the vectors. Similarly, the inner product of any two signals $f(t)$ and $x(t)$ over an interval $t_1 \leq t \leq t_2$ gives the area under the product of $f(t)$ and $x(t)$. The squared magnitude of the signal over an interval $t_1 \leq t \leq t_2$ gives the area under the product of the signal with itself.

That is, the inner product of two signals $f(t)$ and $x(t)$ is

$$\int_{t_1}^{t_2} x(t)f(t)dt, \quad t_1 \leq t \leq t_2, \tag{2.11}$$

and the energy of the signal is

$$E = \int_{t_1}^{t_2} x^2(t)dt. \tag{2.12}$$

Also, if any two signals $f(t)$ and $x(t)$ are orthogonal to each other over an interval $t_1 \leq t \leq t_2$, then $c = 0$ and there is no component of $f(t)$ on $x(t)$.

That is, the condition for orthogonality is

$$\int_{t_1}^{t_2} x(t)f(t)dt = 0. \tag{2.13}$$

For example, the two signals $x_1(t) = \sin t$ and $x_2(t) = \cos t$ over an interval $0 \leq t \leq 2\pi$ are always orthogonal to each other in the two-dimensional signal space since

$$\int_0^{2\pi} \sin t \cos t \, dt = 0. \tag{2.14}$$

Example 2.1

Verify the orthogonality of the following functions: $S_1(t) = 1$ and $S_2(t) = c(1-2t)$ in the interval [0, 1].

Solution

$$\int_0^1 S_1(t)S_2(t)dt = \int_0^1 1 \times c(1-2t)dt$$
$$= c\left[t - t^2\right]_{t=0}^{1}$$
$$= c[(1-1)-(0-0)] = 0.$$

Since $\int_0^1 S_1(t)S_2(t)dt = 0$, $S_1(t)$ and $S_2(t)$ are orthogonal over [0, 1].

2.3 Orthogonal Vector and Signal Space

2.3.1 Orthogonal Vector Space

If the unit vectors along the axes in the three-dimensional Cartesian coordinate system are represented by $\vec{a}_x, \vec{a}_y$ and $\vec{a}_z$, then any vector $\vec{A}$ in the coordinate system can be represented as

$$\vec{A} = A_x\vec{a}_x + A_y\vec{a}_y + A_z\vec{a}_z. \tag{2.15}$$

Since the unit vectors $\vec{a}_x, \vec{a}_y$ and $\vec{a}_z$ are orthogonal,

$$\vec{a}_x \cdot \vec{a}_y = \vec{a}_y \cdot \vec{a}_z = \vec{a}_z \cdot \vec{a}_x = 0 \quad \text{(the axes are orthogonal to each other)}$$

$$\vec{a}_x \cdot \vec{a}_x = \vec{a}_y \cdot \vec{a}_y = \vec{a}_z \cdot \vec{a}_z = 1 \quad \text{(the angle is zero).}$$

This can be expressed as

$$\vec{a}_m \cdot \vec{a}_n = \begin{cases} 0 & m \neq n \\ 1 & m = n. \end{cases} \tag{2.16}$$

It is observed that if a coordinate system has either one or two axes, it is not possible to express the vector in this coordinate system without the error vector. Therefore, the vector should be represented with the complete set of coordinate components, i.e., all the three coordinates in a three-dimensional system.

To understand this concept, let us take two basis vectors $\vec{x}_1$ and $\vec{x}_2$ orthogonal to each other. Any vector $\vec{A}$ can be expressed in the three-dimensional space in terms of these basis vectors as

$$\vec{A} = c_1\vec{x}_1 + c_2\vec{x}_2 + \vec{e}, \tag{2.17}$$

where c_1, c_2 are real variables and $\vec{e}$ is the error vector. For the best approximation,

$$\vec{e} = \vec{A} - (c_1\vec{x}_1 + c_2\vec{x}_2)$$

should be minimum, or from Eq.(2.17) the error vector $\vec{e}$ should be mutually orthogonal to $\vec{x}_1$ and $\vec{x}_2$.

Let $\vec{e} = c_3\vec{x}_3$ be a projection of $\vec{A}$ along another basis vector $\vec{x}_3$. Then,

$$\vec{A} = c_1\vec{x}_1 + c_2\vec{x}_2 + c_3\vec{x}_3. \tag{2.18}$$

By taking the dot product with basis vectors,

$$\vec{A}.\vec{x}_1 = \vec{x}_1.(c_1\vec{x}_1 + c_2\vec{x}_2 + c_3\vec{x}_3) = c_1\vec{x}_1 \cdot \vec{x}_1 + c_2\vec{x}_1 \cdot \vec{x}_2 + c_3\vec{x}_1 \cdot \vec{x}_1 = c_1\vec{x}_1 \cdot \vec{x}_1 + 0 + 0.$$

The constants are: $c_1 = \frac{\vec{A}.\vec{x}_1}{\vec{x}_1.\vec{x}_1}, \quad c_2 = \frac{\vec{A}.\vec{x}_2}{\vec{x}_2.\vec{x}_2} \quad \text{and} \quad c_3 = \frac{\vec{A}.\vec{x}_3}{\vec{x}_3.\vec{x}_3}.$ (2.19)

Thus, in the three dimensional space any vector $\vec{A}$ can be expressed in terms of three mutually orthogonal basis vectors $\vec{x}_1$, $\vec{x}_2$ and $\vec{x}_3$ with zero error, and the basis vectors $\vec{x}_1$, $\vec{x}_2$ and $\vec{x}_3$ represent the complete set of orthogonal vectors in three-dimensional space. We can extend this to an N-dimensional system, whereby a vector can be expressed completely in terms its N mutually perpendicular coordinate components.

Therefore, in the N-dimensional vector space, any vector $\vec{A}$ can be expressed in terms of N mutually orthogonal basis vectors $\vec{x}_1, \vec{x}_2, \vec{x}_3 \ldots\ldots \vec{x}_N$ with zero error. That is,

$$\vec{A} = c_1\vec{x}_1 + c_2\vec{x}_2 + c_3\vec{x}_3 + \cdots\cdots .. + c_N\vec{x}_N. \tag{2.20}$$

The basis vectors $\vec{x}_1, \vec{x}_2, \vec{x}_3 \ldots\ldots \vec{x}_N$ represent the complete set of orthogonal vectors in the N dimensional vector space, and any vector $\vec{A}$ can be expressed as a linear sum of these independent vectors called the basis vectors. The set of basis vectors define the orthogonal vector space. In general, the constants are

$$c_i = \frac{\vec{A}\cdot\vec{x}_i}{\vec{x}_i\cdot\vec{x}_i} = \frac{1}{x_i^2}\left(\vec{A}\cdot\vec{x}_i\right), \quad i = 1,2,3\cdots N. \tag{2.21}$$

The condition for the basis vectors $\vec{x}_1$, $\vec{x}_2, \vec{x}_3 \ldots\ldots \vec{x}_N$ to be mutually orthogonal is

$$\vec{x}_m \cdot \vec{x}_n = \begin{cases} 0 & m \neq n \\ K_m & m = n, \end{cases} \tag{2.22}$$

where $K_m = x_m^2$ is a constant.

If basis vectors $\vec{x}_1, \vec{x}_2, \vec{x}_3 \ldots\ldots \vec{x}_N$ are mutually orthonormal or normalised orthogonal vectors and represent an orthonormal vector space, then

$$\vec{x}_m \cdot \vec{x}_n = \begin{cases} 0 & m \neq n \\ 1 & m = n. \end{cases} \tag{2.23}$$

For example, in the 3-dimensional vector space, the vector (3, 4, 6) can be represented as

$$(3,4,6) = 3(1,0,0) + 4(0,1,0) + 6(0,0,1).$$

where (1, 0, 0), (0, 1, 0), and (0, 0, 1) are the complete set of basis vectors.

If the basis vectors $\vec{x}_1$, $\vec{x}_2, \vec{x}_3 \ldots\ldots \vec{x}_N$ are not a complete set of orthogonal vectors, then the error vector is

$$\vec{e} = \vec{A} - \sum_{i=0}^{N} c_n\vec{x}_n. \tag{2.24}$$

2.3.2 Orthogonal Signal Space

The concept of vector space can be extended to signals. In analogy with orthogonal vector space where any vector can be expressed as a sum of its components along the N-dimensional space, any function $f(t)$ can be expressed as a linear sum of mutually orthogonal functions if these functions form a complete set. Therefore, we can approximate any function $f(t)$ by a set of N mutually orthogonal functions.

2.4 Signal Approximation Using Orthogonal Functions

Consider N real signals $x_1(t), x_2(t), x_3(t) \cdots\cdots x_N(t)$ as a set of mutually orthogonal signals in the N-dimensional signal space over an interval $t_1 \le t \le t_2$. The condition for orthogonality is

$$\int_{t_1}^{t_2} x_m(t) x_n(t) dt = \begin{cases} 0 & m \ne n \\ E_n & m = n \end{cases} \tag{2.25}$$

where $E_n = \int_{t_1}^{t_2} x_n^2(t) dt$ (2.26)

represents the energy of the signal for all n.

Now any function $f(t)$ can be approximated in terms of mutually orthogonal signals $x_1(t), x_2(t), x_3(t) \cdots\cdots x_N(t)$, called the basis functions, over the interval $t_1 \le t \le t_2$.

$$f(t) \cong c_1 x_1(t) + c_2 x_2(t) + c_3 x_3(t) + \cdots\cdots + c_N x_N(t) = \sum_{n=1}^{N} c_n x_n(t), \tag{2.27}$$

where $c_1, c_2, \cdots, c_n$ are constants.
If the error signal is $f_e(t)$,

then $f(t) = \sum_{n=1}^{N} c_n x_n(t) + f_e(t)$ (2.28)

or $f_e(t) = f(t) - \sum_{n=1}^{N} c_n x_n(t).$ (2.29)

For the best approximation, the error signal should be minimised. That is, minimise:

$$e = \frac{1}{(t_2 - t_1)} \int_{t_1}^{t_2} f_e^2(t) dt = \frac{1}{(t_2 - t_1)} \int_{t_1}^{t_2} \left| f(t) - \sum_{n=1}^{N} c_n x_n(t) \right|^2 dt \quad t_1 \le t \le t_2 \tag{2.30}$$

where e is called the mean square error or power of the error signal.

We know that the necessary condition to minimise the mean square error with respect to the constant c is $\frac{\partial e}{\partial c_j} = 0,\ j = 1, 2, 3, \cdots, N.$

$$\frac{\partial e}{\partial c_j} = \frac{\partial}{\partial c_j}\left[\frac{1}{(t_2 - t_1)}\left(\int_{t_1}^{t_2} f_e^2(t) dt\right)\right] = \frac{\partial}{\partial c_j}\left[\frac{1}{(t_2 - t_1)} \int_{t_1}^{t_2} \left| f(t) - \sum_{n=1}^{N} c_n x_n(t) \right|^2 dt\right] = 0. \tag{2.31}$$

While expanding the integration terms, it is noticed that due to orthogonality the cross product of orthogonal functions are zero. This means that all the terms other than the jth term (terms without c_j) are zero. Hence

$$\frac{\partial}{\partial c_j}\left(\int_{t_1}^{t_2} f^2(t) dt\right) = \frac{\partial}{\partial c_j}\left(\int_{t_1}^{t_2} c_n^2 x_n^2(t) dt\right) = \frac{\partial}{\partial c_j}\left(\int_{t_1}^{t_2} c_n f(t) x_n(t) dt\right) = 0.$$

The remaining two non-zero terms are:

$$\frac{\partial}{\partial c_j} \int_{t_1}^{t_2} \left[-2c_j f(t) x_j(t) + c_j^2 x_j^2(t)\right] dt = 0.$$

Changing the order of differentiation and integration,

$$\int_{t_1}^{t_2}\left[-2\frac{\partial}{\partial c_j}(c_j f(t)x_j(t))+\frac{\partial}{\partial c_j}(c_j^2x_j^2(t))\right]dt=0$$

$$\int_{t_1}^{t_2} 2f(t)x_j(t)dt=\int_{t_1}^{t_2} 2c_jx_j^2(t)dt$$

$$\text{or}\quad c_j=\frac{\int_{t_1}^{t_2} f(t)x_j(t)dt}{\int_{t_1}^{t_2} x_j^2(t)dt}=\frac{\int_{t_1}^{t_2} f(t)x_j(t)dt}{E_j}. \tag{2.32}$$

Thus, for the best approximation of the function given in Eq. (2.27), the constants should be chosen from Eq. (2.32).

Example 2.2

Obtain the condition under which two signals $f_1(t)$ and $f_2(t)$ are orthogonal to each other. Hence prove that $\cos(n\omega_0 t)$ and $\cos(m\omega_0 t)$ are orthogonal over any interval $\left(t_0, t_0+\frac{2\pi}{\omega_0}\right)$ for integer values of n and m.

Solution

Given that $f_1(t)=\cos(n\omega_0 t)$ and $f_2(t)=\cos(m\omega_0 t)$ over the interval $[t_0,\ t_0,+2\pi/\omega_0]$.

The condition for orthogonality between two signals $f_1(t)$ and $f_2(t)$ in this case is:

$$\int_{t_0}^{t_0+\frac{2\pi}{\omega_0}} f_1(t).f_2(t)dt=0$$

$$\int_{t_0}^{t_0+\frac{2\pi}{\omega_0}} \cos n\omega_0 t.\cos m\omega_0 t dt=\frac{1}{2}\int_{t_0}^{t_0+\frac{2\pi}{\omega_0}}[\cos(n+m)\omega_0 t+\cos(n-m)\omega_0 t]dt$$

$$=\frac{1}{2}\left[\frac{\sin(n+m)\omega_0 t}{(n+m)\omega_0}+\frac{\sin(n-m)\omega_0 t}{(n-m)\omega_0}\right]_{t_0}^{t_0}+\frac{2\pi}{\omega_0}$$

$$=\frac{1}{2}\left[\left(\frac{\sin(n+m)(\omega_0 t_0+2\pi)}{(n+m)\omega_0}+\frac{\sin(n-m)(\omega_0 t_0+2\pi)}{(n-m)\omega_0}\right)\right.$$

$$\left.-\left(\frac{\sin(n+m)\omega_0 t_0}{(n+m)\omega_0}+\frac{\sin(n-m)\omega_0 t_0}{(n-m)\omega_0}\right)\right]$$

$$=0$$

$$[\because \sin(2\pi N+\theta)=\sin\theta].$$

Therefore, $\cos(n\omega_0 t)$ and $\cos(m\omega_0 t)$ are orthogonal over any interval $\left(t_0, t_0+\frac{2\pi}{\omega_0}\right)$ for all integer values of n and m.

2.5 Mean Square Error

The mean square error (MSE) in signal approximation can be evaluated from the orthogonal functions when the optimum values of $c_1, c_2, c_3, \cdots\cdots c_N$ are obtained from the Eq. (2.32).

$$c_j = \frac{1}{E_j}\int_{t_1}^{t_2} f(t)x_j(t)dt.$$

Now, the mean square error

$$e = \frac{1}{(t_2-t_1)}\int_{t_1}^{t_2}\left|f(t)-\sum_{n=1}^{N} c_n x_n(t)\right|^2 dt \tag{2.33}$$

$$e = \frac{1}{(t_2-t_1)}\left[\int_{t_1}^{t_2} f^2(t)dt + \sum_{n=1}^{N} c_n^2\int_{t_1}^{t_2} x_n^2(t)dt - 2\sum_{n=1}^{N} c_n\int_{t_1}^{t_2} f(t)x_n(t)dt\right]. \tag{2.34}$$

From Eq. (2.32), $\int_{t_1}^{t_2} f(t)x_n(t)dt = c_n E_n.$

Therefore, the MSE is

$$e = \frac{1}{(t_2-t_1)}\left[\int_{t_1}^{t_2} f^2(t)dt + \sum_{n=1}^{N} c_n^2 E_n - 2\sum_{n=1}^{N} c_n c_n E_n\right] = \frac{1}{(t_2-t_1)}\left[\int_{t_1}^{t_2} f^2(t)dt - \sum_{n=1}^{N} c_n^2 E_n\right]$$

$$= \frac{1}{(t_2-t_1)}\left[\int_{t_1}^{t_2} f^2(t)dt - (c_1^2E_1 + c_2^2E_2 + c_3^2E_3 + \cdots\cdots + c_N^2E_N)\right]. \tag{2.35}$$

2.6 Closed or Complete Set of Mutually Orthogonal Functions

The mean square error for N-dimensional orthogonal space from Eq. (2.35) is

$$e = \frac{1}{(t_2-t_1)}\left[\int_{t_1}^{t_2} f^2(t)dt - (c_1^2E_1 + c_2^2E_2 + c_3^2E_3 + \cdots\cdots + c_N^2E_N)\right].$$

It is observed that as $N \to \infty$, the approximation becomes more accurate with large number of orthogonal functions and the error $e \to 0$ and the orthogonal set becomes a complete set.

When the error becomes zero,

$$\int_{t_1}^{t_2} f^2(t)dt = \sum_{n=1}^{\infty} c_n^2 E_n. \tag{2.36}$$

The energy of $f(t)$is equal to the sum of the energies of the orthogonal components. This equation is called *Parseval's Theorem.*

Therefore the function from Eq. (2.27) becomes

$$f(t) = \sum_{n=1}^{\infty} c_n x_n(t) \quad t_1 \le t \le t_2. \tag{2.37}$$

Hence the infinite series with weighted sum of orthogonal functions converges to the function $f\ (t)$ such that the mean square error is zero.

Therefore a set of functions $x_1(t), x_2(t), x_3(t)\cdots\cdots x_N(t)$ mutually orthogonal over the interval $t_1 \leq t \leq t_2$ is said be a closed or a complete set if there exist no function $f(t)$ for which it is true that

$$\int_{t_1}^{t_2} f(t)x_n(t)dt = 0, \quad \text{for } n = 1,2,3\ldots \tag{2.38}$$

That is, there is no function $f(t)$existing other than the set of functions$x_1(t), x_2(t), x_3(t)\cdots\cdots x_N(t)$. Obviously, this means that the function$f(t)$ should be equal to any one of the basis function in the complete orthogonal set.

Note: In the N-dimensional vector space, any vector can be expressed as a sum of its components resolved along N mutually orthogonal vectors, provided they form a complete set. Similarly any function $f(t)$ can be expressed as a sum of its components resolved along mutually orthogonal functions, provided they form a closed or complete orthogonal set.

2.7 Orthogonality in Complex Signals

Our discussion so far has been restricted to the case where the basis functions $x(t)$ are real valued. Now, let the basis functions $x(t)$ be complex functions of the real variable t. Let us approximate a function $f(t)$ by a complex function $x(t)$ over an interval $t_1 \leq t \leq t_2$.

$$f(t) \approx cx(t); \quad t_1 \leq t \leq t_2. \tag{2.39}$$

The error in approximation is,

$$f_e(t) = f(t) - cx(t). \tag{2.40}$$

The optimum value of c to minimise the mean square error over the interval $[t_1, t_2]$ is,

$$e = \frac{1}{(t_2-t_1)}\left[\int_{t_1}^{t_2} |f(t)-cx(t)|^2 dt\right] = \frac{1}{(t_2-t_1)}\left[\int_{t_1}^{t_2} \{f(t)-cx(t)\}\{f^*(t)-cx^*(t)\}dt\right] \tag{2.41}$$

$$= \int_{t_1}^{t_2} |f(t)|^2 dt + c^2\int_{t_1}^{t_2} |x(t)|^2 dt - c\int_{t_1}^{t_2} f^*(t)x(t)dt - c\int_{t_1}^{t_2} f(t)x^*(t)dt$$

$$= \int_{t_1}^{t_2} |f(t)|^2 dt + c^2 E_x - c\int_{t_1}^{t_2} f^*(t)x(t)dt - c\int_{t_1}^{t_2} f(t)x^*(t)dt$$

$$+\left[\frac{1}{\sqrt{E_x}}\int_{t_1}^{t_2} f(t)x^*(t)dt\right]^2 - \left(\frac{1}{\sqrt{E_x}}\int_{t_1}^{t_2} f(t)x^*(t)dt\right)^2$$

$$= \int_{t_1}^{t_2} |f(t)|^2 dt + \left[c\sqrt{E_x} - \frac{1}{\sqrt{E_x}}\int_{t_1}^{t_2} f(t)x^*(t)dt\right]^2 - \left[\frac{1}{\sqrt{E_x}}\int_{t_1}^{t_2} f(t)x^*(t)dt\right]^2, \tag{2.42}$$

where $x^*(t)$ is the complex conjugate of $x(t)$ and $f^*(t)$ is the complex conjugate of $f(t)$.

$$E_x = \int_{t_1}^{t_2} |x(t)|^2 dt$$

is the energy of the signals.

We have to minimise the error energy by choosing suitable value of c. It is seen from the above equation that the first and the last terms are not functions of c. We can minimise e by setting the second term to zero, which yields,

$$c\sqrt{E_x} - \frac{1}{\sqrt{E_x}}\int_{t_1}^{t_2} f(t)x^*(t)dt = 0$$

$$c = \frac{1}{E_x}\int_{t_1}^{t_2} f(t)x^*(t)dt \text{ or } c = \frac{\int_{t_1}^{t_2} f(t)x^*(t)dt}{\int_{t_1}^{t_2} x(t)x^*(t)dt}. \tag{2.43}$$

If $f(t)$ and $x(t)$ are orthogonal, $c = 0$. Thus, if the two signals $x_1(t)$ and $x_2(t)$ are orthogonal over an interval $[t_1, t_2]$, then,

$$\int_{t_1}^{t_2} x_1(t)x_2^*(t)dt = 0. \tag{2.44}$$

Similarly, consider N complex signals $x_1(t), x_2(t), x_3(t), \cdots, x_n(t)$ as a set of mutually orthogonal signals in the N-dimensional signal space over an interval $t_1 \le t \le t_2$. The condition for orthogonality is

$$\int_{t_1}^{t_2} x_m(t)\, x_n^*(t)\; dt = \begin{cases} 0 & m \neq n \\ E_n & m = n \end{cases} \tag{2.45}$$

where $x_n^*(t)$ is the complex conjugate of $x_n(t)$. The set of basis functions $x_n(t)$ is said to be normalised if the energy

$$E_n = \int_{t_1}^{t_2} |x_n(t)|^2\, dt = 1 \quad \text{for all } n. \tag{2.46}$$

Now, any function $f(t)$ can be approximated over on interval $t_1 \le t \le t_2$ as

$$f(t) \cong c_1x_1(t) + c_2x_2(t) + c_3x_3(t) + \cdots + c_nx_n(t) = \sum_{n=1}^{N} c_nx_n(t).$$

The error function is $f_e(t) = f(t) - \sum_{n=1}^{N} c_nx_n(t)$. For best approximation, the minimum error function contants $c_1,\ c_2,\ c_3 \cdots c_n$ are

$$c_j = \frac{\int_{t_1}^{t_2} f(t)x_j^*(t)dt}{\int_{t_1}^{t_2} x_j(t)dt} = \frac{1}{E_j}\int_{t_1}^{t_2} f(t)x_j^*(t)dt. \tag{2.47}$$

Note 1: For real basis functions, $x_n^*(t) = x_n(t)$.
Note 2: Energy of orthogonal signals:
If $x(t) = x_1(t) \pm x_2(t)$; $\quad t_1 \le t \le t_2$ where $x_1(t)$ and $x_2(t)$ are orthogonal, then the energy of orthogonal signal is

$$E_x = E_{x_1} + E_{x_2}. \tag{2.48}$$

Thus the energy of the sum or difference of two orthogonal signals is equal to the sum of the energies of the two signals.

Note 3: Sinusoidal functions $\sin(n\omega_0 t)$, cosine functions $\cos(n\omega_0 t)$, and complex exponential functions $e^{\pm jn\omega_0 t}$ for $n = 1,2,3\cdots..$ are mutually orthogonal over any interval $\left(t_0, t_0 + \frac{2\pi}{\omega_0}\right)$ for all integer values of n and m.

Example 2.3

Prove that complex exponential functions are orthogonal functions.

Solution

Let $f_1(t) = e^{jn\omega_0 t}$ and $f_2(t) = e^{jm\omega_0 t}$ be two complex functions over any interval $\left(t_0, t_0 + \frac{2\pi}{\omega_0}\right)$ for all integer values of n and m. Then the condition for orthogonality is

$$\int_{t_0}^{t_0+\frac{2\pi}{\omega_0}} f_1(t).f_2^*(t)dt = \int_{t_0}^{t_0+\frac{2\pi}{\omega_0}} f_1^*(t).f_2(t)dt = 0$$

Now,

$$\int_{t_0}^{t_0+\frac{2\pi}{\omega_0}} f_1(t)f_2^*(t)dt = \int_{t_0}^{t_0+\frac{2\pi}{\omega_0}} e^{jn\omega_0 t}(e^{jm\omega_0 t})^* dt = \int_{t_0}^{t_0+\frac{2\pi}{\omega_0}} e^{jn\omega_0 t}e^{-jm\omega_0 t}dt$$

$$= \int_{t_0}^{t_0+\frac{2\pi}{\omega_0}} e^{j(n-m)\omega_0 t}dt = \left[\frac{e^{j(n-m)\omega_0 t}}{j(n-m)\omega_0}\right]_{t_0}^{t_0+\frac{2\pi}{\omega_0}}$$

$$= \frac{e^{j(n-m)\omega_0 t_0}e^{j(n-m)2\pi} - e^{j(n-m)\omega_0 t_0}}{j(n-m)\omega_0} = \frac{e^{j(n-m)\omega_0 t_0} - e^{j(n-m)\omega_0 t_0}}{j(n-m)\omega_0} = 0$$

since $e^{j(n-m)2\pi} = 1$.

Hence the complex exponential functions are orthogonal over any interval $\left(t_0, t_0 + \frac{2\pi}{\omega_0}\right)$ for all integer values of n and m.

2.8 Exponential and Sinusoidal Signals

From the previous sections it is evident that an arbitrary function can be expressed in terms of exponential or sinusoidal functions. Exponential signals are extremely important in the study of signals and in the analysis of systems. They are used as a driving function for most of the applications of analysing linear systems. Some of their unique properties are:

(i) Exponential signals are invariant under the action of linear time-invariant systems. That is, the response of an LTI system to an exponential signal is also an exponential signal.

(ii) Any arbitrary signal can be expressed as a sum of various exponential functions.

(iii) Transform tools like Fourier, Laplace and Z-transform use complex exponential signals for analysis.

(iv) Many physical signals such as sound heat and light that occur in nature can be expressed in terms of some basic signal types such as sinusoids and exponential signals.

2.8.1 Exponential Signal

An exponential signal can be categorised into the following signal types:

1. *Continuous time complex exponential signal*

   The continuous-time complex exponential signal is defined as

   $$f(t) = ce^{\alpha t} \qquad -\infty < t < \infty \tag{2.49}$$

   where c and α are complex numbers. Depending upon the values of these parameters, the complex exponential can take on several different characteristics.

2. *Real exponential signal*

   A real exponential signal is defined as

   $$f(t) = ce^{\alpha t}, \tag{2.50}$$

   where both c and α are real as explained in Section 1.3.

3. *Imaginary exponential signal*

   imaginary exponential signal is defined as

   $$f(t) = e^{j\omega_0 t} \qquad -\infty < t < \infty \tag{2.51}$$

   where $\omega_0 = \dfrac{2\pi}{T}$ is the fundamental frequency. An important property of this signal is that it is periodic with period T since $f(t) = f(t+T)$ for all t. That is,

   $$e^{j\omega_0 t} = e^{j\omega_0 t} \cdot e^{j\omega_0 T}.$$

   If $\omega_0 \neq 0$, then the fundamental period $T = 2\pi/\omega_0$ seconds or $\omega_0 T = 2\pi$ and $e^{j\omega_0 T} = 1$. So, the signals $e^{j\omega_0 t}$ and $e^{-j\omega_0 t}$ both are periodic with the same fundamental period.

   If $\omega_0 = 0$, then $f(t) = 1$, which is constant for any value of T.

2.8.2 Sinusoidal Signal

A continuous sinusoidal signal, in general, can be expressed as

$$f(t) = A\cos(\omega_0 t + \phi), \tag{2.52}$$

where A is the amplitude, ω_0 is the angular frequency in radians per second, ϕ is the phase shift in radians and $\omega_0 = 2\pi f_0 = \dfrac{2\pi}{T}$ as explained in Section 1.3.

An important property of the sinusoidal signal is that it is periodic with single fundamental frequency f_0 and time period T. Since $f(t)$ is periodic with period T,

$$f(t) = f(t+T) \text{ for all } t,$$

$$\text{i.e.,} \quad f(t+T) = A\cos(\omega_0(t+T) + \phi). \tag{2.53}$$

If $\omega_0 \neq 0$, then the fundamental period $T = \frac{2\pi}{\omega_0}$ seconds, or $\omega_0 T = 2\pi$. Then

$$f(t+T) = A\cos(\omega_0(t+T) + \phi) = A\cos(\omega_0 t + 2\pi + \phi)$$

$$= A\cos(\omega_0 t + \phi) = f(t).$$

If $\omega_0 = 0$, then $f(t) = A\cos\phi$, which is a constant for any value of T.

Sinusoidal Signals in Terms of Complex Exponential

The imaginary exponential signal $f(t) = e^{j\omega_0 t}$ can be written in terms of sinusoidal signals with the same fundamental period by using Euler's relation

$$e^{j\omega_0 t} = \cos\omega_0 t + j\sin\omega_0 t. \tag{2.54}$$

Similarly, a sinusoidal signal can be written in terms of periodic complex exponentials, again with the same fundamental period,

$$\cos(\omega_0 t) = \frac{1}{2}\left(e^{j\omega_0 t} + e^{\omega_0 t}\right) \text{ and } \sin(\omega_0 t) = \frac{1}{2j}\left(e^{j\omega_0 t} - e^{-j\omega_0 t}\right) \tag{2.55}$$

With phase angle

$$\cos(\omega_0 t + \phi) = \frac{1}{2}\left(e^{j\phi} e^{j\omega_0} t + e^{-j\omega_0 t}\right) \text{ and } \sin(\omega_0 t + \phi) = \frac{1}{2j}\left(e^{j\phi} e^{j\omega_0 t} - e^{-j\phi} e^{-j\omega_0 t}\right). \tag{2.56}$$

Note that the two exponentials in Eq. (2.56) have complex amplitudes. Alternatively, we can express a sinusoidal signal in term of the complex exponential signal as

$$\cos(\omega_0 t + \phi) = \text{Re}\left\{e^{j(\omega_0 t + \phi)}\right\}$$

$$\sin(\omega_0 t + \phi) = \text{Im}\left\{e^{j(\omega_0 t + \phi)}\right\} \tag{2.57}$$

where Re{.} denotes the real part and Im {.} denotes the imaginary part of the complex signal. Further, with $\phi = 0$,

$$\text{Re}\left\{e^{j\omega_0 t}\right\} = \cos(\omega_0 t) \text{ and } \text{Im}\left\{e^{j\omega_0 t}\right\} = \sin(\omega_0 t). \tag{2.58}$$

The corresponding waveforms are shown in Fig. 2.2.

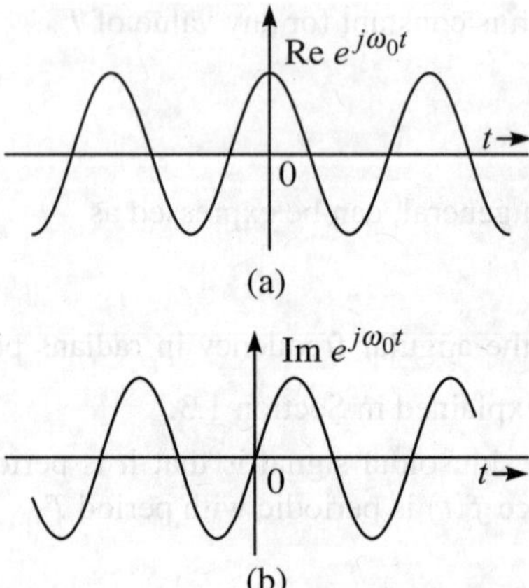

Fig. 2.2 *(a) Real part and (b) imaginary part of the complex exponential*

Damped Sinusoidal Signal

Consider the complex exponential signal

$$f(t) = ce^{\alpha t}, \ -\infty < t < \infty \tag{2.59}$$

where c and α are complex variables. If c is expressed in polar form and α in rectangular form, i.e., $c = |c|e^{j\theta}$ and $\alpha = r + j\omega_0$, then

$$ce^{\alpha t} = |c|\, e^{j\theta} e^{(r+j\omega_0)t} = |c|\, e^{rt} e^{j(\omega_0 t+\theta)}.$$

Using Euler's relation, the expansion of the above is

$$ce^{at} = |c|\, e^{rt}\cos(\omega_0 t+\theta) + j\,|c|\, e^{rt}\sin(\omega_0 t+\theta). \tag{2.60}$$

Thus, for $r = 0$, the real and imaginary parts of a complex exponential are sinusoidal. For $r > 0$ they correspond to sinusoidal signals multiplied by a growing exponential, and for $r < 0$ they correspond to sinusoidal signals multiplied by a decaying exponential. These two cases are shown on Fig. 2.3. The dashed curves show the envelope of the oscillatory waves.

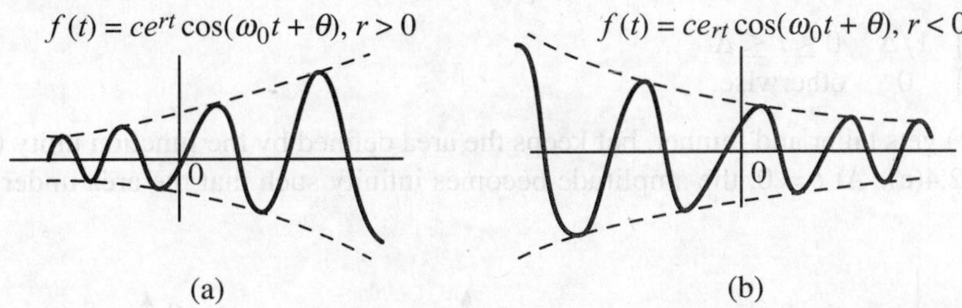

Fig. 2.3 *(a) Growing sinusoidal signal, and (b) decaying sinusoidal signal*

Example 2.4

Determine whether or not each of the following signals is periodic. If periodic, determine the fundamental period.

(a) $x(t) = e^{-j2t}$ (b) $x(t) = e^{-j5t}\cos \pi t$

Solution

(a) Given $x(t) = e^{-j2t} = \cos 2t - j\sin 2t$.

Here, the signals have the period $T = \dfrac{2\pi}{\omega_1} = \dfrac{2\pi}{2} = \pi$ s. Hence the signal is periodic signal.

(b) Given $x(t) = e^{-j5t}\cos \pi t = (\cos 5t - j\sin 5t)\cos \pi t$.

$$= (\cos 5t \cos \pi t - j\sin 5t\cos \pi t)$$

$$= \frac{1}{2}(\cos(5+\pi)t + \cos(5-\pi)t - j\sin(5+\pi)t - j\sin(5-\pi)t).$$

The periods of the two components of the signal are $T_1 = \dfrac{2\pi}{\omega_1} = \dfrac{2\pi}{5+\pi}$ sec and $T_2 = \dfrac{2\pi}{\omega_2} = \dfrac{2\pi}{5-\pi}$ sec.

The ratio $\dfrac{T_1}{T_2} = \dfrac{2\pi/(5-\pi)}{2\pi/(5+\pi)} = \dfrac{5+\pi}{5-\pi}$ is not a rational number and hence the signal is aperiodic.

2.9 Concepts of Impulse Function

The continuous unit impulse function, which was defined in Section 1.3, is

$$\delta(t) = \begin{cases} 1 & t = 0 \\ 0 & \text{otherwise.} \end{cases} \tag{2.61}$$

The delayed unit impulse function is defined as

$$\delta(t-t_0)=\begin{cases}1 & \text{for } t=t_0\\ 0 & \text{otherwise.}\end{cases} \tag{2.62}$$

and

$$\int_{-\infty}^{\infty}\delta(t-t_0)\,dt=1. \tag{2.63}$$

Since at the time instant $t=0$ it is not possible to define the function, the impulse function may be defined for an infinitesimal time period Δ as (Fig. 2.4(b))

$$\delta_\Delta(t)=\begin{cases}1/\Delta & 0\le t\le\Delta\\ 0 & \text{otherwise.}\end{cases} \tag{2.64}$$

As $\Delta\to 0$, $\delta_\Delta(t)$ gets taller and thinner, but keeps the area defined by the function unity (a constant) as shown in Fig. 2.4(c). At $t=0$, the amplitude becomes infinity such that the area under the curve is a finite value.

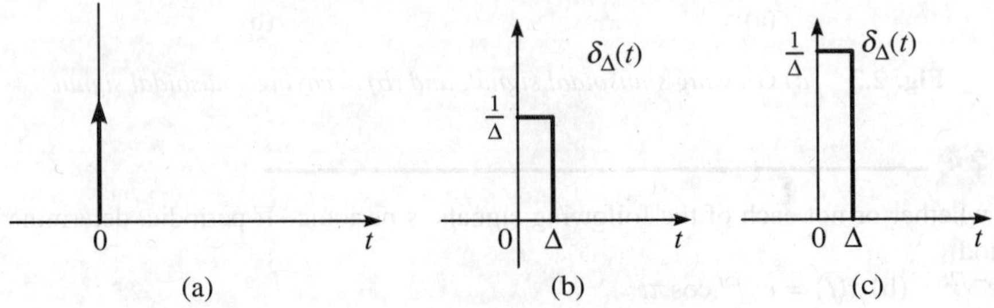

Fig. 2.4 *(a) Unit impulse function and (b) and (c) unit impulse function for the period* Δ

At the limit, the continuous unit impulse function can be expressed as

$$\delta(t)=\lim_{\Delta\to 0}\delta_\Delta(t)\ \text{ or }\ \delta(t)=\begin{cases}1 & t=0\\ 0 & \text{otherwise.}\end{cases} \tag{2.65}$$

The unit impulse function $\delta(t)$, also known as the Dirac delta function, plays a central role in the analysis of the signals and systems. Traditionally, $\delta(t)$ is often defined as the limit of a suitably chosen conventional function having unity area over an infinitesimal time interval.

Some of the conventional functions from which the impulse function can be derived are

(a) Unit step function,

$$\delta(t)=\frac{du(t)}{dt}\ \text{ and }\ u(t)=\int_{-\infty}^{t}\delta(\tau)d\tau. \tag{2.66}$$

(b) The rectangular function,

$$\delta(t)=\lim_{\tau\to 0}\frac{1}{\tau}\left[u(t+\tau/2)-u(t-\tau/2)\right]. \tag{2.67}$$

(c) The exponential function,

$$\delta(t)=\lim_{\tau\to\infty}\frac{1}{\tau}e^{-t/\tau}u(t). \tag{2.68}$$

(d) The unit ramp function,

$$\delta(t) = \lim_{\tau\to\infty} \frac{1}{\tau}\left(1-\frac{|t|}{\tau}\right)u(t), \quad |t| < \tau. \tag{2.69}$$

(e) The sampling function,

$$\delta(t) = \lim_{t\to\infty} \frac{k}{\pi}\ \mathrm{Sa}\,(kt). \tag{2.70}$$

Properties of the Unit Impulse Function

1. The area of an impulse function is always unity.

$$\int_{-\infty}^{\infty} \delta(t)\,dt = \int_{0-}^{0^+} \delta(t)\,dt = 1 \tag{2.71}$$

2. *Sampling property:* The sampling of any signal with impulse function gives the instantaneous value of a signal at given time instant.

$$\int_{-\infty}^{\infty} f(t)\ \delta(t)\ dt = f(0) \tag{2.72}$$

Proof: Consider an arbitrary function $f(t)$ for all time. Then,

$$\begin{aligned}\int_{-\infty}^{\infty} f(t)\,\delta(t)\ dt &= \int_{-\infty}^{0^-} f(t)\,\delta(t)\ dt + \int_{0-}^{0^+} f(t)\,\delta(t)\ dt + \int_{0+}^{\infty} f(t)\,\delta(t)\ dt\\ &= 0 + \int_{0-}^{0^+} f(t)\ \delta(t)\ dt + 0\\ &= f(0)\int_{0-}^{0^+} \delta(t)\ dt = f(0). \qquad \text{(Proved)}\end{aligned}$$

Note: The product of any function with impulse unit function gives

$$f(t)\ \delta(t) = f(0)\ \delta(t). \tag{2.73}$$

Similarly for other time instants,

$$f(t-t_0)\ \delta(t-t_1) = f(t_1-t_0)\ \delta(t-t_1). \tag{2.74}$$

3. *Shifting property*: It gives the instantaneous value of a signal at a given time instant. That is, the sampling of any delayed signal with delayed impulse function given

$$\int_{-\infty}^{\infty} f(t)\,\delta(t-t_0)\,dt = f(t_0) \text{ and } \int_{-\infty}^{\infty} f(t-t_0)\ \delta(t-t_1)\ dt = f(t_1-t_0). \tag{2.75}$$

Proof: Consider an arbitrary function $f(t)$ for all time. Then,

$$\begin{aligned}\int_{-\infty}^{\infty} f(t)\delta(t-t_0)dt &= \int_{-\infty}^{t_0^-} f(t)\delta(t-t_0)dt + \int_{t_0^-}^{t_0^+} f(t)\delta(t-t_0)dt + \int_{t_0^+}^{\infty} f(t)\delta(t-t_0)dt\\ &= 0 + \int_{t_0^-}^{t_0^+} f(t)\delta(t-t_0)dt + 0\\ &= f(t_0)\int_{t_0^-}^{t_0^+} \delta(t-t_0)dt = f(t_0) \qquad \text{(Proved)}\end{aligned}$$

Note: If $f(t)$ is defined in the finite interval $-t_1 < t < t_2$, then

$$\int_{-t_1}^{t_2} f(t)\delta(t-t_0)dt = \begin{cases} f(t_0) & t_1 < t_0 < t_2 \\ 0 & \text{otherwise.} \end{cases} \tag{2.76}$$

4. *Symmetrical property*: The impulse function has even symmetry. That is,

$$\delta(t) = \delta(-t). \tag{2.77}$$

Proof : Consider the integral of impulse function $\int_{-\infty}^{\infty} \delta(-t)\, dt$.

Let $x = -t$, then $dx = -dt$.

$$\int_{-\infty}^{\infty} \delta(-t)\, dt = -\int_{\infty}^{-\infty} \delta(x)\, dx.$$

Since the impulse function has even symmetry, we can interchange the integration limits.

$$\int_{-\infty}^{\infty} \delta(-t)\, dt = \int_{-\infty}^{\infty} \delta(x)\,(-dx) = \int_{-\infty}^{\infty} \delta(t)\, dt.$$

Eliminating the integration, we get $\delta(t) = \delta(-t)$ (Proved).

5. *Scaling property*:

$$\delta(at) = \frac{1}{|a|}\delta(t), \text{where } a \text{ is a constant} \tag{2.78}$$

Proof: Consider an impulse function $\delta(at)$ with scaling factor a.

If $a > 0$, the integral is $\int_{-\infty}^{\infty} \delta(at)\, dt$.

Let $x = at$; then $dx = adt$.

Now, $\int_{-\infty}^{\infty} \delta(at)\, dt = \frac{1}{a}\int_{-\infty}^{\infty} \delta(x)\, dx = \frac{1}{a}$.

If $a < 0$,

let $x = -at$. Then $dx = -adt$.

Now, $\int_{-\infty}^{\infty} \delta(at)\, dt = \frac{-1}{a}\int_{\infty}^{-\infty} \delta(-x)\, dx$.

Interchanging the integration limits,

$$\int_{-\infty}^{\infty} \delta(at)\, dt = \frac{-1}{a}\int_{-\infty}^{\infty} \delta(x)\, dx = \frac{-1}{a}.$$

Therefore, $\frac{1}{|a|}\int_{-\infty}^{\infty} \delta(t)\, dt = \frac{1}{|a|}$. (Proved)

6. *Differentiation property*: For any function $f(t)$ defined in the interval $-\infty < t < \infty$, if its derivative $\dot{f}(t) = \frac{df(t)}{dt}$ is continuous at $t = t_0$, then,

$$\int_{-\infty}^{\infty} \frac{df(t)}{dt}\delta(t-t_0)dt = \left.\frac{df(t)}{dt}\right|_{t=t_0} \tag{2.79}$$

$$\text{and} \quad \int_{-\infty}^{\infty} f(t)\frac{d\delta(t-t_0)}{dt}dt = -\left.\frac{df(t)}{dt}\right|_{t=t_0} \tag{2.80}$$

Note: The derivative of unit impulse function $\dot{\delta}(t) = \frac{d\delta(t)}{dt}$ is called unit doublet.

Proof: Given a function $f(t)$ defined in the interval $-\infty < t < \infty$, if its derivative $\dot{f}(t) = \frac{df(t)}{dt}$ is continuous at $t - t_0$, then,

(a) Evaluating the integration by parts:

$$\int_{-\infty}^{\infty} \delta(t)\frac{df(t-t_0)}{dt}dt = \frac{df(t-t_0)}{dt}\int_{-\infty}^{\infty}\delta(t)dt$$

$$= \left.\frac{df(t)}{dt}\right|_{t=t_0} \quad \text{(Proved)}$$

(b) Evaluating the integration by parts:

$$\int_{-\infty}^{\infty} f(t)\frac{d\delta(t-t_0)}{dt}dt = f(t)\delta(t)\Big|_{-\infty}^{\infty} - \int_{-\infty}^{\infty}\frac{df(t)}{dt}\delta(t)dt$$

Since if $t \to \pm\infty, \quad x(t) \to 0,$

$$\int_{-\infty}^{\infty} f(t)\frac{d\delta(t-t_0)}{dt}dt = 0 - \int_{-\infty}^{\infty}\frac{df(t)}{dt}\delta(t-t_0)dt$$

$$= -\left.\frac{df(t)}{dt}\right|_{t=t_0} \quad \text{(Proved)}$$

2.10 Unit Step Function and Signum Function

2.10.1 Unit Step Function

The unit step function $u(t)$ (defined in Section 1.3) given by

$$u(t) = \begin{cases} 1 & t \geq 0 \\ 0 & t < 0 \end{cases} \tag{2.81}$$

is shown in Fig. 2.5(a). Since the function is discontinuous at $t = 0$ (2.81), the value at $t = 0$ is undefined.

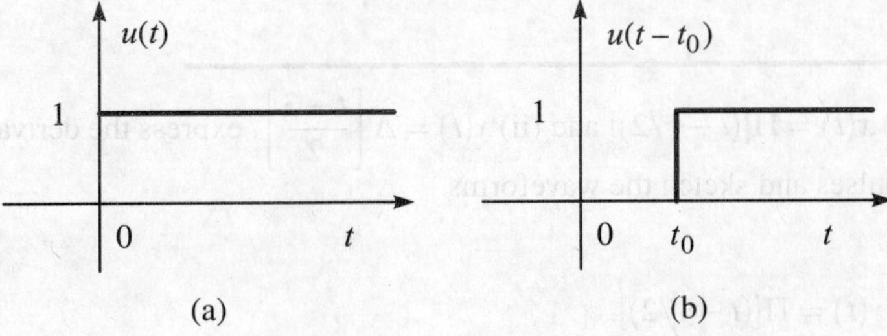

Fig. 2.5 *(a) Unit step function and (b) delayed unit step function*

Note: Any signal when multiplied with a unit step function results in a causal signal.
A delayed unit step function $u(t-t_0)$ as shown in Fig. 2.5 (b) is defined as

$$u(t-t_0) = \begin{cases} 1 & t \geq t_0 \\ 0 & t < t_0. \end{cases} \tag{2.82}$$

The unit step function is related to the unit impulse function as

$$u(t) = \int_{-\infty}^{t} \delta(\tau)\, d\tau. \tag{2.83}$$

Also, $$\delta(t) = \frac{du(t)}{dt}$$

and $$u(t-t_0) = \int_{-\infty}^{t-t_0} \delta(\tau)\, d\tau. \tag{2.84}$$

2.10.2 Signum Function

The signum function $sgn\,(t)$ is another useful function employed in signal analysis and is defined as

$$\text{sgn}\,(t) = \begin{cases} -1 & t < 0 \\ 1 & t > 0 \end{cases}. \tag{2.85}$$

Figure 2.6 shows a sketch of the function. The signum function is related the unit step function, as follows:

$$\text{sgn}\,(t) = 2u\,(t) - 1 \;\text{ or }\; u(t) = \frac{1}{2}[\text{sgn}(t) + 1]. \tag{2.86}$$

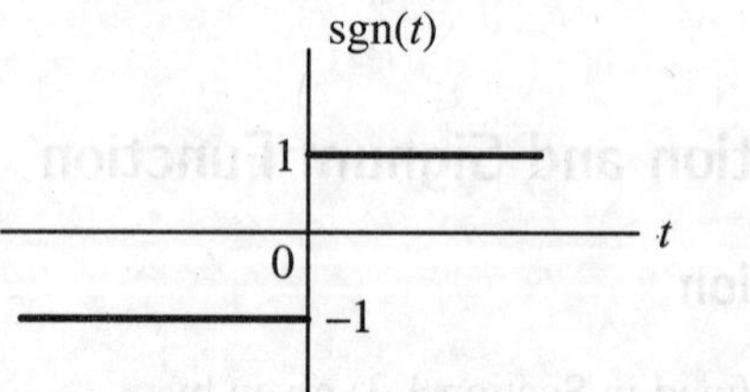

Fig. 2.6 *Signum function*

The signum function is characterised by a sign change at $t = 0$. Its value at $t = 0$ is also undefined and is chosen as zero for practical purposes. Just as the step function (or its folded version) can be used to represent the turning of a signal $x(t)$ on (or off), the signum function can be used to switch the polarity of a signal.

Example 2.5

Given the signals (i) $x(t) = \Pi[(t-1/2)]$ and (ii) $x(t) = \Delta\left[\frac{t-3}{2}\right]$, express the derivatives of the signals in terms of unit impulses and sketch the waveforms.

Solution

Given the signal (i) $x(t) = \Pi[(t-1/2)]$.

(i) The signal is expressed as $x(t) = \begin{cases} 1 & 0 < t < 1 \\ 0 & \text{otherwise} \end{cases}$.

It is a gate function centered at $t = 1/2$ and width 1.

Using unit step functions, $x(t) = u(t) - u(t-1)$.

The derivative of the signal is $\dfrac{dx(t)}{dt} = \dfrac{du(t)}{dt} - \dfrac{du(t-1)}{dt} = \delta(t) - \delta(t-1)$.

The waveforms are shown in Fig. 2.7(a).

(ii) Given the signal $x(t) = \Delta\left[\dfrac{t-3}{2}\right]$. It is a triangular function centered at $t = 3$ and width 4. Using unit ramp functions, the signal is expressed as

$$x(t) = r(t-1) - 2r(t-3) + r(t-5).$$

The derivative of the signal is

$$\frac{dx(t)}{dt} = \frac{dr(t-1)}{dt} - 2\frac{dr(t-3)}{dt} + \frac{dr(t-5)}{dt} = \frac{1}{2}u(t-1) - u(t-3) + \frac{1}{2}u(t-5).$$

And the second derivative is

$$\frac{d^2x(t)}{dt^2} = \frac{1}{2}\frac{du(t-1)}{dt} - \frac{du(t-3)}{dt} + \frac{1}{2}\frac{du(t-5)}{dt} = \frac{1}{2}\delta(t-1) - \delta(t-3) + \frac{1}{2}\delta(t-5).$$

The waveforms are shown in Fig. 2.7(b).

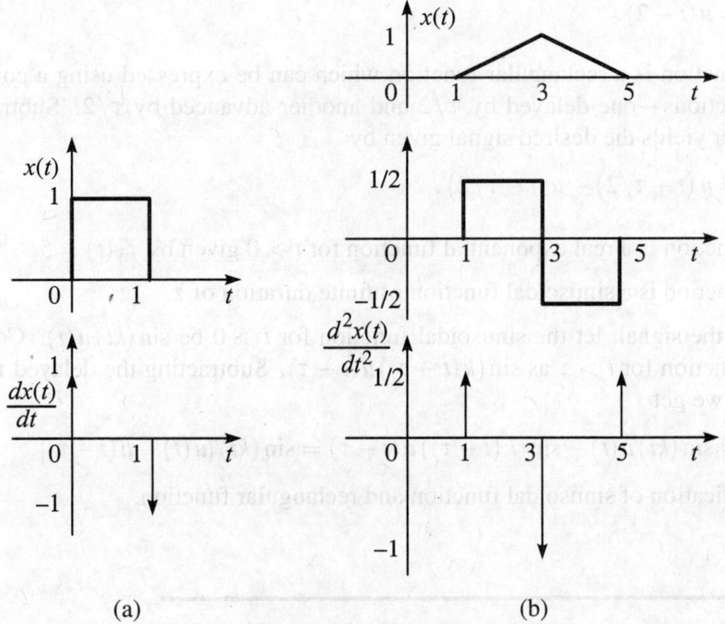

Fig. 2.7 *Gate and triangular signals and their derivatives*

Example 2.6

With the help of unit step function, find the expression for the signals shown in Fig. 2.8.

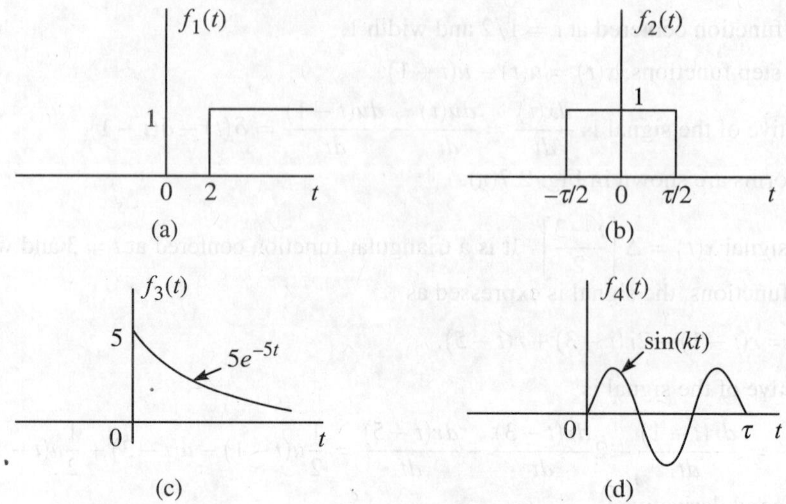

Fig. 2.8 *Examples of continuous signals for Example 2.6*

Solution

(a) The given function is a delayed unit step function with delay 2. The function is

$$f_1(t) = u(t-2).$$

(b) The given function is a rectangular function which can be expressed using a combination of two unit step functions—one delayed by $\tau/2$ and another advanced by $\tau/2$. Subtracting one signal from the other yields the desired signal given by

$$f_2(t) = u(t+\tau/2) - u(t-\tau/2).$$

(c) The given function is a real exponential function for $t > 0$ given by $f_3(t) = 5e^{-3t}u(t)$.

(d) The given function is a sinusoidal function of finite duration of τ.

To construct the signal, let the sinusoidal function for $t > 0$ be $\sin(kt)u(t)$. Consider a delayed sinusoidal function for $t > \tau$ as $\sin(k(t-\tau))u(t-\tau)$. Subtracting the delayed function from the given signal, we get

$$f_4(t) = \sin(kt)u(t) - \sin(k(t-\tau))u(t-\tau) = \sin(kt)[u(t) - u(t-\tau)].$$

It is a multiplication of sinusoidal function and rectangular function.

Example 2.7

Simplify the following expressions:

(a) $\left(\dfrac{\sin t}{t^2+2}\right)\delta(t)$ (b) $\left(\dfrac{j\omega+2}{\omega^2+9}\right)\delta(\omega)$ (c) $\left[e^{-t}\cos\left(3t-\dfrac{\pi}{3}\right)\right]\delta(t)$

Solution

(a) Given $\left(\dfrac{\sin t}{t^2+2}\right)\delta(t)$.

Since $\delta(t)$ is defined only at $t=0$, using the property$f(t)\delta(t)=f(0)\delta(t)$,

$$\left(\frac{\sin t}{t^2+2}\right)\delta(t)=\left(\frac{\sin t}{t^2+2}\right)\Bigg|_{t=0}\delta(t)=\left(\frac{\sin 0}{0+2}\right)\delta(t)=0.$$

(b) Given the function $\left(\frac{j\omega+2}{\omega^2+9}\right)\delta(\omega)$.

Since $\delta(t)$ is defined only at $t=0$, using the property $f(t)\delta(t)=f(0)\delta(t)$,

$$\left(\frac{j\omega+2}{\omega^2+9}\right)\delta(\omega)=\left(\frac{j\omega+2}{\omega^2+9}\right)\Bigg|_{\omega=0}\delta(\omega)=\left(\frac{2}{9}\right)\delta(\omega).$$

(c) Given the function $\left[e^{-t}\cos\left(3t-\frac{\pi}{3}\right)\right]\delta(t)$.

Since $\delta(t)$ is defined only at $t=0$, using the property$f(t)\delta(t)=f(0)\delta(t)$,

$$\left[e^{-t}\cos\left(3t-\frac{\pi}{3}\right)\right]\Bigg|_{t=0}\delta(t)=\left[e^{0}\cos\left(0-\frac{\pi}{3}\right)\right]\delta(t)=\frac{1}{2}\delta(t).$$

Additional Problems

Example 2.8

Verify the orthogonality of the following functions: $S_1(t)=5$ and $S_2(t)=5(1-2t)$ in the interval [0, 1].

Solution

Given the two functions, $S_1(t)=5$ and $S_2(t)=5(1-2t)$ in the interval [1, 0]. The condition for orthogonality is

$$\int_0^1 S_1(t)S_2(t)dt=\int_0^1 5\times 5(1-2t)dt=25\left[t-t^2\right]_{t=0}^{1}=25[(1-1)-(0-0)]=0.$$

Since $\int_0^1 S_1(t)S_2(t)dt=0$, $S_1(t)$ and $S_2(t)$ are orthogonal over [0, 1].

Example 2.9

Verify the orthogonality of the following functions: $S_1(t)=2t$ and $S_2(t)=\left(\frac{1}{t}-3t\right)$ in the interval [0, 1].

Solution

The condition for orthogonality of the given functions is

$$\int_0^1 S_1(t)S_2(t)dt=\int_0^1 2t\times\left(\frac{1}{t}-3t\right)dt$$

$$=2\left[t-t^3\right]_{t=0}^{1}=2[(1-1)-(0-0)]=0.$$

Since $\int_0^1 S_1(t)S_2(t)dt=0$, $S_1(t)$ and $S_2(t)$ are orthogonal over [0, 1].

Example 2.10

Verify the orthogonality of the functions $f_1(t) = \begin{cases} -1 & 0 < t < 2 \\ 1 & 2 < t < 4 \end{cases}$ and $f_2(t) = \begin{cases} -1 & 0 < t < 1 \\ 1 & 1 < t < 3 \\ -1 & 3 < t < 4 \end{cases}$

in the interval [0, 4].

Solution

The condition for orthogonality is

$$\int_0^4 f_1(t)f_2(t)dt = \int_0^1 f_1(t)f_2(t)dt + \int_1^2 f_1(t)f_2(t)dt + \int_2^3 f_1(t)f_2(t)dt + \int_3^4 f_1(t)f_2(t)dt$$

$$= \int_0^1 (-1)(-1)\,dt + \int_1^2 (-1)(1)\,dt + \int_2^3 (1)(1)\,dt + \int_3^4 (1)(-1)\,dt$$

$$= \int_0^1 dt - \int_1^2 dt + \int_2^3 dt - \int_3^4 dt = 0.$$

Since $\int_0^4 f_1(t)f_2(t)dt = 0$, $S_1(t)$ and $S_2(t)$ are orthogonal over [0, 4].

Example 2.11

Prove that $\sin(n\omega_0 t)$and $\cos(m\omega_0 t)$ are orthogonal over any interval $\left(t_0, t_0 + \dfrac{2\pi}{\omega_0}\right)$ for all integer values of n and m.

Solution

Given $f_1(t) = \sin(n\omega_0 t)$ and $f_2(t) = \cos(m\omega_0 t)$, over the interval $(t_0,\ t_0 + 2\pi/\omega_0)$.

The condition for orthogonality is:

$$\int_{t_0}^{t_0+\frac{2\pi}{\omega_0}} f_1(t).f_2(t)dt = 0$$

$$\int_{t_0}^{t_0+\frac{2\pi}{\omega_0}} \sin(n\omega_0 t).\cos(m\omega_0 t)dt = \frac{1}{2}\int_{t_0}^{t_0+\frac{2\pi}{\omega_0}} [\sin(n+m)\omega_0 t + \sin(n-m)\omega_0 t]dt$$

$$= -\frac{1}{2}\left[\frac{\cos(n+m)\omega_0 t}{(n+m)\omega_0} + \frac{\cos(n-m)\omega_0 t}{(n-m)\omega_0}\right]_{t_0}^{t_0+\frac{2\pi}{\omega_0}}$$

$$= \frac{1}{2}\left[\left(\frac{\cos(n+m)(\omega_0 t_0 + 2\pi)}{(n+m)\omega_0} + \frac{\cos(n-m)(\omega_0 t_0 + 2\pi)}{(n-m)\omega_0}\right) - \left(\frac{\cos(n+m)\omega_0 t_0}{(n+m)\omega_0} + \frac{\cos(n-m)\omega_0 t_0}{(n-m)\omega_0}\right)\right]$$

$$= 0$$

$$[\because\ \cos(2\pi N + \theta) = \cos\theta].$$

Therefore, $\cos(n\omega_0 t)$and $\cos(m\omega_0 t)$ are orthogonal over any interval $\left(t_0, t_0 + \dfrac{2\pi}{\omega_0}\right)$ for all integer values of n and m.

Example 2.12

Prove that sinusoidal functions and complex exponential functions are orthogonal functions.

Solution

Let $f_1(t) = \sin(n\omega_0 t)$ or $\cos(n\omega_0 t)$ and $f_2(t) = e^{jm\omega_0 t}$ or $e^{-jm\omega_0 t}$.

If any of the functions $f_1(t)$ and $f_2(t)$ or both are complex, then the condition for orthogonality is

$$\int_{t_0}^{t_0+\frac{2\pi}{\omega_0}} f_1(t).f_2^*(t)dt = \int_{t_0}^{t_0+\frac{2\pi}{\omega_0}} f_1^*(t).f_2(t)dt = 0.$$

Let $f_1(t) = \sin(n\omega_0 t)$and $f_2(t) = e^{jm\omega_0 t}$.

$$\int_{t_0}^{t_0+\frac{2\pi}{\omega_0}} f_1(t)f_2^*(t)dt = \int_{t_0}^{t_0+\frac{2\pi}{\omega_0}} \sin(n\omega_0 t)(e^{jm\omega_0 t})^* dt = \int_{t_0}^{t_0+\frac{2\pi}{\omega_0}} \sin(n\omega_0 t)e^{-jm\omega_0 t}dt$$

$$= \int_{t_0}^{t_0+\frac{2\pi}{\omega_0}} \sin(n\omega_0 t)\cos(m\omega_0 t)dt - \int_{t_0}^{t_0+\frac{2\pi}{\omega_0}} j\sin(n\omega_0 t)\sin(m\omega_0 t)dt$$

$$= 0 - 0 = 0.$$

Since sine and cosine functions are orthogonal over any interval $\left(t_0, t_0 + \frac{2\pi}{\omega_0}\right)$ for all integer values of n and m except at $m \neq n$.

$$\int_{t_0}^{t_0+\frac{2\pi}{\omega_0}} \sin(n\omega_0 t)\cos(m\omega_0 t)dt = \int_{t_0}^{t_0+\frac{2\pi}{\omega_0}} \cos(n\omega_0 t)\cos(m\omega_0 t)dt$$

$$= \int_{t_0}^{t_0+\frac{2\pi}{\omega_0}} \sin(n\omega_0 t)\sin(m\omega_0 t)dt = 0.$$

Therefore, the sinusoidal functions and exponential functions are orthogonal over any interval $\left(t_0, t_0 + \frac{2\pi}{\omega_0}\right)$ for all integer values of n and m except at $m \neq n$.

Example 2.13

A rectangular function is defined as $f(t) = \begin{cases} A, & 0 < t < \dfrac{\pi}{2} \\ -A, & \dfrac{\pi}{2} < t < \dfrac{3\pi}{2} \\ A, & \dfrac{3\pi}{2} < t < 2\pi \end{cases}$.

Approximate the above rectangular function by $A\cos t$ between the intervals (0, 2π) such that the mean square error is minimum.

Solution

$$\text{Given that } f(t) = \begin{cases} A, & 0 < t < \dfrac{\pi}{2} \\ -A, & \dfrac{\pi}{2} < t < \dfrac{3\pi}{2} \\ A, & \dfrac{3\pi}{2} < t < 2\pi \end{cases}$$

For the best approximation, with a given function $A\cos t$ in the interval $(0, 2\pi)$,

$$f(t) \approx CA\cos t, \quad 0 < t < 2\pi,$$

where $C = \dfrac{\int_0^{2\pi} f(t)A\cos t\,dt}{\int_0^{2\pi} (A\cos t)^2 dt}$.

$$\int_0^{2\pi} f(t)A\cos t\,dt = A\left[\int_0^{\frac{\pi}{2}} A\cos t\,dt - \int_{\frac{\pi}{2}}^{\frac{3\pi}{2}} A\cos t\,dt + \int_{\frac{3\pi}{2}}^{2\pi} A\cos t\,dt\right]$$

$$= A^2\left[[\sin t]_0^{\frac{\pi}{2}} - [\sin t]_{\frac{\pi}{2}}^{\frac{3\pi}{2}} + [\sin t]_{\frac{3\pi}{2}}^{2\pi}\right]$$

$$= A^2[(1-0)-(-1-1)+(0-1)] = A^2(1+2-1) = 2A^2.$$

$$\int_0^{2\pi} A^2\cos^2 t\,dt = A^2\int_0^{2\pi} \frac{1+\cos 2\pi t}{2}dt$$

$$= A^2\left[t + \frac{\sin 2t}{2}\right]_0^{2\pi} = A^2[(2\pi+0)-(0+0)] = A^2\pi.$$

Now, $C = \dfrac{2A^2}{A^2\pi} = \dfrac{2}{\pi}$.

$$\therefore \quad f(t) = \frac{2A}{\pi}\cos t; \; 0 < t < 2\pi.$$

Example 2.14

A rectangular function $f(t)$ is approximated by signal $f(t) = \dfrac{4}{\pi}\sin t$. Show that the error function $f_e(t) = f(t) - \dfrac{4}{\pi}\sin t$ is orthogonal to $\sin t$ over $(0, 2\pi)$.

Solution

Given the error $f_e(t) = f(t) - (4/\pi)\sin t$ and $\sin t$ over $(0, 2\pi)$, the condition for orthogonality is

$$\int_0^{2\pi} f_e(t)\sin t\,dt = \int_0^{2\pi}\left[f(t) - \frac{4}{\pi}\sin t\right]\sin t\,dt$$

$$= \int_0^{2\pi} f(t)\sin t\,dt - \frac{4}{\pi}\int_0^{2\pi}\sin^2 t\,dt.$$

Since it is given that $f(t)$ is a rectangular function, assume that $f(t) = \begin{cases} 1, & 0 < t < \pi \\ -1, & \pi < t < 2\pi \end{cases}$.

So, $\displaystyle\int_0^{2\pi} f(t)\sin t\,dt = \int_0^{\pi}\sin t\,dt + \int_{\pi}^{2\pi}(-1)\sin t\,dt$

$$= [-\cos t]_0^{\pi} + [\cos t]_{\pi}^{2\pi} = -(-1-1) + (1-(-1)) = 4.$$

And $\int_0^{2\pi} \sin^2 t dt = \frac{1}{2}\int_0^{2\pi} (1-\cos 2t)dt$

$$= \frac{1}{2}\left[t - \frac{\sin 2t}{2}\right]_0^{2\pi} = \frac{1}{2}.2\pi = \pi.$$

$$\therefore \quad \int_0^{2\pi} f_e(t).\sin t dt = 4 - \frac{4}{\pi}.\pi = 0.$$

Hence $f_e(t) = f(t) - \frac{4}{\pi}\sin t$ and $\sin t$ are orthogonal over $(0, 2\pi)$.

Example 2.15

A rectangular function is defined by

$$f(t) = \begin{cases} 1, & 0 < t < \pi \\ -1, & \pi < t < 2\pi. \end{cases}$$

Approximate the above function by a single sinusoid $\sin t$ over the interval $0 < t < 2\pi$ such that the mean square error is minimum. Evaluate the mean square error in the approximation. Also, show what happens when more number of sinusoids are used for approximations.

Solution

Given $f(t) = \begin{cases} 1, & 0 < t < \pi \\ -1, & \pi < t < 2\pi \end{cases}$ and a single sinusoid $x(t) = \sin t$.

(a) For the best approximation of $f(t)$ with given signal $x(t)$ in the interval $(0, 2\pi)$ is

$f(t) \approx C\sin t,\ 0 < t < 2\pi$. For minimum mean square error, the constant is $C = \dfrac{\int_0^{2\pi} f(t)\sin t dt}{\int_0^{2\pi}\sin^2 t dt}$.

Now, $\int_0^{2\pi} f(t)\sin t dt = \int_0^{\pi}\sin t dt - \int_{\pi}^{2\pi}\sin t dt$

$$= -[\cos t]_0^{\pi} + [\cos t]_{\pi}^{2\pi} = -(-1-1) + (1-(-1)) = 4.$$

And $\int_0^{2\pi}\sin^2 t dt = \int_0^{2\pi}\frac{-\cos 2t + 1}{2}dt$

$$= \frac{1}{2}\left[\frac{-\sin 2t}{2} + t\right]_{t=0}^{2\pi} = \frac{1}{2}[(0+2\pi)-(0+0)] = \pi.$$

Then $C = \dfrac{\int_0^{2\pi} f(t)\sin t dt}{\int_0^{2\pi}\sin^2 t dt} = \dfrac{4}{\pi}.$

$$\therefore \quad f(t) = \frac{4}{\pi}\sin t,\ 0 < t < 2\pi.$$

Mean square error

$$e = \frac{1}{t_2 - t_1}\left[\int_{t_1}^{t_2} f^2(t)dt - C^2E\right].$$

$$\int_0^{2\pi} f^2(t)dt = \int_0^{\pi} 1^2dt + \int_{\pi}^{2\pi} (-1)^2dt = \pi + 2\pi - \pi = 2\pi.$$

Now, and $E = \int_0^{2\pi} \sin^2 tdt = \pi$ joules.

$$\therefore \quad e = \frac{1}{2\pi - 0}\left[2\pi - \left(\frac{4}{\pi}\right)^2 .\pi\right] = \frac{1}{2\pi}\left[2\pi - \frac{16}{\pi}\right]$$

$$= 1 - \frac{8}{\pi^2} = 0.1894.$$

(b) When we approximate with N functions, the best approximation with a given signal $x(t) = \sin nt$ in the interval $(0, 2\pi)$ is

$$f(t) \approx \sum_{n=1}^{N} C_n x_n(t), \; 0 < t < 2\pi.$$

For minimum mean square error, the constants are

$$C_n = \frac{\int_0^{2\pi} f(t)\sin nt dt}{\int_0^{2\pi} \sin^2 nt dt}.$$

$$\text{Now} \int_0^{2\pi} f(t)\sin nt dt = \int_0^{\pi} \sin nt dt - \int_{\pi}^{2\pi} \sin nt dt$$

$$= -\frac{1}{n}[\cos nt]_0^{\pi} + \frac{1}{n}[\cos nt]_{\pi}^{2\pi}$$

$$= -\frac{2}{n}(-1-1) = \frac{4}{n}, \text{ for } n = \text{odd and zero for } n = \text{even}.$$

$$\text{And} \int_0^{2\pi} \sin^2 nt dt = \int_0^{2\pi} \frac{-\cos 2nt + 1}{2} dt$$

$$= \frac{1}{2}\left[\frac{-\sin 2nt}{2} + t\right]_{t=0}^{2\pi} = \frac{1}{2}[(0+2\pi) - (0+0)] = \pi.$$

$$C_n = \begin{cases} \dfrac{4}{\pi n}, & \text{if } n \text{ is odd} \\ 0, & \text{if } n \text{ is even.} \end{cases}$$

For two terms, $f(t) \approx C_1x_1(t) + C_3x_3(t) = \frac{4}{\pi}\sin t + \frac{4}{3\pi}\sin 3t$;

Mean square error $e = \dfrac{1}{(t_2 - t_1)}\left[\int_{t_1}^{t_2} f^2(t)dt - C_1^2E_1 - C_2^2E_2\right]$

$$e = \frac{1}{2\pi}\left[2\pi - \left(\frac{4}{\pi}\right)^2 \pi - \left(\frac{4}{3\pi}\right)^2 \pi\right] = 0.1.$$

For three terms,

$$f(t) = C_1x_1(t) + C_3x_3(t) + C_5x_5(t) = \frac{4}{\pi}\sin t + \frac{4}{3\pi}\sin 3t + \frac{4}{5\pi}\sin 5t.$$

$$\therefore \quad e = \frac{1}{2\pi}\left[2\pi - \left(\frac{4}{\pi}\right)^2\pi - \left(\frac{4}{3\pi}\right)^2\pi - \left(\frac{4}{5\pi}\right)^2\pi\right] = 0.0675.$$

For four terms,

$$f(t) = C_1x_1(t) + C_3x_3(t) + C_5x_5(t) + C_7x_7(t)$$

$$= \frac{4}{\pi}\sin t + \frac{4}{3\pi}\sin 3t + \frac{4}{5\pi}\sin 5t + \frac{4}{7\pi}\sin 7t.$$

$$\therefore \quad e = \frac{1}{2\pi}\left[2\pi - \left(\frac{4}{\pi}\right)^2\pi - \left(\frac{4}{3\pi}\right)^2\pi - \left(\frac{4}{5\pi}\right)^2\pi - \left(\frac{4}{7\pi}\right)^2\pi\right] = 0.051$$

and so on.

Thus as more number of sinusoids are used for approximations, the mean square error decreases.

Example 2.16

For the signals $f(t)$ and $x(t)$ depicted in Fig. 2.9, find the component of the form $x(t)$ contained in $f(t)$. In other words, find (a) the optimum value of c in the approximation $f(t) \approx cx(t)$ so that the error signal energy is minimum. (b) Find the error signal $e(t)$ and its energy E.

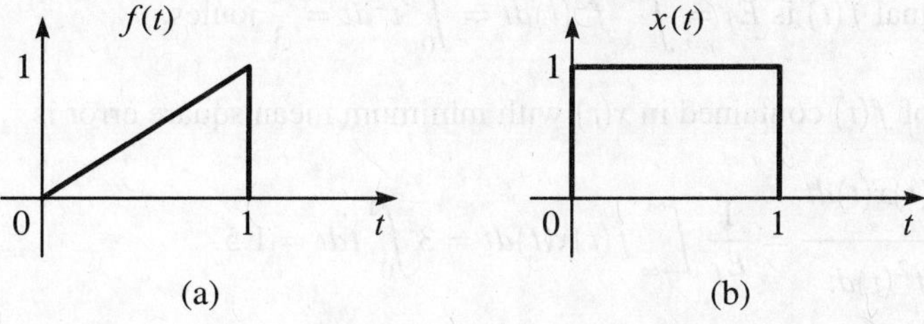

Fig. 2.9 *Signals for x(t) and f(t)*

Solution

(a) Given the signals $x(t) = \begin{cases} 1 & 0 \le t \le 1 \\ 0 & 1 < t \le \infty \end{cases}$ and $f(t) = \begin{cases} t & 0 \le t \le 1 \\ 0 & 1 < t \le \infty \end{cases}$.

Energy of the signal $x(t)$ is $E_x = \int_{-\infty}^{\infty} x^2(t)dt = \int_0^1 1.dt = 1$ joule.

The component of $x(t)$ contained in $f(t)$ with minimum mean square error is

$$c = \frac{\int_{-\infty}^{\infty} f(t)x(t)dt}{\int_{-\infty}^{\infty} x^2(t)dt} = \frac{1}{E_x}\int_{-\infty}^{\infty} f(t)x(t)dt = \frac{1}{1}\int_0^1 tdt = 0.5.$$

(b) Given the approximation $f(t) \approx c.x(t)$.

Therefore $f(t) \approx 0.5x(t)$ or $f(t) = 0.5x(t) + f_e(t)$.

The error signal is $f_e(t) = f(t) - 0.5x(t)$ or $f_e(t) = t - 0.5 \quad 0 \le t \le 1$.

The energy of the error signal is $E = \int_{-\infty}^{\infty} f_e^2(t)dt = \int_0^1 (t-0.5)^2 dt = \frac{1}{12}$ joules.

Example 2.17

For the signals $f(t)$ and $x(t)$ shown in Fig. 2.10, find the component of the form $f(t)$ contained in $x(t)$. In other words, find the optimum value of c in the approximation $x(t) = cf(t)$ so that the error signal energy is minimum. What is the error signal energy?

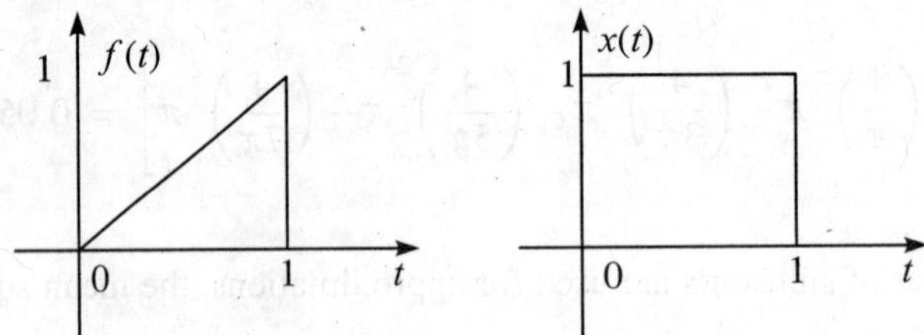

Fig. 2.10 *Signals f(t) and x(t)*

Solution

Given the signals $x(t) = \begin{cases} 1 & 0 \le t \le 1 \\ 0 & 1 < t \le \infty \end{cases}$ and $f(t) = \begin{cases} t & 0 \le t \le 1 \\ 0 & 1 < t \le \infty \end{cases}$.

Energy of the signal $f(t)$ is $E_f = \int_{-\infty}^{\infty} f^2(t)dt = \int_0^1 t^2 dt = \frac{1}{3}$ joules.

(a) The component of $f(t)$ contained in $x(t)$ with minimum mean square error is

$$c = \frac{\int_{-\infty}^{\infty} f(t)x(t)dt}{\int_{-\infty}^{\infty} f^2(t)dt} = \frac{1}{E_f}\int_{-\infty}^{\infty} f(t)x(t)dt = 3\int_0^1 tdt = 1.5.$$

(b) Given the approximation $x(t) \approx cf(t)$.

Therefore, $x(t) \approx 1.5f(t)$ or $x(t) = 1.5f(t) + f_e(t)$.

The error signal is $f_e(t) = x(t) - 1.5f(t)$ or $f_e(t) = 1 - 1.5t \quad 0 \le t \le 1$.

The energy of the error signal is

$$E = \int_{-\infty}^{\infty} f_e^2(t)dt = \int_0^1 (1-1.5t)^2 dt = \frac{1}{4} \text{ joules.}$$

Example 2.18

(a) For the signals $f(t)$ and $x(t)$ depicted in Fig. 2.11(a), find the component of the form $x(t)$ contained in $f(t)$. In other words, find the optimum value of c in the approximation $f(t) \approx cx(t)$ so that the error signal energy is minimum. (b) Find the error signal $e(t)$ and its energy E.

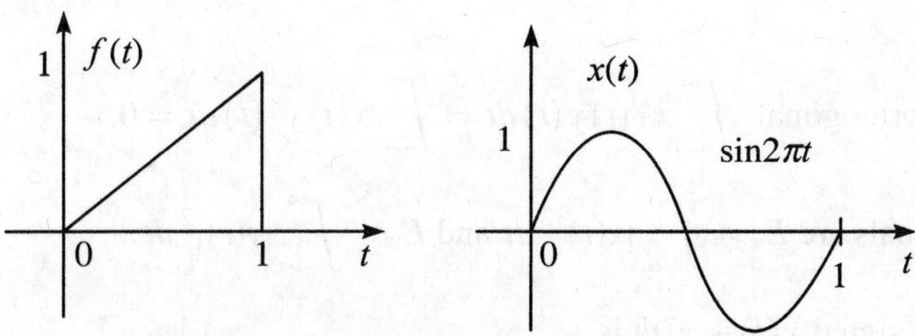

Fig. 2.11 *Signals for Example 2.18*

Solution

Given that the signals are, $f(t) = \begin{cases} t & 0 \le t \le 1 \\ 0 & 1 < t \le \infty \end{cases}$ and $x(t) = \sin 2\pi t;\ \ 0 \le t \le 1.$

The energy of the signal $x(t)$ is $E_x = \int_{-\infty}^{\infty} x^2(t)dt$

$$E_x = \int_0^1 \sin^2 2\pi t dt = 0.5 \text{ joules.}$$

The component of $x(t)$ contained in $f(t)$ with minimum mean square error is

$$c = \frac{\int_{-\infty}^{\infty} f(t)x(t)dt}{\int_{-\infty}^{\infty} x^2(t)dt} = \frac{1}{E_x}\int_{-\infty}^{\infty} f(t)x(t)dt.$$

$$\therefore \quad c = \frac{1}{0.5}\int_0^1 t\sin 2\pi t dt = \frac{-1}{\pi}.$$

Given $f(t) \approx c.x(t)$.

Therefore, $f(t) \approx \left(\frac{-1}{\pi}\right)x(t)$ or $f(t) = \left(\frac{-1}{\pi}\right)x(t) + f_e(t)$.

and the error is $f_e(t) = f(t) - \left(\frac{-1}{\pi}\right)x(t) = t + \left(\frac{1}{\pi}\right)\sin 2\pi t \qquad 0 \le t \le 1.$

The energy of the error signal is $E_e = \int_{-\infty}^{\infty} e^2(t)dt$

$$= \int_0^1 \left(t + \left(\frac{1}{\pi}\right)\sin 2\pi t\right)^2 dt = \frac{1}{3} - \frac{1}{2\pi^2} \text{ joules.}$$

Example 2.19

If $x(t)$ and $y(t)$ are orthogonal, show that the energy of the signal $x(t) + y(t)$ is identical to the energy of the signal $x(t) - y(t)$ and is given by $E_x + E_y$.

Solution

As $x(t)$ and $y(t)$ are orthogonal, $\int_{-\infty}^{\infty} x^*(t)y(t)\,dt = \int_{-\infty}^{\infty} x(t)y^*(t)\,dt = 0.$

The energy of the signals are $E_x = \int_{-\infty}^{\infty} |x(t)|^2\,dt$ and $E_y = \int_{-\infty}^{\infty} |y(t)|^2\,dt.$

The energy of the signal $x(t)+y(t)$ is

$$\int_{-\infty}^{\infty} |x(t)+y(t)|^2\,dt = \int_{-\infty}^{\infty} [x(t)+y(t)][x^*(t)+y^*(t)]dt$$

$$\int_{-\infty}^{\infty} |x(t)|^2\,dt + \int_{-\infty}^{\infty} |y(t)|^2\,dt + \int_{-\infty}^{\infty} x^*(t)y(t)\,dt + \int_{-\infty}^{\infty} x(t)y^*(t)\,dt$$

$$= \int_{-\infty}^{\infty} |x(t)|^2\,dt + \int_{-\infty}^{\infty} |y(t)|^2\,dt = E_x + E_y.$$

Similarly, the energy of the signal $x(t)-y(t)$ is

$$\int_{-\infty}^{\infty} |x(t)-y(t)|^2\,dt = \int_{-\infty}^{\infty} [x(t)-y(t)][x^*(t)-y^*(t)]dt$$

$$\int_{-\infty}^{\infty} |x(t)|^2\,dt + \int_{-\infty}^{\infty} |y(t)|^2\,dt - \int_{-\infty}^{\infty} x^*(t)y(t)\,dt - \int_{-\infty}^{\infty} x(t)y^*(t)\,dt$$

$$= \int_{-\infty}^{\infty} |x(t)|^2\,dt + \int_{-\infty}^{\infty} |y(t)|^2\,dt = E_x + E_y.$$

Therefore, if $x(t)$ and $y(t)$ are orthogonal, then the energy of the signal $x(t)+y(t)$ is equal to the energy of the signal $x(t)-y(t)$ and is given by $E_x + E_y$.

Example 2.20

Determine whether or not each of the following signals is periodic. If periodic, determine its fundamental period.

(a) $x(t) = e^{-j5t}\sin^2 10t$; (b) $x(t) = e^{j2t}\sin \pi t$ and (c) $x(t) = e^{j2\pi t}\cos 4\pi t$.

Solution

(a) Given

$$x(t) = e^{-j5t}\sin^2 10t = \frac{e^{-j5t}}{2}(1-\cos 20t) = \frac{1}{2}(\cos 5t - j\sin 5t)(1-\cos 20t)$$

$$= \frac{1}{2}(\cos 5t - j\sin 5t - \cos 5t\cos 20t + j\sin 5t\cos 20t)$$

$$= \frac{1}{2}(\cos 5t - j\sin 5t - \cos 25t - \cos 15t + j\sin 25t + \sin 15t)$$

$$= \frac{1}{2}(\cos 5t - \cos 25t - \cos 15t + j(\sin 25t + \sin 15t - \sin 5t)).$$

The periods of the three components of the signal are $T_1 = \frac{2\pi}{\omega_1} = \frac{2\pi}{5}$ s,

$$T_2 = \frac{2\pi}{\omega_2} = \frac{2\pi}{15} \text{ sec and } T_3 = \frac{2\pi}{\omega_3} = \frac{2\pi}{25} \text{ s.}$$

The ratio $\frac{T_1}{T_2} = \frac{2\pi/5}{2\pi/15} = 3$ and $\frac{T_1}{T_3} = \frac{2\pi/5}{2\pi/25} = 5$ is a rational number. Hence the signal is periodic.

The time period of the signal is $T = \text{LCM of}\left(\frac{2\pi}{5}, \frac{2\pi}{15}, \frac{2\pi}{25}\right) = \frac{2\pi}{5}$ s.

(b) Given $x(t) = e^{j2t}\sin\pi t = (\cos 2t + j\sin 2t)\sin\pi t$

$$= (\cos 2t\sin\pi t + j\sin 2t\sin\pi t) = \frac{1}{2}(\sin(\pi+2)r + \sin(\pi-2)r + j\cos(\pi-2)t - j\sin(\pi+2)t).$$

The periods of the two components of the signal are $T_1 = \frac{2\pi}{\omega_1} = \frac{2\pi}{\pi-2}$ sec and $T_2 = \frac{2\pi}{\omega_2} = \frac{2\pi}{\pi+2}$ sec.

The ratio $\frac{T_1}{T_2} = \frac{2\pi/(\pi-2)}{2\pi/(\pi+2)} = \frac{\pi+2}{\pi-2}$ is not a rational number hence the signal is aperiodic.

(c) Given $x(t) = e^{j2\pi t}\cos 4\pi t = (\cos 2\pi t + j\sin 2\pi t)\cos 4\pi t$

$$= (\cos 2\pi t\cos 4\pi t + j\sin 2\pi t\sin 4\pi t) = \frac{1}{2}(\cos 6\pi t + \cos 2\pi t + j\ \sin 2\pi t - j\sin 6\pi t).$$

The periods of the two components of the signal are $T_1 = \frac{2x}{\omega_1} = \frac{2x}{2x} = 1$ sec

and $T_2 = \frac{2\pi}{\omega_2} = \frac{2\pi}{6\pi} = \frac{1}{3}$ sec.

The ratio $\frac{T_1}{T_2} = \frac{1}{1/3} = 3$ is a rational number hence the signal is periodic.

The time period of the signal is $T = \text{LCM } of\left(1, \frac{1}{3}\right) = 3$ sec.

Example 2.21

Express the functions given below in terms of impulse functions and sketch the waveforms.

(i) $x(t) = \Pi[(t-6)/4]$ (ii) $x(t) = u(t+2) - u(t-2) + u(t-4) - u(t-6)$

Solution

(i) Given function $x(t) = \Pi[(t-6)/4]$ is a rectangular function which is centred at 6 and having a width of 4.

Therefore, $x(t) = u(t-4) - u(t-8)$.

The derivative is $\frac{dx(t)}{dt} = \delta(t-4) - \delta(t-8)$.

(ii) Given function $x(t) = u(t+2) - u(t-2) + u(t-4) - u(t-6)$.

The derivative is $\frac{dx(t)}{dt} = \delta(t+1) - \delta(t-2) + \delta(t-4) - \delta(t-6)$.

The waveforms are shown in Fig. 2.12.

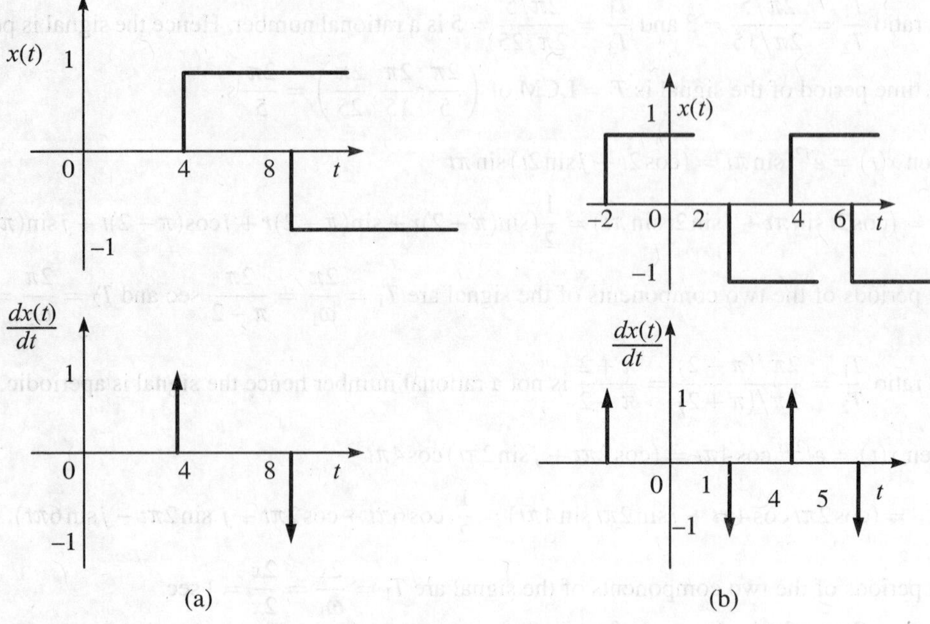

Fig. 2.12 *Signals and their derivatives*

Example 2.22

Express the functions given below in terms of step function and sketch the waveforms.

(i) $x(t) = \Delta[(t+6)/4]$ (ii) $x(t) = \Delta[(t+3)/2] + \Delta[(t-3)/2]$

Solution

(i) Given rectangular function $x(t) = \Delta[(t+6)/4]$ which is centered at -6 and having a width of 8. In terms of ramp functions, $x(t)$ can be expressed as $x(t) = r(t+10) - 2r(t+6) + r(t+2)$.

The derivative is $\dfrac{dx(t)}{dt} = \dfrac{1}{4}u(t+10) - \dfrac{1}{2}u(t+6) + \dfrac{1}{4}u(t+2)$.

The second derivative is $\dfrac{dx^2(t)}{dt^2} = \dfrac{1}{4}\delta(t+10) - \dfrac{1}{2}\delta(t+6) + \dfrac{1}{4}\delta(t+2)$.

The waveforms are as shown in Fig. 2.13(a).

(ii) The given rectangular function $x(t) = \Delta[(t+3)/2] + \Delta[(t-3)/2]$ has two triangular functions. One is centered at -3 and has a width of 4 and the other is centered at $+3$ and has a width of 4.

Let $x_1(t) = \Delta[(t+3)/2]$ and $x_2(t) = \Delta[(t-3)/2]$.

In terms of ramp functions, $x_1(t)$ can be expressed as $x_1(t) = r(t+5) - 2r(t+3) + r(t+1)$ and $x_2(t)$ can be expressed as $x_2(t) = r(t-1) - 2r(t-3) + r(t-5)$.

Therefore $x(t) = x_1(t) + x_2(t) = r(t+5) - 2r(t+3) + r(t+1) + r(t-1) - 2r(t-3) + r(t-5)$.

The first derivatives is

$$\frac{dx(t)}{dt} = \frac{1}{2}u(t+5) - u(t+3) + 1u(t+1) + \frac{1}{2}u(t-1) - 1u(t\ -3) + \frac{1}{2}u(t-5).$$

The second derivative is

$$\frac{dx^2(t)}{dt^2} = \frac{1}{2}\delta(t+5) - 1\delta(t+3) + \frac{1}{2}\delta(t+1) + \frac{1}{2}\delta(t-1) - 1\delta(t-3) + \frac{1}{2}\delta(t-5)$$

The waveforms are as shown in Fig. 2.13(b).

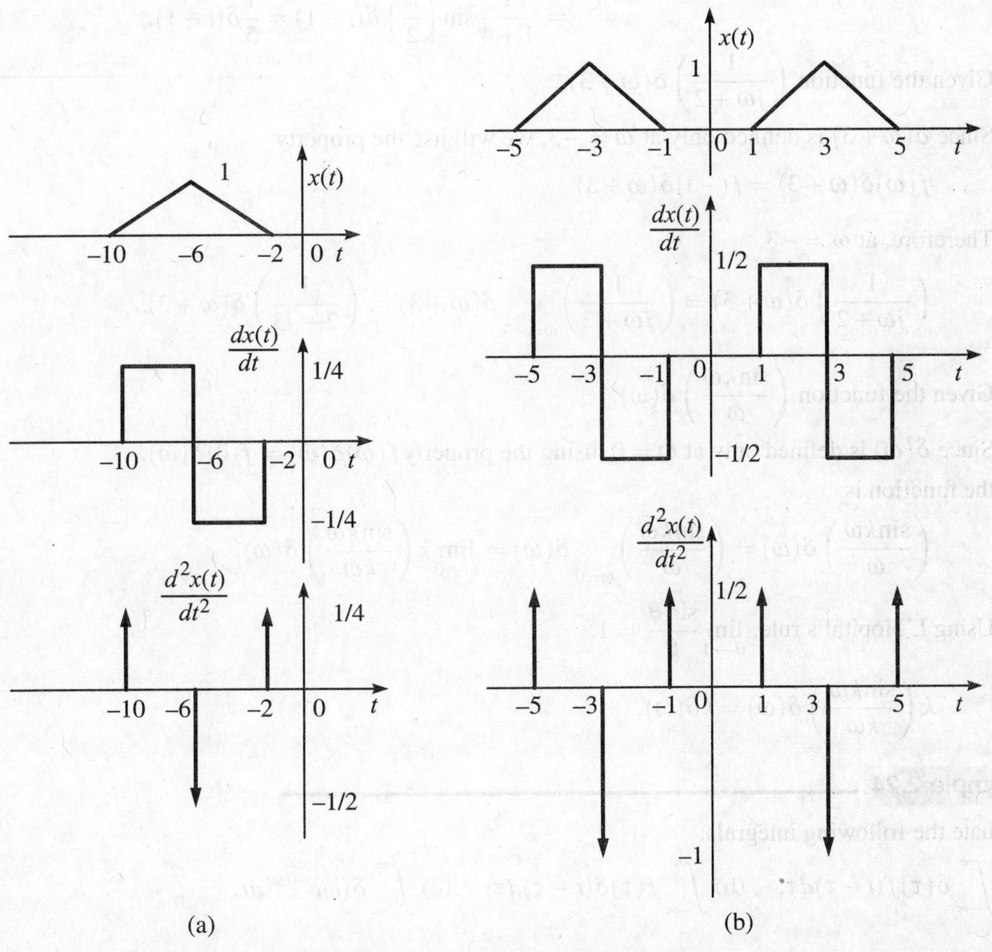

Fig. 2.13 *Signals and their derivatives*

Example 2.23

Simplify the following expressions:

(a) $\left(\dfrac{1}{t^2+4}\right)\sin\left(\dfrac{\pi}{2}(t-2)\right)\delta(t-1)$ (b) $\left(\dfrac{1}{j\omega+2}\right)\delta(\omega+3)$ (c) $\left(\dfrac{\sin k\omega}{\omega}\right)\delta(\omega)$

Solution

(a) Given the function $\left(\dfrac{1}{t^2+4}\right)\sin\left(\dfrac{\pi}{2}(t-2)\right)\delta(t-1)$.

Since $\delta(t-1)$ is defined only at $t=1$, using the property $f(t)\delta(t-1)=f(1)\delta(t-1)$, the function is

$$\left(\frac{1}{t^2+4}\right)\sin\left(\frac{\pi}{2}(t-2)\right)\delta(t-1)=\frac{1}{t^2+4}\sin\left(\frac{\pi}{2}(t-2)\right)\bigg|_{t=1}\delta(t-1)$$

$$=\frac{1}{1+4}\sin\left(\frac{\pi}{2}\right)\delta(t-1)=\frac{1}{5}\delta(t-1).$$

(b) Given the function $\left(\frac{1}{j\omega+2}\right)\delta(\omega+3)$.

Since $\delta(\omega+3)$ is defined only at $\omega=-3$, we will use the property

$$f(\omega)\delta(\omega+3)=f(-3)\delta(\omega+3).$$

Therefore, at $\omega=-3$,

$$\left(\frac{1}{j\omega+2}\right)\delta(\omega+3)=\left(\frac{1}{j\omega+2}\right)_{\omega=-3}\delta(\omega+3)=\left(\frac{1}{2-j3}\right)\delta(\omega+3).$$

(c) Given the function $\left(\frac{\sin k\omega}{\omega}\right)\delta(\omega)$.

Since $\delta(\omega)$ is defined only at $\omega=0$, using the property $f(\omega)\delta(\omega)=f(0)\delta(\omega)$, the function is

$$\left(\frac{\sin k\omega}{\omega}\right)\delta(\omega)=\left(\frac{\sin k\omega}{\omega}\right)_{\omega=0}\delta(\omega)=\lim_{\omega\to 0}k\left(\frac{\sin k\omega}{k\omega}\right)\delta(\omega).$$

Using L'Hopital's rule, $\lim_{\theta\to 0}\frac{\sin\theta}{\theta}=1$,

$$k\left(\frac{\sin k\omega}{k\omega}\right)\delta(\omega)=k\delta(\omega).$$

Example 2.24

Evaluate the following integrals:

(a) $\int_{-\infty}^{\infty}\delta(\tau)f(t-\tau)d\tau$; (b) $\int_{-\infty}^{\infty}f(\tau)\delta(t-\tau)d\tau$, (c) $\int_{-\infty}^{\infty}\delta(t)e^{-j\omega t}dt$.

Solution

(a) Given $\int_{-\infty}^{\infty}\delta(\tau)f(t-\tau)d\tau$.

Since the impulse $\delta(\tau)$ is located at $\tau=0$,

$$\int_{-\infty}^{\infty}\delta(\tau)f(t-\tau)d\tau=\int_{0^-}^{0^+}f(t-\tau)d\tau f(t).$$

(b) Given $\int_{-\infty}^{\infty}f(\tau)\delta(t-\tau)d\tau$.

Since the impulse $\delta(t-\tau)$ is located at $\tau=t$,

$$\int_{-\infty}^{\infty}f(\tau)\delta(t-\tau)d\tau=\int_{t^-}^{t^+}f(\tau)d\tau=f(t).$$

(c) $\int_{-\infty}^{\infty} \delta(t)e^{-j\omega t}dt$.

Since the impulse $\delta(t)$ is located at $t = 0$, hence

$$\int_{-\infty}^{\infty} \delta(t)e^{-j\omega t}dt = e^{-j\omega(0)} = 1.$$

Example 2.25

Evaluate the following integrals:

(a) $\int_{0}^{\infty} e^{-2t}\delta(t+5)dt$, (b) $\int_{0}^{\infty} e^{-2t}\delta(t-5)dt$ and (c) $\int_{-\infty}^{\infty} e^{-2t}\frac{d\delta(t-5)}{dt}dt$.

Solution

(a) Given $\int_{0}^{\infty} e^{-2t}\delta(t+5)dt$.

Since the impulse function $\delta(t+5)$ is located at $t = -5$, $\int_{0}^{\infty} e^{-5t}\delta(t+5)dt$ is zero.

(b) Given $\int_{0}^{\infty} e^{-2t}\delta(t-5)dt$

Since the impulse function $\delta(t-5)$ is located at $t = 5$,

$$\int_{0}^{\infty} e^{-2t}\delta(t-5)dt = e^{-2(5)} = e^{-10}.$$

(c) Given $\int_{-\infty}^{\infty} e^{-2t}\frac{d\delta(t-5)}{dt}dt$.

We know that $\int_{-\infty}^{\infty} x(t)\frac{d\delta(t-t_0)}{dt}dt = -\frac{dx(t)}{dt}\bigg|_{t=t_0}$

$$\int_{-\infty}^{\infty} e^{-2t}\frac{d\delta(t-5)}{dt}dt = -\frac{d}{dt}(e^{-2t})\bigg|_{t=5} = -2e^{-2(5)} = 2e^{-10}.$$

Example 2.26

Evaluate the integrals:

(a) $\int_{-\infty}^{\infty} \frac{de^{(t-1)}}{dt}\delta(t-4)dt$ and (b) $\int_{-\infty}^{\infty} \frac{df(t)}{dt}\delta(t-4)dt$, where $f(t) = 2t^2+5$

Solution

(a) Given the integral $\int_{-\infty}^{\infty} \frac{de^{(t-1)}}{dt}\delta(t-1)dt$.

Using the property $\int_{-\infty}^{\infty} \frac{df(t)}{dt}\delta(t-t_0)dt = \frac{df(t)}{dt}\bigg|_{t=t_0}$

$$\int_{-\infty}^{\infty} \frac{de^{(t-1)}}{dt}\delta(t-1)dt = \frac{de^{(t-1)}}{dt}\bigg|_{t=4} = e^{(4-1)} = e^3.$$

(b) Given the integral $\int_{-\infty}^{\infty} \frac{df(t)}{dt}\delta(t-4)dt$

Using the property $\int_{-\infty}^{\infty} \frac{df(t)}{dt}\delta(t-t_0)dt = \left.\frac{df(t)}{dt}\right|_{t=t_0}$,

$$\int_{-\infty}^{\infty} \frac{df(t)}{dt}\delta(t-4)dt = \left.\frac{df(t)}{dt}\right|_{t=4}.$$

Given $f(t) = 2t^2 + 5$,

then $\left.\frac{df(t)}{dt}\right|_{t=4} = 4t\Big|_{t=4} = 16.$

Summary

1. A signal is a quantity that has both magnitude and direction. Therefore, all types of signals are vectors.
2. Since vectors can be expressed as a sum of its components resolved along the direction of coordinates, signals too can be represented as a sum of its components oriented along the direction of the defined coordinate system.
3. In the two-dimensional vector space, if two vectors $\vec{A}$ and $\vec{B}$ represent orthogonal vectors, then $\vec{A}\cdot\vec{B} = 0$. These vectors are called basis vectors.
4. If two signals $e(t)$ and $x(t)$ are orthogonal to each other over an interval $t_1 \le t \le t_2$, then $\int_{t_1}^{t_2} x(t)e(t)dt = 0$. These signals are called basis signals in the two-dimensional space. This is also the condition for orthogonality.
5. In the three-dimensional system, any vector $\vec{A}$ can be represented as $\vec{A} = A_x\vec{a}_x + A_y\vec{a}_y + A_z\vec{a}_z$. where the unit vectors, $\vec{a}_x, \vec{a}_y$ and $\vec{a}_z$ are mutually orthogonal.
6. In the N-dimensional vector space, any vector $\vec{f}$ can be expressed in terms of N mutually orthogonal basis vectors $\vec{x}_1$, $\vec{x}_2$, $\vec{x}_3$......$\vec{x}_N$. Therefore $\vec{f} = c_1\vec{x}_1 + c_2\vec{x}_2 + c_3\vec{x}_3 + \cdots\cdots\cdots + c_N\vec{x}_N$.
7. The condition for basis vectors $\vec{x}_1$, $\vec{x}_2$, $\vec{x}_3$......$\vec{x}_N$ to be mutually orthogonal is
$$\vec{x}_m \cdot \vec{x}_n = \begin{cases} 0 & m \neq n \\ K_m & m = n, \end{cases}$$
where $K_m = x_m^2$ is a constant.
8. The condition for basis vectors $\vec{x}_1$, $\vec{x}_2$, $\vec{x}_3$......$\vec{x}_N$ to be mutually orthonormal or normalised orthogonal vectors is
$$\vec{x}_m \cdot \vec{x}_n = \begin{cases} 0 & m \neq n \\ 1 & m = n. \end{cases}$$
9. If the basis vectors $\vec{x}_1$, $\vec{x}_2$, $\vec{x}_3$......$\vec{x}_N$ are not the complete set of orthogonal vectors, then the error vector is $\vec{e} = \vec{A} - \sum_{n=0}^{N} c_n\vec{x}_n$.

10. For real signals $x_1(t), x_2(t), x_3(t) \cdots\cdots x_N(t)$ as a set of mutually orthogonal signals in the N-dimensional signal space over an interval $t_1 \leq t \leq t_2$, the condition for orthogonality is $\int_{t_1}^{t_2} x_m(t)x_n(t)dt = \begin{cases} 0 & m \neq n \\ E_n & m = n. \end{cases}$

11. The energy of the signal is $E_n = \int_{t_1}^{t_2} x_n^2(t)dt$.

12. Any function $f(t)$ can be approximated in terms of the basis functions $x_1(t), x_2(t), x_3(t) \cdots\cdots x_N(t)$ over an interval $t_1 \leq t \leq t_2$ as

$$f(t) \cong c_1x_1(t) + c_2x_2(t) + c_3x_3(t) + \cdots\cdots + c_Nx_N(t) = \sum_{n=1}^{N} c_nx_n(t).$$

13. The error signal is $f_e(t) = f(t) - \sum_{n=1}^{N} c_nx_n(t)$ and the mean square error (MSE) or power of the error signal is

$$e = \frac{1}{(t_2 - t_1)}\left[\int_{t_1}^{t_2} f^2(t)dt - (c_1^2E_1 + c_2^2E_2 + c_3^2E_3 + \cdots\cdots + c_N^2E_N)\right].$$

For minimum MSE, the constant are obtained from

$$c_j = \frac{\int_{t_1}^{t_2} f(t)x_j(t)dt}{\int_{t_1}^{t_2} x_j^2(t)dt} = \frac{\int_{t_1}^{t_2} f(t)x_j(t)dt}{E_j}.$$

14. In the N-dimensional signal space, as $N \to \infty$, the error $e \to 0$ and the energy of $f(t)$ is equal to the sum of the energies of the orthogonal components. That is,

$$\int_{t_1}^{t_2} f^2(t)dt = \sum_{n=1}^{\infty} c_n^2E_n.$$

This equation is called Parseval's theorem, and the function $f(t)$ is

$$f(t) = \sum_{n=1}^{\infty} c_nx_n(t); \quad t_1 \leq t \leq t_2.$$

15. To approximate the function $f(t)$ by a complex function $x(t)$ over an interval $t_1 \leq t \leq t_2$, the orthogonality condition in the N-dimensional space is given by

$$\int_{t_1}^{t_2} x_m(t)x_n^*(t)dt = \begin{cases} 0 & m \neq n \\ E_n & m = n, \end{cases}$$

where $x_n^*(t)$ is the complex conjugate of $x_n(t)$.

16. The energy of the sum or difference of two orthogonal signals is equal to the sum of the energies of the two signals.

17. The sinusoidal functions namely $\sin(n\omega_0t)$, $\cos(n\omega_0t)$ and the complex exponential functions $e^{\pm jn\omega_0t}$ for $n = 1, 2, 3 \cdots\cdots$, are mutually orthogonal over any interval $\left(t_0, t_0 + \frac{2\pi}{\omega_0}\right)$ for all integer values of n and m.

18. Exponential signals are extremely important in signals and systems analysis. They are used as a driving function for most of the applications in linear system analysis. Some of the unique properties of this class of signal are:

    (i) Exponential signals are invariant under the action of linear time-invariant systems. That is, the response of LTI system to an exponential signal is also an exponential signal.

    (ii) Any arbitrary signal can be expressed as a sum of various exponential functions.

    (iii) Transform tools like Fourier, Laplace and Z-transform use complex exponential signals for analysis.

19. A continuous-time complex exponential signal is defined as $f(t) = ce^{at}$, where c and a, are complex numbers. If both c and a are real, then $f(t) = ce^{at}$ is called a real exponential signal. The imaginary exponential signal is defined as $f(t) = e^{j\omega_0 t}$.

20. A continuous sinusoidal signal in general can be expressed as $f(t) = A\cos(\omega_0 t + \phi)$.

21. Complex exponential and sinusoidal signals are always periodic with single fundamental frequency f_0 or time period T, where $T = 2\pi/\omega_0$ seconds or $\omega_0 T = 2\pi$.

22. A sinusoidal signal can be expressed in term of the complex exponential signal as

    (i) $A\cos(\omega_0 t + \phi) = A\text{Re}\left\{e^{j(\omega_0 t+\phi)}\right\}$ and $A\sin(\omega_0 t + \phi) = A\text{Im}\left\{e^{j(\omega_0 t+\phi)}\right\}$

    (ii) $\text{Re}\{e^{j\omega_0 t}\} = \cos(\omega_0 t)$ and $\text{Im}\{e^{j\omega_0 t}\} = \sin(\omega_0 t)$.

23. A damped sinusoidal signal is defined as $f(t) = ce^{at}$, where if c and a are complex numbers in the form $c = |c|e^{j\theta}$ and $a = r + j\omega_0$, then $ce^{at} = |c|e^{rt}\cos(\omega_0 t + \theta) + j|c|e^{rt}\sin(\omega_0 t + \theta)$. If $r = 0$, the real and imaginary parts of a complex exponential are sinusoidal. If $r > 0$, the signal is a sinusoidal signal multiplied by a growing exponential, and if $r < 0$, the signal is a sinusoidal signal multiplied by a decaying exponential.

24. The continuous time impulse function is given by $\delta(t) = \begin{cases} 1 & t = 0 \\ 0 & t \neq 0 \end{cases}$.

    It is also known as the Dirac delta function.

25. The unit step function $u(t)$ is defined as $u(t) = \begin{cases} 1 & t \geq 0 \\ 0 & t < 0. \end{cases}$

26. The unit impulse function $\delta(t)$ and the unit step function are related as

$$u(t) = \int_{-\infty}^{t} \delta(\tau)d\tau \quad \text{and} \quad \delta(t) = \frac{du(t)}{dt}.$$

27. The properties of unit impulse function are:

    (i) Area under the function is $\int_{-\infty}^{\infty} \delta(t)\,dt = 1$.

    (ii) Sampling property: $\int_{-\infty}^{\infty} f(t)\,\delta(t)\,dt = f(0)$.

    (iii) The product of any function with unit impulse function gives

$$f(t)\delta(t) = f(0)\delta(t) \quad \text{and} \quad f(t-t_0)\delta(t-t_1) = f(t_1-t_0)\delta(t-t_1).$$

    (iv) *Shifting property*:

$$\int_{-\infty}^{\infty} f(t)\delta(t-t_0)dt = f(t_0) \text{ and } \int_{-\infty}^{\infty} f(t-t_0)\delta(t-t_1)dt = f(t_1-t_0)$$

Note: If $f(t)$ is defined in the finite interval $-t_1 < t < t_2$, then

$$\int_{-t_1}^{t_2} f(t)\delta(t-t_0)dt = \begin{cases} f(t_0) & t_1 < t_0 < t_2 \\ 0 & \text{otherwise.} \end{cases}$$

(v) *Symmetrical property*: The impulse function is an even symmetry. That is, $\delta(t) = \delta(-t)$.

(vi) *Scaling property*: $\delta(at) = \frac{1}{|a|}\delta(t)$, where a is a constant.

(vii) *Differentiation property*: For any function $f(t)$ defined in the interval $-\infty < t < \infty$, if its derivative $\dot{f}(t) = \frac{df(t)}{dt}$ is continuous at $t = t_0$, then

$$\int_{-\infty}^{\infty} \frac{df(t)}{dt}\delta(t-t_0)dt = \left.\frac{df(t)}{dt}\right|_{t=t_0} \quad \text{and} \quad \int_{-\infty}^{\infty} f(t)\frac{d\delta(t-t_0)}{dt}dt = -\left.\frac{df(t)}{dt}\right|_{t=t_0}.$$

28. The derivative of the unit impulse function is called unit doublet.
29. The signum function is defined as sgn $(t) = \begin{cases} -1 & t < 0 \\ 1 & t > 0, \end{cases}$ and $\text{sgn}(t) = 2u(t) - 1$.

Review Questions

1. Bring out the analogy between vectors and signals. (Nov 2015)
2. Define orthogonal signal space and bring out clearly its application in representing a signal.
3. Obtain the condition under which two signals $f_1(t)$ and $f_2(t)$ are said to be orthogonal to each other. Hence, prove that $\sin n\omega_o t$ and $\cos n\omega_o t$ are orthogonal to each other for all integer values of (m,n). (May 2015)
4. Discuss how an unknown function $f(t)$ can be expressed using infinite number of mutually orthogonal functions. (May 2015)
5. Define and discuss the conditions for orthogonality of functions. (Nov 2015)
6. Prove that sinusoidal functions are orthogonal functions. (Nov 2015)
7. Prove that complex exponential functions are orthogonal functions.
8. Write a short note on orthogonal functions. (May 2014)
9. Explain the analogy between vectors and signals in terms of the orthogonality condition and how the constant is evaluated.
10. Define mean square error and derive the expression of the same in N-dimensional signal space.
11. What is an orthonormal vector and orthonormal set of vectors. (May 2011)
12. Find the component of a waveform $\sin\omega_2 t$ contained in another waveform $\sin\omega_1 t$ over the interval $(-T,\ T)$ for all real values of ω_1 and $\omega_2 (\omega_1 \neq \omega_2)$. How does this component change with T? (May 2011)
13. Explain the orthogonality property between two complex functions $f_1(t)$ and $f_2(t)$ for a real variable t. (Nov 2015)

14. Verify that the signals $\sin n\omega_0 t$ and $\sin m\omega_0 t$ are orthogonal over the interval $\left(t_0,\, t_0 + \frac{2\pi}{\omega_0}\right)$. (May 2015)
15. Derive the expression for the component vector of the approximating function $f_1(t)$ over $f_2(t)$ and also prove that the component vector becomes zero if the $f_1(t)$ and $f_2(t)$ are orthogonal.
16. Approximate this function by a waveform $\sin t$ over the interval $(0, 2\pi)$ such that the mean square error is minimum. (Dec 2011)
17. Define a complete set and hence show that the error can be minimised when the function $f(t)$ is approximated using n set of orthogonal functions. Approximate this function by a waveform with single term, $\sin t$, two terms, $\sin t$ and $\sin 3t$, and three terms, $\sin t$, $\sin 3t$ and $\sin 5t$ over the interval $(0, 2\pi)$ and show that the mean square error is minimum when the function is approximated by three terms rather than by a single term. (May 2013)
18. Discuss the concept of orthogonality in complex functions and derive the expression for component vector of approximating the function $f_1(t)$ over $f_2(t)$ in case of complex functions. (Dec 2011)
19. Derive the expression for mean square error in approximating a function $f(t)$ by a set of n orthogonal functions. (Dec 2011)
20. Define orthogonal signal space and a closed or complete set of orthogonal functions and hence represent a function by a closed or complete set of mutually orthogonal functions. (May 2013)
21. Derive the expression for approximating a given function $f(t)$ by a set of an orthogonal functions $g_1(t), g_2(t), \cdots\cdots g_n(t)$ and also show that the mean square error can be reduced when a function is approximated by more number of orthogonal functions. (May 2013)
22. Define and sketch the following elementary continuous-time signals.
    (i) Unit impulse signal
    (ii) Signum function
    (iii) Unit step function
23. Define and sketch the following signals:
    (i) Impulse
    (ii) Double-sided exponential
    (iii) Step function
    (iv) Signum function
24. What is the relationship between unit step function and unit impulse function? Prove it.
25. Show that the derivative of unit-step function is an impulse function.
26. Define the following basic signals and give graphical representations of the same.
    (i) Unit impulse signal
    (ii) Unit step signal
    (iii) Ramp signal
    (iv) Sinusoidal signal
27. Define impulse function and explain clearly how impulse function can be used for analysing any system.
28. Define and sketch real exponential signal in different cases.
29. Define complex exponential signal..

30. List and explain the properties of impulse function.
31. Define and sketch (i) complex exponential, (ii) sinusoidal signal.
32. Derive the relation between unit step function and signum function along with their appropriate definitions.
33. Explain how

    (i) signum function is expressed in terms of unit step function,
    (ii) step function is expressed in terms of impulse function, and
    (iii) impulse function is expressed in terms of step function.

Problems

1. Show that $f(t)$ is orthogonal to signals $\cos t$, $\cos 2t$, $\cos 3t, \ldots, \cos mt$ for all integer values of m, $m \neq 0$, over the interval $(0, 2\pi)$ if

$$f(t) = \begin{cases} 1, & 0 < t < \pi \\ -1, & \pi < t < 2\pi \end{cases}.$$

2. Prove that the signals $x(t)$ and $y(t)$ defined as below are orthogonal over the interval [0, 4].

$$x(t) = \begin{cases} 1 & 0 < t < 2 \\ -1 & 2 < t < 4 \end{cases}$$

$$y(t) = \begin{cases} 1 & 0 < t < 1 \\ -1 & 1 < t < 3 \\ 1 & 3 < t < 4 \end{cases}$$

3. Verify the orthogonality of the following functions: $S_1(t) = t^2$ and $S_2(t) = \left(\frac{1}{t^2} - 3\right)$ in the interval [0, 1].
4. If the signals $x_e(t)$ and $x_o(t)$ are respectively the even and odd functions of $x(t)$, show that they are orthogonal over the interval $[-T, T]$.
5. Show that the signals $[1, \cos\omega_0 t, \cos 2\omega_0 t, \cdots\cdots \cos n\omega_0 t, \sin\omega_0 t, \sin 2\omega_0 t, \cdots . \sin n\omega_0 t]$ are orthogonal over the interval $T = \frac{2\pi}{\omega_0}$.
6. Show that the signals $\left[1, e^{\pm j\omega_0 t}, e^{\pm j2\omega_0 t}, e^{\pm j3\omega_0 t}, \ldots e^{\pm jn\omega_0 t}\right]$ are orthogonal over the interval $T = \frac{2\pi}{\omega_0}$.
7. For the signals $f(t)$ and $x(t)$ depicted in Fig. P2.7, (a) find the component of the form $f(t)$ contained in $x(t)$. In other words, find the optimum value of c in the approximation $x(t) \approx cf(t)$ so that the error signal energy is minimum. (b) Find the error signal $e(t)$ and its energy E.

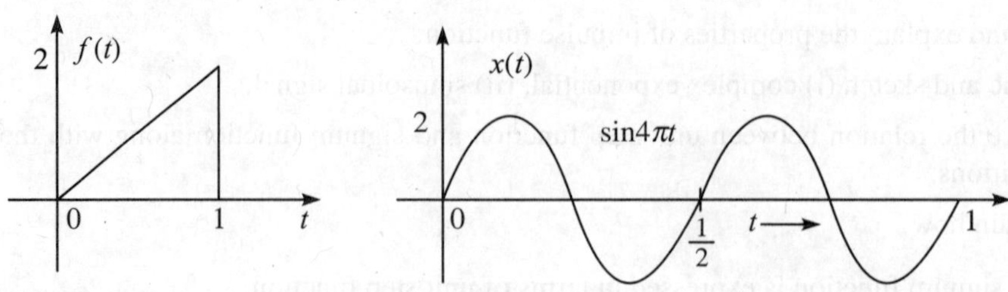

Fig. P2.7

8. Consider the rectangular pulse $x(t)$ as shown in Fig. P2.8. Approximate the above function by A sin 2πt and show that the mean square error is minimum. (May 2011)

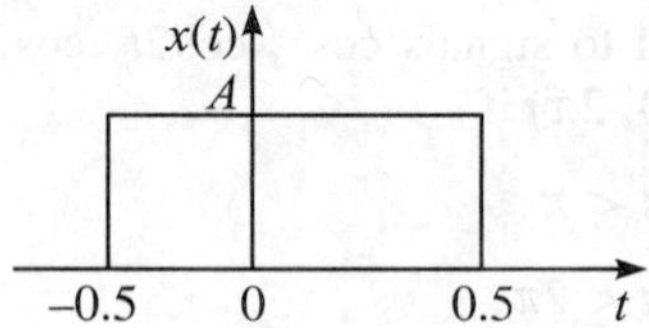

Fig. P2.8

9. Which of the following signals or functions are periodic and if periodic, what is the fundamental period?

   (i) $g(t) = e^{-j60\pi t}$

   (ii) $g(t) = e^{-j60\pi t} + e^{j60\pi t}$

10. Which of the following signals or functions are periodic and if periodc, what is the fundamental period?

   (i) $g(t) = 1 + e^{-j6t} + e^{-j6\pi t}$

   (ii) $g(t) = e^{-j2\pi t} \sin 3t$

11. Which of the following signals or functions are periodic and if periodic, what is the fundamental period?

   (i) $g(t) = e^{j3\pi t} + e^{-j4\pi t}$

   (ii) $g(t) = e^{-j4\pi t}(\cos 2\pi t + j \sin 2\pi t)$

12. Determine energy of the signal $x(t) = e^{j4t} \sin 7t$.

13. Compute the following integrals:

   (i) $\int_{-\infty}^{\infty} \delta(t) \sin 2\pi dt$

   (ii) $\int_{-\infty}^{\infty} [\delta(t)\cos t + \delta(t-1)\sin t] dt$

14. Evaluate the following integrals:

   (a) $\int_{-\infty}^{\infty} \delta(t-2) \sin \pi t dt$

(b) $\int_{-\infty}^{\infty} \delta(t+3)e^{-t}dt$

(c) $\int_{-\infty}^{\infty} (t^3+4)\,\delta(1-t)dt$

15. Evaluate the following integrals:

(a) $\int_{0}^{\infty} \delta(t-2)\sin\frac{\pi}{4}tdt$

(b) $\int_{0}^{\infty} \delta(t+3)e^{-t}dt$

(c) $\int_{0}^{\infty} (t^3+4)\delta(1+t)dt$

16. Evaluate the following integrals:

(a) $\int_{-\infty}^{\infty} f(2-t)\delta(3-t)dt$

(b) $\int_{-\infty}^{\infty} e^{(t-1)}\cos\left(\frac{\pi}{2}(t-5)\right)\delta(t-3)dt$

17. Show that the function $g(t) = \frac{1}{\sqrt{2\pi\sigma^2}}e^{-t^2/2\sigma^2}$ has the properties of unit step function as $\sigma \to 0$.

18. Evaluate the following integrals:

(a) $\int_{-\infty}^{\infty} \sin\pi t\frac{d\delta(t-2)}{dt}dt$

(b) $\int_{-\infty}^{\infty} e^{-5t^2}\frac{d\delta(t)}{dt}dt$

(c) $\int_{-\infty}^{\infty} (t+4)\frac{d\delta(1-t)}{dt}dt$

19. Evaluate the following integrals:

(a) $\int_{-\infty}^{\infty} \frac{df(t)}{dt}\delta(3-t)dt$

(b) $\int_{-\infty}^{\infty} \frac{de^{(x-1)}}{dt}\delta(x-1)dx$

20. Express the functions given below in terms of step functions and sketch the waveforms.

(i) $x(t) = \Pi[(t-6)/4]$

(ii) $x(t) = \Pi[(t-2)/4] + \Pi[(2-t)/4]$

21. Express the functions given below in terms of step functions and sketch the waveforms.

(i) $x(t) = \Delta[(t+6)/4]$

(ii) $x(t) = \Delta[(t+3)/2] + \Delta[(t-3)/2]$

22. Given the signals (i) $x(t) = r(t) - 2r(t-1) + r(t-2)$ and (ii) $x(t) = r(t) - r(t-1) - r(t-3) + r(t-4)$, find in terms of impulse functions $\frac{d^2x(t)}{dt^2}$ and sketch the waveforms.

23. Given the signals (i) $x(t) = \Pi[(t+1)/4]$ and (ii) $x(t) = \Pi[(t-2)/4]$, express the derivative of the signals in terms of unit impulses and sketch the waveforms.
24. Given the signals (i) $x(t) = \Delta[(t+1)/4]$ and (ii) $x(t) = \Delta[(t-2)/4]$, express the derivatives of the signals in terms of unit impulses and sketch the waveforms.
25. If the nth derivative of $x(t)$ exists and is continuous at $t = t_0$, show that

$$\int_{-t_1}^{t_2} x(t)\delta^{(n)}(t-t_0)dt = (-1)^n x^{(n)}(t_0), \text{ for } t_1 < t_0 < t_2$$

where $x^{(n)}(t_0)$ is the nth derivative of $x(t)$ at $t = t_0$ and $\delta^{(n)}(t-t_0)$ is the nth derivative of $\delta(t-t_0)$.

Answers

7. Error signal $e(t) = 2t + \left(\frac{2}{\pi}\right)\sin 4\pi t; \quad 0 \le t \le 1$

   Energy of the error signal $= \frac{4}{3} - \frac{2}{\pi^2}$ joules
8. $f(t) = \left(\frac{4A}{\pi}\right)\sin 4\pi t; \quad -0.5 \le t \le 0.5$
9. (i) Periodic with period $T = \frac{1}{30}$ s; (ii) Periodic with period $T = \frac{1}{30}$ s
10. (i) Non-periodic; (ii) Non-periodic
11. (i) Periodic with period $T = 2$s; (ii) Periodic with period $T = 1$ s
12. ∞
13. (i) 0; (ii) $1 + \sin(1)$
14. (a) 0; (b) e^3 ; (c) 5
15. (a) 1; (b) 0; (c) 0
16. (a) $f(-1)$; (b) $-e^2$
18. (a) $-\pi$; (b) 0; (c) - 1
19. (a) $\frac{\partial f(3)}{\partial t}$; (b) 0
20. (i) $u(t-4) - u(t-8)$; (ii) $u(t) - u(t-1) + u(-t+1) - u(-t)$
21. (i) $1/4(t+10)u(t+10) - 1/2(t+6)u(t+6) + 1/4(t+2)u(t+2)$; (ii) $1/2(t+5)u(t+5) - 1(t+3)u(t+3) + 1/2(t+1)u(t+1) + 1/2(t-1)u(t-1) - 1(t-3)u(t-3) + 1/2(t-5)u(t-5)$
22. (i) $\delta(t) - 2\delta(t-1) + \delta(t-2)$; (ii) $\delta(t) - \delta(t-1) + \delta(t-3) + \delta(t-4)$
23. (i) $\frac{d^2x(t)}{dt^2} = \delta(t+3) - \delta(t-1)$; (ii) $\frac{d^2x(t)}{dt^2} = \delta(t) - \delta(t-4)$
24. (i) $\frac{d^2x(t)}{dt^2} = 4\delta(t+5) - 8\delta(t+1) + 4\delta(t-3)$; (ii) $\frac{d^2x(t)}{dt^2} = 4\delta(t+2) - 8\delta(t-2) + 4\delta(t-6)$

Multiple-choice questions

1. Two vectors are said to be orthogonal if their dot product is

   (a) infinity (b) zero
   (c) one (d) none of these

2. A set of m signals is orthonormal if

   (a) they are orthogonal (b) their norms are all unity
   (c) both (a) and (b) holds (d) none of these

3. In a set of signals, if no signal can be represented as a linear combination of the remaining signals, then such signals are

   (a) linearly independent (b) linearly dependent
   (c) orthogonal (d) orthonarmal

4. In the two-dimensional vector space, if two vectors $\vec{A}$and $\vec{B}$ represent orthogonal vectors, then

   (a) $\vec{A} \cdot \vec{B} = 0$ (b) $\vec{A} \cdot \vec{B} = 1$
   (c) $\vec{A}B = 0$ (d) $A\vec{B} = 0$

5. Basis vectors are

   (a) orthogonal (b) linearly independent
   (c) (a) & (b) (d) only (b)

6. Two signals are orthogonal if the integral of their product is

   (a) infinity (b) zero
   (c) one (d) none of these

7. If we approximate a function by its orthogonal function, the error will be

   (a) infinity (b) zero
   (c) large (d) small

8. For the best approximation of a signal, the mean square error (MSE) over the interval must be

   (a) minimum (b) maximum
   (c) undefined (d) none of these

9. A function $x(t)$ can be exactly represented by a linear combination of orthogonal functions provided they form

   (a) an incomplete set (b) a complete set
   (c) null set (d) none of these

10. Two complex functions $x(t)$ and $y(t)$ are orthogonal over the interval (t_1, t_2) if $\int_{-\infty}^{\infty} x^*(t)\, y(t)\, dt$ is equal to

    (a) zero (b) infinity
    (c) one (d) none of these

11. A set of functions$\{g_i(t)\}$ mutually orthogonal over the interval $[t_1, t_2]$ is called a complete or a closed set if there exists no function $f(t)$ such that

(a) $\int_{t_1}^{t_2} f(t)\, g_i(t)\, dt = 1$ (b) $\int_{t_1}^{t_2} f(t)\, g_i(t)\, dt = 0$

(c) $\int_{t_1}^{t_2} m(t)\, dt = 1$ (d) $\int_{t_1}^{t_2} m(t)\, dt = 0$

12. The energy of the sum of two orthogonal signals is equal to the
(a) sum of the energies of the two signals
(b) difference of the energies of the two signals
(c) product of the energies of the two signals
(d) the energy of any one signal

13. The energy of the difference of two orthogonal signals is equal to the
(a) product of the energies of the two signals
(b) difference of the energies of the two signals
(c) sum of the energies of the two signals
(d) the energy of any one signal

14. The equation $\int_{t_1}^{t_2} f^2(t)dt = \sum_{n=1}^{\infty} c_n^2 E_n$ is called

(a) Fourier's theorem (b) Parseval's theorem
(c) Energy theorem (d) Power theorem

15. The energy of the signal $x(t)$ in the interval $[t_1, t_2]$ is defined as

(a) $E = \frac{1}{(t_2 - t_1)} \int_{t_1}^{t_2} x^2(t)dt$ (b) $E_n = \int_{t_1}^{t_2} x(t)dt$

(c) $E_n = \frac{1}{(t_2 - t_1)} \int_{t_1}^{t_2} x(t)dt$ (d) $E = \int_{t_1}^{t_2} x^2(t)dt$

16. For real signals $x_1(t), x_2(t), x_3(t) \cdots\cdots x_N(t)$ as a set of mutually orthogonal signals in N-dimensional signal space over an interval $t_1 \le t \le t_2$, the condition for orthogonality is

(a) $\int_{t_1}^{t_2} x_m(t)x_n(t)dt = \begin{cases} 0 & m = n \\ E_n & m \neq n \end{cases}$

(b) $\int_{t_1}^{t_2} x_m(t)x_n(t)dt = \begin{cases} 0 & m \neq n \\ E_n & m = n \end{cases}$

(c) $\int_{t_1}^{t_2} x_m(t)x_n(t)dt = \begin{cases} 0 & m \neq n \\ 1 & m = n \end{cases}$

(d) $\int_{t_1}^{t_2} (x_m^2(t) + x_n^2(t))\, dt = \begin{cases} 0 & m \neq n \\ 1 & m = n \end{cases}$

17. For the best approximation of a signal $f(t) \cong c_1x_1(t) + c_2x_2(t) + c_3x_3(t) + \cdots\cdots + c_Nx_N(t)$ in terms of basis functions $x_1(t), x_2(t), x_3(t) \cdots\cdots x_N(t)$ over an interval $t_1 \le t \le t_2$, the minimum mean square error (MSE) over the interval the constant c is

(a) $c_n = \dfrac{\int_{t_1}^{t_2} x_n(t)dt}{\int_{t_1}^{t_2} f(t)x(t)dt}$ (b) $c_n = \dfrac{\int_{t_1}^{t_2} f(t)x_n(t)dt}{\int_{t_1}^{t_2} x_n^2(t)dt}$

(c) $c_n = \dfrac{\int_{t_1}^{t_2} f(t)x_n(t)dt}{\int_{t_1}^{t_2} x(t)dt}$ (d) $c_n = \dfrac{\int_{t_1}^{t_2} f(t)x_n(t)dt}{\int_{t_1}^{t_2} f^2(t)dt}$

18. Consider a set of mutually orthogonal signals $x_1(t), x_2(t), x_3(t) \cdots\cdots x_N(t)$ in the N-dimensional signal space over an interval $t_1 \le t \le t_2$. The condition $\int_{t_1}^{t_2} x_m(t)x_n(t)dt = \begin{cases} 0 & m \ne n \\ 1 & m = n \end{cases}$ is called

(a) orthogonality (b) orthonarmality
(c) mean square error (d) normalisation

19. The relation between unit step function and signum function is

(a) $\text{sgn}(t) = 2u(t) - 1$ (b) $\text{sgn}(t) = 2u(t) + 1$
(c) $\text{sgn}(t) = 2u(t)$ (d) none of these

20. The differentiation of unit step function $u(t)$ is

(a) $\text{sgn}(t)$ (b) $r(t)$
(c) $\delta(t)$ (d) none of these

21. The evaluation of $\int_{-\infty}^{\infty} f(t)\delta(t)dt$ yields

(a) 0 (b) 1 (c) ∞ (d) $f(0)$

22. The evaluation of $\int_{0}^{\infty} t^2\delta(t-3)dt$ yields

(a) 0 (b) 1 (c) 9 (d) ∞

23. The integration of unit step function $u(t)$, $\int u(t)dt$ yields

(a) $\text{sgn}(t)$ (b) $r(t)$ (c) $\delta(t)$ (d) none of these

24. The integration of unit impulse function $\int_{-\infty}^{\infty} \delta(t)dt$ yields

(a) 0 (b) 1 (c) ∞ (d) none of these

25. The even part of the unit step signal is

(a) $\dfrac{1}{2}$ (b) 1 (c) $\dfrac{1}{2}\text{sgn}(t)$ (d) 0

26. Tick the false statement

(a) $t\delta(t) = 0$ (b) $\cos t\delta(t-\pi) = -\delta(t-\pi)$

(c) $\int u(t)dt = \delta(t)$ (d) $t\delta'(t) = \delta(t)$

27. $\int_{-\infty}^{t} \cos \tau u(\tau) d\tau$ is equal to

(a) $+\sin t$ (b) $-\sin t$ (c) $-\cos t$ (d) $+\cos t$

28. For $t > 0$, $\int_{-\infty}^{t} \cos \tau \delta(\tau) d\tau$ is equal to

(a) 1 (b) 0 (c) ∞ (d) none of these

29. For $t < 0$, $\int_{-\infty}^{t} \cos \tau \delta(\tau) d\tau$ is equal to

(a) 1 (b) 0 (c) ∞ (d) none of these

30. $\int_{0}^{2\pi} t \sin\left(\frac{t}{2}\right) \delta(\pi - t) dt$ is equal to

(a) 1 (b) 0 (c) π (d) none of these

Answers

1. (b)	2. (c)	3. (a)	4. (a)	5. (c)	6. (b)	7. (d)	8. (a)	9. (b)	10. (a)
11. (b)	12. (a)	13. (c)	14. (b)	15. (d)	16. (b)	17. (b)	18. (b)	19. (a)	20. (c)
21. (d)	22. (c)	23. (b)	24. (b)	25. (a)	26. (c)	27. (a)	28. (a)	29. (b)	30. (c)

3

Fourier Series

Learning Objectives

The main objective of this chapter is to understand thoroughly the basic concepts of Fourier series. We will learn how periodic signals are expanded using Fourier series. By reading this chapter, students will be able to relate to the following topics.

– Representation of Fourier series
– Continuous-time periodic signals
– Trigonometric and exponential Fourier series
– Dirichlet's conditions
– Properties of Fourier series
– Complex Fourier spectrum

3.1 Introduction

In a previous chapter we had learnt that the basis functions $\sin\omega_0 t$, $\sin 2\omega_0 t, \ldots \sin n\omega_0 t \ldots$ etc. form an orthogonal set over any interval $(t_0, t_0 + 2\pi/\omega_0)$. Since the mean square error is not zero, these basis functions are not a complete set. It was also proved that the basis function $\cos n\omega_0 t$ is orthogonal to $\sin m\omega_0 t$ over the same interval and the basis functions $\cos\omega_0 t$, $\cos 2\omega_0 t, \ldots \cos n\omega_0 t \ldots$ etc., form an orthogonal set over the same interval $(t_0, t_0 + 2\pi/\omega_0)$. Also, the basis functions of both sine and cosine terms form a complete orthogonal set. These functions are called harmonic signals or complex sinusoids with fundamental frequency ω_0.

French mathematician Joseph Fourier (1768–1830) found that any periodic waveform can be expressed as a linear combination of an infinite number of harmonic signals or sinusoids. The representation is called Fourier series. The process of applying signals in the form of Fourier series on systems and their study at the output is called Fourier analysis. This analysis has applications in many branches of engineering.

The advantages of choosing to represent a signal as a set of harmonic signals are that (i) it allows a simple, consistent, and unique method to find out the coefficients (constant weights) for each mean component and (ii) it has only one fundamental frequency. The Fourier series also yields the smallest square error in power between the waveform and its series representation with infinite number of harmonics. It is a least-squares fit to a periodic signal.

This chapter begins with the Fourier series representation of continuous-time periodic signals in trigonometric and exponential forms. It examines the effects of symmetry, consistency and introduces the concept of the complex spectrum, which leads to frequency domain representation of signals. Also, it explains many useful properties of Fourier series and describes some applications of Fourier series in system analysis.

3.2 Representation of Fourier Series

It is noted that any periodic function $f(t)$ with fundamental frequency ω_0 can be represented as Fourier series in terms of a linear weighted sum of orthogonal signals over any interval $(t_0, t_0 + 2\pi/\omega_0)$. For non-periodic signals, the Fourier series does not converge. There are two important representations of Fourier series. They are:

1. Trigonometric Fourier series.
2. Exponential Fourier series.

3.3 Continuous-time Periodic Signals

Trigonometric Fourier Series

In trigonometric Fourier series, a periodic function $f(t)$ is represented in terms of a linear weighted sum of harmonic signals of sine and cosine terms over any interval $(t_0, t_0 + 2\pi/\omega_0)$ with fundamental frequencyω_0.

Thus, for any real-valued and periodic signals the trigonometric Fourier series is expressed as

$$\begin{aligned} f(t) = a_0 + a_1 \cos \omega_0 t + a_2 \cos 2\omega_0 t + a_3 \cos 3\omega_0 t + \ldots\ldots + a_n \cos n\omega_0 t \ldots\ldots \\ + b_1 \sin \omega_0 t + b_2 \sin 2\omega_0 t + b_3 \sin 3\omega_0 t + \ldots\ldots + b_n \sin n\omega_0 t + \ldots\ldots \end{aligned}$$

over the interval $(t_0, t_0 + T)$, or

$$f(t) = a_0 + \sum_{n=1}^{\infty} [a_n \cos n\omega_0 t + b_n \sin n\omega_0 t], \quad (t_0 < t < t_0 + T) \tag{3.1}$$

where n is an integer, T is the period of $f(t)$, $\omega_0 = \dfrac{2\pi}{T}$ is called fundamental frequency, and $n\omega_0$ is the nth harmonic of the sinusoidal frequency ω_0.

Thus a periodic function $f(t)$ is a sum of components comprising the dc component, the fundamental frequency ω_0, second harmonic frequency $2\omega_0$, third harmonic frequency $3\omega_0$ and so on, as shown in Fig. 3.1.

Since the basis functions form a complete orthogonal set, the coefficients are given by

$$a_n = \frac{\displaystyle\int_{t_o}^{t_0+T} f(t) \cos n\omega_0 t \, dt}{\displaystyle\int_{t_o}^{t_0+T} \cos^2 n\omega_0 t \, dt}, \quad b_n = \frac{\displaystyle\int_{t_o}^{t_0+T} f(t) \sin n\omega_0 t \, dt}{\displaystyle\int_{t_o}^{t_0+T} \sin^2 n\omega_0 t \, dt} \tag{3.2}$$

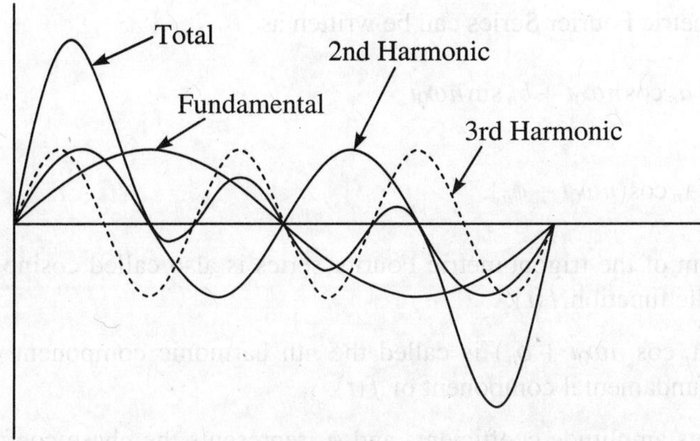

Fig. 3.1 *Frequency components of a periodic function $f(t)$*

where $n = 1, 2, \ldots.\infty$.

Since sine and cosine terms are orthogonal to each other,

$$\int_{t_0}^{t_0+T} \cos m\omega_0 t \cos n\omega_0 t \, dt = \begin{cases} 0 & m \neq n \\ T/2 & m = n \neq 0 \end{cases} \tag{3.3}$$

$$\int_{t_0}^{t_0+T} \sin m\omega_0 t \sin n\omega_0 t \, dt = \begin{cases} 0 & m \neq n \\ T/2 & m = n \neq 0 \end{cases} \tag{3.4}$$

$$\int_{t_0}^{t_0+T} \sin m\omega_0 t \cos n\omega_0 t \, dt = 0, \quad \text{for all } m, n. \tag{3.5}$$

Therefore the Fourier coefficients are:

For $n = 0$,

$$a_0 = \frac{1}{T} \int_{t_o}^{t_0+T} f(t) \, dt, \text{ and } b_0 = 0. \tag{3.6}$$

For $n = 1, 2, \ldots.\infty$,

$$a_n = \frac{2}{T} \int_{t_o}^{t_0+T} f(t) \cos n\omega_0 t \, dt, \quad b_n = \frac{2}{T} \int_{t_o}^{t_0+T} f(t) \sin n\omega_0 t \, dt. \tag{3.7}$$

Note: It is observed that a_0 is the average value of $f(t)$ over the interval $t_0 < t < t_0 + T$. It represents the dc component of the signal $f(t)$.

Compact Form of the Trigonometric Fourier Series

The Fourier series can also be represented in a polar form, which is more compact and easier to manipulate. We know that

$$a_n \cos n\omega_0 t + b_n \sin \omega_0 t = A_n \cos(n\omega_0 t + \phi_n).$$

Or, in polar form

$$A_n \angle \phi_n = a_n - jb_n,$$

where amplitude $A_n = \sqrt{a_n^2 + b_n^2}$ and phase angle $\phi_n = -\tan^{-1}\left(\dfrac{b_n}{a_n}\right)$. (3.8)

Then the trigonometric Fourier Series can be written as

$$f(t) = a_0 + \sum_{n=1}^{\infty} a_n \cos n\omega_0 t + b_n \sin n\omega_0 t$$

$$f(t) = a_0 + \sum_{n=1}^{\infty} A_n \cos(n\omega_0 t + \phi_n). \tag{3.9}$$

Thus the compact form of the trigonometric Fourier series is also called cosine representation or the polar form of a periodic function $f(t)$.

Note 1: The term $A_n \cos(n\omega_0 t + \phi_n)$ is called the nth harmonic component of $f(t)$ and the term $A_1 \cos(\omega_0 t + \phi_1)$ the fundamental component of $f(t)$.

Note 2: A_n represents amplitude coefficients and ϕ_n represents the phase coefficients of the Fourier series.

3.3.1 Evaluation of Fourier Coefficients

In trigonometric Fourier series, the Fourier coefficients a_0, a_n and b_n can be evaluated by taking integration on the function over the interval $t_0 < t < t_0 + T$ and applying the orthogonality properties.

Evaluation of a_0:

$$f(t) = a_0 + \sum_{n=1}^{\infty} a_n \cos n\omega_0 t + b_n \sin n\omega_0 t$$

$$\int_{t_0}^{t_0+T} f(t)dt = \int_{t_0}^{t_0+T} \left[a_0 + \sum_{n=1}^{\infty} [a_n \cos n\omega_0 t + b_n \sin n\omega_0 t] \right] dt$$

$$= \int_{t_0}^{t_0+T} a_0 dt + \int_{t_0}^{t_0+T} \sum_{n=1}^{\infty} a_n \cos n\omega_0 t\, dt + \int_{t_0}^{t_0+T} \sum_{n=1}^{\infty} b_n \sin n\omega_0 t\, dt$$

$$= \int_{t_0}^{t_0+T} a_0 dt + \sum_{n=1}^{\infty} a_n \int_{t_0}^{t_0+T} \cos n\omega_0 t\, dt + \sum_{n=1}^{\infty} b_n \int_{t_0}^{t_0+T} \sin n\omega_0 t\, dt$$

$$= a_0 \int_{t_0}^{t_0+T} dt + \sum_{n=1}^{\infty} a_n(0) + \sum_{n=1}^{\infty} b_n(0) = a_0 T.$$

Therefore,

$$\int_{t_0}^{t_0+T} f(t)dt = a_0 T.$$

$$a_0 = \frac{1}{T} \int_{t_0}^{t_0+T} f(t)dt.$$

Evaluation of a_n:

Take the series $\displaystyle\int_{t_0}^{t_0+T} f(t)dt = \int_{t_0}^{t_0+T} \left[a_0 + \sum_{n=1}^{\infty} a_n \cos n\omega_0 t + b_n \sin n\omega_0 t \right] dt.$

Then, $\int_{t_0}^{t_0+T} f(t)\cos n\omega_0 t\, dt = \int_{t_0}^{t_0+T} a_0 + \sum_{n=1}^{\infty} [a_n \cos n\omega_0 t + b_n \sin n\omega_0 t] \cos n\omega_0 t\, dt.$

$$= \int_{t_0}^{t_0+T} a_0 \cos n\omega_0 t\, dt + \sum_{n=1}^{\infty} a_n \int_{t_0}^{t_0+T} \cos n\omega_0 t \cos n\omega_0 t\, dt + \sum_{n=1}^{\infty} b_n \int_{t_0}^{t_0+T} \cos n\omega_0 t \sin n\omega_0 t\, dt$$

From Eq. (3.3) and Eq.(3.5)

$$\int_{t_0}^{t_0+T} f(t)\cos n\omega_0 t\, dt = a_0(0) + a_n \frac{T}{2} + \sum_{n=1}^{\infty} b_n(0) = a_n \frac{T}{2}.$$

$$\int_{t_0}^{t_0+T} f(t)\cos n\omega_0 t\, dt = a_n \frac{T}{2}.$$

Therefore, $a_n = \frac{2}{T} \int_{t_0}^{t_0+T} f(t)\cos n\omega_0 t\, dt.$

Evaluation of b_n:

Take the series $\int_{t_0}^{t_0+T} f(t)dt = \int_{t_0}^{t_0+T} (a_0 + \sum_{n=1}^{\infty} [a_n \cos n\omega_0 t + b_n \sin n\omega_0 t])dt.$

Then $\int_{t_0}^{t_0+T} f(t)\sin n\omega_0 t\, dt = \int_{t_0}^{t_0+T} (a_0 + \sum_{n=1}^{\infty} [a_n \cos n\omega_0 t + b_n \sin n\omega_0 t] \sin n\omega_0 t\,)dt$

$$= \int_{t_0}^{t_0+T} a_0 \sin n\omega_0 t\, dt + \sum_{n=1}^{\infty} a_n \int_{t_0}^{t_0+T} \cos n\omega_0 t \sin n\omega_0 t\, dt + \sum_{n=1}^{\infty} b_n \int_{t_0}^{t_0+T} \sin n\omega_0 t \sin n\omega_0 t\, dt$$

From Eq.(3.4) and Eq.(3.5),

$$\int_{t_0}^{t_0+T} f(t)\sin n\omega_0 t\, dt = a_0(0) + a_n(0) + b_n \frac{T}{2} = b_n \frac{T}{2}$$

Therefore, $b_n = \frac{2}{T} \int_{t_0}^{t_0+T} f(t)\sin n\omega_0 t\, dt.$

Example 3.1

Evaluate the trigonometric Fourier series for the waveform shown in Fig. 3.2.

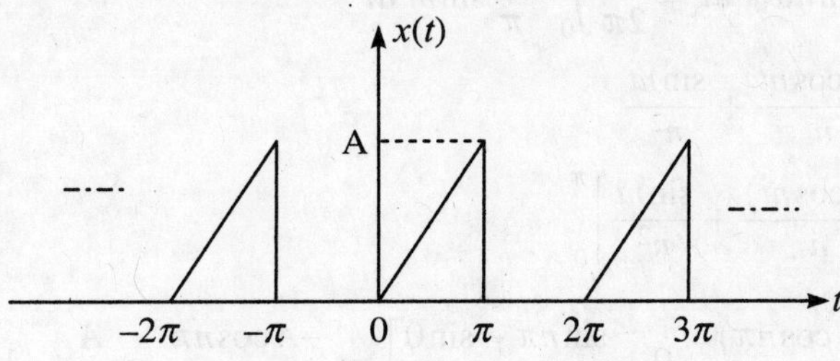

Fig. 3.2 *Triangular waveform*

Solution

From the given figure, the triangular function can be expressed as

$$f(t) = \begin{cases} \dfrac{At}{\pi} & 0 \le t \le \pi \\ 0 & \pi \le t \le 2\pi \end{cases}$$

Time period $T = 2\pi$ sec, fundamental frequency $\omega_0 = 1$ rad/sec.

The trigonometric Fourier series is

$$f(t) = a_0 + \sum_{n=1}^{\infty} (a_n \cos n\omega_0 t + b_n \sin n\omega_0 t)$$

$$\text{where } a_0 = \frac{1}{T}\int_0^T f(t)dt = \frac{1}{2\pi}\left[\int_0^{\pi} \frac{At}{\pi}dt\right]$$

$$= \frac{A}{2\pi^2}\left[\frac{t^2}{2}\right]_0^{\pi} = \frac{A}{2\pi^2}\left[\frac{\pi^2}{2}\right] = \frac{A}{4}.$$

$$a_n = \frac{2}{T}\int_0^T f(t)\cos n\omega_0 t\, dt = \frac{2}{2\pi}\int_0^{\pi} \frac{At}{\pi}\cos nt\, dt$$

$$\text{Since } \int t\cos nt\, dt = \frac{t\sin nt}{n} + \frac{\cos nt}{n^2},$$

$$a_n = \frac{A}{\pi^2}\left[t\frac{\sin nt}{n} + \frac{\cos nt}{n^2}\right]_0^{\pi}$$

$$= \frac{A}{\pi^2}\left[\left(2\pi\frac{\sin \pi n}{n} + \frac{\cos \pi n}{n^2}\right) - \left(0\frac{\sin 0}{n} + \frac{\cos 0}{n^2}\right)\right]$$

$$= \frac{A}{\pi^2 n^2}(\cos \pi n - \cos 0).$$

$$a_n = \begin{cases} \dfrac{-2A}{n^2\pi^2} & \text{for } n = \text{odd} \\ 0 & \text{for } n = \text{even} \end{cases}$$

$$b_n = \frac{2}{T}\int_0^T f(t)\sin n\omega_0 t\, dt = \frac{2}{2\pi}\int_0^{\pi} \frac{At}{\pi}\sin nt\, dt$$

$$\text{Since } \int t\sin nt\, dt = \frac{-t\cos nt}{n} + \frac{\sin nt}{n^2}$$

$$b_n = \frac{A}{\pi^2}\left[t\frac{(-\cos nt)}{n} + \frac{\sin nt}{n^2}\right]_0^{\pi}$$

$$= \frac{A}{\pi^2}\left[\pi\frac{(-\cos n\pi)}{n} + 0 + \frac{\sin n\pi - \sin 0}{n^2}\right] = \frac{-A\cos n\pi}{n\pi} = \frac{A}{n\pi}(-1)^{n+1}$$

The trigonometric Fourier series is

$$f(t) = \frac{A}{4} - \frac{2A}{\pi^2}\sum_{n=\text{odd}}^{\infty}\frac{\cos nt}{n^2} + \frac{A}{\pi}\sum_{n=1}^{\infty} -1^{n+1}\frac{\sin nt}{n}$$

$$f(t) = \frac{A}{4} - \frac{2A}{\pi^2}\left(\cos t + \frac{1}{9}\cos 3t + \frac{1}{25}\cos 5t + ...\right)$$

$$+\frac{A}{\pi}\left(\sin t - \frac{1}{2}\sin 2t + \frac{1}{3}\sin 3t - \frac{1}{4}\sin 4t + \frac{1}{5}\sin 5t....\right).$$

Exponential Fourier Series

Any real valued periodic function $f(t)$ can be expressed as a linear combination of complex exponential functions over any interval $(t_0, t_0 + 2\pi/\omega_0)$. The expression is called the exponential Fourier series of $f(t)$ and is expressed as

$$f(t) = c_0 + c_1 e^{j\omega_0} + c_2 e^{j2\omega_0 t} + c_3 e^{j3\omega_0 t} + ... + c_n e^{jn\omega_0 t} + ...$$
$$+ c_{-1} e^{-j\omega_0 t} + c_{-2} e^{-j2\omega_0 t} + c_{-3} e^{-j3\omega_0 t} + ... + c_{-n} e^{-jn\omega_0 t} + ...$$

Or $\quad f(t) = \sum_{n=-\infty}^{\infty} c_n e^{jn\omega_0 t}$ (3.12)

We know that a set of complex exponential functions $e^{jn\omega_0 t}$, $n = 0, \pm 1, \pm 2$.....etc., form a complete orthogonal set over an interval $(t_0, t_0 + 2\pi/\omega_0)$.

Let $f_1(t) = e^{jn\omega_0 t}$ and $f_2(t) = e^{jm\omega_0 t}$.

The condition for orthogonality is

$$\int_{t_0}^{t_0 + \frac{2\pi}{\omega_0}} f_1(t) f_2^*(t) dt = \int_{t_0}^{t_0 + \frac{2\pi}{\omega_0}} e^{jn\omega_0 t} (e^{jm\omega_0 t})^* dt \tag{3.13}$$

If $n = m$, then $e^{jn\omega_0 t} e^{-jm\omega_0 t} = e^{jn\omega_0 t} e^{-jn\omega_0 t} = 1$.

Hence $\int_{t_o}^{t_o+T} f_1(t) f_2^*(t) dt = \int_{t_o}^{t_o+T} dt = T.$

If $n \neq m$, then

$$\int_{t_o}^{t_o+T} f_1(t) f_2^*(t) dt = \int_{t_o}^{t_o+T} e^{jn\omega_1 t} e^{-jm\omega_1 t} dt$$

$$\int_{t_o}^{t_o+T} e^{j(n-m)\omega_0 t} dt = \int_{t_o}^{t_o+T} \cos(n-m)\omega_0 t\, dt + j\int_{t_o}^{t_o+T} \sin(n-m)\omega_0 t\, dt = 0.$$

Therefore,

$$\int_{t_0}^{t_0+T} e^{jn\omega_0 t} e^{-jm\omega_0 t} dt = \begin{cases} T & m = n \\ 0 & m \neq n \end{cases}. \tag{3.14}$$

where n is an integer $(n = 0, \pm 1, \pm 2, \cdots)$ and $\omega_1 = \frac{2\pi}{T}$. The value of n is called the harmonic number or harmonic.

Evaluation of c_n

The exponential Fourier series is

$$f(t) = \sum_{n=-\infty}^{\infty} c_n e^{jn\omega_0 t}.$$

Taking integration and multiplying with $e^{jn\omega_0 t}$,

$$\int_{t_o}^{t_o+T} f(t)e^{-jn\omega_0 t}\,dt = \int_{t_o}^{t_o+T} \sum_{n=-\infty}^{\infty} c_n e^{jn\omega_0 t} e^{-jn\omega_0 t}\,dt$$

$$= \sum_{n=-\infty}^{\infty} c_n \int_{t_o}^{t_o+T} e^{jn\omega_0 t} e^{-jn\omega_0 t}\,dt = c_n T$$

$$\int_{t_o}^{t_o+T} f(t)e^{-jn\omega_0 t}\,dt = c_n T.$$

Or, $c_n = \dfrac{1}{T}\displaystyle\int_{t_0}^{t_0+T} f(t)e^{-jn\omega_0 t}\,dt.$

Note:

1. It is observed that for $n = 0$, $c_0 = \dfrac{1}{T}\displaystyle\int_{t_0}^{t_0+T} f(t)dt$ is the average value of $f(t)$ over the interval $(t_0 < t < t_0 + T)$. It gives the dc component of $f(t)$ as a_0.
2. Since $f(t)$is a periodic signal with period T, i.e., $f(t) = f(t+T)$,the Fourier series over the interval $t_0 < t < t_0 + T$ holds for any value of t, that is, over the entire interval $-\infty < t < \infty$.
3. The value of c_n is independent of time t_0. In general, it can be taken as either $t_0 = 0$ or $t_0 = -T/2$. If $t_0 = 0$ is chosen, then

   $$c_n = \frac{1}{T}\int_0^T f(t)e^{-jn\omega_0 t}\,dt \qquad -\infty < t < \infty.$$

   If $t_0 = \dfrac{-T}{2}$ is chosen, then

   $$c_n = \frac{1}{T}\int_{-T/2}^{T/2} f(t)e^{-jn\omega_0 t}\,dt \qquad -\infty < t < \infty. \tag{3.15}$$

4. Any periodic function $f(t)$ with period T can be expressed in terms of a trigonometric or exponential Fourier series over the entire interval $-\infty < t < \infty$. This statement is called *Fourier theorem.*
5. Fourier series of a periodic function $f(t)$ of period T separates the function into sinusoidal components of harmonic frequencies of $\omega_0, 2\omega_0, 3\omega_0, ...n\omega_0...$ etc., where $\omega_0 = 2\pi/T$ is the fundamental frequency. This analysis is hence also called harmonic analysis.

Relationship between the Trigonometric and Exponential Fourier Series

The trigonometric Fourier series $f(t) = a_0 + \sum_{n=1}^{\infty} [a_n \cos n\omega_0 t + b_n \sin n\omega_0 t]$

where $a_0 = \dfrac{1}{T}\displaystyle\int_{-T/2}^{T/2} f(t)dt$; $a_n = \dfrac{2}{T}\displaystyle\int_{-T/2}^{T/2} f(t)\cos n\omega_0 t dt$, and $b_n = \dfrac{2}{T}\displaystyle\int_{-T/2}^{T/2} f(t)\sin \omega_0 t\, dt$

and the exponential Fourier series $c_n = \frac{1}{T}\int_{-T}^{t} f(t)e^{-jn\omega_0 t}dt$ are related as follows.

From Euler's formula we know that $e^{-jn\omega_0 t} = \cos n\omega_0 t - j\sin n\omega_0 t$. Hence

$$c_n = \frac{1}{T}\int_{-T}^{t} f(t)\,(\cos n\omega_0 t - j\sin n\omega_0 t)\,dt$$

$$c_n = \frac{1}{T}\int_{-T/2}^{T/2} f(t)\cos n\omega_0 t\;dt - \frac{1}{T}j\int_{-T/2}^{T/2} f(t)\sin n\omega_0 t\;dt$$

$$c_n = \frac{1}{2}\frac{2}{T}\int_{-T/2}^{T/2} f(t)\cos n\omega_0 t\;dt - \frac{1}{2}\frac{2}{T}j\int_{-T/2}^{T/2} f(t)\sin n\omega_0 t\;dt$$

$$= \frac{1}{2}a_n - \frac{1}{2}jb_n = \frac{1}{2}(a_n - jb_n) = \frac{1}{2}A_n e^{j\phi_n}.$$

That is, $c_n = \frac{1}{2}(a_n - jb_n)$ and $a_n = 2\text{Re}\,\{c_n\}$, $\quad b_n = -2\text{Im}\,\{c_n\}$. (3.16)

Example 3.2

Find the exponential Fourier series expansion for the rectangular wave shown in Fig. 3.3.

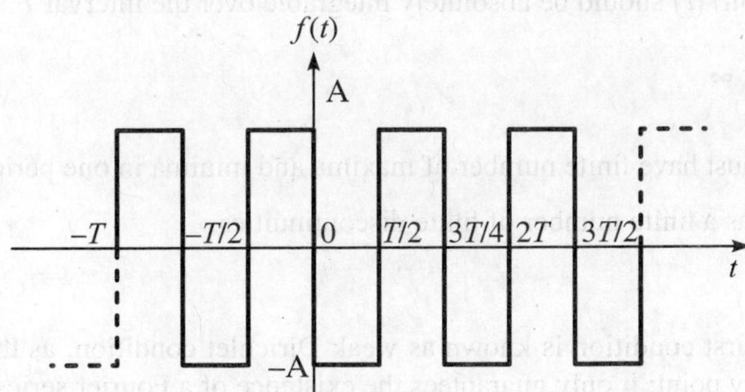

Fig. 3.3 *Rectangular wave signal*

Solution

From Fig. 3.3 the expression for rectangular wave is

$$f(t) = \begin{cases} -A; & -\frac{T}{2} \le t \le 0 \\ +A; & 0 \le t \le +\frac{T}{2} \end{cases}.$$

Time period $= T$, frequency $\omega_0 = \frac{2\pi}{T}$ rad/s.

The exponential Fourier series is $f(t) = \sum_{n=-\infty}^{\infty} c_n e^{jn\omega_0 t}$, where

$$c_n = \frac{1}{T}\int_{-T/2}^{T/2} f(t)e^{-jn\omega_0 t}dt = \frac{1}{T}\left[\int_{-T/2}^{0} -A\,e^{-jn\omega_0 t}\,dt + \int_{0}^{T/2} Ae^{-jn\omega_0 t}\,dt\right]$$

$$= \frac{-A}{jn\omega_0 T}\left[e^{\frac{-jn\omega_0 T}{2}} - 1\right] - \frac{A}{jn\omega_0 T}\left[e^{\frac{-jn\omega_0 T}{2}} - 1\right]$$

$$= \frac{A}{jn\pi}\left[1 - e^{-jn\pi}\right] = \frac{A}{jn\pi}\left[1 - \cos n\pi\right].$$

$$c_n = \begin{cases} \dfrac{2A}{jn\pi} & \text{for } n = \text{odd} \\ 0 & \text{for } n = \text{even} \end{cases}.$$

Therefore, $f(t) = \sum_{n=-\infty}^{\infty} c_n e^{jn\omega_0 t} = \sum_{n=-\infty}^{\infty} \frac{A}{jn\pi}\left[1 - \cos n\pi\right] e^{jn\omega_0 t}$.

3.4 Dirichlet's Conditions

For any periodic function $f(t)$, the Fourier series over the interval $-\infty < t < \infty$ is valid if the series converges at every point. If the function has infinite frequency component at some point, it will not converge at that point. Therefore the conditions imposed on the periodic function $f(t)$ is such that the Fourier series should converge at every point over the interval $(-\infty < t < \infty)$. The conditions are called Dirichlet conditions, given as:

1. The periodic function $f(t)$ should be absolutely integrable over the interval T sec. i.e.,

$$\int_0^T |f(t)|\, dt < \infty \tag{3.17}$$

2. The function $f(t)$ must have finite number of maxima and minima in one period.
3. The function $f(t)$ has a finite number of finite discontinuities.

Note:

1. Compliance to the first condition is known as weak Dirichlet condition, as the Fourier series may not converge at every point; it only guarantees the existence of a Fourier series.
2. Compliance of all three conditions of Dirichlet is called strong Dirichlet conditions. These necessary conditions provide both existence and convergence of Fourier series.
3. It is observed that at the point of discontinuity (finite jump) the function has two values. The Fourier series converges to the mean value of these two values. For example, if a step function has a jump at $t = 0$ and the values at the discontinuity are $f(0^-) = -1$ and $f(0^+) = 1$, then the Fourier series converges at $\frac{1}{2}\left(f(0^-) + f(0^+)\right) = 0$.

Gibbs Phenomenon

It has been observed that while reconstructing a function with truncated Fourier series (i.e., series limited to finite number of components) the peak amplitudes of ripples (overshoots) appear closer to the discontinuities in the signal. Further, the number of peaks in the ripples is equal to the number of nonzero harmonics N used in the reconstruction. The overshoots decay as damped oscillations away from the discontinuities. This is called the Gibbs phenomenon, after the mathematical physicist who first provided an explanation to this phenomenon.

An interesting observation is that the peak amplitude of these ripples does not diminish when more terms are added to the truncated Fourier series. In fact, as N grows larger, the first overshoot on both sides of the discontinuity remains at 9% of the height of the discontinuity.

As an example, let us consider the spectral coefficients up to $N = 7$ of a square wave signal. Figure 3.4 shows the plot of the real approximation to the square wave signal over one period. As N increases, intuitively, we expect the oscillations to die out and lead to perfect reconstruction, i.e., the energy in these ripples vanishing as $N \to \infty$. But it is observed form the Fig. 3.4(b) that the approximation gets closer to a square wave, but still rather large, but narrow ripples around the discontinuities can be seen.

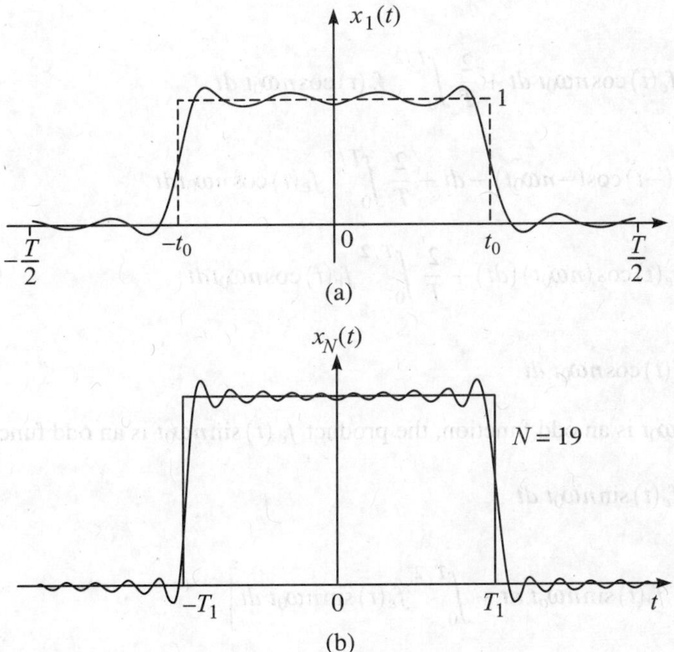

Fig. 3.4 *Gibbs phenomenon*

Note: The Gibbs effect occurs only for signals with discontinuities whose convergence is $1/n$. For signals that converge as $1/n^2$ or faster, the Gibbs effect is absent. For such signals, we get better reconstruction with fewer terms and an exact reconstruction as $N = \infty$.

3.5 Symmetry Conditions of Fourier Series

The Fourier series of a periodic signal $f(t)$ with no symmetry contains both odd (sine) and even (dc and cosine) components. It is observed that depending on the symmetrical conditions of the function, some components of the series become zero. Therefore, if we consider the symmetrical conditions of a function, the analysis of Fourier series will be simplified. Some of the symmetries that we can consider of a function with Fourier series are

1. Even symmetry
2. Odd symmetry
3. Rotation or half-wave symmetry
4. Hidden symmetry

Even Symmetry

A function $f(t)$ is said to be an even function or having even symmetry if

$$f(t) = f(-t). \tag{3.18}$$

Now, let us investigate the effect that even symmetry produces in a Fourier series.

If $f_e(t) = f(t)$ is an even function and since $\cos n\omega_0 t$ is also even function, then the product $f_e(t)\cos n\omega_0 t$ is an even function. The Fourier coefficients are:

$$a_n = \frac{2}{T}\int_{-T/2}^{T/2} f_e(t)\cos n\omega_0 t\, dt$$

$$= \frac{2}{T}\int_{-T/2}^{0} f_e(t)\cos n\omega_0 t\, dt + \frac{2}{T}\int_{0}^{T/2} f_e(t)\cos n\omega_0 t\, dt$$

$$= \frac{2}{T}\int_{T/2}^{0} f_e(-t)\cos(-n\omega_0 t) - dt + \frac{2}{T}\int_{0}^{T/2} f_e(t)\cos n\omega_0 t dt$$

$$= \frac{2}{T}\int_{0}^{T/2} f_e(t)\cos(n\omega_0 t)\,(dt) + \frac{2}{T}\int_{0}^{T/2} f_e(t)\cos n\omega_0 t dt$$

$$a_n = \frac{4}{T}\int_{0}^{T/2} f(t)\cos n\omega_0 t\, dt. \tag{3.19}$$

Similarly, since $\sin n\omega_0 t$ is an odd function, the product $f_e(t)\sin n\omega_0 t$ is an odd function. So,

$$b_n = \frac{2}{T}\int_{-T/2}^{T/2} f_e(t)\sin n\omega_0 t\, dt$$

$$= \frac{2}{T}\left[\int_{-T/2}^{0} f_e(t)\sin n\omega_o t\, dt + \int_{0}^{T/2} f_e(t)\sin n\omega_0 t\, dt\right]$$

$$= \frac{2}{T}\left[\int_{T/2}^{0} f_e(-t)\sin(-n\omega_o t)\;(-dt) + \int_{0}^{T/2} f_e(t)\sin n\omega_0 t\, dt\right]$$

$$= \frac{2}{T}\left[-\int_{0}^{T/2} f_e(t)\sin n\omega_o t\, dt + \int_{0}^{T/2} f_e(t)\sin n\omega_0 t\, dt\right] = 0$$

$$b_n = 0.$$

And $a_0 = \frac{1}{T}\int_{-T/2}^{T/2} f_e(t)dt = \frac{1}{T}\int_{-T/2}^{0} f_e(t)dt + \frac{1}{T}\int_{0}^{T/2} f_e(t)dt$

$$= \frac{1}{T}\int_{0}^{T/2} f_e(t)dt + \frac{1}{T}\int_{0}^{T/2} f_e(t)dt = \frac{2}{T}\int_{0}^{T/2} f_e(t)dt$$

$$a_0 = \frac{2}{T}\int_{0}^{T/2} f(t)dt. \tag{3.20}$$

Therefore, if $f(t)$ has even symmetry, then

$$b_n = 0; \quad a_n = \frac{4}{T}\int_0^{T/2} f(t)\cos n\omega_0 t\, dt, \text{ and } a_0 = \frac{2}{T}\int_0^{T/2} f(t)dt.$$

Odd Symmetry

A function $f(t)$ is said to be an odd function or possessing odd symmetry if

$$f(t) = -f(-t). \tag{3.21}$$

Now, let us investigate the effect that odd symmetry produces in a Fourier series.

If $f_o(t) = f(t)$ is an odd function and since $\sin n\omega_0 t$ is also odd function, then the product $f_o(t)\sin n\omega_0 t$ is even function.

$$b_n = \frac{2}{T}\int_{-T/2}^{T/2} f_o(t)\sin n\omega_0 t\, dt$$

$$= \frac{2}{T}\left[\int_{-T/2}^{0} f_o(t)\sin n\omega_o t\, dt + \int_0^{T/2} f_o(t)\sin n\omega_0 t\, dt\right]$$

$$= \frac{2}{T}\left[\int_{T/2}^{0} f_o(-t)\sin(-n\omega_o t)\ (-dt) + \int_0^{T/2} f_o(t)\sin n\omega_0 t\, dt\right].$$

Since $\sin n\omega_o t = -\sin(-n\omega_o t)$ and $f_o(t) = -f_o(-t)$,

$$b_n = \frac{2}{T}\left[\int_0^{T/2} f_O(t)\sin n\omega_o t\, dt + \int_0^{T/2} f_O(t)\sin n\omega_0 t\, dt\right]$$

$$b_n = \frac{4}{T}\int_0^{T/2} f_o(t)\sin n\omega_0 t\, dt. \tag{3.22}$$

Similarly, since $\cos n\omega_0 t$is even function, the product $f_o\,(t)\cos n\omega_0 t$ is an odd function. Then,

$$a_n = \frac{2}{T}\int_{-T/2}^{T/2} f_o(t)\cos n\omega_0 t\, dt$$

$$= \frac{2}{T}\int_{-T/2}^{0} f_o(t)\cos n\omega_0 t\, dt + \frac{2}{T}\int_0^{T/2} f_o(t)\cos n\omega_0 t\, dt$$

$$= \frac{2}{T}\int_{T/2}^{0} f_o(-t)\cos(-n\omega_0 t) - dt + \frac{2}{T}\int_0^{T/2} f_o(t)\cos n\omega_0 t\, dt$$

$$= \frac{-2}{T}\int_0^{T/2} f_o(t)\cos n\omega_0 t\, dt + \frac{2}{T}\int_0^{T/2} f_o(t)\cos n\omega_0 t\, dt = 0$$

$$a_n = 0$$

and $a_0 = \frac{1}{T}\int_{-T/2}^{T/2} f_o(t)dt = \frac{1}{T}\int_{-T/2}^{0} f_o(t)dt + \frac{1}{T}\int_0^{T/2} f_o(t)dt$

$$= \frac{-1}{T}\int_0^{T/2} f_o(t)dt + \frac{1}{T}\int_0^{T/2} f_o(t)dt = 0.$$

Therefore, if $f(t)$ has odd symmetry, then

$$a_0 = 0,\ a_n = 0 \text{ and } b_n = \frac{4}{T}\int_0^{T/2} f(t)\sin n\omega_0 t\, dt.$$

Hidden Symmetry

Consider a periodic function which does not satisfy either of the symmetry conditions but the Fourier series consists of a dc term (A) and sine terms alone.

The reason why the symmetry of the function is obscured is because of the presence of the dc term. If we subtract half of the dc term ($A/2$) from the function, that is, shift the function horizontally up to a value $A/2$, the new function is seen to have an odd symmetry.

If a function $f(t)$ has hidden symmetry, then

$$f_0(t) = f(t) - \frac{A}{2} \tag{3.23}$$

where $f_0(t)$ is an odd function.

Thus a hidden symmetry is exhibited when a dc term and sine terms appear together as shown in Fig. 3.5.

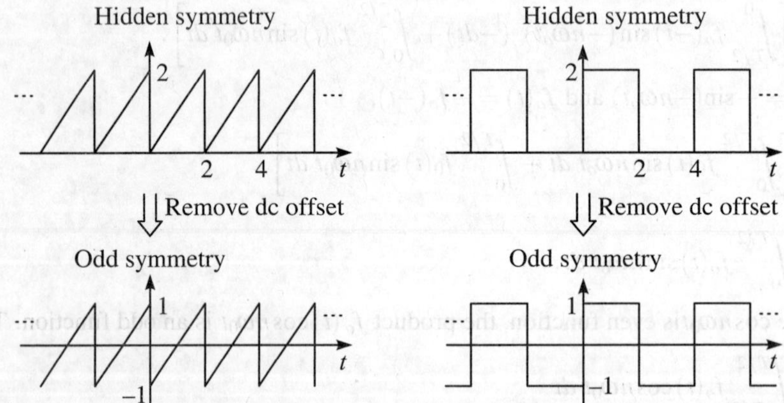

Fig. 3.5 *Hidden symmetry and dc offset of (a) a sawtooth wave, (b) a rectangular pulse wave*

Rotation or Half-wave Symmetry

A function is said to be a rotation or half-wave symmetry of t if

$$f(t) = -f\left(t \pm \frac{T}{2}\right). \tag{3.24}$$

In half-wave symmetry, unlike even and odd symmetry, the shape of the signal is identical, with opposite sign at every alternate half cycles. Figure 3.6 illustrates two half-wave symmetry signals.

It may be shown that the Fourier series of any function which has half-wave symmetry contains only odd harmonics of both sine and cosine terms.

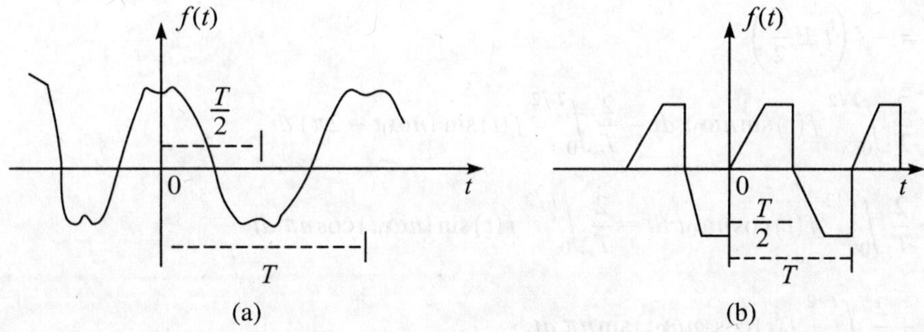

Fig. 3.6 *Half-wave symmetry signals*

Proof: Consider a periodic function $f(t)$. For rotation symmetry the Fourier coefficients are

$$a_n = \frac{2}{T}\int_0^T f(t)\cos n\omega_0 t\,dt = \frac{2}{T}\int_0^{T/2} f(t)\cos n\omega_0 t\,dt + \frac{2}{T}\int_{T/2}^{T} f(t)\cos n\omega_0 t\,dt$$

By changing t to $t+\dfrac{T}{2}$ in the second term to get

$$a_n = \frac{2}{T}\int_0^{T/2} f(t)\cos n\omega_0 t\,dt + \frac{2}{T}\int_0^{T/2} f\left(t+\frac{2}{T}\right)\cos n\omega_0\left(t+\frac{2}{T}\right)dt.$$

Since $f(t) = -f\left(t \pm \dfrac{T}{2}\right)$,

$$a_n = \frac{2}{T}\int_0^{T/2} f(t)\cos n\omega_0 t\,dt - \frac{2}{T}\int_0^{T/2} f(t)\cos\left(n\omega_0 t - 2\pi\right)dt$$

$$= \frac{2}{T}\int_0^{T/2} f(t)\cos n\omega_0 t\,dt - \frac{2}{T}\int_0^{T/2} f(t)\cos\left(n\omega_0 t\right)\cos n\pi\,dt$$

$$-\frac{2}{T}\int_0^{T/2} f(t)\sin\left(n\omega_0 t\right)\sin n\pi\,dt.$$

Since $\sin n\pi = 0$ and $\cos n\pi = \begin{cases} 1 & \text{for } n = \text{even} \\ -1 & \text{for } n = \text{odd.} \end{cases}$

Then, $a_n = \dfrac{2}{T}\displaystyle\int_0^{T/2} f(t)\cos n\omega_0 t\,dt - \dfrac{2}{T}\displaystyle\int_0^{T/2} f(t)\cos n\omega_0 t\cos n\pi\,dt$

$$\text{or } a_n = \begin{cases} \dfrac{4}{T}\displaystyle\int_0^{T/2} f(t)\cos n\omega_0 t\,dt & \text{for } n = \text{odd} \\ 0 & \text{for } n = \text{even.} \end{cases} \tag{3.25}$$

$$b_n = \frac{2}{T}\int_0^T f(t)\sin n\omega_0 t\,dt = \frac{2}{T}\int_0^{T/2} f(t)\sin n\omega_0 t\,dt + \frac{2}{T}\int_{T/2}^{0} f(t)\sin n\omega_0 t\,dt.$$

By changing t to $t+\frac{T}{2}$ in second term, we get

$$a_n = \frac{2}{T}\int_0^{T/2} f(t)\sin n\omega_0 t\,dt + \frac{2}{T}\int_0^{T/2} f\left(t+\frac{2}{T}\right)\sin n\omega_0\left(t+\frac{2}{T}\right)dt.$$

Since $f(t) = -f\left(t \pm \frac{T}{2}\right)$

$$a_n = \frac{2}{T}\int_0^{T/2} f(t)\sin n\omega_0 t\, dt - \frac{2}{T}\int_0^{T/2} f(t)\sin\,(n\omega_0 t - 2\pi)\, dt$$

$$= \frac{2}{T}\int_0^{T/2} f(t)\cos n\omega_0 t\, dt - \frac{2}{T}\int_0^{T/2} f(t)\sin\,(n\omega_0 t)\cos n\pi\, dt$$

$$+ \frac{2}{T}\int_0^{T/2} f(t)\cos\,(n\omega_0 t)\sin n\pi\, dt.$$

Since $\sin n\pi = 0$ and $\cos n\pi = \begin{cases} 1 & \text{for } n = \text{even} \\ -1 & \text{for } n = \text{odd,} \end{cases}$

then, $b_n = \frac{2}{T}\int_0^{T/2} f(t)\sin n\omega_0 t\, dt - \frac{2}{T}\int_0^{T/2} f(t)\sin\,(n\omega_0 t)\cos n\pi\, dt$

$$\text{or } b_n = \begin{cases} \frac{4}{T}\int_0^{T/2} f(t)\sin n\omega_0 t\, dt & \text{for } n = \text{odd} \\ 0 & \text{for } n = \text{even} \end{cases}. \tag{3.26}$$

Therefore, if $f(t)$ has rotation symmetry, then

$$a_n = \begin{cases} 0 & \text{for } n = \text{even} \\ \frac{4}{T}\int_0^{T/2} f(t)\cos n\omega_0 t\, dt & \text{for } n = \text{odd} \end{cases} \quad \text{and}$$

$$b_n = \begin{cases} 0 & \text{for } n = \text{even} \\ \frac{4}{T}\int_0^{T/2} f(t)\sin n\omega_0 t\, dt & \text{for } n = \text{odd.} \end{cases}$$

Note: If the rotation symmetry is hidden by a dc term, remove the dc component from the signal by shifting the function horizontally up by a value A/2; the new function now reveals the hidden symmetry. The Fourier series contains the dc term and odd harmonics of both sine and cosine terms.

Example 3.3

For the periodic signals shown in Fig. 3.7, find the trigonometric Fourier series.

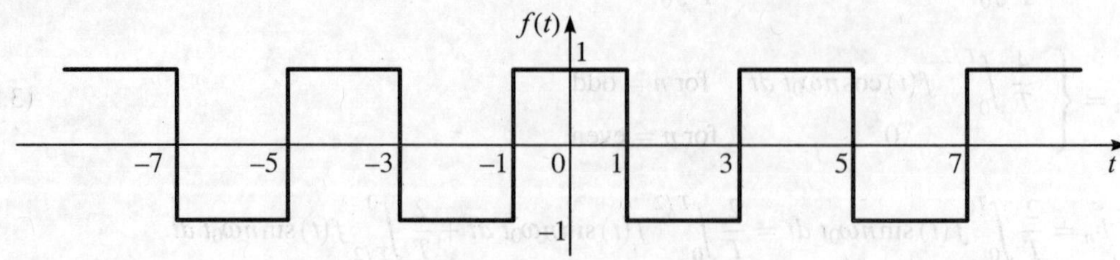

Fig. 3.7 *Square wave signal*

Solution:

From Fig. 3.7 given, the expression for the square wave is

$$f(t) = \begin{cases} +1; & -1 \le t \le +1 \\ -1; & +1 \le t \le +3 \end{cases}.$$

Time period $T = 4$; fundamental frequency $\omega_0 = \frac{2\pi}{T} = \frac{\pi}{2}$.

The trigonometric Fourier series is $f(t) = a_0 + \sum_{n=1}^{\infty} (a_n \cos n\omega_0 t + b_n \sin n\omega_0 t)$.

As the waveform follows even symmetry, all the sine terms are zero; $b_n = 0$; and

$$a_n = \frac{4}{T}\int_0^{T/2} f(t) \cos n\omega_0 t \, dt.$$

And since the time axis is positioned at the middle of the signal amplitude, the dc component. $a_0 = 0$.

$$a_n = \frac{2}{T}\int_{-T/2}^{T/2} f(t) \cos n\omega_0 t \, dt$$

$$= \frac{2}{4}\left[\int_{-1}^{+1} (1)\cos\left(\frac{n\pi}{2}t\right) dt + \int_{+1}^{+3} (-1)\cos\left(\frac{n\pi}{2}t\right) dt\right]$$

$$= \frac{2}{4}\left[\int_{-1}^{1} \cos(n\pi t/2)\, dt - \int_{1}^{3} \cos(n\pi t/2)\, dt\right]$$

$$= \frac{1}{2}\left[\left(\frac{\sin(n\pi t/2)}{n\pi/2}\right)_{-1}^{+1} - \left(\frac{\sin(n\pi t/2)}{n\pi/2}\right)_{+1}^{+3}\right]$$

$$= \frac{1}{n\pi}[\sin(\pi n/2) + \sin(\pi n/2) - (\sin(3\pi n/2) - \sin(\pi n/2))]$$

$$a_n = \frac{4}{n\pi}\sin\left(\frac{n\pi}{2}\right) = \begin{cases} 4/n\pi & \text{for } n = 1, 5, 9.. \\ -4/n\pi & \text{for } n = 3, 7, 11.. \\ 0 & \text{for } n = \text{even}. \end{cases}$$

Therefore the trigonometric Fourier series for $f(t)$ is

$$f(t) = \frac{4}{\pi}\left[\cos\left(\frac{\pi t}{2}\right) - \frac{1}{3}\cos\left(\frac{3\pi t}{2}\right) + \frac{1}{5}\cos\left(\frac{5\pi t}{2}\right) - \frac{1}{7}\cos\left(\frac{7\pi t}{2}\right) + \ldots\ldots\ldots\ldots\right].$$

Example 3.4

For the periodic signal shown in Fig. 3.8, find the trigonometric Fourier series.

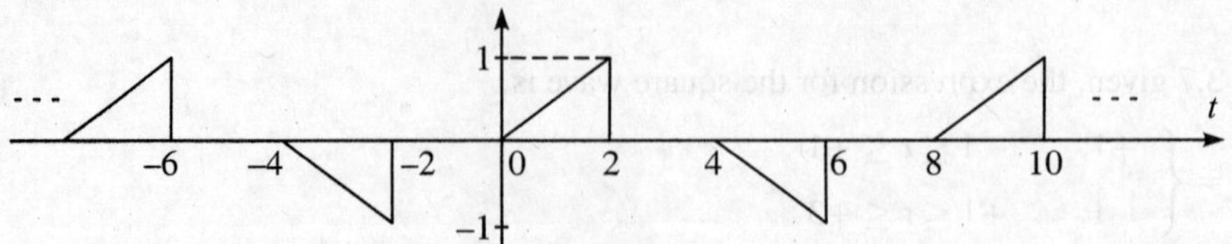

Fig. 3.8 *Triangular waveform*

Solution

From the given figure, the expression for the triangular wave is

$$f(t) = \begin{cases} \dfrac{t}{2} & 0 \le t \le 2 \\ 0 & 2 \le t \le 4 \\ \dfrac{-t}{2} & 4 \le t \le 6 \\ 0 & 6 \le t \le 8 \end{cases}$$

and time period $T = 8$; fundamental frequency $\omega_0 = \dfrac{2\pi}{8} = \dfrac{\pi}{4}$ rad/s.

The given waveform satisfies half-wave or rotation symmetry. Then $f(t) = -f\left(t \pm \dfrac{T}{2}\right)$ and the average value $a_0 = 0$.

The Fourier coefficients are:

$$a_n = \begin{cases} \dfrac{4}{T}\displaystyle\int_0^{T/2} f(t)\cos n\omega_0 t\, dt & \text{for } n = \text{odd} \\ 0 & \text{for } n = \text{even} \end{cases} \quad \text{and}$$

$$b_n = \begin{cases} \dfrac{4}{T}\displaystyle\int_0^{T/2} f(t)\sin n\omega_0 t\, dt & \text{for } n = \text{odd} \\ 0 & \text{for } n = \text{even.} \end{cases}$$

$$a_n = \frac{4}{T}\int_0^{T/2} f(t)\cos n\omega_0 t dt = \frac{1}{4}\int_0^2 t\cos\left(\frac{n\pi}{4}t\right) dt = \frac{1}{4}\left(\frac{t\sin(n\pi t/4)}{n\pi/4} + \frac{\cos(n\pi t/4)}{(n\pi/4)^2}\right)_0^2$$

$$= \frac{1}{4}\left(\frac{2\sin(n\pi/2)}{n\pi/4} + \frac{\cos(n\pi/2)}{(n\pi/4)^2} - \frac{1}{(n\pi/4)^2}\right)$$

$$= \frac{1}{4(n\pi/4)^2}\left(\frac{n\pi}{2}\sin(n\pi/2) + \cos(n\pi/2) - 1\right)$$

$$= \frac{4}{n^2\pi^2}\left(\cos(n\pi/2) + \frac{n\pi}{2}\sin(n\pi/2) - 1\right) = \frac{4}{n^2\pi^2}\left(\frac{n\pi}{2}\sin(n\pi/2) - 1\right).$$

For $n =$ odd, since $\cos(n\pi/2) = 0$ and $\sin(n\pi/2) = \begin{cases} 1 & n = 1,5,9... \\ -1 & n = 3,7,11... \end{cases}$

$$a_n = \begin{cases} \dfrac{4}{n^2\pi^2}\left(\dfrac{n\pi}{2} - 1\right) & n = 1,5,9,13,.... \\ \dfrac{-4}{n^2\pi^2}\left(\dfrac{n\pi}{2} + 1\right) & n = 3,7,11,15,.... \end{cases}$$

$$b_n = \frac{4}{T}\int_0^{T/2} f(t)\sin n\omega_0 t\, dt = \frac{4}{8}\int_0^2 \frac{t}{2}\sin\left(\frac{n\pi}{4}t\right) dt$$

$$= \frac{1}{4}\left(\frac{t(-\cos(n\pi t/4))}{n\pi/4} + \frac{\sin(n\pi t/4)}{(n\pi/4)^2}\right)_0^2 = \frac{1}{4}\left(\frac{-2\cos(n\pi/2)}{n\pi/4} + \frac{\sin(n\pi/2)}{(n\pi/4)^4}\right)$$

$$= \frac{4}{n^2\pi^2}\left(\sin\frac{n\pi}{2} - \frac{n\pi}{2}\cos\frac{n\pi}{2}\right).$$

$$b_n = \frac{4}{n^2\pi^2}\sin\frac{n\pi}{2};\quad n \text{ is odd.}$$

$$b_n = \begin{cases} \dfrac{4}{n^2\pi^2} & n = 1,5,9,.... \\ \dfrac{-4}{n^2\pi^2} & n = 3,7,11,.... \end{cases}$$

The Fourier series is $f(t) = \sum_{n=1}^{\infty}(a_n \cos n\omega_0 t + b_n \sin n\omega_0 t)$.

As the waveform follows half-wave symmetry,

$$f(t) = \sum_{n=1,3,5,...}^{\infty}\left(a_n \cos\frac{n\pi}{4}t + b_n \sin\frac{n\pi}{4}t\right).$$

$$f(t) = \sum_{n=1,5,7...}^{\infty}\left(\frac{4}{n^2\pi^2}\left(\frac{n\pi}{4} - 1\right)\cos\frac{n\pi}{4}t + \frac{4}{n^2\pi^2}\sin\frac{n\pi}{4}t\right)$$

$$- \sum_{n=3,9,11...}^{\infty}\left(\frac{4}{n^2\pi^2}\left(\frac{n\pi}{4} - 1\right)\cos\frac{n\pi}{4}t + \frac{4}{n^2\pi^2}\sin\frac{n\pi}{4}t\right).$$

3.6 The Complex Fourier Spectrum

The interpretation of the Fourier series for periodic signals is that a signal $f(t)$ with period T is decomposed in terms of its harmonics or frequency components $\omega_0, 2\omega_0, 3\omega_0......n\omega_0$...etc., where $\omega_0 = 2\pi/T$. The range of frequencies that a signal contains is called frequency spectrum or spectrum of frequencies. Thus the Fourier series of a periodic signal represent its frequency spectrum.

We know that the waveform representation of the function $f(t)$ as a function of time t is called its time domain representation. Similarly, the representation of frequency components at harmonic frequencies $\omega_0, 2\omega_0, 3\omega_0......n\omega_0$... etc., is called its frequency domain representation. In the discrete function representation, also called line spectrum, only the amplitude of the components are specified at equally spaced frequency values.

Exponential Fourier series is more preferable to represent frequency spectrum than trigonometric Fourier series because of its simplicity. The complex Fourier spectrum or the frequency spectrum of $f(t)$ is equal to the representation of Fourier coefficients c_n at $n = \pm 1, \pm 2, \ldots$ That is, the coefficients c_n, in general, are complex numbers, expressed as

$$c_n = |c_n|\, e^{j\phi_n} \tag{3.27}$$

where $|c_n|$ is the amplitude (or magnitude) spectrum and ϕ_n the phase spectrum. In the case of discrete time periodic signals this spectrum is discrete; that is, the spectrum has a non-zero value only for the frequencies$\{n\omega_0\}$, $n = \pm 1, \pm 2, \cdots$. It is to be noted that along the negative axis for $n = -1, -2, ...$, the frequencies are indicated as negative frequency. Since negative frequency does not exist, these frequencies provide index for the exponential function. Also the sum of positive and negative frequency complex exponentials at a particular frequency results a real function as

$$e^{jn\omega_0 t} + je^{-jn\omega_0 t} = 2\cos n\omega_0 t. \tag{3.28}$$

The spectrum for positive frequencies is $c_n = |c_n| e^{j\phi_n} = \dfrac{2}{T}\displaystyle\int_{-T/2}^{T/2} f(t)e^{-jn\omega_0 t}\,dt.$

The spectrum for negative frequencies is $c_{-n} = |c_n| e^{-j\phi_n} = \dfrac{2}{T}\displaystyle\int_{-T/2}^{T/2} f(t)e^{jn\omega_0 t}\,dt$

Also, $c_n = c_{-n}^*$ and $|c_n| = |c_{-n}|$. (3.29)

Therefore, the amplitude spectrum is symmetrical about the origin (even symmetry), while the phase spectrum is antisymmetric (odd function). Since the spectrum is extended from $-\infty$ to ∞, it is a double sided spectrum. For trigonometric Fourier series the amplitude spectrum is represented with either a_n or b_n or both, depending upon the symmetry.

Since the amplitude spectrum for both coefficients extend from 0 to ∞, it is a one-sided spectrum and has no negative frequencies. Also, no phase spectrum exists.

Notation: Consider a periodic signal $x(t)$ with period T and exponential Fourier coefficients c_n. The notation to represent the Fourier series of $x(t)$ with c_n is

$$x(t) \overset{\text{FS}}{\longleftrightarrow} c_n \text{ or } \text{FS}[x(t)] = c_n \text{ and } \text{FS}^{-1}[c_n] = x(t).$$

Example 3.5

(a) Evaluate the trigonometric Fourier series expansion of a full-wave rectified cosine function shown in Fig. 3.9; (b) derive the corresponding exponential Fourier series, and (c) draw the complex Fourier spectrum.

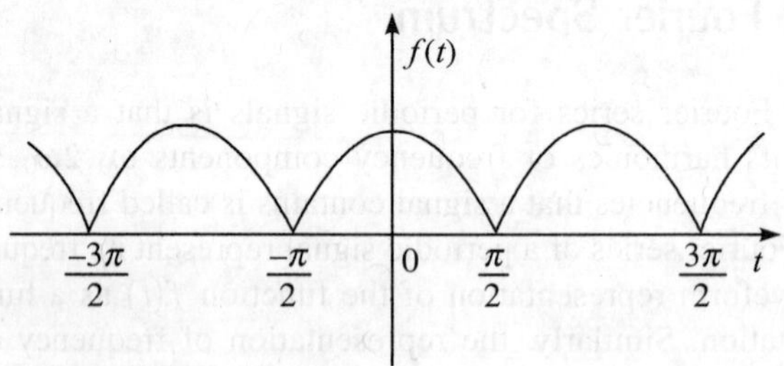

Fig. 3.9 *Full-wave rectified cosine signal*

Solution

Given that $f(t) = \cos t$; $-\dfrac{\pi}{2} \le t \le \dfrac{+\pi}{2}$; time period $T = \pi$; fundamental frequency $\omega_0 = \dfrac{2\pi}{\pi} = 2$.

(a) The trigonometric Fourier series:

$$f(t) = a_0 + \sum_{n=1}^{\infty} (a_n \cos n\omega_0 t + b_n \sin n\omega_0 t).$$

As the waveform follows even symmetry, all the sine terms are zero and $b_n = 0$.

$$f(t) = a_0 + \sum_{n=1}^{\infty} a_n \cos n\omega_0 t.$$

$$a_0 = \frac{1}{T}\int_{-T/2}^{T/2} f(t)dt = \frac{1}{\pi}\int_{-\pi/2}^{\pi/2} \cos t\, dt = \frac{1}{\pi}(\sin t)\Big|_{\frac{-\pi}{2}}^{\frac{+\pi}{2}} = \frac{2}{\pi}.$$

$$a_n = \frac{2}{T}\int_{-T/2}^{T/2} f(t)\cos n\omega_0 t\, dt$$

$$= \frac{2}{\pi}\left(\int_{\frac{-\pi}{2}}^{\frac{+\pi}{2}} \cos t \cos(2nt)dt\right) = \frac{1}{\pi}\int_{\frac{-\pi}{2}}^{\frac{+\pi}{2}} (\cos(2n-1)t + \cos(2n-1)t)\, dt$$

$$= \frac{1}{\pi}\left(\left[\frac{\sin(2n-1)t}{2n\quad 1}\right]_{\frac{-\pi}{2}}^{\frac{\pi}{2}} + \left[\frac{\sin(2n+1)t}{2n+1}\right]_{\frac{-\pi}{2}}^{\frac{\pi}{2}}\right)$$

$$\therefore \quad \sin\left((2n-1)\frac{\pi}{2}\right) = (-1)^{n+1} \text{ and } \sin\left((2n+1)\frac{\pi}{2}\right) = (-1)^n.$$

$$a_n = \frac{2}{\pi}\left(\frac{(-1)^{n+1}}{2n-1} + \frac{(-1)^n}{2n+1}\right)$$

Therefore the trigonometric Fourier series for $f(t)$ is

$$f(t) = \frac{2}{\pi} + \sum_{n=1}^{\infty}\frac{2}{\pi}\left\{\frac{2}{\pi}\left(\frac{(-1)^{n+1}}{2n-1} + \frac{(-1)^n}{2n+1}\right)\cos(2nt)\right\}$$

$$f(t) = \frac{2}{\pi}\left[1 + \frac{2}{3}\cos 2t - \frac{2}{15}\cos 4t + \frac{2}{35}\cos 6t.........\right].$$

(b) Exponential Fourier series:

$$f(t) = \sum_{n=-\infty}^{\infty} c_n e^{jn\omega_0 t} = \sum_{n=-\infty}^{\infty} c_n e^{j2nt}$$

From trigonometric Fourier series, we know that

$$c_n = \frac{1}{2}(a_n - jb_n) = \frac{a_n}{2}$$

$$c_n = \frac{1}{\pi}\left(\frac{(-1)^n}{2n+1} + \frac{(-1)^{n+1}}{2n-1}\right)$$

$$f(t) = \sum_{n=-\infty}^{\infty} c_n e^{jn\omega_0 t} = \sum_{n=-\infty}^{\infty}\frac{1}{\pi}\left(\frac{(-1)^n}{2n+1} + \frac{(-1)^{n+1}}{2n-1}\right)e^{j2nt}.$$

$$c_0 = \frac{2}{\pi};\quad c_1 = c_{-1} = \frac{1}{2}a_1 = \frac{2}{3\pi};\quad c_2 = c_{-2} = \frac{1}{2}a_2 = \frac{-2}{15\pi};\ ...$$

$$f(t) = \frac{2}{\pi} + \frac{2}{3\pi}e^{j2t} - \frac{2}{15\pi}e^{j4t} + ...+ \frac{2}{3\pi}e^{-j2t} - \frac{2}{15\pi}e^{-j4t} + ...$$

(c) The frequency spectrum

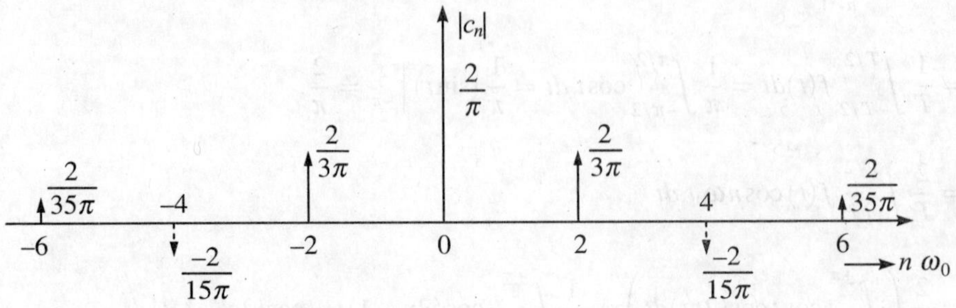

Fig. 3.10 *Amplitude spectrum of full-wave rectified cosine signal*

As c_n is real, only amplitude plot is required. The spectrum shown in Fig. 3.10 is symmetrical about the vertical axis passing through the origin, which is true for every periodic real function. The negative amplitude signifies the phase opposition of the corresponding harmonic.

3.7 Properties of Fourier Series

Fourier series of continuous time signals have many properties that are useful for solving problems. The important properties are explained below.

Linearity

Let two periodic signals $x_1(t)$ and $x_2(t)$ with period T have exponential Fourier coefficients c_n and d_n respectively.

The linearity property states that if

$$x_1(t) \overset{\text{FS}}{\longleftrightarrow} c_n, \quad x_2(t) \overset{\text{FS}}{\longleftrightarrow} d_n,$$

then the linear combination of the two signals with same time period T satisfy the relation

$$Ax_1(t) + Bx_2(t) \overset{\text{FS}}{\longleftrightarrow} Ac_n + Bd_n, \tag{3.30}$$

where A and B are arbitrary constants.

Proof: Let two periodic signals $x_1(t)$ and $x_2(t)$ with period T have exponential Fourier coefficients c_n and d_n respectively.

$$\begin{aligned}\text{FS }[Ax_1(t) + Bx_2(t)] &= \frac{2}{T}\int_0^T [Ax_1(t) + Bx_2(t)]\, e^{-jn\omega_0 t}\, dt \\ &= \frac{2}{T}\int_0^T Ax_1(t)e^{-jn\omega_0 t}\, dt + \frac{2}{T}\int_0^T Bx_2(t)e^{-jn\omega_0 t}\, dt \\ &= A\frac{2}{T}\int_0^T x_1(t)e^{-jn\omega_0 t}\, dt + B\frac{2}{T}\int_0^T x_2(t)e^{-jn\omega_0 t}\, dt \\ &= Ac_n + Bd_n \qquad \text{(Proved).}\end{aligned}$$

Time Shifting

The time shifting property states that if $x(t) \overset{FS}{\longleftrightarrow} c_n$, then

$$x(t-t_0) \overset{FS}{\longleftrightarrow} [e^{-jk\omega_0 t_0}]c_n. \tag{3.31}$$

Proof: Let a periodic signal $x(t)$ with period T have exponential Fourier coefficient c_n.

Then, $x(t) = \sum_{n=-\infty}^{\infty} c_n e^{jn\omega_0 t}$.

For time delay t_0, $x(t-t_0) = \sum_{n=-\infty}^{\infty} c_n e^{jn\omega_0 (t-t_0)}$

$$x(t-t_0) = \sum_{n=-\infty}^{\infty} \left[c_n e^{-jn\omega_0 t_0}\right] e^{jn\omega_0 t}.$$

Then, $x(t-t_0) \overset{FS}{\longleftrightarrow} [e^{-jk\omega_0 t_0}]c_n$. (Proved).

Time Reversal

The time reversal property states that if $x(t) \overset{FS}{\longleftrightarrow} c_n$, then

$$x(-t) \overset{FS}{\longleftrightarrow} c_{-n}. \tag{3.32}$$

Proof: Let a periodic signal $x(t)$ with period T have its exponential Fourier coefficient as c_n.

Then, $x(t) = \sum_{n=-\infty}^{\infty} c_n e^{jn\omega_0 t}$.

Now, $x(-t) = \sum_{n=-\infty}^{\infty} c_n e^{-jn\omega_0 t} = \sum_{n=-\infty}^{\infty} c_n e^{j-n\omega_0 t} = \sum_{n=-\infty}^{\infty} c_{-n} e^{jn\omega_0 t}$

$$x(-t) = \sum_{n=-\infty}^{\infty} c_{-n} e^{jn\omega_0 t}.$$

Therefore, $x(-t) \overset{FS}{\longleftrightarrow} c_{-n}$ (Proved).

Time Scaling

The time scaling property states that if $x(t) \overset{FS}{\longleftrightarrow} c_n$, then

$$x(at) \overset{FS}{\longleftrightarrow} c_n. \qquad \text{with frequency } a\omega_0. \tag{3.33}$$

Proof: Let a periodic signal $x(t)$ with period T have its exponential Fourier coefficient as c_n.

Then $x(t) = \sum_{n=-\infty}^{\infty} c_n e^{jn\omega_0 t}$.

Now $x(at) = \sum_{n=-\infty}^{\infty} c_n e^{-jn\omega_0 at} = \sum_{n=-\infty}^{\infty} c_n e^{jn(a\omega_0)t} = c_n.$

Therefore, $x(at) \overset{FS}{\longleftrightarrow} c_n$ with frequency $a\omega_0$ (Proved).

Conjugation and Conjugate Symmetry

The conjugation property states that if $x(t) \overset{FS}{\longleftrightarrow} c_n$ then for complex $x(t)$,

$$x^*(t) \overset{FS}{\longleftrightarrow} c^*_{-n} \tag{3.34}$$

and the conjugate symmetry for real $x(t), a_{-n} = a_n^*$, or $a_n = a_{-n}^*$ (3.35)

Proof: Let a periodic signal $x(t)$ with period T have its exponential Fourier coefficient as c_n.

Then, $x(t) = \sum_{n=-\infty}^{\infty} c_n e^{jn\omega_0 t}$.

$$x^*(t) = \left[\sum_{n=-\infty}^{\infty} c_n e^{jn\omega_0 t}\right]^* = \sum_{n=-\infty}^{\infty} c_n e^{jn\omega_0 t^*}$$

$$= \sum_{n=-\infty}^{\infty} c_n^* e^{-jn\omega_0 t} = \sum_{n=-\infty}^{\infty} c_n^* e^{j(-n)\omega_0 t} = \sum_{n=-\infty}^{\infty} c_{-n}^* e^{jn\omega_0 t}$$

$$x^*(t) \overset{\text{FS}}{\longleftrightarrow} c_{-n}^*$$ (Proved).

Time Differentiation

The time differentiation property states that if $x(t) \overset{\text{FS}}{\longleftrightarrow} c_n$, then

$$\frac{dx(t)}{dt} \overset{\text{FS}}{\longleftrightarrow} jn\omega_0 c_n. \tag{3.36}$$

Proof: Let a periodic signal $x(t)$ with period T have its exponential Fourier coefficient as c_n. Then $x(t) = \sum_{n=-\infty}^{\infty} c_n e^{jn\omega_0 t}$.

Differentiating both sides, we get

$$\frac{dx(t)}{dt} = \frac{d}{dt}\left[\sum_{n=-\infty}^{\infty} c_n e^{jn\omega_0 t}\right] = \sum_{n=-\infty}^{\infty} \frac{d}{dt}\left[c_n e^{jn\omega_0 t}\right]$$

$$= \sum_{n=-\infty}^{\infty} c_n \frac{d}{dt}\left[e^{jn\omega_0 t}\right] = \sum_{n=-\infty}^{\infty} (jn\omega_0 c_n)\, e^{jn\omega_0 t}.$$

$$\frac{dx(t)}{dt} \overset{\text{FS}}{\longleftrightarrow} jn\omega_0 c_n$$ (Proved).

Time Integration

The time integration property states that if $x(t) \overset{\text{FS}}{\longleftrightarrow} c_n$, then

$$\int_{-\infty}^{t} x(\tau)\, d\tau \overset{\text{FS}}{\longleftrightarrow} \frac{c_n}{jn\omega_0}, \quad \text{with } c_0 = 0. \tag{3.37}$$

Proof: Let a periodic signal $x(t)$ with period T have its exponential Fourier coefficient as c_n.

Then $x(t) = \sum_{n=-\infty}^{\infty} c_n e^{jn\omega_0 t}$.

Differentiating both sides, we get

$$\int_{-\infty}^{t} x(\tau)\, d\tau = \int_{-\infty}^{t} \left[\sum_{n=-\infty}^{\infty} c_n e^{jn\omega_0 t}\right] d\tau = \sum_{n=-\infty}^{\infty} \int_{-\infty}^{t} \left[c_n e^{jn\omega_0 t}\right] d\tau$$

$$= \sum_{n=-\infty}^{\infty} c_n \int_{-\infty}^{t} e^{jn\omega_0 t}\, d\tau = \sum_{n=-\infty}^{\infty} \left(\frac{c_n}{jn\omega_0}\right) e^{jn\omega_0 t}.$$

$$\frac{dx(t)}{dt} \overset{FS}{\longleftrightarrow} \frac{c_n}{jn\omega_0} \text{ and at } n = 0,\ c_0 = 0.$$ (Proved)

Convolution

Consider two periodic signals $x_1(t)$ and $x_2(t)$ with period T having exponential Fourier coefficients c_n and d_n respectively.

The convolution property states that if

$$x_1(t) \overset{FS}{\longleftrightarrow} c_n, \quad x_2(t) \overset{FS}{\longleftrightarrow} d_n,$$

then $x_1(t) * x_2(t) \overset{FS}{\longleftrightarrow} Tc_n d_n$,

where $x_1(t) * x_2(t) = \int_0^T x_1(\tau)x_2(t-\tau)d\tau$.

Proof: Let two periodic signals $x_1(t)$ and $x_2(t)$ with period T have exponential Fourier coefficients c_n and d_n respectively.

$$\int_0^T x_1(\tau)x_2(t-\tau)d\tau = \int_0^T \sum_{n=-\infty}^{\infty} c_n e^{jn\omega_0 t} \sum_{n=-\infty}^{\infty} d_n e^{jn\omega_0 (t-\tau)} d\tau$$

Interchanging integration and summations,

$$= \int_0^T \sum_{n=-\infty}^{\infty} \sum_{n=-\infty}^{\infty} c_n d_n e^{jn\omega_0 t} e^{jn\omega_0 (t-\tau)} d\tau$$

$$= \sum_{n=-\infty}^{\infty} \sum_{n=-\infty}^{\infty} c_n d_n e^{jn\omega_0 t} \int_0^T d\tau = T \sum_{n=-\infty}^{\infty} \sum_{n=-\infty}^{\infty} c_n d_n e^{jn\omega_0 t}$$

$$= Tc_n d_n.$$ (Proved)

Multiplication

Let two periodic signals $x_1(t)$ and $x_2(t)$ with period T have exponential Fourier coefficients c_n and d_n respectively.

The multiplication property states that if

$$x_1(t) \overset{FS}{\longleftrightarrow} c_n, \quad x_2(t) \overset{FS}{\longleftrightarrow} d_n,$$

then $x_1(t)x_2(t) \overset{FS}{\longleftrightarrow} \sum_{m=-\infty}^{\infty} c_m d_{n-m}$

Proof: Let two periodic signals $x_1(t)$ and $x_2(t)$ with period T have exponential Fourier coefficients c_n and d_n respectively.

$$\text{FS } [x_1(t)x_2(t)] = \frac{1}{T}\int_0^T [x_1(t)x_2(t)]\, e^{-jn\omega_0 t}$$

Now substitute the Fourier series $x_1(t) = \sum_{m=-\infty}^{\infty} c_m e^{jm\omega_0 t}$. Then

$$\text{FS } [x_1(t)x_2(t)] = \frac{1}{T}\int_0^T \left[x_1(t) \sum_{m=-\infty}^{\infty} c_m e^{jm\omega_0 t} \right] e^{-jn\omega_0 t} = \frac{1}{T}\int_0^T \left[x_1(t) \sum_{m=-\infty}^{\infty} c_m e^{-j(n-m)\omega_0 t} \right]$$

$$= \sum_{m=-\infty}^{\infty} c_m \frac{1}{T}\int_0^T \left[x_1(t) e^{-j(n-m)\omega_0 t} \right] = c_m d_{n-m}.$$

3.8 Power Spectrum of Periodic Signals (Parseval's Theorem)

The power components of a periodic function $f(t)$ at harmonic frequencies is called its power spectrum.

Let $f(t) = \sum_{n=-\infty}^{\infty} c_n e^{jn\omega_0 t}$ be the exponential Fourier series of $f(t)$.

The average power of a periodic signal across a one-ohm resistance is defined as

$$P = \frac{1}{T}\int_{-T/2}^{T/2} f^2(t)\,dt = \frac{1}{T}\int_{-T/2}^{T/2} \left|\sum_{n=-\infty}^{\infty} c_n e^{-jn\omega_0 t}\right|^2 dt.$$

Interchanging the summation and integration,

$$P = \sum_{n=-\infty}^{\infty} \frac{1}{T}\int_{-T/2}^{T/2} |c_n e^{jn\omega_0 t}|^2\,dt = \sum_{n=-\infty}^{\infty} |c_n|^2 \frac{1}{T}\int_{-T/2}^{T/2} |e^{jn\omega_0 t}|^2\,dt.$$

since $|e^{jn\omega_0 t}| = 1$,

$$P = \sum_{n=-\infty}^{\infty} |c_n|^2 \frac{1}{T}(T) = \sum_{n=-\infty}^{\infty} |c_n|^2. \tag{3.38}$$

Hence $P = \frac{1}{T}\int_{-T/2}^{T/2} f^2(t)\,dt = \sum_{n=-\infty}^{\infty} |c_n|^2.$

This statement is called Parseval's theorem. The Parseval's theorem or Parseval's equality simply says that the total average power of a periodic signal $f(t)$is equal to the sum of the average powers in all of its harmonic components. The magnitude of the power component is

$$|c_{-n}|^2 = |c_n|^2. \tag{3.39}$$

Therefore the average power of the periodic signal is equal to the sum of the powers of its components. The power spectrum is discrete and exists only at harmonic frequencies with even symmetry.

Note: For $n = 0$ the power component is the dc power.

For trigonometric Fourier series, the average power can be obtained as

$$P = \frac{1}{T}\int_{-T/2}^{T/2} f^2(t)\,dt = \sum_{n=-\infty}^{\infty} |c_n|^2$$

$$= c_0^2 + \sum_{n=-\infty}^{-1} |c_n|^2 + \sum_{n=1}^{\infty} |c_n|^2$$

$$= c_0^2 + 2\sum_{n=1}^{\infty} |c_n|^2 = c_0^2 + 2\sum_{n=+1}^{\infty} \left|\frac{1}{2}a_n - \frac{1}{2}jb_n\right|^2.$$

$$P = c_0^2 + \frac{1}{2}\sum |a_n - jb_n|^2. \tag{3.40}$$

In polar form, $f(t) = a_0 + \sum_{n=1}^{\infty} A_n \cos(n\omega_0 t + \phi_n)$, the average power is

$$P = a_0^2 + \frac{1}{2}\sum_{n=1}^{\infty} A_n^2. \tag{3.41}$$

Example 3.6

Evaluate the Fourier series coefficients of the continuous time periodic signal $f(t) = 1 + \sin\omega_0 t + 2\cos\omega_0 t + \cos\left(2\omega_0 t + \frac{\pi}{4}\right)$.

Solution

Given $f(t) = 1 + \sin\omega_0 t + 2\cos\omega_0 t + \cos\left(2\omega_0 t + \frac{\pi}{4}\right)$.

Expanding $f(t)$ in terms of exponentials,

$$f(t) = 1 + \frac{e^{j\omega_0 t} - e^{-j\omega_0 t}}{2j} + 2\frac{e^{j\omega_0 t} + e^{-j\omega_0 t}}{2} + \frac{e^{j(2\omega_0 t + \frac{\pi}{4})} + e^{-j(2\omega_0 t + \frac{\pi}{4})}}{2}$$

$$f(t) = 1 + \left(1 + \frac{1}{2j}\right)e^{j\omega_0 t} + \left(1 - \frac{1}{2j}\right)e^{-j\omega_0 t} + \left(\frac{1}{2}e^{j\frac{\pi}{4}}\right)e^{j2\omega_0 t} + \left(\frac{1}{2}e^{-j\frac{\pi}{4}}\right)e^{-j2\omega_0 t}.$$

Comparing with the exponential Fourier series

$$f(t) = \sum_{-\infty}^{\infty} c_n e^{jn\omega_0 t} = \cdots + c_{-2}e^{-2j\omega_0 t} + c_{-1}e^{-j\omega_0 t} + c_0 + c_1 e^{j\omega_0 t} + c_2 e^{2j\omega_0 t} + \ldots.$$

the Fourier series coefficients of $f(t)$ are

$$c_0 = 1;$$

$$c_1 = 1 + \frac{1}{2j} = 1 - j\frac{1}{2}$$

$$c_{-1} = 1 - \frac{1}{2j} = 1 + j\frac{1}{2}$$

$$c_2 = \frac{1}{2}e^{j\frac{\pi}{4}} = \frac{1}{2}\left[\cos\frac{\pi}{4} + j\sin\frac{\pi}{4}\right] = \frac{1}{2\sqrt{2}}(1+j)$$

$$c_{-2} = \frac{1}{2}e^{-j\frac{\pi}{4}} = \frac{1}{2}\left[\cos\frac{\pi}{4} - j\sin\frac{\pi}{4}\right] = \frac{1}{2\sqrt{2}}(1-j)$$

$$c_n = 0 \quad \text{for } |n| > 2.$$

Example 3.7

Determine the Fourier series representation for $x(t) = 2\sin(2\pi t - 3) + \sin 6\pi t$.

Solution

Exponential Fourier series: From the given equation, we have

$$x(t) = 2\sin(2\pi t - 3) + \sin(6\pi t)$$

$$= \frac{2\left[e^{j2\pi t - 3} - e^{-j2\pi t - 3}\right]}{j2} + \frac{e^{j6\pi t} - e^{-j6\pi t}}{j2}$$

$$x(t) = \left(-\frac{1}{j2}\right)e^{-j32\pi t} + \left(-\frac{e^{j3}}{j}\right)e^{-j2\pi t} + \left(\frac{e^{-j3}}{j}\right)e^{j2\pi t} + \left(\frac{1}{j2}\right)e^{j32\pi t}.$$

Comparing with the exponential Fourier series and considering $\omega_o = 2\pi$,

$$x(t) = \sum_{n=-\infty}^{\infty} c_n e^{jn\omega_0 t} = \cdots + c_{-3}e^{-3j\omega_0 t} + c_{-1}e^{-j\omega_0 t} + c_1 e^{j\omega_0 t} + c_3 e^{3j\omega_0 t} + \ldots.$$

the exponential Fourier series coefficients are:

$$c_{-3} = j0.5;\quad c_{-1} = je^{j3};\quad c_1 = -je^{-j3};\quad c_3 = -j0.5;\ \text{and}\ c_m = 0,\quad \text{except}\,|m| \neq 1\ \text{and}\ 3.$$

Trigonometric Fourier series:

$$x(t) = 2\sin(2\pi t - 3) + \sin(6\pi t) = -2\sin(3)\cos(2\pi t) + 2\cos(3)\sin(2\pi t) + \sin(6\pi t).$$

We know that

$$x(t) = a_0 + \sum_{n=1}^{\infty}(a_n \cos n\omega_0 t + b_n \sin n\omega_0 t) = \cdots + a_1\cos(\omega_0 t) + b_1 \sin(\omega_0 t) + b_3 \sin(3(\omega_0 t)) + \ldots$$

Consider $\omega_0 = 2\pi$. Comparing the coefficients, the trigonometric Fourier series coefficients are
$a_1 = 2\sin 3$ and $a_m = 0$ except $|m| \neq 1$
$b_1 = 2\cos 3$; $b_3 = 1$; $b_m = 0$ except $|m| \neq 1$ and 3.

Example 3.8

For the periodic signal shown in Fig. 3.11, find the trigonometric Fourier series.

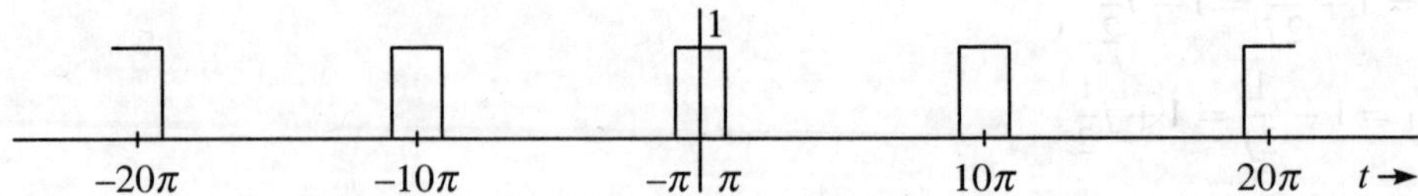

Fig. 3.11 *Pulse waveform*

Solution

From Fig. 3.11, the expression for the triangular waveform is

$$f(t) = \begin{cases} 1 & -\pi \leq t \leq +\pi \\ 0 & +\pi \leq t \leq \dfrac{9\pi}{2} \end{cases}$$

Time period $T = 10\pi$. The fundamental frequency $\omega_0 = \dfrac{2\pi}{10\pi} = \dfrac{1}{5}$ rad/s.

The trigonometric Fourier series is given by

$$f(t) = a_0 + \sum_{n=1}^{\infty}(a_n \cos n\omega_0 t + b_n \sin n\omega_0 t).$$

The given waveform satisfies even symmetry; hence all the sine terms are zero and $b_n = 0$.

$$a_0 = \frac{1}{T}\int_{-T/2}^{+T/2} f(t)dt$$

$$= \frac{1}{10\pi}\left[\int_{-\pi}^{+\pi}(1)dt + \int_{+\pi}^{+\frac{9\pi}{2}}(0)dt\right] = \frac{1}{10\pi}[t]_{-\pi}^{+\pi} = \frac{2\pi}{10\pi} = \frac{1}{5}.$$

$$a_n = \frac{2}{T}\int_{-T/2}^{T/2} f(t)\cos n\omega_0 t\, dt$$

$$= \frac{2}{10\pi}\int_{-\pi}^{+\pi} \cos n\omega_0 t\, dt + 0 = \frac{2}{10\pi}\left[\int_{-\pi}^{+\pi} \cos\left(\frac{n}{5}t\right) dt\right] + 0$$

$$= \frac{1}{5\pi}\left[\frac{\sin(nt/5)}{nt/5}\right]_{-\pi}^{+\pi} = \frac{2}{n\pi}\sin\left(\frac{n\pi}{5}\right).$$

Therefore, the trigonometric Fourier series for $f(t)$is

$$f(t) = \frac{1}{5} + \sum_{n=1}^{\infty} \frac{2}{n\pi}\sin\left(\frac{n\pi}{5}\right)\cos\left(\frac{n}{5}t\right).$$

Example 3.9

Evaluate the trigonometric Fourier series expansion of a triangular function shown in Fig. 3.12.

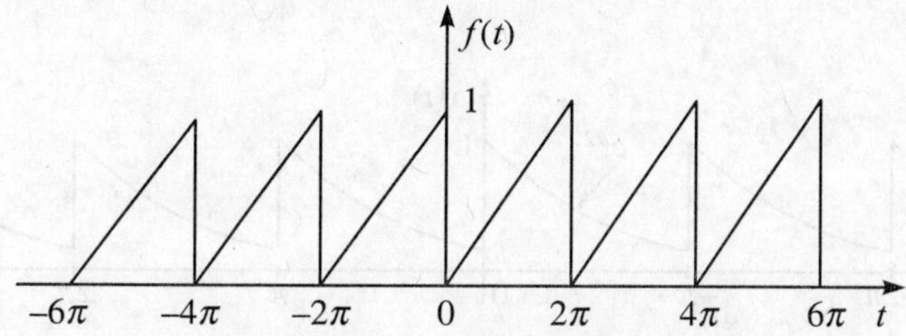

Fig. 3.12 *Triangular function*

Solution

From Fig. 3.12, the triangular function is expressed as

$$f(t) = \frac{t}{2\pi}, \qquad 0 \le t \le 2\pi.$$

Time period $T = 2\pi$, fundamental frequency $\omega_0 = 1$.

We know that the trigonometric Fourier series is

$$f(t) = a_0 + \sum_{n=1}^{\infty} (a_n \cos n\omega_0 t + b_n \sin n\omega_0 t)$$

Since the waveform follows odd symmetry with dc term, all the cosine terms are zero, i.e., $a_n = 0$.

So, $f(t) = a_0 + \sum_{n=1}^{\infty} b_n \sin n\omega_0 t.$

$$a_0 = \frac{1}{T}\int_0^T f(t)dt = \frac{1}{2\pi}\left[\int_0^{2\pi} \frac{t}{2\pi}dt\right]$$

$$= \frac{1}{4\pi^2}\left[\frac{t^2}{2}\right]_0^{2\pi} = \frac{1}{4\pi^2}\left[\frac{4\pi^2}{2}\right] = \frac{1}{2}.$$

$$b_n = \frac{2}{T}\int_0^T f(t)\sin n\omega_0 t\, dt$$

$$= \frac{2}{2\pi}\int_0^{2\pi} \frac{t}{2\pi} \sin nt \, dt = \frac{1}{2\pi^2}\left[t\frac{-\cos nt}{n} + \frac{\sin nt}{n^2}\right]_0^{2\pi}$$

$$= \frac{1}{2\pi^2}\left(\frac{-2\pi}{n} - 0\right) = \frac{-1}{n\pi}.$$

Therefore the trigonometric Fourier series for $f(t)$is

$$f(t) = \frac{1}{2} + \sum_{n=1}^{\infty}\left(\frac{-1}{n\pi}\right)\sin nt$$

$$f(t) = \frac{1}{2} - \frac{1}{\pi}\left[\sin t + \frac{1}{2}\sin 2t + \frac{1}{3}\sin 3t + \frac{1}{4}\sin 4t + \ldots\ldots\ldots\ldots\right].$$

Example 3.10

For the periodic signal shown in Fig. 3.13, find the trigonometric Fourier series.

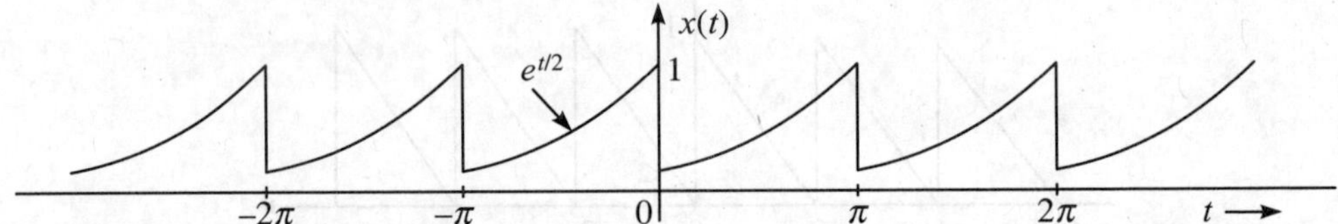

Fig. 3.13 *Exponential signal*

Solution

From Fig. 3.13, the exponential function is expressed as

$$x(t) = e^{-t/2}; \qquad 0 \le t \le \pi;$$

time period $T = \pi$; frequency $\omega_0 = \frac{2\pi}{\pi} = 2.$

We know that the trigonometric Fourier series is

$$x(t) = a_0 + \sum_{n=1}^{\infty}(a_n \cos n\omega_0 t + b_n \sin n\omega_0 t).$$

$$a_0 = \frac{1}{T}\int_0^T f(t)dt = \frac{2}{\pi}\int_0^{\pi} e^{-t/2}dt$$

$$= \frac{1}{4\pi^2}\left[\frac{e^{-t/2}}{-1/2}\right]_0^{\pi} = \frac{-1}{2\pi^2}\left[e^{-\pi/2} - e^0\right] = \frac{1}{2\pi^2}\left[1 - e^{-\pi/2}\right]$$

$$a_n = \frac{2}{T}\int_0^T f(t)\cos(n\omega_0 t)\, dt = \frac{2}{\pi}\int_0^{\pi} e^{-t/2}\cos(2nt)dt$$

$$\therefore \int e^{ax}\cos(bx)dx = \frac{e^{-ax}}{a^2+b^2}(-a\cos(bx) + b\sin(bx)),$$

$$a_n = \frac{2}{\pi}\left[\frac{e^{-t/2}}{\frac{1}{4}+4n^2}\left[\frac{-1}{2}\cos(2nt)+2n\sin(2nt)\right]\right]_0^{\pi}$$

$$a_n = \frac{2}{\pi}\left(\frac{4}{1+16n^2}\right)\left[e^{-\pi/2}\left(\frac{-1}{2}+0\right)-1\left(\frac{-1}{2}\right)\right] = \frac{4}{\pi(1+16n^2)}\left(1-e^{-\pi/2}\right).$$

$$b_n = \frac{2}{T}\int_0^T f(t)\sin n\omega_0 t\, dt = \frac{2}{\pi}\int_0^{\pi} e^{-\pi/2}\sin(2nt)\, dt$$

$$\therefore \int e^{-ax}\sin(bx)dx = \frac{e^{-ax}}{a^2+b^2}\left(a\sin(bx)+b\cos(bx)\right),$$

$$= \frac{2}{\pi}\left[\frac{e^{-t/2}}{\frac{1}{4}+4n^2}\left[\frac{1}{2}\sin(2nt)+2n\cos(2nt)\right]\right]_0^{\pi}$$

$$b_n = \frac{2}{\pi}\frac{4}{1+16n^2}\left[e^{-\pi/2}(2n)-2n\right] = \frac{16n}{\pi(1+16n^2)}\left(e^{-\pi/2}-1\right).$$

Therefore the trigonometric Fourier series for $f(t)$is

$$x(t) = \sum_{n=0}^{\infty}\frac{4\left(1-e^{-\pi/2}\right)}{\pi(1+16n^2)}\left(\cos(2nt)-4n\sin(nt)\right).$$

Example 3.11

Find the exponential Fourier series expansion for the half-wave rectified sine wave shown in Fig. 3.14.

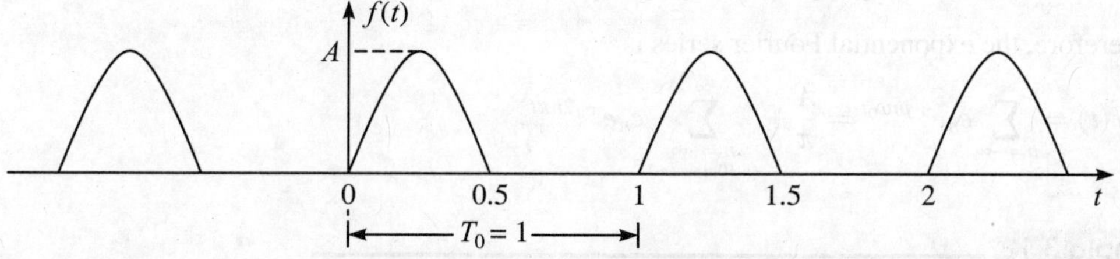

Fig. 3.14 *Half-wave rectified sine wave*

Solution

From Fig. 3.14, the half-wave rectified sine wave is expressed as

$$f(t) = \begin{cases} A\sin 2\pi t; & 0 \le t \le 0.5 \\ 0; & 0.5 \le t \le 1 \end{cases}.$$

Time period; fundamental frequency $\omega_0 = \dfrac{2\pi}{1} = 2\pi.$

The exponential Fourier series is

$$f(t) = \sum_{n=-\infty}^{\infty} c_n e^{jn\omega_0 t} = \sum_{n=-\infty}^{\infty} c_n e^{j2\pi nt}.$$

Now, $c_0 = \frac{1}{T}\int_0^T f(t)dt = \frac{1}{1}\left[\int_0^{0.5} A\sin 2\pi t dt + 0\right]$

$$= A\left[\frac{-\cos 2\pi t}{2\pi}\right]_0^{0.5} = A\left[-\frac{-1-1}{2\pi}\right] = \frac{A}{\pi} \text{ and}$$

$$c_n = \frac{1}{T}\int_0^T f(t)e^{-jn\omega_0 t}dt$$

$$= \frac{1}{1}\left[\int_0^{0.5}(A\sin 2\pi t)\,e^{-j2\pi nt}dt + \int_{0.5}^{1}(0)\,e^{-j2\pi nt}dt\right] = \int_0^{0.5}(A\sin 2\pi t)e^{-j2\pi nt}dt$$

$$= A\left[\frac{e^{-j2\pi nt}}{2\pi^2 + -j2\pi n^2}[(-j2\pi n)\sin(2\pi t) - (2\pi)\cos(2\pi t)]\right]_0^{0.5}$$

$$= \frac{Ae^{-j2\pi n/2}}{4\pi^2(1-n^2)}((-j2\pi n)\sin\pi - (2\pi)\cos\pi) + \frac{A2\pi}{4\pi^2(1-n^2)}$$

$$= A\left[\frac{(-1)^n}{4\pi^2(1-n^2)}[(-j2\pi n)(0) - (2\pi)(-1)] + \frac{2\pi}{4\pi^2(1-n^2)}\right]$$

$$= \frac{A[(-1)^n + 1]}{2\pi(1-n^2)}.$$

If n is odd, then $c_n = \frac{A[-1+1]}{2\pi[1-n^2]} = 0.$

If n is even, then $c_n = \frac{A[1+1]}{2\pi(1-n^2)} = \frac{A}{\pi(1-n^2)}.$

Therefore, the exponential Fourier series is

$$f(t) = \sum_{n=-\infty}^{\infty} c_n e^{+jn\omega_0 t} = \frac{A}{\pi} + \sum_{\substack{n=-\infty \\ (n\neq 0, n=even)}}^{\infty} c_n e^{+j2n\pi t}.$$

Example 3.12

Determine the Fourier series of the repetitive waveform shown in Fig. 3.15 up to 5th harmonic when repetition time T=1 second.

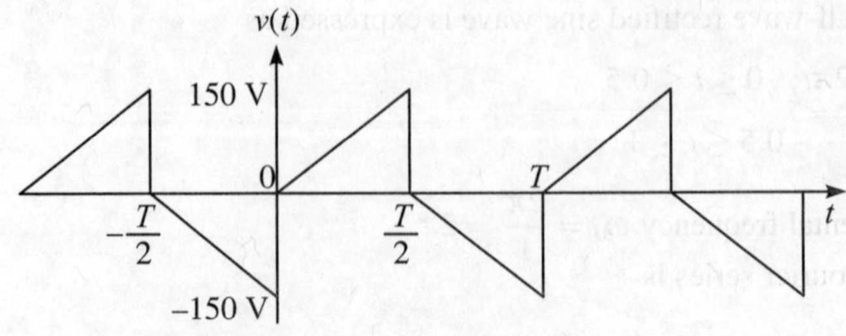

Fig. 3.15 *Triangular waveform*

Solution

From Fig. 3.15 the triangular wave is expressed as

$$v(t) = \begin{cases} \dfrac{300t}{T} & 0 \le t \le T/2 \\ 150 - \dfrac{300t}{T} & T/2 \le t \le T. \end{cases}$$

Time period $T = 1$; frequency $\omega_0 = \dfrac{2\pi}{1} = 2\pi$.

We know that the trigonometric Fourier series is

$$v(t) = a_0 + \sum_{n=1}^{\infty} (a_n \cos n\omega_0 t + b_n \sin n\omega_0 t).$$

As $v(t) = -v(t \pm T/2)$, the given waveform satisfies half-wave or rotation symmetry.

So, the trigonometric Fourier coefficients are:

average value $a_0 = 0$.

$$a_n = \begin{cases} 0 & \text{for } n = \text{even} \\ \dfrac{4}{T}\displaystyle\int_0^{T/2} v(t)\cos n\omega_0 t\, dt & \text{for } n = \text{odd} \end{cases}$$

$$b_n = \begin{cases} 0 & \text{for } n = \text{even} \\ \dfrac{4}{T}\displaystyle\int_0^{T/2} v(t)\sin n\omega_0 t\, dt & \text{for } n = \text{odd} \end{cases}$$

So, $$a_n = \frac{4}{T}\int_0^{T/2} \frac{300t}{T}\cos n\omega_0 t\, dt = \frac{1200}{T^2}\int_0^{T/2} t\cos n\omega_0 t\, dt$$

$$= \frac{1200}{T^2}\left[\frac{t\sin n\omega_0 t}{n\omega_0} + \frac{\cos n\omega_0 t}{n^2\omega_0^2}\right]_0^{T/2} = \frac{1200}{T^2}\left[\frac{(T/2)\sin n\pi}{n\omega_0} + \frac{\cos n\pi - 1}{n^2\omega_0^2}\right]$$

$$= \frac{1200}{T^2}\left[\frac{\cos n\pi - 1}{n^2\omega_0^2}\right] = \frac{-2400}{n^2(2\pi)^2} = \frac{-600}{n^2\pi^2} \quad \text{for } n \text{ odd.}$$

$$a_n = \begin{cases} \dfrac{-600}{n^2\pi^2} & \text{for } n = \text{ odd} \\ 0 & \text{for } n = \text{ even} \end{cases}$$

Similarly, $$b_n = \frac{4}{T}\int_0^{T/2} v(t)\sin n\omega_0 t\, dt$$

$$b_n = \frac{4}{T}\int_0^{T/2} \frac{300t}{T}\sin n\omega_0 t\, dt = \frac{1200}{T^2}\int_0^{T/2} t\sin n\omega_0 t\, dt$$

$$= \frac{1200}{T^2}\left[\frac{t(-\cos n\omega_0 t)}{n\omega_0} - \frac{\sin \omega_0 t}{n^2\omega_0^2}\right]_0^{T/2} = \frac{300}{n\pi}(-1)^{n+1} = \frac{300}{n\pi} \quad \text{for } n \text{ odd.}$$

$$b_n = \begin{cases} \dfrac{300}{n\pi} & \text{for } n = \text{ odd} \\ 0 & \text{for } n = \text{ even} \end{cases}$$

$$v(t) = a_0 + \sum_{n=1}^{\infty}(a_n \cos n\omega_0 t + b_n \sin n\omega_0 t)$$

$$v(t) = \frac{-600}{\pi^2}\cos\omega_0 t - \frac{600}{9\pi^2}\cos 3\omega_0 t - \frac{600}{25\pi^2}\cos 5\omega_0 t - \ldots\ldots$$

$$+\frac{300}{\pi}\sin\omega_0 t + \frac{300}{3\pi}\sin 3\omega_0 t + \frac{300}{5\pi}\sin 5\omega_0 t + \ldots\ldots$$

Example 3.13

For the periodic signal shown in Fig. 3.16, find the trigonometric Fourier series.

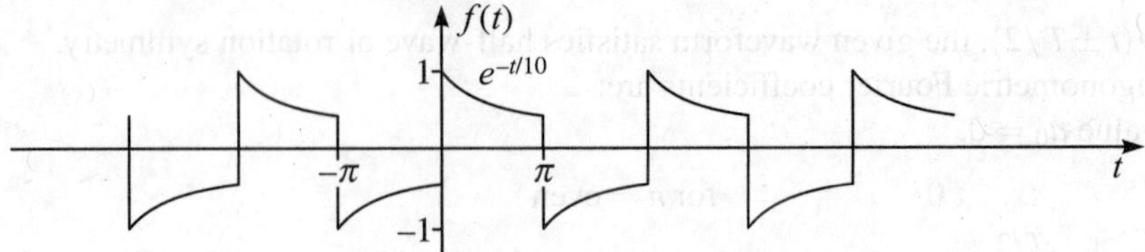

Fig. 3.16 *Periodic waveform*

Solution

From Fig. 3.16, the exponential periodic wave is expressed as

$$f(t) = e^{-t/10}; \qquad 0 \le t \le \pi;\ \text{time period } T = 2\pi;\ \text{fundamental frequency } \omega_0 = \frac{2\pi}{2\pi} = 1.$$

We know that the trigonometric Fourier series is

$$f(t) = a_0 + \sum_{n=1}^{\infty}(a_n \cos n\omega_0 t + b_n \sin n\omega_0 t)$$

As $f(t) = -f(t \pm T/2)$, the given waveform satisfies half-wave symmetry.
So, the trigonometric Fourier coefficients are:
Average value $a_0 = 0$.

$$a_n = \begin{cases} 0 & \text{for } n = \text{even} \\ \dfrac{4}{T}\displaystyle\int_0^{T/2} f(t)\cos n\omega_0 t\, dt & \text{for } n = \text{odd.} \end{cases}$$

$$b_n = \begin{cases} 0 & \text{for } n = \text{even} \\ \dfrac{4}{T}\displaystyle\int_0^{T/2} f(t)\sin n\omega_0 t\, dt & \text{for } n = \text{odd} \end{cases}$$

$$= \frac{2}{\pi}\left[\frac{e^{-\pi/10}}{n^2+0.01}(-0.1\cos n\pi + n\sin n\pi) + \frac{0.1}{n^2+0.01}\right].$$

For n odd, $a_n = \dfrac{0.2}{\pi}\left[\dfrac{e^{\pi/10}+1}{n^2+0.01}\right].$

Example 3.14

Find the Fourier series representation of a periodic train of impulses, called an impulse train, defined by $f(t) = \sum_{n=-\infty}^{\infty} \delta(t-nT)$ and plots its spectrum.

Solution

A train of impulses does not meet the Dirichlet's conditions as it has infinite energy over one period. We can nevertheless calculate the Fourier series coefficients of the impulse train by using the formula

$$c_n = \frac{1}{T}\int_{-T/2}^{T/2} f(t)e^{-jn\omega_0 t}\,dt = \frac{1}{T}\int_{-T/2}^{T/2} \delta(t)e^{-jn\omega_0 t}\,dt$$

$$= \frac{1}{T} \quad \text{for } n = \pm 1, \pm 2, \cdots$$

Thus c_n is a real function since phase is zero. The graph of spectrum is shown in Fig. 3.17.

The spectrum of an impulse train is a real, contains harmonics of equal strength at all frequencies, up to infinity.

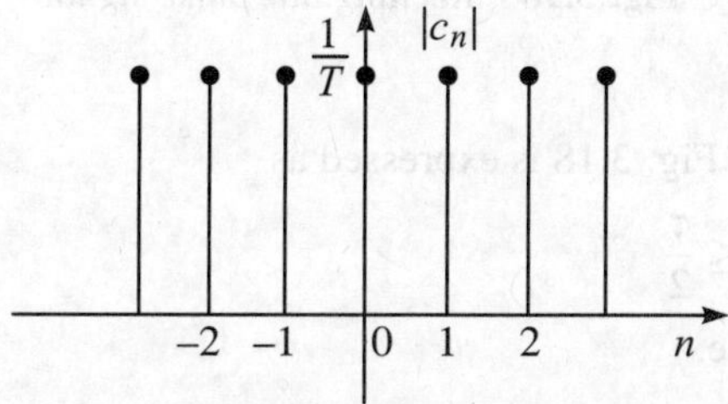

Fig. 3.17 *Amplitude spectrum of impulse train*

Example 3.15

Determine the average power of $f(t) = 2\sin 100t$ from time domain and frequency domain.

Solution

From time domain, we know the average power is

$$P = \frac{1}{T}\int_{-T/2}^{T/2} 4\sin^2 100t\,dt.$$

$$= \frac{4}{T}\int_{-T/2}^{T/2}\left(\frac{1-\cos 2(100t)}{2}\right)dt = \frac{4}{T}\left(\frac{T}{2}\right) - 0 = 2 \text{ watts.}$$

In frequency domain, $f(t) = 2\sin 100t$ can represented in Fourier series as

$$f(t) = 2\cdot\frac{e^{j100t} - e^{-j100t}}{2j} = -je^{j100t} + je^{-j100t}$$

so that the Fourier coefficient of $f(t)$ are

$$c_1 = -j, \quad c_{-1} = j$$

$$c_n = 0 \quad \text{for all } n \neq \pm 1.$$

In frequency domain, we know the average power is

$$P = \sum_{n=-\infty}^{\infty} |c_n|^2 = |-j|^2 + |j|^2 = 1 + 1 = 2 \text{ watts.}$$

Example 3.16

A periodic rectangular pulse signal is plotted in Fig. 3.18 whose width, amplitude and period are τ, A and T respectively. Find the exponential and trigonometric Fourier series. Plot and discuss its spectrum for different cases of τ and T.

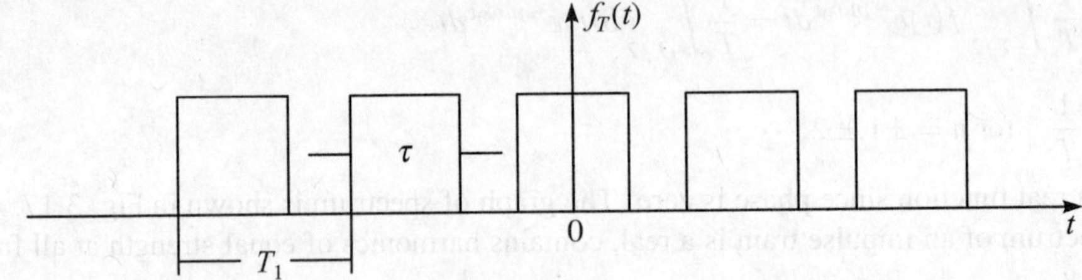

Fig. 3.18 *Rectangular pulse signal*

Solution

The rectangular pulse signal of Fig. 3.18 is expressed as

$$f(t) = \begin{cases} A & -\dfrac{\tau}{2} \le t \le \dfrac{\tau}{2} \\ 0 & \text{otherwise.} \end{cases}$$

Its time period $= T$, fundamental frequency $\omega_0 = \dfrac{2\pi}{T}$. We know that exponential Fourier series is

$$f(t) = \sum_{n=-\infty}^{\infty} c_n e^{jn\omega_0 t}, \text{ where}$$

$$c_n = \frac{1}{T}\int_{-T/2}^{T/2} f(t) e^{-jn\omega_0 t} dt$$

$$c_n = \frac{1}{T}\int_{-\tau/2}^{\tau/2} A e^{-jn\omega_0 t} dt = \frac{A}{-jn\omega_0 T}\left(e^{-jn\omega_0\tau/2} - e^{jn\omega_0\tau/2}\right)$$

$$= \frac{A}{n\omega_0 T} = \frac{\left(e^{jn\omega_0\tau/2} - e^{-jn\omega_0\tau/2}\right)}{2j} = \frac{2A}{Tn\omega_0}\sin\left(n\omega_0\frac{\tau}{2}\right)$$

$$c_n = \frac{A\tau}{T}\frac{\sin(n\omega_0\tau/2)}{n\omega_0\tau/2} = \frac{A\tau}{T}\text{Sa}\left(\frac{n\omega_1\tau}{2}\right),$$

where $\text{Sa}\left(\dfrac{n\omega_1\tau}{2}\right) = \dfrac{\sin(n\omega_1\tau/2)}{n\omega_1\tau/2}$ is the sampling function.

The exponential Fourier series is

$$f(t) = \sum_{n=-\infty}^{\infty} \frac{A\tau}{T}\text{Sa}\left(\frac{n\omega_0\tau}{2}\right) e^{jn\omega_0 t}.$$

For trigonometric Fourier series, since $f(t)$ is an even function, $b_n = 0$.

From the relation

$$c_n = \frac{1}{2}a_n - \frac{1}{2}jb_n,$$

$$a_n = 2c_n = \frac{2A\tau}{T}\text{Sa}\left(\frac{n\omega_0\tau}{2}\right),$$

$$a_0 = \frac{1}{T}\int_{-T/2}^{T/2} f(t)dt = \frac{1}{T}\int_{-\tau/2}^{\tau/2} Adt = \frac{A\tau}{T},$$

so that the trigonometric Fourier series is

$$f(t) = \frac{A\tau}{T} + \frac{2A\tau}{T}\sum_{n=1}^{\infty}\text{Sa}\left(\frac{n\omega_0\tau}{2}\right)\cos n\omega_0 t.$$

To plot the spectrum, we consider three cases for τ and T.

Case 1: Suppose $\tau = 1/10$ and $T = 1/2$. In this case the spectrum is $c_n = \frac{A}{5}\text{Sa}\left(\frac{n\pi}{5}\right)$ and a plot is shown in Fig. 3.19. The fundamental frequency is $\omega_0 = 2\pi/T = 4\pi$ rad/s, and so the harmonic spacing $\Delta\omega_1$ is 4π.

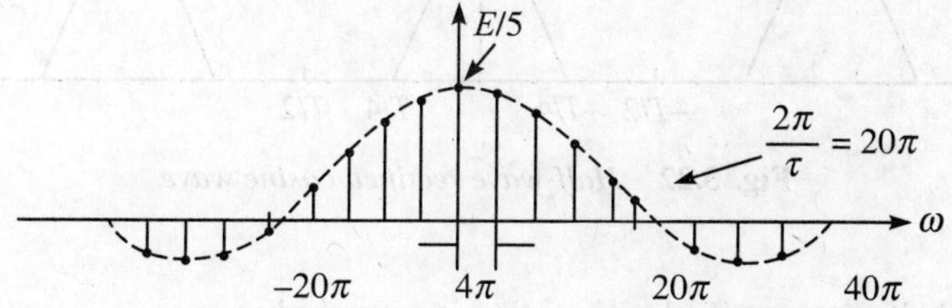

Fig. 3.19 *Amplitude spectrum for* $\tau = \frac{1}{10}$ *and* $T = \frac{1}{2}$

Case 2: Assume $\tau = \frac{1}{10}$ as before, but we increase the period to $T = 1$. Now the spectrum is $c_n = \frac{A}{10}\text{Sa}\left(\frac{n\pi}{10}\right)$, which is depicted in Fig. 3.20. The fundamental frequency is decreased to $\omega_0 = 2\pi$ and so is the harmonic spacing to 2π.

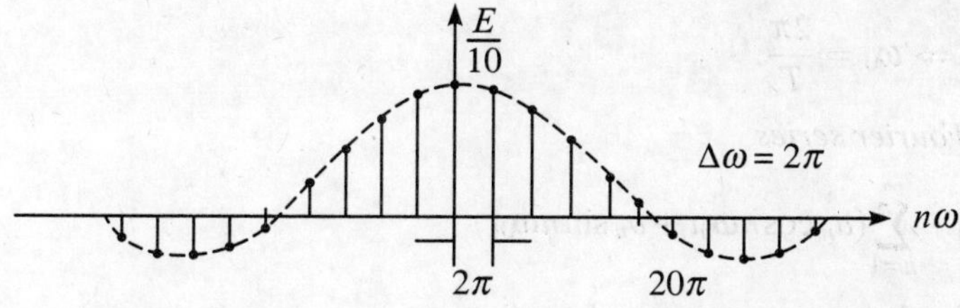

Fig. 3.20 *Amplitude spectrum for* $\tau = \frac{1}{10}$ *and* $T = 1$

Case 3: Assume $\tau = \frac{1}{20}$ and we retain $T = 1$. The spectrum is $c_n = \frac{A}{20}\text{Sa}\left(\frac{n\pi}{20}\right)$ as shown in Fig. 3.21.

In this example we observe that as the period T increases, two qualitative features of the spectrum are changed: both the spacing between the frequency components and the amplitudes of the frequency components of $f(t)$ decrease. Also, it is noticed that as τ decreases the fundamental frequency or spacing of the frequency components of $f(t)$ does not change, but the bandwidth is increased.

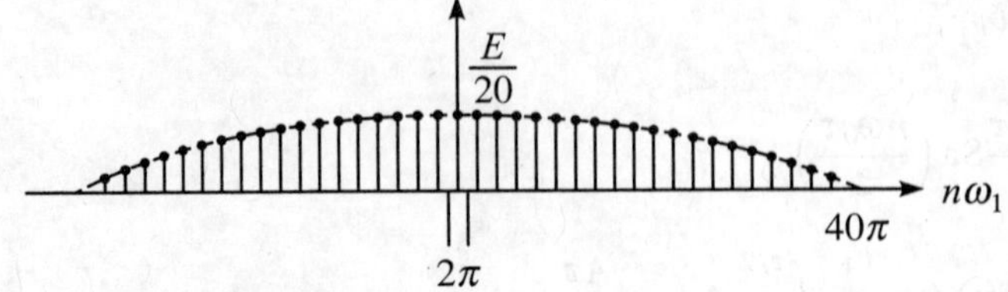

Fig. 3.21 *Amplitude spectrum for* $\tau = \frac{1}{20}$ *and* $T = 1$

Example 3.17

Obtain the trigonometric Fourier series and exponential Fourier series expansion for the half-wave rectified cosine wave shown in Fig. 3.22.

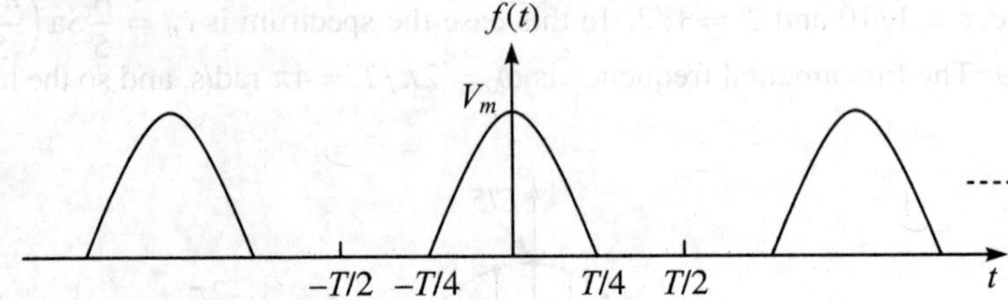

Fig. 3.22 *Half-wave rectified cosine wave*

Solution

From Fig. 3.22, the half-wave rectified cosine wave is expressed as

$$f(t) = \begin{cases} 0 & \text{for } \frac{-T}{2} \le t \le \frac{-T}{4} \\ V_m \cos \omega_0 t & \text{for } \frac{-T}{4} \le t \le \frac{T}{4} \\ 0 & \text{for } \frac{T}{4} \le t \le \frac{T}{2} \end{cases}$$

Time period $T_0 = T \Rightarrow \omega_0 = \dfrac{2\pi}{T}$.

(a) *Trigonometric Fourier series*

$$f(t) = a_0 + \sum_{n=1}^{\infty} (a_n \cos n\omega_0 t + b_n \sin n\omega_0 t)$$

Since the function has even symmetry, $b_n = 0$.

$$f(t) = a_0 + \sum_{n=1}^{\infty} (a_n \cos n\omega_0 t).$$

$$a_0 = \frac{1}{T}\int_{-T/2}^{T/2} f(t)dt = \frac{1}{T}\int_{-T/4}^{T/4} V_m \cos \omega_0 t \, dt = \frac{V_m}{2\pi}\left[\sin \frac{\omega_0 T}{4} - \sin\left(-\frac{\omega_0 T}{4}\right)\right] = \frac{V_m}{\pi}.$$

$$a_n = \frac{1}{T}\int_{-T/2}^{T/2} f(t)\cos n\omega_0 t\, dt = \frac{1}{T}\int_{-T/4}^{T/4} V_m \cos\omega_0 t \cos n\omega_0 t\, dt$$

$$= \frac{V_m}{2T}\int_{-T/4}^{T/4} (\cos(n+1)\omega_0 t + \cos(n-1)\omega_0 t)\, dt$$

$$= \frac{V_m}{2T}\left(\frac{\sin(n+1)\omega_0 t}{\omega_0(n+1)} + \frac{\sin(n-1)\omega_0 t}{\omega_0(n-1)}\right)_{-T/4}^{T/4}$$

$$= \frac{2v_m}{4\pi}\left(\frac{\sin((n+1)\pi/2)}{n+1} + \frac{\sin((n-1)\pi/2)}{n-1}\right)$$

$$= \frac{V_m}{2\pi}\left(\frac{\sin\left(\frac{n\pi}{2}+\frac{\pi}{2}\right)}{n+1} + \frac{\sin\left(\frac{n\pi}{2}-\frac{\pi}{2}\right)}{n-1}\right)$$

$$= \frac{V_m}{\pi}\left(\frac{\cos(n\pi/2)}{n+1} + \frac{-\cos(n\pi/2)}{n-1}\right)$$

$$= \frac{2V_m\cos(n\pi/2)}{\pi(1-n^2)}, \text{ for } n \neq 1.$$

For $n = 1$,

$$a_1 = \frac{1}{T}\int_{-T/4}^{T/4} V_m \cos^2 \omega_0 t\, dt = \frac{1}{T}\int_{-T/4}^{T/4} V_m \frac{1-\cos 2\omega_0 t}{2}\, dt = \frac{V_m}{2}.$$

$$\therefore \quad f(t) = \frac{V_m}{\pi} + \frac{V_m}{2}\cos\omega t + \frac{2V_m}{3\pi}\cos 2\omega t - \frac{2V_m}{15\pi}\cos 4\omega t + \frac{2V_m}{35\pi}\cos 6\omega t - \ldots\ldots\ldots\ldots$$

(b) *Exponential Fourier series:*

$$f(t) = \sum_{n=-\infty}^{\infty} c_n e^{jn\omega_0 t}$$

where, $c_n = \frac{1}{2}(a_n - jb_n)$; $\quad c_{-n} = \frac{1}{2}(a_n + jb_n)$

$$c_0 = a_0 = \frac{V_m}{\pi}.$$

$$c_n = c_{-n} = \frac{1}{2}a_n = \frac{V_m}{\pi}\left(\frac{\cos\left(\frac{n\pi}{2}\right)}{1-n^2}\right), \text{ provided } n \neq 1.$$

$$c_2 = c_{-2} = \frac{1}{2}a_2 = \frac{V_m}{3\pi}; \quad c_3 = c_{-3} = \frac{1}{2}a_3 = 0; \quad c_4 = c_{-4} = \frac{1}{2}a_4 = \frac{-V_m}{15\pi}.$$

The exponential Fourier series expansion for the given signal is:

$$f(t) = \frac{V_m}{\pi} + \frac{V_m}{3\pi}e^{j2\omega_0 t} - \frac{V_m}{15\pi}e^{j4\omega_0 t} + \ldots\ldots\ldots\ldots + \frac{V_m}{3\pi}e^{-j2\omega_0 t} - \frac{V_m}{15\pi}e^{-j4\omega_0 t} + \ldots\ldots$$

Example 3.18

Obtain the exponential Fourier series expansion for the full-wave rectified sine wave shown in Fig. 3.23.

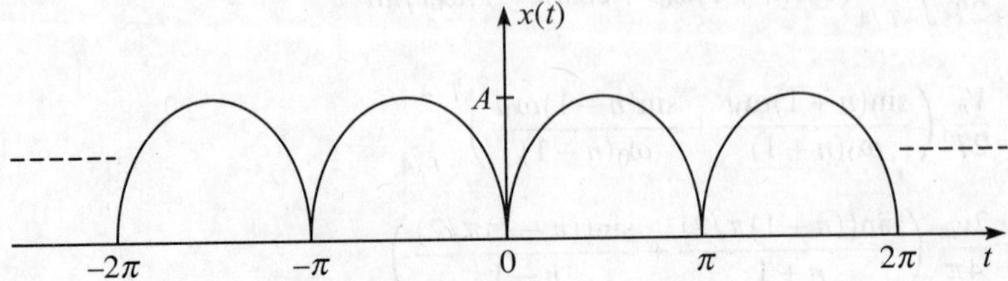

Fig. 3.23 *Full-wave rectified sine wave*

Solution

From Fig. 4.23, the full-wave rectified sine wave is expressed as

$$x(t) = A\sin t; \quad 0 \le t \le \pi.$$

Time period $T_0 = \pi$, fundamental frequency $\omega_0 = \dfrac{2\pi}{T} = 2\text{rad/s}$.

The exponential Fourier series is $x(t) = \sum\limits_{n=-\infty}^{\infty} c_n e^{jn\omega_0 t} = \sum\limits_{n=-\infty}^{\infty} c_n e^{j2\pi nt}$

$$c_0 = \frac{1}{T}\int_0^T f(t)dt = \frac{1}{2}\left[\int_0^{\pi} A\sin t dt\right] = \frac{A}{\pi}[-\cos t]_0^{\pi}$$

$$= \frac{A}{\pi}[-\cos\pi - \cos 0] = \frac{2A}{\pi}.$$

$$c_n = \frac{1}{T}\int_0^T f(t)e^{-jn\omega_0 t}dt = \frac{1}{\pi}\left[\int_0^{\pi} A\sin t e^{-j2nt}dt\right]$$

$$= \frac{A}{j2\pi}\left[\int_0^{\pi}\left(e^{j(1-2n)t} - e^{-j(1+2n)t}\right)dt\right] = \frac{A}{j2\pi}\left[\frac{e^{j(1-2n)t}}{j1-2n} - \frac{e^{-j(1+2n)t}}{-j1+2n}\right]_0^{\pi}$$

$$= \frac{A}{j2\pi}\left[\frac{e^{j(1-2n)\pi} - e^0}{j(1-2n)} - \frac{e^{-j(1+2n)\pi} - e^0}{-j(1+2n)}\right]$$

$$c_n = \frac{2A}{\pi(1-4n^2)}.$$

The exponential Fourier series is:

$$x(t) = \sum_{n=-\infty}^{\infty}\frac{2A}{\pi(1-4n^2)}e^{j2nt} = \frac{2A}{\pi} + \frac{2A}{\pi}\sum_{\substack{n=-\infty \\ n\neq 0}}^{\infty}\frac{2A}{\pi(1-4n^2)}.$$

$$c_0 = \frac{2A}{\pi}$$

$$c_1 = c_{-1} = \frac{2A}{\pi(1-4(1)^2)} = -\frac{2A}{3\pi}$$

$$c_2 = c_{-2} = \frac{2A}{\pi(1-4(2)^2)} = -\frac{2A}{15\pi}$$

$$c_3 = c_{-3} = \frac{2A}{\pi(1-4(3)^2)} = -\frac{2A}{35\pi}$$

Figure 3.24 shows the amplitude spectrum of the full-wave rectified sine wave.

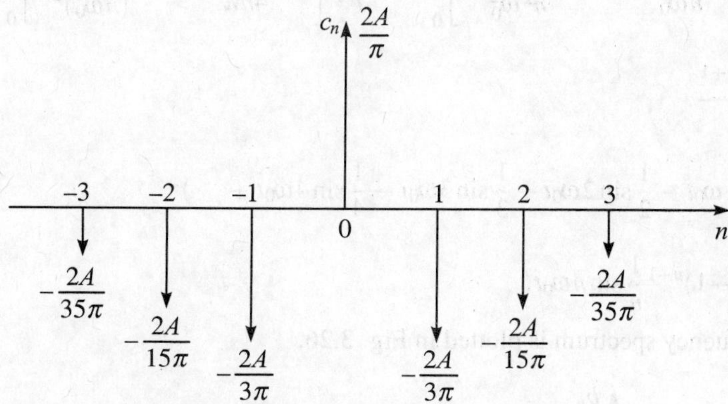

Fig. 3.24 *Amplitude spectrum of full-wave rectified sine wave*

Example 3.19

A periodic saw tooth pulse signal with period T is depicted in Fig. 3.25. Determine the trigonometric Fourier series and draw its spectrum.

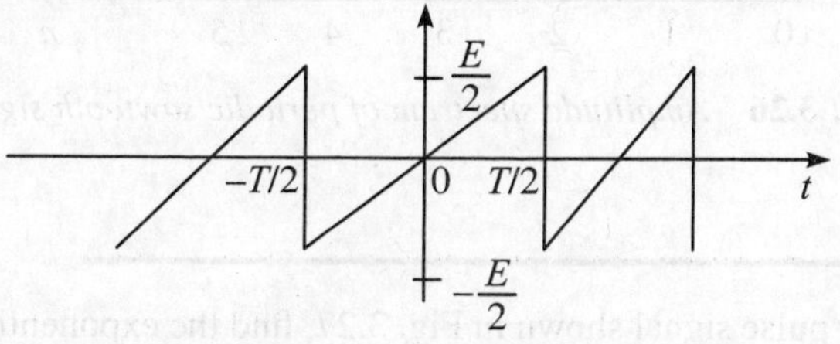

Fig. 3.25 *Saw tooth pulse signal*

Solution

From Fig. 4.25, the sawtooth pulse signal is expressed as

$$f(t) = \frac{Et}{T} \qquad \frac{-T}{2} \le t \le \frac{T}{2}.$$

Time period $= T$; fundamental frequency $\omega_0 = \dfrac{2\pi}{T}$ rad/s.

The trigonometric Fourier series is

$$f(t) = a_0 + \sum_{n=1}^{\infty}(a_n \cos n\omega_0 t + b_n \sin n\omega_0 t).$$

It is obvious that the signal is odd function and hence $a_0 = a_n = 0$.

$$f(t) = \sum_{n=1}^{\infty} b_n \sin n\omega_0 t$$

$$b_n = \frac{4}{T}\int_0^{T/2} f(t)\sin n\omega_0 t\, dt = \frac{4}{T}\int_0^{T/2} \frac{E}{T} t \sin n\omega_0 t\, dt$$

$$= \frac{4E}{T^2}\left[\frac{-t\cos\left(\frac{n\omega_0 t}{2}\right)}{n\omega_0} + \frac{\sin\left(\frac{n\omega_0 t}{2}\right)}{n^2\omega_0^2}\right]_0^{T/2} = \frac{4E}{T^2}\left[\frac{(-1)^{n+1}T^2}{4n\pi} + \frac{\sin\left(\frac{n\omega_0 t}{2}\right)}{(n\omega_0)^2}\right]_0^{T/2}.$$

$$= \frac{E(-1)^{n+1}}{n\pi}.$$

$$f(t) = \frac{E}{\pi}(\sin\omega_0 t - \frac{1}{2}\sin 2\omega_0 t + \frac{1}{3}\sin 3\omega_0 t - \frac{1}{4}\sin 4\omega_0 t + \cdots)$$

$$= \frac{E}{\pi}\sum(-1)^{n+1}\frac{1}{n}\sin n\omega_0 t.$$

The amplitude frequency spectrum is plotted in Fig. 3.26.

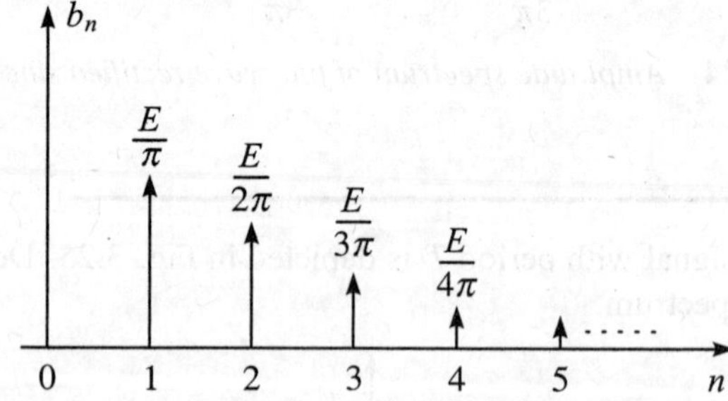

Fig. 3.26 *Amplitude spectrum of periodic sawtooth signal*

Example 3.20

For the periodic triangular pulse signal shown in Fig. 3.27, find the exponential Fourier series and draw its spectrum.

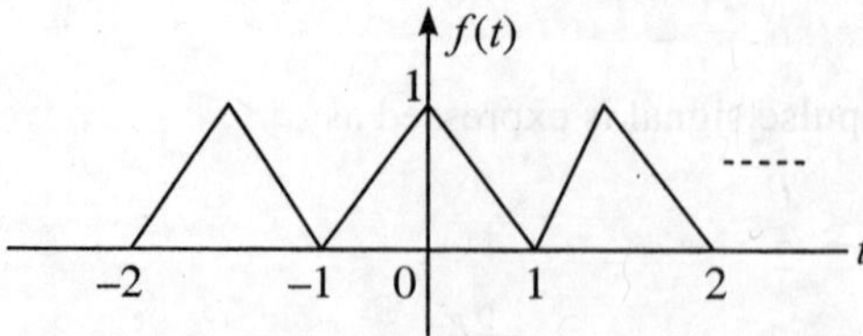

Fig. 3.27 *Periodic triangular pulse signal*

Solution

From Fig. 3.27, the periodic triangular pulse signal is expressed as

$$f(t) = \begin{cases} 1+t & -1 \le t \le 0 \\ 1-t & 0 \le t \le 1. \end{cases}$$

Time period $T = 2$; fundamental frequency $\omega_0 = \pi$.

The exponential Fourier series is $x(t) = \sum_{n=-\infty}^{\infty} c_n e^{jn\omega_0 t} = \sum_{n=-\infty}^{\infty} c_n e^{j2\pi nt}$, where

$$c_n = \frac{1}{T}\int_{-T/2}^{T/2} f(t)e^{-jn\omega_0 t}dt$$

$$c_n = \frac{1}{2}\int_{-1}^{0}(1+t)e^{-jn\omega_0 t}dt + \frac{1}{2}\int_{0}^{1}(1-t)e^{-jn\omega_0 t}dt$$

$$c_n = \frac{1}{n^2\pi^2}(1-\cos n\pi) = \frac{1}{n^2\pi^2}(1-(-1)^n)$$

$$c_n = \begin{cases} \frac{2}{n^2\pi^2} & n\text{, odd} \\ 0 & n \ne 0\text{, even.} \end{cases}$$

For $n = 0, c_n = \frac{1}{2}\int_{-1}^{0}(1+t)dt + \frac{1}{2}\int_{0}^{1}(1-t)dt = \frac{1}{2}$.

The exponential Fourier series expansion for the given signal is:

$$f(t) = \sum_{n=-\infty}^{\infty} c_n e^{jn\omega_0 t} = \sum_{n=-\infty}^{\infty} \frac{2}{\pi^2 n^2} e^{jn\pi t}$$

$$f(t) = \frac{1}{2} + \frac{2}{\pi^2}\left[e^{j\pi t} + \frac{1}{3^2}e^{j3\pi t} + \frac{1}{5^2}e^{j5\pi t} + \cdots\right].$$

The amplitude spectrum of that signal is plotted in Fig. 3.28.

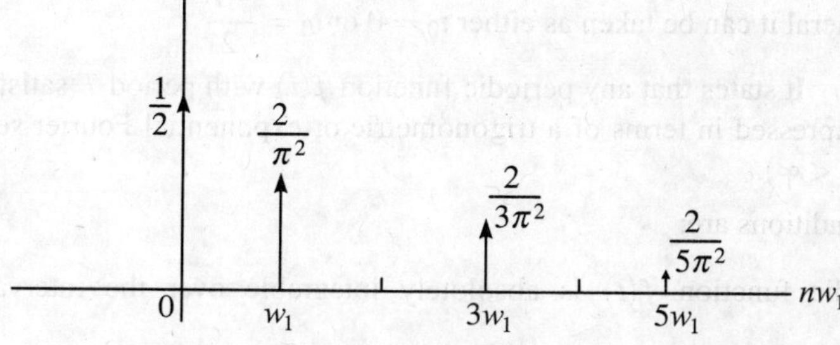

Fig. 3.28 *Spectrum of the triangular impulse signal*

Summary

1. *Fourier series:* Any periodic function $f(t)$ can be represented in terms of a linear weighted sum of orthogonal signals over an interval $(t_0, t_0 + 2\pi/\omega_0)$ with fundamental frequency ω_0.
2. In trigonometric Fourier series, any periodic function $f(t)$ can be represented in terms of a linear weighted sum of harmonic signals of sine and cosine terms over an interval $(t_0, t_0 + 2\pi/\omega_0)$ with fundamental frequency ω_0. The expression is

$$f(t) = a_0 + \sum_{n=1}^{\infty} [a_n \cos n\omega_0 t + b_n \sin n\omega_0 t] \ (-\infty < t < \infty),$$

where $n = 1, 2, \ldots.\infty$ is an integer, a_n and b_n are Fourier coefficients. T is the period of $f(t)$, $\omega_0 = 2\pi/T$ is called the fundamental frequency, and $n\omega_0$ is called the nth harmonic of frequency ω_0.

3. Since the Fourier series separates a periodic function into sinusoidal frequency components at harmonic frequencies $n\omega_0$; the Fourier analysis is also called harmonic analysis.
4. The trigonometric Fourier coefficients or the constant values of Fourier series are given by

$$a_n = \frac{\int_{t_o}^{t_0+T} f(t) \cos n\omega_0 t \, dt}{\int_{t_o}^{t_0+T} \cos^2 n\omega_0 t \, dt}, \quad b_n = \frac{\int_{t_o}^{t_0+T} f(t) \sin n\omega_0 t \, dt}{\int_{t_o}^{t_0+T} \sin^2 n\omega_0 t \, dt}$$

5. The constant value $a_0 = \frac{1}{T} \int_{t_o}^{t_0+T} f(t) \, dt$ is the average value or the dc component of $f(t)$.
6. The compact form of the trigonometric Fourier series is $f(t) = a_0 + \sum_{n=1}^{\infty} A_n \cos(n\omega_0 t + \phi_n)$, where $A_n = \sqrt{a_n^2 + b_n^2}$ and $\phi_n = -\tan^{-1}\left(\frac{b_n}{a_n}\right)$.
7. The exponential Fourier series is given by $f(t) = \sum_{n=-\infty}^{\infty} c_n e^{jn\omega_0 t}$, where $c_n = \frac{1}{T} \int_{t_o}^{t_0+T} f(t) e^{-jn\omega_0 t} dt$ for $n = -\infty \ldots. -2, -1, 0, 1, 2, \ldots.\infty$ are the Fourier coefficients. Also, it is evident that $c_{-n} = c_n^*$ and $|c_{-n}|^2 = |c_n|^2$.
8. Since $f(t)$ is a periodic signal with period T, i.e., $f(t) = f(t+T)$, the Fourier series over the interval $(t_0 < t < t_0 + T)$ holds for the entire interval $(-\infty < t < \infty)$. The value of c_n is independent of time t_0. In general it can be taken as either $t_0 = 0$ or $t_0 = \frac{-T}{2}$.
9. *Fourier theorem*: It states that any periodic function $f(t)$ with period T satisfying Dirichlet conditions can be expressed in terms of a trigonometric or exponential Fourier series over the entire interval $(-\infty < t < \infty)$.
10. The Dirichlet conditions are:

    (i) The periodic function $f(t)$ is absolutely integrable over the interval T seconds. i.e., $\int_0^T |f(t)| dt < \infty$.

    (ii) The function $f(t)$must have finite number of maxima and minima in one period.

    (iii) The function $f(t)$ has a finite number of finite discontinuities.

11. The relationship between the coefficients of the two form of Fourier series is $c_n = \frac{1}{2}(a_n - jb_n)$ and $a_n = 2\text{Re}\,[c_n],\quad b_n = -2\text{Im}\,[c_n]$.
12. The symmetrical properties of $f(t)$ are given as follows.

S. No	Symmetry of $f(t)$	Expression for Fourier coefficients
1	Even symmetry $f(t) = f(-t)$	$b_n = 0;\ a_n = \frac{4}{T}\int_0^{T/2} f(t)\cos n\omega_0 t\, dt$ and $a_0 = \frac{2}{T}\int_0^{T/2} f(t)dt$
2	Odd symmetry $f(t) = -f(-t)$	$a_0 = 0,\ a_n = 0$ and $b_n = \frac{4}{T}\int_0^{T/2} f(t)\sin n\omega_0 t\, dt$
3	Hidden symmetry $f_0(t) = f(t) - \frac{A}{2}$ where $f_0(t)$ is odd symmetry of $f(t)$.	$a_0 = \frac{A}{2}$, where A is the maximum amplitude of $f(t)$, $a_n = 0$ and $b_n = \frac{4}{T}\int_0^{T/2} f(t)\sin n\omega_0 t\, dt$
4.	Rotation symmetry $f(t) = -f\left(t \pm \frac{T}{2}\right)$	$a_n = \begin{cases} 0 & \text{for } n = \text{even} \\ \frac{4}{T}\int_0^{T/2} f(t)\cos n\omega_0 t\, dt & \text{for } n = \text{odd} \end{cases}$ $b_n = \begin{cases} 0 & \text{for } n = \text{even} \\ \frac{4}{T}\int_0^{T/2} f(t)\sin n\omega_0 t\, dt & \text{for } n = \text{odd} \end{cases}$

13. The representation of Fourier series frequency components at harmonic frequencies $\omega_0, 2\omega_0, 3\omega_0 \ldots\ldots n\omega_0 \ldots$, is called its frequency spectrum.
14. The amplitude spectrum is symmetrical about the origin (even symmetric), whereas the phase spectrum is anti-symmetrical (odd function).
15. The magnitude of the power contained in each frequency components at harmonic frequencies is $|c_n|^2$. Also $|c_{-n}|^2 = |c_n|^2$.
16. The representation of frequency components $|c_n|^2$at harmonic frequencies $\omega_0, 2\omega_0, 3\omega_0 \ldots\ldots n\omega_0 \ldots$ etc., is called power spectrum.
17. For any periodic function $f(t)$ with period T, if the Fourier series pair are $x_1(t) \overset{\text{FS}}{\longleftrightarrow} c_n,\quad x_2(t) \overset{\text{FS}}{\longleftrightarrow} d_n$ and $x(t) \overset{\text{FS}}{\longleftrightarrow} c_n$, then the properties of the Fourier series are

S. No	Property	Periodic function	Expression for Fourier coefficients
1	Linearity	$Ax_1(t)+Bx_2(t)$	Ac_n+Bd_n
2	Time shifting	$x(t-t_0)$	$e^{-jn\omega_0 t_0}c_n$
3	Time reversal	$x(-t)$	c_{-n}
4	Time scaling	$x(at)$	$c_n$with frequency $a\omega_0$
5	Conjugation	For complex $x(t), x^*(t)$	c^*_{-n}
		For real $x(t)$	$a_{-n}=a^*_n,\ a_n=a^*_{-n}$
		For imaginary $x(t)$	$a_{-n}=-a^*_n\ a_n=-a^*_{-n}$
6	Parseval's theorem	Average power is $P=\frac{1}{T}\int_{-T/2}^{T/2} f^2(t)\,dt$	$\sum_{n=-\infty}^{\infty}\lvert c_n\rvert^2$ and $\lvert c_{-n}\rvert^2=\lvert c_n\rvert^2$

Review Questions

1. State and explain trigonometric Fourier series. (June 2014)
2. State and explain exponential Fourier series. (June 2014)
3. Derive the expressions for coefficients of trigonometric Fourier series. (June 2014)
4. Derive the expression for coefficients of exponential Fourier series. (June 2014)
5. (a) Discuss the compact form of Fourier series.
   (b) State the necessary and sufficient conditions for the existence of Fourier series representation of a periodic signal. (June 2016)
6. State and prove Parseval's theorem. (June 2015)
7. State and explain Dirichlet's conditions. (June 2014)
8. Explain Gibb's phenomenon. (Dec 2016)
9. Derive the relationship between coefficients of the trigonometric Fourier series and the exponential Fourier series. (June 2014)
10. Define and explain complex frequency spectrum. (June 2016)
11. Define amplitude and phase spectra. (June 2013)
12. Define even symmetry and derive the trigonometric Fourier series coefficients for even function. (June 2015)
13. Define odd symmetry and derive the trigonometric Fourier series coefficients for odd function. (Dec 2015)
14. Define rotation or half-wave symmetry and derive the trigonometric Fourier series coefficients for half-wave symmetrical function. (Dec 2014)
15. Explain hidden symmetry. (June 2014)
16. State and derive the properties of Fourier series. (Dec 2014)
17. Define continuous-time Fourier series. List some of its properties. (May 2012)
18. Find the cosine Fourier series of a half-wave rectified sine function. (June 2015)

19. State and prove the following properties of Fourier series: (i) time differentiation property, (ii) scaling property. (June 2013)
20. Find the line spectrum of a half-wave rectified rise wave with period of 2 seconds. (June 2014)
21. Justify the following with respect to Fourier series: (i) odd functions have only sine terms, (ii) even functions have only cosine terms. (June 2013, June 2014)
22. Explain the symmetry of waveform and comment on Fourier coefficient.
23. Explain the concept of negative frequency. (June 2011)
24. Define line spectrum and power spectrum. (June 2016)
25. What is the effect of half-wave symmetry on Fourier coefficients of a signal? (June 2013)
26. Write the Parseval's relation for continuous time periodic signals. (June 2015)
27. Prove that discrete magnitude spectrum is symmetrical about vertical axis, whereas phase spectrum is anti-symmetrical about vertical axis. (June 2016)

Problems

1. Find the trigonometric Fourier series of the periodic impulse train shown in Fig. P3.1.

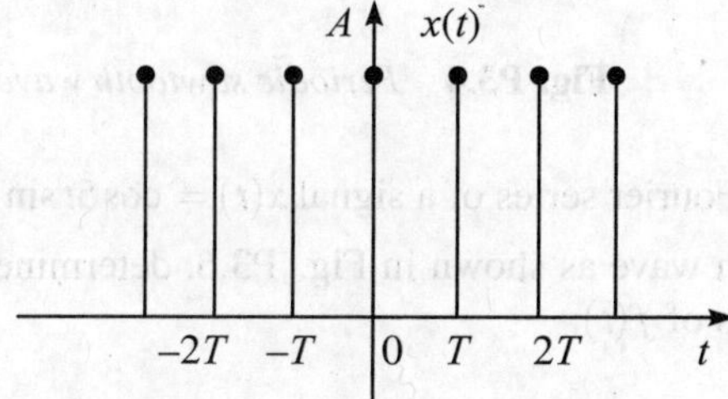

Fig. P3.1 *Periodic impulse train*

2. For the periodic rectangular signal as shown in Fig. P3.2, find the trigonometric and exponential Fourier series and plot the amplitude spectrum.

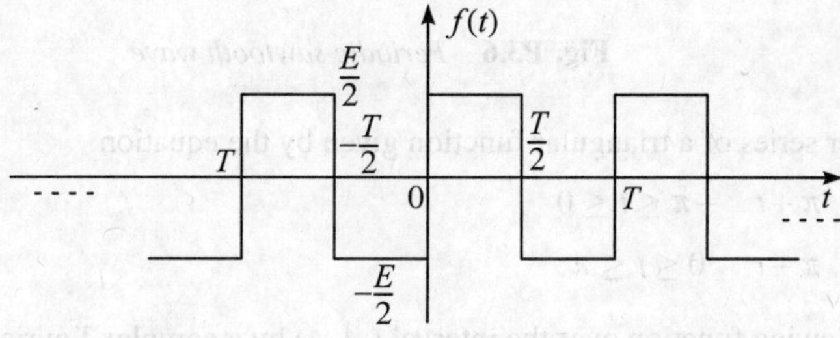

Fig. P3.2 *Periodic rectangular signal*

3. A periodic square wave is shown in Fig. P3.3. Determine the trigonometric and complex exponential Fourier series of $f(t)$.

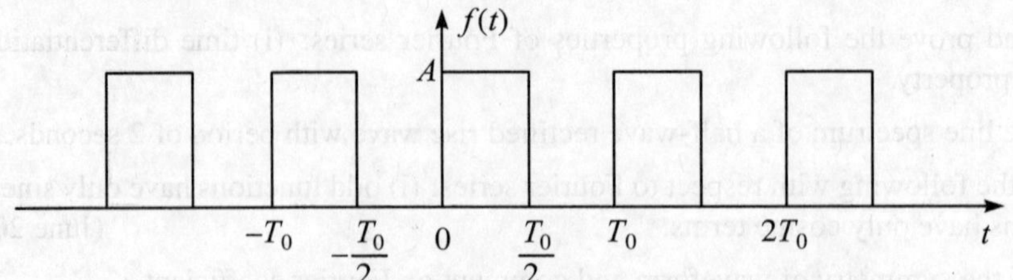

Fig. P3.3 *Periodic square wave*

4. Find the trigonometric Fourier series of the periodic sawtooth wave shown in Fig. P3.4.

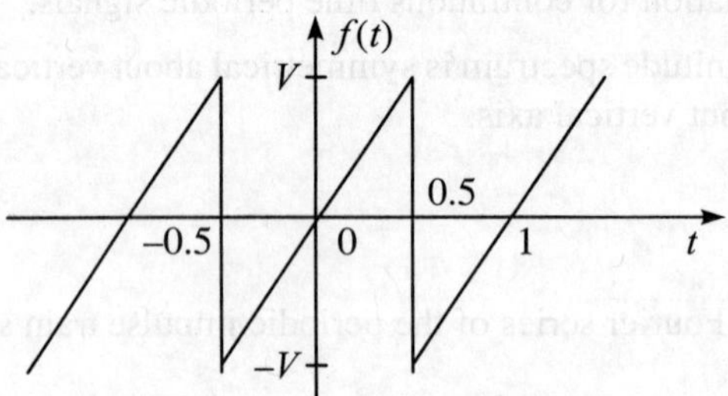

Fig. P3.4 *Periodic sawtooth wave*

5. (a) Find the exponential Fourier series of a signal $x(t) = \cos 5t \sin 3t$.

6. For the periodic sawtooth wave as shown in Fig. P3.6, determine the complex and trigonometric exponential Fourier series of $f(t)$.

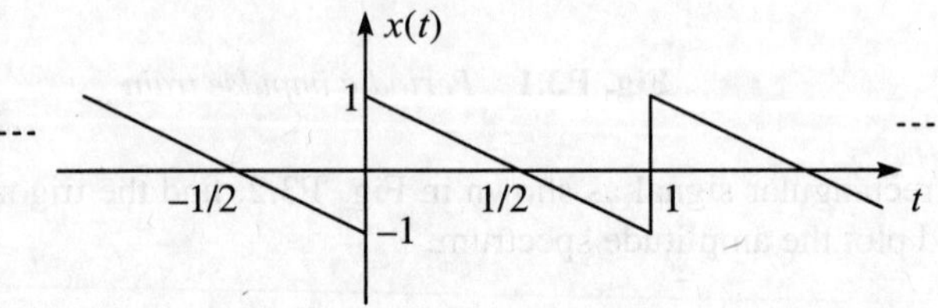

Fig. P3.6 *Periodic sawtooth wave*

7. Find the Fourier series of a triangular function given by the equation

$$f(t) = \begin{cases} \pi + t & -\pi \le t \le 0 \\ \pi - t & 0 \le t \le \pi \end{cases}.$$

8. Expand the following function over the interval (-4, 4) by a complex Fourier series

$$f(t) = \begin{cases} 1 & -2 < t < 2 \\ 0 & \text{elsewhere} \end{cases}$$

9. Obtain the trigonometric Fourier series for the waveform shown in Fig. P3.9.

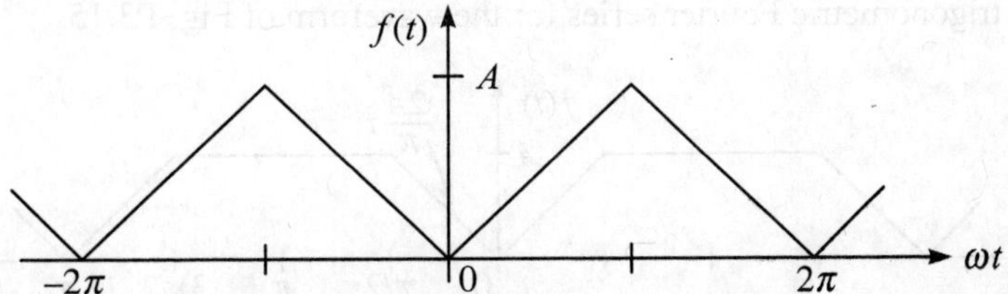

Fig. P3.9 *Triangular waveform*

10. Find the Fourier series expansion of the half-wave rectified sine wave shown Fig. P3.10.

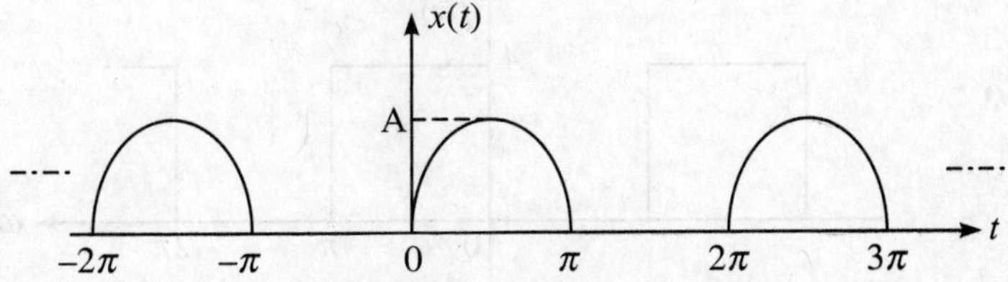

Fig. P3.10 *Half-wave rectified sine wave*

11. Obtain the trigonometric Fourier series coefficients for the signal shown in Fig. P3.11. (Dec 2011)

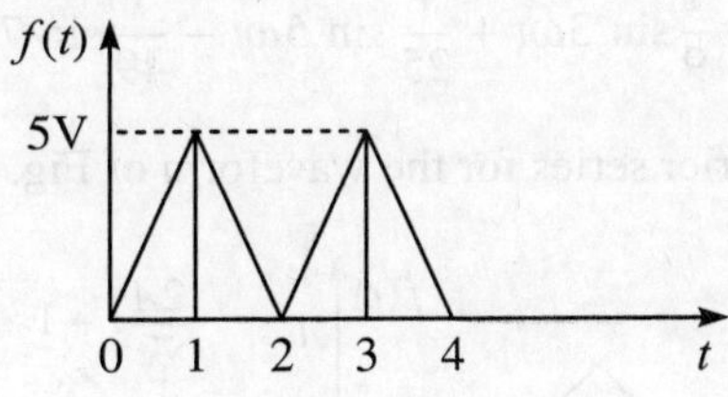

Fig. P3.11

12. Expand the following function over the interval −4, 4 using complex Fourier series.

$$f(t) = \begin{cases} 1, & -2 \leq t \leq 2 \\ 0, & \text{elsewhere} \end{cases}$$ (Dec 2011)

13. Find the exponential Fourier series coefficients for the periodic function $x(t) = \sin \pi x$ (June 2014)
14. For the periodic function given below, derive the exponential form of the Fourier series. (June 2016)

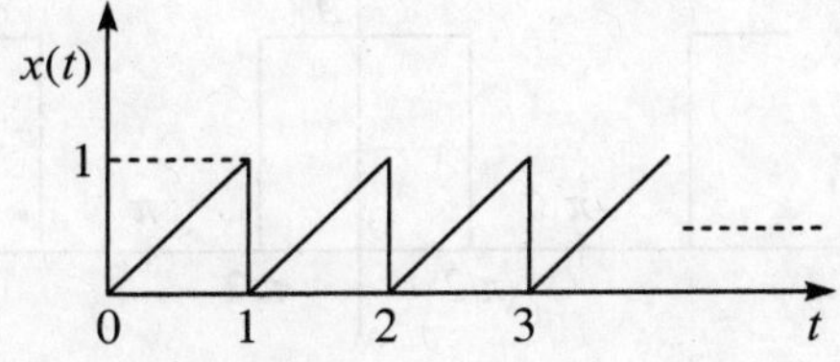

Fig. P3.14 *Sawtooth function*

15. Find out the trigonometric Fourier series for the waveform of Fig. P3.15.

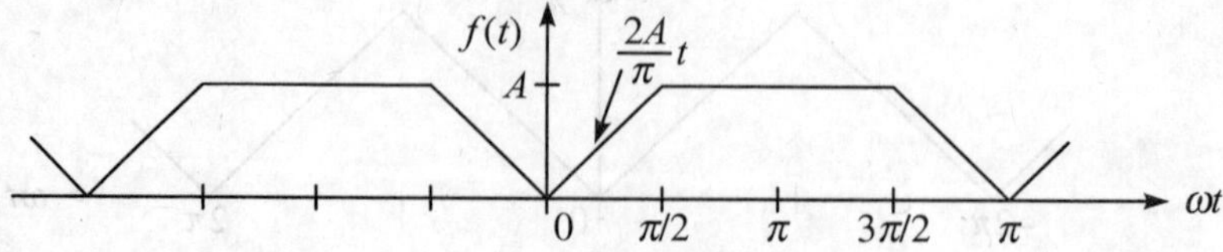

Fig. P3.15 *Trapezoidal wave*

16. Find the trigonometric Fourier series for the waveform of Fig. P3.16.

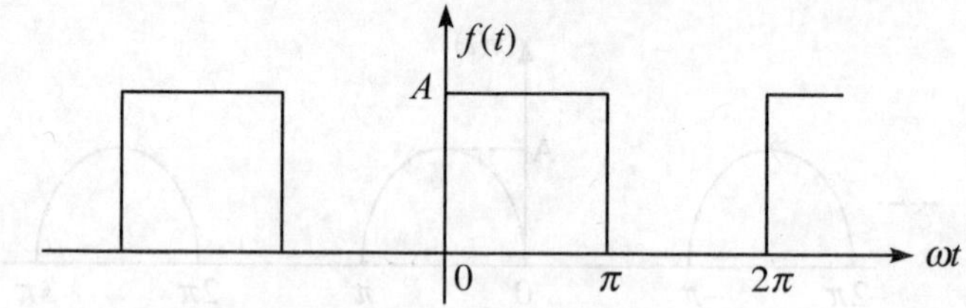

Fig. P3.16 *Square waveform*

17. Prove that the trigonometric Fourier series for the triangular waveform with odd symmetry is

$$f(t) = \frac{8A}{\pi^2}\left(\sin\,\omega t - \frac{1}{9}\sin\,3\omega t + \frac{1}{25}\sin\,5\omega t - \frac{1}{49}\sin\,7\omega t + \ldots\right).$$

18. Find the trigonometric Fourier series for the waveform of Fig. P3.18.

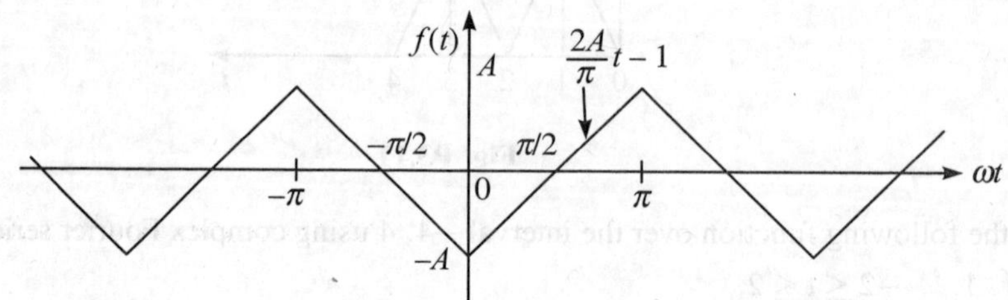

Fig. P3.18 *Triangular waveform*

19. Find the trigonometric Fourier series for the waveform of Fig. P3.19.

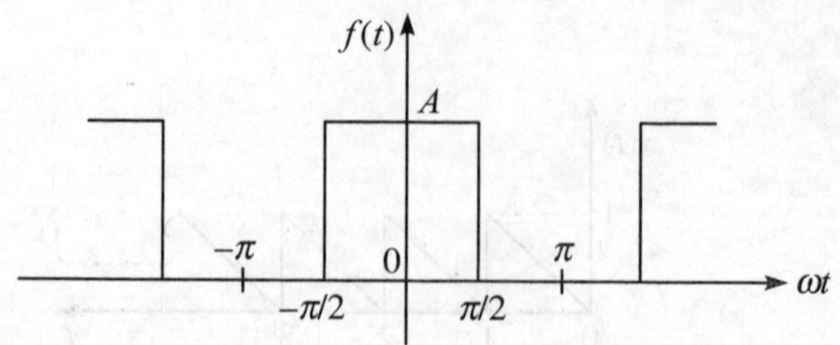

Fig. P3.19 *Rectangular waveform*

20. Prove that the trigonometric Fourier series for the sawtooth waveform with odd symmetry is

$$f(t) = \frac{2A}{\pi}\left(\sin\,\omega t - \frac{1}{2}\sin\,2\omega t + \frac{1}{3}\sin\,3\omega t - \frac{1}{4}\sin\,4\omega t + \ldots\right).$$

21. Compute the trigonometric Fourier series of the sawtooth waveform of Fig. P3.21.

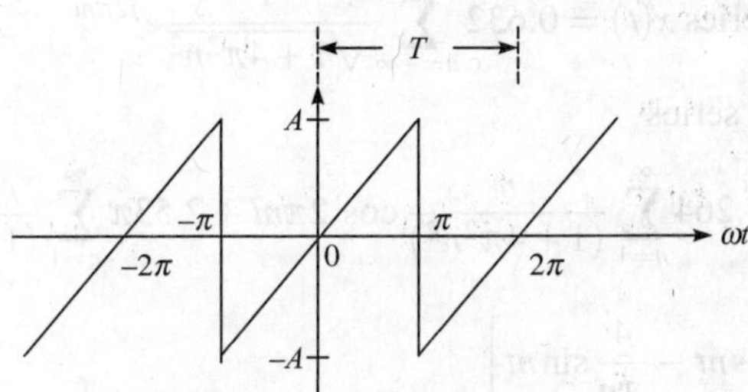

Fig. P3.21 *Sawtooth waveform*

22. Find the exponential Fourier series for the piroidic signal $x(t) = e^{-t} \quad 0 < t < 1$ shown in Fig. P3.22.

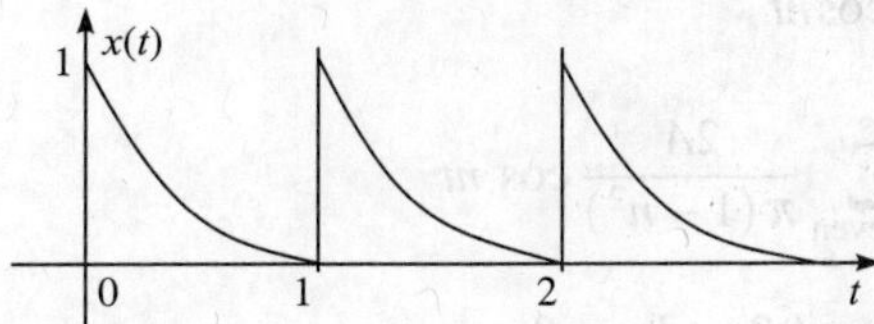

Fig. P3.22 *Exponential waveform*

23. Determine the complex exponential Fourier series representation for each of the following signals:

(a) $f(t) = \cos\,4t + \sin\,6t$

(b) $f(t) - \cos\left(3t + \frac{\pi}{5}\right)$

(c) $f(t) = 4 + 2\cos\left(\frac{2\pi t}{3} + \frac{\pi}{6}\right) + 4\sin\left(\frac{7\pi t}{3}\right)$

Answers

1. $x(t) = \frac{A}{T}\left[1 + \sum_{n=1}^{\infty} 2\cos\frac{2\pi nt}{T}\right]$

2. $f(t) - \sum_{n=-\infty}^{\infty} \frac{E}{jn\pi} e^{jn\omega_0 t}$

3. Trigonometric Fourier series

$$f(t) = \frac{A}{2} + \sum_{n=1}^{\infty} \frac{2A}{n\pi}\sin\left(\frac{2\pi n}{T}t\right) \quad \text{for odd } n$$

Exponential Fourier series $f(t) = \frac{A}{2} + \sum_{n=\text{odd}}^{\infty}\left(\frac{A}{jn\pi}\right)e^{-\frac{j2\pi n}{T}}$

4. $f(t) = \sum_{n=-\infty}^{\infty} \left(\frac{-2V}{n\pi} \cos n\pi \right) \sin(2\pi nt)$

5. $x(t) = \frac{1}{j4}\left[e^{j8t} - e^{-j8t} - e^{j2t} + e^{-j2t}\right]$

6. Exponential Fourier series $x(t) = 0.632 \sum_{n=-\infty}^{\infty} \frac{1}{\sqrt{1+4\pi^2 n^2}} e^{j2\pi nt}$

Trigonometric Fourier series

$$x(t) = 0.632 + 1.264 \sum_{n=1}^{\infty} \frac{n}{(1+4\pi^2 n^2)} \cos 2\pi nt + 2.53\pi \sum_{n=1}^{\infty} \frac{n}{(1+4\pi^2 n^2)} \sin 2\pi nt$$

7. $f(t) = \frac{\pi}{2} + \sum_{n=1}^{\infty} \left[\frac{4}{n^2} \cos nt - \frac{4}{\pi n} \sin nt \right]$

8. $f(t) = \frac{\pi}{2} + \sum_{n=-\infty}^{\infty} \frac{j}{2n\pi} \left[e^{-jn\pi t} - e^{jn\pi t}\right],\ n \neq 0$

9. $f(t) = \frac{A}{2} - \frac{4A}{\pi} \sum_{n=\text{odd}}^{\infty} \frac{1}{n^2} \cos nt$

10. $x(t) = \frac{A}{\pi} + \frac{A}{2\pi} \sin t + \sum_{n=\text{even}}^{\infty} \frac{2A}{\pi(1-n^2)} \cos nt$

11. $a_n = \left(\frac{-5}{2n^2}\right)[1-(-1)^n]\, n \neq 0, \quad b_n = 0$

12. $c_n = \begin{cases} \frac{1}{2} & n = 0 \\ \frac{1}{n\pi} \sin\left(\frac{n\pi}{2}\right) & n \neq 0 \end{cases}$

13. $c_n = \begin{cases} \frac{-j}{2} & n = 1 \\ \frac{j}{2} & n = -1 \end{cases}$

14. $c_n = \begin{cases} 0 & n = 0 \\ \frac{-1}{j2n\pi} & n \neq 0 \end{cases}$

15. $f(t) = \frac{3A}{4} - \frac{4A}{\pi^2} \left(\cos t + \frac{1}{2} \cos 2t + \frac{1}{9} \cos 3t + \ldots \right)$

16. $f(t) = \frac{A}{2} + \frac{A}{j\pi} \left(\cdots - \frac{1}{3} e^{-j3\omega t} - e^{-j\omega t} + e^{j\omega t} + \frac{1}{3} e^{j3\omega t} + \ldots \right)$

18. $f(t) = \frac{4A}{\pi^2} \left(\cdots - \frac{1}{9} e^{-j3\omega t} - e^{-j\omega t} + e^{j\omega t} + \frac{1}{9} e^{j3\omega t} + \ldots \right)$

19. $f(t) = \frac{A}{2} + \sum_{n=\text{odd}} (-1)^{(n-1)/2} \frac{A}{n\pi} e^{jn\omega t}$

21. $= \frac{2A}{\pi} \sum (-1)^{n-1} \frac{1}{n} \sin n\omega t$

22. $x(t) = \sum_{n=-\infty}^{\infty} \frac{0.6321}{1+j2\pi n} e^{j2\pi nt}$

23. (a) $f(t) = \frac{1}{j2}e^{-j6t} + \frac{1}{2}e^{-j4t} + \frac{1}{2}e^{j4t} \frac{1}{j2}e^{j6t}$

(b) $f(t) = \frac{1}{2}\left[e^{j\pi/5}\, e^{j3t} + e^{-j\pi/5}\, e^{-j3t}\right]$

(c) $f(t) = 4 + e^{j\pi/6} e^{j2\pi t/3} + e^{-j\pi/6} e^{-j2\pi t/3} + 2/je^{j7\pi t/3} - 2/je^{-j7\pi t/3}$

Multiple-choice Questions

1. The Fourier series of an odd periodic function contains only

(a) Odd harmonics (b) Even harmonics

(c) Cosine terms (d) Sine terms

2. The trigonometric Fourier series of an even function of time does not have

(a) the dc term (b) cosine terms

(c) sine terms (d) odd harmonic terms

3. The Fourier series of a real periodic function has only

(a) Cosine terms if it is even (b) Sine terms if it is even

(c) Cosine terms if it is odd (d) Sine terms if it is odd

Which of the above statements are correct?

(i) (a) and (d) (ii) (a) and (c) (iii) (b) and (d) (iv) (b) and (c)

4. The trigonometric Fourier series for the waveform $f(t)$ shown below in Fig. M3.4 contains

(a) Only cosine terms and zero value for the dc component

(b) Only cosine terms and a positive value for the dc component

(c) Only cosine terms and a negative value for the dc component

(d) Only sine terms and a negative value for the dc component

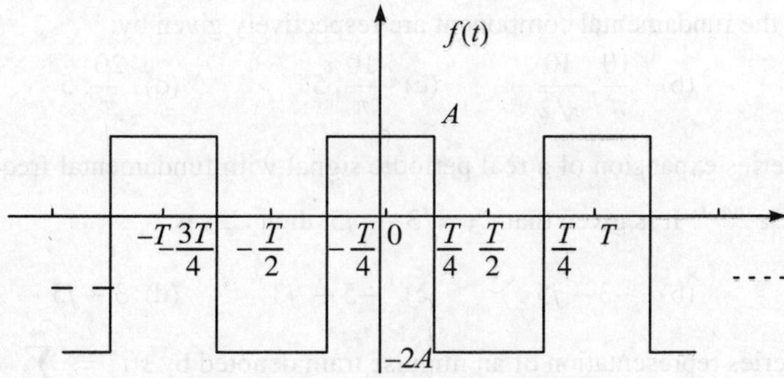

Fig. M3.4 *Rectangular waveform*

5. The trigonometric Fourier series of an even function does not have the

(a) dc term (b) cosine terms

(c) sine terms (d) odd harmonic terms

6. One period $(0,T)$ each of two periodic waveforms W_1 and W_2 are shown in Fig. M3.6 (a) and Fig. M3.6(b). The magnitudes of the nth Fourier series coefficients of W_1 and W_2 for $n \geq 1, n$ odd, are respectively proportional to

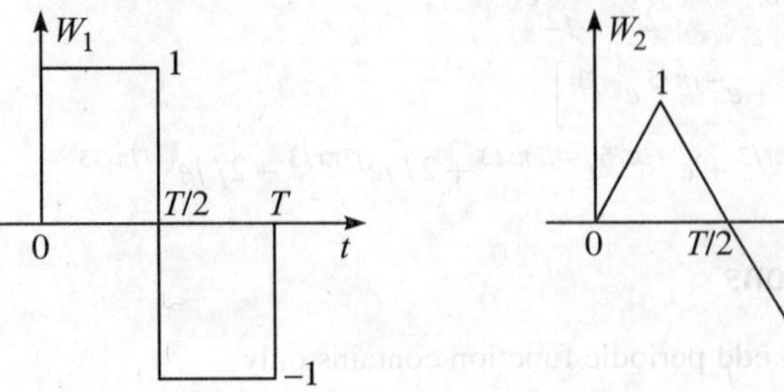

Fig. M3.6 *(a) Rectangular pulse; (b) Triangular pulse*

(a) $|n_{-3}|$ and $|n_{-2}|$ (b) $|n_{-2}$ and $|n_{-3}|$

(c) $|n_{-1}|$ and $|n_{-2}|$ (d) $|n_{-4}|$ and $|n_{-2}|$

7. The trigonometric Fourier series of a periodic time function can have only

(a) cosine terms (b) sine terms

(c) cosine and sine terms (d) dc and cosine terms

8. A periodic signal $x(t)$ of period T_0 is given by

$$x(t) = \begin{cases} 1 & |t| < T_1 \\ 0 & T_1 < |t| < \dfrac{T_0}{2} \end{cases}$$

The dc component of $x(t)$ is

(a) $\dfrac{T_1}{T_0}$ (b) $\dfrac{T_1}{2T_0}$ (c) $\dfrac{2T_1}{T_0}$ (d) $\dfrac{T_0}{T_1}$

9. A half-wave rectified sinusoidal waveform has a peak voltage of 10 V. The average value and the peak value of the fundamental component are respectively given by:

(a) $\dfrac{20}{\pi}, \dfrac{10}{\sqrt{2}}$ (b) $\dfrac{10}{\pi}, \dfrac{10}{\sqrt{2}}$ (c) $\dfrac{10}{\pi}, 5$ (d) $\dfrac{20}{\pi}, 5$

10. The Fourier series expansion of a real periodic signal with fundamental frequency ω_0 is given by $f(t) = \sum_{n=-\infty}^{\infty} c_n e^{jn\omega_0 t}$. It is given that $C_3 = 3 + j5$, then C_{-3} is

(a) $5 + j3$ (b) $-3 - j5$ (c) $-5 + j3$ (d) $3 - j5$

11. The Fourier series representation of an impulse train denoted by $s(t) = \sum_{n=-\infty}^{\infty} \delta(t - nt_0)$ is given by

(a) $\left(\dfrac{1}{T_0}\right) \sum_{n=-\infty}^{\infty} \exp\left(\dfrac{-j2\pi nT}{T_0}\right)$ (b) $\left(\dfrac{1}{T_0}\right) \sum_{n=-\infty}^{\infty} \exp\left(\dfrac{-j\pi nT}{T_0}\right)$

(c) $\left(\dfrac{1}{T_0}\right) \sum_{n=-\infty}^{\infty} \exp\left(\dfrac{j\pi nT}{T_0}\right)$ (d) $\left(\dfrac{1}{T_0}\right) \sum_{n=-\infty}^{\infty} \exp\left(\dfrac{j2\pi nT}{T_0}\right)$

12. Which of the following cannot be the Fourier series expansion of a periodic signal?
   (a) $x(t) = 2\cos \pi t + 7\cos t$ (b) $x(t) = 2\cos t + 3\cos 3t$
   (c) $x(t) = \cos t + 0.5$ (d) $x(t) = 2\cos 1.5\pi t + \sin 3.5\pi t$
13. Choose the function $f(t)$, $-\infty < t < +\infty$, for which a Fourier series cannot be defined.
   (a) $3\sin 25t$ (b) $4\cos 20t + 3 + 2\sin 10t$
   (c) $e^{-|t|}\sin 25t$ (d) 1
14. A function is given by $f(t) = \sin^2 t + \cos 2t$. Which of the following is true?
   (a) $f(t)$ has frequency components at 0 and $1/2\pi$ Hz
   (b) $f(t)$ has frequency components at 0 and $1/\pi$ Hz
   (c) $f(t)$ has frequency components at $1/2\pi$ and $1/\pi$ Hz
   (d) $f(t)$ has frequency components at 0, $1/2\pi$ and $1/\pi$ Hz
15. For a periodic signal $x(t)$ with period T, condition for half wave symmetry is
   (a) $x(t) = x(t - T/2)$ (b) $x(t) = x(t \pm T)$
   (c) $x(t) = -x(t \pm T/2)$ (d) $x(t) = -x(t \pm T)$
16. Dirichlet's condition is
   (a) The function $f(t)$ has a finite number of finite discontinuities.
   (b) The function $f(t)$ has a finite number of maxima.
   (c) The function $f(t)$ has a finite number of minima.
   (d) The function $f(t)$ is singularity function.
17. The convolution of two periodic signals (period T) having exponential Fourier coefficients c_n and d_n is
   (a) $c_n * d_n$ (b) $c_n d_n/T$ (c) $c_n d_n$ (d) $T c_n d_n$
18. The convolution of two periodic signals (period T) having exponential Fourier coefficients c_n and d_n is
   (a) $c_n * d_n$ (b) $c_n d_n/T$ (c) $c_n d_n$ (d) $T c_n d_n$
19. If $f(t) \overset{FS}{\longleftrightarrow} c_n$, then
   (a) $f(-t) \overset{FS}{\longleftrightarrow} c_{-n}$ (b) $f(t) \overset{FS}{\longleftrightarrow} c_{-n}$
   (c) $f(-t) \overset{FS}{\longleftrightarrow} c_n$ (d) $f(-t) \overset{FS}{\longleftrightarrow} -c_n$
20. $f^*(t) = f(t)$, then
   (a) $c_n^* = c_n$ (b) $c_n^* = -c_n$ (c) $c_{-n}^* = c_n$ (d) $c_n^* = c_{-n}$
21. Fourier series of any periodic signal $x(t)$ with fundamental period T can be obtained if
   (a) $x(t)$ has rational symmetry
   (b) $x(t)$ has infinite number of discontinuities
   (c) $\int_{-T/2}^{T/2} x(t)dt < \infty$
   (d) none of the above

22. Spectrum of a periodic pulse is
    (a) continuous sync function (b) discrete Ssync function
    (c) pulse train (d) impulse function
23. Fourier series could be applied to
    (a) power signal (b) energy signal
    (c) aperiodic signal (d) unit step signal
24. Harmonics of a periodic signal are equal to
    (a) integer multiple of fundamental frequency
    (b) half of fundamental frequency
    (c) fundamental frequency
    (d) none of the above
25. The exponential Fourier series coefficient c_{-n} in terms of trigonometric fourier series coefficients is
    (a) $c_{-n} = \frac{1}{2}(a_n + jb_n)$ (b) $c_{-n} = \frac{1}{2}(a_n - jb_n)$
    (c) $c_{-n} = (a_n - jb_n)$ (d) $c_{-n} = (a_n + jb_n)$
26. The trigonometric Fourier series coefficient a_n in terms of exponential fourier series coefficients is
    (a) $a_n = c_n + c_{-n}$ (b) $a_n = c_n - c_{-n}$
    (c) $a_n = c_n + jc_{-n}$ (d) $a_n = c_n - jc_{-n}$
27. The trigonometric Fourier series coefficient b_n in terms of exponential fourier series coefficients is
    (a) $b_n = j(c_n + c_{-n})$ (b) $b_n = j(c_n - c_{-n})$
    (c) $b_n = c_n + c_{-n}$ (d) $b_n = c_n - c_{-n}$
28. Even symmetry is also called
    (a) rotation symmetry (b) mirror symmetry
    (c) full symmetry (d) half symmetry
29. Half wave symmetry is also called
    (a) rotation symmetry (b) mirror symmetry
    (c) full symmetry (d) even symmetry
30. The coefficient a_n is zero for ____________ functions.
    (a) even (b) odd
    (c) both (a) and (b) (d) none of these

Answers

1. (d)	2. (c)	3. (a)	4. (c)	5. (c)	6. (c)	7. (c)	8. (c)	9. (c)	10. (d)
11. (d)	12. (c)	13. (c)	14. (b)	15. (c)	16. (a)	17. (d)	18. (c)	19. (a)	20. (d)
21. (c)	22. (b)	23. (a)	24. (a)	25. (a)	26. (a)	27. (b)	28. (b)	29. (a)	30. (b)

4

The Fourier Transform

Learning Objectives

The main objective of this chapter is to understand thoroughly the basic concepts of Fourier transform. We will leam how periodic signals can be represented by using Fourier transform. By reading this chapter, students will be able to relate to the following topics:

- Concepts of Fourier transform
- Fourier transform of signals
- Properties of Fourier transforms
- Fourier transforms involving impulse function and signum function

4.1 Introduction

We are familiar with the basic concepts of the Fourier series representation of a periodic function which can be expressed as a linear combination of complex exponential functions over the interval $-\infty < t < \infty$. That is, the Fourier series is a spectra of amplitude lines spaced at harmonics of the fundamental frequency of a continuous-time periodic signal. The Fourier series also represents a non-periodic signal which is a continuous sum of complex exponential functions over a finite interval $-t_0 < t < t_0 + T$. The fundamental requirement of any function to be represented as a Fourier series is that it should satisfy Dirichlet's conditions. For non-periodic signals, the Fourier transform can be derived from the Fourier series by taking the time period as tending to infinity. That is, the frequency domain representation of any arbitrary function over the entire interval $-\infty < t < \infty$ is called the Fourier transform.

It is to be noted that for a large value of time period, the harmonic frequencies in the discrete spectrum of a periodic signal become more closely spaced. That is, a larger number of frequency components will appear closer to each other in a given range. As it is known that any non-periodic signal, of finite or infinite duration, can be constructed from a periodic signal with a single period tending to infinity. Therefore, a non-periodic function is a periodic function that has only one period in the entire interval, and thus its spectrum becomes continuous, leading to a frequency-domain representation for any arbitrary signal in the form of the Fourier transform.

This chapter explains the derivation of the Fourier transform from Fourier series. Also, Fourier transforms of different types of functions and the properties of the Fourier transform will be discussed. The Fourier transform discussed here for continuous-time functions is called continuous-time Fourier transform. Likewise, for discrete time functions the Fourier transform is called discrete Fourier transform (DFT).

Note: Just as any periodic signal represented by a linear combination of harmonics at discrete frequencies is called a Fourier series, any non-periodic or periodic signal that can be represented by a continuous frequency in the frequency domain is called Fourier transform.

4.2 Fourier Transform from Fourier Series

We will develop the concept of Fourier transform by beginning with a periodic function and then letting the period become infinite. Consider a periodic function $f_T(t)$ over the interval $-\infty < t < \infty$. The exponential form of the Fourier series is

$$f_T(t) = \sum_{-\infty}^{\infty} c_n e^{jn\omega_0 t} \tag{4.1}$$

where the Fourier coefficient

$$c_n = \frac{1}{T}\int_{-T/2}^{T/2} f_T(t)e^{-jn\omega_0 t}\,dt \text{ and } \omega_0 = \frac{2\pi}{T}. \tag{4.2}$$

The Fourier coefficient c_n represents the amplitude of the frequency components.

Let $f(t)$ be a non-periodic signal of finite duration, i.e.,

$$f(t) = 0 \quad |t| > T. \tag{4.3}$$

Now, the non-periodic function can be expressed in terms of a periodic function with infinite time period as

$$f(t) = \lim_{T\to\infty} f_T(t). \tag{4.4}$$

As $T \to \infty$, the fundamental frequency ω_0 becomes vanishingly small. Also, c_n becomes very small. But the shape of the spectrum remains unaltered and becomes continuous with infinite number of frequency components. Therefore, the fundamental frequency ω_0 can be written as $\omega_0 \to d\omega$ and

$$\frac{1}{T} = \frac{\omega_0}{2\pi} \to \frac{d\omega}{2\pi}. \tag{4.5}$$

Thus the frequency of any "harmonic" $n\omega_0$ should correspond to the frequency variable ω which describes the continuous spectrum. In other words, n should tend to infinity as ω_0 approaches zero, so that the product $n\omega_0 \to \omega$ is finite.

Now, let the function c_n of Eq.(4.2) with T be denoted as $Tc_n = F(\omega)$, a function of ω. Then the equation becomes

$$c_nT = \int_{-T/2}^{T/2} f_T(t)e^{-j\omega t}\,dt. \tag{4.6}$$

From Eq.(4.4), as $T \to \infty$,

$$F(\omega) = c_nT = \lim_{T\to\infty}\int_{-T/2}^{T/2} f_T(t)e^{-j\omega t}\,dt$$

or $F(\omega) = \int_{-\infty}^{\infty} f(t)e^{-j\omega t}dt$ (4.7)

Equation (4.7) is a function of ω and represents the Fourier transform of $f(t)$. Now, from Eq. (4.1), the Fourier series of a periodic function $f_T(t)$, is

$$f_T(t) = \sum_{-\infty}^{\infty} c_n e^{jn\omega_0 t}.$$

By multiplying and dividing the summation by time period T,

$$f_T(t) = \frac{1}{T}\sum_{-\infty}^{\infty} c_n T e^{jn\omega_0 t}$$

and by replacing $c_n T$ by the function $F(\omega)$, the summation becomes an integral. Therefore,

$$f_T(t) = \int_{-\infty}^{\infty} F(\omega)\frac{1}{T}e^{j\omega t}d\omega.$$

Substituting the value of T from Eq.(4.5),

$$f_T(t) = \frac{1}{2\pi}\int_{-\infty}^{\infty} F(\omega)e^{j\omega t}d\omega. \tag{4.8}$$

As $T \to \infty$, the function $f_T(t) \to f(t)$. Thus,

$$f(t) = \frac{1}{2\pi}\int_{-\infty}^{\infty} F(\omega)e^{j\omega t}d\omega. \tag{4.9}$$

Equations (4.7) and (4.9) are collectively called the Fourier transform pair.

4.3 Fourier Transform of an Arbitrary Signal

For any arbitrary signal $f(t)$ the function $F(\omega)$ is its Fourier transform; conversely, $f(t)$ is the inverse Fourier transform of $F(\omega)$.

The Fourier transform pair can be written as:

$$F(\omega) = \mathcal{F}[f(t)] = \int_{-\infty}^{\infty} f(t)e^{-j\omega t}dt,$$

$$f(t) = \mathcal{F}^{-1}[f(\omega)] = \frac{1}{2\pi}\int_{-\infty}^{\infty} F(\omega)e^{j\omega t}d\omega.$$

Also, the notation $f(t) \longleftrightarrow F(\omega)$ is to used to show Fourier transform. The Fourier transform pair can also be represented with the continuous frequency f Hz instead of ω rad/s by substituting $\omega = 2\pi f$. That is,

$$\mathcal{F}(f) = \int_{-\infty}^{\infty} f(t)e^{-j2\pi ft}dt, \tag{4.10}$$

$$f(t) = \int_{-\infty}^{\infty} \mathcal{F}(f)e^{j2\pi ft}df. \tag{4.11}$$

The Fourier transform $F(\omega)$ is, in general, complex-valued and it can be expressed as

$$F(\omega) = |F(\omega)|e^{J\phi(\omega)}. \tag{4.12}$$

If $f(t)$ is real,

$$F(-\omega) = \int_{-\infty}^{\infty} f(t)e^{j\omega t}dt = F^*(\omega). \tag{4.13}$$

Thus, $|F(-\omega)| = |F(\omega)|$ (4.14)

and $\phi(-\omega) = -\phi(\omega).$ (4.15)

Therefore, for real signals the amplitude spectrum $|X(\omega)|$ is an even function and the phase spectrum $\emptyset(\omega)$ is an odd function of ω. The plots of amplitude spectrum and phase spectrum are collectively called the frequency spectrum or spectral density of the signal. The condition for the convergence of Fourier transform is the same as that of Fourier series.

Existence of a Fourier Transform

The Dirichlets conditions which are necessary for the existence of Fourier series of periodic signals can be extended to the existence of Fourier transforms of non-periodic signals as follows.

1. The function $f(t)$ should be absolutely integrable over the interval $-\infty < t < \infty$.

$$\int_{-\infty}^{\infty} |f(t)|dt < \infty. \tag{4.16}$$

2. The function $f(t)$ has a finite number of maxima and minima with any finite interval.
3. The function $f(t)$ has a finite number of finite discontinuities with in any finite interval.

A sufficient condition for the existence of $F(\omega)$ is that the function must be absolutely integrable over the interval $-\infty < t < \infty$ and that it has a finite number of discontinuities.

Although these conditions are sufficient, they are not necessary since some periodic functions which do not meet the conditions still have a Fourier transform; this is possible by considering impulse functions in the Fourier transforms. For example, the step function which does not meet the Dirichlets conditions has a Fourier transform. Thus the Fourier series representation of a periodic time function is just a part of a more general Fourier transform representation. Hence, it is necessary to emphasise that the Fourier transform-pair relationship is unique, that is, for a given $f(t)$ there exist only one $F(\omega)$ and vice versa.

4.4 Fourier Transform of Standard Signals

The Fourier transform of some standard non-periodic functions like exponential function and pulse function can be derived directly from the Fourier transform pair as follows.

4.4.1 Fourier Transform of Exponential Function

Consider an exponential function

$$f(t) = e^{-at}u(t) \quad a > 0. \tag{4.17}$$

Its Fourier transform is

$$F(\omega) = \int_{-\infty}^{\infty} f(t)e^{-j\omega t}dt = \int_{-\infty}^{\infty} e^{-at}u(t)e^{-j\omega t}dt = \int_{0}^{\infty} e^{-(a+j\omega)t}dt.$$

$$\therefore \quad F(\omega) = \frac{1}{a + j\omega}, \quad a > 0. \tag{4.18}$$

Since the Fourier transform is complex-valued, to plot it as a function of ω, $F(\omega)$ should be expressed in term of its magnitude and phase. The magnitude and phase spectrum of $F(\omega)$ as shown in Fig. 4.1 are given by

$$|F(\omega)| = \frac{1}{\sqrt{a^2 + \omega^2}} \text{ and } \angle F(\omega) = -\tan^{-1}\left(\frac{\omega}{a}\right). \tag{4.19}$$

Note: It is observed that if $a < 0$, the expression goes to infinity and the Fourier transform does not exist.

The inverse Fourier transform of $F(\omega)$ is given by

$$f(t) = F^{-1}\left(\frac{1}{a + j\omega}\right)$$

$$= \frac{1}{2\pi}\int_{-\infty}^{\infty} \frac{1}{a + j\omega} e^{j\omega t} d\omega = e^{-at}u(t) \tag{4.20}$$

Or, this can be represented as a Fourier transform pair

$$e^{-at}u(t) \longleftrightarrow \frac{1}{a + j\omega}. \tag{4.21}$$

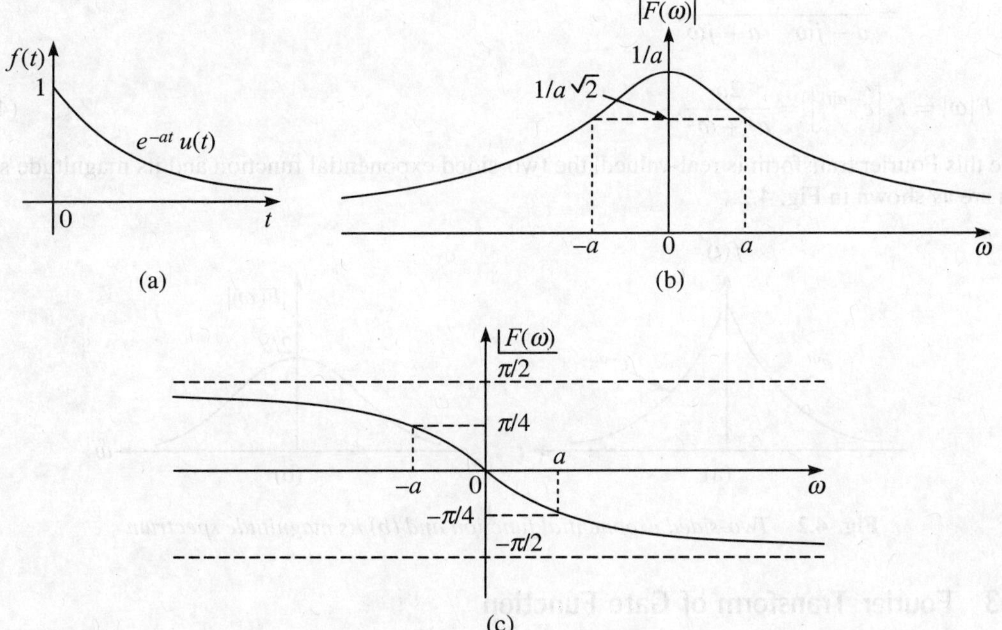

Fig. 4.1 *(a) The exponential function, (b) the magnitude and (c) the phase spectrum of its Fourier transform*

4.4.2 Fourier Transform of Two-sided Exponential Function

Consider a two-sided exponential function $f(t) = e^{-a|t|}$, $a > 0$. The function can be expressed as

$$f(t) = \begin{cases} e^{-at} & \text{for } t \geq 0 \\ e^{at} & \text{for } t \leq 0. \end{cases} \tag{4.22}$$

The Fourier transform is

$$\begin{aligned}
\mathcal{F}[f(t)] = F(\omega) &= \int_{-\infty}^{+\infty} f(t)e^{-j\omega t}dt \\
&= \int_{-\infty}^{0} f(t)e^{-j\omega t}dt + \int_{0}^{+\infty} f(t)e^{-j\omega t}dt \\
&= \int_{-\infty}^{0} e^{at}e^{-j\omega t}dt + \int_{0}^{+\infty} e^{-at}e^{-j\omega t}dt \\
&= \int_{-\infty}^{0} e^{(a-j\omega)t}dt + \int_{0}^{+\infty} e^{-(a+j\omega)t}dt \\
&= \left(\frac{e^{(a-j\omega)t}}{a-j\omega}\right)_{-\infty}^{0} + \left(\frac{e^{(a+j\omega)t}}{-(a+j\omega)}\right)_{0}^{+\infty} \\
&= \frac{1}{a-j\omega} + \frac{1}{a+j\omega}.
\end{aligned}$$

$$\therefore \quad F[\omega] = F\left[e^{-a|t|}\right] = \frac{2a}{a^2+\omega^2}. \tag{4.23}$$

Since this Fourier transform is real-valued, the two-sided exponential function and its magnitude spectrum are as shown in Fig. 4.2.

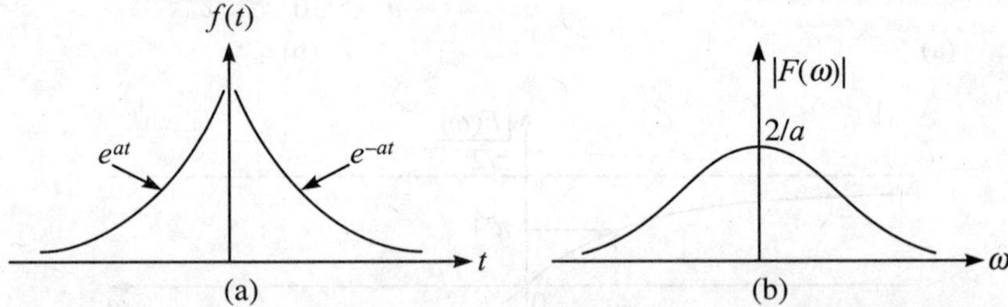

Fig. 4.2 *Two-sided exponential function and (b) its magnitude spectrum*

4.4.3 Fourier Transform of Gate Function

A gate function is a rectangular function defined as

$$G_\tau(t) = rect\left(\frac{t}{\tau}\right) = \Pi\left(\frac{t}{\tau}\right) = \begin{cases} 1 & |t| < \frac{\tau}{2} \\ 0 & \text{elsewhere} \end{cases} \tag{4.24}$$

where τ is pulse width.

It can be expressed in terms of unit step functions as

$$G_\tau(t) = u(t+\tau/2) - u(t-\tau/2). \qquad (4.25)$$

The Fourier transform is

$$G_\tau(\omega) = \int_{-\infty}^{+\infty} f(t)e^{-j\omega t}dt = \int_{-\infty}^{+\infty} (u(t+\tau/2) - u(t-\tau/2))e^{-j\omega t}dt$$

$$= \int_{-\tau/2}^{\tau/2} e^{-j\omega t}dt = \frac{1}{j\omega}\left(\frac{j\omega\tau}{e^2} - \frac{-j\omega\tau}{e^2}\right)$$

$$= \frac{2}{\omega}\left(\frac{e^{\frac{j\omega\tau}{2}} - e^{\frac{-j\omega\tau}{2}}}{j2}\right) = \frac{2}{\omega}\sin\left(\frac{\omega\tau}{2}\right).$$

Multiplying and dividing by τ we get

$$G_\tau(\omega) = \tau\frac{\sin\left(\frac{\omega\tau}{2}\right)}{\frac{\omega\tau}{2}} = \tau\,\mathrm{Sa}\left(\frac{\omega\tau}{2}\right).$$

In terms of linear frequency f,

$$G_\tau(f) = \tau\frac{\sin\left(\frac{2\pi\tau}{2}\right)}{\frac{2\pi\tau}{2}} = \tau\frac{\sin(\pi\tau f)}{\pi\tau f} = \tau\,\mathrm{sinc}(\tau f)$$

$$G_\tau(f) = \tau\,\mathrm{sinc}(\tau f).$$

Thus the Fourier transform of a Gate function is a sampling function that is real and an even function of ω. Gate function and its spectrum are shown in Fig. 4.3. It is observed that as τ increases $G_\tau(t)$ becomes broader, while the first lobe of its spectrum becomes narrower with increasing peak. As $\tau \to \infty$, the spectrum converges to an impulse function.

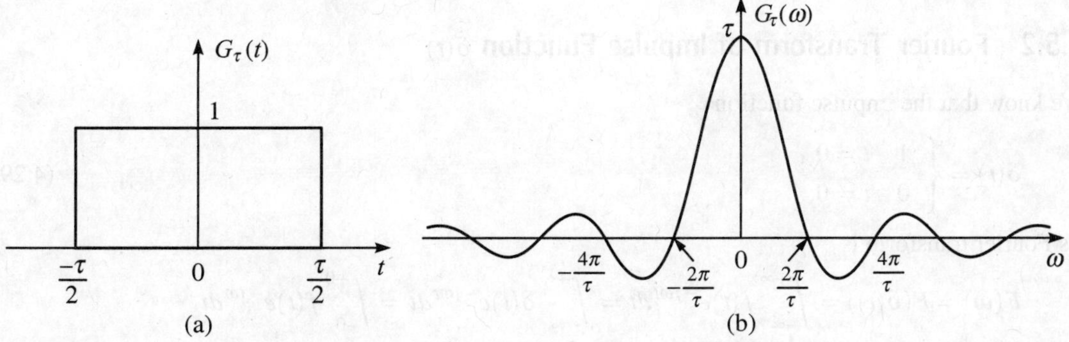

Fig. 4.3 *(a) Gate function and (b) its spectrum*

The Fourier transform pair for gate function is

$$G_\tau(t) \longleftrightarrow \tau\mathrm{Sa}\left(\frac{\omega\tau}{2}\right) \text{ or } G_\tau(t) \longleftrightarrow \tau\,\mathrm{sin}c(\tau f). \qquad (4.26)$$

The practical bandwidth of the gate function corresponds to the first zero crossing in the spectrum. Therefore, the bandwidth of the pulse or the gate function is $\frac{2\pi}{\tau}$ radians/s or $\frac{1}{\tau}$ Hz. Thus, we can write

$$\text{Bandwidth of the gate function} = B = \frac{1}{\tau} = \frac{1}{\text{Pulse width}}\text{Hz}. \tag{4.27}$$

4.5 Fourier Transforms Involving Impulse Function and Signum Function

4.5.1 Singularity Functions

Singularity functions are a class of discontinuous functions or functions whose derivatives are discontinuous at its points of singularity. A singularity is a point at which a function does not possess a derivative. Some of the common singularity functions are:

1. Impulse function
2. Step function
3. Ramp function and its higher derivatives

If a function has a jump discontinuity at a point $t = t_0$, then the function does not possess a derivative at that point. For example, a unit step function has no derivative at the origin. To overcome this problem, an impulse function designated as $\delta(t)$, also called the Dirac delta function, of strength equal to the amount of discontinuity is introduced as a derivative of the function at the discontinuity point. Then the derivative of the unit step function at the origin is an impulse function of unit strength, i.e., the unit impulse function.

$$\frac{du(t)}{dt} = \delta(t). \tag{4.28}$$

Hence at the point of discontinuity an impulse function is used to represent the Fourier transform. (The impulse, step and ramp functions have already been dealt with in detail in Chapters 1 and 2.)

4.5.2 Fourier Transform of Impulse Function $\delta(t)$

We know that the impulse function

$$\delta(t) = \begin{cases} 1 & t = 0 \\ 0 & t \neq 0. \end{cases} \tag{4.29}$$

Its Fourier transform is

$$F(\omega) = F(\delta(t)) = \int_{-\infty}^{+\infty} f(t)e^{-j\omega t}dt = \int_{-\infty}^{+\infty} \delta(t)e^{-j\omega t}dt = \int_{-0}^{+0} f(t)e^{-j\omega}dt$$

Using the sifting property of impulse function,

$$\delta(\omega) = [e^{-j\omega t}]_{t=0} = 1. \tag{4.30}$$

Hence, the Fourier transform of an impulse function is unity. Figure 4.4 shows an impulse function and its Fourier transform. It can be see that the impulse function contains the entire frequency component ($-\infty$ to $+\infty$) of unity magnitude and zero phase, i.e., the bandwidth of the delta function is infinite.

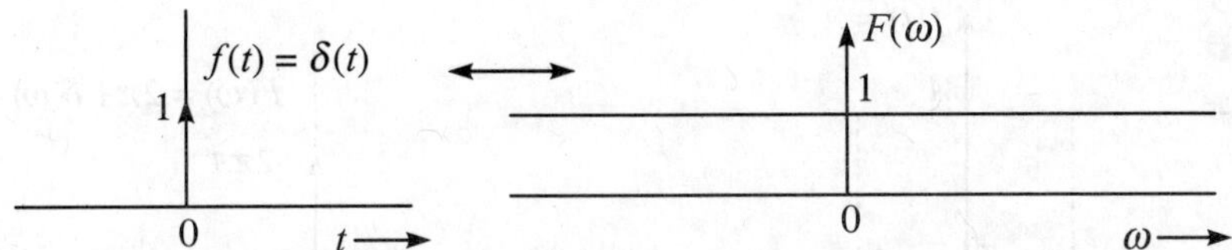

Fig. 4.4 *Impulse function and its Fourier transform*

4.5.3 Fourier Transform of a Constant Function or the DC Signal

Consider a constant function

$$f(t) = A \quad -\infty < t < \infty \tag{4.31}$$

where A is a constant over an entire interval $(-\infty, \infty)$ as shown in Fig. 4.5. This function is not absolutely integrable, but its Fourier transform can be determined using the concept of delta (impluse) function.

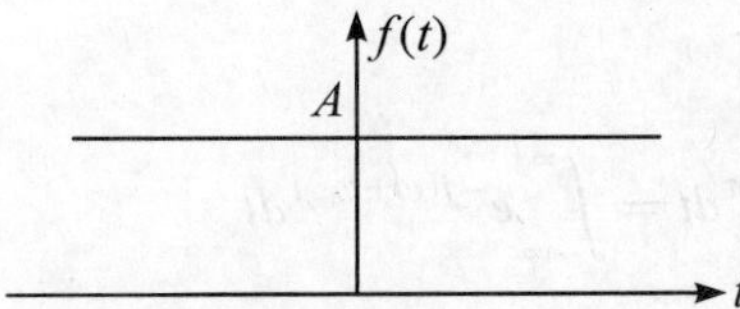

Fig. 4.5 *Constant function*

Consider a gate function of height A and width τ as shown in Fig. 4.6(a). The Fourier transform of the gate function is given as

$$G_\tau(t) \longleftrightarrow A\tau\,\mathrm{Sa}\left(\frac{\omega\tau}{2}\right).$$

A constant function can be derived from the gate function as the width τ tends to infinity. Therefore, the Fourier transform of a constant function is,

$$\begin{aligned}\mathcal{F}[A] &= \lim_{\tau\to\infty} \mathcal{F}\left[AG_\tau(t)\right] = \lim_{\tau\to\infty} A\tau\,\mathrm{Sa}\left(\frac{\omega\tau}{2}\right)\\ &= 2\pi A \lim_{\tau\to\infty} \frac{\tau}{2\pi}\,\mathrm{Sa}\left(\frac{\omega\tau}{2}\right).\end{aligned}$$

As in the limit $\tau \to \infty$, the sampling function approaches a delta function. That is,

$$\lim_{\tau\to\infty} \frac{\tau}{2\pi}\mathrm{Sa}\left(\frac{\omega\tau}{2}\right) = \delta(\omega). \tag{4.32}$$

Therefore,

$$\mathcal{F}[A] = 2\pi A\delta(\omega) \quad \text{or} \quad A \longleftrightarrow 2\pi A\delta(\omega) \tag{4.33}$$

The Fourier transform plotted in Fig. 4.6(b) shows that the delta function contains a single frequency component at $\omega = 0$, which is a pure dc signal. When plotted as a function of angular frequency ω, the magnitude of the dc signal is $2\pi A$. Considering it with respect to linear frequency f, the magnitude would be A.

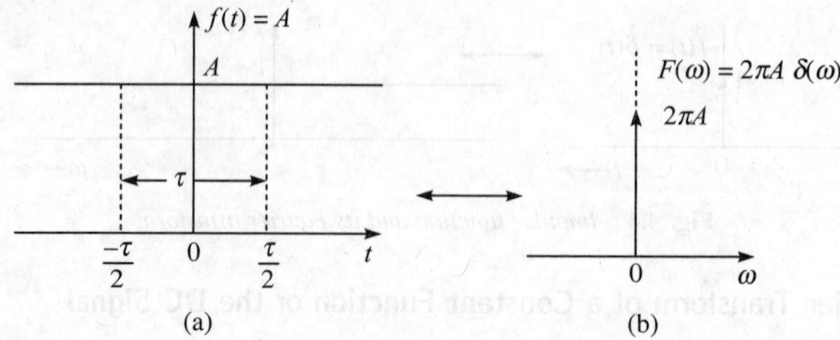

Fig. 4.6 *(a) Deriving a constant form gate function and (b) its Fourier transform*

4.5.4 Fourier Transform of Complex Exponential Function

Consider a complex exponential function given as

$$f(t) = e^{\mp j\omega_0 t} \quad -\infty < t < \infty \tag{4.34}$$

The Fourier transform is

$$\mathcal{F}[e^{\pm j\omega_0 t}] = \int_{-\infty}^{\infty} e^{\pm j\omega_0 t} e^{-j\omega t}\, dt = \int_{-\infty}^{\infty} e^{-j(\omega \pm \omega_0)t}\, dt$$

$$\mathcal{F}[e^{\pm j\omega_0 t}] = \int_{-\infty}^{\infty} e^{\pm j\omega_0 t} e^{-j\omega t}\, dt$$

$$= \int_{-\infty}^{\infty} e^{-j(\omega \mp \omega_0)}\, dt.$$

Since the function is not absolutely integrable in the interval $-\infty < t < \infty$, using limits as $\tau \to \infty$,

$$F(\omega) = \lim_{\tau\to\infty} \int_{-\tau/2}^{\tau/2} e^{-j(\omega\pm\omega_0)t}\, dt =$$

$$= \lim_{\tau\to\infty} \frac{2}{(\omega\pm\omega_0)} \left(\frac{e^{\frac{j(\omega\pm\omega_0)}{2}} - e^{\frac{-j(\omega\pm\omega_0)}{2}}}{2j} \right) = \lim_{\tau\to\infty} \frac{2}{(\omega\pm\omega_0)} \sin\left(\frac{(\omega\pm\omega_0)\tau}{2}\right)$$

$$= \lim_{\tau\to\infty} \tau \frac{\sin\left(\frac{(\omega\pm\omega_0)\tau}{2}\right)}{(\omega\pm\omega_0)\tau/2} = \lim_{\tau\to\infty} \tau \,\mathrm{Sa}\left(\frac{(\omega\pm\omega_0)\tau}{2}\right)$$

$$= 2\pi \lim_{\tau\to\infty} \frac{\tau}{2\pi} \,\mathrm{Sa}\left(\frac{(\omega\pm\omega_0)\tau}{2}\right).$$

We know that as the limit $\tau \to \infty$, sampling function approaches a delta function. That is,

$$\lim_{\tau\to\infty} \frac{\tau}{2\pi} \,\mathrm{Sa}\left(\frac{\omega\tau}{2}\right) = \delta(\omega).$$

Therefore, $F(\omega) = 2\pi\delta(\omega \pm \omega_0)$.

The Fourier transform pair is

$$e^{\mp j\omega_0 t} \longleftrightarrow 2\pi\delta(\omega \pm \omega_0) \text{ or } e^{\pm jf_0 t} \longleftrightarrow \delta(f \pm f_0). \tag{4.35}$$

Example 4.1

Find the Fourier transform of $\cos(\omega_0 t)$.

Solution

Given the function $f(t) = \cos\omega_0 t$, the Fourier transform is

$$F[\cos(\omega_0 t)] = \int_{-\infty}^{\infty} \cos(\omega_0 t)e^{-j\omega t}dt = \frac{1}{2}\int_{-\infty}^{\infty} \left(e^{j\omega_0 t} + e^{-j\omega_0 t}\right)e^{-j\omega t}dt$$

$$F(\omega) = \int_{-\infty}^{\infty} \cos(\omega_0 t)dt = \frac{1}{2}\int_{-T/2}^{T/2} (e^{j\omega_0 t} + e^{-j\omega_0 t})e^{-j\omega t}dt$$

$$= \frac{1}{2}\int_{-\infty}^{\infty} e^{j\omega_0 t}e^{-j\omega_0 t}dt + \frac{1}{2}\int_{-\infty}^{\infty} e^{-j\omega_0 t}e^{-j\omega t}dt$$

$$= \frac{1}{2}F[e^{j\omega_0 t}] + \frac{1}{2}F[e^{-j\omega_0 t}] = \frac{2\pi}{2}[\delta(\omega - \omega_0) + \delta(\omega + \omega_0)].$$

$$F(\omega) = \pi[\delta(\omega - \omega_0) + \delta(\omega + \omega_0)]$$

$$\therefore \quad \cos(\omega_0 t) \longleftrightarrow \pi[\delta(\omega - \omega_0) + \delta(\omega + \omega_0)]$$

or

$$\cos(\omega_0 t) \longleftrightarrow \frac{1}{2}[\delta(f - f_0) + \delta(f + f_0)].$$

Figure 4.7 shows a cosine function and its Fourier transform.

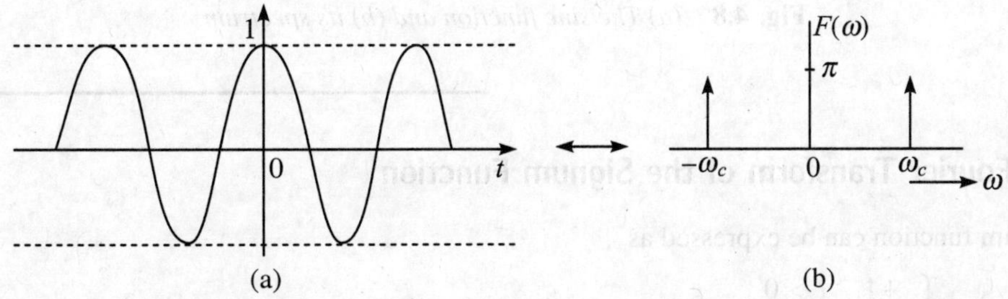

Fig. 4.7 *(a) Cosine wave signal and (b) its spectrum*

Example 4.2

Find the Fourier transform of $\sin(\omega_0 t)$.

Solution

Given the function $f(t) = \sin \omega_0 t$, the Fourier transform is

$$F[\sin(\omega_0 t)] = \int_{-\infty}^{\infty} \sin(\omega_0 t) e^{-j\omega t} dt = \frac{1}{2j} \int_{-\infty}^{\infty} \left(e^{j\omega_0 t} - e^{-j\omega_0 t}\right) e^{-j\omega t} dt$$

$$= \frac{1}{2j} \int_{-\infty}^{\infty} e^{j\omega_0 t} e^{-j\omega t} dt - \frac{1}{2j} \int_{-\infty}^{\infty} e^{-j\omega_0 t} e^{-j\omega t} dt$$

$$= \frac{1}{2j} \mathcal{F}\left[e^{j\omega_0 t}\right] - \frac{1}{2j} \mathcal{F}\left[e^{-j\omega_0 t}\right] = \frac{2\pi}{2j} [\delta(\omega - \omega_0) - \delta(\omega + \omega_0)]$$

$$= \frac{\pi}{j} [\delta(\omega - \omega_0) - \delta(\omega + \omega_0)].$$

Therefore, $\sin(\omega_0 t) \longleftrightarrow \frac{\pi}{j} [\delta(\omega - \omega_0) - \delta(\omega + \omega_0)]$

or $\sin(\omega_0 t) \longleftrightarrow \frac{1}{2j} [\delta(f - f_0) - \delta(f + f_0)]$.

Figure 4.8 shows a sine function and its Fourier transform.

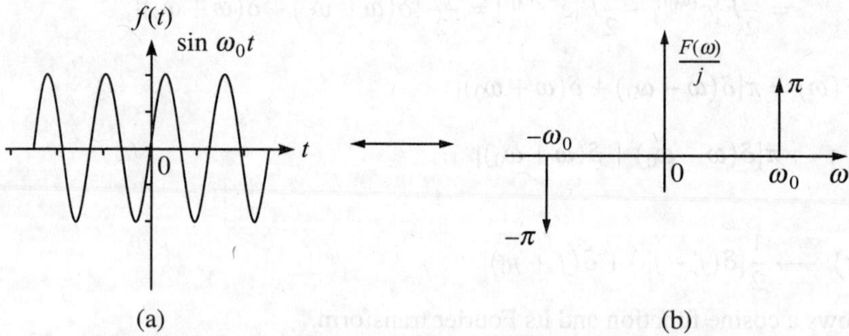

Fig. 4.8 *(a) The sine function and (b) its spectrum*

4.5.5 Fourier Transform of the Signum Function

The signum function can be expressed as

$$\operatorname{sgn}(t) = \begin{cases} +1, & t > 0 \\ -1, & t < 0 \end{cases}. \tag{4.36}$$

The Fourier transform is

$$\mathcal{F}[\operatorname{sgn}(t)] = \int_{-\infty}^{\infty} \operatorname{sgn}(t) e^{-j\omega t} dt = -\int_{-\infty}^{0} e^{-j\omega t} dt + \int_{0}^{\infty} e^{-j\omega t} dt.$$

Since the function is not absolutely integrable in the interval $-\infty < t < \infty$, the Fourier transform can be determined by using step function. We know that

$$\operatorname{sgn}(t) = 2u(t) - 1 \quad \text{or} \quad u(t) = \frac{1}{2}[\operatorname{sgn}(t) + 1]. \tag{4.37}$$

Consider the positive half of the signum function expressed as $\lim_{a\to 0} e^{-at}u(t) = 1$ and the negative half expressed as $\lim_{a\to 0} e^{at}u(-t) = 1$ as shown in Fig. 4.9. The total function can be expressed as,

$$\text{sgn}(t) = \lim_{a\to 0}[e^{-at}u(t) - e^{at}u(-t)] \tag{4.38}$$

Now the Fourier transform is

$$\mathcal{F}[\text{sgn}(t)] = \int_{-\infty}^{\infty} \text{sgn}(t)e^{-j\omega t}\,dt = \int_{-\infty}^{\infty} \lim_{a\to 0}\left[e^{-at}u(t) - e^{at}u(-t)\right]e^{-j\omega t}\,dt$$

$$= \lim_{a\to 0}\left[\int_{0}^{\infty} e^{-at}e^{-j\omega t}\,dt - \int_{-\infty}^{0} e^{at}e^{-j\omega t}\,dt\right] = \lim_{a\to 0}\left[\int_{0}^{\infty} e^{-(a+j\omega)t}\,dt - \int_{-\infty}^{0} e^{(a-j\omega)t}\,dt\right]$$

$$= \lim_{a\to 0}\left[\frac{-2j\omega}{a^2+\omega^2}\right] = \frac{2}{j\omega}. \tag{4.39}$$

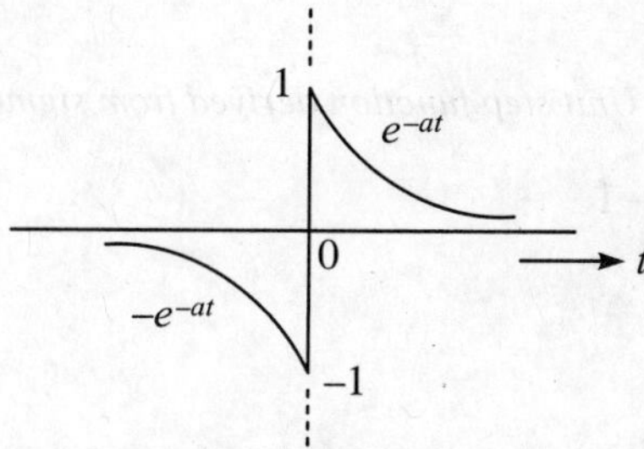

Fig. 4.9 *Deriving a signum function from exponential function*

The signum function and its spectrum is shown in Fig. 4.10.

The Fourier transform pair is

$$\text{sgn}(t) \longleftrightarrow \frac{2}{j\omega} \quad \text{or} \quad \text{sgn}(t) \longleftrightarrow \frac{1}{j\pi f}. \tag{4.40}$$

Fig. 4.10 *Signum function and its Fourier transform*

4.5.6 Fourier Transform of the Unit Step Function

The unit step function can be expressed as

$$u(t) = \begin{cases} 1, & t > 0 \\ 0, & t < 0. \end{cases} \tag{4.41}$$

The Fourier transform is

$$\mathcal{F}[u(t)] = \int_{-\infty}^{\infty} u(t)e^{-j\omega t}dt.$$

Since the function is not absolutely integrable, in the interval $-\infty < t < \infty$, the Fourier transform can be determined by using signum function. The step function can be taken as the sum of a signum function and a constant function, as shown in Fig. 4.11.

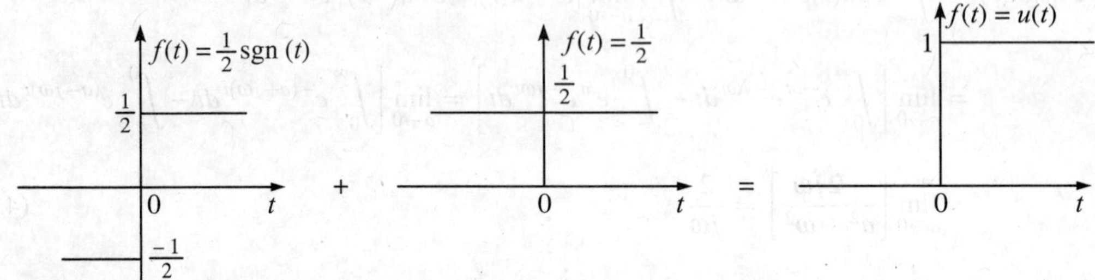

Fig. 4.11 *Unit step function derived from signum function*

We know that $\text{sgn}(t) = 2u(t) - 1$.

$$u(t) = \frac{1}{2}[\text{sgn}(t) + 1] \tag{4.42}$$

Its Fourier transform is

$$\mathcal{F}[u(t)] = \frac{1}{2}\mathcal{F}[1] + \frac{1}{2}\mathcal{F}[\text{sgn}(t)].$$

As the Fourier transforms $\mathcal{F}[1] = 2\pi\delta(\omega)$ and $\mathcal{F}[\text{sgn}(t)] = \dfrac{2}{j\omega}$, hence

$$\mathcal{F}[u(t)] = \pi\delta(\omega) + \frac{1}{j\omega}.$$

The unit step function and its spectrum are shown in Fig. 4.12.

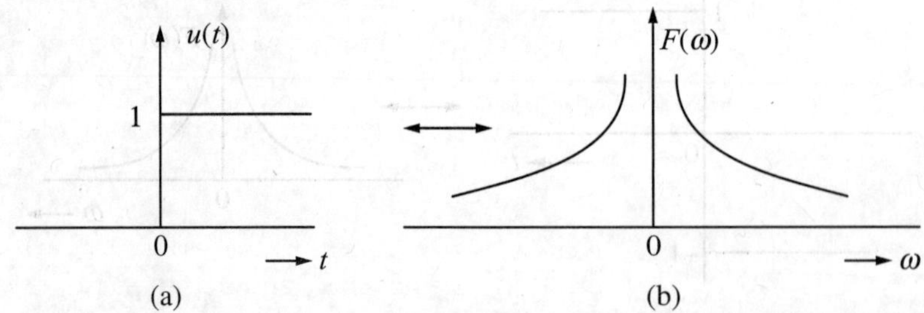

Fig. 4.12 *(a) The unit step function and (b) its spectrum*

The Fourier transform pair is

$$u(t) \longleftrightarrow \pi\delta(\omega) + \frac{1}{j\omega} \quad \text{or} \quad u(t) \longleftrightarrow \frac{1}{2}\delta(f) + \frac{1}{j2\pi f} \tag{4.43}$$

Figure 4.12(b) shows the spectrum which consists of a frequency component at $\omega = 0$ (equivalent to dc signal), along with other frequency components extending up to infinity. The unit step function appears to be a dc signal for $t > 0$, and hence the spectrum has an impulse at $\omega = 0$. However, it is not a pure dc signal because it is zero for $t < 0$ and abruptly discontinuous at $t = 0$, which is the source for other high frequency components in the spectrum.

4.6 Fourier Transform of Periodic Signals

Here, the relationship between the Fourier exponential series and Fourier transform will be discussed, which will be useful for finding the periodic signal expression in terms of the exponential Fourier series through the evaluation of the Fourier transform. Actually, the Fourier transform of a periodic function does not exist since it fails to satisfy Dirichlet's conditions, but using impulse functions at discontinuity, the Fourier transform for periodic functions can be expressed as a sum of Fourier transforms of its individual components.

Fourier Transform Using Fourier Series

Consider an arbitrary periodic signal $f_T(t)$ with period T. The Fourier series is $f_T(t) = \sum_{-\infty}^{\infty} c_n e^{jn\omega_0 t}$, where $c_n = \frac{1}{T}\int_{-T/2}^{T/2} f_T(t)e^{-jn\omega_0 t}dt$ and $T = \frac{2\pi}{\omega_0}$.

Applying Fourier transform on both sides,

$$\mathcal{F}[f_T(t)] = \mathcal{F}\left[\sum_{n=-\infty}^{\infty} c_n e^{jn\omega_0 t}\right] = \sum_{n=-\infty}^{\infty} c_n \mathcal{F}[e^{jn\omega_0 t}].$$

We know that the Fourier transform of exponential signal is

$$\mathcal{F}\left[e^{jn\omega_0 t}\right] = 2\pi\delta(\omega - n\omega_0).$$

$$\therefore \quad F_T(\omega) = 2\pi \sum_{n=-\infty}^{\infty} c_n \delta(\omega - n\omega_0) \tag{4.44}$$

Equation (4.44) represents the Fourier transform of a periodic function consisting of infinite sum of impulses at harmonics with a strength equal to 2π times the Fourier coefficient c_n as shown in Fig. 4.13. Also, we know that the Fourier transform of a signal $f(t)$ for the period T is

$$F(\omega) = \int_{-\infty}^{\infty} f(t)e^{-j\omega t}dt = \int_{-T/2}^{T/2} f(t)e^{-j\omega t}dt.$$

Also, the coefficient of Fourier exponential series is

$$c_n = \frac{1}{T}\int_{-T/2}^{T/2} f(t)e^{-jn\omega_0 t}dt. \tag{4.45}$$

Substituting the expression for $F(\omega)$, c_n can be expressed as

$$c_n = \frac{1}{T}\int_{-T/2}^{T/2} f(t)e^{-j\omega t}dt\bigg|_{\omega=n\omega_0}$$

$$\text{Or} \quad c_n = \frac{1}{T}F(\omega)\bigg|_{\omega=n\omega_0}. \tag{4.46}$$

This result provides us an another way to find the coefficients of the Fourier exponential series from Fourier transform.

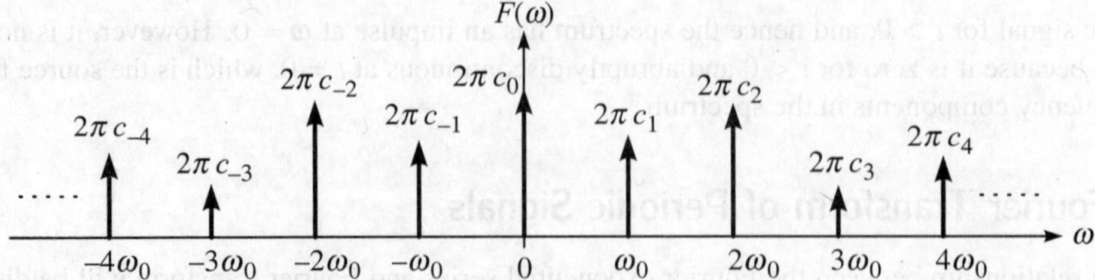

Fig. 4.13 *The line spectrum of c_n or the Fourier transform of the periodic function*

Fourier Transform Using Convolution Theorem

Fourier transform for a periodic function can be obtained by applying the convolution theorem. Consider an arbitrary signal $f(t)$ with duration $|t| < T/2$, where $T = \dfrac{2\pi}{\omega_0}$.

The periodic signal $f_T(t)$ with period T can be obtained by taking the convolution of the impulse train with $f(t)$. That is,

$$f_T(t) = \sum_{m=-\infty}^{\infty} \delta(t - mT) * f(t) = \sum_{m=-\infty}^{\infty} f(t - mT).$$

Applying the Fourier transform,

$$F_T(\omega) = F\left(\sum_{m=-\infty}^{\infty} \delta(t - mT)\right) F(\omega) = \omega_0 F(\omega)\left(\sum_{m=-\infty}^{\infty} \delta(\omega - m\omega_0)\right)$$

$$F_T(\omega) = \omega_0\left(\sum_{m=-\infty}^{\infty} F(m\omega_0)\delta(\omega - m\omega_0)\right).$$

The Fourier transform pair is

$$\sum_{m=-\infty}^{\infty} f(t - mT) \longleftrightarrow \omega_0\left(\sum_{m=-\infty}^{\infty} F(m\omega_0)\delta(\omega - m\omega_0)\right), \tag{4.47}$$

where $F(n\omega_0)$ is the amplitude of the Fourier transform at harmonic frequencies.

Example 4.3

Given a periodic signal shown in Fig. 4.14, find out c_n from the Fourier transform of the pulse. Also, find out the Fourier transform of the periodic signal.

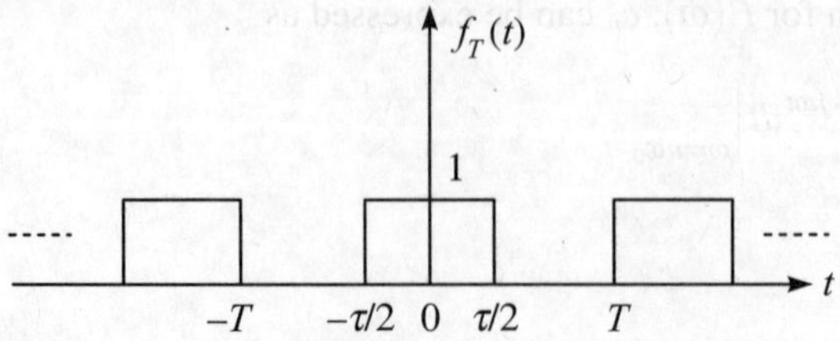

Fig. 4.14 *Periodic signal of rectangular pulse*

Solution

Given that the signal is a periodic rectangular pulse or a gate function, its Fourier transform is

$$F(\omega) = \int_{-T/2}^{T/2} f(t)e^{-j\omega t}dt = \int_{-\tau/2}^{\tau/2} e^{-j\omega t}dt$$

$$= \frac{2}{\omega}\sin\frac{\omega\tau}{2} = \tau\,\text{Sa}\left(\frac{\omega\tau}{2}\right).$$

The Fourier coefficient c_n from the Fourier transform is

$$c_n = \frac{1}{T}F(\omega)\bigg|_{\omega=n\omega_0} = \frac{1}{T}\tau\,\text{Sa}\left(\frac{n\omega_o\tau}{2}\right).$$

We know that the Fourier transform a periodic signal is

$$F_T(\omega) = 2\pi\sum_{n=-\infty}^{\infty} c_n\delta(\omega - n\omega_0)$$

$$F_T(\omega) = \frac{2\pi\tau}{T}\sum_{n=-\infty}^{\infty}\text{Sa}\left(\frac{n\omega_0\tau}{2}\right)\delta(\omega - n\omega_0).$$

4.7 Properties of Fourier Transform

The properties of continuous Fourier transform are useful in solving problems. The important properties are explained below. For the convenience of expression, we will use the notation $f(t) \longleftrightarrow F(\omega)$, or $F(\omega) = \mathcal{F}[f(t)]$, or $f(t) = \mathcal{F}^{-1}[F(\omega)]$ to denote transform pair.

1. Linearity (Superposition)

Consider two signals $f_1(t)$ and $f_2(t)$ with Fourier transforms $F_1(\omega)$ and $F_2(\omega)$ respectively. The linearity property states that the linear combination of the two signals satisfy

$$F[a_1f_1(t) + a_2f_2(t)] = a_1F_1(\omega) + a_2F_2(\omega)$$

$$\text{or}\quad a_1f_1(t) + a_2f_2(t) \longleftrightarrow a_1F_1(\omega) + a_2F_2(\omega), \tag{4.48}$$

where a_1 and a_2 are arbitrary constants.

Proof: Consider two signals $f_1(t)$ and $f_2(t)$ with Fourier transforms $F_1(\omega)$ and $F_2(\omega)$ respectively.

$$\text{Then } F[a_1f_1(t) + a_2f_2(t)] = \int_{-\infty}^{\infty}[a_1f_1(t) + a_2f_2(t)]e^{-j\omega t}dt$$

$$= \int_{-\infty}^{\infty} a_1f_1(t)e^{-j\omega t}dt + \int_{-\infty}^{\infty} a_2f_2(t)e^{-j\omega t}dt$$

$$= a_1F_1(\omega) + a_2F_2(\omega).$$

Example 4.4

Find the Fourier transform of $x(t)$ shown in Fig. 4.15.

Solution

The given function $x(t)$ is a sum of two gate functions

$$x(t) = G_4(t) + G_2(t),$$

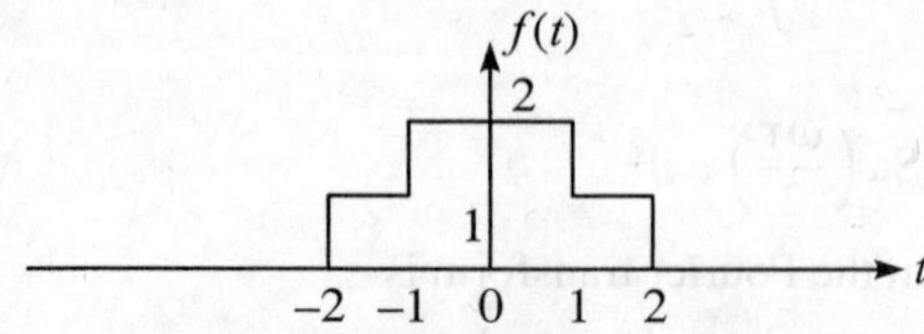

Fig. 4.15 *Sum of gate functions*

where $G_2(t) = u(t+1) - u(t-1)$; and $G_4(t) = u(t+2) - u(t-2)$.

The Fourier transform is

$$X(\omega) = \mathcal{F}[g_4(t) + g_2(t)] = \mathcal{F}[g_4(t)] + \mathcal{F}[g_2(t)].$$

We know that $G_\tau(t) \longleftrightarrow \tau \mathrm{Sa}(\frac{\omega\tau}{2})$.

$$X(\omega) = 4\mathrm{Sa}(2\omega) + 2\mathrm{Sa}(\omega).$$

---

2. Complex Conjugate Property

For any complex-valued signal $f(t)$, the complex conjugate property states that

$$\mathcal{F}[f^*(t)] = F^*(-\omega). \tag{4.49}$$

Proof: Consider any complex-valued signal $f(t)$. The Fourier transform is

$$\mathcal{F}[f^*(t)] = \int_{-\infty}^{\infty} f^*(t)e^{-j\omega t}dt = [\int_{-\infty}^{\infty} f(t)e^{j\omega t}dt]^* = F^*(-\omega).$$

An important consequence of this property is that if the signal $f(t)$ is real valued, then $f^*(t) = f(t)$ and $F^*(-\omega) = F(\omega)$.

3. Time Shifting (Delay) Property

If a signal $f(t)$ with period T has Fourier transform $F(\omega)$, then the time shifting property states that

$$F[f(t \pm t_0)] = F(\omega)e^{\pm j\omega t_0}. \tag{4.50}$$

Changing the variable of integration, let $x = t - t_0$, and $dx = dt$. Then,

$$F[f(t - t_0)] = \int_{-\infty}^{\infty} f(x)e^{-j\omega(x+t_0)}dx$$

$$= e^{-j\omega t_0} \int_{-\infty}^{\infty} f(x)e^{-j\omega t}dx = e^{-j\omega t_0}F(\omega).$$

Similarly, we can prove

$$F[f(t + t_0)] = e^{j\omega t_0}F(\omega). \tag{4.51}$$

Note: Thus if a signal $f(t)$ is delayed in time by t_0, the magnitude spectral density remains unchanged but for a negative phase $(-\omega t_0)$ is added to each frequency component. If t_0 is negative, the time function is advanced in time and the phase spectral density added has a positive slope.

Example 4.5

Determine the Fourier transform of three rectangular pulsesshown in Fig. 4.16.

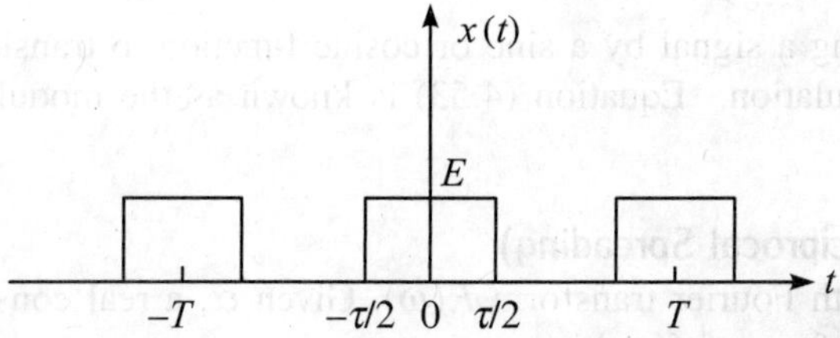

Fig. 4.16 *Pulse train of three rectangular pulses*

Solution

Given $G(\tau)=\left[u\left(t+\frac{\tau}{2}\right)-u\left(t-\frac{\tau}{2}\right)\right]$.

Then, for the three rectangular pulses,

$$x(t)=EG_\tau(t+T)+EG_\tau(t)+EG_\tau(t-T).$$

We know that $G_\tau(t)\longleftrightarrow\tau\,\text{Sa}\left(\frac{\omega\tau}{2}\right)$.

Using time shifting property, the Fourier transform is

$$X(\omega)=EF[G_\tau(t+T)+G_\tau(t)+G_\tau(t-T)].$$

$$=E\left[\tau\,\text{Sa}\left(\frac{\omega\tau}{2}\right)e^{j\omega T}+\tau\,\text{Sa}\left(\frac{\omega\tau}{2}\right)+\tau\,\text{Sa}\left(\frac{\omega\tau}{2}\right)e^{-j\omega T}\right]$$

$$=E\tau\left[\text{Sa}\left(\frac{\omega\tau}{2}\right)\left[1+e^{j\omega T}+e^{-j\omega T}\right]\right]=E\tau\,\text{Sa}\left(\frac{\omega\tau}{2}\right)[1+2\cos(\omega T)].$$

4. Frequency Shifting (Modulation) Property

The dual of the delay property is the frequency-translation property. Let $F(\omega)$ be the Fourier transform of a signal $f(t)$ with period T. Then the frequency shifting property states that,

$$\mathcal{F}\left[f(t)e^{\pm j\omega_0 t}\right]=F(\omega\mp\omega_0). \tag{4.52}$$

Consider a signal $f(t)$ with period T, and $f(t)\longleftrightarrow F(\omega)$.Then the Fourier transform of $f(t)e^{\pm j\omega_0 t}$ is

$$=\int_{-\infty}^{\infty}f(t)e^{\pm j\omega_0 t}e^{-j\omega t}dt=\int_{-\infty}^{\infty}f(t)e^{-j(\omega\mp\omega_0)t}dt$$

$$=F(\omega\mp\omega_0).$$

Therefore, multiplying a time function by $e^{j\omega_0 t}$ causes its spectral density to be translated in frequency by ω_0 radians/s. For real-valued $f(t)$, it is now a relatively simple matter to find the Fourier transform.

$$\text{Re}\{f(t)e^{j\omega_0 t}\}=f(t)\cos\omega_0 t=\frac{1}{2}f(t)\left[e^{j\omega_0 t}+e^{-j\omega_0 t}\right].$$

Using the frequency translation property, we get

$$\mathcal{F}[f(t)\cos\omega_0 t] = \frac{1}{2}[\mathcal{F}(\omega+\omega_0)+\mathcal{F}(\omega-\omega_0)]. \tag{4.53}$$

This process of multiplying a signal by a sine or cosine function to translate the spectral density is known as amplitude modulation. Equation (4.53) is known as the modulation property of Fourier transform.

5. Scaling Property (Reciprocal Spreading)

Consider a signal $f(t)$ with Fourier transform $F(\omega)$. Given α, a real constant, the scaling property states that the Fourier transform of $f(\alpha t)$ is

$$\mathcal{F}[f(\alpha t)] = \frac{1}{|\alpha|}F\left[\frac{\omega}{a}\right]. \tag{4.54}$$

Proof: For positive α, changing the variable of integration to $x = \alpha t$ and $dx = dt$,

$$\mathcal{F}[f(\alpha t)] = \int_{-\infty}^{\infty} f(x)e^{-j\omega x/\alpha}\frac{dx}{\alpha} = \frac{1}{\alpha}\int_{-\infty}^{\infty} f(x)e^{-jx\omega/\alpha}dx$$

$$= \frac{1}{\alpha}F\left[\frac{\omega}{\alpha}\right], \quad \text{for } \alpha > 0.$$

When α is negative,

$$F[f(\alpha t)] = \int_{-\infty}^{\infty} f(x)e^{-j\omega x/\alpha}\frac{dx}{\alpha} = \frac{-1}{\alpha}\int_{-\infty}^{\infty} f(x)e^{-j\omega x/\alpha}dx$$

$$F[f(\alpha t)] = -\frac{1}{\alpha}F\left[\frac{\omega}{\alpha}\right], \quad \text{for } \alpha < 0.$$

These two cases can be combined into the more compact form

$$\mathcal{F}[f(\alpha t)] = \frac{1}{|\alpha|}F\left[\frac{\omega}{\alpha}\right].$$

Note:

1. When $\alpha = -1, \mathcal{F}[f(-t)] = F(-\omega)$. (4.55)
2. If α is positive and greater than unity, $f(\alpha t)$ is a compressed version of $f(t)$ and its spectral density is expanded in frequency by $1/\alpha$. The magnitude of the spectral density also changes, an effect necessary to maintain an energy balance between the two domains.
3. If α is positive but less than unity, $f(\alpha t)$ is an expanded version of $f(t)$ and its spectral density is compressed.
4. If α is negative, $f(\alpha t)$ is reversed in time compared to $f(t)$ and is expanded or compressed depending on whether $|\alpha|$ is less than or greater than unity.

Example 4.6

If $\mathcal{F}[e^{-at}u(t)] = \dfrac{1}{a+j\omega}$, find the $F(\omega)$ of $e^{-a|t|}$.

Solution

Given $F\left[e^{-at}u(t)\right] = \dfrac{1}{a+j\omega}$, the given function $e^{-a|t|}$ can be written as

$$e^{-a|t|} = e^{-at}u(t) + e^{at}u(-t)$$

The Fourier transform is

$$F\left[e^{-a|t|}\right] = F_1(\omega) + F_1(-\omega).$$

since $F[f(-t)] = F(-\omega)$,

therefore $F\left[e^{-a|t|}\right] = \dfrac{1}{a+j\omega} + \dfrac{1}{a-j\omega} = \dfrac{2a}{a^2+\omega^2}$.

---

6. Duality (Symmetry) Property

One of the features of the transform is the symmetry in variables t and ω. This symmetry can be used to great advantage in the table of transform pair. If $\mathcal{F}[f(t)] = F(\omega)$, then

$$\mathcal{F}[F(t)] = 2\pi f(-\omega). \tag{4.56}$$

Proof: Consider a signal $f(t)$ with period T and Fourier transform $F(\omega)$. We we know that the inverse Fourier transform is $f(t) = \dfrac{1}{2\pi}\displaystyle\int_{-\infty}^{\infty} F(\omega)e^{j\omega t}\,d\omega$.

Then $2\pi f(-t) = \displaystyle\int_{-\infty}^{\infty} F(vt)e^{-jvt}dv$,

where we have replaced the variable ω by v and t by $-t$. Now if we replace t by ω, we have

$$2\pi f(-\omega) = \int_{-\infty}^{\infty} F(vt)e^{jvt}dv.$$

Finally, replace v by t. Thus,

$$2\pi f(-\omega) = \int_{-\infty}^{\infty} F(t)e^{-jt\omega}dt$$

or $\mathcal{F}[F(t)] = 2\pi f(-\omega)$.

Note: If $f(t)$ is an even function, $f(t) = f(-t)$. Then the above result reduces to

$$\mathcal{F}[F(t)] = 2\pi f(\omega). \tag{4.57}$$

Example 4.7

Given $f(t) = 1$, find the inverse Fourier transform of $X(\omega)$ using duality property.

Solution

We know that the Fourier transform of $x(t) = 1$ is $X(\omega) = 2\pi\delta(\omega)$. Now replacing t by ω we have

$$X(t) = 2\pi x(-\omega) = 2\pi \times 1.$$

Therefore, $2\pi\delta(t) \longleftrightarrow 2\pi$

or $\delta(t) \longleftrightarrow 1$.

Thus, $\mathcal{F}^{-1}[X(\omega)] = \mathcal{F}^{-1}[1] = \delta(t)$.

---

7. Symmetry Property

Consider a signal $f(t)$ with period T and Fourier transform $F(\omega)$. Then the symmetry property states that for real and even signals, the Fourier transforms are real and even functions of frequency and for real and odd signals, the Fourier transforms are imaginary and odd functions of frequency.

If $f(t) = f_e(t) + f_o(t)$ and $F(\omega) = F_R(\omega) + jF_I(\omega)$, then

$$\mathcal{F}[f_e(t)] = F_R(\omega) \text{ and } \mathcal{F}[f_o(t)] = jF_I(\omega) \tag{4.58}$$

where $f_e(t)$ and $f_o(t)$ are even and odd parts of $f(t)$ respectively, and $F_R(\omega)$ and $F_I(\omega)$ are real and imaginary parts of $F(\omega)$ respectively.

Proof: Let a signal $f(t)$ be real and has Fourier transform $F(\omega)$.

$$\mathcal{F}[f(t)] = F(\omega) = F_R(\omega) + jF_I(\omega)$$

and $\mathcal{F}[f(-t)] = F^*(\omega) = F_R(\omega) - jF_I(\omega)$

(a) Even part of $f(t)$: We know that $f_e(t) = \frac{1}{2}(f(t) + f(-t))$. Then

$$\mathcal{F}[f_e(t)] = \mathcal{F}\left[\frac{1}{2}(f(t) + f(-t))\right] = \frac{1}{2}(\mathcal{F}[f(t)] + \mathcal{F}[f(-t)])$$

$$= \frac{1}{2}(F_R(\omega) + jF_I(\omega) + F_R(\omega) - jF_I(\omega))$$

$$\mathcal{F}[f_e(t)] = F_R(\omega).$$

(b) Odd part of $f(t)$: We know that $f_o(t) = \frac{1}{2}(f(t) - f(-t))$. Then

$$\mathcal{F}[f_o(t)] = \mathcal{F}\left[\frac{1}{2}(f(t) - f(-t))\right] = \frac{1}{2}(\mathcal{F}[f(t)] - \mathcal{F}[f(-t)])$$

$$= \frac{1}{2}(F_R(\omega) + jF_I(\omega) - F_R(\omega) + jF_I(\omega))$$

$$\mathcal{F}[f_o(t)] = jF_I(\omega) \quad \text{(Proved).}$$

8. Differentiation Property in Time Domain

Let $F(\omega)$ be the Fourier transform of a signal $f(t)$ with period T. If $\frac{df(t)}{dt}$ is absolutely integrable, then,

$$\mathcal{F}\left[\frac{d}{dt}f(t)\right] = j\omega F(\omega) \quad and \quad \mathcal{F}\left[\frac{d^n}{dt^n}f(t)\right] = (j\omega)^n F(\omega). \tag{4.59}$$

Proof: Assume that $\frac{df(t)}{dt}$ is absolutely integrable. Using inverse Fourier transform, we have

$$f(t) = \frac{1}{2\pi}\int_{-\infty}^{\infty} F(\omega)e^{j\omega t}\,d\omega.$$

Differentiating both sides with respect to t,

$$\frac{df(t)}{dt} = \frac{1}{2\pi}\int_{-\infty}^{\infty} F(\omega)\frac{d}{dt}e^{j\omega t}\,d\omega = \frac{1}{2\pi}\int_{-\infty}^{\infty} j\omega F(\omega)e^{j\omega t}\,d\omega.$$

$$= j\omega \frac{1}{2\pi}\int_{-\infty}^{\infty} F(\omega)e^{j\omega t}d\omega = j\omega f(t)$$

$$\frac{df(t)}{dt} = j\omega f(t)$$

Taking Fourier transform,

$$\mathcal{F}\left[\frac{df(t)}{dt}\right] = j\omega F(\omega).$$

We can extend this result to the nth order derivative by repeated differentiation within the integral.

$$F\left[\frac{d^n f(t)}{dt^n}\right] = (j\omega)^n F(\omega)$$

or the Fourier transform pair is

$$\frac{d^n f(t)}{dt^n} \longleftrightarrow (j\omega)^n F(\omega). \tag{4.60}$$

Thus, the differentiation of a function in time domain corresponds to multiplication of its Fourier transform by $(j\omega)$.

9. Differentiation Property in Frequency Domain

Consider next the effect of differentiating the frequency-domain representation of a signal.

Let $f(t)$ be a signal with period T and Fourier transform $F(\omega)$. If $\dfrac{dF(\omega)}{d\omega}$ is absolutely integrable, then by differentiation property in the frequency domain,

$$F[(-jt)f(t)] = \frac{dF(\omega)}{d\omega}$$

And $$F[(-jt)^n f(t)] = \frac{d^n F(\omega)}{d\omega^n} \tag{4.61}$$

Proof: For signal $f(t)$, assume that $\dfrac{df(t)}{dt}$ is absolutely integrable. The Fourier transform is

$$F(\omega) = \int_{-\infty}^{\infty} f(t)e^{-j\omega t}dt$$

Differentiating both sides with respect to ω,

$$\frac{dF}{d\omega}F(\omega) = \int_{-\infty}^{\infty} -jtf(t)e^{-j\omega t}dt.$$

$$\frac{dF(\omega)}{d\omega} = (-jt)\int_{-\infty}^{\infty} f(t)e^{-j\omega t}dt = (-jt)F(\omega).$$

From this it follow that

$$tf(t) \longleftrightarrow j\frac{d}{d\omega}F(\omega). \tag{4.62}$$

We can extend to nth-order derivatives by repeated differentiations within the integral. Thus,

$$F[(-jt)^n f(t)] = \frac{d^n F(\omega)}{d\omega^n}$$

or the Fourier transform pair is

$$(-jt)^n f(t) \longleftrightarrow \frac{d^n F(\omega)}{d\omega^n} \tag{4.63}$$

Thus, differentiation of a Fourier transform in frequency domain corresponds to multiplication of the signal by $(-jt)$ in the time domain.

Example 4.8

Calculate the Fourier transform $F(\omega)$ for the signal $f(t)$ displayed in Fig. 4.17 using differentiation property.

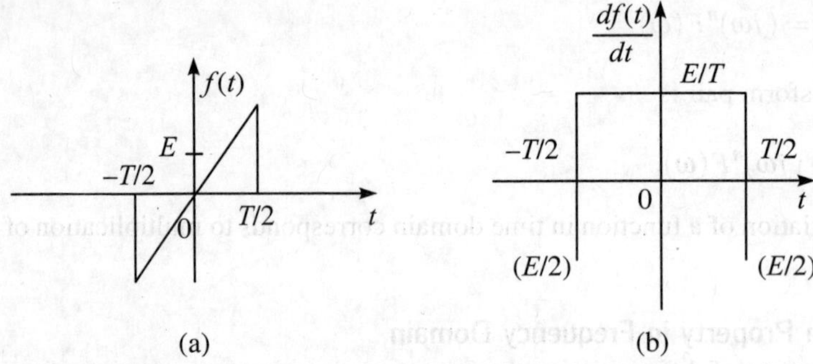

Fig. 4.17 *(a) Triangular function and (b) its differentiation*

Solution

Using differentiation, the $\frac{df(t)}{dt}$ displayed in Fig. 4.17 is represented as

$$\frac{df(t)}{dt} = \frac{E}{T}G_T(t) - \frac{E}{2}\delta(t+T/2) - \frac{E}{2}\delta(t-T/2).$$

Taking the Fourier transform of both sides, we get

$$j\omega F(\omega) = \frac{E}{T}T\ \text{Sa}\left(\frac{T\omega}{2}\right) - \frac{E}{2}e^{j\omega T/2} - \frac{E}{2}e^{-j\omega T/2} = E\ \text{Sa}\left(\frac{\omega T}{2}\right) - E\cos\left(\frac{\omega T}{2}\right)$$

so that $F(\omega) = \frac{E}{j\omega}\left[\text{Sa}\left(\frac{\omega T}{2}\right) - \cos\left(\frac{\omega T}{2}\right)\right]$.

Example 4.9

Use the frequency-differentiation property to find the Fourier transform of $f(t) = te^{-at}u(t)$.

Solution

Given $f(t) = te^{-at}u(t)$,

Let $f_1(t) = e^{-at}u(t)$. The Fourier transform is

$$F_1(\omega) = \frac{1}{a+j\omega}.$$

Using differentiation property in frequency domain,

$$F[f(t)] = F[tf_1(t)] = j\frac{dF_1(\omega)}{d\omega}.$$

Now, $j\dfrac{dF_1(\omega)}{d\omega} = j\dfrac{-j}{(a+j\omega)^2} = \dfrac{1}{(a+j\omega)^2}$.

Therefore, $\mathcal{F}\left[te^{-\alpha t}u(t)\right] = \dfrac{1}{(a+j\omega)^2}$.

10. Integration Property

Consider the Fourier transform $F(\omega)$ of a signal $f(t)$ with period T. The Fourier transform of the integration of the function $f(t)$ is

$$\mathcal{F}\left[\int_{-\infty}^{t} f(\tau)d\tau\right] = \pi F(0)\delta(\omega) + \frac{F(\omega)}{j\omega}. \tag{4.64}$$

Proof: Let $f(t)$ be a signal with period T. Then

$$\mathcal{F}\left[\int_{-\infty}^{t} f(\tau)d\tau\right] = \int_{-\infty}^{\infty}\left[\int_{-\infty}^{t} f(\tau)d\tau\right] e^{-j\omega t}dt$$

$$= \int_{-\infty}^{\infty} [f(t) * u(t)]e^{-j\omega t}dt = \int_{-\infty}^{\infty}\left[\int_{-\infty}^{\infty} f(\tau)u(t-\tau)d\tau\right] e^{-j\omega t}dt$$

$$= \int_{-\infty}^{\infty} f(\tau)\left[\int_{-\infty}^{\infty} u(t-\tau)e^{-j\omega(t-\tau)}dt\right] e^{-j\omega\tau}d\tau$$

$$= \int_{-\infty}^{\infty} f(\tau)\left[\pi\delta(\omega) + \frac{1}{j\omega}\right] e^{-j\omega\tau}d\tau.$$

$$= F(\omega)\left[\pi\delta(\omega) + \frac{1}{j\omega}\right] = \pi F(0)\delta(\omega) + \frac{F(\omega)}{j\omega}.$$

Example 4.10

Consider a pulse signal having gate function $f(t) = \dfrac{1}{\tau}G_\tau\left(t - \dfrac{\tau}{2}\right)$. Find the spectrum of $y(t) = \int_{-\infty}^{t} f(\tau)d\tau$ using the integration property in the time domain.

Solution

Given g function $f(t) = \dfrac{1}{\tau}G_\tau\left(t - \dfrac{\tau}{2}\right)$, we know that

$$G_\tau(t) = u\left(t + \frac{\tau}{2}\right) - u\left(t - \frac{\tau}{2}\right).$$

The Fourier transform is

$$G_\tau(\omega) = \tau \mathrm{Sa}\left(\frac{\omega\tau}{2}\right).$$

Using time shifting property,

$$F\left[G_\tau\left(t - \frac{\tau}{2}\right)\right] = e^{-j\omega\tau/2}G_\tau(\omega).$$

Therefore, $F(\omega) = \frac{1}{\tau} e^{-j\omega\tau/2}\, G_\tau(\omega) e^{-j\omega\tau/2}\, \mathrm{Sa}\left(\frac{\omega\tau}{2}\right)$.

Using the integration property in time domain:

If $y(t) = \int_{-\infty}^{t} f(\tau)d\tau$, the Fourier transform is

$$Y(\omega) = \pi F(0)\delta(\omega) + \frac{F(\omega)}{j\omega}.$$

Now, $F(0) = e^{j(0)\tau/2}\, \mathrm{Sa}\left(\frac{\omega\tau}{2}\right) = 1$.

Therefore, $Y(\omega) = \pi\delta(\omega) + \frac{1}{j\omega}\, \mathrm{Sa}\left(\frac{\omega\tau}{2}\right) e^{-j\omega\tau/2}$.

9. Convolution Properties

Convolution is a particularly powerful way of characterising the input-output relationship of time-invariant linear systems. As the convolution integral is not always simple to evaluate, transform domain offers a convenient method of avoiding the convolution operation. There are two convolution theorems, one for the time domain and the other for the frequency domain.

Time convolution property: Let $F_1(\omega)$ and $F_2(\omega)$ respectively be the Fourier transforms of two signals $f_1(t)$ and $f_2(t)$. The Fourier transform of their time convolution is

$$\mathcal{F}[f_1(t) * f_2(t)] = F_1(\omega)F_2(\omega). \tag{4.65}$$

Proof: The Fourier transform of the time convolution of $f_1(t)$ and $f_2(t)$ is

$$\mathcal{F}[f_1(t) * f_2(t)] = \int_{-\infty}^{\infty}\left[\int_{-\infty}^{\infty} f_1(\tau)f_2(t-\tau)d\tau\right] e^{-j\omega t} dt$$

$$= \int_{-\infty}^{\infty} f_1(\tau)\left[\int_{-\infty}^{\infty} f_2(t-\tau)e^{-j\omega t} dt\right] d\tau$$

$$= \int_{-\infty}^{\infty} f_1(\tau)\cdot f_2(\omega)e^{-j\omega\tau} d\tau = F_1(\omega)F_2(\omega)$$

For example, the convolution of two identical gate functions is a triangular function given by $G_\tau(t) * G_\tau(t) = \tau\Delta(1/\tau)$. Then the Fourier transform of the triangular function is

$$\mathcal{F}[\tau\Delta(t/\tau)] = F[G_\tau(t) * G_\tau(t)] = G_\tau(\omega)G_\tau(\omega).$$

Since $G_\tau(t) \longleftrightarrow \tau\, \mathrm{Sa}\left(\frac{\omega\tau}{2}\right)$,

therefore, $F[\tau\Delta(t/\tau)] = \tau^2\, \mathrm{Sa}^2(\omega\tau/2)$.

Frequency convolution property: Let $F_1(\omega)$ and $F_2(\omega)$ be the Fourier transforms of two signals $f_1(t)$ and $f_2(t)$. Then the frequency convolution property states that

$$\mathcal{F}[f_1(t)f_2(t)] = \frac{1}{2\pi}F_1(\omega) * F_2(\omega). \tag{4.66}$$

Proof: The Fourier transform of the product of $f_1(t)$ and $f_2(t)$ is

$$\mathcal{F}[f_1(t)f_2(t)] = \int_{-\infty}^{\infty} f_1(t)f_2(t)e^{-j\omega t} dt.$$

Using inverse Fourier transform to represent $f_1(t)$,

$$\mathcal{F}[f_1(t)f_2(t)] = \int_{-\infty}^{\infty}\left[\frac{1}{2\pi}\int_{-\infty}^{\infty} F_1(\lambda)e^{j\lambda t}d\lambda\right] f_2(t)e^{-j\omega t}dt$$

$$= \frac{1}{2\pi}\int_{-\infty}^{\infty} F_1(\lambda)\left[\int_{-\infty}^{\infty} f_2(t)e^{-j(\omega-\lambda)t}dt\right]d\lambda$$

$$= \frac{1}{2\pi}\int_{-\infty}^{\infty} F_1(\lambda)F_2(\omega-\lambda)d\lambda$$

$$= \frac{1}{2\pi}F_1(\omega) * F_2(\omega).$$

$$f_1(t)f_2(t) \longleftrightarrow \frac{1}{2\pi}F_1(\omega) * F_2(\omega) \quad \text{or} \quad f_1(t)f_2(t) \longleftrightarrow F_1(f) * F_2(f).$$

Thus the convolution operation in one domain is transformed as a product operation in the other domain.

12. Parseval's Theorem

Consider $F(\omega)$, the Fourier transform of signal $f(t)$. The energy of the signal can be expressed in the frequency domain using Parseval's theorem. It states that the normalised energy of the signal is

$$E = \int_{-\infty}^{\infty} |f(t)|^2 dt = \frac{1}{2\pi}\int_{-\infty}^{\infty} |F(\omega)|^2 d\omega \text{ joules} \tag{4.67}$$

Proof: Let $F(\omega)$ be the Fourier transform of the signal $f(t)$. The total or normalised energy of the signal is

$$E = \int_{-\infty}^{\infty} |f(t)|^2 dt = \int_{-\infty}^{\infty} f(t)f^*(t)dt.$$

We know that the inverse Fourier transform is

$$f(t) = \frac{1}{2\pi}\int_{-\infty}^{\infty} F(\omega)e^{j\omega t}d\omega.$$

$$E = \int_{-\infty}^{\infty} f^*(t)\frac{1}{2\pi}\int_{-\infty}^{\infty} F(\omega)e^{j\omega t}d\omega\, dt$$

$$= \frac{1}{2\pi}\int_{-\infty}^{\infty} F(\omega)\left[\int_{-\infty}^{\infty} f^*(t)e^{j\omega t}dt\right]d\omega = \frac{1}{2\pi}\int_{-\infty}^{\infty} F(\omega)\left[\int_{-\infty}^{\infty} f(t)e^{-j\omega t}dt\right]^* d\omega$$

$$= \frac{1}{2\pi}\int_{-\infty}^{\infty} F(\omega)F^*(\omega)d\omega = \frac{1}{2\pi}\int_{-\infty}^{\infty} |F(\omega)|^2 d\omega$$

Therefore, $E = \int_{-\infty}^{\infty} |f(t)|^2 dt = \frac{1}{2\pi}\int_{-\infty}^{\infty} |F(\omega)|^2 d\omega$ or

$$E = \int_{-\infty}^{\infty} |f(t)|^2 dt = \int_{-\infty}^{\infty} |F(f)|^2 dt. \tag{4.68}$$

This statement is called Parseval's theorem for Fourier transforms.

Energy Spectral Density

For a signal $f(t)$, the squared magnitude of the Fourier transform $|F(\omega)|^2$ is called energy spectral density. It is denoted as

$$S(\omega) = |F(\omega)|^2 \text{ joules/Hz.} \tag{4.69}$$

Example 4.11

Find the inverse Fourier transform of $\delta(\omega)$.

Solution

Given the function $F(\omega) = \delta(\omega)$, the inverse Fourier transform is

$$f(t) = \frac{1}{2\pi}\int_{-\infty}^{\infty} F[\omega]e^{j\omega t}d\omega.$$

$$= \frac{1}{2\pi}\int_{-\infty}^{\infty} \delta(\omega)e^{j\omega t}d\omega = \frac{1}{2\pi}.$$

Thus,

$$\frac{1}{2\pi} \longleftrightarrow \delta(\omega) \text{ or, } 1 \longleftrightarrow 2\pi\delta(\omega).$$

Additional Problems

Example 4.12

Find the Fourier transform of a two-sided exponential $f(t) = \text{sgn}(t)e^{-a|t|}$.

Solution

Given that $f(t) = \text{sgn}(t)e^{-a|t|}$, the function can be expressed as

$$f(t) = \begin{cases} e^{-at} & \text{for } t \geq 0 \\ -e^{at} & \text{for } t \leq 0 \end{cases}$$

The Fourier transform is

$$F[f(t)] = \int_{-\infty}^{+\infty} f(t)e^{-j\omega t}dt = \int_{-\infty}^{0} f(t)e^{-j\omega t}dt + \int_{0}^{+\infty} f(t)e^{-j\omega t}dt$$

$$= -\int_{-\infty}^{0} e^{at}e^{-j\omega t}dt + \int_{0}^{+\infty} e^{-at}e^{-j\omega t}dt = -\int_{-\infty}^{0} e^{(a-j\omega)t}dt + \int_{0}^{+\infty} e^{-(a+j\omega)t}dt$$

$$= -\left(\frac{e^{(a-j\omega)t}}{a-j\omega}\right)_{-\infty}^{0} + \left(\frac{e^{(a+j\omega)t}}{-(a+j\omega)}\right)_{0}^{+\infty}$$

$$= \frac{-1}{a-j\omega} + \frac{1}{a+j\omega}$$

$$F[f(t)] = F[e^{-a|t|}] = -\frac{j2\omega}{a^2+\omega^2}.$$

The two-sided exponential function and its magnitude spectrum is shown in Fig 4.18.

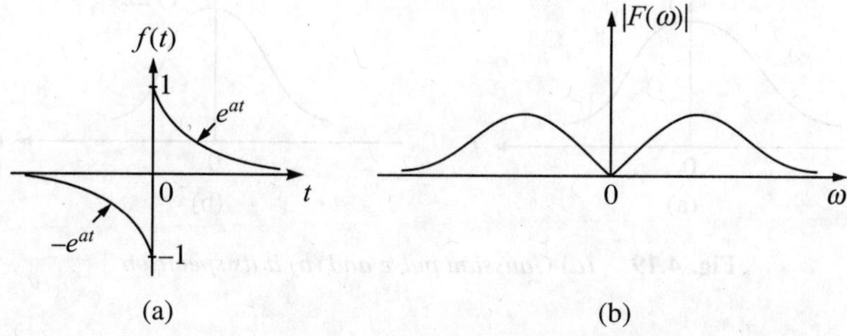

Fig. 4.18 *(a) Two-sided exponential function and (b) its spectrum*

Example 4.13

Find the Fourier transform of the Gaussian pulse $x(t) = e^{-t^2/2\tau^2}$

Solution

Given $f(t) = e^{-t^2/2\tau^2}$. The Fourier transform of the Gaussian pulse is

$$X(\omega) = \int_{-\infty}^{\infty} f(t)e^{-j\omega t}dt = \int_{-\infty}^{\infty} e^{-t^2/2\tau^2}e^{-j\omega t}dt.$$

Let $\frac{t}{\tau} = x; \quad dt = \tau dx$. Then,

$$X(\omega) = \tau\int_{-\infty}^{\infty} e^{-x^2/2}e^{-j\omega\tau x}dx = \tau\int_{-\infty}^{\infty} e^{-\left(\frac{x^2}{2}+j\omega\tau x\right)}dt$$

Now, let $\frac{x^2}{2} + j\omega\tau x = \frac{x^2}{2} + j\omega\tau x + \left(\frac{j\omega\tau}{\sqrt{2}}\right)^2 - \left(\frac{j\omega\tau}{\sqrt{2}}\right)^2$

$$= \left(\frac{x}{\sqrt{2}} + \frac{j\omega\tau}{\sqrt{2}}\right)^2 - \left(\frac{j\omega\tau}{\sqrt{2}}\right)^2$$

The Fourier transform becomes

$$X(\omega) = \tau\int_{-\infty}^{\infty} e^{-\left(\frac{x}{\sqrt{2}}+\frac{1}{\sqrt{2}}j\omega\tau\right)^2+\left(\frac{j\omega\tau}{\sqrt{2}}\right)^2}dx = \tau e^{\left(\frac{j\omega\tau}{\sqrt{2}}\right)^2}\int_{-\infty}^{\infty} e^{-\left(\frac{x}{\sqrt{2}}+\frac{1}{\sqrt{2}}j\omega\tau\right)^2}.$$

Again, let $u = \frac{x}{\sqrt{2}} + \frac{1}{\sqrt{2}}j\omega\tau$ and $du = \frac{dx}{\sqrt{2}}$.

$$X(\omega) = \tau e^{\frac{\omega^2\tau^2}{2}}\int_{-\infty}^{\infty} e^{-u^2}\sqrt{2}du$$

We know that $\int_{-\infty}^{\infty} e^{-u^2}du = \sqrt{\pi}$.

Therefore, $X(\omega) = \tau\sqrt{2\pi}e^{\frac{\omega^2\tau^2}{2}}$. Thus, the Fourier transform of a Gaussian pulse is also a Gaussian pulse as shown in Fig. 4.19.

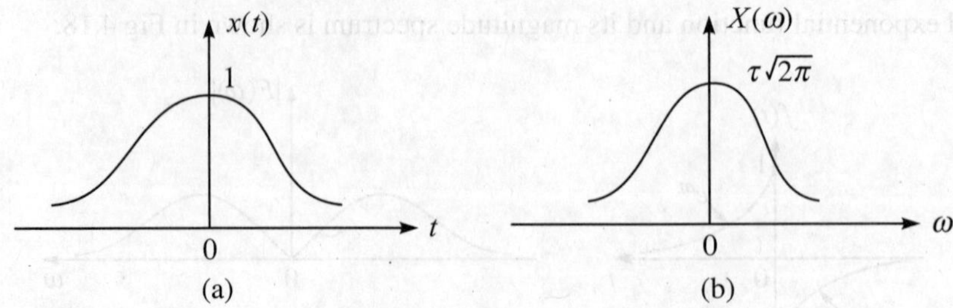

Fig. 4.19 *(a) Gaussian pulse and (b) it its spectrum*

Example 4.14

If $G(t) \longleftrightarrow E\tau \ \text{Sa}(\omega\tau)$, find the spectrum of $G\left(\frac{t}{2}\right)$, $G(2t)$.

Solution

Given $G(t) \longleftrightarrow E\tau \ \text{Sa}(\omega\tau)$. From the scaling property,

$$\mathcal{F}[f(\alpha t)] = \frac{1}{|\alpha|} F\left[\frac{\omega}{a}\right].$$

The Fourier transforms are $\mathcal{F}[G(t/2)] = 2E\tau \ \text{Sa}(\omega\tau)$

And $\mathcal{F}[G(2t)] = \frac{E\tau}{2} \ \text{Sa}\left(\frac{\omega\tau}{4}\right)$.

The three rectangular pulses and their spectra are shown in Fig. 4.20.

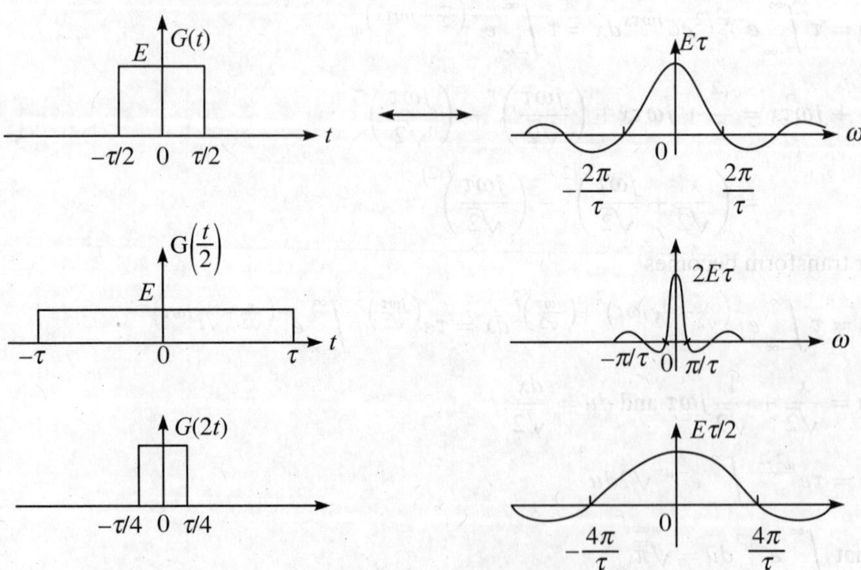

Fig. 4.20 *Effect of scaling: rectangular pulses and their spectra*

Example 4.15

Find the Fourier transform of the signal shown in Fig. 4.21.

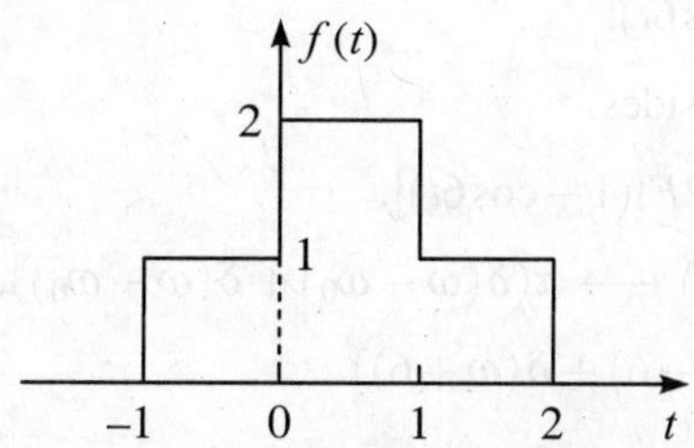

Fig. 4.21 *Sum of two gated signals*

Solution

The signal $f(t)$ is a sum of two gated signals. Figure 4.22 shows the gated signal and its components.

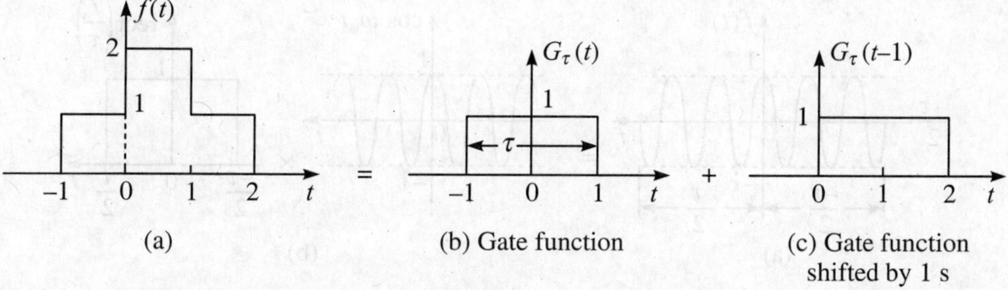

Fig. 4.22 *Gated signal and its components*

It can be represented as

$$f(t) = G_2(t) + G_2(t-1).$$

We know that the Fourier transform of the gate function is

$$G_\tau(t) \longleftrightarrow \tau \text{ Sa}\left(\frac{\omega\tau}{2}\right).$$

Here, $\tau = 2$ such that $G_2(t) \longleftrightarrow 2 \text{ Sa}(\omega)$.

The Gate function shown in Fig. 4.23(c) is $G_2(t-1)$.

Using time shifting property, the Fourier transform

$$\mathcal{F}[G_2(t-1)] = 2 \text{ Sa}(\omega)e^{-j\omega}$$

Now, $f(t) = G_2(t) + G_2(t-1)$.

Using the linearity property of Fourier transform,

$$\mathcal{F}[f(t)] = 2 \text{ Sa}(\omega) + 2 \text{ Sa}(\omega)e^{-j\omega}$$

$$f(t) \longleftrightarrow 2 \text{ Sa}(\omega)[1 + e^{-j\omega}].$$

Example 4.16

Determine the Fourier transform for a signal given by $x(t) = 4\sin^2(3t)$.

Solution

Given $x(t) = 4\sin^2(3t)$,

$$x(t) = 4\sin^2(3t) = 2(1-\cos 6t).$$

Taking Fourier transform on both sides,

$$X(\omega) = 2F[(1-\cos 6t)] = 2F[(1-\cos 6t)].$$

Since $1 \longleftrightarrow 2\pi\delta(\omega)$ and $\cos(\omega_0 t) \longleftrightarrow \pi(\delta(\omega-\omega_0)+\delta(\omega+\omega_0))$,

$$X(\omega) = 4\pi\delta(\omega) - 2\pi(\delta(\omega-6)+\delta(\omega+6)).$$

Example 4.17

Find the Fourier transform a radio frequency pulse shown in Fig. 4.23 for $\tau = \dfrac{8}{f_0}$.

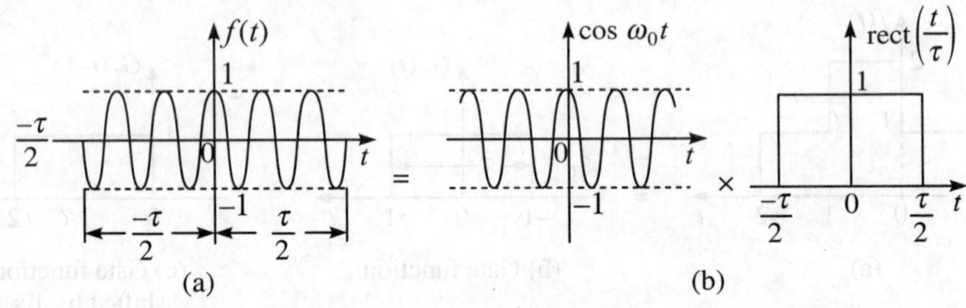

Fig. 4.23 *(a) Radio frequency pulse and (b) its components*

Solution

The radiofrequency pulse shown in the Fig. 4.23 is a sinusoidal signal having a frequency ω_0 extending from $-\dfrac{\tau}{2}$ to $\dfrac{\tau}{2}$. The function $f(t)$ may be considered as multiplication of a sinusoid $\cos\omega_0 t$ and Gate function $G_\tau(t)$ as shown in Fig. 4.23(b).

Given $f(t) = G_\tau(t)\cos\omega_0 t$, for $\tau = \dfrac{8}{f_0}$ and $\omega_0 = \dfrac{16\pi}{\tau}$,

$$\therefore \quad f(t) = \begin{cases} \cos\omega_0 t & |t| < \dfrac{8}{f_0} \\ 0 & |t| > \dfrac{8}{f_0}. \end{cases}$$

We know that the Gate function $G_\tau(t) = \left[u\left(t+\dfrac{\tau}{2}\right) - u\left(t-\dfrac{\tau}{2}\right)\right]$.

Its Fourier transform is $G_\tau(\omega) = \tau\,\mathrm{Sa}\left(\dfrac{\omega\tau}{2}\right)$.

Using modulation property,

$$F(\omega) = F[G_\tau(t)\cos\omega_0 t] = \frac{1}{2}\tau\left[\mathrm{Sa}\left(\frac{\omega+\omega_0}{2}\tau\right) + \mathrm{Sa}\left(\frac{\omega-\omega_0}{2}\tau\right)\right].$$

The spectrum is shown in Fig. 4.24 for $\tau = \frac{8}{f_0}$ corresponding to a function $\cos\omega_0 t$ existing only for 8 cycles, i.e., $\omega_0 = \frac{16\pi}{\tau}$. The peaks of the sampling function exist at $|\omega_0| = \left(\frac{16\pi}{\tau}\right)$ and the first zero occurs at $\frac{2\pi}{\tau}$ from the peaks. Hence the origin of coordinate axis will touch the 8th zero from each peak of the sampling function.

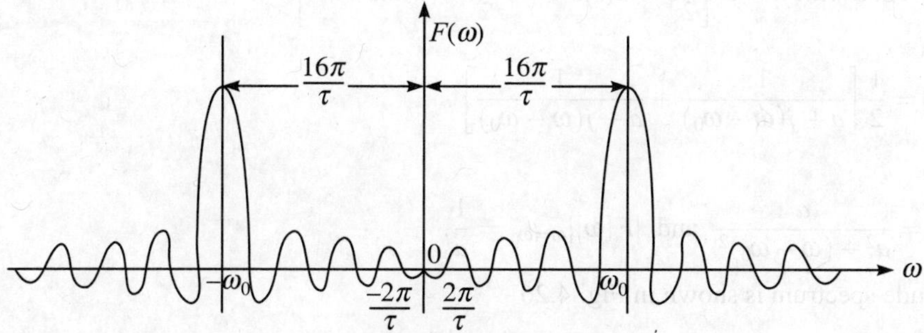

Fig. 4.24 *Spectrum of radio frequency pulse*

Example 4.18

Find the Fourier transform of the damped sinusoidal $f(t) = e^{-at}\cos\omega_0 t u(t)$.

Solution

Given function $f(t) = e^{-at}\cos\omega_0 t u(t)$. By multiplying $\cos\omega_0 t$, e^{-at} with $u(t)$, we will get the damped sinusoid shown in Fig. 4.25 (a).

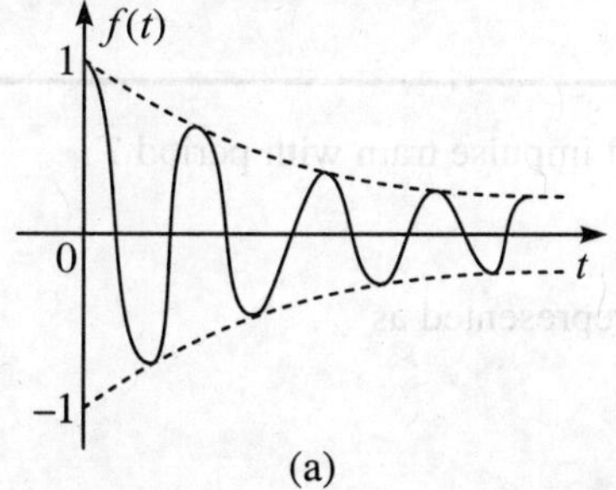

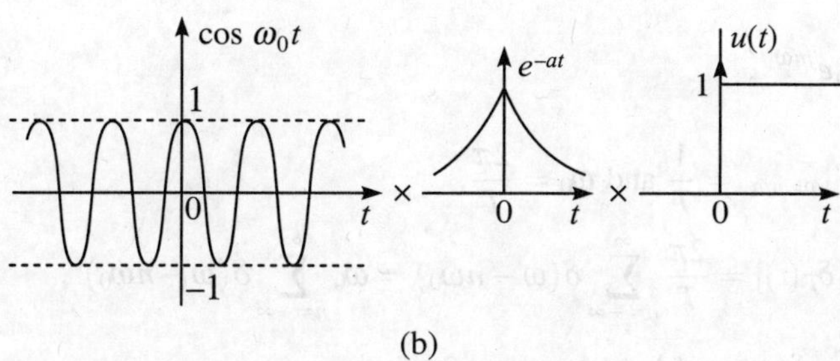

Fig. 4.25 *(a) Damped sinusoid and (b) its components*

We know that the Fourier transform of single-sided exponential $e^{-at}u(t)$ is $\frac{1}{a+j\omega}$.

$$e^{-at}u(t) \longleftrightarrow \frac{1}{a+j\omega}.$$

Using the frequency shifting property,

$$F[e^{-at}\cos\omega_0 t u(t)] = F\left[\frac{1}{2}e^{-at}u(t)e^{j\omega_0 t} + \frac{1}{2}e^{-at}u(t)e^{-j\omega_0 t}\right]$$

$$F[\omega] = \frac{1}{2}\left[\frac{1}{a+j(\omega-\omega_0)} + \frac{1}{a-j(\omega-\omega_0)}\right].$$

Therefore,

$$F[\omega] = \frac{a}{a^2+(\omega-\omega_0)^2} \text{ and } |F[\omega]|_{\omega=\omega_0} = \frac{1}{a}.$$

The magnitude spectrum is shown in Fig. 4.26

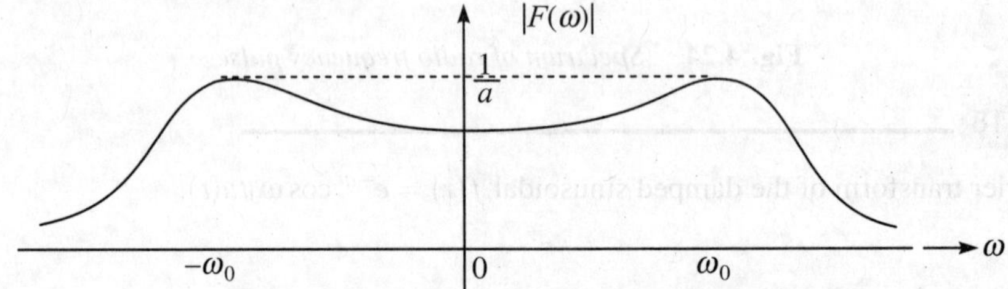

Fig. 4.26 *Spectrum of the damped sinusoid*

Example 4.19

Find the Fourier transform of a unit impulse train with period T.

Solution

The periodic impulse train can be represented as

$$\delta_T(t) = \sum_{n=-\infty}^{\infty} \delta(t-nT).$$

Expanding $\delta_T(t)$ to Fourier series, we get

$$\delta_T(t) = \sum_{n=-\infty}^{\infty} c_n e^{jn\omega_s t},$$

where $c_n = \frac{1}{T}\mathcal{F}[\delta(t)]|_{\omega=n\omega_0} = \frac{1}{T}$ and $\omega_0 = \frac{2\pi}{T}$.

Thus $\delta_T(\omega) = \mathcal{F}[\delta_T(t)] = \frac{2\pi}{T}\sum_{n=-\infty}^{\infty}\delta(\omega-n\omega_0) = \omega_0\sum_{n=-\infty}^{\infty}\delta(\omega-n\omega_0)$

or, $\delta_T(f) = f_s\sum_{n=-\infty}^{\infty}\delta(\omega-nf_s)$, where $\omega_0 = 2\pi f_s$.

Figure 4.27 shows the periodic impulse train and its spectrum.

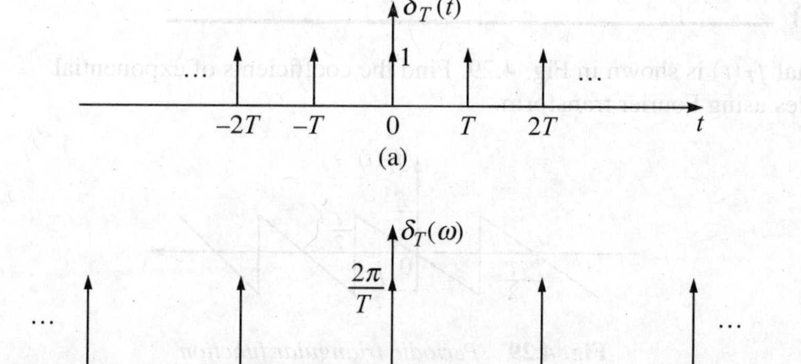

Fig. 4.27 *(a) Periodic Impulse train and (b) its spectrum*

It shows that the Fourier transform of a periodic impulse train with period T, in the time domain, is also a periodic train of equally spaced delta functions with period ω_0 in the frequency domain.

Example 4.20

Find the Fourier coefficients of the periodic signal shown in Fig. 4.28 from its Fourier transform.

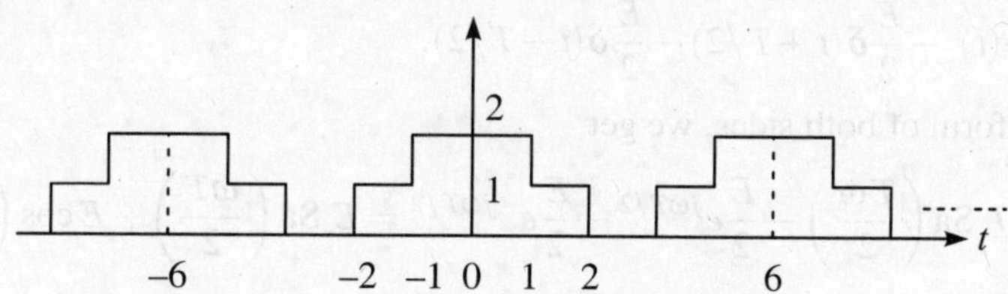

Fig. 4.28 *Periodic signal of gatedfunction*

Solution

The given the periodic function $f(t)$ is a sum of two gate functions

$$f(t) = G_4(t) + G_2(t),$$

where $G_2(t) = u(t+1) - u(t-1)$ and $G_4(t) = u(t+2) - u(t-2)$.

Period $T = 6$ and $\omega_o = \dfrac{2\pi}{T} = \dfrac{2\pi}{6} = \dfrac{\pi}{3}$.

The Fourier transform, using linearity property, is

$$F(\omega) = F[G_4(t) + G_2(t)] = F[G_4(t)] + F[G_2(t)].$$

We know that $G_\tau(t) \longleftrightarrow \tau \text{ Sa}\left(\dfrac{\omega\tau}{2}\right)$.

Therefore, $F(\omega) = 4 \text{ Sa}(2\omega) + 2 \text{ Sa}(\omega)$.

We know that the Fourier coefficients $c_n = \dfrac{1}{T} F(\omega)\Big|_{\omega = n\omega_0}$

Then, $c_n = \dfrac{1}{6}[4 \text{ Sa}(2n\omega_0) + 2 \text{ Sa}(n\omega_0)] = \dfrac{2}{3} \text{ Sa}\left(\dfrac{2\pi n}{3}\right) + \dfrac{1}{3} \text{ Sa}\left(\dfrac{\pi n}{3}\right)$.

Example 4.21

A periodic signal $f_T(t)$ is shown in Fig. 4.29. Find the coefficients of exponential Fourier series using Fourier transform.

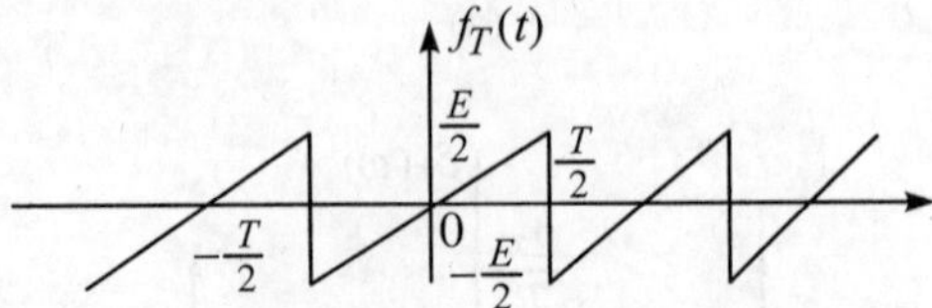

Fig. 4.29 *Periodic triangular function*

Solution

Given a periodic triangular function with period T.

We know that the coefficient of an exponential Fourier series is

$$c_n = \frac{1}{T}F(\omega)\bigg|_{\omega=n\omega_0}.$$

Let $f(t)$ be the triangular function in the interval $-T/2 < t < T/2$. The differentiation $\frac{df(t)}{dt}$ can be represented as

$$\frac{df(t)}{dt} = \frac{E}{T}G_T(t) - \frac{E}{2}\delta(t+T/2) - \frac{E}{2}\delta(t-T/2).$$

Taking Fourier transform of both sides, we get

$$j\omega F(\omega) = \frac{E}{T}T\ \mathrm{Sa}\left(\frac{T\omega}{2}\right) - \frac{E}{2}e^{j\omega T/2} - \frac{E}{2}e^{-j\omega T/2} = E\ \mathrm{Sa}\left(\frac{\omega T}{2}\right) - E\cos\left(\frac{\omega T}{2}\right)$$

so that $F(\omega) = \frac{E}{j\omega}\left[\mathrm{Sa}\left(\frac{\omega T}{2}\right) - \cos\left(\frac{\omega T}{2}\right)\right]$.

Then $c_n = \frac{1}{T}F(\omega)\Big|_{\omega=n\omega_0}$ where $\omega_o = \frac{2\pi}{T}$.

$$c_n = \frac{E}{jTn\omega_0}\left[\mathrm{Sa}\frac{n\omega_0 T}{2} - \cos\frac{n\omega_0 T}{2}\right]$$

$$= \frac{ET}{jTn2\pi}\left[\mathrm{Sa}\frac{n2\pi T}{2T} - \cos\frac{n2\pi T}{2T}\right] = \frac{E}{jn2\pi}(\ \mathrm{Sa}(n\pi) - \cos(n\pi)).$$

Therefore, $c_n = j\frac{E}{2\pi n}\cos n\pi = (-1)^n\frac{E}{2\pi n}j.$

We know that the Fourier transform of a periodic function is

$$F_T(\omega) = 2\pi\sum_{n=-\infty}^{\infty} c_n\delta(\omega - n\omega_0) = \sum_{n=-\infty}^{\infty}(-1)^n\frac{jE}{n}\delta(\omega - n\omega_0).$$

Example 4.22

Find the Fourier transform of the rectangular pulse shown in Fig. 4.30 using differentiation property.

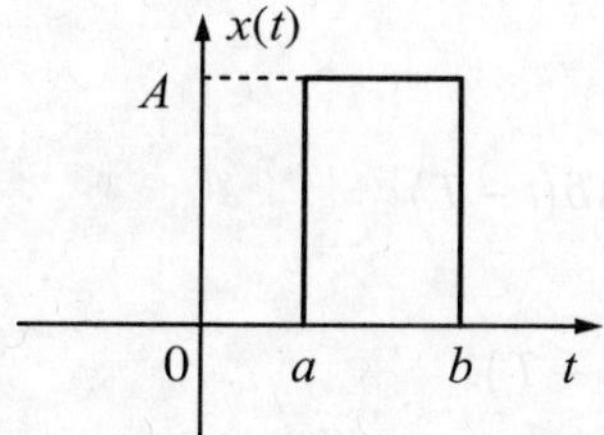

Fig. 4.30 *Rectangular pulse*

Solution

The given rectangular pulse can be expressed as

$$x(t) = u(t-a) - u(t-b).$$

Using differentiation property,

$$\frac{dx(t)}{dt} = A(\delta(t-a) - \delta(t-b))$$

Taking the Fourier transform of both sides we get

$$j\omega X(\omega) = A\left(e^{-j\omega a} - e^{-j\omega b}\right)$$

so that $X(\omega) = \dfrac{A}{j\omega}\left(e^{-j\omega a} - e^{-j\omega b}\right)$.

Example 4.23

Find the Fourier transform of a triangular waveform $f(t)$ shown in Fig. 4.31.

Solution

The given triangular waveform $f(t)$ can be synthesised from the ramp functions shown in Fig. 4.31.

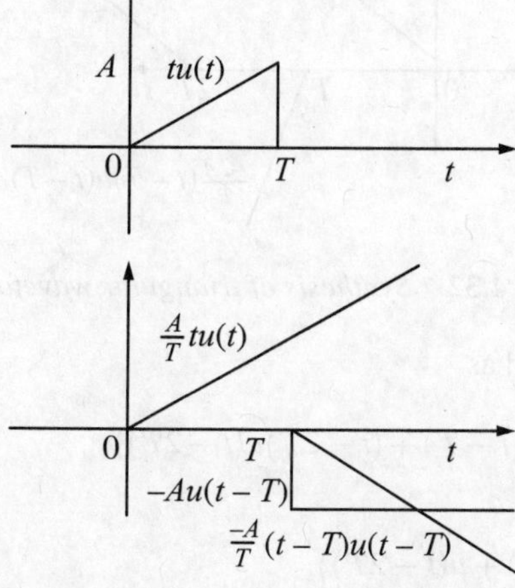

Fig. 4.31 *Synthesis of triangular waveform*

Thus the signal can be represented as:

$$f(t)=\frac{A}{T}[tu(t)-(t-T)u(t-T)]-Au(t-T).$$

Using differentiation,

$$\frac{df(t)}{dt}=\frac{A}{T}[u(t)-u(t-T)]-A\delta(t-T)$$

$$=\frac{A}{T}G_T(t-T/2)-A\delta(t-T).$$

Taking the Fourier transform of both sides, we get

$$(j\omega)F(\omega)=\frac{A}{T}Te^{-j\omega T/2}\ \mathrm{Sa}(\omega T/2)-Ae^{-j\omega T}$$

$$F(\omega)=\frac{A}{j\omega}e^{-j\omega T/2}\ \mathrm{Sa}(\omega T/2)-\frac{A}{j\omega}e^{-j\omega T}$$

Example 4.24

Find the Fourier transform of a triangular waveform $f(t)$ shown in Fig. 4.32.

Solution

The given triangular waveform can be synthesised from ramp functions shown in Fig. 4.32.

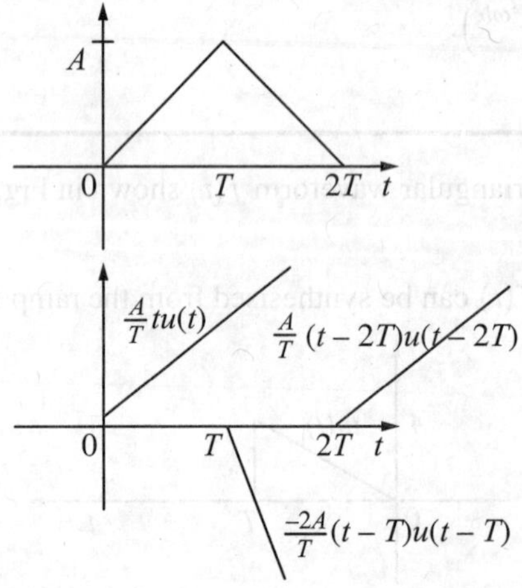

Fig. 4.32 *Synthesis of triangular waveform*

Thus the signal can be represented as

$$x(t)=\frac{A}{T}(tu(t)-2(t-T)u(t-T)+(t-2T)u(t-2T)).$$

$$\frac{dx(t)}{dt}=\frac{A}{T}(u(t)-2u(t-T)+u(t-2T)).$$

$$\frac{d^2x(t)}{dt^2} = \frac{A}{T}(\delta(t) - 2\delta(t-T) + \delta(t-2T)).$$

Taking the Fourier transform of both sides, we get

$$(j\omega)^2 F(\omega) = \frac{A}{T}(1 - 2e^{-j\omega T} + e^{-j2\omega T}).$$

$$F(\omega) = \frac{-A}{\omega^2 T}(1 - 2e^{-j\omega T} + e^{-j2\omega T}) = Te^{-j\omega T}\,\mathrm{Sa}^2(\omega T/2).$$

Example 4.25

Find the Fourier transform of a trapezoidal waveform shown in Fig. 4.33.

Solution

The given trapezoidal waveform can be synthesised from ramp functions as shown in Fig. 4.33.

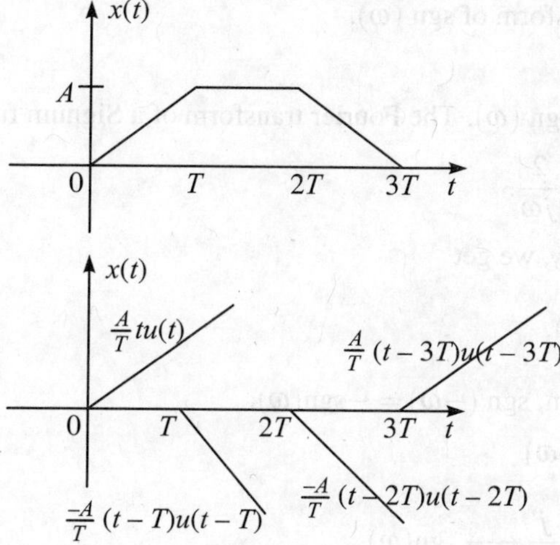

Fig. 4.33 *Synthesis of tropezoidol waveform*

Thus the signal can be represented as

$$x(t) = \frac{A}{T}(tu(t) - (t-T)u(t-T) - (t-2T)u(t-2T) + (t-3T)u(t-3T)).$$

Taking differentiation on both sides,

$$\frac{dx(t)}{dt} = \frac{A}{T}(u(t) - u(t-T) - u(t-2T) + u(t-3T))$$

$$\frac{d^2x(t)}{dt^2} = \frac{A}{T}(\delta(t) - \delta(t-T) - \delta(t-2T) - \delta(t-3T)).$$

Taking the Fourier transform of both sides, we get

$$(j\omega)^2 F(\omega) = \frac{A}{T}\left(1 - e^{-j\omega T} - e^{-j2\omega T} + e^{-j3\omega T}\right).$$

$$F(\omega) = \frac{-A}{\omega^2 T}\left(1 - e^{-j\omega T} - e^{-j2\omega T} + e^{-j3\omega T}\right).$$

Example 4.26

Find the Inverse Fourier transform of $\delta(\omega-\omega_0)$.

Solution

Given the function $F(\omega)=\delta(\omega-\omega_0)$, the inverse Fourier transform of a signal $F(\omega)$ can be expressed as

$$f(t)=\frac{1}{2\pi}\int_{-\infty}^{\infty}F(\omega)e^{j\omega t}d\omega=\frac{1}{2\pi}\int_{-\infty}^{\infty}\delta(\omega-\omega_0)e^{j\omega t}d\omega=\frac{1}{2\pi}e^{j\omega_0 t}$$

Therefore, the Fourier transform pair is

$$\frac{1}{2\pi}e^{j\omega_0 t}\longleftrightarrow\delta(\omega-\omega_0).$$

Example 4.27

Find the inverse Fourier transform of sgn (ω).

Solution

Given the function $F(\omega)=$ sgn (ω). The Fourier transform of a Signum function is

$$\text{sgn}(\omega)=F[\text{sgn}(t)]=\frac{2}{j\omega}.$$

Applying the duality property, we get

$$F\left(\frac{2}{jt}\right)=2\pi\text{sgn}(-\omega).$$

As sgn (ω) is an odd function, sgn $(-\omega)=-\text{sgn}(\omega)$.

$$\text{Hence, } \frac{2}{jt}\longleftrightarrow-2\pi\text{sgn}(\omega)$$

$$\frac{-1}{jt}\longleftrightarrow\pi\ \text{sgn}(\omega)\ \text{or}\ \frac{j}{\pi t}\longleftrightarrow\text{sgn}(\omega).$$

Therefore, the inverse Fourier transform of sgn (ω) is $\frac{j}{\pi t}$.

Example 4.28

Find the inverse Fourier transform of $u(\omega)$.

Solution

Given $F(\omega)=u(\omega)$. We know that

$$u(\omega)=\frac{1}{2}[1+\text{sgn}(\omega))].$$

Applying linearity property,

$$u(\omega)=F^{-1}\left[\frac{1}{2}\right]+F^{-1}\left[\frac{1}{2}\text{sgn}(\omega)\right]$$

$$u(\omega)=\frac{1}{2}\delta(t)+\frac{j}{2\pi t}.$$

Example 4.29

Using convolution property in time domain, find the inverse Fourier transform of $F(\omega)=\frac{1}{(\alpha+j\omega)^2}$.

Solution

Given $F(\omega)=\frac{1}{(\alpha+j\omega)^2}$, let $F_1(\omega)=\frac{1}{\alpha+j\omega}$. The inverse Fourier transform is

$$f_1(t)=e^{-\alpha t}u(t)$$

Now, using convolution property in time domain,

$$\frac{1}{(\alpha+j\omega)^2}=F_1(\omega)F_1(\omega)\longleftrightarrow f_1(t)*f_1(t)$$

$$f(t)=f_1(t)*f_1(t)=\int_{-\infty}^{\infty}f_1(\tau)f_1(t-\tau)d\tau$$

$$=\int_{-\infty}^{\infty}e^{-a\tau}e^{-a(t-\tau)}u(\tau)u(t-\tau)d\tau=\int_0^t e^{-at}d\tau=te^{-at}u(t).$$

Therefore the inverse Fourier transform of $\frac{1}{(\alpha+j\omega)^2}$ is $te^{-\alpha t}u(t)$.

Example 4.30

Find the energy of the signal $f(t)=e^{-at}u(t)$, $a>0$ in the frequency range, $-\infty<\omega<\infty$.

Solution

Given an exponential function $f(t)=e^{-at}u(t)$, $a>0$.

The Fourier transform is

$$F(\omega)=\int_{-\infty}^{\infty}f(t)e^{-j\omega t}dt=\int_{-\infty}^{\infty}e^{-at}u(t)e^{-j\omega t}dt=\int_0^{\infty}e^{-(a+j\omega)t}dt$$

$$F(\omega)=\frac{1}{a+j\omega},\ a>0.$$

The energy spectral density is $|F(\omega)|^2=\frac{1}{a^2+\omega^2}$

The energy is $E=\frac{1}{2\pi}\int_{-\infty}^{\infty}|F(\omega)|^2d\omega=\frac{1}{2\pi}\int_{-\infty}^{\infty}\frac{1}{a^2+\omega^2}d\omega$

Since $\int_{-\infty}^{\infty}\frac{a}{a^2+\omega^2}d\omega=\pi$,

$$E=\frac{1}{2\pi a}\int_{-\infty}^{\infty}\frac{a}{a^2+\omega^2}d\omega=\frac{\pi}{2\pi a}=\frac{1}{2a}\text{ joules.}$$

Example 4.31

Find the energy of the signal $f(t)=e^{-at}u(t)$, $a>0$ in the frequency range $|\omega|<W$.

Solution

The Fourier transform of the given an exponential function is

$$F(\omega) = \int_{-\infty}^{\infty} f(t)e^{-j\omega t}dt = \int_{-\infty}^{\infty} e^{-at}u(t)e^{-j\omega t}dt = \int_{0}^{\infty} e^{-(a+j\omega)t}dt.$$

$$F(\omega) = \frac{1}{a+j\omega}, \ a > 0.$$

The energy spectral density is $|F(\omega)|^2 = \dfrac{1}{a^2+\omega^2}, \quad |\omega| < W$

The energy is $E = \dfrac{1}{2\pi}\displaystyle\int_{-\infty}^{\infty} |F(\omega)|^2 d\omega = \dfrac{1}{2\pi}\int_{-W}^{W} \dfrac{1}{a^2+\omega^2}d\omega$

$$E = \frac{1}{\pi a}\tan^{-1}(W/a) \text{ joules.}$$

Summary

1. Any periodic signal with period T can be represented by a combination of harmonics at discrete frequencies called Fourier series, and any aperiodic signal be represented by harmonics of continuous frequency called Fourier transform.
2. A non-periodic function can be expressed in terms of periodic function as $f(t) = \lim_{T\to\infty} f_T(t)$.

   For any periodic signal with period T, as the period T increases, the harmonics becomes closer, and as T tends to infinity, the periodic signal becomes an aperiodic signal and the frequency becomes continuous, $n\omega_0 \to \omega$.
3. The coefficients of the Fourier exponential series related to the Fourier transform $F(\omega)$ are

   $$c_n = \frac{1}{T}F(\omega)\bigg|_{\omega=n\omega_0}.$$

   For aperiodic signal, as $T\to\infty$,

   $$F(\omega) = c_n T = \lim_{T\to\infty}\int_{-T/2}^{T/2} f_T(t)e^{-j\omega t}dt \quad \text{or} \quad F(\omega) = \int_{-\infty}^{\infty} f(t)e^{-j\omega t}dt.$$

   Thus the Fourier transform of a periodic function is $F_T(\omega) = 2\pi\sum_{n=-\infty}^{\infty} c_n\delta(\omega - n\omega_0)$. It consists of an infinite sum of impulses at harmonics with a strength equal to 2π times of the Fourier coefficient c_n.
4. Fourier transforms pair are:

   With continuous frequency ω rad/s,

   $$F(\omega) = \mathcal{F}[f(t)] = \int_{-\infty}^{\infty} f(t)e^{-j\omega t}dt$$

   $$f(t) = \mathcal{F}^{-1}[F(\omega)] = \frac{1}{2\pi}\int_{-\infty}^{\infty} F(\omega)e^{j\omega t}d\omega$$

With continuous frequency f Hz,

$$F(f) = \mathcal{F}[f(t)] = \int_{-\infty}^{\infty} f(t)e^{-jft}dt$$

$$f(t) = \mathcal{F}^{-1}[F(f)] = \int_{-\infty}^{\infty} F(f)e^{jft}df$$

5. The notation for Fourier transform pair is $f(t) \longleftrightarrow F(\omega)$.
6. If $f(t)$ is real, $F(-\omega) = \int_{-\infty}^{\infty} f(t)e^{j\omega t}dt = F^*(\omega)$.
7. For real signal, the amplitude spectrum $|F(\omega)|$ is an even function and the phase spectrum $\phi(\omega)$ is an odd function of ω.
8. For real and even signals, the Fourier transforms are real and even functions of frequency and for real and odd signals, the Fourier transforms are imaginary and odd functions of frequency. That is, if $f(t) = f_e(t) + f_o(t)$ and $F(\omega) = F_R(\omega) + jF_I(\omega)$, then

   $F[f_e(t)] = F_R(\omega)$ and $F[f_e(t)] = F_R(\omega)$.
9. A sufficient condition for the existence of $F(\omega)$ is that the signal $f(t)$ should be absolutely integrable over the entire interval, i.e., $\int_{-\infty}^{\infty} |f(t)|dt < \infty$.
10. Singularity ffinctions are a class of discontinuous functions, or their derivatives are discontinuous at its singular points. Fourier transform does not exists for singularity functions.
11. To overcome the singularity problem, an impulse function, designated as $\delta(t)$, also called the Dirac delta function, of strength equal to the amount of discontinuity is introduced as a derivative of the function at the discontinuity point.
12. The unit impulse function $\delta(t)$ exists only at $t = 0$ and is zero elsewhere so that $\int_{-\infty}^{\infty} \delta(t)dt = 1$.
13. Using impulse functions at discontinuity, it is possible to express the Fourier transform for periodic functions as the sum of Fourier transform of its individual components.
14. The commonly used singularity functions are: impulse function, step function, and ramp function and its higher derivatives.
15. The derivative of $u(t)$ is $\dfrac{du(t)}{dt} = \delta(t)$; also, $u(t) = \int_{-\infty}^{t} \delta(t)dt$.
16. The Fourier transform of a periodic function consists of infmite sum of harmonics with a strength equal to 2π times the Fourier coefficient impulses at c_n. That is, $F(\omega) = 2\pi \sum_{n=-\infty}^{\infty} c_n\delta(\omega - n\omega_0)$.
17. The linearity property states that if

   $f_1(t) \longleftrightarrow F_1(\omega)$, $f_2(t) \longleftrightarrow F_2(\omega)$,

   then the linear combination of the two signals satisfy $\mathcal{F}[a_1 f_1(t) + a_2 f_2(t)] = a_1 F_1(\omega) + a_2 F_2(\omega)$, where a_1 and a_2 are arbitrary constants.
18. For any complex-valued signal, if $f(t) \longleftrightarrow F(\omega)$, then the complex conjugate property is $\mathcal{F}[f^*(t)] = F^*(-\omega)$.
19. If a signal $f(t)$ with period T has Fourier transform $F(\omega)$, then $\mathcal{F}[f(t \pm t_0)] = F(\omega)e^{\pm j\omega t_0}$ and $\mathcal{F}[f(t)e^{\pm j\omega_0 t}] = F(\omega \mp \omega_0)$.

20. Parseval's theorem for Fourier transforms is $E = \int_{-\infty}^{\infty} |f(t)|^2\,dt = \frac{1}{2\pi}\int_{-\infty}^{\infty} |F(\omega)|^2\,dt.$

21. The Fourier transform for some standard aperiodic signals are given below.

Function name	**Time domain signal $x(t)$**	**Frequency domain signal $F(\omega)$**
Impulse function	$\delta(t)$	1
DC or constant signal	1	$2\pi\delta(\omega)$
Unit step function	$u(t) = \begin{cases} 1 & t \geq 0 \\ 0 & t < 0 \end{cases}$	$\frac{1}{j\omega} + \pi\delta(\omega)$
Unit ramp function	$r(t) = tu(t)$	$\frac{1}{(j\omega)^2}$
Signum function	$\mathrm{sgn}(t) \equiv \begin{cases} 1 & t \geq 0 \\ -1 & t < 0 \end{cases}$	$\frac{2}{j\omega}$
Linear time-decaying signal	$\frac{1}{t}u(t)$	$-j\pi\,\mathrm{sgn}(\omega)$
Rectangle pulse	$\mathrm{rect}\left(\frac{t}{\tau}\right) = \Pi\left(\frac{t}{\tau}\right) = \begin{cases} 1 & \|t\| \leq \tau/2 \\ 0 & \text{elsewhere} \end{cases}$	$\tau\,\mathrm{Sa}\left(\frac{\tau\omega}{2}\right)$
Sampling pulse	$\mathrm{Sa}(Wt) \equiv \frac{\sin(Wt)}{Wt}$	$\frac{1}{W}\,\mathrm{rect}\left(\frac{\omega}{2W}\right)$
Triangle pulse	$\mathrm{tri}\left(\frac{t}{W}\right) = \Delta\left(\frac{t}{W}\right) = \begin{cases} 1 - \|t\|/W & \|t\| \leq W \\ 0 & \text{elsewhere} \end{cases}$	$W\,\mathrm{Sa}^2\left(\frac{W\omega}{2}\right)$
Squared sampling pulse	$\mathrm{Sa}^2(Wt)$	$\frac{1}{W}\mathrm{tri}\left(\frac{\omega}{2W}\right)$
Exponential decaying signal	$e^{-at}u(t),\ \mathrm{Re}\{a\} > 0$	$\frac{1}{a + j\omega}$
Ramp with exponential decaying signal	$e^{-at}u(t),\ \mathrm{Re}\{a\} > 0$	$\frac{1}{(a + j\omega)^2}$
Two-sided exponential signal	$e^{-a\|t\|},\ a > 0$	$\frac{2a}{a^2 + \omega^2}$
Gaussian pulse	$\exp\left(\frac{t^2}{(2\sigma)^2}\right)$	$(\sigma\sqrt{2\pi})\exp\left(-\frac{\sigma^2\omega^2}{2}\right)$

22. The Fourier transform of some standard periodic signals:

Function name	**Time domain signal $f(t)$**	**Frequency domain signal $F(\omega)$**
Cosine signal	$\cos(\omega_0 t)$	$\pi[\delta(\omega-\omega_0)+\delta(\omega+\omega_0)]$
Sinusoidal signal	$\sin(\omega_0 t)$	$-j\pi[\delta(\omega-\omega_0)-\delta(\omega+\omega_0)]$
Cosine signal with phase shift	$\cos(\omega_0 t+\theta)$	$\pi[e^{j\theta}\delta(\omega-\omega_0)+e^{-j\theta}\delta(\omega+\omega_0)]$
Sine signal with phase shift	$\sin(\omega_0 t+\theta)$	$-j\pi[e^{j\theta}\delta(\omega-\omega_0)-e^{-j\theta}\delta(\omega+\omega_0)]$
Complex exponential signal	$e^{j\omega_0 t}$	$2\pi\delta(\omega-\omega_0)$
Impulse train	$\sum_{n=-\infty}^{\infty}\delta(t-nT)$	$\frac{2\pi}{T}\sum_{n=-\infty}^{\infty}\delta\left(\omega-\frac{2\pi n}{T}\right)$
Pulse waveform	$x(t)=\begin{cases}1 & \|t\|\le\tau \\ 0 & \tau<\|t\|\le T/2\end{cases}$ $x(t+T)=x(t)$	$\frac{4\pi}{\tau}\sum_{n=-\infty}^{\infty}\mathrm{Sa}(n\omega\tau)\delta\left(\omega-\frac{2\pi n}{\tau}\right)$
Fourier series	$\sum_{n=-\infty}^{\infty}c_n e^{jn\omega_0 t}$, where $c_n=\frac{1}{T}\int_{-T/2}^{T/2}x(t)e^{-jn\omega_0 t}dt$	$2\pi\sum_{n=-\infty}^{\infty}c_n\delta(\omega-n\omega_0)$ $c_n=\frac{1}{T}F\omega\Big\|_{\omega=n\omega_0}$

23. If signals $f(t), f_1(t)$ and $f_2(t)$ have Fourier transforms $F(\omega), F_1(\omega)$ and $F_2(\omega)$ respectively. That is, $f(t)\longleftrightarrow F(\omega)$, $f_1(t)\longleftrightarrow F_1(\omega)$, $f_2(t)\longleftrightarrow F_2(\omega)$, the following properties hold.

Property	**Expression in time domain**	**Expression in frequency domain**
Linearity	$a_1 f_1(t)+a_2 f_2(t)$	$a_1F_1(\omega)+a_2F_2(\omega)$
Time shifting	$f(t-t_0)$	$e^{-j\omega t_0}F(\omega)$
Time scaling	$f(at)$	$\frac{1}{\|a\|}F\left(\frac{\omega}{a}\right)$
Conjugation	$f^*(t)$	$F^*(-\omega)$
Frequency shifting	$f(t)e^{j\omega_0 t}$	$F(\omega-\omega_0)$
Duality	$F(t)$	$2\pi f(-\omega)$
Symmetry	$f(t)=f_e(t)+f_o(t)$	$F(\omega)=F_R(\omega)+jF_1(\omega)$ $\mathcal{F}[f_e(t)]=F_R(\omega)$ and $\mathcal{F}[f_o(t)]=jF_1(\omega)$
Modulation	$f(t)\cos(\omega_0 t)$	$\frac{1}{2}[F(\omega-\omega_0)+F(\omega+\omega_0)]$
Time differentiation	$\frac{df(t)}{dt}$ and $\frac{d^n f(t)}{dt^t}$	$(j\omega)F(\omega)$ and $(j\omega)^n F(\omega)$

Frequency differentiation	$(-jt)f(t)$ and $(-jt)^n f(t)$	$\frac{dF(\omega)}{d\omega}$, $\frac{d^nF(\omega)}{d\omega^n}$
Time integration	$\int_{-\infty}^{t} f(\tau)d\tau$	$\pi F(0)\delta(\omega)+\frac{F(\omega)}{j\omega}$
Frequency integration	$\frac{f(t)}{(-jt)}$	$\int F(\omega)d\omega$
Convolution	$f_1(t)*f_2(t)$ $=\int_{-\infty}^{\infty} f_1(\tau)f_2(t-\tau)d\tau$	$F_1(\omega)F_2(\omega)$
Multiplication	$f_1(t)f_2(t)$	$F_1(\omega)*F_2(\omega)$
Symmetry	$f(t)$ is real function	$F(-\omega)=F^*(\omega)$ and $\|F(-\omega)\|$ $=\|F(\omega)\|\angle F(-\omega)=-\angle G(\omega))$

Review Questions

1. State and prove the time shifting and frequency shifting properties of Fourier transforms. (Nov 2015)
2. Find the Fourier transform of symmetrical gate pulse and sketch the spectrum. (May 2015)
3. State and prove following properties of Fourier transform: (May 2015)
   (i) Time shifting, (ii) Differentiation in time domain.
4. Prove that the time shift in time domain is equal to phase shift in frequency domain. (Dec 2011)
5. State all the properties of the Fourier transform. (Dec 2011)
6. State and prove the symmetry and duality properties of the Fourier transform. (June 2014)
7. State and prove any four properties of the Fourier transform. (Dec 2011)
8. Compute the Fourier transform of standard signals like the unit step, unit impulse, single-sided exponential, double sided exponential and gate function. (May 2013)

Problems

1. Find the Fourier transform for the signals given in Fig. P4.1.

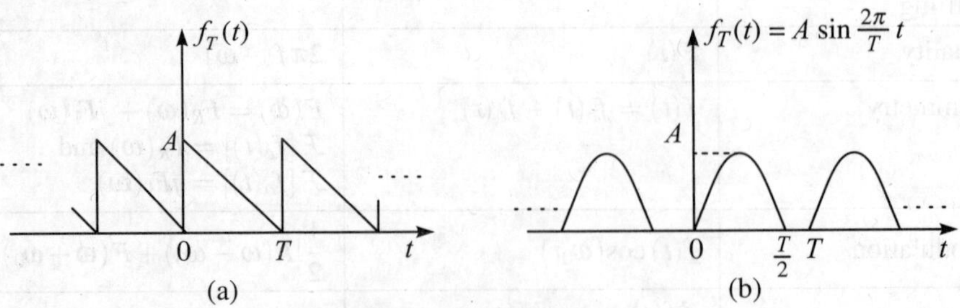

Fig. P4.1 *Periodic waveforms*

2. Find the inverse Fourier transform for following $F(\omega)$ as shown as Fig. P4.2.

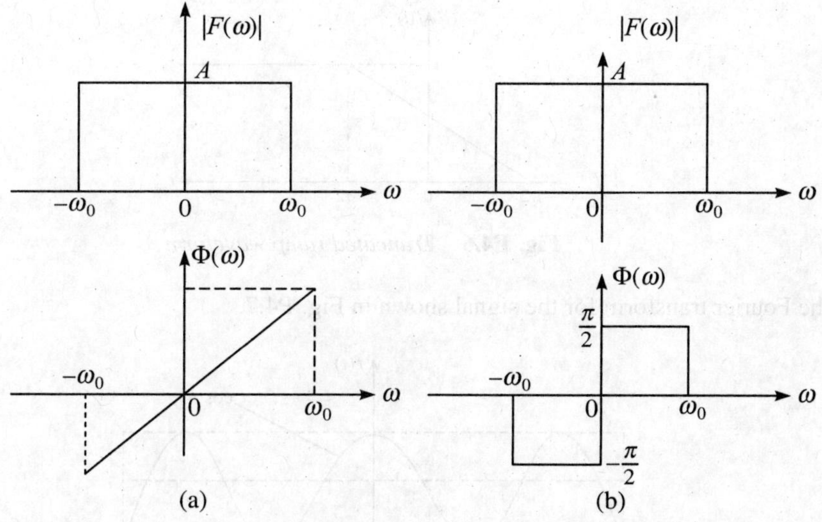

Fig. P4.2 *Pulse waveforms*

3. Find the inverse Fourier transform for each of the following $F(\omega)$.

   a. $F(\omega) = \delta(\omega - \omega_0)$

   b. $F(\omega) = \begin{cases} e^{-2\omega} & \omega \geq 0 \\ 0 & \text{otherwise} \end{cases}$

   c. $F(\omega) = u(\omega + \omega_0) - u(\omega - \omega_0)$

4. If $f(t) \longleftrightarrow F(\omega)$, find the Fourier transforms for the following signals using the properties of the Fourier transform.

   (a) $tf(2t)$ (b) $(t-1)f(t)$ (c) $(t-2)f(-2t)$

   (d) $f(2t-5)$ (e) $t\dfrac{df(t)}{dt}$ (f) $\displaystyle\int_{-\infty}^{t} \tau f(\tau)d\tau$

5. Find the $F(\omega)$ using properties of the Fourier transform for the $f(t)$ shown in Fig. P4.5.

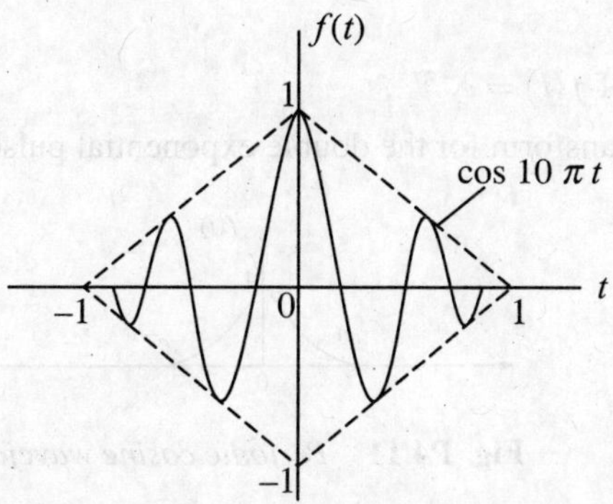

Fig. P4.5 *Modulated waveform*

6. Determine the spectrum function for signal shown in Fig. P4.6. (Use integral property.)

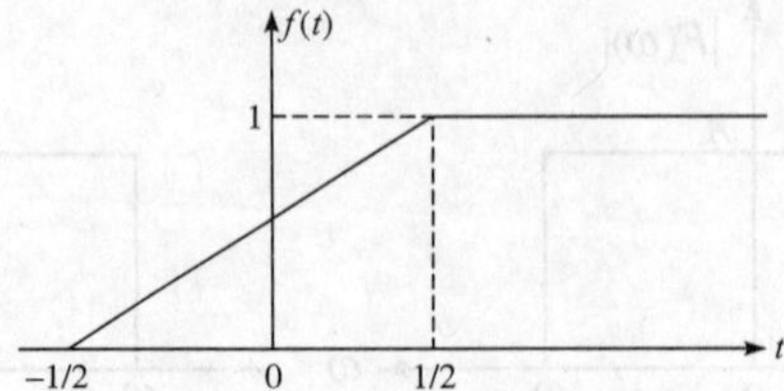

Fig. P4.6 *Truncated ramp waveform*

7. Find the Fourier transform for the signal shown in Fig. P4.7.

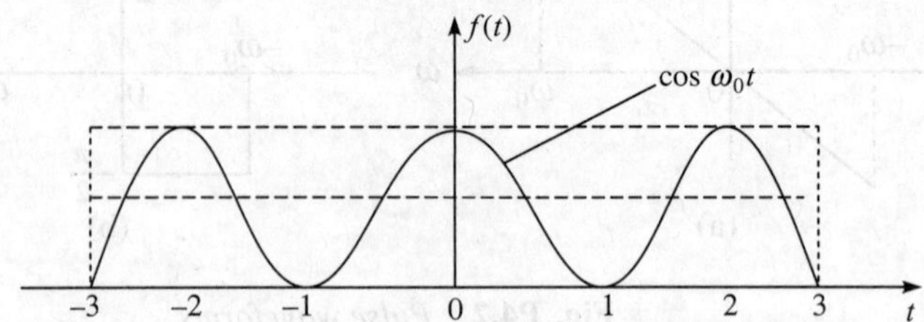

Fig. P4.7 *Periodic cosine waveform*

8. Plot the spectrum of a signal that has been sampled by an impulse train. The signals are given in Fig. P4.8.

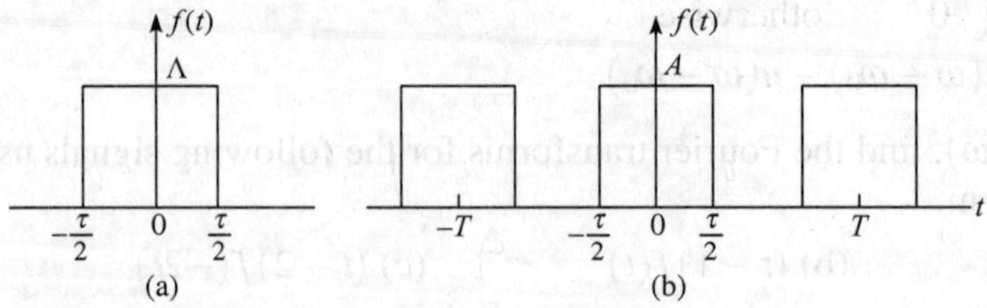

Fig. P4.8 *Pulse train*

9. Determine the Fourier transforms for a function is given by $f(t) = 4e^{2t}u(-t)$. Also draw its magnitude and phase spectra.

10. Find Fourier transform of $f(t) = e^{-\pi t^2}$. (Nov 2015)

11. Determine the Fourier transform for the double exponential pulse shown in Fig. P4.11.

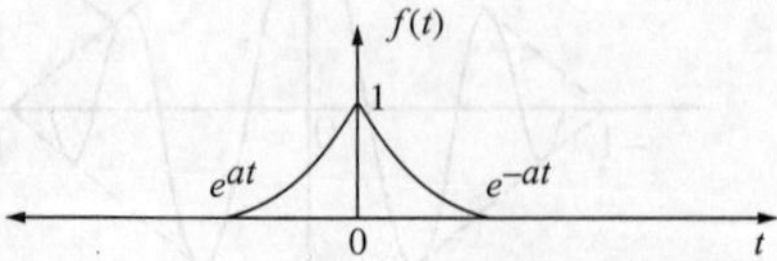

Fig. P4.11 *Periodic cosine waveform*

12. Find the Fourier transform of the following functions: (Dec 2011)
(i) $f(t) = e^{-a|t|}\sin(t)$, (ii) $f(t) = \cos at^2$, (iii) $f(t) = \sin at^2$

13. Find the Fourier transform of (Dec 2011)
(i) $f(t) = e^{-at}\cos(bt)$, (ii) $f(t) = t\cos(at)$
14. Compute the Fourier transform of (Dec 2011)
(i) $f(t) = e^{4t}u(-t-1)$, (ii) $f(t) = \sin(\pi t)u(t)$.
15. Determine the Fourier transform for double exponential pulse whose function is given by $y(t) = e^{-2|t|}$. Also draw its magnitude and phase spectra. (Dec 2015)
16. Find the Fourier transform of: (i) triangular pulse with period $T = 8$ s and amplitude $A = 10$ V, (ii) one cycle of sine wave (Dec 2015)
17. Find the Fourier transform of the following functions: (i) a single symmetrical triangular pulse, (ii) a single symmetrical gate pulse. (Jan 2015)
18. Determine the Fourier transform of the signal $x(t) = 5G_4(t-2)\cos(8t)$.
19. Determine the Fourier transform for a symmetrical triangular pulse with period $T = 4$s and amplitude $A = 10$.
20. Determine the Fourier transform of the signal $x(t) = 5 + 2\cos(8\pi t)$.
21. Determine the Fourier transform of (a) $x(t)\cos(\omega_0 t + \theta)$ and (b) $x(t)\sin(\omega_0 t + \theta)$.

Answers

1. (a) $F(\omega) = 2\pi A \sum_{\substack{n=-\infty \\ n\neq 0}}^{\infty} c_n \delta(\omega - 2\pi n/T)$, where $c_n = \begin{cases} 1/2 & n = 0 \\ \dfrac{1}{j2\pi n} + \dfrac{1}{4\pi^2 n^2}\left(1 - e^{-j2\pi n}\right) & n \neq 0 \end{cases}$

(b). $F(\omega) = 2\pi A \sum_{n=-\infty}^{\infty} c_n \delta(\omega - 2\pi n/T)$, where $c_n = \dfrac{(1 - e^{-j\pi n})}{2(1-n^2)}$.

2. (a) $f(t) = \dfrac{A\omega_0}{\pi}\,\text{Sa}(\omega_0(t+t_0))$ (b) $f(t) = \dfrac{-A}{\pi t}(1 - \cos\omega_0 t)$

3. (a) $\dfrac{1}{2\pi}e^{j\omega_0 t}$ (b) $\dfrac{1}{2\pi(2-jt)}$ (c) $\dfrac{\omega_0}{\pi}\,\text{Sa}(\omega_0 t)$

4. (a) $\dfrac{j}{2}\dfrac{d}{\omega}\left[F\left(\dfrac{\omega}{2}\right)\right]$ (b) $j\dfrac{d}{d\omega}[F(\omega)] - F(\omega)$

(c) $\dfrac{j}{2}\dfrac{d}{d\omega}\left[F\left(-\dfrac{\omega}{2}\right)\right] - F\left(-\dfrac{\omega}{2}\right)$ (d) $\dfrac{e^{-j5\omega}}{2}F\left(\dfrac{\omega}{2}\right)$

(e) $j\dfrac{d}{d\omega}[\omega F(\omega)]$ (f) $\dfrac{F(\omega)}{d\omega}$

5. $\dfrac{1}{2}\left[\text{Sa}^2\left(\dfrac{\omega+10\pi}{2}\right) + \text{Sa}^2\left(\dfrac{\omega-10\pi}{2}\right)\right]$

6. $\dfrac{1}{j\omega}\,\text{Sa}^2(\omega/2)$

7. $2[\,\text{Sa}(3(\omega-\pi)) + 2\,\text{Sa}(3(\omega+\pi))]$

8. $F(\omega) = \dfrac{2\pi A\tau}{T}\sum_{n=-\infty}^{\infty}\text{Sa}(\omega - n\omega_0\tau/2)\delta(\omega - n\omega_0)$

9. $|F(\omega)| = \dfrac{16}{\sqrt{4+\omega^2}}, \quad \theta(\omega) = \tan^{-1}(\omega/2)$

10. $F(\omega) = e^{-\omega^2/4\pi}$

11. $\dfrac{2a}{a^2+\omega^2}$

12. (i) $\dfrac{1}{j}\left[\dfrac{a}{a^2+(\omega+1)^2} - \dfrac{a}{a^2+(\omega-1)^2}\right]$

(ii) $\sqrt{\dfrac{\pi}{a}}\cos\left(\dfrac{1}{4a}(\omega^2 - a\pi)\right)$

(iii) $-\sqrt{\dfrac{\pi}{a}}\sin\left(\dfrac{1}{4a}(\omega^2 - a\pi)\right)$

13. (i) $\dfrac{a+j\omega}{(a+j\omega)^2+b^2}$ (ii) $= \dfrac{a^2+\omega^2}{(a^2-\omega^2)^2}$

14. (i) $\dfrac{e^{-(4-j\omega)}}{4-j\omega}$ (ii) $F(\omega) = \dfrac{\pi}{2j}[\delta(\omega-\pi) - \delta(\omega+\pi)] + \dfrac{\pi}{\pi^2+\omega^2}$

15. $Y(\omega) = \dfrac{4}{4+\omega^2}$

16. (i) $40\ \text{Sa}^2\, 2\omega$ (ii) $\dfrac{A\omega_0}{\omega_0^2-\omega^2}\left(1+e^{-j\omega T}\right)$

17. (i) $A\tau\ \text{Sa}^2\left(\dfrac{\omega\tau}{2}\right)$ (ii) $A\tau\ \text{Sa}\left(\dfrac{\omega\tau}{2}\right)$

18. $F(\omega) = 10e^{-j2\omega}(\text{Sa}(2\omega+16) + \text{Sa}(2\omega-16))$

19. $X(\omega) = \dfrac{20}{\omega^2}(1-\cos 2\omega)$

20. $F(\omega) = 2\pi\delta(\omega+10\pi) + 10\pi\delta(\omega) + 2\pi\delta(\omega-10\pi)$

21. (a) $\pi\left[e^{j\theta}X(\omega-\omega_0) + e^{-j\theta}X(\omega+\omega_0)\right]$ (b) $-j\pi\cdot\left[e^{j\theta}X(\omega-\omega_0) - e^{-j\theta}X(\omega+\omega_0)\right]$

Multiple-choice Questions

1. The Fourier transform of a real valued time signal has
   (a) odd symmetry (b) even symmetry
   (c) conjugate symmetry (d) no symmetry

2. The function $f(t)$ has the Fourier transform $g(\omega)$. The Fourier transform of $g(t) = \left(\int_{-\infty}^{\infty} g(t)e^{-j\omega t}\,dt\right)$ is
   (a) $\dfrac{1}{2\pi}f(\omega)$ (b) $\dfrac{1}{2\pi}f(-\omega)$
   (c) $2\pi t f(-\omega)$ (d) none of the these

3. The amplitude spectrum of a Gaussian pulse is
   (a) uniform (b) a sine function
   (c) Gaussian (d) an impulse function
4. The Fourier transform of a function $x(\omega)$ is $X(\omega)$. The Fourier transform of $\frac{dx(t)}{dt}$ will be
   (a) $\frac{dX(f)}{dt}$ (b) $j2\pi fX(f)$ (c) $jfX(f)$ (d) $\frac{X(f)}{jf}$
5. The Fourier transform of a voltage signal $x(\omega)$ is $X(\omega)$. The unit of $X(f)$ is
   (a) volt (b) volt-second (c) volt/second (d) volt^2
6. A signal $x(t)$ has a Fourier transform $X(\omega)$. If $x(t)$ is a real and odd function of t, then $X(\omega)$ is
   (a) a real and even function of ω (b) an imaginary and odd function of ω
   (c) an imaginary and even function of ω (d) a real and odd function of ω
7. The Fourier transform of the signal $x(t) = e^{-3t^2}$ is of the following form, where A and B are constants:
   (a) $Ae^{-B|f|}$ (b) Ae^{-Bf} (c) $A+B|f|^2$ (d) Ae^{-Bf^2}
8. If a signal $f(t)$ has energy E, the energy of the signal $f(2t)$ is equal to
   (a) E (b) $E/2$ (c) $2E$ (d) $4E$
9. The Fourier transform $\mathcal{F}\{e^{-t}u(t)\}$ is equal to $\frac{1}{1+j2\pi f}$. Therefore, $\mathcal{F}\left(\frac{1}{1+j2\pi t}\right)$ is
   (a) $e^{f}u(f)$ (b) $e^{-f}u(f)$ (c) $e^{f}u(-f)$ (d) $e^{-f}u(-f)$
10. The Fourier transform of a conjugate symmetric function is always
    (a) imaginary (b) conjugate anti-symmetric
    (c) real (d) conjugate symmetric
11. Let $x(t) \longleftrightarrow X(j\omega)$ be a Fourier transform pair. The Fourier transform of the signal $x(5t-3)$ in terms of $X(j\omega)$ is given as
    (a) $\frac{1}{5}e^{-\frac{j3\omega}{5}}X\left(\frac{j\omega}{5}\right)$ (b) $\frac{1}{5}e^{\frac{j3\omega}{5}}X\left(\frac{j\omega}{5}\right)$
    (c) $\frac{1}{5}e^{-j3\omega}X\left(\frac{j\omega}{5}\right)$ (d) $\frac{1}{5}e^{j3\omega}X\left(\frac{j\omega}{5}\right)$
12. If $G(f)$ represents the Fourier transform of a signal $g(t)$ which is real and odd symmetric in time, then
    (a) $G(f)$ is complex (b) $G(f)$ is imaginary
    (c) $G(f)$ is real (d) $G(f)$ is real and non-negative
13. If the Fourier transform of the deterministic signal $g(t)$ is $G(f)$, then
    (1) The Fourier Transform of $g(t-2)$ is (A) $G(f)e^{-j4\pi f}$
    (2) The Fourier Transform of $g(t/2)$ is (B) $G(2f)$
    (C) $2G(2f)$
    (D) $G(f-2)$

Match each of the items 1, 2 on the left with the most appropriate item A, B, C or D on the right.

(a) 1-B, 2-C (b) 1-A, 2-C (c) 1-D, 2-C (d) 1-C, 2-A

14. A signal $x(t)$ has a Fourier transform $X(\omega)$. If $x(t)$ is a real and odd function of t, then $X(\omega)$ is
    - (a) a real and even function of ω
    - (b) an imaginary and odd function of ω
    - (c) an imaginary and even function of ω
    - (d) a real and odd function of ω

15. If signals $x(t)$ and $y(t)$ with Fourier transforms $X(f)$ and $Y(f)$ respectively are related as $y(t) = -x(2(t-1))$, that is, $y(t)$ is compressed, shifted and inverted form of $x(t)$ as shown in Fig. P4.15(a) and (b), then $Y(f)$ is

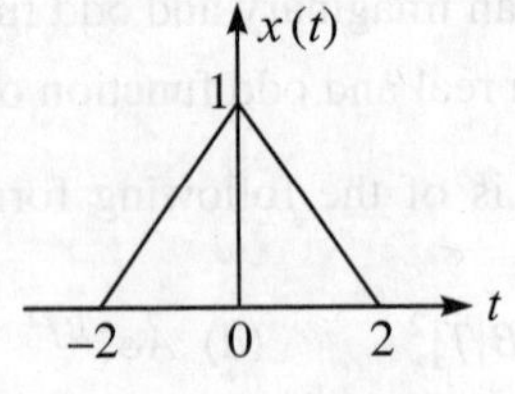

Fig. P4.15(a)

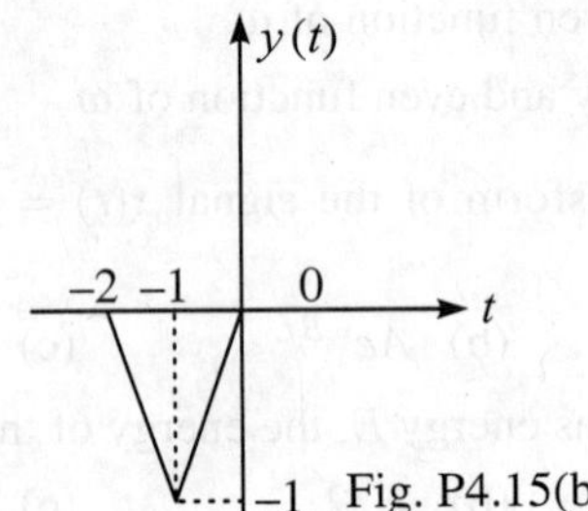

Fig. P4.15(b)

(a) $-\frac{1}{2}X\left(\frac{f}{2}\right)e^{-j2\pi f}$ (b) $-\frac{1}{2}X\left(\frac{f}{2}\right)e^{j2\pi f}$

(c) $-X\left(\frac{f}{2}\right)e^{j2\pi f}$ (d) $-X\left(\frac{f}{2}\right)e^{-j2\pi f}$

16. For a signal $x(t)$, the Fourier transform is $X(f)$. Then the inverse Fourier transform of $X(3f+2)$ is given by

(a) $\frac{1}{2}x\left(\frac{t}{2}\right)e^{j3\pi t}$ (b) $\frac{1}{2}x\left(\frac{t}{2}\right)e^{-\frac{j4\pi t}{3}}$

(c) $3x(3t)e^{-j4\pi t}$ (d) $x(3t+2)$

17. The signal $x(t)$ is described by $x(t) = \begin{cases} 1 & -1 \le t \le +1 \\ 0 & \text{otherwise} \end{cases}$. Two of the angular frequencies at which its Fourier transform becomes zero are

(a) $\pi, 2\pi$ (b) $0.5\pi, 1.5\pi$ (c) $0, \pi$ (d) $2\pi, 2.5\pi$

18. The frequency domain of a periodic triangular function is a
    - (a) discrete sampling function
    - (b) discrete sync function
    - (c) continuous sampling function
    - (d) discrete sampling square function

19. An impulse function consists of
    - (a) the entire frequency range with same relative phase
    - (b) infinite bandwidth with linear phase variations
    - (c) pure dc
    - (d) large dc along with weak harmonics.

20. Sinusoidal function is used as basis function in electrical communication because
    (a) of convenience (b) response to linear system is sine wave
    (c) both a and b (d) none of the these
21. The derivative of an ideal step function is
    (a) an impulse function (b) zero
    (c) sync function (d) undefined
22. A signum function is
    (a) zero for t greater than zero, (b) zero for t less than zero
    (c) unity for t greater than zero (d) $2u(t) - 1$
23. A sine wave generator starts at $t = 0$. It consists of
    (a) a pure sine wave (b) a pure cosine wave
    (c) both a. and b. (d) sine wave plus other frequencies
24. The Fourier transform of a signum function
    (a) does not exist (b) is another unit step function
    (c) contains an impulse function (d) is $2/j\omega$
25. The Fourier transform of a voltage signal $x(t)$ is $X(f)$. The unit of $|X(f)|$ is
    (a) volt (b) volt second (c) volt/second (d) volt Hz
26. The Fourier transform of $x(t)$ is $X(f)$. The Fourier transform of $\frac{dx}{dt}$ will be
    (a) $\frac{dX(f)}{df}$ (b) $j2\pi fX$ (c) $jtx(t)$ (d) $\frac{X(f)}{jf}$
27. The magnitude spectrum of a Fourier transform of a real valued time signal has
    (a) odd symmetry (b) even symmetry
    (c) conjugate symmetry (d) no symmetry
28. The magnitude squared of the Fourier transform $|X(\omega)|^2$ is
    (a) Energy (b) Energy spectral density
    (c) Power (d) Power density
29. The Fourier transform of a voltage signal $x(t)$ is $X(\omega)$. The unit of $|X(\omega)|^2$ is
    (a) volt (b) volt-sec (c) joule/Hz (d) joule-Hz
30. The convolution of two gate functions gives
    (a) Gate function (b) Exponential function
    (c) Impulse function (d) Triangular function

Answers

1. (c)	2. (c)	3. (c)	4. (b)	5. (b)	6. (b)	7. (d)	8. (b)	9. (c)	10. (c)
11. (a)	12. (b)	13. (b)	14. (a)	15. (b)	16. (b)	17. (b)	18. (d)	19. (a)	20. (b)
21. (a)	22. (d)	23. (d)	24. (d)	25. (a)	26. (b)	27. (b)	28. (b)	29. (c)	30. (d)

5

Signal Transmission through Linear Systems

Learning Objectives

The main objective of this chapter is to get a clear understanding of the basic concepts of linear systems. We will learn how signals are transmitted through linear systems and in the process relate to the following topics.

— Linear systems and their response
— Transfer function of an LTI system
— Distortionless transmission through a system
— Ideal filter characteristics of linear systems
— Causality and Paley–Wiener criterion for physical realisation
— Signals, system bandwidth and rise time
— Convolution in time domain
— Convolution in frequency domain with Fourier transforms
— Graphical representation of convolution

5.1 Introduction

The modelling of linear systems has great advantage in real time applications since several physical systems can be easily modelled as LTI systems in contrast to non-LTI systems where no general procedures exist for such modelling. There are several methods to describe the behaviour of the output response when an arbitrary input signal is applied to an LTI system. In the earlier chapter we have seen that the unit impulse can be used as a building block to represent various signals, which together with superposition and time invariance properties can be used to completely characterise an LTI system in terms of its response to unit impulse. Thus, just as any arbitrary signal can be approximated as the superposition of weighted and time shifted impulses, the impulse response of an LTI system too can be expressed as the superposition of weighted and time shifted impulse responses.

Many physical processes can be represented by linear time-invariant (LTI) systems as models. For example, RLC circuits, filter circuits, IC amplifier circuits, motor circuits, etc., which can be expressed by linear differential equations, are usually modelled as LTI systems for analysis. LTI models also serve well in the analysis of simple systems design in industrial applications.

5.2 Linear System

A system is said to be linear if it satisfies the principle of superposition. This implies that the response of a linear system to a weighted sum of input signals is equal to the same weighted sum of output signals, each output signal being associated with particular input signal acting on the system independently of all other input signals. If a system violates the principle of superposition, it is called nonlinear.

Let $x(t) = \sum_{i=1}^{N} a_i x_i(t)$ be the sum of N number of input signals, where a_i are weighted constants. Then the output signal is

$$y(t) = T\left[x(t)\right] = T\left[\sum_{i=1}^{n} a_i x_i(t)\right] \tag{5.1}$$

where $T[\ ]$ is a transformation operator.

If the system is linear, then,

$$y(t) = \sum_{i=1}^{N} a_i y_i(t) = T\left[\sum_{i=1}^{n} a_i x_i(t)\right], \tag{5.2}$$

where $y_i(t) = T\left[x_i(t)\right]$ is the output of the system in response to the inputs $x_i(t)$ acting alone.

5.3 Linear Time Invariant Systems

A system is said to be a linear time invariant (LTI) system if it satisfies linear and time invariance properties. A system is time-invariant if a time-delay or time-advance of the input signal leads to an identical time-shift in the output signal. Stated in another way, the characteristics of a time-invariant system do not change with time.

For an LTI system, if the input is delayed by t_0 seconds, the system satisfies superposition and homogeneity principles. Also, the output is delayed by the same time t_0 seconds.

If $y(t) = T\left[x(t)\right]$, then if $y(t-t_0) = T\left[x(t-t_0)\right]$ (5.3)

it is a time invariant system.

Also, $y(t,t_0) = y(t-t_0)$. (5.4)

Note 1: The linear and time variant properties for signals explained in chapter1 are also valid for systems.

Note 2: The coefficients of the differential equation describing LTI systems are always constant.

5.4 Linear Time Variant Systems

A system is said to be linear time variant (LTV) if it satisfies the linear property but not the time invariant property. For an LTV system, if the input is delayed by t_0 seconds, the system satisfies superposition and homogeneity principles but the output varies with t_0. The coefficients in the differential equation are time variant.

Therefore, $y(t,t_0) \neq y(t-t_0)$. (5.5)

Also, the coefficients in the differential equation describing an LTV system are time variant, unlike the LTI system characterised by constant coefficients.

5.5 Response of a Linear System

Now we shall consider the various methods for describing the relationship between the input and the output of a linear time invariant (LTI) system. The response of a continuous-time LTI system can be computed by the convolution of the system's impulse response with the input signal using a convolution integral. The methods we shall discuss are given in the following subsections.

5.5.1 Impulse Response

When the input signal to a system is the impulse signal, that is $x(t) = \delta(t)$, then the response of the systems is known as impulse response denoted by $h(t)$. The impulse response of an LTI system due to an impulse input applied at time $t = 0$ is

$$h(t) = T[\delta(t)]. \tag{5.6}$$

Note:

1. The zero state response of the system can be deduced by the impulse response of the system.
2. The impulse response completely characterises the behaviour of an LTI system.

Response to An Arbitrary Input

A continuous-time signal can be expressed as the weighted superposition of the time-shifted impulses:

$$x(t) = \int_{-\infty}^{\infty} x(\tau)\delta(t-\tau)d\tau \tag{5.7}$$

This result can be obtained by dividing the signal $x(t)$ into sections of width Δ as shown in Fig. 5.1.

Then the signal approximation can be expressed as a staircase function

$$\hat{x}(t) := \sum_{k=-\infty}^{\infty} x(k\Delta)\delta_\Delta(t-k\Delta)\Delta, \tag{5.8}$$

$$\delta_\Delta = \begin{cases} \frac{1}{\Delta} & 0 \leq t \leq \Delta \\ 0 & \text{otherwise} \end{cases} \tag{5.9}$$

is a unit pulse function and $x(k\Delta)$ is the signal height.

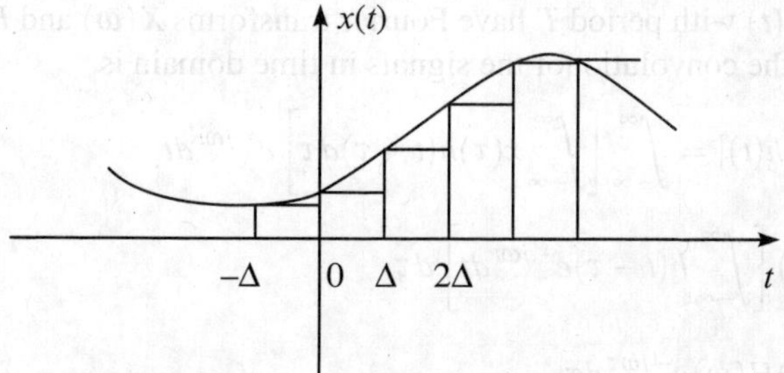

Fig. 5.1 *Signal approximation*

Now, taking limits, Eq. (5.8) becomes

$$x(t) = \lim_{\Delta \to 0} \sum_{k=-\infty}^{\infty} x(k\Delta)\delta_{\Delta}(t - k\Delta)\Delta. \tag{5.10}$$

As $\Delta \to 0$, the summation approaches the integral, and

$$k\Delta \to \tau, \quad \Delta \to d\tau, \quad \delta_{\Delta}(t - k\Delta) \to \delta(t - \tau).$$

Then, $x(t) = \int_{-\infty}^{\infty} x(\tau)\delta(t-\tau)d\tau$.

Since the system is linear, the response $y(t)$ of the system to an arbitrary input $x(t)$ can be expressed as,

$$y(t) = T\{x(t)\} = T\int_{-\infty}^{\infty} x(\tau)\delta(t-\tau)d\tau = \int_{-\infty}^{\infty} x(\tau)T\{\delta(t-\tau)\}d\tau.$$

For time-invariant system, $h(t-\tau) = T\{\delta(t-\tau)\}$.

$$y(t) = \int_{-\infty}^{\infty} x(\tau)h(t-\tau)d\tau. \tag{5.11}$$

Equation (5.11) indicates that a continuous-time LTI system is completely characterised by its impulse response. This equation is called a convolution integral and it is denoted as $x(t) * h(t)$.

Thus the system response is

$$y(t) = x(t) * h(t) = \int_{-\infty}^{\infty} x(\tau)h(t-\tau)d\tau. \tag{5.13}$$

A continuous-time LTI system with response is shown in Fig. 5.2.

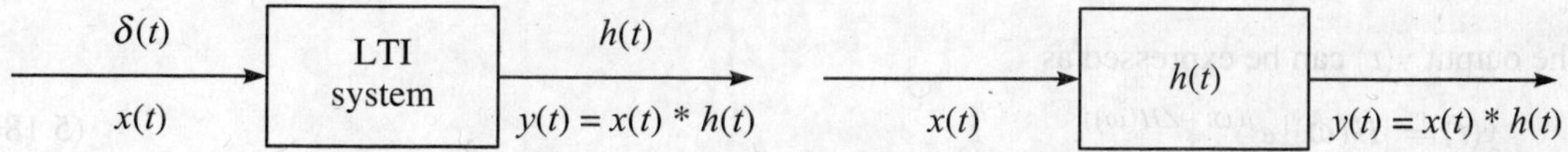

Fig. 5.2 *Continuous-time LTI system*

5.5.2 Frequency Response of an LTI System

If $y(t)$ is the response of an LTI system with impulse response $h(t)$ to an arbitrary input $x(t)$, then convolution is a particularly powerful way of characterizing the input–output relationship of the time-invariant linear system. The transform domain offers a convenient method of evaluating the convolution operation.

Let signals $x(t)$ and $h(t)$ with period T have Fourier transforms $X(\omega)$ and $H(\omega)$ respectively. Then the Fourier transform of the convolutionof the signals in time domain is

$$\mathcal{F}[y(t)] = \mathcal{F}[x(t) * h(t)] = \int_{-\infty}^{\infty}\left[\int_{-\infty}^{\infty} x(\tau)h(t-\tau)d\tau\right] e^{-j\omega t}dt$$

$$= \int_{-\infty}^{\infty} x(\tau)\left[\int_{-\infty}^{\infty} h(t-\tau)e^{-j\omega t}dt\right] d\tau$$

$$= \int_{-\infty}^{\infty} x(\tau)H(\omega)e^{-j\omega\tau}d\tau.$$

$$Y(\omega) = X(\omega)H(\omega). \tag{5.14}$$

Thus, a convolution operation in the time domain results in a multiplication operation in the frequency domain.

The output response of the system is

$$y(t) = \mathcal{F}^{-1}[Y(\omega)]$$

$$y(t) = \frac{1}{2\pi}\int_{-\infty}^{\infty} X(\omega)H(\omega)\cdot e^{j\omega t}d\omega. \tag{5.15}$$

$H(\omega)$ represents the frequency response of the system. It is a complex valued function and can be expressed as

$$H(\omega) = |H(\omega)|\angle H(\omega), \tag{5.16}$$

where $|H(\omega)|$ is called the magnitude response of the system and $\angle H(\omega)$ the phase response of the system.

Note: The magnitude response is symmetric and phase response is anti-symmetric.

Response to Eigen Functions

If the input to the system is an exponential function $x(t) = e^{j\omega t}$, then its output is

$$y(t) = x(t) * h(t) = \int_{-\infty}^{\infty} x(t-\tau)h(\tau)d\tau$$

$$= \int_{-\infty}^{\infty} e^{j\omega(t-\tau)}h(\tau)d\tau = e^{j\omega t}\int_{-\infty}^{\infty} e^{-j\omega\tau}h(\tau)d\tau$$

$$y(t) = e^{j\omega t}H(\omega). \tag{5.17}$$

The output $y(t)$ can be expressed as

$$y(t) = |H(\omega)|\,e^{j(\omega t+\angle H(\omega))}. \tag{5.18}$$

The output is a complex exponential of the same frequency as input multiplied by the complex constant $H(\omega)$. Similarly, for $\sin\omega t$ and $\cos\omega t$ inputs. An input signal is called an eigenfunction of the system if the corresponding output is a constant multiple of the input signal. Thus, the functions $e^{j\omega t}$, $\sin\omega t$ and $\cos\omega t$ are all eigenfunctions as we get the same function in the output as in the input.

Example 5.1

A system has a transfer function $H(\omega) = K$, where K is a constant. Find the impulse response of this system.

Solution

Given that the transfer function of the systems is $H(\omega) = K$. The magnitude of the transfer function $|H(\omega)| = |K| = K$ is a constant independent of frequency. The phase of the transfer function $\angle H(\omega) = 0$ is a zero for all the frequencies.

The impulse response of the system is

$$h(t) = \mathcal{F}^{-1}[H(\omega)] = \mathcal{F}^{-1}[K] = K\delta(t).$$

Example 5.2

The impulse response of a continuous time RC filter is given by $h(t) = \frac{1}{RC}e^{-(1/RC)t}u(t)$. Find the frequency response of the system.

Solution

Given the impulse response of the system $h(t) = \frac{1}{RC}e^{-(1/RC)t}u(t)$.

The frequency response of the system $H(\omega)$ is the Fourier transform of the impulse response of the system.

$$\therefore\ H(\omega) = \mathcal{F}[h(t)] = \mathcal{F}\left[\frac{1}{RC}e^{-(1/RC)t}u(t)\right]$$

$$= \frac{1}{RC}\frac{1}{j\omega + (1/RC)} = \frac{1}{1 + j\omega RC}.$$

Magnitude response $|H(\omega)| = \frac{1}{\sqrt{1 + (\omega RC)^2}}$.

Phase response $\angle H(\omega) = -\tan^{-1}(\omega RC)$.

5.5.3 Properties of LTI System

Commutative Property

The output of an LTI system with input $x(t)$ and impulse response $h(t)$ is identical to the output of an LTI system with input $h(t)$ and impulse response $x(t)$. That is, $x(t) * h(t) = h(t) * x(t)$.

$$\int_{-\infty}^{\infty} x(t-\tau)h(\tau)d\tau = \int_{-\infty}^{\infty} h(t-\tau)x(\tau)d\tau \qquad (5.19)$$

Associative Property

This implies that a cascade of two (or more) LTI systems (Fig. 5.3) can be reduced to a single system with impulse response equal to the convolution of the impulse responses of the cascaded systems.

$$\{x(t) * h_1(t)\} * h_2(t) = x(t) * \{h_1(t) * h_2(t)\}. \qquad (5.20)$$

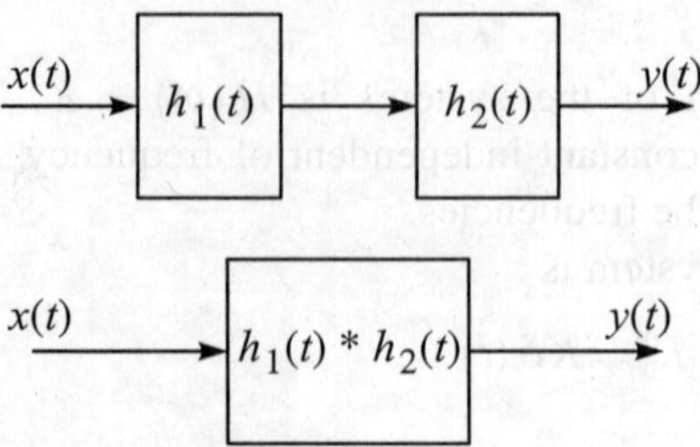

Fig. 5.3 *Cascade of two LTI systems*

Distributive Property

This property gives that summing the outputs of two systems subjected to the same input is equivalent to a system whose impulse response equals to the sum of the impulse responses of the two individual systems as shown in Fig. 5.4.

$$x(t) * \{h_1(t)\} + h_2(t)\} = x(t) * h_1(t) + x(t)h_2(t)$$

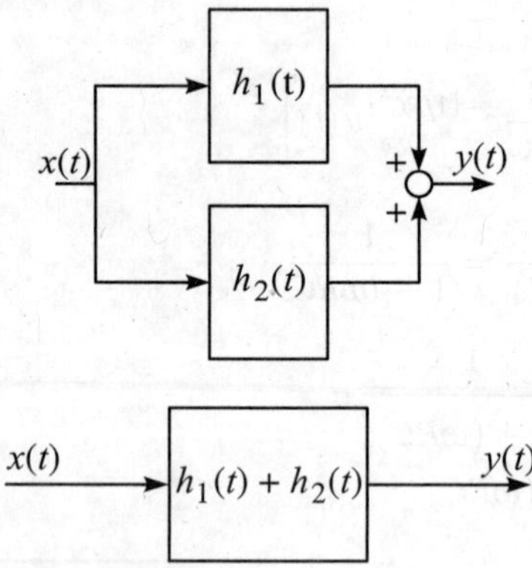

Fig. 5.4 *Demonstrating distributive property of LTI systems*

Static and Dynamic Systems

A system S is static or memoryless if its output at any time depends only on the value of its input at that instant of time. For an LTI system, this property can only hold if its impulse response is itself an impulse. But from the convolution property, we know that the output depends on the previous samples of the input. Therefore an LTI system has memory and hence it is a dynamic system.

Causality

A continuous-time LTI system is said to be causal if and only if its impulse response is $h(t) = 0$, $t < 0$. Then the integral becomes,

$$y(t) = \int_0^{\infty} h(\tau)x(t-\tau)d\tau = \int_{-\infty}^{t} x(\tau)h(t-\tau)d\tau. \tag{5.22}$$

If the signal $x(t)$ is also causal, then

$$y(t) = \int_0^{t} h(\tau)x(t-\tau)d\tau = \int_0^{t} x(\tau)h(t-\tau)d\tau. \tag{5.23}$$

Note: All physical systems are causal since it is impossible to predict their future.

Stability

A continuous-time LTI system is bounded input, bounded output (BIBO) stable if and only if the impulse response is absolutely integrable.

Proof: Consider an LTI system with impulse response $h(t)$. The output response is

$$y(t) = \int_{-\infty}^{\infty} h(\tau)x(t-\tau)d\tau.$$

Therefore, $|y(t)| \le \int_{-\infty}^{\infty} |h(\tau)|\,|x(t-\tau)|\,d\tau.$

If $x(t)$ is bounded and $|x(t-\tau)| < K_1 < \infty$, then

$$|y(t)| \le K_1 \int_{-\infty}^{\infty} |h(\tau)|\,d\tau.$$

Also, for the response, $y(t) = \int_{-\infty}^{\infty} x(\tau)h(t-\tau)d\tau$

$$|y(t)| \le K_1 \int_{-\infty}^{\infty} |h(t-\tau)|d\tau.$$

For bounded output, i.e., $|y(t)| < \infty$, the impulse response should be absolutely integrable. Hence,

$$\int_{-\infty}^{\infty} |h(\tau)|d\tau < \infty \quad \text{or} \quad \int_{-\infty}^{\infty} |h(t-\tau)|d\tau < \infty. \tag{5.24}$$

Therefore, for LTI systems Eq. (5.24) gives the necessary and sufficient condition for BIBO stability.

Invertibility

We have already seen that a system S is invertible if and only if there exists an inverse system S^{-1} for it such that SS^{-1} is an identity system. For an LTI system with impulse response $h_1(t)$, this is equivalent to the existence of another system with impulse response $h_2(t)$ such that $h_1(t) * h_2(t) = \delta(t)$, as shown in Fig. 5.5.

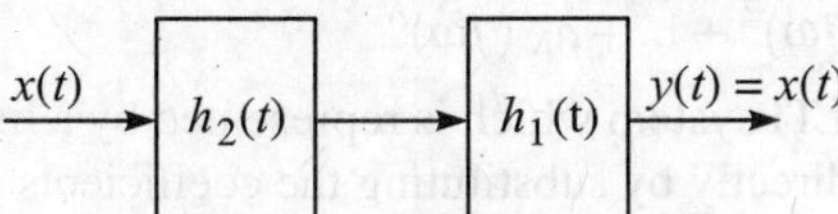

Fig. 5.5 *Inverse of two LTI systems*

5.6 Transfer Function of an LTI System

The ratio of the Fourier transform of the output signal $Y(\omega)$ to the Fourier transform of the input signal $X(\omega)$ of an LTI system is called its transfer function. It is expressed as

$$H(\omega) = \frac{Y(\omega)}{X(\omega)}. \tag{5.25}$$

The inverse Fourier transform of $H(\omega)$ gives the impulse response of the system. That is

$$h(t) = \mathcal{F}^{-1}[H(\omega)]. \tag{5.26}$$

Solution of Differential Equation Representation of LTI System

A general Nth-order linear constant-coefficient differential equation with zero initial conditions is given by

$$\sum_{k=0}^{N} a_k \frac{d^k y(t)}{dt^k} = \sum_{k=0}^{M} b_k \frac{d^k x(t)}{dt^k}, \tag{5.27}$$

where the coefficients a_k and b_k are real constants. The order N refers to the highest derivative of $y(t)$ in the above equation.

Using the differentiation property of Fourier transform,

$$\sum_{k=0}^{N} a_k (j\omega)^k Y(\omega) = \sum_{k=0}^{M} b_k (j\omega)^k X(\omega)$$

$$Y(\omega) \sum_{k=0}^{N} a_k (j\omega)^k = X(\omega) \sum_{k=0}^{M} b_k (j\omega)^k$$

$$\frac{Y(\omega)}{X(\omega)} = \frac{\sum_{k=0}^{M} b_k (j\omega)^k}{\sum_{k=0}^{N} a_k (j\omega)^k}. \tag{5.28}$$

The system's transfer function is

$$H(\omega) = \frac{\sum_{k=0}^{M} b_k (j\omega)^k}{\sum_{k=0}^{N} a_k (j\omega)^k} \tag{5.29}$$

$$= \frac{b_0 + j\omega b_1 + b_2 (j\omega)^2 + \ldots + b_M (j\omega)^M}{a_0 + j\omega a_1 + a_2 (j\omega)^2 + \ldots + a_N (j\omega)^N}. \tag{5.30}$$

Thus the transfer function of an LTI system which is represented by a linear constant-coefficient differential equation can be obtained directly by substituting the coefficients of the differential equation.

Example 5.3

A system produces an output $y(t) = e^{-t}u(t)$ for an input $x(t) = e^{-2t}u(t)$. Determine the impulse response and frequency response of the system.

Solution

Given output $y(t) = e^{-t}u(t)$ and input $x(t) = e^{-2t}u(t)$, their Fourier transforms are:

$$Y(\omega) = \mathcal{F}\left[e^{-t}u(t)\right] = \frac{1}{j\omega + 1} \text{ and } X(\omega) = \mathcal{F}\left[e^{-2t}u(t)\right] = \frac{1}{j\omega + 2}.$$

The transfer function of the system is

$$H(\omega) = \frac{Y(\omega)}{X(\omega)} = \frac{1/(j\omega+1)}{1/(j\omega+2)} = \frac{j\omega+2}{j\omega+1}.$$

$$H(\omega) = \frac{j\omega+1+1}{j\omega+1} = 1 + \frac{1}{j\omega+1}.$$

The impulse response of the system is

$$h(t) = \mathcal{F}^{-1}[H(\omega)] = \mathcal{F}^{-1}\left[1+\frac{1}{j\omega+1}\right] = \delta(t) + e^{-t}u(t).$$

The frequency response of the system is

$$H(\omega) = \frac{j\omega+2}{j\omega+1}.$$

Magnitude response is $|H(\omega)| = \dfrac{\sqrt{4+\omega^2}}{\sqrt{1+\omega^2}}$

Phase response is $\angle H(\omega) = \tan^{-1}\dfrac{\omega}{2} - \tan^{-1}\omega.$

Example 5.4

Consider a causal LTI system with frequency response $H(\omega) = \dfrac{1}{j\omega+4}$. For a particular input $x(t)$, the system is observed to produce an output $y(t) = e^{-2t}u(t) - e^{-4t}u(t)$. Find the input signal $x(t)$.

Solution

Given the frequency response $H(\omega) = \dfrac{1}{j\omega+4}$ and the output $y(t) = e^{-2t}u(t) - e^{-4t}u(t)$. The Fourier transform is

$$Y(\omega) = \mathcal{F}\left[e^{-2t}u(t)\right] - \mathcal{F}\left[e^{-4t}u(t)\right]$$

$$= \frac{1}{j\omega+2} - \frac{1}{j\omega+4} = \frac{2}{(j\omega+2)(j\omega+4)}.$$

The transfer function of the system is $H(\omega) = \dfrac{Y(\omega)}{X(\omega)}$.

From this, the input can be obtained as follows.

$$X(\omega) = \frac{Y(\omega)}{H(\omega)} = \frac{2/(j\omega+2)(j\omega+4)}{1/(j\omega+4)} = \frac{2}{(j\omega+2)}.$$

Taking the inverse Fourier transform,

Input $x(t) = \mathcal{F}^{-1}[X(\omega)] = \mathcal{F}^{-1}\left[\dfrac{2}{j\omega+2}\right]$

$$x(t) = 2e^{-2t}u(t).$$

Example 5.5

Consider the system described by $y''(t) + 5y'(t) + 6y(t) = -x'(t)$. Find the impulse response $h(t)$ of the system.

Solution

Given the differential equation $y''(t)+5y'(t)+6y(t) = -x'(t)$. Taking Fourier transform on both sides,

$$(j\omega)^2 Y(\omega)+5(j\omega)Y(\omega)+6Y(\omega) = -j\omega X(\omega)$$

$$\left((j\omega)^2+5(j\omega)+6\right)Y(\omega) = -j\omega X(\omega)$$

$$H(\omega) = \frac{Y(\omega)}{X(\omega)} = \frac{-j\omega}{(-\omega^2+5j\omega+6)} = \frac{-j\omega}{(j\omega+2)(j\omega+3)} H(\omega) = \frac{2}{(j\omega+2)} - \frac{3}{(j\omega+3)}.$$

Taking inverse Fourier transform, we obtain the impulse response $h(t)$of the system:

$$h(t) = 2e^{-2t}u(t) - e^{-3t}u(t).$$

5.7 Distortionless Transmission through a System

Distortionless transmission through an LTI system requires the exact input signal shape be reproduced at the output although its amplitude may vary and the signal may be delayed in time.

Therefore, for any arbitrary input $x(t)$, if the output is

$$y(t) = Kx(t-t_0) \tag{5.31}$$

and its Fourier transform is

$$Y(\omega) = Ke^{-j\omega t_0}X(\omega), \tag{5.32}$$

where K is the amplitude and t_0 is the time delay in seconds. Then the system has distortionless transmission.

Thus, for distortionless transmission, the conditions is that the system transfer function is

$$H(\omega) = Ke^{-j\omega t_0}.$$

Or $|H(\omega)| = K$ and

$$\angle H(\omega) = -j\omega t_0 \quad \text{or} \quad \angle H(\omega) = n\pi - j\omega t_0 (n \text{ integer}). \tag{5.33}$$

This implies that the amplitude of $H(\omega)$ must be constant over the entire frequency range and the phase of $H(\omega)$ must be linear with frequency as shown in Fig. 5.6.

Note: For distortionless transmission, an LTI system should have a flat amplitude response and linear phase variation over the given frequency range.

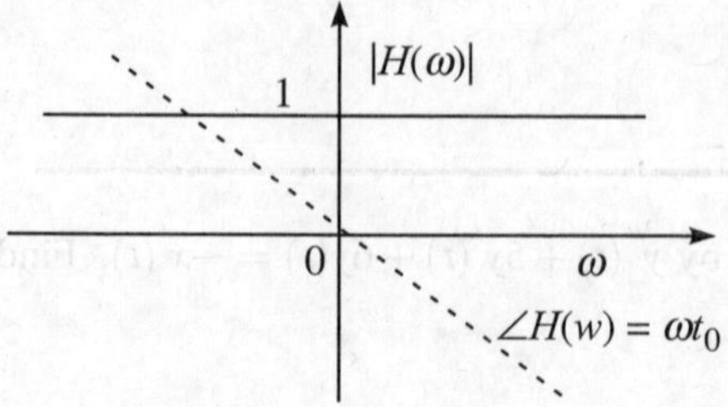

Fig. 5.6 *Magnitude and phase response of $H(\omega)$*

5.8 Filter Characteristics of Linear Systems

In general, a linear system performs filtering action by modifying the input spectrum. It is observed that some frequency components of the input signal are amplified, some are attenuated and some remain the same. That is, the frequency components of the input signal vary according to the system filter characteristics. An LTI system acts as a filter depending on the transfer function $H(\omega)$ of the system. The $H(\omega)$ acts as a weighting function to different frequency components of the input signal.

LTI systems may be classified in to five types of filters.

1. *Lowpass filter (LPF):* The system allows low frequency components to pass and attenuates high frequency components above a certain frequency of the input signal.
2. *High-pass filter (HPF):* The system allows high frequency components to pass and attenuates low frequency components below a certain frequency of the input signal.
3. *Bandpass filter (BPF):* The system allows frequency components within certain frequency band to pass and attenuates frequency components outside that band of the input signal.
4. *Band-rejection filter (BRF):* The system attenuates frequency components within certain frequency band and allows frequency components outside of that band of the input signal to pass..
5. *All-pass filter:* System allows all frequency components of the input signal to pass without distortion.

The *passband* of a filter is the range of frequencies that is allowed by the system without distortion. The *stopband* of a filter is the range of frequencies that is attenuated by the system.

Ideal Filter

An ideal filter-passes all signal frequency components in its pass band without distortion and completely blocks frequency components outside its pass band. There is a discontinuity in frequency spectrum between pass band and stop band. But in practical cases filters have a gradual transition gap between pass band and stop band. The range of frequencies over which there is a gradual attenuation between the passband and the stopband is called the *transition band.* Filters with small transition gap (sharp cutoff) is very difficult to design.

5.9 Ideal LPF, HPF, BPF and BRF Characteristics

Ideal LPF

An ideal lowpass filter transmits all signal frequency components below a certain frequency W rad/s, called cutoff frequency, without distortion. The signal above this frequency is filtered completely.

The transfer function of an ideal LPF is given by

$$H(\omega) = \begin{cases} e^{-j\omega t_0} & |\omega| < W \\ 0 & |\omega| > W. \end{cases} \tag{5.34}$$

The magnitude and phase characteristics of this filter are:

$$|H(\omega)| = 1 \text{ and } \angle H(\omega) = -j\omega t_0 \text{ for } |\omega| < W. \tag{5.35}$$

The response is shown in Fig. 5.7.

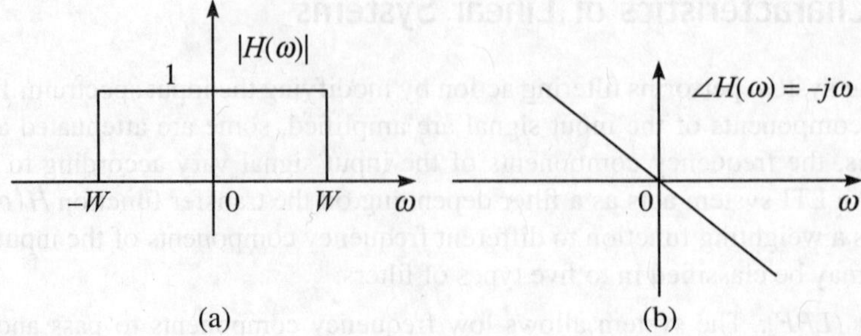

Fig. 5.7 *Ideal lowpass filter characteristics: (a) magnitude response and (b) phase response*

Ideal HPF

An ideal high-pass filter transmits all signal frequency components above a certain frequency W rad/s without distortion. The signal below this frequency is stopped completely. The transfer function of an ideal HPF is given by

$$H(\omega) = \begin{cases} 0 & |\omega| < W \\ e^{-j\omega t_0} & |\omega| > W. \end{cases} \tag{5.36}$$

The magnitude and phase characteristics are:

$$|H(\omega)| = 1 \quad \text{and} \quad \angle H(\omega) = -j\omega t_0 \quad \text{for} \quad |\omega| > W. \tag{5.37}$$

The response is shown in Fig. 5.8.

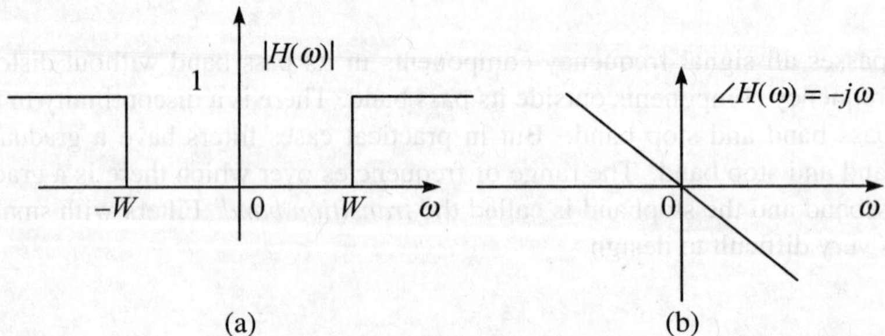

Fig. 5.8 *Ideal high-pass filter characteristics: (a) magnitude response and (b) phase response*

Ideal BPF

An ideal bandpass filter transmits all signal frequency components within a certain frequency band $W_2 - W_1$ rad/s without distortion. The signal with frequency outside this band is stopped completely. The transfer function of an ideal BPF is given by

$$H(\omega) = \begin{cases} e^{-j\omega t_0} & W_1 < |\omega| < W_2 \\ 0 & \text{otherwise.} \end{cases} \tag{5.38}$$

The magnitude and phase characteristics are:

$$|H(\omega)| = 1 \quad \text{and} \quad \angle H(\omega) = -j\omega t_0 \quad \text{for} \quad W_1 < |\omega| < W. \tag{5.39}$$

The response is shown in Fig. 5.9.

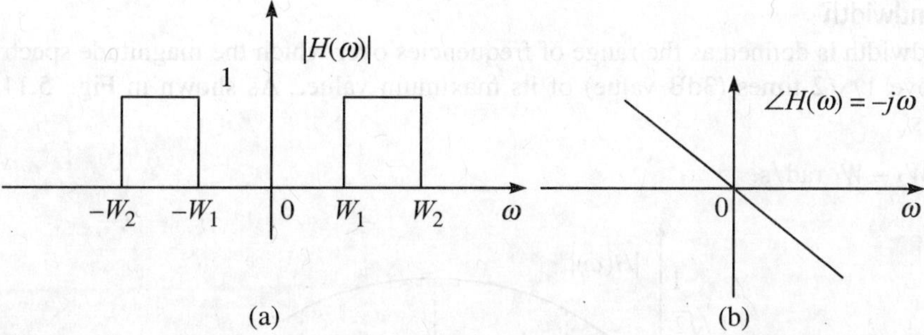

Fig. 5.9 *Ideal bandpass filter characteristics: (a) magnitude response and (b) phase response*

Ideal BRF

An ideal band-rejection filter rejects or stops all signal frequency components within certain frequency band of $W_2 - W_1$ rad/s. The signal outside this band is transmitted without distortion. The transfer function of an ideal BRF is given by

$$H(\omega) = \begin{cases} 0 & W_1 < |\omega| < W_2 \\ e^{-j\omega t_0} & \text{otherwise} \end{cases} . \tag{5.40}$$

The magnitude and phase characteristics of an ideal filter are:

$$|H(\omega)| = 1 \text{ and } \angle H(\omega) = -j\omega t_0 \text{ for } |\omega| < W_1 \text{ and } |\omega| > W_2.$$

The response is shown in Fig. 5.10.

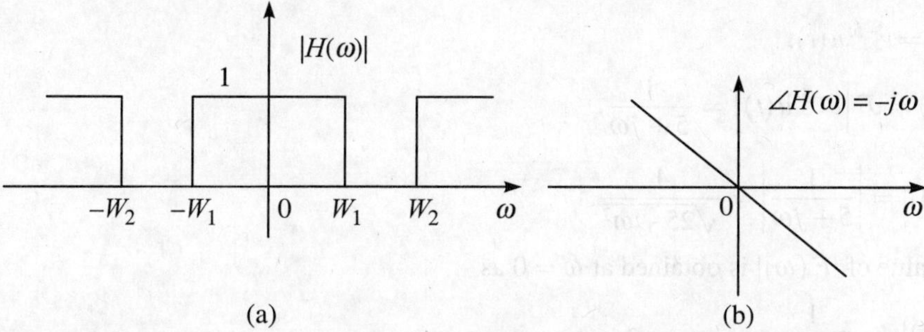

Fig. 5.10 *Ideal rejection-pass filter characteristics a) Magnitude response b) phase response*

Note: All ideal filters are non-causal and distortionless transmission systems.

Filters are used to eliminate the unwanted frequency components of a signal or the noise components from a signal; they also are used to recover the input signal from a noisy signal.

5.10 Bandwidth of Signals and Systems

Bandwidth of A Signal

It is the range of significant frequency components present in a signal. A signal may have frequency components in the entire frequency range from $-\infty$ to ∞. However, since all practical signals have finite energy, only some frequency components of the signals have significant amplitude within a certain frequency band. Outside this band the signals have negligible amplitude. The amplitude of the significant frequency components is within the $1/\sqrt{2}$ times (3dB value) of maximum signal amplitude.

System Bandwidth

System bandwidth is defined as the range of frequencies over which the magnitude spectrum of $H(\omega)$ remains above $1/\sqrt{2}$ times (3dB value) of its maximum value. As shown in Fig. 5.11, the system bandwidth is

$$W = W_2 - W_1 \text{ rad/s.} \tag{5.41}$$

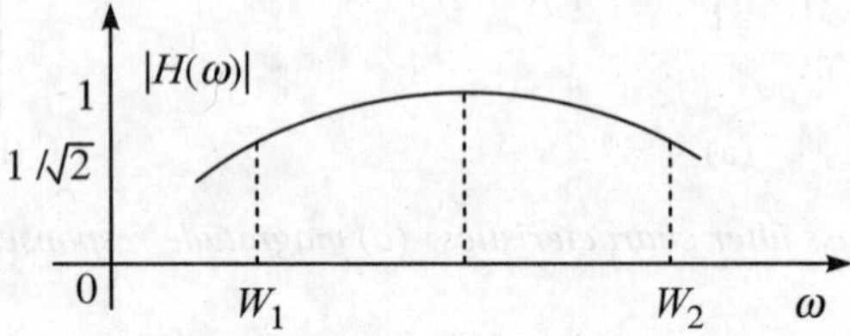

Fig. 5.11 *System bandwidth*

For distortionless transmission, a system should have infinite bandwidth. But due to physical limitations it is impossible to design an ideal filter having infinite bandwidth.

For satisfactory distortionless transmission, therefore, an LTI system should have high bandwidth compared to the signal bandwidth.

Example 5.6

For a low-pass filter, the impulse response is $h(t) = e^{-5t}u(t)$. Determine the bandwidth of the system.

Solution

Given $h(t) = e^{-5t}u(t)$.

$$F(\omega) = \mathcal{F}\left[e^{-5t}u(t)\right] = \frac{1}{5+j\omega}.$$

$$|F(\omega)| = \left|\frac{1}{5+j\omega}\right| = \frac{1}{\sqrt{25+\omega^2}}.$$

The peak value of $|F(\omega)|$ is obtained at $\omega = 0$ as

$$|F(\omega)|_{\text{peak}} = \frac{1}{5}.$$

Given that at $\omega = \omega_c$,

$$|F(\omega)| = \frac{1}{\sqrt{2}} \text{of } |F(\omega)|_{\text{peak}} = \frac{1}{\sqrt{2}}\left(\frac{1}{5}\right) = 0.1414 \quad \text{or} \quad 0.1414 = \frac{1}{\sqrt{25+\omega_c^2}}.$$

The system bandwidth is $\omega_c = 5$ rad/s.

5.11 Causality and Physical Realisability

The Paley–Wiener Criterion

For real time applications, all systems should be physically realisable. For this to be true, in the *time domain* approach the impulse response $h(t)$ of a physically realisable system must be causal. That is,

$h(t) = 0$ for $t < 0$. This condition, called the causal condition, ensures that the system produces an output response only after an input signal is applied. In the *frequency domain* approach, the necessary and sufficient condition for the magnitude response function $|H(\omega)|$ to be physically realisable is:

$$\int_{-\infty}^{\infty} \frac{|\ln H(\omega)|}{1+\omega^2} d\omega < \infty. \tag{5.42}$$

This condition is known as the Paley–Wiener Criterion. To satisfy this condition the function $|H(\omega)|$ must be square integrable, i.e.,

$$\int_{-\infty}^{\infty} |H(\omega)|^2 \, d\omega < \infty. \tag{5.43}$$

Note: All causal systems that satisfy Paley-Wiener Criterion are physically realisable.

Ideal filters are not physically realisable. But it is possible to construct physically realisable filters close to the ideal filter characteristics. As an example, a low-pass filter having transfer function

$$H(\omega) = \begin{cases} e^{-j\omega t_0} & |\omega| < W \\ \in & |\omega| > W, \end{cases} \tag{5.44}$$

where ε is an arbitrarily small value, produces very nearly ideal filter characteristics. The magnitude response is shown in Fig. 5.12.

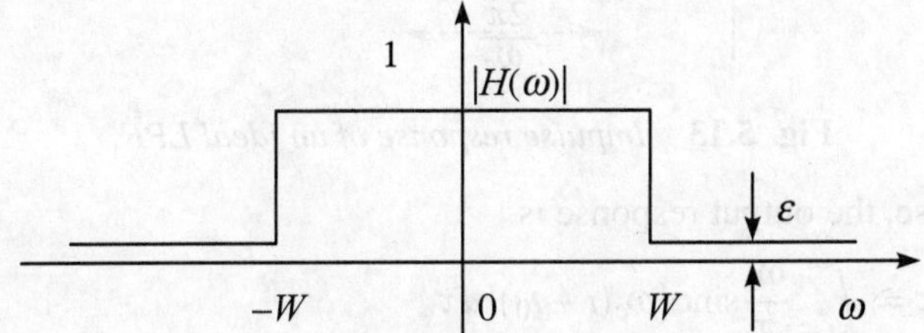

Fig. 5.12 *Magnitude response of a physically realisable LPF*

5.12 Bandwidth and Risetime

The system bandwidth can be obtained from the rise time, which can be derived from the output response of the system.

Rise Time

The rise time t_r of the output response is defined as the time the response takes to reach from 10% to 90% of the maximum value of the signal, or, in general, it is the time of the response to reach from zero to the final value of the signal.

$$\left.\frac{dy(t)}{dt}\right|_{t_0} = \frac{1}{t_r}. \tag{5.45}$$

Relationship between Bandwidth and Rise Time

Consider an ideal LPF. Its transfer function is given by

$$H(\omega) = \begin{cases} e^{-j\omega t_0} & |\omega| < \omega_c \\ 0 & |\omega| > \omega_c \end{cases}, \tag{5.46}$$

where ω_c (rad/s) is the cutoff frequency or 3-dB bandwidth of the filter.

By applying inverse Fourier transform, the impulse response is

$$h(t) = \frac{1}{2\pi}\int_{-\infty}^{\infty} H(\omega)\, e^{j\omega t}\, d\omega = \frac{1}{2\pi}\int_{-\omega_c}^{\omega_c} e^{-j\omega t_0} e^{j\omega t}\, d\omega$$

$$= \frac{1}{2\pi}\int_{-\omega_c}^{\omega_c} e^{j\omega(t-t_0)}\, d\omega = \frac{1}{2\pi}\frac{e^{j\omega(t-t_0)}}{j(t-t_0)}\Bigg|_{-\omega_c}^{\omega_c} = \frac{1}{\pi}\frac{\sin\omega_c(t-t_0)}{(t-t_0)}$$

$$h(t) = \frac{\omega_c}{\pi}\text{sinc}\left[\omega_c(t-t_0)\right]. \tag{5.47}$$

The impulse response is shown in Fig. 5.13.

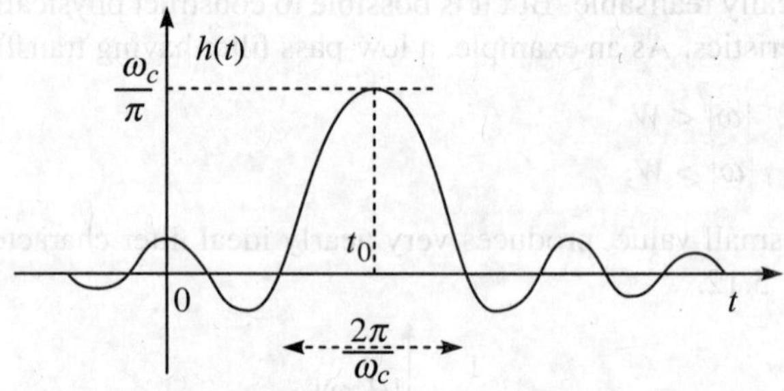

Fig. 5.13 *Impulse response of an ideal LPF*

If the input is impulse, the output response is

$$y(t) = \int_{-\infty}^{\infty} h(\tau) d\tau = \int_{-\infty}^{\infty} \frac{\omega_c}{\pi}\text{sinc}\left[\omega_c(t-t_0)\right] d\tau.$$

The derivative is

$$\frac{dy(t)}{dt} = \frac{\omega_c}{\pi}\text{sinc}\left[\omega_c(t-t_0)\right]$$

$$\frac{dy(t)}{dt} = \frac{\omega_c}{\pi}\frac{\text{sinc}\left[\omega_c(t-t_0)\right]}{\omega_c(t-t_0)} = \frac{\omega_c}{\pi}\frac{\text{sinc}\left[\omega_c(t-t_0)\right]}{\omega_c(t-t_0)}.$$

At $t = t_0$, take limits $\left.\frac{dy(t)}{dt}\right|_{t_0} = \frac{\omega_c}{\pi} = \frac{1}{t_r}$, or

$$\text{Rise Time is } t_r = \frac{\pi}{\omega_c} \text{ seconds,} \tag{5.48}$$

where the filter bandwidth is ω_c rad/s.

Therefore, the product of bandwidth and rise time is π (a constant).

Note: It is observed that the rise time of the output is inversely proportional to the system bandwidth.

Output Response

The output response is $y(t) = \int_{-\infty}^{\infty} h(\tau) d\tau$.

$$y(t) = \int_{-\infty}^{\infty} \frac{\omega_c}{\pi}\text{sinc}\left[\omega_c(t-t_0)\right] d\tau = \frac{1}{\pi}\text{Si}\left[\omega_c(t-t_0)\right], \tag{5.49}$$

where Si $[\omega_c(t-t_0)]$ is a sine integral function as shown in Fig. 5.14.

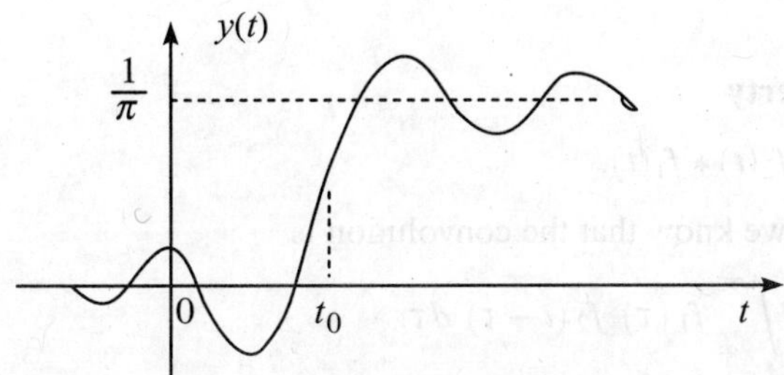

Fig. 5.14 *Sine integral function*

5.13 Concept of Convolution in Time Domain

The process of expressing the output signal in terms of the superposition of weighted and time shifted impulse responses is called convolution. Convolution is a particularly powerful way of characterising the input–output relationship of time-invariant linear systems. The mathematical tool for evaluating the convolution of continuous-time signals is called convolution integral; for discrete-time signals, it is called convolution sum. The convolution integral plays an important role in system analysis in time and frequency domains. It is an important process for signal processing and detection in communication systems.

5.13.1 The Convolution Integral

Let $x_1(t)$ and $x_2(t)$ be two continuous-time signals. Then convolution of $x_1(t)$ and $x_2(t)$ can be expressed as

$$y(t) = x_1(t) * x_2(t) = \int_{-\infty}^{\infty} x_1(\tau)\, x_2(t-\tau)\, d\tau, \tag{5.50}$$

where τ is an independent time variable. The symbol $*$ denotes convolution process. Equation (5.50) is called the convolution integral. Thus the output of any continuous-time LTI system is the convolution of the input $x(t)$ with the impulse response $h(t)$ of the system.

The convolution integral can be applied to any LTI system. However, the convolution integral can be simplified if we assume the following:

First, if $x(t)$ is a positive signal i.e., $x(t) = 0$ for $t < 0$, then

$$y(t) = \int_{0}^{\infty} x(\tau)\, h(t-\tau)\, d\tau. \tag{5.51}$$

Second, if the system is causal, i.e., $h(t) = 0$ for $t < 0$, then $h(t-\tau) = 0$ for $t-\tau < 0$ or $\tau > t$ and the convolution is

$$y(t) = \int_{-\infty}^{t} x(\tau)\, h(t-\tau)\, d\tau. \tag{5.52}$$

Third, if $x(t)$ is a positive signal and the system $h(t)$ is causal, then

$$y(t) = \int_{0}^{t} x(\tau)\, h(t-\tau)\, d\tau. \tag{5.53}$$

5.13.2 Properties of the Convolution Integral

The convolution integral has the following properties. Let $f_1(t)$ and $f_2(t)$ be two continuous-time signals.

(i) Commutative property

$$f_1(t) * f_2(t) = f_2(t) * f_1(t). \tag{5.54}$$

Proof: By definition we know that the convolution is

$$f_1(t) * f_2(t) = \int_{-\infty}^{\infty} f_1(\tau)\, f_2(t-\tau)\, d\tau.$$

Let $t - \tau = x$. Then

$$f_1(t) * f_2(t) = \int_{-\infty}^{\infty} f_1(t-x)\, f_2(x)\, dx$$

$$= \int_{-\infty}^{\infty} f_1(t-x)\, f_2(x)\, dx = f_2(t) * f_1(t)$$

(ii) Distributive property

$$f_1(t) * [f_2(t) + f_3(t)] = f_1(t) * f_2(t) + f_1(t) * f_3(t) \tag{5.55}$$

(iii) Associative property

$$[f_1(t) * f_2(t)] * f_3(t) = f_1(t) * [f_2(t) * f_3(t)] = f_1(t) * f_2(t) * f_3(t) \tag{5.56}$$

(iv) Shift property

If the signal $f_2(t)$ is shifted by t_0 and $f_1(t) * f_2(t) = f(t)$, then the convolution is

$$f_1(t) * f_2(t-t_0) = f(t-t_0). \tag{5.57}$$

Proof: By definition we know that

$$f_1(t) * f_2(t-t_0) = \int_{-\infty}^{\infty} f_1(\tau)\, f_2(t-\tau-t_0)\, d\tau$$

$$= \int_{-\infty}^{\infty} f_1(\tau)\, f_2(t-t_0-\tau)\, d\tau = f(t-t_0)$$

Similarly, $f_1(t-t_0) * f_2(t) = f(t-t_0)$ and

$$f_1(t-t_1) * f_2(t-t_2) = f(t-t_1-t_2). \tag{5.58}$$

(v) Convolution of a function with impulse

Let for an arbitrary function $f(t)$, the convolution with an impulse function $\delta(t-t_0)$ delayed by t_0 be given by

$$f(t) * \delta(t-t_0) = f(t-t_0). \tag{5.59}$$

Proof: By definition we know that

$$f(t) * \delta(t-t_0) = \int_{-\infty}^{\infty} f(\tau)\, \delta(t-\tau-t_0)\, d\tau$$

$$= \int_{-\infty}^{\infty} f(\tau)\, \delta(t - t_0 - \tau)\, d\tau = f(t - t_0).$$

If $t_0 = 0$ then, the convolution with an impulse function is

$$f(t) * \delta(t) = f(t).$$

Also,

$$f(t - t_0) * \delta(t) = f(t - t_0) \quad \text{and} \quad f(t) * \delta'(t - t_0) = f'(t - t_0).$$

(vi) Convolution of a function with unit step

For an arbitrary function $f(t)$ the convolution with an unit step function $u(t - t_0)$ delayed by t_0 is given by

$$f(t) * u(t - t_0) = \int_{-\infty}^{t - t_0} f(\tau)\, d\tau \quad \text{or} \quad f(t) * u(t) = \int_{-\infty}^{t} f(\tau)\, d\tau. \tag{5.60}$$

Proof: The unit step function $u(t - t_0)$ can be represented as

$$u(t - t_0) = \int_{-\infty}^{t} \delta(\tau - t_0) d\tau = \int_{-\infty}^{t - t_0} \delta(\tau)\, d\tau.$$

Its convolution with $f(t)$ is

$$f(t) * u(t - t_0) = f(t) * \int_{-\infty}^{t - t_0} \delta(t)\, dt$$

$$= \int_{-\infty}^{t - t_0} f(t) * \delta(t)\, dt = \int_{-\infty}^{t - t_0} f(\tau)\, d\tau.$$

When $t_0 = 0$, then

$$f(t) * u(t) = \int_{-\infty}^{t} f(\tau)\, d\tau.$$

Similarly, $u(t) * \delta'(t) = u'(t) = \delta(t)$,

where the symbol prime ($'$) indicates differentiation.

(vii) Width property

An important property of the convolution integral is the width property. If the durations of finite signals $x(t)$ and $h(t)$ are T_1 and T_2, respectively, then the duration of $y(t) = x(t) * h(t)$ is equal to the sum of the durations of $x(t)$ and $h(t)$, i.e.,

$$T = T_1 + T_2 \tag{5.61}$$

Also, if the areas under finite signals $x(t)$ and $h(t)$ are A_1 and A_2 respectively, then the area under $y(t) = x(t) * h(t)$ is the product of both areas, i.e.,

$$A = \text{Area under } y(t) = \text{Area under } x(t) \times \text{Area under } h(t) = A_1 A_2. \tag{5.62}$$

The property is illustrated in Fig. 5.15.

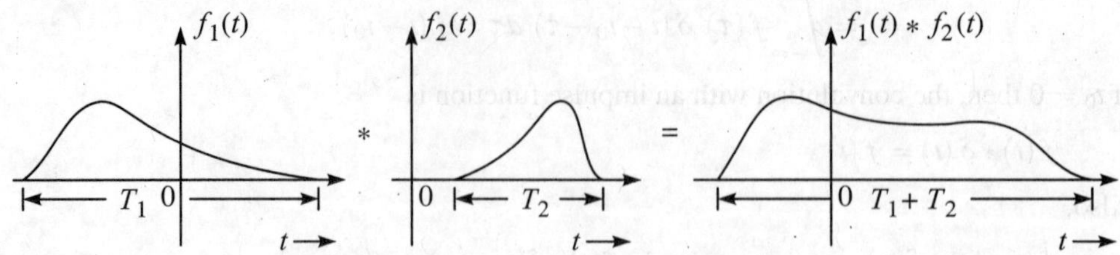

Fig. 5.15 *Illustrating width property*

Note: The width property does not hold for signals extending to $\pm\infty$.

5.13.3 Evaluation of Convolution Integral

The convolution integral can be evaluated in three different ways:

1. Analytical method, which involves performing the integration by mathematical procedure when $x(t)$ and $h(t)$ are specified analytically.
2. Graphical method, which is useful when $x(t)$ and $h(t)$ are provided in graphical form.
3. Numerical method, where we approximate $x(t)$ and $h(t)$ by numerical sequences and obtain $y(t)$ by discrete convolution using a digital computer.

Example 5.7

The impulse response of an LTI system is $h(t) = e^{-2t}u(t)$. Determine the response $y(t)$ for the input $x(t) = e^{-t}u(t)$.

Solution

Given $h(t) = e^{-2t}u(t)$ and $x(t) = e^{-t}u(t)$.

The output response $y(t) = x(t) * h(t) = \int_{-\infty}^{\infty} x(\tau)h(t-\tau)d\tau.$

$$y(t) = \int_{-\infty}^{\infty} e^{-\tau}e^{-2(t-\tau)}u(\tau)u(t-\tau)d\tau$$

Since $u(\tau)u(t-\tau) = 1$ for $\tau > 0$ and $\tau < t$, therefore,

$$y(t) = \int_{0}^{t} e^{-\tau}e^{-2(t-\tau)}d\tau.$$

Integrating, $y(t) = e^{-2t}\int_0^t e^{\tau}d\tau = e^{-2t}(e^t - 1) \quad t > 0.$

Also, $\quad y(t) = 0 \quad$ for $t < 0$.

Thus, $\quad y(t) = (e^{-t} - e^{-2t})u(t).$

Example 5.8

The impulse response of a LTI system is $h(t) = u(t-2)$. Determine the response $y(t)$ for the input, $x(t) = u(t+1)$.

Solution

Given $h(t) = u(t-2)$ and $x(t) = u(t+1)$.

The output response $y(t) = x(t) * h(t) = \int_{-\infty}^{\infty} x(\tau)h(t-\tau)d\tau$.

$$y(t) = \int_{-\infty}^{\infty} u(\tau+1)u(t-\tau-2)d\tau.$$

$$u(\tau+1)u(t-\tau-2) = 1 \text{ for } \tau > -1 \text{ and } \tau < t-2.$$

The integration becomes $y(t) = \int_{-1}^{t-2} e^{-\tau} d\tau$.

For $t < 1 \quad y(t) = 0$.

For $t > 1$, $\quad y(t) = \int_{-1}^{t-2} d\tau = t|_{-1}^{t-2} = t-1$.

Therefore, $y(t) = \begin{cases} 0 & t < 1 \\ t-1 & t > 1 \end{cases}$ or $y(t) = (t-1)u(t-1)$.

The output is a delayed ramp signal drawn in Fig. 5.16.

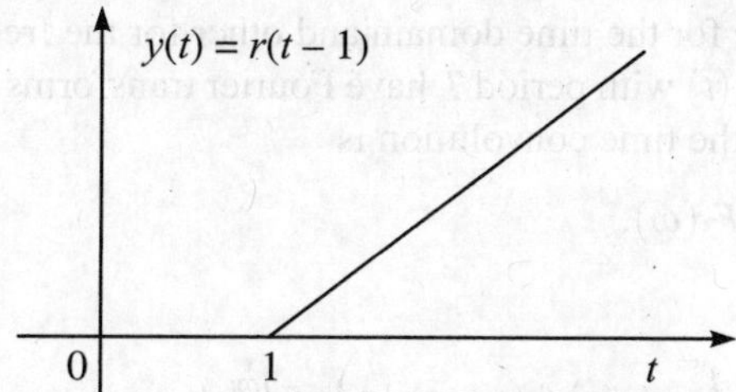

Fig. 5.16 *Delayed ramp signal*

Example 5.9

Find the convolution of the two signals $x_1(t) = e^{-3t}u(t)$ and $x_2(t) = u(t-3) - u(t-5)$.

Solution

Given the two signals $x_1(t) = e^{-3t}u(t)$ and $x_2(t) = u(t-3) - u(t-5)$.

The output response $y(t) = x(t) * h(t) = \int_{-\infty}^{\infty} x(\tau)h(t-\tau)d\tau$.

$$y(t) = \int_{-\infty}^{\infty} e^{-3\tau}u(\tau)u(t-\tau-3)d\tau - \int_{-\infty}^{\infty} e^{-3\tau}u(\tau)u(t-\tau-5)d\tau.$$

$$\because \quad u(\tau)u(t-\tau-3) = 1 \text{ for } \tau > 0 \text{ and } \tau < t-3$$

$$u(\tau)u(t-\tau-5) = 1 \text{ for } \tau > 0 \text{ and } \tau < t-5.$$

The integration becomes $y(t) = \int_{0}^{t-3} e^{-3\tau} d\tau - \int_{0}^{t-5} e^{-3\tau} d\tau$.

For $t < 3, \quad y(t) = 0$

For $3 < t \le 5, \quad y(t) = \int_{0}^{t-3} e^{-3\tau} d\tau = \frac{1}{3}\left(1 - e^{-3(t-3)}\right)$.

$$\text{For } t > 5, \quad y(t) = \int_0^{t-3} e^{-3\tau} d\tau - \int_0^{t-5} e^{-3\tau} d\tau$$

$$= \frac{1}{3}\left(1 - e^{-3(t-3)}\right) - \frac{1}{3}\left(1 - e^{-3(t-5)}\right) = \frac{1}{3}\left(e^{-3(t-5)} - e^{-3(t-3)}\right).$$

$$\text{Therefore, } y(t) = \begin{cases} 0 & t < 3 \\ \frac{1}{3}\left(1 - e^{-3(t-3)}\right) & 3 \le t < 5 \\ \frac{1}{3}\left(e^{-3(t-5)} - e^{-3(t-3)}\right) & t > 5. \end{cases}$$

5.14 Convolution in Frequency Domain

The transform domain offers a convenient method of avoiding the convolution operation. There are two-convolution theorems, one for the time domain and other for the frequency domain.

Let two signals $f_1(t)$ and $f_2(t)$ with period T have Fourier transforms $F_1(\omega)$ and $F_2(\omega)$ respectively. Then the Fourier transform of the time convolution is

$$\mathcal{F}[f_1(t) * f_2(t)] = F_1(\omega)F_2(\omega). \tag{5.63}$$

Proof:

$$\mathcal{F}[f_1(t) * f_2(t)] = \int_{-\infty}^{\infty} \left(\int_{-\infty}^{\infty} f_1(\tau) f_2(t-\tau) d\tau \right) e^{-j\omega t} dt$$

$$= \int_{-\infty}^{\infty} f_1(\tau) \left(\int_{-\infty}^{\infty} f_2(t-\tau) e^{-j\omega t} dt \right) d\tau$$

$$= \int_{-\infty}^{\infty} f_1(\tau) \cdot f_2(\omega) e^{-j\omega\tau} d\tau = F_1(\omega)F_2(\omega).$$

The inverse Fourier transform gives the convolution. This property is called time convolution.

5.15 Convolution Property of Fourier Transforms

Let two signals $f_1(t)$ and $f_2(t)$ with period T have Fourier transforms $F_1(\omega)$ and $F_2(\omega)$respectively then the convolution of the Fourier transforms is given by

$$F_1(\omega) * F_2(\omega) = 2\pi \mathcal{F}[f_1(t) f_2(t)]. \tag{5.64}$$

Proof: We know that

$$\mathcal{F}[f_1(t) f_2(t)] = \int_{-\infty}^{\infty} f_1(t) f_2(t) e^{-j\omega t} dt.$$

Using inverse of Fourier transform to represent $f_1(t)$, then

$$\mathcal{F}[f_1(t) f_2(t)] = \int_{-\infty}^{\infty} \left[\frac{1}{2\pi} \int_{-\infty}^{\infty} F_1(\lambda) e^{j\lambda t} d\lambda \right] f(t) e^{-j\omega t} dt$$

$$= \frac{1}{2\pi}\int_{-\infty}^{\infty} F_1(\lambda)\left[\int_{-\infty}^{\infty} f_2(t)e^{-j(\omega-\lambda)t}dt\right]d\lambda$$

$$= \frac{1}{2\pi}\int_{-\infty}^{\infty} F_1(\lambda)F_2(\omega-\lambda)d\lambda$$

$$= \frac{1}{2\pi}F_1(\omega) * F_2(\omega).$$

$$F_1(\omega) * F_2(\omega) = 2\pi F\left[f_1(t)f_2(t)\right].$$

Thus the convolution operation in one domain is transformed a product operation in the other domain.

5.16 Graphical Representation of Convolution

When two signals $f_1(t)$ and $f_2(t)$ are provided in graphical form, the convolution can be performed by graphical method. It involves the following steps.

1. For the given signals $f_1(t)$ and $f_2(t)$, draw the signals $f_1(\tau)$ and $f_2(\tau)$ as a function of independent time variable τ.
2. Draw the function $f_2(-\tau)$, which is a time-reversed function of $f_2(\tau)$ (folding at the origin). Then shift the function by time t to form $f_2(t-\tau)$.
3. Draw both signals $f_1(t)$ and $f_2(t-\tau)$ on the same time τ- axis with large time shift t along the negative axis (no overlap for finite time signals).
4. Increase the time shift t along the positive time axis. Multiply the signals $f_1(\tau)f_2(t-\tau)$ and integrate over the overlapping period of the two signals to obtain convolution at t.
5. Increase the time shift t step by step and obtain convolution using step 4.
6. Draw the convolution function $f(t)$ with the values obtained in steps 4 and 5 as a function of t.

Example 5.10

Consider two finite duration functions $f_1(t)$ and $f_2(t)$ is defined by

$$f_1(t) = u(t) - u(t-2); \text{ and } f_2(t) = u(t) - u(t-1).$$

Find the convolution of $f_1(t)$with$f_2(t)$ using graphical method.

Solution

Given the two functions $f_1(t) = u(t) - u(t-2)$ and $f_2(t) = u(t) - u(t-1)$ as shown in Fig. 5.17(a).

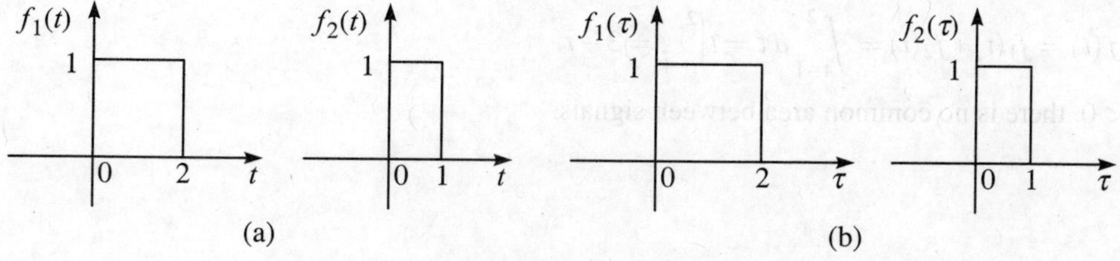

Fig. 5.17 *Signals with (a) time t and (b) time τ*

1. Replace t in $f_1(t)$ and $f_2(t)$ by τ and get $f_1(\tau)$ and $f_2(\tau)$ as shown in Fig. 5.17(b).
2. Fold $f_2(\tau)$ that is, replace τ by $-\tau$ in the $f_2(\tau)$ and get $f_2(-\tau)$ as shown in Fig. 5.18 (a) and shift it by t to get $f_2(t-\tau)$. Draw the signals $f_1(t)$ and $f_2(t-\tau)$ on the same time axis. Initially, t should be a large value on the negative axis such that no overlap of signals take place. The graph of Fig. 5.18(b) illustrates the result.

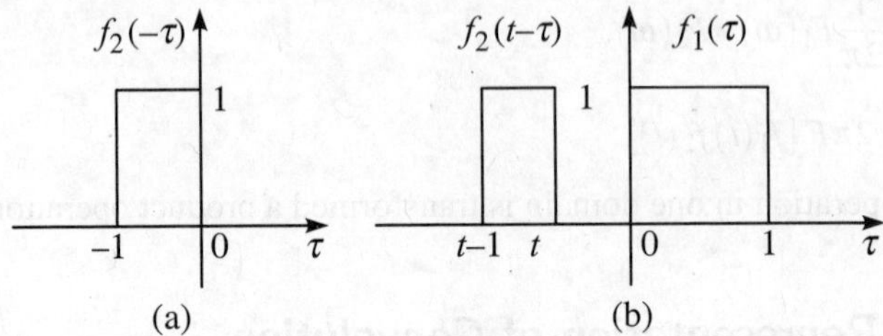

Fig. 5.18 *(a) Folded signal $f_2(-\tau)$, (b) Signals $f_1(t)$ and $f_2(t-\tau)$ on the same time axis*

3. Now shift the time t along the positive axis, step by step, and obtain the common area of the two signals. The convolution at different time intervals is given by

   For $t<0$, there is no common area between signals.
   For $0<t<1$, the common area is shown in Fig. 5.19(a).

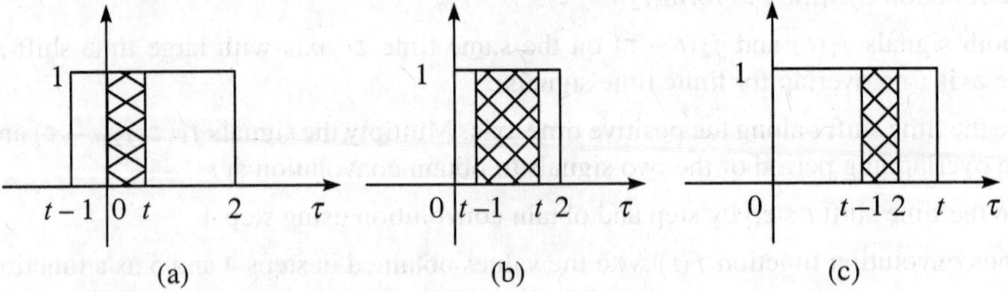

Fig. 5.19 *Shifting of signal $f_2(t-\tau)$ on $f_1(t)$ at (a) $0<t<1$, (b) $1<t<2$, (c) $2<t<3$*

$$f(t) = f_1(t) * f_2(t) = \int_0^t d\tau = t.$$

For $1<t<2$, the common area is shown in Fig. 5.19(b).

$$f(t) = f_1(t) * f_2(t) = \int_1^2 d\tau = t\Big|_1^2 = 1.$$

For $2<t<3$, the common area is shown in Fig. 5.19(c).

$$f(t) = f_1(t) * f_2(t) = \int_{t-1}^2 d\tau = t\Big|_{t-1}^2 = 3-t.$$

For $t<0$, there is no common area between signals.

Therefore the convolution is given by

$$f(t)=\begin{cases}0 & t<0\\ t, & 0\leq t<1\\ 1, & 1\leq t<2\\ 3-t, & 2\leq t<3\\ 0, & t\geq 5.\end{cases}$$

The graph of the convolution of the two signals is depicted in Fig. 5.20.

$$y(t)=tu(t)-(t-1)u(t-1)-(t-2)u(t-2)+(t-3)u(t-3).$$

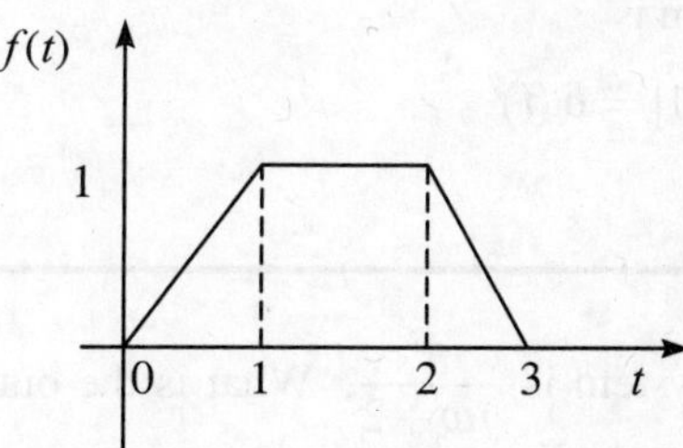

Fig. 5.20 *Graph of convolution*

Additional Problems

Example 5.11

The output of a system in response to an input $x(t)=e^{-2t}u(t)$ is $y(t)=e^{-t}u(t)$. Find the frequency response and the impulse response of the system.

$$X(\omega)=\frac{1}{2+j\omega},\quad Y(\omega)=\frac{1}{1+j\omega}=X(\omega)H(\omega).$$

$$\therefore\quad H(\omega)=\frac{2+j\omega}{1+j\omega}=1+\frac{1}{1+j\omega}.$$

Taking inverse Fourier transform, we get

$$h(t)=\delta(t)+e^{-t}u(t).$$

Example 5.12

For the system shown below,

$$f(t)=\begin{cases}e^{-at} & t\geq 0, |a|\geq 1\\ 0 & \text{otherwise}\end{cases}$$

$$Y(\omega)=\begin{cases}\dfrac{1}{a+j\omega} & -a<\omega<a\\ 0 & \text{otherwise,}\end{cases}$$

find the transfer function and impulse response.

Solution

Given $f(t) = e^{-at}$ for $t \geq 0$, i.e., $f(t) = e^{-at}u(t)$.

$$\therefore\ F(\omega) = \mathcal{F}\left[e^{-at}u(t)\right] = \frac{1}{a + j\omega}.$$

Given $Y(\omega) = \dfrac{1}{a + j\omega}$.

The transfer function of the system is

$$H(\omega) = \frac{Y(\omega)}{F(\omega)} = \frac{1/(a + j\omega)}{1/(a + j\omega)} = 1.$$

The impulse response of the system is

$$h(t) = \mathcal{F}^{-1}[H(\omega)] = \mathcal{F}^{-1}[1] = \delta(t).$$

Example 5.13

The transfer function of an LTI system is $\dfrac{1}{j\omega + 2}$. What is the output of the system for an input of $(0.8)^t u(t)$?

Solution

Given the transfer function of the LTI system as $H(\omega) = \dfrac{1}{j\omega + 2}$ and input $x(t) = (0.8)^t u(t)$, the impulse response of the system is

$$h(t) = \mathcal{F}^{-1}[H(\omega)] = \mathcal{F}^{-1}\left[\frac{1}{j\omega + 2}\right] = e^{-2t}u(t).$$

We know that the output $y(t)$ is

$$y(t) = x(t) * h(t) = \int_{-\infty}^{\infty} h(\tau)x(t-\tau)\,d\tau$$

$$= \int_{-\infty}^{\infty} e^{-2\tau}u(\tau)(0.8)^{t-\tau}u(t-\tau)\,d\tau = \int_{-\infty}^{\infty} e^{-2\tau}(0.8)^t(0.8)^{-\tau}\,d\tau$$

$$= (0.8)^t \int_{-\infty}^{\infty} \left(0.8e^2\right)^{-\tau} d\tau.$$

Let $\left(0.8e^2\right)^{-1} = a$.

$$y(t) = (0.8)^t \int_{-\infty}^{\infty} a^\tau d\tau = (0.8)^t \left[\frac{a^\tau}{\ln a}\right]_0^\infty = (0.8)^t \left[\frac{\left(0.8e^2\right)^{-\tau}}{\ln\left[\left(0.8e^2\right)^{-1}\right]}\right]_0^\infty = \frac{(0.8)^t}{-\ln\left(0.8e^2\right)}[0-1]$$

$$= \frac{(0.8)^t}{\ln\left(0.8e^2\right)} = \frac{(0.8)^t}{(\ln 0.8 + 2)}.$$

$\therefore$ The output of the system $y(t) = \dfrac{(0.8)^t}{(\ln 0.8 + 2)}$.

Example 5.14

Find the impulse response for the *RL* filter shown in Fig. 5.21. Also find the transfer function. What would be its frequency response?

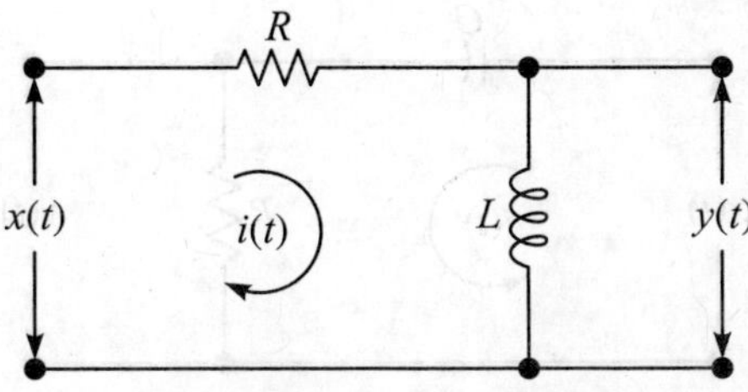

Fig. 5.21 *An RL filter*

Solution

The transfer function of the *RL* filter can be determined by transforming the network into frequency domain as shown in Fig. 5.22.

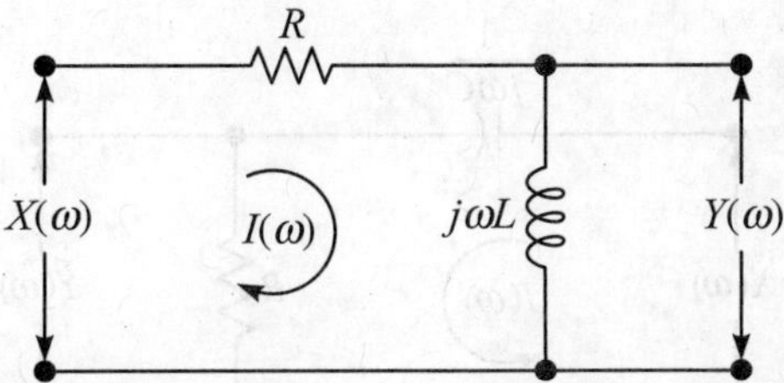

Fig. 5.22 *The RL filter in frequency domain*

From the network, the transfer function is

$$H(\omega) = \frac{Y(\omega)}{X(\omega)} = \frac{j\omega L}{j\omega + R} = \frac{j\omega}{j\omega + (R/L)}.$$

The impulse response of the system is $h(t) = \mathcal{F}^{-1}[H(\omega)] = \mathcal{F}^{-1}\left[\frac{j\omega}{j\omega + (R/L)}\right]$

$$= \mathcal{F}^{-1}\left[1 - \frac{R}{L}\frac{1}{j\omega + (R/L)}\right].$$

$$h(t) = \delta(t) - \frac{R}{L}e^{-(R/L)t}u(t).$$

The frequency response of the system is $H(\omega) = \dfrac{j\omega}{j\omega + (R/L)}$.

Magnitude response $|H(\omega)| = \dfrac{\omega}{\sqrt{\omega^2 + (R/L)^2}}$.

Phase response $\angle H(\omega) = \dfrac{\pi}{2} - \tan^{-1}\dfrac{L}{R}\omega$.

Example 5.15

Find the impulse response for the *RC* filter shown in Fig. 5.23. Also find the transfer function. What would be its frequency response?

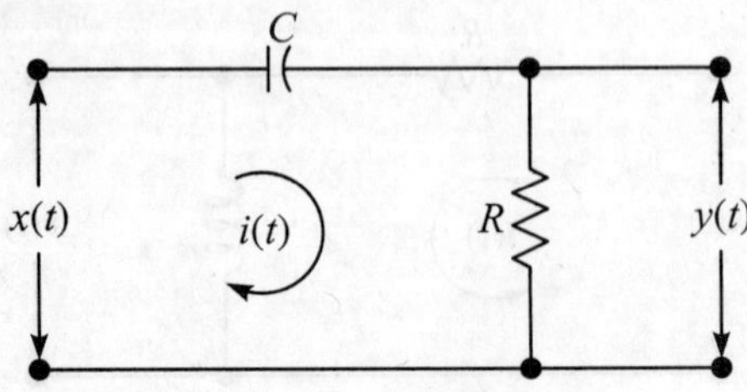

Fig. 5.23 *RC filter*

Solution

The transfer function of the *RC* filter can be determined by transforming the network into frequency domain as shown in Fig. 5.24.

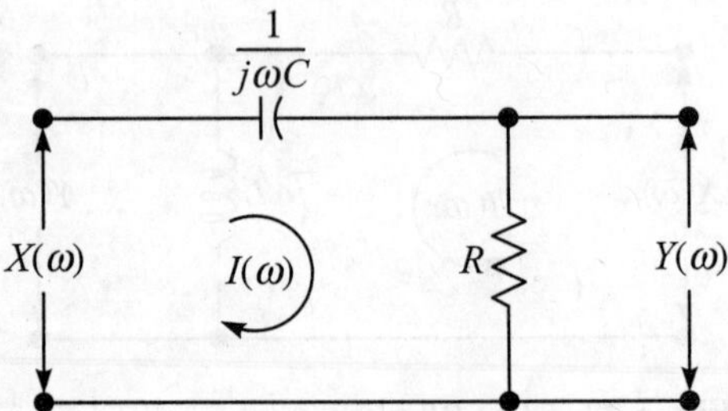

Fig. 5.24 *RC filter in frequency domain*

From the Fourier transformed network, the frequency response (transfer function) is

$$H(\omega) = \frac{Y(\omega)}{X(\omega)} = \frac{R}{R + (1/j\omega C)} = \frac{j\omega RC}{1 + j\omega RC} = \frac{j\omega}{j\omega + (1/RC)}.$$

The impulse response of the system is

$$h(t) = \mathcal{F}^{-1}[H(\omega)] = \mathcal{F}^{-1}\left[\frac{j\omega}{j\omega + (1/RC)}\right]$$

$$= \mathcal{F}^{-1}\left[1 - \frac{(1/RC)}{j\omega + (1/RC)}\right].$$

$$h(t) = \delta(t) - \frac{1}{RC} e^{-(1/RC)t} u(t).$$

The frequency response of the system is $H(\omega) = \dfrac{j\omega RC}{1 + j\omega RC}$.

Magnitude response $|H(\omega)| = \dfrac{\omega RC}{\sqrt{1+(\omega RC)^2}}$.

Phase response $\angle H(\omega) = \dfrac{\pi}{2} - \tan^{-1}(\omega RC)$.

Example 5.16

Find the impulse response of the system with time constant $RC = 0.125$ seconds as shown in Fig. 5.25. Also find the transfer function. What would be its frequency response?

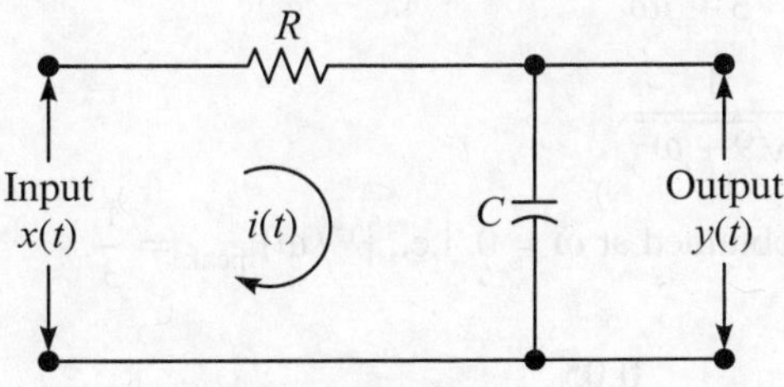

Fig. 5.25 *RC filter*

Solution

The transfer function of the *RC* filter can be determined by transforming the network into frequency domain as shown in Fig. 5.26.

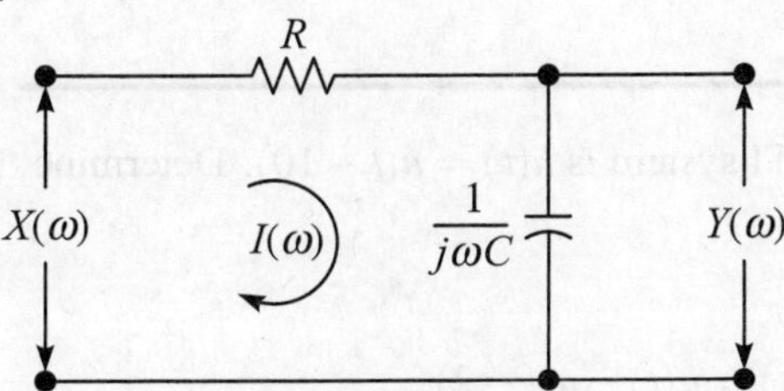

Fig. 5.26 *RC filter in frequency domain*

From the Fourier transformed network, the transfer function is

$$H(\omega) = \frac{Y(\omega)}{X(\omega)} = \frac{(1/j\omega C)}{R+(1/j\omega C)}.$$

Given $RC = 0.125$, the frequency response of the system is $H(\omega) = \dfrac{1}{1+j\omega RC} = \dfrac{8}{j\omega+8}$.

The impulse response of the system is

$$h(t) = \mathcal{F}^{-1}[H(\omega)] = \mathcal{F}^{-1}\left[\frac{8}{j\omega+8}\right].$$

$$h(t) = 8e^{-8t}u(t).$$

Magnitude response $|H(\omega)| = \dfrac{1}{\sqrt{1+(0.125\omega)^2}}$.

Phase response $\angle H(\omega) = -\tan^{-1}(0.125\omega)$.

Example 5.17

For a low-pass signal, the essential bandwidth may be chosen as the frequencies where the amplitude spectrum of the signal decays to 5% of its peak value. Determine the essential bandwidth of $f(t) = e^{-3t}u(t)$.

Solution

Given $f(t) = e^{-3t}u(t)$

$$F(\omega) = \mathcal{F}\left[e^{-3t}u(t)\right] = \frac{1}{3+j\omega}.$$

$$|F(\omega)| = \left|\frac{1}{3+j\omega}\right| = \frac{1}{\sqrt{9+\omega^2}}.$$

The peak value of $|F(\omega)|$ is obtained at $\omega = 0$, i.e., $|F(\omega)|_{\text{peak}} = \dfrac{1}{3}$.

Given that at $\omega = \omega_c$,

$$|F(\omega)| = 5\% \text{ of } |F(\omega)|_{\text{peak}} = \frac{0.05}{1/3} = 0.15$$

or $0.15 = \dfrac{1}{\sqrt{9+\omega_c^2}}$.

Therefore, the essential bandwidth is $\omega_c = 5.953$ rad/s.

Example 5.18

The impulse response of an LTI system is $h(t) = u(t-10)$. Determine the response $y(t)$ for the input, $x(t) = u(t) - u(t-2)$.

Solution

Given $h(t) = u(t-10)$ and $x(t) = u(t) - u(t-2)$,

the output response $y(t) = x(t) * h(t) = \displaystyle\int_{-\infty}^{\infty} x(\tau)\, h(t-\tau) d\tau$

$$y(t) = \int_{-\infty}^{\infty} [u(\tau) - u(\tau-2)]\, u(t-\tau-10) d\tau$$

$$= \int_{-\infty}^{\infty} u(\tau)u(t-\tau-10) d\tau - \int_{-\infty}^{\infty} u(\tau-2)u(t-\tau-10) d\tau.$$

From properties of unit step function,

$$u(\tau)u(t-\tau-10) = 1 \text{ for } \tau > 0 \text{ and } \tau < t-10$$

$$u(\tau-2)u(t-\tau-10) = 1 \text{ for } \tau > 2 \text{ and } \tau < t-10,$$

the integration becomes

$$y(t) = \int_{0}^{t-10} d\tau - \int_{2}^{t-10} d\tau.$$

For $t < 10$ $y(t) = 0$.

For $10 < t < 12$, $y(t) = \int_0^{t-10} d\tau = t|_0^{t-10} = t - 10.$

For $t \geq 12$, $y(t) = \int_0^{t-10} d\tau - \int_2^{t-10} d\tau$

$$= t\Big|_0^{t-10} - t\Big|_2^{t-10} = t - 10 - t + 12 = 2.$$

Hence, $$y(t) = \begin{cases} 0 & t < 10 \\ t - 10 & 10 < t \leq 12 \\ 2 & t > 12. \end{cases}$$

or $y(t) = (t - 10)u(t - 10) - (t - 12)u(t - 12).$

The output is a delayed truncated ramp signal drawn as shown in Fig. 5.27.

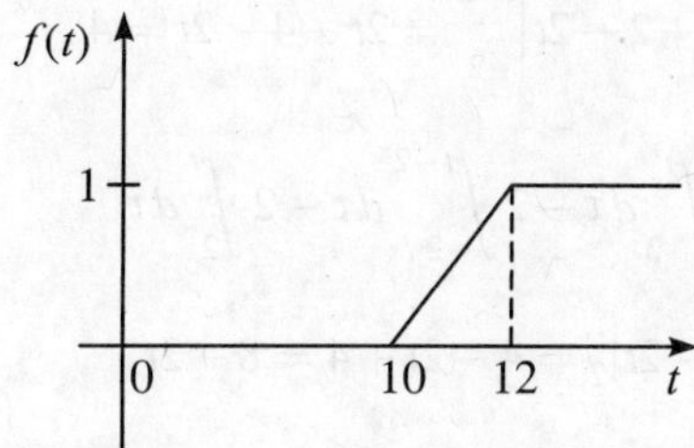

Fig. 5.27 *Delayed truncated ramp signal*

Example 5.19

Obtain the convolution of two signals $h(t) = 2[u(t) - u(t-2)]$ and $x(t) = u(t+2) - u(t-2)$.

Solution

Given $h(t) = 2[u(t) - u(t-2)]$ and $x(t) = u(t+2) - u(t-2)$, the convolution is

$$y(t) = x(t) * h(t) = \int_{-\infty}^{\infty} x(\tau)h(t-\tau)d\tau$$

$$y(t) = \int_{-\infty}^{\infty} [u(\tau+2) - u(\tau-2)][u(t-\tau) - u(t-\tau-2)]\,d\tau$$

$$y(t) = 2\int_{-\infty}^{\infty} u(\tau+2)u(t-\tau)d\tau - 2\int_{-\infty}^{\infty} u(\tau+2)u(t-\tau-2)d\tau$$

$$-2\int_{-\infty}^{\infty} u(\tau-2)u(t-\tau)d\tau + 2\int_{-\infty}^{\infty} u(\tau-2)u(t-\tau-2)d\tau.$$

From the properties of unit step function,

$u(\tau+2)u(t-\tau) = 1$ for $\tau > -2$ and $\tau < t$.

$u(\tau+2)u(t-\tau-2) = 1$ for $\tau > -2$ and $\tau < t-2$.

$$u(\tau-2)u(t-\tau)=1 \text{ for } \tau>2 \text{ and } \tau<t.$$

$$u(\tau-2)u(t-\tau-2)=1 \text{ for } \tau>2 \text{ and } \tau<t-2.$$

The integration becomes

$$y(t)=2\int_{-2}^{t} d\tau-2\int_{-2}^{t-2} d\tau-2\int_{2}^{t} d\tau+2\int_{2}^{t-2} d\tau.$$

For $t<-2 \quad y(t)=0.$

For $-2<t<0, \quad y(t)=2\int_{-2}^{t} d\tau=2\,t|_{-2}^{t}=2t+2.$

For $0<t<2, \quad y(t)=2\int_{-2}^{t} d\tau-2\int_{-2}^{t-2} d\tau$

$$=2t+2-2t\Big|_{-2}^{t-2}=2t+4-2t=4.$$

For $2<t<4, \quad y(t)=2\int_{-2}^{t} d\tau-2\int_{-2}^{t-2} d\tau-2\int_{2}^{t} d\tau$

$$=4-2t|_{2}^{t}=4-2t+4=8-2t.$$

For $t>4, \quad y(t)=2\int_{-2}^{t} d\tau-2\int_{-2}^{t-2} d\tau-2\int_{2}^{t} d\tau+2\int_{2}^{t-2} d\tau$

$$=8-2t-2t|_{2}^{t-2}=8-2t+2t-8=0.$$

$$\text{Hence } y(t)=\begin{cases} 0 & t\le -2 \\ 2t+4 & -2<t\le 0 \\ 4 & 0<t\le 2 \\ 8-2t & 2<t\le 4 \\ 0 & t>4 \end{cases}$$

or $\quad y(t)=(t+2)u(t+2)-tu(t)-(t-2)u(t-2)+(t-4)u(t-4).$

The output is a delayed trepizoidal signal drawn as shown in Fig. 5.28.

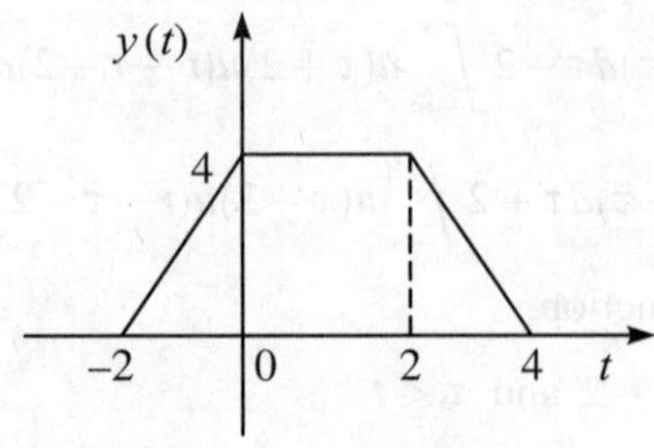

Fig. 5.28 *Delayed trepizoidal signal*

Example 5.20

Consider the system described by $y'(t)+2y(t)=x(t)+x'(t)$. Find the impulse response $h(t)$ of the system.

Solution

Given the differential equation $y'(t)+2y(t)=x(t)+x'(t)$.

Taking Fourier transform of both sides,

$$j\omega Y(\omega)+2Y(\omega)=X(\omega)+j\omega X(\omega)$$

$$Y(\omega)(j\omega+2)=X(\omega)(1+j\omega)$$

$$H(\omega)=\frac{Y(\omega)}{X(\omega)}=\frac{(1+j\omega)}{(j\omega+2)} \quad \text{or} \quad H(\omega)=1-\frac{1}{(2+j\omega)}.$$

Taking inverse Fourier transform,

$$h(t)=\delta(t)-e^{-2t}u(t).$$

Example 5.21

Consider the system described by $y'(t)+2y(t)=x(t)$. Find the output response $h(t)$ of the system if the excitation is $x(t)=e^{-t}u(t)$.

Solution

Given the differential equation $y'(t)+2y(t)=x(t)$.

Taking Fourier transform on both sides,

$$j\omega Y(\omega)+2Y(\omega)=X(\omega)$$

$$Y(\omega)(j\omega+2)=X(\omega)$$

Given $X(\omega)=\dfrac{1}{(j\omega+1)}$

$$Y(\omega)=\frac{1}{(j\omega+2)(j\omega+1)} \quad \text{or} \quad Y(\omega)=\frac{1}{j\omega+1}-\frac{1}{j\omega+2}.$$

Taking inverse Fourier transform,

$$y(t)=e^{-t}u(t)-e^{-2t}u(t).$$

Example 5.22

Consider two functions $x(t)=u(t+1)$ and $h(t)=u(t-2)$. Find the convolution of $x(t)$ and $h(t)$ using graphical method.

Solution

The two functions $x(t)=u(t+1)$ and $h(t)=u(t-2)$ are depicted in Fig. 5.29(a).

1. Replace t in the $x(t)$ and $h(t)$ by τ to get $x(\tau)$ and $h(\tau)$ as shown in Fig. 5.29(b).

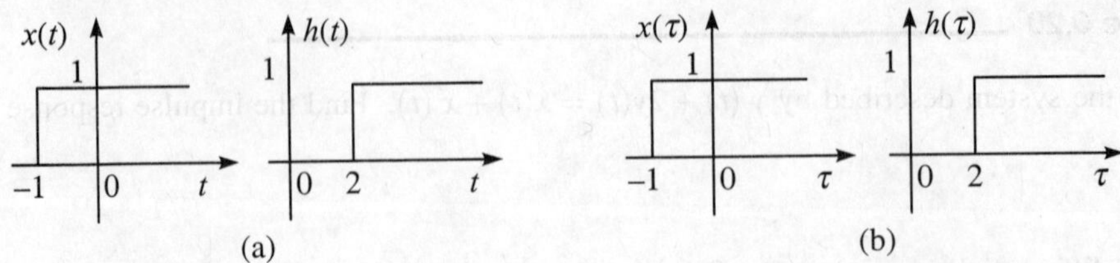

Fig. 5.29 *Signals with (a) time t and (b) time τ*

2. Fold $h(\tau)$, that is, replace τ by $-\tau$ in the $h(\tau)$ get $h(-\tau)$ as shown in Fig. 5.30(a) and shift it by t to get $h(t-\tau)$. Draw the signals $x(t)$ and $h(t-\tau)$ on the same time axis. Initially, t should be a large value on the negative axis such that no overlap of signals takes place. The graph of Fig. 5.30(b) illustrates the result.

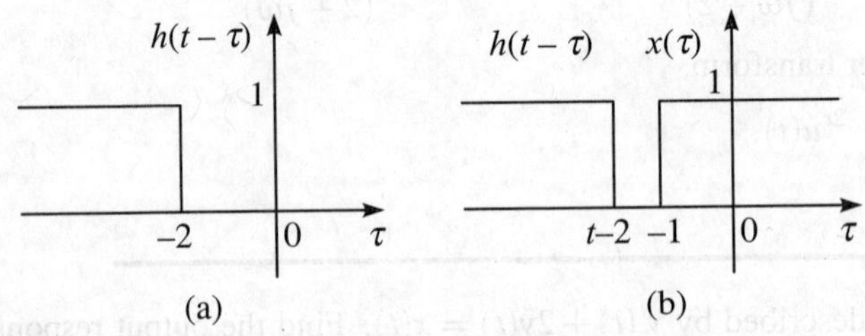

Fig. 5.30 *(a) Folded signal $h(-\tau)$ and (b) signals $x(t)$ and $h(t-\tau)$ on the same time axis*

3. Now shift the time t along positive axis step-by-step and obtain the common area of the two signals. The convolution at different time intervals is given by:

   For $t-2<-1$ or $t<1$, there is no common area between signals.

   For $t \geq 1$, the common area is shown in Fig. 5.31(a).

$$f(t)=x(t)*h(t)=\int_{-1}^{t-2} d\tau = t-1.$$

Hence the convolution is given by

$$f(t)=\begin{cases} 0 & t<1 \\ t-1, & t \geq 1. \end{cases}$$

The graph of convolution depicted in Fig. 5.31(b).

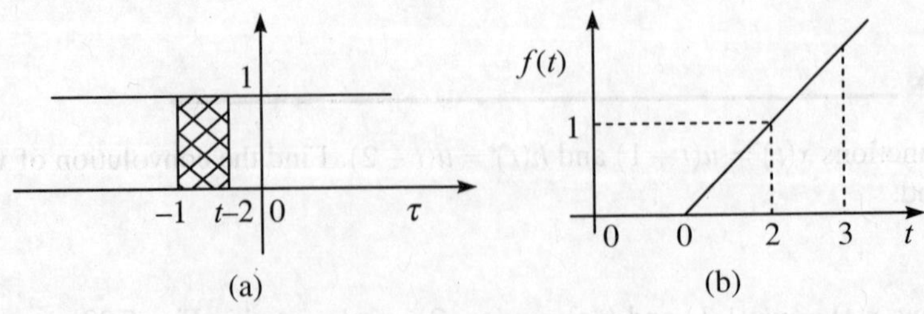

Fig. 5.31 *(a) Shifting of signal $h(t-\tau)$ on $x(t)$ for t $\geq$ 1; (b) convolution*

Example 5.23

Consider two finite duration functions $f_1(t)$ and $f_2(t)$ defined by $f_1(t) = 2e^{-t}u(t)$ and $f_2(t) = u(t) - u(t-2)$. Find the convolution of $f_1(t)$ with $f_2(t)$ using graphical method.

Solution

Given the two functions $f_1(t) = 2e^{-t}u(t)$ and $f_2(t) = u(t) - u(t-2)$ as shown in Fig. 5.32(a).

1. Replace t in $f_1(t)$ and $f_2(t)$ by τ and get $f_1(\tau)$ and $f_2(\tau)$ as shown in Fig. 5.32(b).

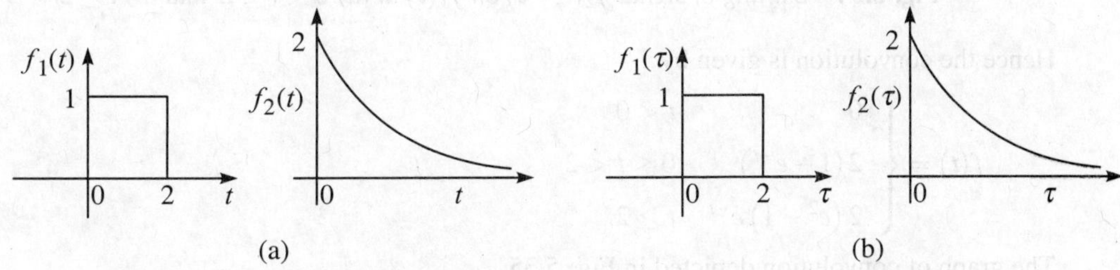

Fig. 5.32 *Signals with (a) time t and (b) timeτ*

2. Fold $f_2(\tau)$, that is, replace τ by $-\tau$ in the $f_2(\tau)$ get $f_2(-\tau)$ as shown in Fig. 5.33(a) and shift it by t to get $f_2(t-\tau)$. Draw the signals $f_1(t)$ and $f_2(t-\tau)$ on the same time axis. Initially, t should be a large value on the negative axis such that no overlap of signals takes place. The graph of Fig. 5.33(b) illustrates the result.

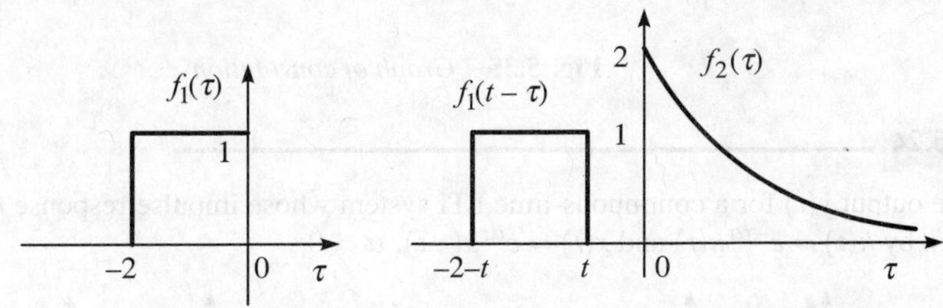

Fig. 5.33 *(a) Folded signal $f_2(-\tau)$, (b) signals $f_1(t)$ and $f_2(t-\tau)$ on the same time axis*

3. Now shift the time t along positive axis step by step and obtain the common area of the two signals. The convolution at different time intervals is given by:

   For $t < 0$, there is no common area between signals.

   For $0 < t < 2$, the common area is shown in Fig. 5.34(a).

$$f(t) = f_1(t) * f_2(t) = \int_0^t 2e^{-(t-\tau)}d\tau = 2\left(1 - e^{-t}\right).$$

   For $t > 2$, the common area is shown in Fig. 5.34(b).

$$f(t) = f_1(t) * f_2(t) = \int_0^2 2e^{-(t-\tau)}d\tau = 2e^{-(t-\tau)}\Big|_0^2 = 2\left(e^2 - 1\right)e^{-t}$$

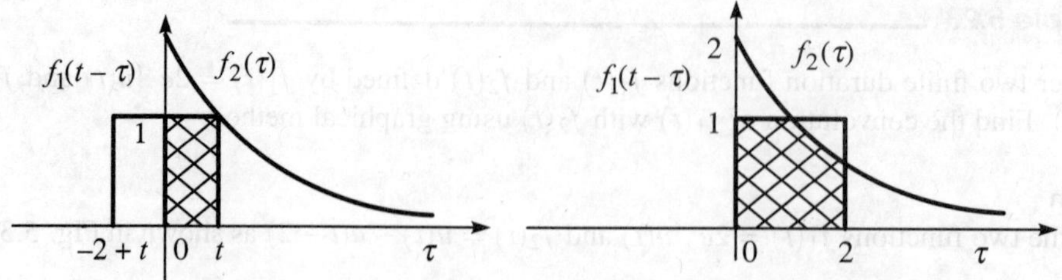

Fig. 5.34 *Shifting of signal $f_2(t-\tau)$ on $f_1(t)$ at (a) $0 < t < 2$ and (b) $t \geq 2$*

Hence the convolution is given by

$$f(t) = \begin{cases} 0 & t < 0 \\ 2\left(1 - e^{-t}\right) & 0 \leq t < 2 \\ 2\left(e^2 - 1\right)e^{-t} & t \geq 2. \end{cases}$$

The graph of convolution depicted in Fig. 5.35.

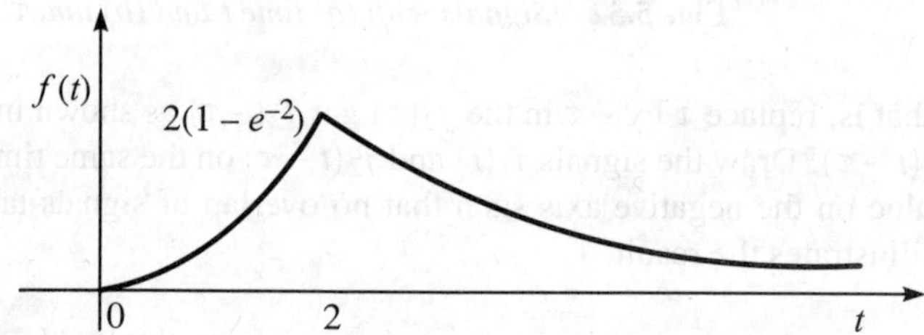

Fig. 5.35 *Graph of convolution*

Example 5.24

Compute the output $y(t)$ for a continuous-time LTI system whose impulse response $h(t)$ and the input $x(t)$ are given by $h(t) = e^{-\alpha t}u(t)$ and $x(t) = e^{\alpha t}u(-t)$, $\alpha > 0$.

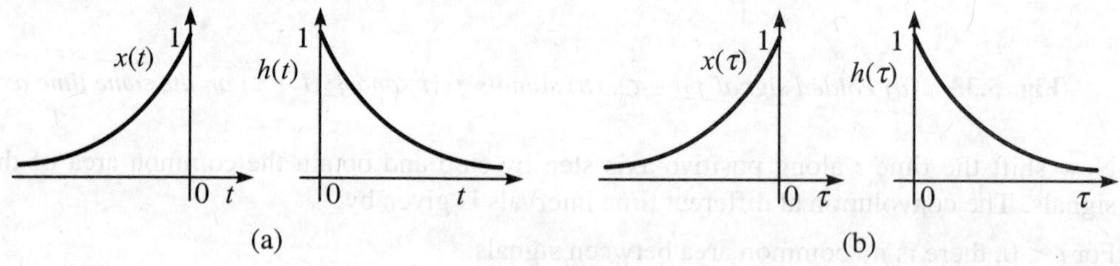

Fig. 5.36

For $t < 0$, $\quad y(t) = \int_{-\infty}^{t} e^{\alpha\tau} e^{-\alpha(t-\tau)} d\tau = e^{-\alpha t} \int_{-\infty}^{t} e^{2\alpha\tau} d\tau,$

we get, $\quad y(t) = \dfrac{1}{2\alpha} e^{\alpha t}.$

For $\quad t > 0, \quad y(t) = \int_{-\infty}^{0} e^{\alpha\tau} e^{-\alpha(t-\tau)} d\tau = e^{-\alpha t} \int_{-\infty}^{0} e^{2\alpha\tau} d\tau,$

we get, $y(t) = \dfrac{1}{2\alpha}e^{-\alpha t}$.

Combining the two, we can write $y(t)$ as

$$y(t) = \frac{1}{2\alpha}e^{-\alpha|t|}, \quad \alpha > 0.$$

The graph of convolution depicted in Fig. 5.38

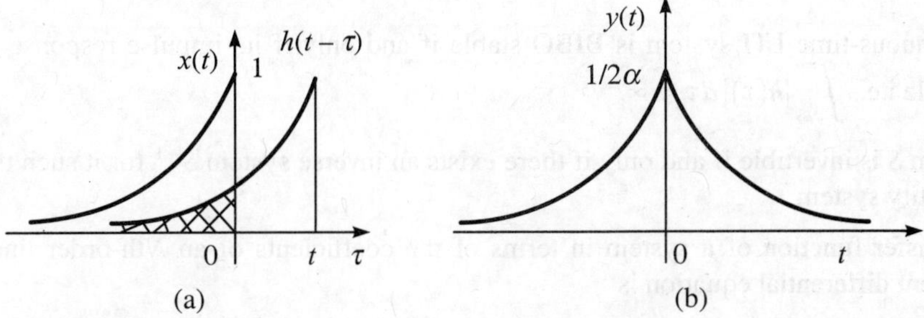

Fig. 5.38 *Shifting of signal $h(t-\tau)$ on $x(t)$ for $t > 0$, (b) graph of convolution*

Summary

1. An LTI system is completely characterised by its impulse response.
2. The input–output relationship of an LTI continuous-time system is given by the convolution integral of the system's impulse response with the input signal.
3. Given the impulse response of an LTI system and a specific input signal, the convolution giving the output signal can be computed using either direct integration, or a graphical approach or a numerical approach.
4. The main properties of an LTI system are derived in terms of its impulse response.
5. If the system is linear, then $y(t) = \sum_{i=1}^{N} a_i y_i(t)$, where, $y_i(t) = T[x_i(t)]$, is the output of the system in response to the input $x_i(t)$ acting alone.
6. A system is said to be linear time invariant (LTI) system if it satisfies linear and time invariant properties.
7. For a time invariant system, if $y(t) = H[x(t)]$; then$y(t-t_0) = H[x(t-t_0)]$. Also, $y(t,t_0) = y(t-t_0)$.
8. The coefficients in the differential equation describing an LTI system are always constant.
9. A system is said to be linear time variant (LTV) system if it satisfies linear property but does not satisfy time invariant properties, i.e., $y(t,t_0) \neq y(t-t_0)$.
10. The impulse response is the output of a LTI system due to an impulse input applied at time $t = 0$, i.e., $h(t) = H[\delta(t)]$.
11. For an LTI system, the response $y(t)$ of the system to an arbitrary input $x(t)$ can be expressed as $y(t) = x(t) * h(t) = \int_{-\infty}^{\infty} x(\tau)h(t-\tau)d\tau$, $Y(\omega) = X(\omega)H(\omega)$.

12. The frequency response of the system is $H(\omega) = |H(\omega)| \angle H(\omega)$.
13. The magnitude response $|H(\omega)|$ is symmetric and phase response $\angle H(\omega)$ is anti-symmetric.
14. A system is static or memoryless if its output at any time depends only on the value of its input at that same time.
15. A continuous-time LTI system is said to be causal if and only if its impulse response $h(t) = 0$, $t < 0$.
16. A continuous-time LTI system is BIBO stable if and only if its impulse response is absolutely integrable i.e., $\int_{-\infty}^{\infty} |h(\tau)|\, d\tau < \infty$.
17. A system S is invertible if and only if there exists an inverse system S^{-1} for it such that is SS^{-1} is the identity system.
18. The transfer function of a system in terms of the coefficients of an Nth-order linear constant-coefficient differential equation is

$$H(\omega) = \frac{b_0 + j\omega b_1 + b_2 (j\omega)^2 + ... + b_M (j\omega)^M}{a_0 + j\omega a_1 + a_2 (j\omega)^2 + ... + a_N (j\omega)^N}.$$

19. The conditions for distortionless transmission are that the system transfer function should be

$H(\omega) = Ke^{-j\omega t_0}$ or $|H(\omega)| = K$ and

$\angle H(\omega) = -j\omega t_0$ or $\angle H(\omega) = n\pi - j\omega t_0$ (n integer).

The amplitude of $H(\omega)$ must be constant over the entire frequency range and the phase of $H(\omega)$ must be linear with frequency.

20. An ideal LPF filter transmits all signal frequency components without distortion below a certain frequency W rad/s, called the cutoff frequency. The signal above this frequency is stopped completely. The transfer function of an ideal LPF is given by $H(\omega) = \begin{cases} e^{-j\omega t_0} & |\omega| < W \\ 0 & |\omega| > W \end{cases}$.
21. An Ideal HPF filter transmits all signal frequency components above a certain frequency W rad/s without distortion. The signal below this frequency is stopped completely.

The transfer function of an ideal HPF is given by $H(\omega) = \begin{cases} 0 & |\omega| < W \\ e^{-j\omega t_0} & |\omega| > W \end{cases}$.

22. An ideal BPF filter transmits all signal frequency components within a certain frequency band, $W_2 - W_1$ rad/s, without distortion. The signal outside this band is stopped completely. The transfer function of an ideal BPF is given by $H(\omega) = \begin{cases} e^{-j\omega t_0} & W_1 < |\omega| < W_2 \\ 0 & \text{otherwise} \end{cases}$
23. An ideal BRF filter rejects or stops all signal frequency components within certain frequency band of $W_2 - W_1$ rad/s. The signal outside this band is transmitted without distortion. The transfer function of an ideal BRF is given by

$$H(\omega) = \begin{cases} 0 & W_1 < |\omega| < W_2 \\ e^{-j\omega t_0} & \text{otherwise.} \end{cases}$$

24. All ideal filters are non-causal and are used to eliminate unwanted frequency components of a signal or the noise components from a signal. It is also used to recover the input signal from a noisy signal.

25. The bandwidth of a signal is a range of significant frequency components present within the $1/\sqrt{2}$ times (3dB value) of the maximum signal amplitude present in the signal.
26. System bandwidth is defined as the range of frequencies over which the magnitude spectrum of $H(\omega)$ remains above $1/\sqrt{2}$ times (3dB value) of its maximum value. As the system bandwidth is $W = W_2 - W_1$ rad/sec, for satisfactory distortionless transmission an LTI system should have high bandwidth compared to the signal bandwidth.
27. In the time domain approach, the impulse response $h(t)$of a physically realisable system must be a causal. That is, $h(t) = 0$ for $t < 0$.
28. In the frequency domain approach, the necessary and sufficient condition for the magnitude response function $|H(\omega)|$ to be physically realisable is that $\int_{-\infty}^{\infty} \frac{|\ln H(\omega)|}{1+\omega^2} d\omega < \infty$. This condition is called Paley–Wiener Criterion.
29. The impulse response of an ideal LPF is $h(t) = \frac{\omega_c}{\pi} \text{sinc}\,[\omega_c(t-t_0)]$.
30. The rise time t_r of the output response is defined as the time of the response to reach the final value of the signal from zero value.
31. The rise time is given by $t_r = \frac{\pi}{\omega_c}$ s, where ω_c rad/s is the bandwidth.
32. The product of bandwidth and rise time is π (a constant).
33. The process of expressing the output signal in terms of the superposition of weighted and time shifted impulse responses is called convolution.
34. Let $x_1(t)$ and $x_2(t)$ be two continuous-time signals. Then convolution of $x_1(t)$and $x_2(t)$ can be expressed as $y(t) = x_1(t) * x_2(t) = \int_{-\infty}^{\infty} x_1(\tau)\, x_2(t-\tau)\, d\tau$, where τ is an independent time variable. The symbol * denotes the convolution process.
35. Commutative property: $f_1(t) * f_2(t) = f_2(t) * f_1(t)$
36. Distributive property: $f_1(t) * [f_2(t) + f_3(t)] = f_1(t) * f_2(t) + f_1(t) * f_3(t)$
37. Associative property: $[f_1(t) * f_2(t)] * f_3(t) = f_1(t) * [f_2(t) * f_3(t)] = f_1(t) * f_2(t) * f_3(t)$
38. Shift property: $f_1(t) * f_2(t-t_0) = f(t-t_0)$. Similarly $f_1(t-t_0) * f_2(t) = f(t-t_0)$ and $f_1(t-t_1) * f_2(t-t_2) = f(t-t_1-t_2)$.
39. Convolution of a function with impulse: $f(t) * \delta(t-t_0) = f(t-t_0)$, $f(t) * \delta'(t-t_0) = f'(t-t_0)$.
40. Convolution of a function with unit step:

$$f(t) * u(t) = \int_{-\infty}^{t} f(\tau)\, d\tau$$

$$f(t) * u(t-t_0) = \int_{-\infty}^{t-t_0} f(\tau)\, d\tau$$

$$u(t) * \delta'(t) = u'(t) = \delta(t)$$

41. Width property: If the durations of finite signals $x(t)$ and $h(t)$ are T_1 and T_2 respectively, then the duration of $y(t) = x(t) * h(t)$ is equal to the sum of the durations of $x(t)$ and $h(t)$, that is, $T = T_1 + T_2$. And if the areas under finite signals $x(t)$ and $h(t)$ are A_1 and A_2 respectively, then the area under $y(t) = x(t) * h(t)$is A= area under $y(t)$= area under $x(t)$;and area under $h(t) = A_1\, A_2$.

42. Convolutions of some important functions are shown in Table 5.1.

Table 5.1

No.	$x_1(t)$	$x_2(t)$	$x_1(t) * x_2(t)$
1.	$x(t)$	$\delta(t)$	$x(t)$
2.	$u(t)$	$u(t)$	$tu(t) = r(t)$
3.	$e^{at}u(t)$	$u(t)$	$\frac{e^{at}-1}{a}u(t)$
4.	$e^{-at}u(t)$	$e^{-bt}u(t)$	$\frac{1}{a-b}\left(e^{-at}-e^{-bt}\right)u(t)$
5.	$e^{at}u(t)$	$e^{bt}u(t)$	$\frac{1}{b-a}\left(e^{bt}-e^{at}\right)u(t)$
6.	$e^{at}u(t)$	$e^{bt}u(t)$	$te^{at}u(t)$
7.	$te^{at}u(t)$	$e^{bt}u(t)$	$\frac{1}{2}t^2e^{at}u(t)$

43. The Fourier transform of the time convolution is $\mathcal{F}[f_1(t) * f_2(t)] = F_1(\omega)F_2(\omega)$

44. The convolution of the Fourier transforms is $F_1(\omega) * F_2(\omega) = 2\pi\mathcal{F}[f_1(t)f_2(t)]$.

45. When two signals $f_1(t)$ and $f_2(t)$ are provided in graphical form, the convolution can be performed by graphical method.

Review Questions

1. Explain the frequency response of a linear system.
2. Explain impulse response of a linear system.
3. What is an LTI system? Derive an expression for the transfer function of an LTI system.
4. What is an LTI system. Show that an LTI system combined with time scaling property may result in time variant system. (May 2013)
5. Define linearity and time-invariant properties of a system.
6. Derive the expression for transfer function of an LTI system.
7. Define an LTI system and state its properties.
8. Obtain the conditions for distortionless transmission through a system.
9. Explain signal bandwidth and system bandwidth.
10. Explain ideal filter characteristics of a linear system.
11. What is Paley–Winear criterion? Explain.
12. Derive the expression for the step response of an LTI system.
13. Define causality and stability of a linear system.
14. Verify the BIBO stability condition for a continuous-time LTI systems.
15. Explain ideal filters. (Nov 2015)

16. Derive the relationship between rise time and bandwidth. (May 2015)
17. Sketch the frequency response of ideal LPF, HPF and BPF. (May 2015)
18. What is a distortionless system? Explain. (May 2015)
19. What are the requirements to be satisfied by an LTI system to provide distortionless transmission of a signal? (Dec 2011)
20. Define the following terms: (Dec 2011)

    (i) Signal bandwidth
    (ii) System bandwidth
    (iii) Linear time variant system
    (iv) Paley–Wiener criteria for physical realisability.

21. Draw the ideal characteristics of low-pass, high-pass, band-pass and bandstop filters. (Dec 2011)
22. Show that the output of an LTI system is given by the linear convolution of the input signal and the impulse response of the system.
23. Derive an expression for convolution integral and state its properties.
24. Explain the graphical approach for computing the convolutional integral.

Problems

1. Show that for a continuous-time LTI system, $T\{e^{st}\} = \lambda_1 e^{st}$ and $T\{z^n\} = \lambda_2 z^n$ where s and z are complex variables and λ_1 and λ_2 are complex constants.
2. Show that the system described by the differential equation $\dfrac{dy(t)}{dt} + ty(t) = x(t)$ is linear.
3. A system is characterised by the equation $y(t) = x(t-2) + x(2-t)$. Determine whether the system is LTI system or not.
4. A system is characterised by the equation $y(t) = x(t/3)$. Determine whether the system is LTI system or not.
5. For an LTI system with impulse response $h(t) = 6e^{-t}u(t)$, determine the system response to the inputs: (a) $2u(t)$, and (b) $3e^{-3t}u(t)$.
6. For an LTI system with impulse response $h(t) = 4e^{-4t}u(t)$, determine the system response to the input rect $[2t - 0.5] = u(t + 0.75) - u(t + 0.25)$.
7. Find the impulse response of the RC low-pass filter shown in Fig. P5.7. Also find the transfer function. What would be its frequency response?

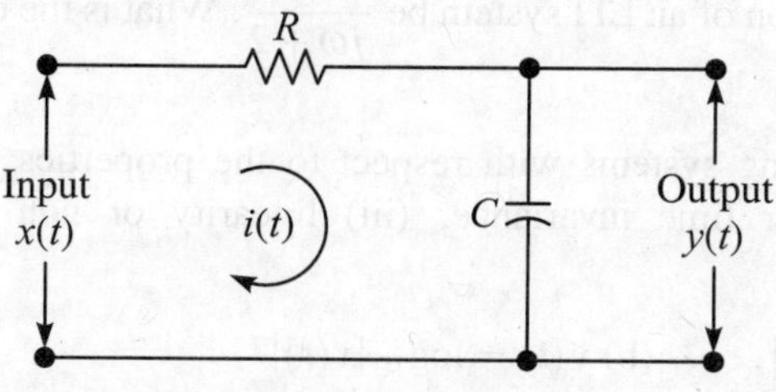

Fig. P5.7

8. Find the impulse response of the system $y''(t)+3y'(t)+2y(t)=2x'(t)+x(t)$.
9. Consider the system described by $y'(t)+2y(t)=x(t)$. Find the output response $y(t)$ of the system if the excitation is $x(t)=u(t)$.
10. Consider two functions $x(t)=u(t-1)+u(t-3)$ and $h(t)=u(t)-u(t-2)$. Find the convolution of $x(t)$ and $h(t)$ using the graphical method.
11. Compute and sketch the output y(t) of the continuous-time LTI system with impulse response $h(t)$ for an input signal $x(t)$ are given by $h(t)=e^{-\alpha t}u(t)$ and $x(t)=e^{\alpha t}u(-t)$.
12. The impulse response of a continuous-time system is expressed as: $h(t)=e^{-2t}u(t)$. Find the frequency response of the system. Plot the frequency response. (Nov 2015)
13. Check whether the following system is linear or not. (May 2015)

$$\frac{dy(t)}{dt}+2y(t)=x^2(t)$$

14. Check the five basic system properties (invertibility, causality, stability, time invariance and linearity) for the following systems. (June 2014)

(i) $y(t)=\int_{-\infty}^{t}x(\tau)\,d\tau$ (ii) $y(t)=x^2(t)$ (iii) $y(t)=\sin(x(t))$

15. Consider an LTI system with an input and output related through the equation

$$y(t)=\int_{-\infty}^{\infty}e^{-(t-\tau)}x(\tau-2)\,d\tau.$$

What is the impulse response $h(t)$ for this system? (Nov 2015)

16. A system has an impulse response $h(t)=4e^{-4t}u(t)$. Find the response of the system to $x(t)=\text{rect}\left[2\left(t-\frac{1}{4}\right)\right]$. (Nov 2009)
17. There are several possible ways of estimating an essential bandwidth of a non-bandlimited signal. For a low-pass signal, for example, the essential bandwidth may be chosen as a frequency where the amplitude spectrum of the signal decays to k per cent of its peak value. The choice of k depends on the nature of application. Choosing $k=5$, determine the essential bandwidth of $g(t)=e^{-at}u(t)$.
18. A stable LTI system is characterised by the differential equation $\frac{d^2y(t)}{dt^2}+6\frac{dy(t)}{dt}+8y(t)=2x(t)$. Find the frequency response and impulse response using Fourier transform. (Dec 2015)
19. A stable LTI system is characterised by the differential equation $\frac{d^2y(t)}{dt^2}+6\frac{dy(t)}{dt}+8y(t)=2x(t)$. What is the response of this system if $x(t)=t\,e^{-2t}u(t)$. (Dec 2015)
20. Let the system function of an LTI system be $\frac{1}{j\omega+2}$. What is the output of the system for an input $(0.5)^t u(t)$. (Jan 2015)
21. Examine the following systems with respect to the properties: (i) causality or non causality, (ii) time variance or time invariance, (iii) linearity or non linearity and (iv) stability or instability.

(a) $y(t)=\cos[x(t)]$, (b) $y(t)=\log_{10}|x(t)|$

Answers

3. Linear time variant system
4. Linear time variant system
5. (a) $y(t) = 12\left[1 - e^{-t}\right]u(t)$ (b) $y(t) = 9\left[e^{-t} - e^{-3t}\right]u(t)$
6. $y(t) = 1.6683e^{-4t}\left[u(t+0.75) - u(t+0.25)\right]$
7. $H(j\omega) = \dfrac{1}{1+j\omega cR}; h(t) = \dfrac{1}{RC}e^{-\frac{t}{RC}}u(t)$
8. $h(t) = -e^{-t}u(t) + 3e^{-2t}u(t)$
9. $y(t) = -\dfrac{e^{-2t}}{2}u(t) + \dfrac{1}{2}u(t)$
11. $y(t) = \dfrac{1}{2\alpha}e^{\alpha|t|},\ \alpha > 0$
12. $H(j\omega) = \dfrac{1}{2+j\omega}$
13. Nonlinear
14. (i) Linear, Time invariant, and non-causal; (ii) Nonlinear, time variant and causal; (iii) Nonlinear, time invariant and causal
15. $h(t) = e^{-(t-2)}u(t)$
16. $$y(t) = \begin{cases} 1 - 0.049e^{-4t} & \text{for } t < 0 \\ e^{-1} - e^{-3} & \text{for } t < 0 \\ (e^{-1} - e^{-3}) & \text{for } t < 0 \end{cases}$$
17. $0.71129a$
18. Frequency response $H(\omega) = \dfrac{1}{j\omega+2} - \dfrac{1}{j\omega+4}$.

    Impulse response $h(t) = \left[e^{-2t} - e^{-4t}\right]u(t)$
19. $y(t) = \left[\dfrac{1}{4}e^{-2t} - \dfrac{1}{2}e^{-2t} + t^2e^{-2t} - \dfrac{1}{4}e^{-4t}\right]u(t)$
20. $y(t) = \dfrac{(0.5)^t}{2+\log 0.5}u(t)$
21. (a) Causal, time invariant, nonlinear and stable

    (b) Causal, time invariant, nonlinear and unstable

Multiple-choice Questions

1. Let h(t) be the impulse response of a linear time invariant system. Then the response of the system for any input u(t) is

   (a) $\displaystyle\int_0^t u(t-\tau)d\tau$ (b) $\displaystyle\frac{d}{dt}\int_0^t h(\tau)u(t-\tau)d\tau$

   (c) $\displaystyle\int_0^t \left[\int_0^t h(\tau)u(t-\tau)d\tau\right] dt$ (d) $\displaystyle\int_0^t h^2(\tau)u(t-\tau)d\tau$

2. The unit impulse response of a linear time invariant system is the unit0 step function $u(t)$. For $t > 0$, the response of the system to an excitation $e^t u(t)$, $a > 0$ will be

   (a) ae^{-at} (b) $(1/a)(1 - e^{-at})$

   (c) $a(l - e^{-at})$ (d) $1 - e^{-at}$

3. A system with an input $x(t)$ and an output $y(t)$ is described by the relation: $y(t) = tx(t)$. This system is

   (a) linear and time-invariant (b) linear and time varying

   (c) nonlinear and time-invariant (d) nonlinear and time-varying

4. Convolution of $x(t+5)$ with impulse function $\delta(t-7)$ is equal to

   (a) $x(t-12)$ (b) $x(t+12)$ (c) $x(t-2)$ (d) $x(t+2)$

5. Let $x(t)$ be the input to a linear, time-invariant system. The required output is $4x(t-2)$. The transfer function of the system should be

   (a) $4e^{j2\omega}$ (b) $2e^{-j4\omega}$ (c) $4e^{-j2\omega}$ (d) $2e^{j4\omega}$

6. Which of the following can be the impulse response of a causal system?

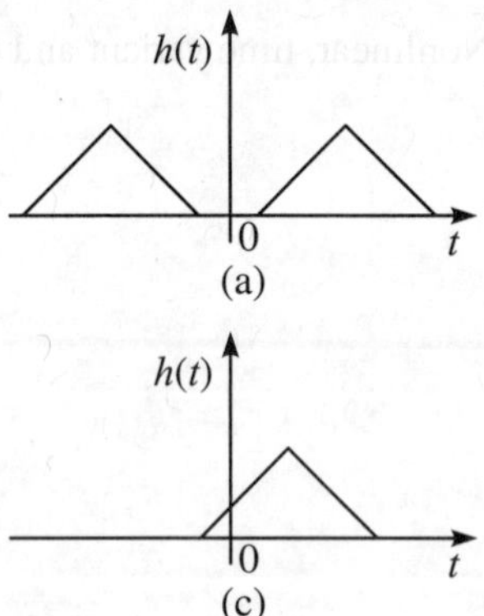

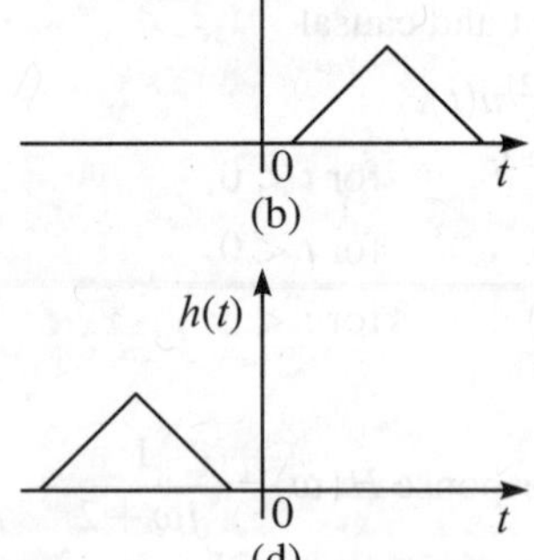

7. The input and output of a continuous time system are respectively denoted by $x(t)$ and $y(t)$. Which of the following descriptions corresponds to a causal system?

   (a) $y(t) = x(t-2) + x(t+4)$ (b) $y(t) = (t-4)x(t+1)$

   (c) $y(t) = (t+4)x(t-1)$ (d) $y(t) = (t+5)x(t+5)$

8. The impulse response $h(t)$ of a linear time-invariant continuous time system is described by $h(t) = \exp(\alpha t)u(t) + \exp(\beta t)u(-t)$, where $u(t)$ denotes the unit step function, and α and β are real constants. This system is stable if

   (a) α is positive and β is positive (b) α is negative and β is negative

   (c) α is positive and β is negative (d) α is negative and β is positive

9. The differential equation $100\frac{d^2y}{dt^2} - 20\frac{dy}{dt} + y = x(t)$ describes a system with an input $x(t)$ and an output $y(t)$. The system, which is initially relaxed, is excited by a unit step input. The output $y(t)$ can be represented by the waveform

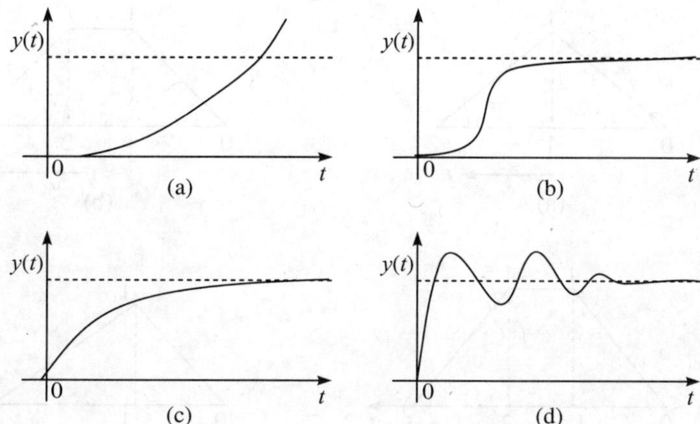

10. The response of an initially relaxed linear constant parameter network to a unit impulse applied at $t = 0$ is $4e^{-2t}5u(t)$. The response of this network to a unit step function will be:

(a) $2[1-e^{-2t}]$ (b) $\sin 2t$

(c) $4[e^{-t}-e^{-2t}]u(t)$ (d) $(1-4e^{-4t})u(t)$

11. The impulse response and the excitation function of a linear time invariant causal system are shown in Fig. M5.11(a) and (b) respectively. The output of the system at $t = 2$ seconds is equal to

(a) 0 (b) 1/2 (c) 3/2 (d) 1

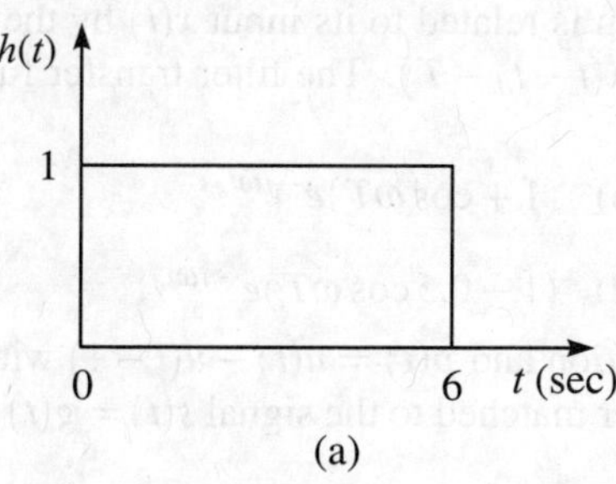

(a)

Fig. M5.11(a)

x(t)
1/2
0 2 6 t (sec)

(b)

Fig. M5.11(b)

12. An excitation is applied to a system at $t = T$ and its response is zero for $-\infty < t < T$. Such a system is a

(a) non-causal system (b) stable system

(c) causal system (d) unstable system

13. A linear time invariant system has an impulse response e^{2t}, $t > 0$. If the initial conditions are zero and the input is e^{3t}, the output for $t > 0$ is

(a) $e^{3t}-e^{2t}$ (b) e^{5t}

(c) $e^{3t}+e^{2t}$ (d) None of the above

14. Let $u(t)$ be the unit step function. Which of the waveforms in Fig. M5.14 (a)-(d) corresponds to the convolution of $[u(t)-u(t-1)]$ with $[u(t)-u(t-2)]$?

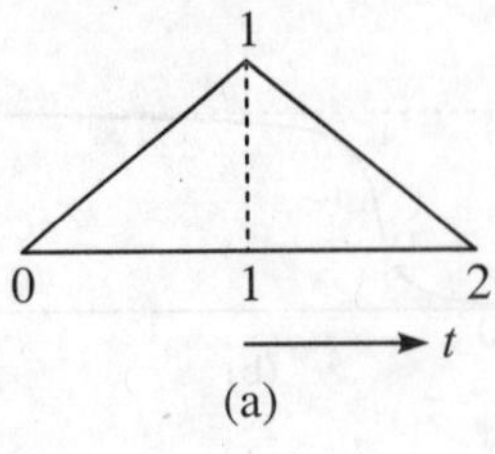

(a)

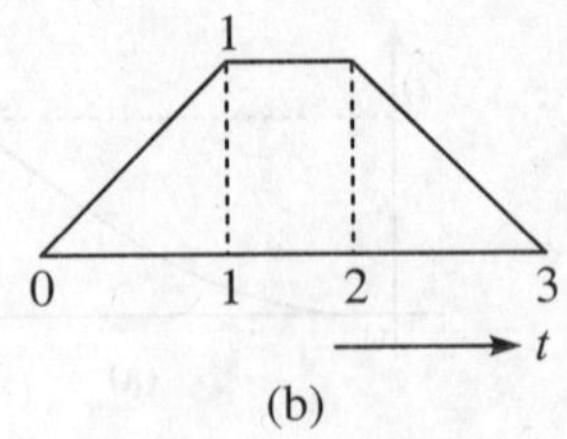

(b)

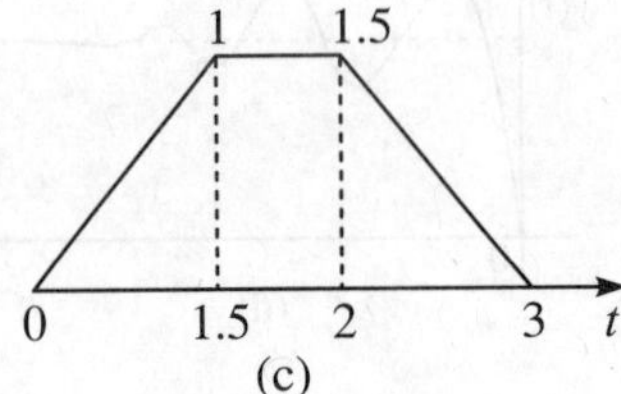

(c)

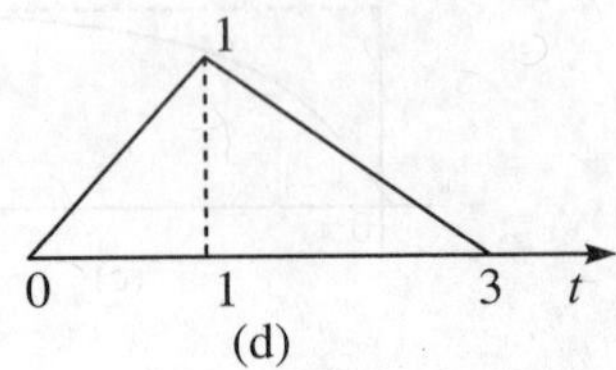

(d)

(a) Fig. M5.14a (b) Fig. M5.14 b (c) Fig. M5.14c (d) Fig. M5.14d

15. A system described by the differential equation $\frac{d^2y}{dt^2}+3\frac{dy}{dt}+2y=x(t)$ is initially at rest. For input $x(t)=2u(t)$, the output $y(t)$ is

(a) $(1-2e^{-t}+e^{-2t})u(t)$ (b) $(1+2e^{-t}-2e^{-2t})u(t)$

(c) $(0.5+e^{-t}+1.5e^{-2t})u(t)$ (d) $(0.5+2e^{-t}+2e^{-2t})u(t)$

16. The output $y(t)$ of a linear time invariant system is related to its input $x(t)$ by thefollowing equation: $y(t)=0.5\,x(t-t_d+T)+x(t-t_d)+0.5\,x(t-t_d-T)$. The filter transfer function $H(\omega)$ of such a system is given by

(a) $(1+\cos\omega T)e^{-j\omega t_d}$ (b) $(1+\cos\omega T)e^{-j\omega t_d}$

(c) $(1+0.5\cos\omega T)e^{-j\omega t_d}$ (d) $(1-0.5\cos\omega T)e^{-j\omega t_d}$

17. Let $g(t)=p(t)*p(t)$, where $*$ denotes convolution and $p(t)=u(t)-u(t-1)$ with $u(t)$ being the unit step function. The impulse response of filter matched to the signal $s(t)=g(t)-8(t-2)*g(t)$ is given as

(a) $s(1-t)$ (b) $-s(1-t)$ (c) $-s(t)$ (d) $s(t)$

18. The frequency response of a linear, time-invariant system is given by $H(f)=\frac{5}{1+j10\pi f}$. The step response of the system is

(a) $5(1-e^{-5t})u(t)$ (b) $5(1-e^{-(t/5)})u(t)$

(c) $(1/5)(1-e^{-5t})u(t)$ (d) $(1/5)(1-e^{-(t/5)})u(t)$

19. A continuous time LTI system is described by $\frac{d^2y(t)}{dt^2}+4\frac{dy(t)}{dt}+3y(t)=\frac{dx(t)}{dt}+4x(t)$. Assuming zero initial conditions, the response $y(t)$ of the above system for the input $x(t)=e^{-2t}u(t)$ is given by

(a) $(e^{t}-e^{3t})u(t)$ (b) $(e^{-t}-e^{-3t})u(t)$

(c) $(e^{-t}+e^{-3t})u(t)$ (d) $(e^{t}+e^{3t})u(t)$

20. An input $x(t) = exp(-2t)u(t) + \delta(t-6)$ is applied to an LTI system with impulse response $h(t) = u(t)$. The output is

(a) $[1 - \exp(-2t)]u(t) + u(t+6)$ (b) $[1 - \exp(-2t)]u(t) + u(t-6)$

(c) $0.5[1 - \exp(-2t)]u(t) + u(t+6)$ (d) $0.5[1 - \exp(-2t)]u(t) + u(t-6)$

21. For a distortionless transmission, the amplitude response $|H(\omega)|$ is

(a) constant (b) zero (c) linear (d) infinite

22. For a distortionless transmission, the phase response is

(a) constant (b) zero

(c) non-linear (d) linear with frequency

23. Rise time of a system is defined as the difference of the time taken from

(a) 0.2 to 0.9 of the final time (b) 0.1 to 0.8 of the final time

(c) 0.1 to 0.9 of the final time (d) 0.5 to 0.9 of the final time

24. For defining a transfer function, the initial conditions must be taken as

(a) zero (b) infinite (c) finite (d) none of these

25. The characteristics of an LTI system are completely characterised by its

(a) impulse response (b) step response

(c) transfer function (d) none of these

26. The output of an LTI system is equal to unit impulse response when the input is

(a) $u(t)$ (b) $\delta(t)$ (c) $r(t)$ (d) none of these

27. The relation between rise time t_r and bandwidth ω_c is

(a) $t_r = \dfrac{\pi}{\omega_c}$ (b) $t_r = \dfrac{\omega_c}{\pi}$ (c) $t_r = \dfrac{1}{\omega_c}$ (d) $t_r = \omega_c$

28. For distortionless transmission,

(a) $y(t) = kx(t)$ (b) $y(t) = x(t - t_d)$

(c) $y(t) = kx(t - t_d)$ (d) none of these

29. For distortionless transmission, system bandwidth must be equal to

(a) signal bandwidth (b) two times signal bandwidth

(c) infinite (d) half of signal bandwidth

30. For distortionless transmission, the bandwidth of the system is

(a) finite (b) infinite (c) zero (d) none of these

Answers

1. (a)	2. (b)	3. (b)	4. (c)	5. (c)	6. (b)	7. (c)	8. (d)	9. (a)	10. (a)
11. (b)	12. (c)	13. (a)	14. (b)	15. (a)	16. (a)	17. (d)	18. (b)	19. (b)	20. (d)
21. (a)	22. (d)	23. (c)	24. (a)	25. (a)	26. (b)	27. (a)	28. (c)	29. (a)	30. (b)

6

Sampling

Learning Objectives

The main objective of this chapter is to understand thoroughly the basic concepts of sampling process. By reading this chapter, students will be able to relate to the following topics.

- Sampling theorem
- Graphical and analytical proof for band-limited signals
- Impulse sampling
- Natural and flat-top sampling
- Reconstruction of signal from its samples
- Effect of undersampling – aliasing

6.1 Introduction

In the previous chapters we looked at continuous-time signals, linear systems, Fourier series and Fourier transforms. It was observed that the Fourier series expansion of a continuous-time signal is an infinite sum of discrete coefficients taken at regular intervals. Mathematically, we know that analog or continuous signals are functions of continuous quantities such as time, space and frequency, whereas discrete signals are functions defined at integer values. Discrete signals are called sequences as the signal is defined only at discrete intervals. The process of converting a continuous signal to a discrete signal is called sampling; a reconstruction process is applied on the samples to get the continuous signal back. Since most of the signals that are naturally available at the system input are continuous-time signals, such as speech, sonar, images, video and seismic, it is necessary to convert these signals into discrete time signals which are needed as inputs for digital devices such as microprocessors and microcomputers. The analog-to-digital converters (ADC) and digital-to-analog converters (DAC) are used to convert analog signals into digital signals and vice-versa. The first step in ADC is to convert a continuous-time signal into a discrete-time signal by using the sampling process.

In digital communications, information is processed in the form of samples taken at regular intervals. At the destination the samples are converted back into the original information using a reconstruction process. Sampling process is also required for the multiplexing of signals. Sampling on discrete time signals is also used to change data rate by adding or removing some samples.

Discrete sequences are gaining widespread attention in many applications as they are easily convertible to digital numbers or data formats that can be processed directly by logic circuits of digital computers. Advances in computer communications, telephony, telemetry, etc., has further made it easy for implementing discrete sequence algorithms on personal computers (PCs).

A discrete-time signal $f(nT)$, $n = 0, \pm 1, \pm 2, \pm 3, ...$, can be obtained from its continuous-time signal $f(t)$ by taking samples at $n = 0, \pm 1, \pm 2, \pm 3, ...$ with amplitude equal to the instantaneous signal level; here T is called the sampling period. The reciprocal of the sampling period T is called the sampling frequency f_s (samples/second). Figure 6.1(a) shows a continuous-time signal and Fig. 6.1(b) the discrete-time signal obtained by sampling it.

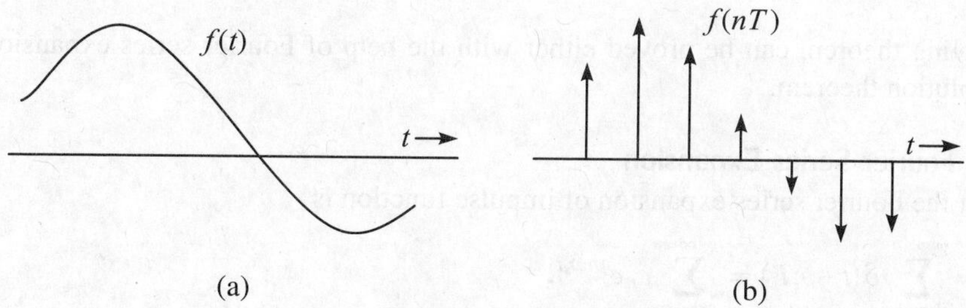

Fig. 6.1 *(a) Continuous-time signal and (b) discrete-time signal*

6.2 Sampling Theorem

The sampling theorem, also known as Shannon's sampling theorem, states that *a band-limited continuous-time signal $f(t)$ which has no frequency component higher than f_m Hz is uniquely represented by its sampling values taken at uniform intervals separated by less than $\frac{1}{2f_m}$ seconds.* For reconstruction of signals from samples, the sampling theorem in frequency domain can be stated as: *a band-limited finite energy signal $f(t)$ which has no frequency component higher than f_m Hz may be completely recovered from the knowledge of its sampled values taken at the sampling frequency equal to or greater than twice of its highest frequency component, i.e., $f_s \geq 2f_m$ Hz.*

The theorem is also called uniform sampling theorem or lowpass sampling theorem because in this process lowpass type signals whose frequency is less than f_m Hz are sampled at regular time intervals. The minimum sampling frequency $f_s = 2f_m$ Hz at which the signal can be completely recovered is called Nyquist rate. The reciprocal of the Nyquist rate is called Nyquist interval.

Proof: Consider $f(t)$ be a band-limited continuous-time signal which has no frequency component higher than f_m Hz and that its Fourier transform is $F(\omega) = 0, \quad f \geq f_m$.

Let $\delta_T(t)$ be a periodic impulse function with period T seconds equal to the sampling interval. The sampling process is the product of the continuous-time signal $f(t)$ and the periodic impulse function $\delta_T(t)$. The resultant discrete time signal is the samples of $f(t)$ taken at uniform interval T sec. The sampling process and the corresponding signals are as shown in Fig. 6.2. Therefore the discrete time signal is

$$f(nT) = f_s(t) = f(t)\delta_T(t), \tag{6.1}$$

where $\delta_T(t) = \sum_{n=-\infty}^{\infty} \delta(t - nT)$.

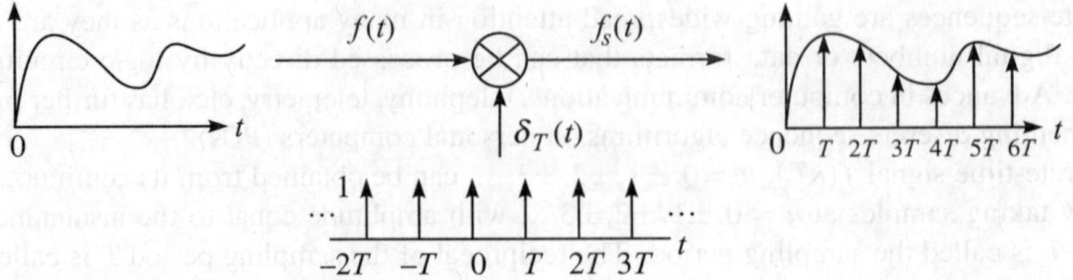

Fig. 6.2 *Sampling process of a continuous-time signal*

The sampling theorem can be proved either with the help of Fourier series expansion or the frequency convolution theorem.

Proof Using Fourier Series Expansion

We know that the Fourier series expansion of impulse function is

$$\delta_T(t) = \sum_{n=-\infty}^{\infty} \delta(t-nT) = \sum_{n=-\infty}^{\infty} c_n e^{jn\omega_s t}, \tag{6.2}$$

where $\omega_s = \frac{2\pi}{T} = 2\pi f_s$ is the sampling frequency and the Fourier coefficient is

$$c_n = \frac{1}{T}\int_{-\infty}^{\infty} \delta(t) e^{-jn\omega_s t} = \frac{1}{T}. \tag{6.3}$$

Therefore,

$$\delta_T(t) = \frac{1}{T}\sum_{n=-\infty}^{\infty} e^{jn\omega_s t}. \tag{6.4}$$

Substituting this in Eq. (6.1) we get

$$f_s(t) = \frac{1}{T}\sum_{n=-\infty}^{\infty} f(t) e^{jn\omega_s t}. \tag{6.5}$$

Now, taking the Fourier transform on both sides,

$$\mathcal{F}[f_s(t)] = \frac{1}{T}\sum_{n=-\infty}^{\infty} \int_{-\infty}^{\infty} f(t) e^{jn\omega_s t} e^{-j\omega t} dt.$$

Let $F(\omega)$ be the Fourier transform of $f(t)$ and $F_s(\omega)$ be the Fourier transform of $f_s(t)$. Then,

$$F_s(\omega) = \frac{1}{T}\sum_{n=-\infty}^{\infty} \int_{-\infty}^{\infty} f(t) e^{-j(\omega - n\omega_s)t} dt$$

$$F_s(\omega) = \frac{1}{T}\sum_{n=-\infty}^{n=\infty} F(\omega - n\omega_s). \tag{6.6}$$

Proof Using Frequency Convolution Theorem

Let $F(\omega)$ be the Fourier transform of $f(t)$, $F_s(\omega)$ the Fourier transform of $f_s(t)$ and $\delta_T(\omega)$ the Fourier transform of $\delta_T(t)$. Using convolution theorem, the Fourier transform of sampled signal is the convolution of $F(\omega)$ and $\delta_T(\omega)$.

$$\mathcal{F}[f_s(t)] = \mathcal{F}[f(t)\delta_T(t)] \tag{6.7}$$

or $F_s(\omega) = \frac{1}{2\pi}[F(\omega) * \delta_T(\omega)].$ (6.8)

We know that the Fourier transform of impulse function is

$$\delta_T(\omega) = \omega_s \sum_{n=-\infty}^{\infty} \delta(\omega - n\omega_s). \tag{6.9}$$

Then $F_s(\omega) = \frac{1}{2\pi} F(\omega) * \omega_s \sum_{n=-\infty}^{\infty} \delta(\omega - n\omega_s).$

Substituting the value $\frac{2\pi}{T} = \omega_s$,

$$F_s(\omega) = \frac{1}{T} \sum_{n=-\infty}^{\infty} F(\omega) * \delta(\omega - n\omega_s)$$

$$F_s(\omega) = \frac{1}{T} \sum_{n=-\infty}^{\infty} F(\omega - n\omega_s).$$

$$F_s(\omega) = \frac{1}{T}[\cdots + F(\omega + 2\omega_s) + F(\omega + \omega_s) + F(\omega) + F(\omega - \omega_s) + F(\omega + 2\omega_s) + \ldots]. \tag{6.10}$$

Note that the Eq. (6.6) and Eq. (6.10) are the same.

Thus the Fourier transform of the sampled signal is equal to the infinite sum of shifted versions of the Fourier transform of the continuous signal $f(t)$. From Eq. (6.10), it is observed that the shifted Fourier transforms $F(\omega)$ are integer multiples of ω_s. If the sampling frequency ω_s is greater than or equal to the frequency band of $F(\omega)$, i.e., $-\omega_m \leq \omega < \omega_m$ or $\omega_s \geq 2\omega_m$ and $T \leq 1/2\pi f_m$, then the shifted spectra of $F(\omega)$ do not overlap with each other and $f(t)$ can be uniquely represented by its sampling values. If $\omega_s \geq 2\omega_m$, the shifted spectrum of $F(\omega)$ overlaps with each other and the basic shape of the spectrum is distorted. This proves the sampling theorem.

6.3 Graphical Proof for Band-limited Signals

Let $f(t)$ be a band-limited continuous-time signal with no frequency component higher than f_m Hz. The signal and its Fourier transform $F(\omega)$ are shown in Fig. 6.3 (a) and (b) respectively, an impulse train and its Fourier transform are shown in Fig. 6.3 (c) and (d) respectively, and a discrete signal and its spectrum $F(\omega)$ are shown in Fig. 6.3 (e) and (f).

It is observed that the function $F(\omega)$ repeats itself every ω_s rad/s interval. If the sampling frequency of $F_s(\omega)$ is greater than or equal to the bandwidth of $F(\omega)$, i.e., $\omega_s \geq 2\omega_m$, then the shifted spectra of $F(\omega)$ do not overlap with each other as shown in Fig. 6.4(a). That is, $F(\omega)$ will repeat periodically in the frequency domain without overlap, provided $\omega_s \geq 2\omega_m$. That is,

$$\frac{2\pi}{T} \geq 2(2\pi f_m) \text{ or } T \leq \frac{1}{2f_m} \text{ seconds.} \tag{6.11}$$

Therefore, if the samples of $f(t)$ are taken at regular intervals less than $1/2\pi f_m$ seconds apart, the continuous signal can be recovered without loss of information by passing the sampled signal through a lowpass filter with cutoff frequency f_m. The filter allows transmission of all frequencies below f_m and attenuates all frequencies above f_m.

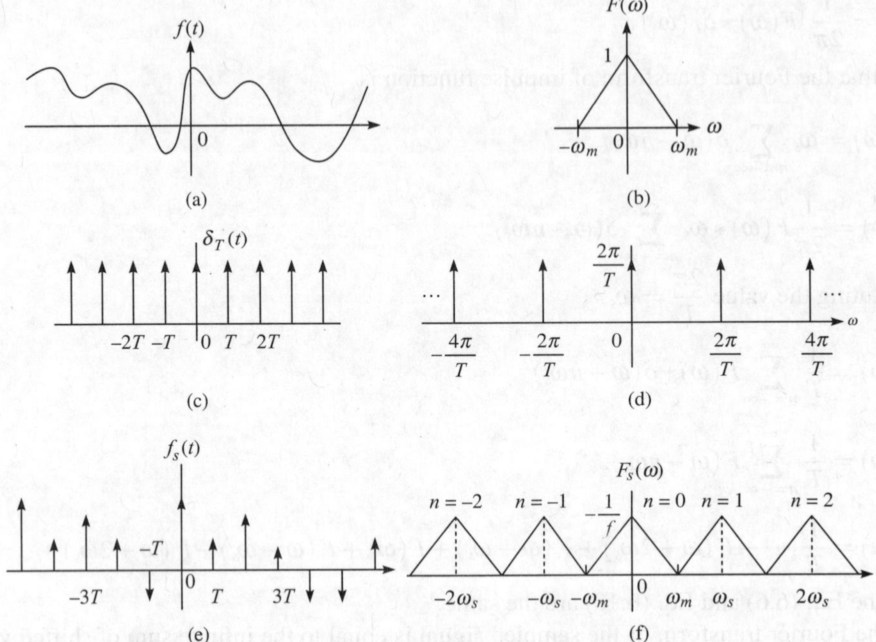

Fig. 6.3 *(a) A continuous-time signal and (b) its spectrum; (c) impulse train and (d) its spectrum; (e) a discrete time signal and (f) its spectrum*

If the sampling frequency of $F_s(\omega)$ is less than the bandwidth of $F(\omega)$, i.e., $\omega_s < 2\omega_m$, then the shifted spectra of $F(\omega)$ overlaps with each other, resulting in a distorted spectrum as shown in Fig. 6.4 (b). It is not possible then to recover the continuous signal without loss of information. Thus, the sampling theorem is graphically proved.

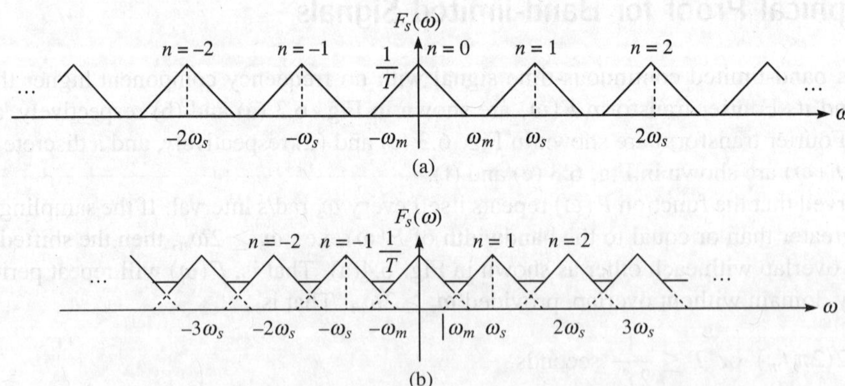

Fig. 6.4 *Spectrum of $F(\omega)$ for (a) $\omega_s > 2\omega_m$ and (b) $\omega_s < 2\omega_m$*

For example, if the highest frequency component in a voice signal is 3.4 kHz, this signal should be sampled at or higher than 6.8 kHz so that the voice signal can be completely specified by its sampled values. The minimum sampling frequency of 6.8 kHz is called Nyquist rate and the Nyquist interval is $T = 0.147$ ms. If the sampling frequency is taken at 6 kHz, the signal is distorted and cannot be recovered. In practice, the voice signal is sampled at 8 kHz for satisfactory reconstruction.

6.4 Impulse Sampling or Ideal Sampling

Consider $f(t)$ to be a band-limited continuous-time signal which has no frequency component higher than f_m Hz. The Fourier transform of the signal $f(t)$ is $F(\omega)$. Let $\delta_T(t)$ be a periodic impulse function with period T seconds. The impulse sampling is the product of the continuous-time signal $f(t)$ and the periodic impulse function or impulse train $\delta_T(t)$ as shown in Fig. 6.2. We know that

$$\delta_T(t) = \sum_{n=-\infty}^{\infty} \delta(t - nT).$$

The impulse sampling is given by

$$f(nT) = f_s(t) = f(t)\delta_T(t). \tag{6.12}$$

$$f_s(t) = f(t)\delta_T(t) = f(t) \sum_{n=-\infty}^{\infty} \delta(t - nT)$$

$$f_s(t) = \sum_{n=-\infty}^{\infty} f(nT)\delta(t - nT). \tag{6.13}$$

Taking the Fourier transform,

$$F_s(\omega) = \frac{1}{T} \sum_{n=-\infty}^{\infty} F(\omega - n\omega_s),$$

where $\omega_s = 2\pi/T = 2\pi f_s$ is the sampling frequency. The process of multiplying the continuous-time signal $f(t)$ with a periodic impulse function or impulse train $\delta_T(t)$ is called ideal or impulse sampling. It is also called instantaneous sampling.

The impulse sampling process in frequency domain as shown in Fig. 6.5 is the convolution of the frequency domain signal $F(\omega)$ and the periodic impulse function or impulse train $\delta_T(\omega)$ as

$$F_s(\omega) = \frac{1}{2\pi} F(\omega) * \delta_T(\omega) = \frac{1}{T} \sum_{n=-\infty}^{\infty} F(\omega) * \delta(\omega - n\omega_s). \tag{6.14}$$

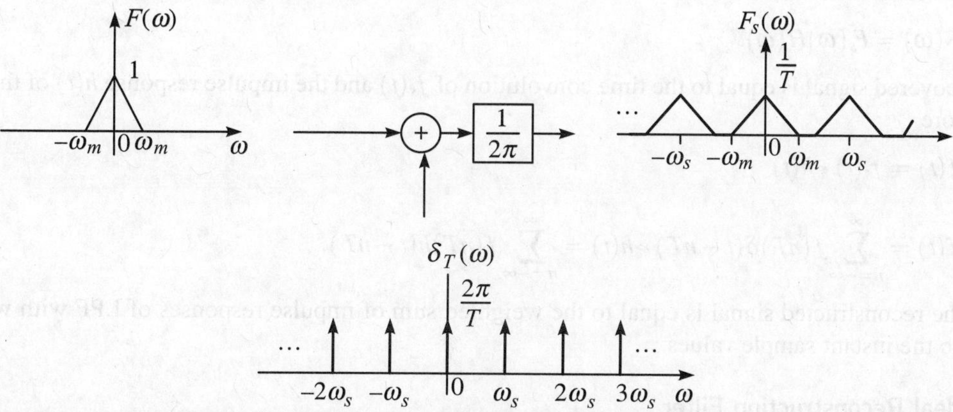

Fig. 6.5 *Impulse sampling in frequency domain*

6.5 Reconstruction of a Signal from Its Samples

Consider a band-limited signal $f(t)$ that is sampled at $\omega_s \geq 4\pi f_m \geq 2\omega_m$. The process of recovering $f(t)$ from its sampled signal $f_s(t)$ is accomplished by passing $f_s(t)$ through a reconstruction filter, which is a lowpass filter (LPF) with cutoff frequency f_m (Fig. 6.7). Mathematically, we can show this by using the time convolution theorem.

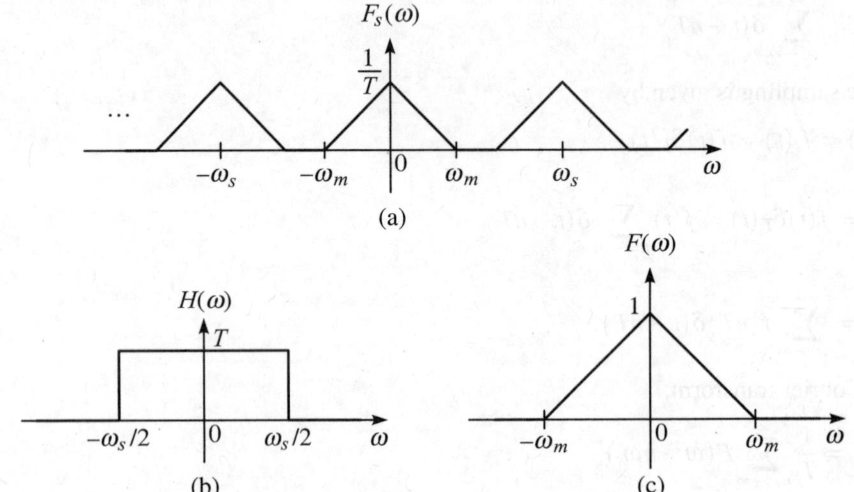

Fig. 6.7 *The reconstruction process*

The spectrum of $f_s(t)$ is given by

$$F_s(\omega) = \frac{1}{T}\sum_{n=-\infty}^{\infty} F(\omega - n\omega_s) = \frac{1}{T}\sum_{n=-\infty}^{\infty} F(\omega - 2n\omega_m). \tag{6.15}$$

The process of lowpass filtering is equivalent to multiplying $F_s(\omega)$ by the transfer function $H(\omega)$ of a reconstruction filter. Thus

$$F(\omega) = F_s(\omega)H(\omega). \tag{6.16}$$

The recovered signal is equal to the time convolution of $f_s(t)$ and the impulse response $h(t)$ of the LPF. Therefore

$$f(t) = f_s(t) * h(t)$$

$$f(t) = \sum_{n=-\infty}^{\infty} f(nT)\delta(t-nT) * h(t) = \sum_{n=-\infty}^{\infty} f(nT)h(t-nT). \tag{6.17}$$

Thus the reconstructed signal is equal to the weighted sum of impulse responses of LPF with weights equal to the instant sample values.

The Ideal Reconstruction Filter

Let a band-limited continuous signal $f(t)$ is sampled at a frequency greater than the Nyquist rate. If the sampled signal $f_s(t)$ is passed through an ideal low-pass filter with cutoff frequency ω_m, the transfer function of an ideal low-pass filter is a gate function given by

$$H(\omega) = \begin{cases} 1 & |\omega| < \omega_m \\ 0 & |\omega| > \omega_m. \end{cases} \tag{6.18}$$

By taking the inverse Fourier transform of $H(\omega)$,

$$h(t) = \int_{-\omega_m}^{\omega_m} e^{j\omega t}\,dt$$

$$h(t) = \frac{\sin(\omega_m t)}{\omega_m t} = 2\omega_m \mathrm{Sa}(\omega_m t).$$

$$f(t) = f_s(t) * h(t) = f_s(t) * \mathrm{Sa}(\omega_m t) \tag{6.19}$$

$$= f_s(t) * 2\omega_m \mathrm{Sa}(\omega_m t) = \sum_{n=-\infty}^{\infty} f(nT)\delta(t-nT) * 2\omega_m \mathrm{Sa}(\omega_m t)$$

$$f(t) = \sum 2\omega_m f(nT)\mathrm{Sa}(\omega_m(t-nT)). \tag{6.20}$$

The function $\mathrm{Sa}(\omega_m(t-nT))$ is called an interpolation function because it allows one to interpolate between the sampled values $f(nT)$ to find $f(t)$ for all t. Graphically, the reconstructed signal is shown in Fig. 6.7.

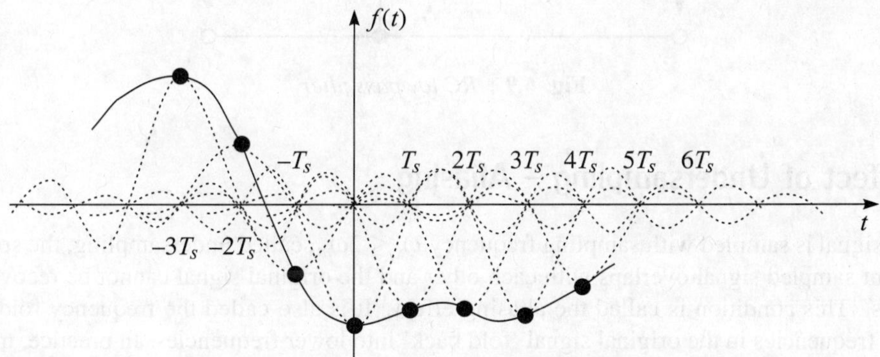

Fig. 6.8 *Reconstruction signal*

Each sample value is multiplied by the sample functions centered on the sampled value. These functions are then added to give the original waveform. Since the ideal filter is non-causal, it is not physically realisable. Also, the impulse function $h(t)$ is not time-limited—an infinite number of impulse responses are needed to interpolate the exact signal values between samples.

Reconstruction with RC Lowpass Filter

For practical real-time applications, an RC filter can be used to reconstruct the continuous signal from samples. It is a causal reconstruction filter and therefore physically realisable. A simple first-order RC lowpass filter is shown in Fig. 6.9. The transfer function of the filter is

$$H(\omega) = T\frac{1/j\omega c}{R + 1/j\omega c}$$

$$= \frac{T}{1 + j\omega CR} = \frac{T}{1 + j\omega/\omega_c}, \tag{6.21}$$

where $\omega_c = 1/RC$ is the 3-dB cut off frequency of the filter and T is the amplitude response of the filter at zero frequency equal to the sampling interval.

The impulse response of the filter is

$$h(t) = F^{-1}[H(\omega)] = \omega_c T e^{-t/RC} u(t). \tag{6.22}$$

From Eq. (6.17), the reconstructed signal is

$$f(t) = \sum_{n=-\infty}^{\infty} f(nT)h(t-nT)$$

$$= \sum_{n=-\infty}^{\infty} f(nT)\omega_c T e^{-(t-nT)/RC} u(t-nT). \tag{6.23}$$

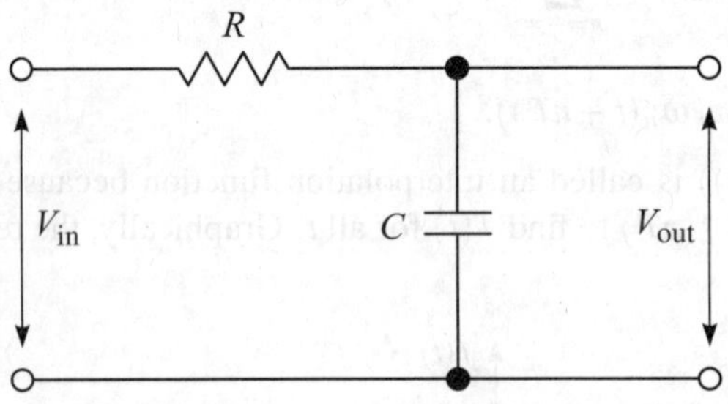

Fig. 6.9 *RC lowpass filter*

6.6 Effect of Undersampling – Aliasing

When the signal is sampled with sampling frequency $\omega_s \leq 2\omega_m$, called undersampling, the spectrum of the adjacent sampled signal overlaps with each other and the original signal cannot be recovered from its samples. This condition is called the aliasing effect. It is also called the frequency folding effect since high frequencies in the original signal 'fold back' into lower frequencies. In practice, most of the band-limited signals have considerable spectral components within a certain frequency level. Above this frequency, as frequency increases, the spectral components rapidly decreases to negligible values. Sampling such signals produce aliasing and therefore distorts the base spectrum of the sampled signal with a resulting loss of information. It also occurs if the signal is not band-limited. To avoid aliasing effect, the signal should be band-limited and the sampling frequency should be increased to greater than Nyquist rate, i.e., $\omega_s \geq 2\omega_m$.

Anti-aliasing Filter

Consider a continuous band-limited signal $f(t)$ which is corrupted by high frequency noise. Even though the signal is sampled with sampling frequency $\omega_s \geq 2\omega_m$, aliasing effect may occur due to high frequency noise components. To avoid this condition, first the continuous-time signal is band-limited to a maximum frequency ω_m by passing it through a lowpass filter whose cutoff frequency is ω_m. The low-pass filter eliminates all high frequency components greater than when the signal is applied before sampling. This filter is called anti-aliasing filter. For example, when the audio signal is sampled consists of frequency components up to 15 kHz, the signal has to pass through an LPF having a cutoff frequency of 15 kHz before sampling, and be sampled with $f_s \geq 30$ kHz in order to avoid aliasing effect. The process is shown in Fig. 6.10.

Note: Aliasing filter ensures that the signal to is band-limited to the highest frequency.

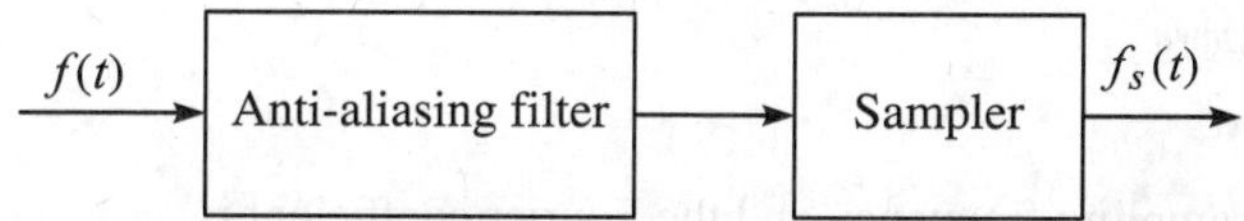

Fig. 6.10 *Sampling process with anti-aliasing filter*

6.6.1 Practical Considerations for Sampling

The following points are to be considered in the practical sampling process:

1. In practice it is impossible to sample a signal and reconstruct it from the samples without error because no practical signal is strictly band limited. All signals in nature have some frequency components at high frequencies with negligible energy. These components cause aliasing effect.
2. *Non existence of ideal reconstruction filters:* Very sharp cutoff lowpass filters are needed to reconstruct the original signal. Since practical filters have a finite spectral slope beyond the cutoff frequency, there is a possibility of entering some frequency components of the adjacent spectrum.

To avoid these difficulties, oversampling, where the sample rate greater than the Nyquist rate, is needed to provide some spectral space between adjacent spectra. This spectral space is called the guard-band and is equivalent to $\omega_s - 2\omega_m$.

Therefore, in the practical sampling process, band-limited signals are usually oversampled at sampling frequency greater than the Nyquist rate. For non-band-limited signals, a pre-aliasing filter can be used at a sampling frequency very much greater than the Nyquist rate.

6.7 Real Sampling or Natural and Flat-top Sampling

In real-time applications, the sampling process can be done using a narrow-width pulse train instead of an impulse train since it is not possible to generate an impulse train function for ideal sampling. In practice, two types of sampling techniques are considered:

1. Natural or pulse sampling
2. Flat-top sampling

Natural or Pulse Sampling

In natural sampling process the baseband signal $f(t)$ is multiplied with a pulse train function $P_T(t)$. The pulse train is a rectangular pulse of unit amplitude with a narrow pulsewidth τ sec repeating every T seconds $(T \gg \tau)$. The sampled signal consists of discrete time pulses with same pulse width and signal amplitude at the same time instants repeating at every T seconds.

Mathematically, the sampled signals $f_s(t)$ can be written as

$$f_s(t) = f(t)P_T(t), \tag{6.24}$$

$$\text{where } P_T(t) = \begin{cases} 1 & \dfrac{\tau}{2} \le t \le \dfrac{\tau}{2} \\ 0 & \dfrac{\tau}{2} \le t \le T - \dfrac{\tau}{2}. \end{cases} \tag{6.25}$$

is a periodic function. Its Fourier series is

$$P_T(t) = \sum_{n=-\infty}^{\infty} c_n e^{jnw_s t}, \tag{6.26}$$

where $\omega_s = \dfrac{2\pi}{T}$ is the sampling frequency and the Fourier coefficient is

$$c_n = \frac{1}{T}\int_{-T/2}^{T/2} P_T(t)e^{-jn\omega_s t}\,dt = \frac{1}{T}\int_{-\tau/2}^{\tau/2} e^{-jn\omega_s t}\,dt$$

$$= \frac{1}{T}\left[\frac{e^{-jn\omega_s t}}{-jn\omega_s}\right]_{-\tau/2}^{\tau/2} = \frac{1}{T}\left[\frac{e^{jn\omega_s\tau/2} - e^{-jn\omega_s\tau/2}}{jn\omega_s}\right]$$

$$= \frac{1}{T}\frac{\sin(n\omega_s\tau/2)}{n\omega_s/2} = \frac{\tau}{T}\mathrm{Sa}\left(\frac{n\omega_s\tau}{2}\right).$$

Substituting in Eq. (6.24),

$$f_s(t) = f(t)\sum_{n=-\infty}^{\infty} c_n e^{jn\omega_s t} = \sum_{n=-\infty}^{\infty} c_n f(t) e^{jn\omega_s t}.$$

Taking the Fourier transform of both sides,

$$F_s(\omega) = \int_{-\infty}^{\infty} f_s(t)e^{-j\omega t}\,dt = \int_{-\infty}^{\infty}\sum_{n=-\infty}^{\infty} c_n f(t)e^{jn\omega_s t}e^{-j\omega t}\,dt.$$

Interchanging the summation and integration functions,

$$F_s(\omega) = \sum_{n=-\infty}^{\infty} c_n \int_{-\infty}^{\infty} f(t)e^{-j(\omega-n\omega_s)t}\,dt.$$

Thus the Fourier transform of the sampled signal is

$$F_s(\omega) = \sum_{n=-\infty}^{\infty}\frac{\tau}{T}\mathrm{Sa}\left(\frac{n\omega_s\tau}{2}\right)F(\omega - n\omega_s). \tag{6.27}$$

This shows that the spectrum of the sampled signal is equal to the infinite sum of the spectra of a continuous signal at harmonic sampling frequencies $n\omega_s$. Each spectra is multiplied with a Fourier coefficient of $P_T(t)$. As n increases, the amplitude of the spectra decreases without distortion. The sampling process is shown in Fig. 6.11.

Reconstruction

Assume that the signal $f(t)$ is band-limited to maximum frequency ω_m. From the Fig. 6.11 it is observed that the spectrum $F(\omega)$ can be recovered from $F_s(\omega)$ for $n = 0$ i.e., $\omega_s \geq 2\omega_m$ using lowpass reconstruction filter with cutoff frequency ω_m.

Flat-top Sampling

In natural sampling the amplitude of the pulses vary along with the input signal in the duration of the pulse width τ. But in most of the cases, it is required to keep pulse amplitude constant during the pulse period. After pulse sampling, a sample and hold (S/H) circuit is used to make the pulse amplitude constant, equal to the instantaneous value of the input signal.

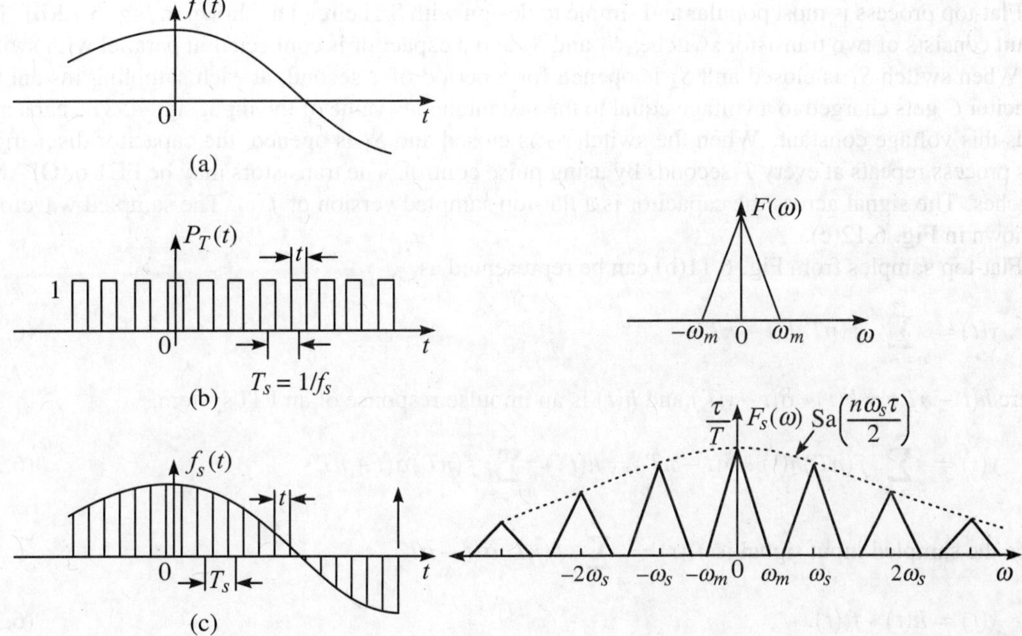

Fig. 6.11 *The natural sampling process*

The sample-and-hold (SH) circuit keeps signal sample up until the following sampling instant. It basically produces a "staircase" signal. It consist of a series switch and shunt capacitor as shown in Fig. 6.12(a). The switch closes only for an infinitesimal period of time at the sampling instants to charge the capacitor to its new amplitude.

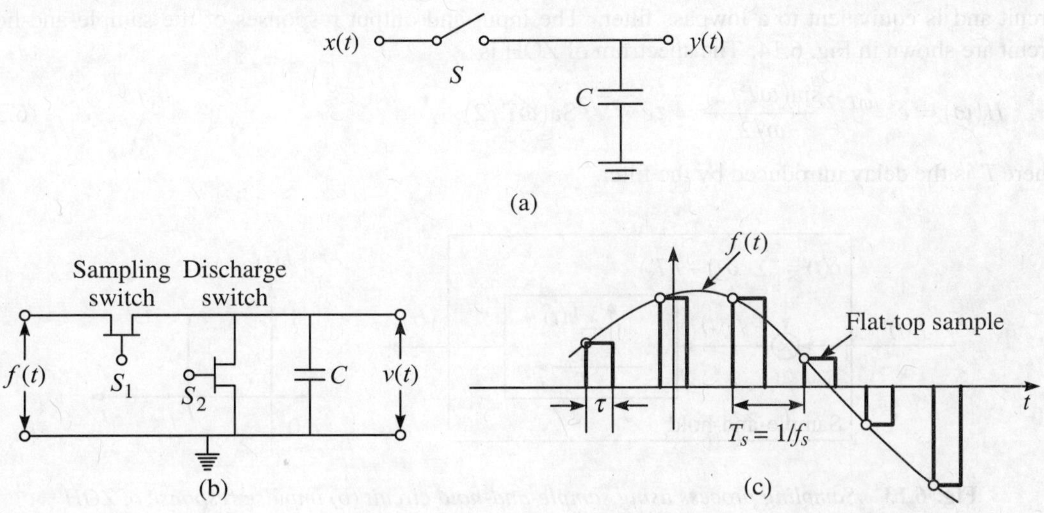

Fig. 6.12 *Sample-and-hold circuit*

Flat-top process is most popular and simple to design with S/H circuit as shown in Fig. 6.12(b). The circuit consists of two transistor switches S_1 and S_2 and a capacitor is connected in parallel with switch S_2. When switch S_1 is closed and S_2 is opened for a period of τ second, at each sampling instant the capacitor C gets charged to a voltage equal to the instantaneous value of the input signal. The capacitor holds this voltage constant. When the switch S_2 is closed and S_1 is opened, the capacitor discharges. This process repeats at every T seconds by using pulse control. The transistors may be FET or OPAMP switches. The signal across the capacitor is a flat-top sampled version of $f(t)$. The sampled waveform is shown in Fig. 6.12(c).

Flat-top samples from Fig. 6.11(b) can be represented as

$$y(t) = \sum_{n=-\infty}^{\infty} f(nT)h(t-nT), \tag{6.28}$$

where $h(t-nT) = h(t) * \delta(t-nT)$ and $h(t)$ is an impulse response of an LTI system.

$$y(t) = \sum_{n=-\infty}^{\infty} f(nT)h(t) * \delta(t-nT) = h(t) * \sum_{n=-\infty}^{\infty} f(nT)\delta(t-nT) \tag{6.29}$$

Since the sampled input signal is $f_s(t) = \sum_{n=-\infty}^{\infty} f(nT)\delta(t-nT)$,

$$y(t) = h(t) * f_s(t). \tag{6.30}$$

Taking Fourier transform, the output spectrum is

$$Y(\omega) = F_s(\omega)H(\omega) = f_s \sum_{n=-\infty}^{\infty} F(\omega - n\omega_s)H(\omega). \tag{6.31}$$

Therefore the flat-top samples can be obtained by convolution of the sampled signal $f_s(t)$ and the impulse response of an LTI system $h(t)$. The sampling process using the sample-and-hold circuit and the filter having unit pulse response is shown in Fig. 6.13. The filter is called zero-order hold (ZOH) circuit and is equivalent to a lowpass filter. The input and output responses of the sample-and-hold circuit are shown in Fig. 6.14. The spectrum of ZOH is

$$H(\omega) = e^{-j\omega T/2}\frac{\sin(\omega T/2)}{\omega/2} = \tau e^{-\omega T/2}\mathrm{Sa}(\omega T/2), \tag{6.32}$$

where T is the delay introduced by the filter.

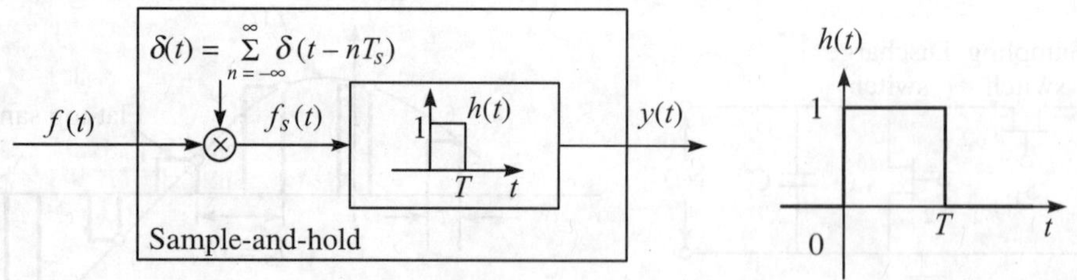

Fig. 6.13 *Sampling process using sample-and-hold circuit (b) impulse response of ZOH*

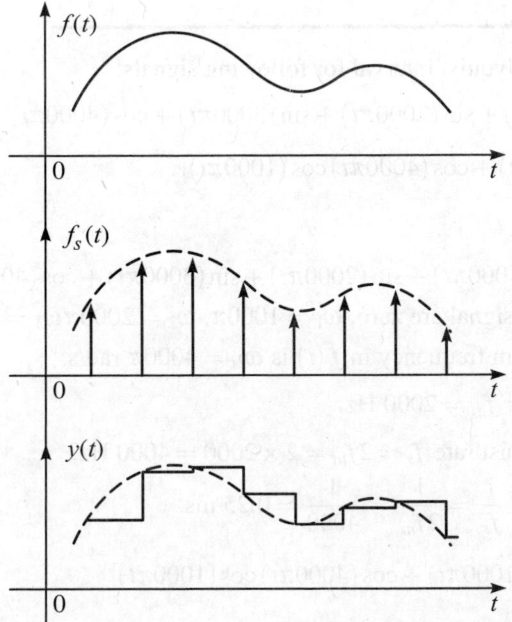

Fig. 6.14 *Input and output responses of sample-and-hold circuit*

In the ZOH process the magnitudes of high frequency components of the spectrum are attenuated more compared to the magnitudes of the low frequency components due to the sampling function. This phenomenon is called aperture effect. To reduce this effect, an equalizer with transfer function

$$H_e(\omega) = \frac{1}{H(\omega)} \quad \text{for } \omega \leq \omega_m \tag{6.33}$$

can be used along with the reconstruction filter as shown in Fig. 6.15.

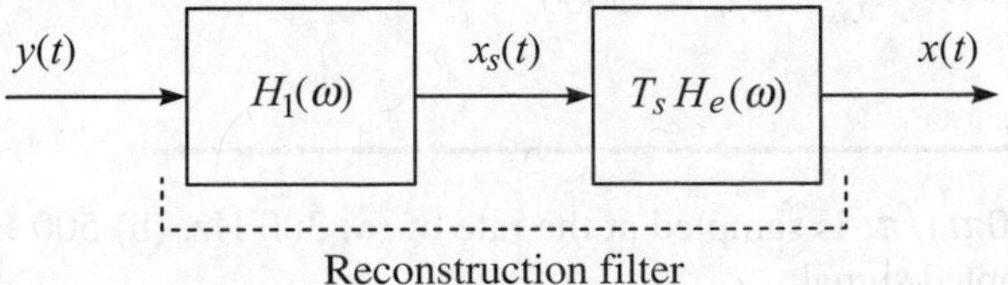

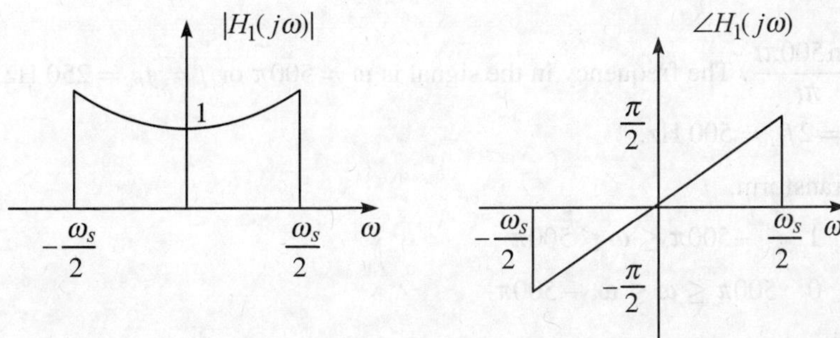

Fig. 6.15 *(a) Reconstruction filter and (b) magnotude and phase spectrum*

Example 6.1

Find the Nyquist rate and Nyquist interval for following signals:

(a) $f(t) = 1 + \sin(1000\pi t) + \sin(2000\pi t) + \sin(3000\pi t) + \cos(4000\pi t)$

(b) $f(t) = \frac{1}{2\pi}[\sin(1000\pi t) + \cos(4000\pi t)\cos(1000\pi t)]$.

Solution

(a) Given $f(t) = 1 + \sin(1000\pi t) + \sin(2000\pi t) + \sin(3000\pi t) + \cos(4000\pi t)$.

The frequencies in the signal are zero, $\omega_1 = 1000\pi$, $\omega_2 = 2000\pi\omega_3 = 3000\pi$ and $\omega_4 = 4000\pi$.

Therefore, the maximum frequency in $f(t)$ is $\omega_m = 4000\pi$ rad/s.

$$2\pi f_m = 4000\pi \text{ or } f_m = 2000 \text{ Hz}.$$

We know that the Nyquist rate $f_s = 2f_m = 2 \times 2000 = 4000$ Hz.

$$\text{The Nyquist interval} = \frac{1}{f_s} = \frac{1}{2f_m} = \frac{1}{4000} = 0.25 \text{ ms}.$$

(b) Given $f(t) = \frac{1}{2\pi}[\sin(1000\pi t) + \cos(4000\pi t)\cos(1000\pi t)]$

$$= \frac{1}{4\pi}[2\sin(1000\pi t) + \cos(5000\pi t) + \cos(3000\pi t)].$$

The frequencies in the signal are $\omega_1 = 1000\pi$, $\omega_2 = 3000\pi$ and $\omega_3 = 5000\pi$. Therefore, the maximum frequency in $f(t)$ is $\omega_3 = 5000\pi$ rad/s.

$$2\pi f_m = 5000\pi, \quad f_m = 2500 \text{ Hz}.$$

Hence the highest frequency is $f_m = 2500$ Hz.

We know that the Nyquist rate $f_N = 2f_m = 5000$ Hz.

$$\text{The Nyquist interval} = \frac{1}{f_s} = \frac{1}{2f_m} = \frac{1}{5000} = 0.2 \text{ ms}.$$

Example 6.2

A signal $f(t) = (\sin 500\pi t)/\pi t$ is sampled at the rate of (a) 300 Hz, (b) 500 Hz and (c) 700 Hz. Find out the spectrum of sampled signal.

Solution

Given $f(t) = \frac{\sin 500\pi t}{\pi t}$. The frequency in the signal is $\omega = 500\pi$ or $f = f_m = 250$ Hz.

Nyquist rate $f_N = 2f_m = 500$ Hz.

Taking Fourier transform,

$$X(\omega) = \begin{cases} 1 & -500\pi \le \omega \le 500\pi \\ 0 & 500\pi \le \omega \le \omega_s - 500\pi. \end{cases}$$

$$X(f) = \begin{cases} 1 & -250 \le f \le 250 \\ 0 & 250\pi \le f \le f_s - 250\pi. \end{cases}$$

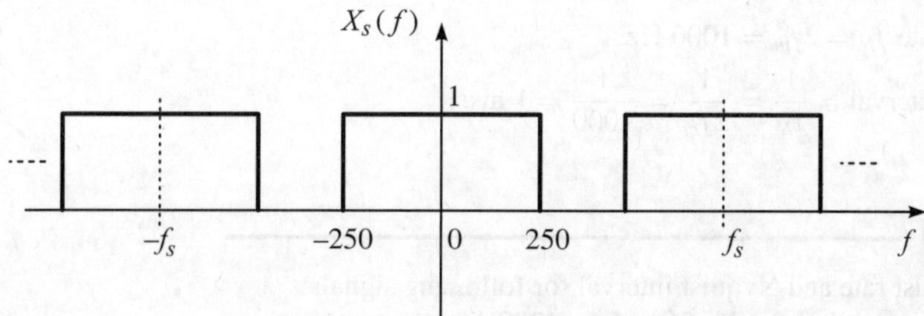

Fig. 6.16 *Spectrum of the sampled signal*

The spectrum of the signal is a rectangular pulse with a bandwidth of 250 Hz. The spectrum of the sampled signal is shown in Fig. 6.16 and is given by

$$X_s(\omega) = \frac{1}{T}\sum_{n=-\infty}^{\infty} X(\omega - n\omega_s) \quad = f_s \sum_{n=-\infty}^{\infty} X(f - nf_s).$$

(a) Given $f_s = 300$ Hz, the above equation becomes

$$X_s(f) = 300 \sum_{n=-\infty}^{\infty} X(f - 300n).$$

Since $f_s < 2f_m$, the spectrum has overlaps.

(b) Given $f_s = 500$ Hz,

$$X_s(f) = 500 \sum_{n=-\infty}^{\infty} X(f - 500n).$$

Since $f_s < 2f_m$, the spectrum has no overlaps and no gap between adjacent spectrum.

(c) Given $f_s = 700$ Hz,

$$X_s(f) = 700 \sum_{n=-\infty}^{\infty} X(f - 700n).$$

The spectrum has no overlaps and there is a gap between adjacent spectrum.

Example 6.3

Find the Nyquist rate and Nyquist interval for the signals (a) $f(t) = \dfrac{\sin 500\pi t}{\pi t}$ and (b) $f(t) = \left[\dfrac{\sin 500\pi t}{\pi t}\right]^2$.

Solution

(a) Given $f(t) = \dfrac{\sin 500\pi t}{\pi t}$, the frequency in the signal is $\omega = 500\pi$ or $f = f_m = 250$ Hz.

Nyquist rate $f_N = 2f_m = 500$ Hz.

The Nyquist Interval $= \dfrac{1}{f_N} = \dfrac{1}{2f_m} = \dfrac{1}{500} = 2$ ms.

(b) Given $f(t) = \left[\dfrac{\sin 500\pi t}{\pi t}\right]^2 = \left[\dfrac{(\sin 500\pi t)^2}{(\pi t)^2}\right] = \left[\dfrac{1 - \cos 10000\pi t}{(\pi t)^2}\right]$, the frequency in the signal is $\omega = 10000\pi$ or $f_m = 5000$ Hz.

Nyquist rate $f_N = 2f_m = 1000$ Hz.

Nyquist interval $= \dfrac{1}{f_N} = \dfrac{1}{2f_m} = \dfrac{1}{1000} = 1$ ms.

Example 6.4

Find the Nyquist rate and Nyquist interval for following signals:
(a) $f(t) = 1 + \text{sinc}300\pi t$, (b) $f(t) = \text{rect}(200t)$.

(a) Given $f(t) = 1 + \text{sinc}300\pi t$, $f(t) = 1 + \dfrac{\sin 300\pi t}{\pi t}$.
The frequency of the signal is $\omega = 300\pi$ or $f = f_m = 150$ Hz.
$\therefore$ Nyquist rate $f_s = 2f_m = 300$ Hz.
The Nyquist interval $= \dfrac{1}{f_s} = \dfrac{1}{2f_m} = \dfrac{1}{300} = 3.33$ ms.

(b) Given $f(t) = \text{rect}(200t)$, which is a rectangular pulse signal having pulse width $\tau = 1/200$ seconds. Since the signal is a finite duration signal, it is not band-limited. The signal spectrum consists of infinite frequencies.
$\therefore$ Nyquist rate is infinity and Nyquist interval is zero.

Example 6.5

Consider a signal $f(t) = \left[\dfrac{\sin 40\pi t}{\pi t}\right]^2$ which is sampled with a sampling frequency of $f_s = 60$ Hz to obtain a signal $g(t)$ with a Fourier transform $G(\omega)$. Determine the maximum value of ω_0 for which it is guaranteed that $G(\omega) = 60F(\omega)$ for $|f| \leq f_0$, where $F(\omega)$ is the Fourier transform of $f(t)$.

Solution

Given $f(t) = \left[\dfrac{\sin 40\pi t}{\pi t}\right]^2 = \left[\dfrac{(\sin 40\pi t)^2}{(\pi t)^2}\right] = \left[\dfrac{1 - \cos 80\pi t}{(\pi t)^2}\right]$.

The frequency in the signal is $\omega = 80\pi$ or $f_m = 40$ Hz.
Given the sampling frequency $f_s = 60$ Hz, since $f_s < 2f_m$, the sampling theorem is not satisfied and the original signal cannot be recovered from its samples. The resultant signal will be distorted due to aliasing effect.
Given $G(\omega) = 60F(\omega)$.
Thus the maximum value of f_0 will be obtained when the signal $G(\omega)$ is perfectly extracted from $F(\omega)$.

$$G(\omega) = 60F(\omega) = \frac{1}{T_s}F(\omega), \text{ or } f_s = \frac{1}{T_s} = 60 \text{ Hz.}$$

Hence to recover $f(t)$, the sampling frequency must be $f_s = 60$ Hz. Thus the signal $f(t)$ is perfectly recovered if $\omega_0 = 60\pi$ rad/s or $f_0 = 60$ Hz.

Example 6.6

Given a continuous-time signal $x(t)$ with Nyquist rate ω_s, determine the Nyquist rate for the following continuous-time signals:
(a) $y(t) = x^2(t)$, and (b) $y(t) = x(t)\cos\omega_0 t$.

Solution

Given $x(t)$ has a Nyquist rate of ω_s. Therefore, $X(\omega) = 0$ for $|\omega| > \dfrac{\omega_s}{2}$.

(a) Given $y(t) = x^2(t)$.

Therefore, $Y(\omega) = \dfrac{1}{2\pi}[X(\omega) * X(\omega)]$.

Hence, $Y(\omega) = 0$ for $\omega > \dfrac{\omega_s}{2} + \dfrac{\omega_s}{2} = \omega_s$.

Therefore, the Nyquist rate for $y(t)$ is $2\omega_s$.

(b) Given $y(t) = x(t)\cos\omega_0 t$.

Therefore, $Y(\omega) = \dfrac{1}{2\pi}[X(\omega - \omega_0) * X(\omega + \omega_0)]$.

Hence $Y(\omega) = 0$ for $|\omega| > \omega_0 + \dfrac{\omega_s}{2}$.

Therefore, the Nyquist rate for $y(t)$ is $2\left(\omega_0 + \dfrac{\omega_s}{2}\right) = 2\omega_0 + \omega_s$.

Example 6.7

Figure 6.17(a) shows the spectrum of a message signal. The signal is sampled at the rate of $f_s = 1.5 f_{\max}$, where $f_{\max} = 1$ Hz is the maximum signal frequency. Sketch the spectrum of the sampled version of the signal. If the sampled signal is passed through an ideal lowpass filter of bandwidth $f_{\max}$, sketch the spectrum of the output signal from this filter.

Solution

When $x(t)$ is sampled instantaneously, its spectrum is given as

$$X_s(f) = f_s X(f) + \sum_{\substack{n=-\infty \\ n\neq 0}}^{\infty} f_s X(f - nf_s).$$

Here $f_s = 1.5 f_{\max}$. Then the above equation becomes

$$X_s(f) = 1.5 f_{\max}\, X(f) + \sum_{\substack{n=-\infty \\ n\neq 0}}^{\infty} 1.5 f_{\max} X(f - 1.5 n f_{\max}).$$

Also, given $f_{\max} = 1$ Hz, the above equation becomes

$$X_s(f) = 1.5 X(f) + \sum_{\substack{n=-\infty \\ n\neq 0}}^{\infty} 1.5 X(f - 1.5 n_s).$$

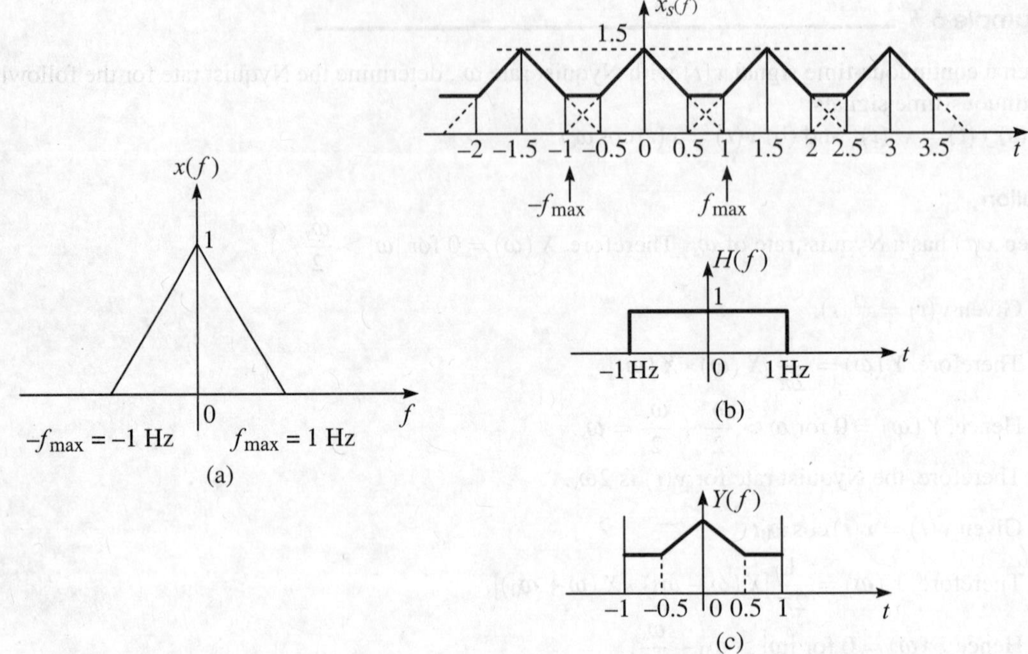

Fig. 6.17 *The spectrum of a message signal, sampled signal and reconstruction*

Because of the undersampling ($f_s = 1.5 f_{\max}$), there is aliasing effect and it is visible in Fig. 6.16 (a). When the sampled signal is passed through an ideal lowpass filter of bandwidth $f_{\max}$ as shown in Fig. 6.16(b), the spectrum of the corresponding output signal is as shown in Fig. 6.16(c). It shows clearly that the spectrum of Fig. 6.16 (c) is not the same as the original signal because of aliasing.

Example 6.8

A signal $g(t) = \cos 6280t + \cos 31400t$ is sampled at 8000 samples per second. Determine the spectrum of the sampled signal. Assume instantaneous sampling. The sampled signal is passed through an LPF with constant unity gain from 0 Hz to 2 kHz, with zero gain at 4 kHz. Determine the output.

Solution

Given $g(t) = \cos 6280t + \cos 31400t$

$$= \cos(2\pi \times 1000t) + \cos(2\pi \times 5000t).$$

The signal consists of two frequencies: $f_1 = 1000$ Hz and $f_2 = 5000$ Hz with amplitude one. We know that the Fourier transform of cosine signal is given as,

$$\mathcal{F}[\cos(\omega_c t)] = \frac{1}{2}[\delta(\omega - \omega_c) + \delta(\omega + \omega_c)].$$

Hence the Fourier transform of the given signal becomes

$$G(\omega) = \frac{1}{2}[\delta(\omega - 6280) + \delta(\omega + 6280)] + \frac{1}{2}[\delta(\omega - 31400) + \delta(\omega + 31400)].$$

$$G(f) = \frac{1}{2}[\delta(\omega - 1000) + \delta(\omega + 1000)] + \frac{1}{2}[\delta(\omega - 5000) + \delta(\omega + 5000)].$$

Spectrum of the sampled signal is given as

$$G_s(\omega) = \frac{1}{T}\sum_{n=-\infty}^{\infty} G(\omega - n\omega_s),$$

$$\text{or} \quad G_s(f) = \frac{1}{T}\sum_{n=-\infty}^{\infty} G(f - nf_s) = f_s \sum_{n=-\infty}^{\infty} G(f - nf_s).$$

Since $f_s = 8000$ Hz, the above equation becomes

$$G_s(f) = 8000 \sum_{n=-\infty}^{\infty} G(f - 8000n)$$

$$= 8000\,[G(f) + G(f+8000) + G(f-8000) + G(f+16000) + G(f-16000) + ...].$$

$G(f)$ repeats at ± 8000, ± 16000, ± 24000, ... and so on. The spectra are shown in Fig. 6.18.

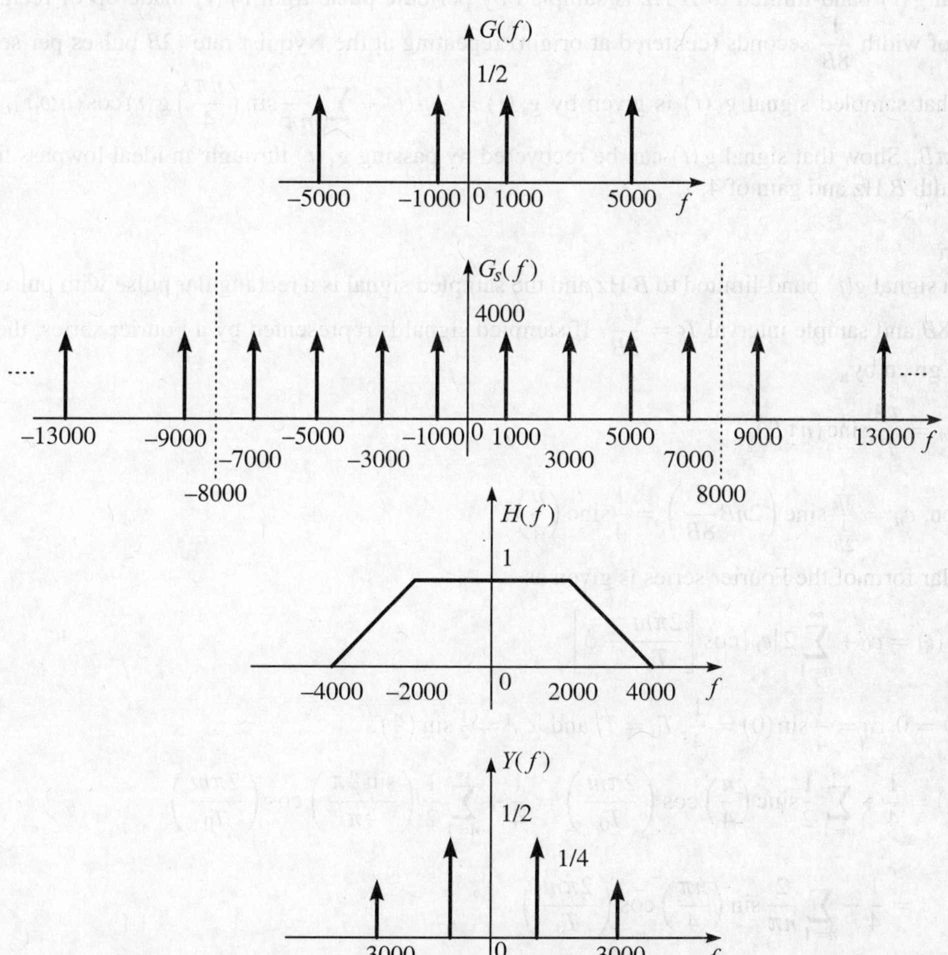

Fig. 6.18 *Spectrum of a given signal $G(f)$, sampled signal $G_s(f)$, lowpass filter $H(f)$ and output signal $Y(f)$*

The sampled signal is passed through the lowpass filter having unity gain from 0 to 2 kHz and linear roll off from 2 to 4 kHz. Hence, the frequency components at only 1 kHz and 3 kHz are passed at the output. Observe that the 3 kHz frequency is attenuated to half strength due to roll-off of lowpass filter. Hence the output spectrum can be expressed as

$$Y(f) = \frac{1}{2}\left[\delta(\omega - 1000) + \delta(\omega + 1000)\right] + \frac{1}{4}\left[\delta(\omega - 3000) + \delta(\omega + 3000)\right].$$

Taking inverse Fourier transform the output signal is

$$y(t) = \cos(2\pi \times 1000t) + \cos(2\pi \times 3000t).$$

The new 3 kHz has appeared and the 5 kHz frequency component has been eliminated due to aliasing. The highest frequency content in output signal is 3000 Hz.

Example 6.9

A signal $g(t)$ band-limited to B Hz is sampled by periodic pulse train $P_T(t)$ made up of rectangular pulses of width $\frac{1}{8B}$ seconds (centered at origin) repeating at the Nyquist rate ($2B$ pulses per second). Show that sampled signal $g_s(t)$ is given by $g_s(t) = \frac{1}{4}g(t) + \sum_{n=1}^{\infty} \frac{2}{n\pi} \sin\left(\frac{n\pi}{4}\right) g(t) \cos(n\omega_s t)$, where $\omega_s = 4\pi B$. Show that signal $g(t)$ can be recovered by passing $g_s(t)$ through an ideal lowpass filter of bandwidth B Hz and gain of 4.

Solution

Given a signal $g(t)$ band-limited to B Hz and the sampled signal is a rectangular pulse with pulse width $\tau = 1/8B$ and sample interval $T_s = \frac{1}{2B}$. If sampled signal is represented by a Fourier series, the value of c_n is given by

$$c_n = \frac{\tau A}{T_s} \operatorname{sinc}(n\tau T_s).$$

$$\text{Then, } c_n = \frac{\frac{1}{8B}}{\frac{1}{2B}} \operatorname{sinc}\left(2nB\frac{1}{8B}\right) = \frac{1}{4}\operatorname{sinc}\left(\frac{n}{4}\right).$$

The polar form of the Fourier series is given as

$$x(t) = c_0 + \sum_{n=1}^{\infty} 2|c_n| \cos\left[\frac{2\pi nt}{T_0} + \phi\right].$$

$$\text{Here, } \phi = 0, c_0 = \frac{1}{4}\sin(0) = \frac{1}{4}, T_0 = T_s \text{ and } |c_n| = \tfrac{1}{4}\sin\left(\tfrac{n}{4}\right).$$

$$\therefore \; x(t) = \frac{1}{4} + \sum_{n=1}^{\infty} \frac{1}{2}\operatorname{sinc}\left(\frac{n}{4}\right)\cos\left(\frac{2\pi nt}{T_0}\right) = \frac{1}{4} + \sum_{n=1}^{\infty}\frac{1}{2}\left(\frac{\sin\frac{n}{4}\pi}{\frac{n}{4}\pi}\right)\cos\left(\frac{2\pi nt}{T_0}\right)$$

$$= \frac{1}{4} + \sum_{n=1}^{\infty} \frac{2}{n\pi}\sin\left(\frac{n\pi}{4}\right)\cos\left(\frac{2\pi nt}{T_0}\right).$$

The sampled signal $g_s(t)$ is related to $x(t)$ as,

$$g_s(t) = x(t).g(t)$$

$$= \frac{1}{4}g(t) + \sum_{n=1}^{\infty} \frac{4}{n\pi} \sin\left(\frac{n\pi}{4}\right) \cos\left[\frac{2\pi nt}{T_0}\right] g(t)$$

$$= \frac{1}{4}g(t) + \sum_{n=1}^{\infty} \frac{2}{n\pi} \sin\left(\frac{n\pi}{4}\right) \cos(n\omega_s t)\, g(t).$$

Since $\omega_s = 4\pi B$, the above equation becomes

$$g_s = \frac{1}{4}g(t) + \sum_{n=1}^{\infty} \frac{2}{n\pi} \sin\left(\frac{n\pi}{4}\right) \cos(n\omega_s t)\, g(t).$$

Here, note that first term $g(t)$ is multiplied by $1/4$. Hence the signal can be recovered by a filter of gain 4 and bandwidth B.

Example 6.10

The signal $x(t) = 2\cos 200\pi t + 6\cos 180\pi t$ is ideally sampled at a frequency of 150 samples per second. The sampled version $x_s(t)$ is passed through a unit-gain ideal LPF with a cut-off frequency of 110 Hz. What frequency components will be present at the output of the LPF? Write down the expression for its output signal.

Solution:

Given $x(t) = 2\cos 200\pi t + 6\cos 180\pi t$.

Taking Fourier transform on both sides, we have

$$X(\omega) = [\delta(\omega - 200\pi) + \delta(\omega + 200\pi)] + 3[\delta(\omega - 180\pi) + \delta(\omega + 180\pi)]$$

or

$$X(f) = [\delta(f - 100) + \delta(f + 100)] + 3[\delta(f - 90) + \delta(f + 90)].$$

This spectrum is depicted in Fig. 6.19.

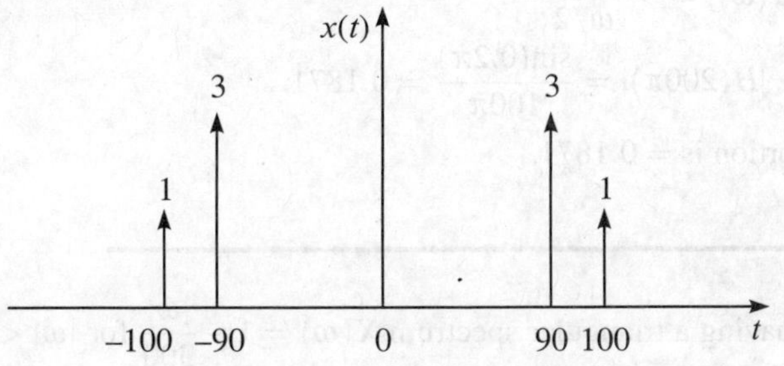

Fig. 6.19 *Spectrum of the signal*

The spectrum of $x_s(t)$, the ideally sampled version of $x(t)$, is an aperiodic repetition of $X(f)$ at regular intervals of f_s, i.e., 150 Hz, and will be as shown in Fig. 6.20.

From this figure, it is clear that the output of the LPF contains frequency components at 50 Hz, 60 Hz, 90 Hz and 100 Hz, although the original analog signal contains only 90 Hz and 100 Hz components. As it is undersampled, aliasing is taking place and the 50 Hz component is the alias of the 100 Hz component and the 60 Hz component is the alias of the 90 Hz component.

The expression for the output of the LPF is given by

$$y(t) = 2[\cos 2\pi(50)t + \cos 2\pi(100)t] + 6[\cos 2\pi(60)t + \cos 2\pi(90)t], \text{ or}$$

$$y(t) = 2\cos 100\pi t + 2\cos 200\pi t + 6\cos 120\pi t + 6\cos 180\pi t.$$

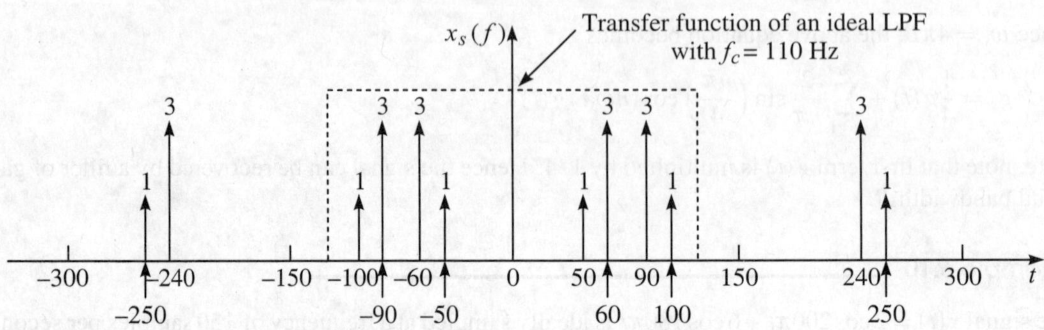

Fig. 6.20 *Spectrum of the sampled signal*

Example 6.11

A signal band-limited to 100 Hz is processed through flat-top sampling system with sampling frequency 200 Hz. The pulse duration is 0.2 seconds. Find out the amplitude distortion due to aperture effect.

Solution

Given sampling frequency $f_s = 200$ Hz; maximum signal frequency $f_m = 100$ Hz and pulse duration $\tau = 2$ ms.

The transfer function of the filter is $H(\omega) = e^{-j\omega\tau/2}\dfrac{\sin(\omega\tau/2)}{\omega/2}$.

The magnitude is $|H(\omega)| = \dfrac{\sin(\omega\tau/2)}{\omega/2}$.

At signal frequency, $|H(200\pi)| = \dfrac{\sin(0.2\pi)}{100\pi} = 0.1871$.

The amplitude distortion is $= 0.1871$.

Example 6.12

Consider a signal $x(t)$ having a triangular spectrum $X(\omega) = 1 - \dfrac{|\omega|}{300}$ for $|\omega| < 300$. Assume that $x(t)$ is ideally sampled at (a) $\omega_s = 500$ rad/s and (b) $\omega_s = 700$ rad/s. Sketch the spectrum of sampled signal.

Solution

Given the spectrum $X(\omega) = 1 - \dfrac{|\omega|}{300}$ for $|\omega| < 300$, the triangular spectrum is as shown in Fig. 6.21(a). It is observed that the highest frequency in the signal is $\omega_m = 300$ rad/s. The Nyquist rate is $\omega_N = 600$ rad/s.

The spectrum of the sampled signal is $X_s(\omega) = \dfrac{\omega_s}{2\pi}\sum_{n=-\infty}^{\infty} X(\omega - n\omega_s)$.

(a) For $\omega_s = 500$ rad/s, since $\omega_s < 2\omega_s$, aliasing effect occurs. The spectrum is as shown in Fig. 6.21 (b) overlap with each other.

(b) For $\omega_s = 700$ rad/s, since $\omega_s > 2\omega_s$, there is no aliasing effect. The spectrum is as shown in Fig. 6.21(c) .

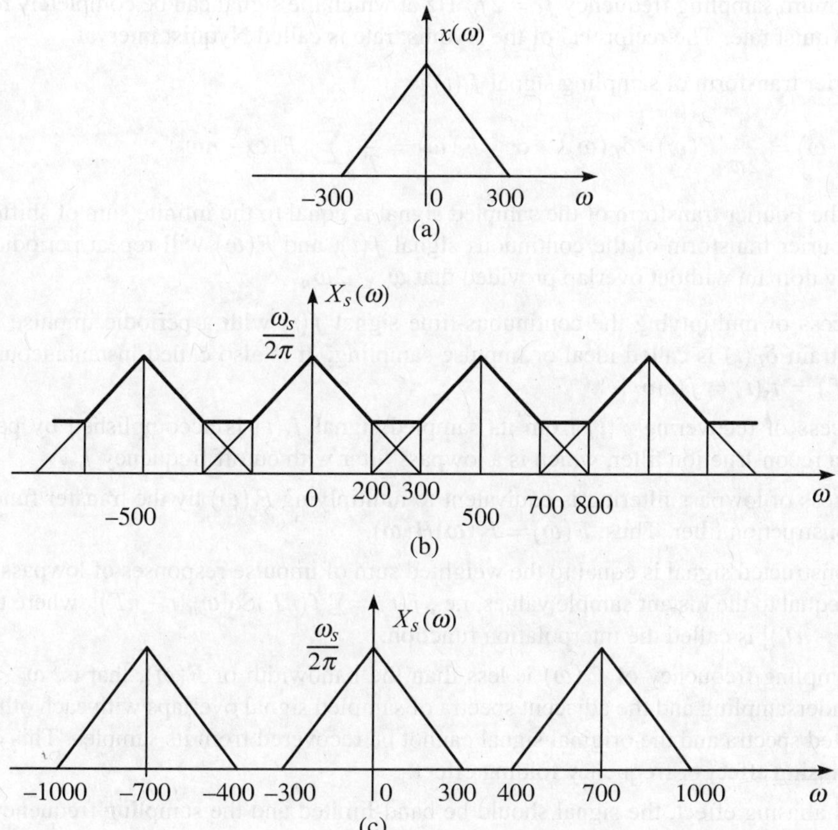

Fig. 6.21 *(a) Spectrum of a given signal $x(t)$ and the spectrum of the sampled signal at (b) $\omega_s = 500$ rad/s and (c) $\omega_s = 700$ rad/s*

Summary

1. The process of converting a continuous-time signal into a discrete time signal is called sampling.
2. A discrete-time signal $f(nT), n = 0, \pm1, \pm2, \pm3, \ldots$ can be obtained from its continuous-time signal $f(t)$ by taking samples at $n = 0, \pm1, \pm2, \pm3, \ldots$ with amplitude equal to the instantaneous signal level. T is called the sampling period.
3. The reciprocal of the sampling period T is called sampling frequency f_s (samples/s).
4. The sampling theorem, also known as Shannon's sampling theorem, states that a band-limited continuous-time signal $f(t)$ which has no frequency component higher than f_m Hz is uniquely represented by its sampling values taken at uniform intervals separated by less than $\frac{1}{2f_m}$ s.

5. For reconstruction of signals from samples, the sampling theorem in frequency domain can be stated as: a band-limited finite energy signal $f(t)$ which has no frequency component higher than f_m Hz maybe completely recovered from the knowledge of its sampled values taken at the sampling frequency equal to or greater than twice its highest frequency component, i.e., $f_s \geq 2f_m$ Hz.
6. The minimum sampling frequency $f_s = 2f_m$ Hz at which the signal can be completely recovered is called Nyquist rate. The reciprocal of the Nyquist rate is called Nyquist interval.
7. The Fourier transform of sampling signal $f_s(t)$ is

$$F_s(\omega) = \frac{1}{2\pi}[F(\omega) * \delta_T(\omega)] \quad \text{or} \quad F_s(\omega) = \frac{1}{T}\sum_{n=-\infty}^{\infty} F(\omega - n\omega_s).$$

That is, the Fourier transform of the sampled signal is equal to the infinite sum of shifted versions of the Fourier transform of the continuous signal $f(t)$, and $F(\omega)$ will repeat periodically in the frequency domain without overlap provided that $\omega_s \geq 2\omega_m$.
8. The process of multiplying the continuous-time signal $f(t)$ with a periodic impulse function or impulse train $\delta_T(t)$ is called ideal or impulse sampling. It is also called instantaneous sampling, i.e., $f(nT) = f_s(t) = f(t)\delta_T(t)$.
9. The process of recovering $f(t)$ from its sampled signal $f_s(t)$ is accomplished by passing $f_s(t)$ through a reconstruction filter, which is a lowpass filter with cutoff frequency f_m.
10. The process of lowpass filtering is equivalent to multiplying $F_s(\omega)$ by the transfer function $H(\omega)$ of a reconstruction filter. Thus, $F(\omega) = F_s(\omega)H(\omega)$.
11. The reconstructed signal is equal to the weighted sum of impulse responses of lowpass filters with weights equal to the instant sample values, i.e., $f(t) = \sum f(nT)\text{Sa}[\omega_m(t-nT)]$, where the function $\text{Sa}[\omega_m(t-nT)]$ is called the interpolation function.
12. If the sampling frequency of $F_s(\omega)$ is less than the bandwidth of $F(\omega)$, that is, $\omega_s < 2\omega_m$, it is called undersampling and the adjacent spectra of sampled signal overlaps with each other resulting in distorted spectra and the original signal cannot be recovered from its samples. This condition is called aliasing effect or frequency folding effect.
13. To avoid aliasing effect, the signal should be band-limited and the sampling frequency should be increased to greater than Nyquist rate, i.e., $\omega_s \geq 2\omega_m$.
14. To avoid aliasing effect due to high frequency noise components, the continuous-time signal is band-limited to a maximum frequency ω_m before sampling by passing it through a lowpass filter whose cutoff frequency is ω_m. This filter is called anti-aliasing filter.
15. To avoid difficulties with noise components and reconstruction filters, in practical sampling process, the continuous-time signal is usually oversampled at sampling frequency greater than the Nyquist rate, $\omega_s > 2\omega_m$.
16. In natural sampling process, the baseband signal $f(t)$ is multiplied with a pulse train function $P_T(t)$. The pulse train is a rectangular pulse of unit amplitude with a narrow pulsewidth τ seconds that repeats at every T seconds ($T \gg \tau$). The sampled signal is $f_s(t) = f(t)P_T(t)$.
17. After the pulse sampling, a sample-and-hold (S/H) circuit is used to make the pulse amplitude constant, equal to the instantaneous value of the input signal. The process is called flat-top sampling.
18. For practical, real-time applications, RC filter can be used to reconstruct the continuous signal from samples. It is a causal reconstruction filter and therefore physically realisable.
19. Analog-to-digital converters (ADC) and digital-to-analog converters (DAC) are used to convert analog signals into digital signals and vice-versa.

Review Questions

1. State and prove sampling theorem for lowpass signals. (June 2013)
2. State uniform sampling theorem and prove it both graphically and analytically. Also, mention all the results used. (June 2015)
3. With the help of a graphical example, explain sampling theorem for band-limited signals. (May 2011)
4. State and prove the sampling theorem for band-limited signals. (Dec 2011)
5. Write short notes on different types of sampling techniques.
6. What is aliasing? Sketch and explain the spectrum of naturally-sampled signal for the following cases : (i) $\omega_0 = 2\omega_m$, (ii) $\omega_0 > 2\omega_m$, (iii) $\omega_0 < 2\omega_m$, where ω_0 is frequency corresponding to the sampling interval and ω_m is maximum frequency in the spectrum of the baseband signal.
7. Define sampling theorem and discuss the method of performing sampling using impulse sampling technique. (Dec 2011)
8. What are the disadvantages of undersampling? (June 2016)
9. Write short notes on: (a) Aliasing effect, and (b) Nyquist sampling theorem. (June 2016)
10. Explain the effect of undersampling with an example and neat diagrams. (Nov 2015)
11. Discuss the effect of aliasing due to undersampling. (Dec 2011)
12. Discuss the process of reconstructing the signal from its samples. (May 2013)
13. With the necessary derivation, explain the operation of a reconstruction filter. (June 2014)
14. Explain the reconstruction of a signal from its sample values using interpolation technique.
15. Explain the reconstruction of signal from its sampled signal. (June 2015)
16. The Fourier transform of a sampled signal is given by $X(f) = \sum_{m=0}^{N-1} x(m)e^{-j2\pi fm}$.

    Using the above equation, prove that the spectrum of a sampled signal is periodic, and hence state the sampling theorem. (June 2014)
17. Explain how an original signal can be recovered from a sampled signal. (June 2014)
18. Define sampling theorem and discuss the method of performing sampling using natural sampling technique. Compare it with the impulse sampling technique. (June 2014)
19. Define Nyquist rate. Compare the merits and demerits of performing sampling using impulse, natural and flat-top sampling techniques. (May 2013)
20. Explain natural and flat-top sampling. (June 2014)
21. Derive the expression for the transfer function of flat-top-sampled signal. (June 2015)
22. Discuss in detail the generation of finite-width-sampled signals. (June 2014)
23. Describe the sample-and-hold operation with the help of neat sketches. What is its advantage? (June 2014)
24. Explain briefly bandpass sampling. (May 2011, 2015)

Problems

1. For a signal $x(t) = 3\cos 25\pi t - 10\sin 200\pi t + \cos 200\pi t x(t)$, calculate the Nyquist rate and Nyquist interval.
2. A continuous-time signal is given as $x(t) = 8\cos 200\pi t$. (i) Determine the minimum sampling rate. (ii) If $f_s = 400$ Hz, what is discrete time signal obtained after sampling? (iii) If $f_s = 800$ Hz, what is discrete time signal obtained after sampling?
3. Consider the analog signal $x(t) = 3\cos 100\pi t$. (i) Determine the minimum sampling rate required to avoid aliasing. (ii) Suppose the signal is sampled at two different sampling rates of $f_s = 200$ Hz, and $f_s = 75$ Hz. Find the discrete time signals obtained after sampling in each case separately.
4. A complex signal $x(t)$ with a Fourier transform $X(j\omega)$ is zero everywhere except for the interval $-5 > \omega > 2$. Using $X_p(j\omega) = \frac{1}{T}\sum_{k=-\infty}^{\infty} X(j(\omega - k\omega_s))$, determine the minimum sampling frequency ω_s to ensure a reconstruction of the original signal $x(t)$ without losing information.
5. Find the Nyquist rate and Nyquist interval for the continuous-time signal
$$x(t) = \frac{1}{2\pi}\cos(5000\pi t)\cos(1000\pi t).$$
6. A continuous-time signal is given as $x(t) = 10\cos(500\pi t)$.
   (i) Determine minimum sampling rate,
   (ii) If $f_s = 1100$ Hz, what is the discrete time signal obtained after sampling?
   (iii) If $f_s = 850$ Hz, what is the discrete time signal obtained after sampling?
7. Obtain the Nyquist rate of the signal $x(t) = \cos 2000\pi t + 10\sin 10000\pi t + 10\cos 5000\pi t$.
8. Determine the minimum sampling rate and Nyquist interval of signals given by (i) $x(t) = \sin(200\pi t) + \sin(1000\pi t)$, and (ii) $x(t) = \cos(200\pi t) + \cos(2000\pi t)$.
9. Given a continuous-time signal $x(t)$ with Nyquist rate ω_s, determine the Nyquist rate for the following continuous-time signals: (i) $y(t) = x^2(t)$; (ii) $y(t) = x(t)\cos\omega_0 t$. (Nov 2015)
10. Plots the spectrum of the signal shown in Fig. P6.10 which has been sampled by an impulse train.

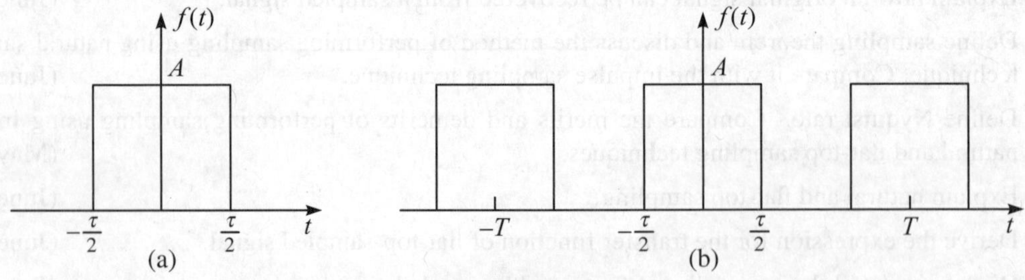

Fig. P6.10

11. Determine Nyquist's frequency and Nyquist's interval for each of the following given signals: (a) sinc$(100t)$; (b) sinc$^2(100t)$; (c) sinc$(100t)$ + sinc$^2(60t)$.
12. Consider the analog signal $x(t) = 6\cos 10\pi t$. Suppose the signal is sampled at two different sampling rates of $f_s = 7$ Hz, and $f_s = 14$ Hz. Find the discrete time signals obtained after sampling in each case separately and draw the spectra.

13. A lowpass signal $x(t)$ has a spectrum given by $X(f) = \begin{cases} 1 - \dfrac{|f|}{200} & |f| < 200 \\ 0 & \text{otherwise} \end{cases}$

    is ideally sampled at $f_s = 300$ Hz. Sketch the spectrum of the sampled signal.
14. Consider the analog signal $x(t) = 2\cos 400\pi t + 6\cos 640\pi t$ sampled at $f_s = 500$ Hz. If it is passed through ideal LPF with cutoff frequency $f_0 = 400$ Hz, find the frequency components in the output signal.
15. Consider the analog signal $x(t) = 4\cos 60\pi t + 5\sin 300\pi t - \cos 120\pi t$. Suppose the signal is sampled at Nyquist rate. Find the discrete time signal obtained and draw the spectra.

Answers

1. 200 Hz, 5 ms　2. 200 Hz　3. 100 Hz　4. 10 Hz
5. 6000 Hz, 0.165 ms　6. 500 Hz　7. 1000 Hz
9. 1000 Hz, 2000 Hz　11. (a) 100 Hz (b) 200 Hz (c) 1200 Hz
14. 180, 200, 300 and 320

Multiple-choice Questions

1. The most commonly used sampling method is
   (a) ideal or impulse sampling　(b) natural sampling using rectangular pulses
   (c) sample-and-hold method　(d) none of the above.
2. Aliasing effect can be reduced if the sampling frequency is
   (a) very high compared to Nyquist rate　(b) slightly more than the Nyquist rate
   (c) less than the Nyquist rate　(d) at Nyquist rate
3. The distortion in the signal arising from aperture effect can be reduced by
   (a) reducing the width of the pulses used for flat-top sampling
   (b) reducing the sampling frequency
   (c) properly band-limiting the signal before sampling it
   (d) using flat-top sampling
4. The impulse response function, $h(t)$, of a zero-order-hold circuit is
   (a) an impulse　(b) a rectangular pulse
   (c) a triangular pulse　(d) none of the above
5. A high rate of sampling will make the design of
   (a) the anti-aliasing filter simpler, but that of the reconstruction filter more difficult
   (b) the anti-aliasing filter more difficult, but that of the reconstruction filter much simpler
   (c) both the filters more difficult
   (d) both the filters much easier

6. A signal $x(t) = e^{\cos(0.02t^2)}\left(\frac{\sin \pi t}{\pi t}\right)$ was sampled at the rate of $\frac{1}{T}$ and it was claimed that $X(e^{j\omega})$, the sampled signal, is equal to 1 for all ω.
   (a) This is possible, but only for $T = 1$.
   (b) This is possible, but only for $T = \frac{1}{2}$.
   (c) This is not possible for any T.
   (d) This is possible, but not for the values of T indicated above.
7. A signal $x(t) = \cos(2t)$ is sampled at the rate of $\frac{1}{T} = 1000$ Hz. The discrete-time signal so obtained has a periodicity of
   (a) 2000 (b) 1000
   (c) infinity, in the sense that it is not periodic (d) none of the above
8. Pre-aliasing filter is used at
   (a) After sampling (b) before sampling
   (c) After reconstruction filter (d) before reconstruction filter
   (e) None of the above statements is correct
9. For a continuous-time signal of frequency 1 kHz, the minimum required sampling frequency is
   (a) 4 kHz (b) 2 kHz (c) 1 kHz (d) 500 kHz
10. The process of reconstruction from discrete-time signal can be aided by the use of
    (a) holding circuit (b) pulse circuit
    (c) sampling circuit (d) none of the above
11. The 'aliasing' process means
    (a) amplitude overlapping (b) phase overlapping
    (c) spectral overlapping (d) peaks overlapping
12. A band-limited lowpass signal is sampled at twice its Nyquist rate with $f_s = 2000$ samples per second. The signal is band-limited to
    (a) 250 Hz (b) 1000 Hz (c) 500 Hz (d) 2000 Hz.
13. A certain lowpass signal $x(t)$ is sampled and the spectrum of the sampled version has guard band from 1500 Hz to 1900 Hz. The sampling frequency used is
    (a) 1500 Hz (b) 1900 Hz (c) 1700 Hz (d) 3400 Hz
14. A lowpass signal band-limited to 1200 Hz was sampled. It was found that the 1000 Hz frequency component was re-appearing in the recovered signal as a 400 Hz component because of aliasing. The sampling frequency used is
    (a) 1400 Hz (b) 1600 Hz (c) 2200 Hz (d) 800 Hz
15. The signal $x(t) = 3\cos^2(250\pi t)$ is sampled at regular intervals of T seconds. The maximum value of T for which $x(t)$ may be recovered from the sampled version without any distortion is equal to
    (a) 1 ms (b) 2 ms (c) 4 ms (d) 0.5 ms
16. A signal $x(t) = 5\cos(240\pi t)$ was sampled at a frequency f_s. The signal recovered from the samples was, however, found to be $3\cos(110\pi t)$. The sampling frequency is equal to
    (a) 175 Hz (b) 350 Hz (c) 130 Hz (d) 65 Hz

17. The aperture effect
    (a) amplifies the high-frequency components
    (b) attenuates the low-frequency components
    (c) amplifies the low-frequency components
    (d) attenuates the high-frequency components
18. If the sampling frequency of a signal is $f_s = 2000$ samples per second, the Nyquist interval is
    (a) 5 ms (b) 0.5 ms (c) 0.5 seconds (d) 5 seconds
19. The process of multiplying the continuous time signal with impulse train is called
    (a) ideal sampling (b) natural sampling
    (c) flat-top sampling (d) none of the above
20. A reconstruction filter used to recover the signals from its samples is a
    (a) low-pass filter (b) band-pass filter
    (c) high-pass filter (d) All-pass filter
21. Due to undersampling, the effect on the sampled signal is
    (a) distortion (b) aliasing
    (c) frequency folding (d) all the above
22. If a sampled signal is processed through ideal LPF having cutoff frequency ω_0 rad/s, the maximum frequency component in the recovered signal is
    (a) $\omega_0 + \omega_0/2$ (b) $\omega_0/2$ (c) $2\omega_0$ (d) ω_0
23. In natural sampling the pulse width τ of the pulse train with period T should be
    (a) $T \gg \tau$ (b) $T \ll \tau$ (c) $T \leq \tau$ (d) $T \geq \tau$
24. A signal $x(t) = 2\sin(100\pi t)$ is sampled with sampling frequency $f_s = 80$ Hz. The frequency components in the recovered signal are
    (a) 50 Hz, 80 Hz (b) 50 Hz, 30 Hz
    (c) 80 Hz, 30 Hz (d) 80 Hz, 100 Hz
25. If a signal $x(t)$ is sampled with sampling frequency $f_s = 80$ Hz, then the Nyquist rate of $x(t)\cos(20\pi t)$ is
    (a) 110 Hz (b) 120 Hz (c) 200 Hz (d) 100 Hz
26. Flat-top sampling of lowpass signals
    (a) give rise to aperture effect (b) implies oversampling
    (c) leads to aliasing (d) introduces delay distortion
27. Sampled output consists of
    (a) sum of original signal and sampling signal
    (b) difference of original signal and sampling signal
    (c) product of original signal and sampling signal
    (d) division of original signal and sampling signal

28. If maximum frequency component in a signal is 50 Hz, then the signal should be sampled with minimum sampling frequency of

(a) 50 Hz (b) 75 Hz (c) 25 Hz (d) 100 Hz

29. The Nyquist sampling rate for the signal $f(t) = \text{sinc}^2(100\pi t)$ is

(a) 50 Hz (b) 100 Hz (c) 200 Hz (d) 400 Hz

30. What is the cutoff frequency of the pre-aliasing filter for sampling the signal $f(t) = 10\cos(50\pi t)$ $\cos^2(150\pi t)$ with out distortion, where t is in seconds is

(a) 150 samples/s (b) 200 samples/s

(c) 300 samples/s (d) 175 samples/s

Answers

1. (c)	2. (a)	3. (a)	4. (b)	5. (c)	6. (a)	7. (b)	8. (b)	9. (b)	10. (d)
11. (c)	12. (b)	13. (b)	14. (b)	15. (b)	16. (c)	17. (d)	18. (b)	19. (a)	20. (a)
21. (d)	22. (d)	23. (a)	24. (b)	25. (b)	26. (a)	27. (c)	28. (d)	29. (c)	30. (d)

7

The Laplace Transform

Learning Objectives

The main objective of this chapter is to understand thoroughly the basic concepts of the Laplace transform. We will learn how signals and systems are represented by Laplace transforms. By reading this chapter, students will be able to relate to the following topics:

- Laplace transform
- Concept of region of convergence (ROC) for Laplace transforms
- Properties of Laplace transforms
- Inverse Laplace transform, partial fraction expansion
- Laplace transform of certain signals using waveform synthesis
- Transfer function and output response of LTI systems

7.1 Introduction

We have come to know from the previous chapters that the analysis of signals and linear systems in time domain involve differential equations and complex integrals that require the evaluation of complex mathematical equations with higher order differential equations. This is often tedious, time-consuming and provide non-unique solutions in most cases. We saw that an alternate approach is to use Fourier transform. In Fourier transforms, the differential equations and complex integrals are transformed into simple algebraic equations in the frequency domain. The solutions are then inverse Fourier transformed to obtain them as functions of time. The Fourier transform of a function $f(t)$ represents a continuous sum of exponential function of form $e^{j\omega t}$, where ω is a real frequency. However, the major limitations of Fourier transforms are that the function $f(t)$ should be absolutely integrable and that the transformation maps the time functions only onto the $j\omega$ axis in the complex frequency plane. The Fourier transform does not exist for most of the functions having complex variable $\sigma + j\omega$ encountered in LTI systems.

The Laplace transform was developed by French mathematician and astronomer Pierre-Simon Laplace in the year 1780 by extending the Fourier transform to complex Fourier transform. It can be used to analyse a wider class of signals and systems that are not absolutely integrable.

7.2 Review of Laplace Transforms

Unilateral Laplace Transform

In the majority of applications involving Laplace transform, the time functions of interest are causal, that is, $f(t) = 0$ for $t < 0$; the Laplace transforms of such functions are termed unilateral or one-sided Laplace transforms.

For any real function $f(t), t \geq 0$, the Laplace transform is defined as

$$F(s) = \int_{0^-}^{\infty} f(t)e^{-st}\,dt, \tag{7.1}$$

where $s = \sigma + j\omega$ is a complex frequency with real numbers σ and ω.

Another notation used for the Laplace transform is $\mathcal{L}[f(t)]$, where $\mathcal{L}[\,]$ is Laplace transform operator.

Note: The lower limit of integration in the Laplace transform is taken as $t = 0^-$. This ensures that the inclusion of initial conditions, impulse functions and its derivatives at $t = 0^-$.

The inverse Laplace transform of a function $F(s)$ is defined as

$$f(t) = \frac{1}{2\pi j}\int_{\sigma-j\infty}^{\sigma+j\infty} F(s)e^{st}\,ds.$$

$$f(t) = \mathcal{L}^{-1}[F(s)], \tag{7.2}$$

where $\mathcal{L}^{-1}[F(s)]$ is inverse Laplace transform operator.

Bilateral Laplace Transform

The Laplace transform can be alternatively defined as the *bilateral Laplace transform* or two-sided Laplace transform by extending the limits of integration to be the entire real axis, i.e., $-\infty \leq t \leq \infty$. Bilateral Laplace transform pair is defined as

$$F(s) = \int_{-\infty}^{\infty} f(t)e^{-st}\,dt$$

$$f(t) = \frac{1}{2\pi j}\int_{\sigma-j\infty}^{\sigma+j\infty} F(s)e^{st}\,ds. \tag{7.3}$$

Note: The inverse Laplace transform involves complex integration and can be solved by either contour integral methods or simply by using standard models of Laplace transform pairs. The notation of Laplace transform pair is $f(t) \leftrightarrow F(s)$. The symbol $\leftrightarrow$ indicates transformation on both sides.

7.3 Convergence of the Laplace Transform

As we know, the Fourier transform of $f(\mathrm{t})$ converges if $f(t)$ is absolutely integrable. Similarly, the necessary condition for convergence of the Laplace transform is the absolute integrability of $f(t)e^{-\sigma t}$. Mathematically, this can be stated as

$$\int_{-\infty}^{\infty} \left|f(t)e^{-\sigma t}\right| dt < \infty. \tag{7.4}$$

Note: Laplace transform exists only for signals which satisfy the Eq. (7.4) in the given region.

Existence of the Laplace Transform

Laplace transform exists if it converges in the given time interval. Therefore, the condition for its existence is that the function $f(t)e^{-\sigma t}$ should be absolutely integrable. That is,

$$\int_{-\infty}^{\infty} \left| f(t)e^{-\sigma t} \right| dt < \infty.$$

Proof: Let $f(t)$ be a causal and an exponential order function. Then it always satisfies the following inequality

$$|f(t)| = Me^{\alpha t} \quad \text{for all } t > 0, \tag{7.5}$$

where M and α are real positive finite constants.

Therefore the condition integral is

$$\begin{aligned} \int_{-\infty}^{\infty} \left| f(t)e^{-\sigma t} \right| dt &= \int_{0}^{\infty} Me^{\alpha t} e^{-\sigma t} dt \\ &= \int_{0}^{\infty} Me^{-(\sigma-\alpha)t} dt = \frac{M}{\sigma-\alpha} e^{-(\sigma-\alpha)t} \Big|_{0}^{\infty} \\ &= \frac{M}{\sigma-\alpha} \quad \text{if } \sigma > \alpha \text{ is a finite value.} \end{aligned} \tag{7.6}$$

Thus, the Laplace transform exists for $\infty > \sigma > \alpha$.

7.3.1 Region of Convergence (ROC)

The region of convergence, known as the ROC, is important to understand because it defines the region where the Laplace transform exists. The range of values of the complex variable $s = \sigma + j\omega$ for which the Laplace transform converges is called the region of convergence (ROC). The region of convergence of $X(s)$ can be represented graphically with a pole–zero plot on the s-plane, s being given by $s = \sigma + j\omega$, where σ is represented along the x-axis and $j\omega$ along the y-axis. The ROC is a shaded area on the pole–zero plot of s-plane; $X(s)$, the Laplace transform, exists for all the values of s in the shaded area.

Poles and Zeros of $X(s)$

The general polynomial form of $X(s)$ can be expressed as

$$X(s) = \frac{a_0 s^m + a_1 s^{m-1} + \ldots + a_m}{b_0 s^n + b_1 s^{n-1} + \ldots + b_n} = \frac{K(s-z_1)(s-z_2)\cdots(s-z_m)}{(s-p_1)(s-p_2)\cdots(s-p_n)}, \tag{7.7}$$

where a_k, $k = 1,2,3\ldots,m$ and b_k, $k = 1,2,3\ldots,n$ are real coefficients, m and n are positive integers and K is a real constant. The roots of the numerator polynomial, z_k, $k = 1,2,3\ldots,$ m are called the zeros of $X(s)$ because $X(s) = 0$ for those values of s. Similarly, the roots of the denominator polynomial, P_k, $k = 1,2,3\ldots,n$ are called the poles of $X(s)$ because $X(s)$ is infinite for those values of s.

$X(s)$ can be represented graphically by locating poles and zeros in the s-plane as shown in Fig. 7.1. Generally, the symbol cross mark ("**x**") is used to indicate single pole location and the symbol circle mark ("**o**") is used to indicate zero location. Multiple poles are indicated with multiple cross symbols and multiple zeros are indicated with multiple circle symbols.

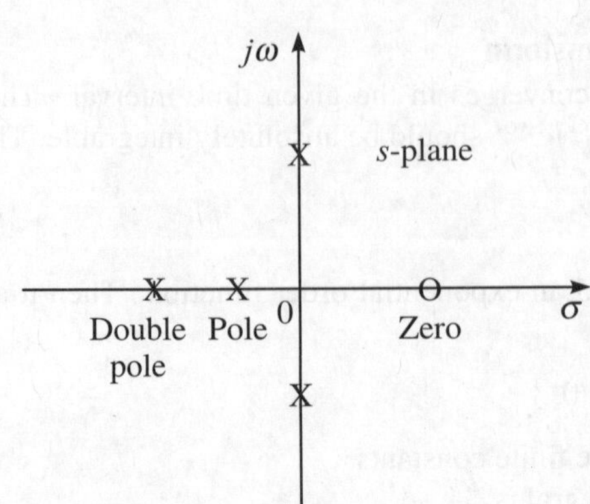

Fig. 7.1 *Pole–zero plot in the s-plane*

7.4 Constraints on ROC for Various Classes of Signals

Assume that a time function $x(t)$ has a Laplace transform $X(s)$ and $X(s)$ is a rational function of s. The ROC depends on the range of time interval of the function $x(t)$. Also, the poles of $X(s)$ should lie outside the ROC, since $X(s)$ does not converge at the poles. Zeros, on the other hand, may lie inside or outside the ROC.

Three classes of signal exist depending on their range of time interval. They are:

1. Right-sided signal, that is, it exists only in the interval $(0 < t < \infty)$
2. Left-sided signal, that is, it exists only in the interval $(-\infty < t < 0)$
3. Two-sided signal, that is, it exists in the time interval $(-\infty < t < \infty)$

7.4.1 Properties of the ROC

The properties of the ROC are given below.

Property 1: The ROC does not contain any poles.

Property 2: If $x(t)$ is a finite duration function, that is, the signal $x(t)$ exist only between two finite time intervals $(t_1 < t < t_2)$, then the ROC is the entire s-plane except at $s = 0$ and $s = \infty$ as shown in Fig. 7.2(a).

Property 3: If $x(t)$ is a right-sided signal, that is, $x(t)$ exists only in interval $(0 < t < \infty)$, then the ROC is

$$\text{Re}\{s\} > \sigma_{\text{max}}, \tag{7.8}$$

where the symbol Re$\{s\}$ represents a real value of s and σ_{max} is the maximum value of all the poles of $X(s)$.

Thus, the ROC covers the entire right side of the half-plane starting from the vertical line Re $\{s\} = \sigma_{\text{max}}$ in the s-plane as shown in Fig. 7.2(b). All poles are located on the left side of the ROC.

Property 4: If $x(t)$ is a left-sided signal, that is, $x(t)$ exists only in interval $(-\infty < t < 0)$, then the ROC is

$$\text{Re}\{s\} < \sigma_{\text{min}}, \tag{7.9}$$

where $\sigma_{\min}$ is the minimum value of all the poles of $X(s)$.

Thus, the ROC covers the entire left side of the half-plane starting from the vertical line $\text{Re}\{s\} = \sigma_{\min}$ in the s-plane as shown in Fig. 7.2(c). All poles are located on the right side of the ROC.

Property 5: If $x(t)$ is a two-sided signal, that is, $x(t)$ an infinite duration signal existing in the time interval $(-\infty < t < \infty)$, then the ROC is of the form

$$\sigma_1 < \text{Re}\{s\} < \sigma_2 \tag{7.10}$$

where σ_1 and σ_2 are the real parts of the two poles of $X(s)$. Thus, the ROC is a vertical strip in the s-plane between the vertical lines $\text{Re}\{s\} = \sigma_1$ and $\text{Re}\{s\} = \sigma_2$ as shown in Fig. 7.2(d).

All poles should be located outside the ROC.

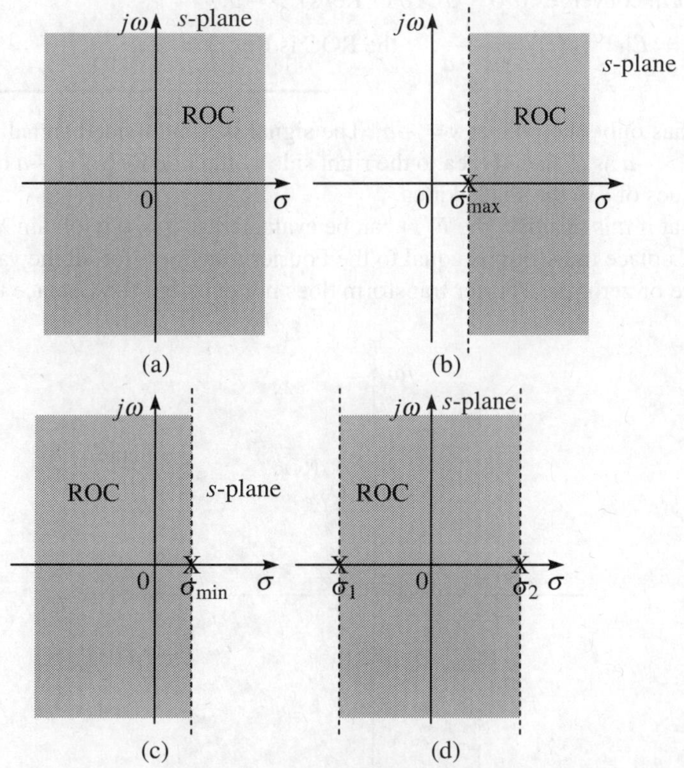

Fig. 7.2 *(a) ROC is the entire s-plane; (b) ROC for right-sided signal; (c) ROC for left-sided signal; d) ROC for both sided signal*

Example 7.1

Obtain the Laplace transform of $x(t) = e^{-at}u(t)$ and compare ROC with its Fourier transform.

Solution

Given signal $x(t) = e^{-at}u(t)$. The Fourier transform $X(\omega)$ is

$$X(\omega) = \int_{-\infty}^{\infty} e^{-at}u(t)e^{-j\omega t} = \int_{0}^{\infty} e^{-at}e^{-j\omega t}\,dt = \frac{1}{a+j\omega} \quad a > 0.$$

The Fourier transform converges for $a > 0$.

Now, the Laplace transform is

$$X(s) = \int_{-\infty}^{\infty} e^{-at}u(t)e^{-st}\,dt = \int_{0}^{\infty} e^{-(s+a)t}\,dt.$$

or, with $s = \sigma + j\omega$,

$$X(\sigma + j\omega) = \int_{0}^{\infty} e^{-(a+\sigma+j\omega)t}\,dt$$

$$= \frac{1}{a+\sigma+j\omega} e^{-(a+\sigma+j\omega)t}\Big|_0^{\infty} = \frac{1}{a+\sigma+j\omega} \quad \sigma + a > 0.$$

The Laplace transform converges if $\sigma + a > 0$ or $\mathrm{Re}\{s\} > -a$.

Therefore, $X(s) = \mathcal{L}[e^{-at}u(t)] = \frac{1}{s+a}$; the ROC is $\mathrm{Re}\{s\} > -a$.

The function $X(s)$ has only one pole at $s = -a$. The signal is a right-sided signal and Fig. 7.3 shows the ROC for $\mathrm{Re}\{s\} > -a$ as a shaded area to the right side of the line $\mathrm{Re}\{s\} = -a$ on the s-plane. $X(s)$ exists for all the values of s in the shaded area.

It is observed that if a is positive, the $F(s)$ can be evaluated at $\sigma = 0$ to obtain $X(0 + j\omega) = \frac{1}{a+j\omega}$. For $\sigma = 0$, the Laplace transform is equal to the Fourier transform for all the values of s on the $j\omega$ *axis*. If a is negative or zero, the Fourier transform does not exist but the Laplace transform exists for all values of a.

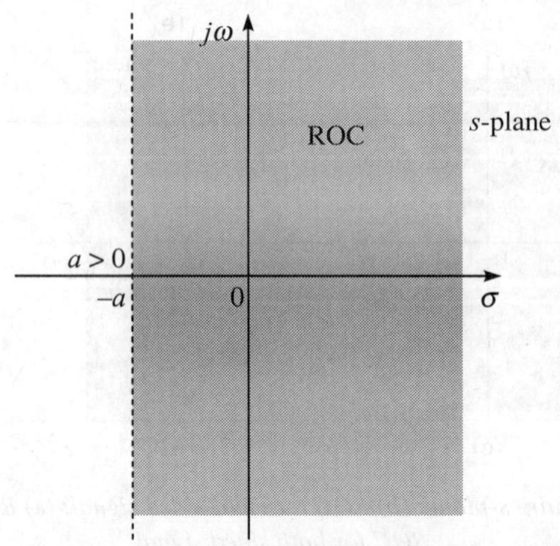

Fig. 7.3 *ROC for* $x(t) = e^{-at}u(t)$

Example 7.2

Obtain Laplace transforms of $x_1(t) = e^{-at}u(t)$ and $x_2(t) = -e^{-at}u(-t)$. Compare their ROCs.

(Nov 2015)

Solution

Given the signal $x_1(t) = e^{-at}u(t)$. Its Laplace transform is

$$X_1(s) = \int_{-\infty}^{\infty} e^{-at}u(t)e^{-st}\,dt = \int_{0}^{\infty} e^{-(s+a)t}\,dt.$$

$$X_1(s) = \frac{1}{s+a}, \quad \text{Re}\{s\} > -a.$$

The Laplace transform converges if $\sigma + a > 0$ or $\text{Re}\{s\} > -a$. Therefore, the ROC is $\text{Re}\{s\} > -a$. The function $X_1(s)$ has only one pole at $s = -a$. Since the signal is a right-sided signal, the ROC for $\text{Re}\{s\} > -a$ is shown as a shaded area right side of the line $\text{Re}\{s\} = -a$ on s-plane (Fig. 7.3).

Now, for the signal $x_2(t) = -e^{-at}u(-t)$, the Laplace transform is

$$X_2(s) = -\int_{-\infty}^{\infty} e^{-at}e^{-st}u(-t)dt = -\int_{-\infty}^{0} e^{-(s+a)t}dt$$

or $\quad X_2(s) = \dfrac{1}{s+a}, \quad \text{Re}\{s\} < -a$

and the Laplace transform converges if $\text{Re}\{s\} < -a$.

$$\mathcal{L}\left[-e^{-at}u(-t)\right] = \frac{1}{s+a}, \quad \text{Re}\{s\} < -a.$$

The function $X_2(s)$ also has only one pole at $s = -a$. Since the signal is left-sided, the ROC, $\text{Re}\{s\} < -a$, is the shaded area on the left side of the line $\text{Re}\{s\} = -a$ on s-plane (Fig. 7.4). It is observed that the algebraic expressions for $X_1(s)$ and $X_2(s)$ for the given two different signals are identical except for the ROCs. Therefore, in order for the Laplace transform to be unique for each signal $x(t)$, the ROC must be specified as part of the transform.

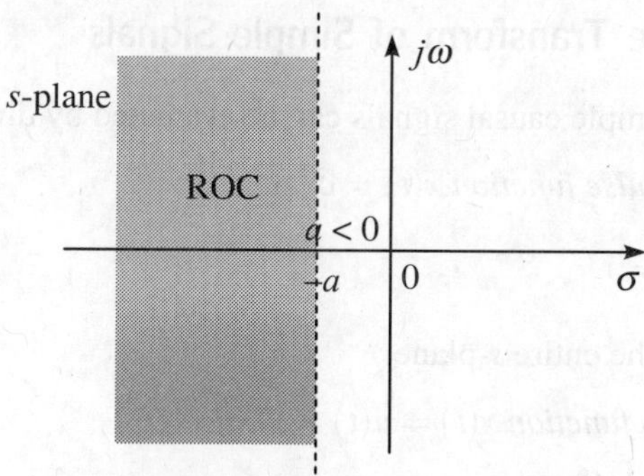

Fig. 7.4 *ROC for* $x(t) = -e^{-at}u(-t)$

Example 7.3

Obtain the Laplace transform of $f(t) = 3e^{-2t}u(t) - 2e^{-t}u(t)$ with its ROC.

Solution

Given the signal $f(t) = 3e^{-2t}u(t) - 2e^{-t}u(t)$

From the basic properties, Laplace transform for signals are

$$e^{-2t}u(t) \leftrightarrow \frac{1}{s+2} \text{ and } e^{-t}u(t) \leftrightarrow \frac{1}{s+1}.$$

Therefore the Laplace transform of $f(t) = 3e^{-2t}u(t) - 2e^{-t}u(t)$ is

$$F(s) = \frac{3}{s+2} - \frac{2}{s+1} = \frac{s-1}{s^2+3s+2}.$$

To determine the ROC, the poles are $p_1 = -1$ and $p_2 = -2$. Since the signal is right-sided signal (Fig. 7.5), the ROC is $\text{Re}\{s\} > -1$.

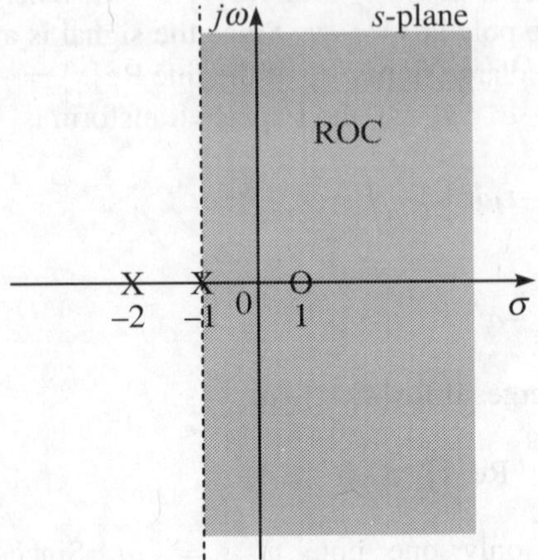

Fig. 7.5 *ROC for $f(t) = 3e^{-2t}u(t) - 2e^{-t}u(t)$ on pole-zero plot*

7.4.2 Unilateral Laplace Transform of Simple Signals

Laplace transform for some simple causal signals can be evaluated by direct integration.

Laplace transform of unit impulse function $x(t) = \delta(t)$

$$X(s) = \int_{-\infty}^{\infty} \delta(t)e^{-st}\,dt = 1 \tag{7.11}$$

Thus, $\delta(t) \leftrightarrow 1$. The ROC is the entire s-plane.

Laplace transform of unit step function $x(t) = u(t)$

$$X(s) = \int_{-\infty}^{\infty} u(t)e^{-st}\,dt = \int_{0}^{\infty} e^{-st}\,dt$$

$$= \int_{0}^{\infty} e^{-st}\,dt = \frac{1}{s}e^{-st}\Big|_{0}^{\infty}$$

$$X(s) = \frac{1}{s}, \quad \text{Re}\{s\} > 0 \tag{7.12}$$

Thus $u(t) \leftrightarrow \dfrac{1}{s}$. The ROC is right half of the s-plane.

Laplace transform of cosine function $x(t) = \cos(\omega_0 t)u(t)$

We know that $\cos(\omega_0 t) = \dfrac{e^{j\omega_0 t} + e^{-j\omega_0 t}}{2}$.

$$X(s) = \int_{-\infty}^{\infty} \cos(\omega_0 t)u(t)e^{-st}\,dt = \int_{0}^{\infty} \frac{e^{j\omega_0 t} + e^{-j\omega_0 t}}{2}e^{-st}\,dt$$

$$= \frac{1}{2}\int_0^{\infty} e^{j\omega_0 t}e^{-st}dt + \frac{1}{2}\int_0^{\infty} e^{-j\omega_0 t}e^{-st}dt = \frac{1}{2}\int_0^{\infty} e^{-(s-j\omega_0)t}dt + \frac{1}{2}\int_0^{\infty} e^{-(s+j\omega_0)t}dt$$

$$X(s) = \frac{1}{2}\left(\frac{1}{s-j\omega_0}\right) + \frac{1}{2}\left(\frac{1}{s+j\omega_0}\right) = \frac{s}{s^2+\omega_0^2}, \quad \text{Re}\{s\} > 0. \tag{7.13}$$

Thus $\cos\omega_0 t u(t) \leftrightarrow \dfrac{s}{s^2+\omega_0^2}, \quad \text{Re}\{s\} > 0.$

Laplace transform of sine function $x(t) = \sin(\omega_0 t)u(t)$

We know that $\sin(\omega_0 t) = \dfrac{e^{j\omega_0 t} - e^{-j\omega_0 t}}{2j}$

$$X(s) = \int_{-\infty}^{\infty} \sin(\omega_0 t)u(t)e^{-st}dt = \int_0^{\infty} \frac{e^{j\omega_0 t} - e^{-j\omega_0 t}}{2j}e^{-st}dt$$

$$= \frac{1}{2j}\int_0^{\infty} e^{j\omega_0 t}e^{-st}dt - \frac{1}{2j}\int_0^{\infty} e^{-j\omega_0 t}e^{-st}dt$$

$$= \frac{1}{2j}\int_0^{\infty} e^{-(s-j\omega_0)t}dt - \frac{1}{2j}\int_0^{\infty} e^{-(s+j\omega_0)t}dt$$

$$X(s) = \frac{1}{2j}\left(\frac{1}{s-j\omega_0}\right) - \frac{1}{2j}\left(\frac{1}{s+j\omega_0}\right) = \frac{s}{s^2+\omega_0^2}, \quad \text{Re}\{s\} > 0. \tag{7.14}$$

$$\sin\omega_0 t u(t) \leftrightarrow \frac{\omega_0}{s^2+\omega_0^2} \quad \text{Re}\{s\} > 0.$$

Laplace transform of unit ramp function $x(t) = tu(t)$

$$X(s) = \int_{-\infty}^{\infty} tu(t)e^{-st}dt = \int_0^{\infty} te^{-st}dt$$

$$= \frac{1}{s^2}e^{-st}\Big|_0^{\infty} = \frac{1}{s^2}, \quad \text{Re}\{s\} > 0. \tag{7.15}$$

7.5 Properties of Laplace Transforms

The properties of unilateral Laplace transform are useful in solving many problems involving integral and differential equations. The important properties are explained below. For the convenience of expression, we use the notation $f(t) \leftrightarrow F(s)$ or $F(s) = \mathcal{L}[f(t)]$, $f(t) = \mathcal{L}^{-1}[F(s)]$ to denote a transform pair.

1. Linearity (Superposition)

Let signals $f_1(t)$ and $f_2(t)\ldots f_n(t)$ for $t > 0$ have Laplace transforms $F_1(s), F_2(s)\ldots.. F_n(s)$ respectively. The linearity property states that if

$$f_1(t) \leftrightarrow F_1(s),\ f_2(t) \leftrightarrow F_2(s)\ldots\ldots\ f_n(t) \leftrightarrow F_n(s), \text{ with ROCs} : R_1, R_2, R_3 \ldots R_n \text{ respectively,}$$

then the linear combination of the signals satisfy

$$\mathcal{L}[a_1f_1(t)+a_2f_2(t)+\ldots+a_nf_n(t)] = a_1F_1(s)+a_2F_2(s)+\ldots+a_nF_n(s)$$

or $$a_1f_1(t)+a_2f_2(t)+\ldots+a_nf_n(t) \leftrightarrow a_1F_1(s)+a_2F_2(s)+\ldots+a_nF_n(s), \tag{7.16}$$

where a_1, $a_2,\ldots,$ a_n are arbitrary constants. The ROC is at least the intersection of $R_1, R_2, R_3\cdots$ and R_n.

Proof: Consider signals $f_1(t)$ and $f_2(t)\ldots f_n(t)$ for $t>0$ have Laplace transforms $F_1(s)$, $F_2(s)\ldots.F_n(s)$ respectively. Then,

$$\begin{aligned}\mathcal{L}[a_1f_1(t)+a_2f_2(t)+\ldots+a_nf_n(t)] &= \int_0^{\infty}[a_1f_1(t)+a_2f_2(t)+\ldots+a_nf_n(t)]e^{-st}dt\\ &= \int_0^{\infty}a_1f_1(t)e^{-st}dt + \int_0^{\infty}a_2f_2(t)e^{-st}dt+\ldots\\ &\quad+\int_0^{\infty}a_nf_n(t)e^{-st}dt\\ &= a_1F_1(s)+a_2F_2(s)+\ldots+a_nF_n(s).\end{aligned}$$

2. Time Shifting Property

Let the signal $f(t)$ for $t>0$ to be delayed (or advanced) in time by some time $t_0>0$ such that $t-t_0>0$. If $f(t)$ has Laplace transform $F(s)$, then the time shifting property states that

$$\mathcal{L}[f(t\pm t_0)] = \mathcal{L}(s)e^{\pm st_0}. \tag{7.17}$$

Proof: Consider a signal $f(t)$ for $t>0$ having Fourier transform $F(s)$. If $f(t)$ is delayed by t_0, then

$$\mathcal{L}[f(t-t_0)u(t-t_0)] = \int_0^{\infty}f(t-t_0)u(t-t_0)e^{-st}dt = \int_{t_0}^{\infty}f(t-t_0)e^{-st}dt.$$

Let $t-t_0=x$. Then $dt=dx$, and $0<x<\infty$.

$$\begin{aligned}\text{Thus } \mathcal{L}[f(t-t_0)u(t-t_0)] &= \int_0^{\infty}f(x)e^{-s(x-t_0)}dx\\ &= e^{-st_0}\int_0^{\infty}f(x)e^{-st}dx = e^{-st_0}F(s).\end{aligned}$$

Similarly, we can prove that $\mathcal{L}[f(t+t_0)] = e^{st_0}F(s)$.

Example 7.4

Find the Laplace transform for $x(t)=u(t+2)+u(t-2)$.

Solution

Given $x(t)=u(t+2)+u(t-2)$.
We know that the Laplace transform

$$u(t)\leftrightarrow\frac{1}{s}.$$

From shifting property,

$$\mathcal{L}[u(t-2)] = \frac{e^{-2s}}{s}; \quad \mathcal{L}[u(t+2)] = \frac{e^{2s}}{s}.$$

$$\therefore \quad X(s) = \mathcal{L}[u(t+2)+u(t-2)] = \frac{e^{2s}+e^{-2s}}{s} = \frac{\cosh 2s}{s}.$$

3. Frequency Shifting Property

The dual of the time delay property is the frequency-shifting property.

Let the signal $f(t)$ for $t > 0$ be delayed (or advanced) in the s-domain by some real value $a > 0$. Then, the frequency shifting property states that

$$e^{-at} f(t) \leftrightarrow F(s+a). \tag{7.18}$$

Proof: Consider $F(s)$, the Laplace transform of the signal $f(t)$. Then

$$\mathcal{L}[e^{-at} f(t)] = \int_0^\infty e^{-at} f(t) e^{-st}\, dt$$

$$= \int_0^\infty f(t) e^{-(s+a)t}\, dt$$

$$= F(s+a).$$

Similarly, $e^{at} f(t) \leftrightarrow F(s-a)$.

Example 7.5

Prove that $\mathcal{L}[e^{-at} \sin \omega_0 t u(t)] = \dfrac{\omega_0}{(s+a)^2+\omega_0^2}$ using the frequency shift property.

Solution

We know that $\mathcal{L}[\sin(\omega_0 t)u(t)] = \dfrac{\omega_0}{s^2+\omega_0^2}$.

Using the frequency shifting property, we have

$$\mathcal{L}[e^{-at} u(t)] = \frac{1}{s+a}.$$

$$\mathcal{L}\left[e^{-at} \sin(\omega_0 t) u(t)\right] = \frac{\omega_0}{(s+a)^2+\omega_0^2}.$$

4 Scaling Property

If a signal $f(t)$ has a Laplace transform $F(s)$, then the scaling property states that

$$f(at) \leftrightarrow \frac{1}{a} F\left(\frac{s}{a}\right); \quad a > 0, \tag{7.19}$$

where $a > 0$ is a real-valued scaling constant.

Proof: Consider a signal $f(t)$ that has a Laplace transform $F(s)$ and $a > 0$ be a real constant. Then,

$$\mathcal{L}[f(at)] = \int_0^\infty f(at)e^{-st}\,dt.$$

Let $at = u$, $adt = du$.

$$\mathcal{L}[f(at)] = \frac{1}{a}\int_0^\infty f(u)e^{su/a}du = \frac{1}{a}F\left(\frac{s}{a}\right).$$

Example 7.6

Find the Laplace transform of the signal $x(t) = \sin(t/2)u(t/2)$.

Solution

The Laplace transform of a sinusoidal signal is

$$\sin(t)u(t) \leftrightarrow \frac{1}{s^2+1}.$$

Using scaling property, the Laplace transform of the signal $x(t) = \sin(t/2)u(t/2)$ is

$$\sin(t/2)u(t/2) \leftrightarrow \frac{2}{(2s)^2+1}.$$

5. Time Differentiation Property

If a signal $f(t)$ has a Laplace transform $F(s)$ and $\frac{df(t)}{dt}$ is absolutely integrable, then time differentiation property states that

$$\mathcal{L}\left[\frac{df(t)}{dt}\right] = sF(s) - f(0^-).$$

$$\text{or} \quad f'(t) \leftrightarrow sF(s) - f(0^-). \tag{7.20}$$

And, in general, we have

$$\frac{d^n f(t)}{dt^n} \leftrightarrow s^n F(s) - s^{n-1}f(0^-) - s^{n-2}f'(0^-) - \cdots - f^{(n-1)}(0^-). \tag{7.21}$$

Proof: Consider that a signal $f(t)$ has a Laplace transform $F(s)$ and $\frac{df(t)}{dt}$ is absolutely integrable. From the definition,

$$\mathcal{L}\left[\frac{df(t)}{dt}\right] = \int_{0^-}^\infty \frac{df(t)}{dt}e^{-st}dt = \int_{0^-}^\infty e^{-st}df(t).$$

Integrate by parts to obtain

$$\mathcal{L}[f'(t)] = f(t)e^{-st}\Big|_{0^-}^\infty + s\int_{0^-}^\infty f(t)e^{-st}dt$$

$$= f(t)e^{-st}\Big|_{t=\infty} - f(0) + sF(s).$$

Now, since $F(s)$ exists, it follows that $f(t)e^{-st}$ is zero at $t=\infty$. Thus,

$$\mathcal{L}[f'(t)] = sF(s) - f(0^-).$$

Similarly, if $y(t) = \dfrac{df(t)}{dt}$ and $\dfrac{d^2 f(t)}{dt^2}$ exists, and

$$\mathcal{L}[y(t)] = Y(s) = sF(s) - f(0^-),$$

then the Laplace transform of the second derivative of $f(t)$ is

$$\mathcal{L}\left[\frac{dy(t)}{dt}\right] = \mathcal{L}\left[\frac{d^2 f(t)}{dt^2}\right] = sY(s) - y(0^-)$$
$$= s(sF(s) - f(0^-)) - y(0^-) = s^2F(s) - sf(0^-) - f'(0^-).$$

The general case is found by repeated application of the above process. That is, the Laplace transform of nth derivative of $f(t)$ is

$$\frac{d^n f(t)}{dt^n} \leftrightarrow s^n F(s) - s^{n-1}f(0^-) - s^{n-2}f'(0^-) - \cdots - sf^{(n-2)}(0^-) - f^{(n-1)}(0^-).$$

For causal function, all initial conditions are zero. That is,

$$\frac{d^n f(t)}{dt^n} \leftrightarrow s^n F(s). \tag{7.22}$$

Example 7.7

Find the transform of $\dfrac{df(t)}{dt}$, where $f(t) = e^{-at}u(t)$.

Solution

Given $f(t) = e^{-at}u(t)$.

$$\mathcal{L}[f'(t)] = \int_{0^-}^{\infty} f'(t)e^{-st}\,dt.$$

$$\int_{0^-}^{\infty} \frac{d(e^{-at}u(t))}{dt} e^{-st}\,dt = \int_0^{\infty} [e^{-at}\delta(t) - ae^{-at}u(t)]e^{-st}\,dt$$
$$= 1 - a\frac{1}{s+a} = \frac{s}{s+a}.$$

Or, using the differentiation property, at $t=0$, $f(0^-)=0$.

The Laplace transform is

$$\mathcal{L}[f'(t)] = sF(s) - f(0^-) = sF(s)$$
$$= s\cdot\frac{1}{s+a} = \frac{s}{s+a}.$$

6. Differentiation in s-Domain Property

If a signal $f(t)$ has a Laplace transform $F(s)$, then the differentiation in s-domain property states that

$$(-t)^n f(t) \leftrightarrow \frac{d^n F(s)}{ds^n}. \tag{7.23}$$

Proof: Consider a signal $f(t)$ that has a Laplace transform $F(s)$. Then

$$F(s) = \mathcal{L}[f(t)] = \int_0^{\infty} f(t)e^{-st}\,dt.$$

Differentiating with respect to s both sides, we have

$$\frac{dF(s)}{ds} = \int_0^{\infty} f(t)\frac{d}{ds}(e^{-st})dt = \int_0^{\infty} -tf(t)e^{-st}\,dt.$$

That is, $\dfrac{dF(s)}{ds} \leftrightarrow -tf(t)$.

Taking second derivative with respect to s on both sides,

$$\frac{d^2F(s)}{ds^2} = \int_0^{\infty} f(t)\frac{d^2}{ds^2}(e^{-st})dt = \int_0^{\infty} (-t)^2 f(t)e^{-st}\,dt.$$

That is, $\dfrac{d^2F(s)}{ds^2} \leftrightarrow (-t)^2 f(t)$.

Similarly, we can get

$$\frac{d^nF(s)}{ds^n} = \int_0^{\infty} (-t)^n f(t)e^{-st}\,dt.$$

Thus, $(-t)^n f(t) \leftrightarrow \dfrac{d^nF(s)}{ds^n}$.

7. Time Integration Property

If a signal $f(t)$ has a Laplace transform $F(s)$, then the time integration property states that

$$\int_0^t f(\tau)d\tau \leftrightarrow \frac{F(s)}{s} \tag{7.24}$$

and $\displaystyle\int_{-\infty}^t f(\tau)d\tau \leftrightarrow \frac{F(s)}{s} + \frac{\int_{-\infty}^0 f(\tau)d\tau}{s}$.

Proof: Consider a signal $f(t)$ with Laplace transform $F(s)$. Then, for causal signals,

$$\mathcal{L}\left[\int_0^t f(\tau)d\tau\right] = \int_0^{\infty}\left[\int_0^t f(\tau)d\tau\right]e^{-st}\,dt.$$

Integrating by parts we obtain

$$\mathcal{L}\left[\int_0^t f(\tau)d\tau\right] = \frac{-e^{-st}}{s}\int_0^t f(\tau)d\tau\Bigg|_0^{\infty} + \frac{1}{s}\int_0^{\infty} f(t)e^{-st}\,dt.$$

If $f(t)$ is an exponential order function, its integration converges at $s=\infty$, that is,

$$\int_0^t f(\tau)d\tau = 0 \quad \text{at } s=\infty.$$

$$\therefore \quad \mathcal{L}\left[\int_0^t f(\tau)d\tau\right] = \frac{1}{s}\int_0^{\infty} f(t)e^{-st}\,dt = \frac{1}{s}F(s).$$

For non-casual signals: If the lower limit is $-\infty$, we obtain

$$\int_{-\infty}^{t} f(\tau)d\tau = \int_{-\infty}^{0} f(\tau)d\tau + \int_{0}^{t} f(\tau)d\tau.$$

Thus,

$$\mathcal{L}\left[\int_{\infty}^{t} f(\tau)d\tau\right] = \frac{\int_{-\infty}^{0} f(\tau)d\tau}{s} + \frac{F(s)}{s} = \frac{f^{-1}(0^-)}{s} + \frac{F(s)}{s}.$$

Example 7.8

Find the Laplace transform of $f(t) = te^{\lambda t}$.

Solution

We know that $\mathcal{L}[e^{-\lambda t}] = \dfrac{1}{s+\lambda}$. By frequency differentiation,

$$\mathcal{L}[te^{-\lambda t}] = -\frac{d}{ds}\left[\frac{1}{s+\lambda}\right] = \frac{1}{(s+\lambda)^2}.$$

8. Integration in *S*-Domain Property

If a signal $f(t)$ has Laplace transform $F(s)$, then the integration in s-domain property states that

$$\int_{s}^{\infty} F(u)du \leftrightarrow \frac{f(t)}{t}. \tag{7.25}$$

Proof: Consider a signal $f(t)$ whose Laplace transform is $F(s)$. Then,

$$\begin{aligned}
\int_{s}^{\infty} F(u)du &= \int_{s}^{\infty}\left(\int_{0}^{\infty} f(t)e^{-jut}dt\right)du \\
&= \int_{0}^{\infty} f(t)\left(\int_{s}^{\infty} e^{-ut}du\right)dt. \\
&= \int_{0}^{\infty} f(t)\left.\frac{e^{-ut}}{t}\right|_{s}^{\infty} dt = \int_{0}^{\infty}\frac{f(t)}{t}e^{-st}dt \\
&= \mathcal{L}\left[\frac{f(t)}{t}\right].
\end{aligned}$$

Example 7.9

Find the Laplace transform for $f(t) = \dfrac{1}{t}[e^{-2t} - e^{-3t}]u(t)$.

Solution

Given $f(t) = \dfrac{1}{t}[e^{-2t} - e^{-3t}]u(t)$. From the properties of Laplace transform we know that

$$\mathcal{L}[e^{-2t}u(t)] = \frac{1}{s+2},$$

$$\mathcal{L}[e^{-3t}u(t)] = \frac{1}{s+3}.$$

By the property of integration in the *s*-domain,

$$\mathcal{L}\left[\frac{1}{t}(e^{-2t} - e^{-3t})u(t)\right] = \int_s^\infty \left(\frac{1}{s+2} - \frac{1}{s+3}\right) ds$$

$$= (\ln(s+2) - \ln(s+3))\Big|_s^\infty = \ln\left(\frac{s+2}{s+3}\right).$$

---

9. Time Convolution Property

If two signals $f_1(t)$ and $f_2(t)$ have Laplace transforms $F_1(s)$ and $F_2(s)$ respectively, then the Laplace transform of the convolution of $f_1(t)$ and $f_2(t)$ is

$$\mathcal{L}[f_1(t) * f_2(t)] = F_1(s)F_2(s). \tag{7.26}$$

Proof: Consider two signals $f_1(t)$ and $f_2(t)$ with Laplace transforms $\mathcal{L}[f_1(t)] = F_1(s)$ and $\mathcal{L}[f_2(t)] = F_2(s)$. Then the convolution of $f_1(t)$ and $f_2(t)$ is

$$f_1(t) * f_2(t) = \int_{-\infty}^{\infty} f_1(\tau)f_2(t-\tau)d\tau.$$

The Laplace transform is

$$\mathcal{L}[f_1(t) * f_2(t)] = \int_0^\infty \left[\int_{-\infty}^{\infty} f_1(\tau)f_2(t-\tau)d\tau\right] e^{-st}dt$$

$$= \int_0^\infty f_1(\tau)\int_0^\infty f_2(t-\tau)e^{-st}dtd\tau$$

$$= \int_0^\infty f_1(\tau)F_2(s)e^{-s\tau}d\tau = F_2(s)\int_0^\infty f_1(\tau)e^{-s\tau}d\tau$$

$$= F_1(s)F_2(s).$$

Example 7.10

If $F_1(s) = \dfrac{1}{s+2}$ and $F_2(s) = \dfrac{1}{s+3}$, find the inverse Laplace transform of $F(s) = F_1(s)F_2(s)$ using convolution property.

Solution

Given $F_1(s) = \dfrac{1}{s+2}$ and $F_2(s) = \dfrac{1}{s+3}$.

$$F(s) = F_1(s)F_2(s) = \left(\frac{1}{s+2}\right)\left(\frac{1}{s+3}\right).$$

Using partial fractions,

$$\left(\frac{1}{s+2}\right)\left(\frac{1}{s+3}\right) = \frac{A}{s+2} + \frac{B}{s+3}.$$

$$A = \frac{1}{(s+3)}\bigg|_{s=-2} = 1; \qquad B = \frac{1}{(s+2)}\bigg|_{-3} = -1.$$

$$F(s) = \frac{1}{s+2} - \frac{1}{s+3}.$$

Apply inverse Laplace transform to obtain

$$f(t) = [e^{-2t} - e^{-3t}]u(t).$$

---

10. Multiplication or Frequency Convolution Property

If two signals $f_1(t)$ and $f_2(t)$ have Laplace transforms $F_1(s)$ and $F_2(s)$ respectively, then the Laplace transforms of the multiplication of $f_1(t)$ and $f_2(t)$ is

$$\mathcal{L}[f_1(t)f_2(t)] = \frac{1}{2\pi j}(F_1(s) * F_2(s)), \tag{7.27}$$

where $F_1(s) * F_2(s) = \int_{\sigma-j\infty}^{\sigma-j\infty} F_1(\lambda)F_2(s-\lambda)d\lambda.$

Proof: Consider two signals $f_1(t)$ and $f_2(t)$ with Laplace transforms $\mathcal{L}[f_1(t)] = F_1(s)$ and $\mathcal{L}[f_2(t)] = F_2(s)$. Then the Laplace transform of the multiplication of $f_1(t)$ and $f_2(t)$ is

$$\mathcal{L}[f_1(t)f_2(t)] = \int_0^{\infty} f_1(t)f_2(t)e^{-st}dt$$

$$= \int_0^{\infty} f_1(t)\frac{1}{2\pi j}\left(\int_{\sigma-j\infty}^{\sigma+j\infty} F_2(\lambda)e^{\lambda t}d\lambda\right)e^{-st}dt$$

$$= \frac{1}{2\pi j}\int_{\sigma-j\infty}^{\sigma+j\infty} F_2(\lambda)\left(\int_0^{\infty} f_1(t)e^{-(s-\lambda)t}dt\right)d\lambda$$

$$= \frac{1}{2\pi j}\int_{\sigma-j\infty}^{\sigma-j\infty} F_2(\lambda)F_1(s-\lambda)d\lambda = \frac{1}{2\pi j}(F_1(s) * F_2(s)).$$

11. Initial Value Theorem

The initial value theorem is used to calculate $f(0)$ directly from the Laplace transform $F(s)$ without the need of inverse transform. It states that if $f(t)$ and its first derivative are Laplace transformable, then the initial value of $f(t)$ is given by

$$f(0^+) = \lim_{t\to 0} f(t) = \lim_{s\to\infty}[sF(s)]. \tag{7.28}$$

Proof: Consider a signal $f(t)$ whose Laplace transform is $F(s)$. Then by the property of differentiation,

$$sF(s) - f(0^-) = \mathcal{L}\left[\frac{df(t)}{dt}\right] = \int_{0^-}^{\infty} \frac{df(t)}{dt}e^{-st}dt$$

$$= \int_{0^-}^{0^+} \frac{df(t)}{dt}e^{-st}dt + \int_{0^+}^{\infty} \frac{df(t)}{dt}e^{-st}dt.$$

If there is a discontinuity in $f(t)$ at $t=0$, the derivative of $f(t)$ is an impulse function of amplitude equal to the value of discontinuity.

So, $$\left.\frac{df(t)}{dt}\right|_{t=0} = (f(0^+)-f(0^-))\delta(t).$$

Therefore, $$\int_{0^-}^{0^+} \frac{df(t)}{dt}e^{-st}\,dt = \int_{0^-}^{0^+} (f(0^+)-f(0^-))\delta(t)e^{-st}\,dt$$
$$= f(0^+)-f(0^-).$$

$$sF(s)-f(0^-) = f(0^+)-f(0^-)+\int_{0^+}^{\infty} \frac{df(t)}{dt}e^{-st}\,dt.$$

Therefore,

$$sF(s) = f(0^+)+\int_{0^+}^{\infty} \frac{df(t)}{dt}e^{-st}\,dt.$$

As $s\to\infty$, the integral $\lim_{s\to\infty}\int_{0^+}^{\infty} \frac{df(t)}{dt}e^{-st}\,dt = 0.$

Thus, $f(0^+) = \lim_{s\to\infty}(sF(s)).$

Note: The theorem is applied only when initial value exists.

Example 7.11

Find the initial value of $f(t)$ if $F(s) = \dfrac{s}{(s+a)^2+\omega^2}$.

Solution

The initial value is $f(0^+) = \lim_{s\to\infty} sF(s) = \lim_{s\to\infty} \dfrac{s^2}{(s+a)^2+\omega^2}$

$$= \lim_{s\to\infty} \frac{1}{(1+\frac{a}{s})^2+(\frac{\omega}{s})^2} = 1.$$

Therefore, $f(0^+) = 1$.

12. Final Value Theorem

The final value theorem permits us to calculate $f(\infty)$ directly from Laplace transform $F(s)$ without the need of inverse transform. It states that if $f(t)$ and its first derivative are Laplace transformable, then the final value of $f(t)$ is given by

$$f(\infty) = \lim_{s\to 0}(sF(s)). \tag{7.29}$$

Proof: Consider a signal $f(t)$ whose Laplace transform is $F(s)$. Then, by the property of differentiation,

$$sF(s)-f(0^-) = \mathcal{L}\left[\frac{df(t)}{dt}\right] = \int_{0^-}^{\infty} \frac{df(t)}{dt}e^{-st}\,dt$$

$$= \int_{0^-}^{0} \frac{df(t)}{dt} e^{-st} dt + \int_{0^+}^{\infty} \frac{df(t)}{dt} e^{-st} dt.$$

If there is a discontinuity in $f(t)$ at $t = 0$, the derivative of $f(t)$ is an impulse function of amplitude equal to the value of discontinuity. So,

$$\left. \frac{df(t)}{dt} \right|_{t=0} = (f(0^+) - f(0^-))\delta(t).$$

Therefore, $\displaystyle\int_{0^-}^{0^+} \frac{df(t)}{dt} e^{-st} dt = \int_{0^-}^{0^+} (f(0^+) - f(0^-))\delta(t) e^{-st} dt$

$$= f(0^+) - f(0^-).$$

$$sF(s) - f(0^-) = f(0^+) - f(0^-) + \int_{0^+}^{\infty} \frac{df(t)}{dt} e^{-st} dt.$$

Therefore, $\displaystyle sF(s) = f(0^+) + \int_{0^+}^{\infty} \frac{df(t)}{dt} e^{-st} dt.$

We take the limit as $s \to 0$.

$$\lim_{s\to 0} [sF(s)] = f(0^+) + \int_{0}^{\infty} \frac{df(t)}{dt} dt$$

$$= f(0^+) + f(t)\Big|_{0^+}^{\infty} = f(0^+) + f(\infty) - f(0^+).$$

Thus, $f(\infty) = \lim_{s\to 0} [sF(s)]$.

Note: The theorem is applied only when final value exists.

Example 7.12

Find the final value of the function $F(s)$ given by $\dfrac{(s-1)}{s(s^2-1)}$. (Dec 2011)

Solution

The final value is

$$f(\infty) = \lim_{s\to 0} sF(s) = \lim_{s\to\infty} \frac{s(s-1)}{s(s^2-1)} = \lim_{s\to 0} \frac{1}{(s+1)} = 1.$$

Therefore, $f(\infty) = 1$.

7.6 Relationship between Laplace Transform and Fourier Transform

The Laplace transform is an extension of Fourier transform in which a function $f(t)$ can be represented as a Fourier transform of $[f(t)e^{-\sigma t}]$. That is,

$$\mathcal{L}[f(t)] = \mathcal{F}[f(t)e^{-\sigma t}]. \tag{7.30}$$

We know that the Fourier transform of a signal $f(t)$ is a representation of the function as a continuous sum of exponential function of the form $e^{j\omega t}$, where ω is a real frequency. The Laplace transform, on the other hand, is an extension of Fourier transform in which a function $f(t)$ is represented by a continuous sum of exponential functions of the form e^{st}, where $s = \sigma + j\omega$ is a complex frequency. Thus the Fourier transform is a special case where $s = j\omega$.

From definition, the Fourier transform pair for $f(t)$ is

$$F(\omega) = \int_{-\infty}^{\infty} f(t)e^{-J\omega t}\,dt,$$

$$f(t) = \frac{1}{2\pi}\int_{-\infty}^{\infty} F(\omega)e^{j\omega t}\,d\omega. \tag{7.31}$$

Now the Fourier transform of $[f(t)e^{-\sigma t}]$ is

$$F[f(t)e^{-\sigma t}] = \int_{-\infty}^{\infty} f(t)e^{-\sigma t}e^{-j\omega t}\,dt$$

$$= \int_{-\infty}^{\infty} f(t)e^{-(\sigma+j\omega)t}\,dt. \tag{7.32}$$

Here, the Fourier transform is a function of $\sigma + j\omega$. Then

$$F(\sigma + j\omega) = \int_{-\infty}^{\infty} f(t)e^{-(\sigma+j\omega)t}\,dt. \tag{7.33}$$

And the inverse Fourier transform $f(t)$ can be expressed as

$$f(t) = \frac{1}{2\pi}\int_{-\infty}^{\infty} F(\sigma + j\omega)e^{(\sigma+j\omega)t}\,d\omega. \tag{7.34}$$

Now, let $s = \sigma + j\omega$ be the complex frequency. Then $ds = jd\omega$. Substituting this in Eq. (7.33) and Eq. (7.34),

$$F(s) = \int_{-\infty}^{\infty} f(t)e^{-st}\,dt; \qquad f(t) = \frac{1}{2\pi j}\int_{\sigma-j\infty}^{\sigma+j\infty} F(s)e^{st}\,ds. \tag{7.35}$$

Equation (7.35) is called the complex Fourier transform pair or the bilateral Laplace transform pair or the two-sided Laplace transform pair.

Note: When a given function $f(t)$ is multiplied by $e^{-\sigma t}$, where σ is any real number, the convergence of the resulting Fourier integral is greatly enhanced by the convergence factor $e^{-\sigma t}$. That is, the Laplace transform exists for many functions for which there is no Fourier transform. This is one of principal advantages of the Laplace transform, that is, its ability to transform functions that are not absolutely integrable.

Thus the Laplace transform is a frequency domain approach for continuous time signals, irrespective of whether the system is stable or unstable.

7.7 Inverse Laplace Transform

Inverse Laplace transform is used to obtain the time function from the Laplace transform. That is, it converts the function from s-domain to time domain. The inverse transformation can be accomplished in several ways.

By Inverse Formula using Residue Methods

The inversion formula can be expressed as a contour integral by the Residue theorem. We know that

$$f(t) = \frac{1}{2\pi j}\oint_c F(s)e^{st}\,dt = \sum_{i=1}^{n} \text{Res}[s_i], \tag{7.36}$$

where $\text{Res}[s_i]$, $i = 1,2,3..$ are the residues of $F(s)e^{st}$ and c is a closed curve.

Consider the inverse of the Laplace transform

$$f(t) = \frac{1}{2\pi j}\int_{\sigma-j\infty}^{\sigma+j\infty} F(s)e^{st}\,dt, \tag{7.37}$$

where $f(t)$ is given by a line integral along a vertical line $\text{Re}\{s\} = \sigma$ in the region of the existence of $F(s)$. In this integral, the real σ is to be selected such that if the ROC of $F(s)$ is $\text{Re}\{s\} > \sigma_1$, then $\infty > \sigma > \sigma_1$.

7.8 Partial Fraction Expansion

An efficient method to evaluate the inverse Laplace transform is by partial fraction expansion. Partial fraction expansion depends on the nature of $F(s)$.

(a) *F(s)* Is A Rational Function

If $F(s)$ is a rational function, the $F(s)$ can be represented as

$$F(s) = \frac{N(s)}{D(s)} = \frac{a_0 s^m + a_1 s^{m-1} + \ldots + a_m}{b_0 s^n + b_1 s^{n-1} + \ldots + b_n} = \frac{K(s-z_1)(s-z_2)\cdots(s-z_m)}{(s-p_1)(s-p_2)\cdots(s-p_n)} \tag{7.38}$$

where a and b are real coefficients, m and n are positive integers, $m < n$, K is a real constant, $z_i =$ roots of the numerator polynomial and $p_i =$ roots of the denominator polynomial. The inversion formula can be evaluated by partial fraction expansion and with the help of the table of Laplace transform pairs.

To find a partial fraction expansion, we first find the roots of the denominator polynomial p_i. These roots are the poles or singularities.

Case 1: Single roots

Assume p_i represents all poles of $F(s)$, where pi, $i = 1,2,3\ldots$ are real roots and distinct. We can write $F(s)$ as

$$F(s) = \frac{N(s)}{(s-p_1)(s-p_2)\cdots(s-p_n)}$$

$$= \frac{k_1}{s-p_1} + \frac{k_2}{s-p_2} + \cdots + \frac{k_n}{s-p_n}.$$

The coefficients k_i can be obtained as

$$k_i = (s-p_i)F(s)\Big|_{s=p_i}. \tag{7.39}$$

Case 2: Repeated roots

If the denominator polynomial $D(s)$ has multiple roots, that is, if it contains factors of the form $(s-p_1)^k$, then the expansion of $F(s)$ will consist of terms of the form

$$F(s) = \frac{k_{11}}{s-p_1} + \frac{k_{12}}{(s-p_1)^2} + \cdots + \frac{k_{1r}}{(s-p_1)^r} + \cdots$$

$$\text{Then } k_{1k} = \frac{1 d^{r-k}}{(r-k)! ds^{r-k}}[(s-p_1)F(s)]_{s=p_1}, \quad k = 1, 2, \cdots, r. \tag{7.40}$$

The inverse of transform of $\dfrac{k_{1r}}{(s-p_1)^r}$ can be obtained from the properties of Laplace transform.

Case 3: The complex conjugate roots

If there is a complex conjugate pair of poles in the denominator polynomial, then $F(s)$ can represented as

$$F(s) = \frac{N(s)}{(s+\alpha-j\beta)(s+\alpha+j\beta)D_1(s)}.$$

The corresponding partial fraction expansion can take the following forms

$$F(s) = \frac{k_1}{s+\alpha-j\beta} + \frac{k_2}{s+\alpha+j\beta} + \ldots$$

where

$$k_1 = (s+\alpha-j\beta)F(s)\Big|_{s=-\alpha+j\beta},$$

$$k_2 = (s+\alpha+j\beta)F(s)\Big|_{s=-\alpha-j\beta}.$$

We see the k_1 and k_2 are complex conjugate pair.

Let $k_1 = A + jB$ and $k_2 = A - jB = k_1^*$. Then the inverse transform of $F(s)$ is

$$f(t) = \mathcal{L}^{-1}\left[\frac{k_1}{s+\alpha-j\beta} + \frac{k_1^*}{s+\alpha+j\beta}\right]$$

$$= k_1 e^{-\alpha t} e^{j\beta t} + k_1^* e^{-\alpha t} e^{-j\beta t}$$

$$= \left[(A+jB)e^{j\beta t} + (A-jB)e^{-j\beta t}\right] e^{-\alpha t}.$$

$$f(t) = 2e^{-\alpha t}[A\cos\beta t - B\sin\beta t]. \tag{7.41}$$

(b) $F(s)$ Is An Improper Rational Function

When $F(s)$ is an improper rational function, that is, when $m \geq n$, we can write $F(s)$ by long division in the form

$$F(s) = \frac{N(s)}{D(s)} = Q(s) + \frac{R(s)}{D(s)}, \tag{7.42}$$

where $N(s)$ and $D(s)$ are the numerator and denominator polynomials respectively of X(s); the quotient $Q(s)$ is a polynomial of the order $m-n$, and the remainder $R(s)$ has a degree strictly less than n.

The inverse Laplace transform of $\frac{R(s)}{D(s)}$ can be evaluated by partial fraction expansion and using tables of Laplace transform pair since it is a proper rational function. The inverse Laplace transform of $Q(s)$ can be computed by using the transform pair

$$\frac{d^k\delta(t)}{dt^k} \leftrightarrow s^k,\ k = 1,2,3\dots. \tag{7.43}$$

Example 7.13

Find the inverse transform or the function

$$F(s) = \frac{5(s+1)}{s(s+2)(s+4)}.$$

Solution

Since $F(s)$ is a proper rational function and has simple poles, using partial fraction expansion,

$$F(s) = \frac{A}{s} + \frac{B}{s+2} + \frac{C}{s+4}.$$

$$A = \left.\frac{5(s+1)}{(s+2)(s+4)}\right|_{s=0} = \frac{5}{8}$$

$$B = \left.\frac{5(s+1)}{s(s+4)}\right|_{s=-2} = \frac{5}{4}$$

$$C = \left.\frac{5(s+1)}{s(s+2)}\right|_{s=-4} = \frac{-15}{8}.$$

Thus $F(s) = \frac{5}{8s} + \frac{5}{4(s+2)} - \frac{15}{8(s+4)}$.

Then $f(t) = \mathcal{L}^{-1}[F(s)] = \frac{5}{8}u(t) + \frac{5}{4}e^{-2t}u(t) - \frac{15}{8}e^{-4t}u(t)$

$$= \frac{5}{8}(1 + 2e^{-2t} - 3e^{-4t})u(t).$$

Example 7.14

Find the inverse transform of the function $F(s) = \frac{s^2+3}{(s+2)(s^2+2s+5)}$.

Solution

Given $F(s) = \frac{s^2+3}{(s+2)(s^2+2s+5)}$. The problem can be solved in two ways:

Method 1: The denominator can be factored in to complex poles as

$$F(s)=\frac{s^2+3}{(s+2)(s+1-2j)(s+1+2j)}$$

$$=\frac{k_1}{s+1-2j}+\frac{k_2}{s+1+2j}+\frac{k_3}{s+2}.$$

$$k_1=\left.\frac{s^2+3}{(s+2)(s+1+2j)}\right|_{s=-1+2j}=-\frac{1}{5}+j\frac{2}{5},$$

$$k_2=k_1^*=-\frac{1}{5}-j\frac{2}{5},\quad k_3=\left.\frac{s^2+3}{s^2+2s+5}\right|_{s=-2}=\frac{7}{5}.$$

Thus, $f(t)=\mathcal{L}^{-1}\left(\frac{1}{5}\left(\frac{-1+2j}{s+1-2j}+\frac{-1-2j}{s+1+2j}\right)+\frac{7}{5(s+2)}\right)$

$$f(t)=\frac{1}{5}\left((-1+2j)e^{-(-1+2j)t}+(-1-2j)e^{(1+2j)t}\right)+\frac{7}{5}e^{-2t}$$

$$f(t)=\left[\frac{7}{5}e^{-2t}-2e^{-t}\left(\frac{1}{5}\cos 2t+\frac{4}{5}\sin 2t\right)\right]u(t).$$

Method 2:

$$F(s)=\frac{s^2+3}{(s+2)(s^2+2s+5)}.$$

$$\frac{s^2+3}{(s+2)(s^2+2s+5)}=\frac{A}{(s+2)}+\frac{Bs+C}{(s^2+2s+5)}$$

$$s^2+3=A(s^2+2s+5)+(Bs+C)(s+2).$$

Substituting $s=-2$, then $A=7/5$, and comparing the s^2 terms,

$$A+B=1.$$

Then $B=1-7/5=-2/5$.
Comparing the constant terms, $5A+2C=3$; $2C=3-5(7/5)$; $C=-2/5$.

Then, $\frac{s^2+3}{(s+2)(s^2+2s+5)}=\frac{1}{5}\left(\frac{7}{(s+2)}+\frac{-2s-10}{(s^2+2s+5)}\right).$

Converting into basic Laplace transforms,

$$F(s)=\frac{1}{5}\left(\frac{7}{(s+2)}+\frac{-2(s+1)-8}{(s+1)^2+4}\right)$$

$$=\frac{1}{5}\left(\frac{7}{(s+2)}+\frac{-2(s+1)}{(s+1)^2+4}+\frac{-8}{(s+1)^2+4}\right).$$

Taking the inverse Laplace transform,

$$f(t)=\left[\frac{7}{5}e^{-2t}-2e^{-t}\left(\frac{1}{5}\cos 2t+\frac{4}{5}\sin 2t\right)\right]u(t).$$

Example 7.15

Find the inverse transform of the function $F(s)=\dfrac{(s+1)(s+3)}{(s+2)(s+4)}$.

Solution

Given $F(s)=\dfrac{(s+1)(s+3)}{(s+2)(s+4)}$. Since $F(s)$ has equal number of poles and zeros, dividing

$$F(s)=\frac{s^2+4s+3}{s^2+6s+8}=\frac{s^2+6s+8-2s-5}{s^2+6s+8}=1-\frac{2s+5}{(s+2)(s+4)}.$$

Using partial fraction expansion,

$$\frac{2s+5}{(s+2)(s+4)}=\frac{A}{(s+2)}+\frac{B}{(s+4)}.$$

$$A=\left.\frac{2s+5}{(s+4)}\right|_{s=-2}=\frac{1}{2};\qquad B=\left.\frac{2s+5}{(s+2)}\right|_{-4}=\frac{3}{2}.$$

Thus, $F(s)=1-\dfrac{1}{2(s+2)}-\dfrac{3}{2(s+4)}$.

Then $f(t)=\mathcal{L}^{-1}[F(s)]=\delta(t)-\dfrac{1}{2}e^{-2t}u(t)-\dfrac{3}{2}e^{-4t}u(t)$.

Example 7.16

Find the inverse transform of the function $F(s)=\dfrac{s+2}{s(s+3)(s+1)^2}$.

Solution

The given $F(s)$ can be expanded as

$$F(s)=\frac{A}{s+1}+\frac{B}{(s+1)^2}+\frac{C}{s}+\frac{D}{s+3}.$$

$$A=\frac{d}{ds}\left[(s+1)^2F(s)\right]_{s=-1}=\frac{d}{ds}\left[\frac{s+2}{s^2+3s}\right]_{s=-1}$$

$$=\left.\frac{(s^2+3s)-(s+2)(2s+3)}{(s^2+3s)}\right|_{s=-1}=\frac{3}{2}$$

$$B=(s+1)^2\left.\frac{s+2}{(s^2+3s)(s+1)^2}\right|_{s=-1}=\left.\frac{s+2}{s^2+3s}\right|_{s=-1}=-\frac{1}{2}.$$

$$C = \frac{s+2}{(s+3)(s+1)^2}\bigg|_{s=0} = \frac{2}{3}; \quad D = \frac{s+2}{s(s+1)^2}\bigg|_{s=-3} = \frac{1}{12}.$$

$$\therefore \quad F(s) = \frac{3}{2(s+1)} - \frac{1}{2(s+1)^2} + \frac{2}{3s} + \frac{1}{12(s+3)}.$$

From the properties of Laplace transform,

$$f(t) = \left[\frac{-3}{4}e^{-t} - \frac{1}{2}te^{-t} + \frac{2}{3} + \frac{1}{12}e^{-3t}\right]u(t).$$

Example 7.17

Find the inverse Laplace transform of $F(s) = \dfrac{-3}{(s+2)(s-1)}$ if the ROC is (a) $-2 < \text{Re}\{s\} < 1$, (b) $\text{Re}\{s\} > 1$, and (c) $\text{Re}\{s\} < -2$.

Solution

Given $F(s) = \dfrac{-3}{(s+2)(s-1)}$. Using partial fraction expansion, $F(s) = \dfrac{1}{s+2} - \dfrac{1}{s-1}$.

(a) If ROC is $-2 < \text{Re}\{s\} < 1$, $\text{Re}\{s\} > -2$ corresponds to the right side or to the causal signal. $\text{Re}\{s\} < 1$ corresponds to left side or to the anti-causal signal. Therefore,

$$f(t) = e^{-2t}u(t) + e^{t}u(-t).$$

(b) Both poles lie to the left of the ROC. So both poles correspond to the right side or to the causal signal. Therefore,

$$f(t) = e^{-2t}u(t) - e^{t}u(t).$$

(c) Both poles lie to the right of ROC. So both poles correspond to the left side or to the anti-causal signal

$$f(t) = (-e^{-2t} + e^{t})u(-t).$$

7.9 Laplace Transform of Certain Signals Using Waveform Synthesis

Laplace transforms can be used to represent any type of signals. Laplace transforms for causal periodic signals can be derived easily from the basic causal waveforms consisting of exponential, ramp, step or delta functions. The basic waveforms are given below.

Pulse Waveform

Consider a causal pulse waveform of width T and amplitude A. The pulse waveform can be synthesized into two step functions as shown in Fig. 7.6. Thus the signal can be represented as

$$x(t) = A(u(t) - u(t-T)).$$

Using the Laplace transform properties,

$$X(s) = L[x(t)] = A\left(\frac{1}{s} - \frac{e^{-sT}}{s}\right) = \frac{A}{s}(1 - e^{-sT}).$$

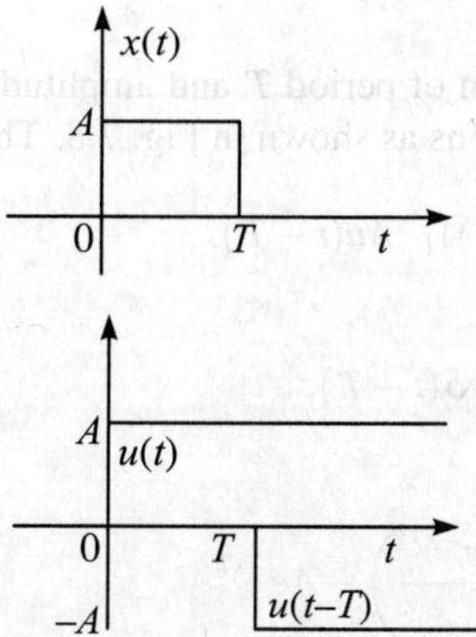

Fig. 7.6 *Synthesis of pulse waveform*

Truncated Ramp Waveform

Consider a causal truncated ramp waveform of period T and amplitude A. The truncated ramp waveform can be synthesised into ramp functions as shown in Fig. 7.7. Thus the signal can be represented as

$$x(t) = \frac{A}{T}(tu(t) - (t-T)u(t-T))$$

$$\frac{dx(t)}{dt} = \frac{A}{T}(u(t) - u(t-T)).$$

Using Laplace transform properties,

$$\mathcal{L}\left[\frac{dx(t)}{dt}\right] = sX(s) = \frac{A}{T}\left(\frac{1}{s} - \frac{e^{-sT}}{s}\right)$$

$$X(s) = \frac{A}{Ts^2}(1 - e^{-sT}).$$

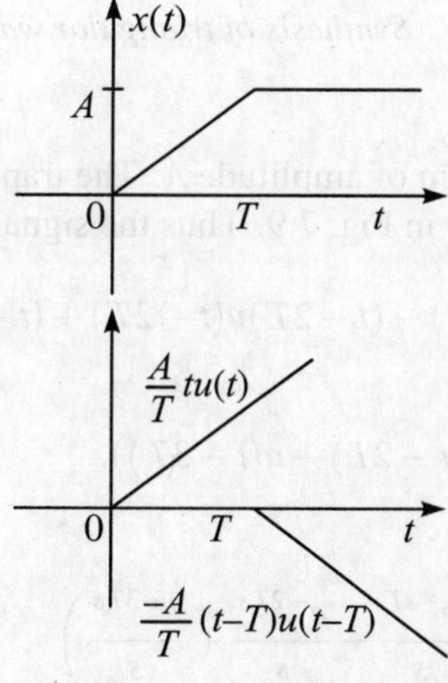

Fig. 7.7 *Synthesis of truncated ramp waveform*

Triangular Waveform

Consider a causal triangular waveform of period T and amplitude A. The triangular waveform can be synthesised into ramp and step functions as shown in Fig. 7.8. Thus the signal can be represented as

$$x(t) = \frac{A}{T}(tu(t) - (t-T)u(t-T)) - Au(t-T).$$

$$\frac{dx(t)}{dt} = \frac{A}{T}(u(t) - u(t-T)) - A\delta(t-T).$$

Using Laplace transform properties,

$$sX(s) = \mathcal{L}\left[\frac{dx(t)}{dt}\right] = \frac{A}{T}\left(\frac{1}{s} - \frac{e^{-sT}}{s}\right) - Ae^{-sT}$$

$$X(s) = \frac{A}{Ts^2}(1 - e^{-sT} - Tse^{-sT}).$$

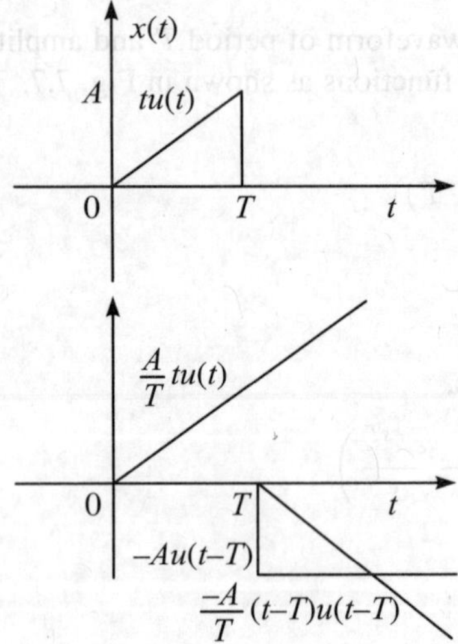

Fig. 7.8 *Synthesis of triangular waveform*

Trapezoidal Waveform

Consider a causal trapezoidal waveform of amplitude A. The trapezoidal waveform can be synthesised into ramp and step functions as shown in Fig. 7.9. Thus the signal can be represented as

$$x(t) = \frac{A}{T}(tu(t) - (t-T)u(t-T) - (t-2T)u(t-2T) + (t-3T)u(t-3T))$$

$$\frac{dx(t)}{dt} = \frac{A}{T}(u(t) - u(t-T) - u(t-2T) + u(t-3T)).$$

Using Laplace transform properties,

$$sX(s) = \mathcal{L}\left[\frac{dx(t)}{dt}\right] = \frac{A}{T}\left(\frac{1}{s} - \frac{e^{-sT}}{s} - \frac{e^{-2Ts}}{s} + \frac{e^{-3Ts}}{s}\right)$$

$$X(s) = \frac{A}{Ts^2}(1 - e^{sT} - e^{-2Ts} + e^{-3Ts}).$$

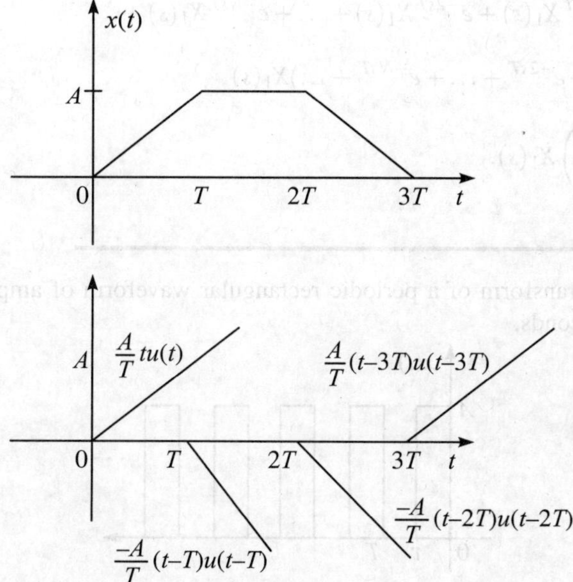

Fig. 7.9 *Synthesis of trapezoidal waveform*

Periodic Waveforms

A periodic waveform can be synthesised using the basic waveforms as follows. Consider a causal periodic waveform $x(t)$ of period T as shown in Fig. 7.10. Let $X(s)$ be the Laplace transform of $x(t)$. The periodic waveform can be expressed as the sum of the basic function delayed every T seconds. Consider a basic waveform $x_1(t)$ for the first period.

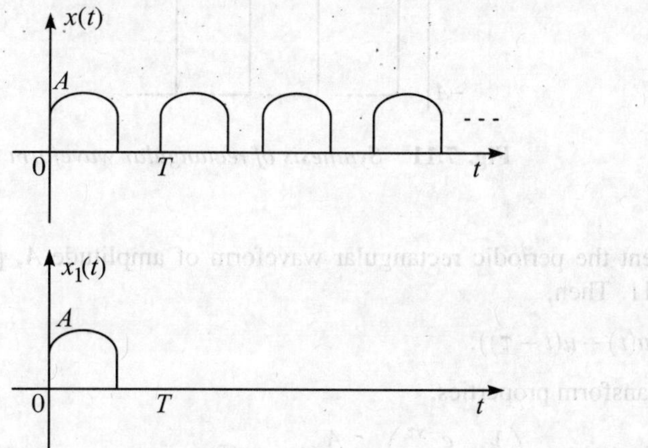

Fig. 7.10 *Periodic waveform*

Let $X_1(s)$ be the Laplace transform of $x_1(t)$. Then,

$$x(t) = x_1(t) + x_1(t-T) + x_1(t-2T) + \ldots + x_1(t-NT) + \ldots$$

Taking Laplace transform of both sides and applying time delay property,

$$x(t) = x_1(t) + x_1(t-T) + x_1(t-2T) + \ldots + x_1(t-NT) + \ldots$$

$$X(s) = X_1(s) + e^{-sT}X_1(s) + e^{-2sT}X_1(s) + \ldots + e^{-NsT}X_1(s) + \ldots$$

$$X(s) = (1 + e^{-sT} + e^{-2sT} + \ldots + e^{-NsT} + \ldots)X_1(s)$$

$$X(s) = \left(\frac{1}{1 - e^{-sT}}\right)X_1(s).$$

Example 7.18

Determine the Laplace transform of a periodic rectangular waveform of amplitude A, pulse width τ seconds and period T seconds.

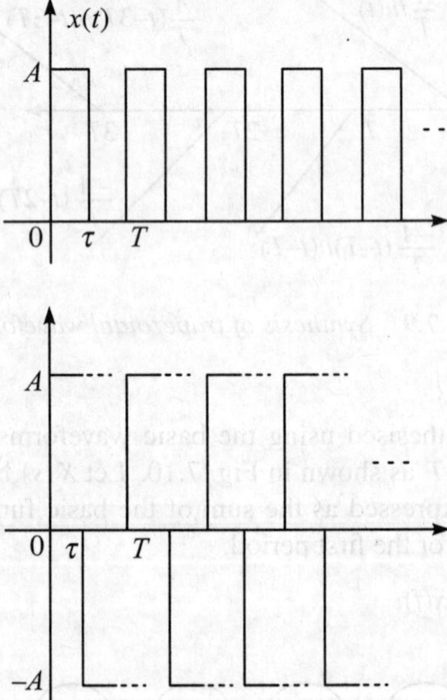

Fig. 7.11 *Synthesis of rectangular waveform*

Solution

Let $x_1(t)$ represent the periodic rectangular waveform of amplitude A, pulse width τ and period T shown in Fig. 7.11. Then,

$$x_1(t) = A(u(t) - u(t - \tau)).$$

Using Laplace transform properties,

$$X_1(s) = \mathcal{L}[x_1(t)] = A\left(\frac{1}{s} - \frac{e^{-s\tau}}{s}\right) = \frac{A}{s}(1 - e^{-s\tau}).$$

Then the Laplace transform of a periodic rectangular waveform is

$$X(s) = \left(\frac{1}{1 - e^{-sT}}\right)X_1(s)$$

$$X(s) = \frac{A}{s}\left(\frac{1}{1 - e^{-sT}}\right)(1 - e^{-s\tau}) = \frac{A}{s}\left(\frac{1 - e^{-s\tau}}{1 - e^{-sT}}\right).$$

Example 7.19

Determine the Laplace transform of a periodic sine waveform of amplitude A, with period T seconds for one cycle.

Solution

Consider one cycle of a sine waveform $x(t)$ of amplitude A, period T as shown in Fig. 7.12. Let $x_1(t)$ be the periodic sine waveform.

$$x(t) = x(t) - x(t-T)$$

$$x(t) = A\left[\sin(\omega_0 t) - \sin(\omega_0(t-T))\right].$$

Using Laplace transform properties, the Laplace transform is

$$X(s) = \mathcal{L}[x(t)] = A\left(\frac{\omega_0}{s^2+\omega_0^2}\right) - A\left(\frac{\omega_0 e^{-sT}}{s^2+\omega_0^2}\right)$$

$$= \frac{A\omega_0(1-e^{-sT})}{s^2+\omega_0^2}.$$

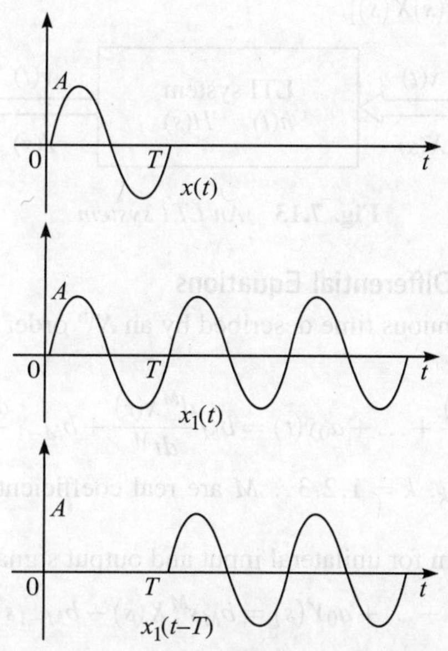

Fig. 7.12 *Synthesis of sinusoidal waveform*

7.10 Applications of Laplace Transform on Linear Systems

The Transfer Function of LTI System

The transfer function of a system completely describes the behaviour of the system with any type of input. Consider an LTI continuous time system with impulse response $h(t)$ as shown in Fig. 7.13. Let $x(t)$, $y(t)$ and $h(t)$ have Laplace transforms $X(s)$, $Y(s)$ and $H(s)$ respectively. The transfer function of a system is defined as the ratio of the Laplace transform of the output signal to the Laplace transform of the input signal when all initial conditions are zero.

$$H(s) = \frac{Y(s)}{X(s)}.$$

We know that if $x(t)$ is the input signal, then the output signal can be obtained from convolution given by $y(t) = x(t) * h(t)$.

The Laplace transform of the output signal is

$$\mathcal{L}[y(t)] = Y(s) = X(s)H(s)$$

$$\text{or,}\quad H(s) = \frac{Y(s)}{X(s)}.$$

The inverse Laplace transform of the transfer function gives the impulse response of the system. That is,

$$h(t) = \mathcal{L}^{-1}[H(s)] = \mathcal{L}^{-1}\left[\frac{Y(s)}{X(s)}\right].$$

Also, the inverse Laplace transform of the output function gives the output signal.

$$y(t) = \mathcal{L}^{-1}[Y(s)] = \mathcal{L}^{-1}[H(s)X(s)].$$

x(t) → LTI system h(t) H(s) → y(t)
X(s) Y(s)

Fig. 7.13 *An LTI system*

Systems Described by Linear Differential Equations

Consider an LTI system in continuous time described by an Nth order linear differential equation with constant coefficients.

$$a_N\frac{d^N y(t)}{dt^N} + a_{N-1}\frac{d^{N-1}y(t)}{dt^{N-1}} + \ldots + a_0 y(t) = b_M\frac{d^M x(t)}{dt^M} + b_{M-1}\frac{d^{M-1}x(t)}{dt^{M-1}} + \ldots + b_0 x(t).$$

where $a_k,\ k = 1,2,3\ldots N$ and $b_k,\ k = 1,2,3\ldots M$ are real coefficients, N = order of the system and $N > M$.

Performing Laplace transform for unilateral input and output signals with zero initial conditions,

$$a_N s^N Y(s) + a_{N-1}s^{N-1}Y(s) + \ldots + a_0 Y(s) = b_M s^M X(s) + b_{M-1}s^{M-1}X(s) + \ldots + b_0 X(s)$$

$$(a_N s^N + a_{N-1}s^{N-1} + \ldots + a_0)Y(s) = (b_M s^M + b_{M-1}s^{M-1} + \ldots + b_0)X(s)$$

$$\text{or,}\quad Y(s) = \frac{(b_M s^M + b_{M-1}s^{M-1} + \ldots + b_0)}{(a_N s^N + a_{N-1}s^{N-1} + \ldots + a_0)}X(s)$$

$$\text{or,}\quad H(s) = \frac{Y(s)}{X(s)} = \frac{(b_M s^M + b_{M-1}s^{M-1} + \ldots + b_0)}{(a_N s^N + a_{N-1}s^{N-1} + \ldots + a_0)} = \frac{\Sigma_{k=1}^{M} b_k s^k}{\Sigma_{k=1}^{N} a_k s^k}.$$

Note:

1. To find out system transfer function, all initial conditions of the input signal should be zero. The transfer function is independent of the input signal.

2. The transfer function is a rational function in s.
3. Frequency response can be obtained by setting $H(\omega) = H(s)\big|_{s=j\omega}$.

Steady-state Frequency Response of LTI System

The steady-state frequency response of a linear discrete time system can be obtained from the Laplace transform by substituting $s = j\omega$ in the system transfer function.

Therefore, the steady-state frequency response is

$$H(\omega) = \left.\frac{Y(s)}{X(s)}\right|_{s=j\omega} = \frac{Y(j\omega)}{X(j\omega)}.$$

$$H(\omega) = |H(\omega)|\angle\phi(\omega).$$

$H(\omega)$ is a continuous complex function with frequency ω, where the magnitude response and is phase response are respectively:

$$|H(\omega)| = \frac{|Y(j\omega)|}{|X(j\omega)|}, \quad \text{and} \quad \angle\phi(\omega) = \tan^{-1}\left(\frac{Y(j\omega)}{X(j\omega)}\right).$$

Causality

It is observed that $h(t) = 0$ for $t < 0$ a causal system and thus is right-sided. Therefore, the ROC associated with the transfer function of a causal system is a right half-plane. The converse, however, is not true. For example, a right-sided signal starting at the negative axis also leads to an ROC that is a right half-plane. However, if we know that the transfer function is rational, then it suffices to check that the ROC is the right half-plane to the right of the rightmost pole in the s-plane to conclude that the system is causal.

Stability

So far we have seen that bounded-input bounded-output (BIBO) stability of a continuous-time LTI system is equivalent to its impulse response, being absolutely integrable, in which case its Fourier transform converges. Also, the stability of an LTI differential system is equivalent to having all the zeros of its characteristic polynomial having a negative real part. For the Laplace transform, the first stability condition translates to the following:

- An LTI system is stable if and only if the ROC of its transfer function contains the $j\omega$ axis.
- A causal system with a proper rational transfer function $H(s)$ is stable if and only if all of its poles are in the left-half of the s-plane. That is, all the poles have negative real parts.

Advantages of Laplace Transform

The advantages of using Laplace transform are as follows:

1. The higher order differential equations can be easily solved by using simple algebraic equations.
2. It transforms higher order differential equations with initial conditions in the time domain into simple algebraic equations in the s-domain, since the initial conditions are automatically included in the solution.
3. The total solution for differential equations can be obtained by using inverse Laplace transform.
4. It is a powerful tool for analysing system properties in the form of transform function.
5. It can be used to analyse many classes of signals and systems which are not absolutely integrable.

6. It provides solutions for many unstable systems such as impulse functions.
7. Fourier transform can be obtained from Laplace transform by substituting $\sigma = 0$, i.e., $s = j\omega$.

Limitations

1. Laplace transform does not converge for some type of signals whose amplitude grows faster than time.
2. The region of convergence is needed to obtain inverse Laplace transform.
3. It is very difficult to solve complex integrals directly in the process of inverse Laplace transform.

Example 7.20

Find the transfer function of an (a) ideal delay T, (b) ideal differentiator and (c) ideal integrator.

Solution

(a) *Ideal delay*: We know that $y(t) = x(t-T)$. Taking Laplace transform,

$$Y(s) = e^{-sT}X(s).$$

Then, $H(s) = e^{-sT}$.

(b) *Ideal differentiator*: We know that $y(t) = dx(t)/dt$. Taking Laplace transform,

$$Y(s) = sX(s).$$

Then $H(s) = s$.

(c) *Ideal integrator*: We know that $y(t) = \int_0^t x(\tau)d\tau$. Taking Laplace transform,

$$Y(s) = X(s)/s.$$

Then, $H(s) = 1/s$.

Example 7.21

Find the transfer function of a system described by $2\frac{d^2y(t)}{dt^2} + 3\frac{dy(t)}{dt} + y(t) = 2\frac{dx(t)}{dt} - 3x(t)$.

Solution

Given $2\frac{d^2y(t)}{dt^2} + 3\frac{dy(t)}{dt} + y(t) = 2\frac{dx(t)}{dt} - 3x(t)$, assume that all initial conditions are zero.

Apply Laplace transform on both sides:

$$2s^2Y(s) + 3Y(s) + Y(s) = 2sX(s) - 3X(s),$$

$$(2s^2 + 3s + 1)Y(s) = (2s-3)X(s).$$

Therefore, the transfer function is

$$H(s) = \frac{Y(s)}{X(s)} = \frac{2s-3}{2s^2+3s+1}.$$

Example 7.22

Solve the following differential equation $\frac{d^2y(t)}{dx^2} + 3\frac{dy(t)}{dx} + 2 = u(t)$ using Laplace transform with the initial conditions $y(0) = 0$, $y'(0) = 1$.

Solution

Given $\frac{d^2y}{dx^2}+3\frac{dy}{dx}+2=u(t)$ with the initial conditions $y(0)=0,\ y'(0)=1$.

Applying Laplace transforms on both sides,

$$s^2Y(s)-sy(0^-)-y'(0^-)+3sY(s)-3y(0^-)+2=\frac{1}{s}$$

$$s^2Y(s)-1+3sY(s)+2=\frac{1}{s}$$

$$(s^2+3s)Y(s)=\frac{1}{s}-1.$$

Therefore the output function is

$$Y(s)=\frac{-s+1}{s^2(s+3)}.$$

Using partial fractions,

$$Y(s)=\frac{-s+1}{s^2(s+3)}=\frac{A}{s}+\frac{B}{s^2}+\frac{C}{s+3}$$

$$-s+1=As(s+3)+B(s+3)+Cs^2.$$

At $s=0$, $3B=1$ or $B=\frac{1}{3}$.

At $s=-3$, $9C=4$ or $C=\frac{4}{9}$.

Comparing the s^2 terms,

$$A+C=0,\quad \text{or}\quad A=\frac{-4}{9}.$$

Substituting the values,

$$Y(s)=\frac{-s+1}{s^2(s+3)}=\frac{1}{9}\left(\frac{-4}{s}+\frac{3}{s^2}+\frac{4}{s+3}\right).$$

Taking inverse Laplace transform,

$$y(t)=\frac{-4}{9}u(t)+\frac{1}{3}tu(t)+\frac{4}{9}e^{-3t}u(t).$$

Additional Problems

Example 7.23

Find the Laplace transform of $x(t)=e^{-2t}u(t)+e^{-t}(\cos 3t)u(t)$.

Solution

Given $x(t)=e^{-2t}u(t)+e^{-t}(\cos 3t)u(t)$, $x(t)$ can be expressed as

$$x(t)=[e^{-2t}+\frac{1}{2}e^{-(1-3j)t}+\frac{1}{2}e^{-(1+3j)t}]u(t).$$

The Laplace transforms of the basic signals are

$$e^{-2t}u(t) \leftrightarrow \frac{1}{s+2}, \quad \text{Re}\{s\} > -2$$

$$e^{-(1-3j)t}u(t) \leftrightarrow \frac{1}{s+(1-3j)}, \quad \text{Re}\{s\} > -1$$

$$e^{-(1+3j)t}u(t) \leftrightarrow \frac{1}{s+(1+3j)}, \quad \text{Re}\{s\} > -1.$$

For all the three Laplace transforms to converge simultaneously, we must have $\text{Re}\{s\} > -1$ as ROC. Consequently, the Laplace transform of $x(t)$ is

$$X(s) = \frac{1}{s+2} + \frac{1/2}{s+(1-3j)} + \frac{1/2}{s+(1+3j)}, \quad \text{Re}\{s\} > -1$$

$$X(s) = \frac{2s^2+5s+12}{(s^2+2s+10)(s+2)}, \quad \text{Re}\{s\} > 1.$$

Example 7.24

Find the Laplace transform of the signal $x(t) = \delta(t) - \frac{4}{3}e^{-t}u(t) + \frac{1}{3}e^{2t}u(t)$.

Solution

The Laplace transform of a signal $x(t)$ can be expressed as

$$X(s) = \mathcal{L}[x(t)] = \int_{-\infty}^{\infty} x(t)e^{-st}dt.$$

The Laplace transform of the unit impulse function is

$$\mathcal{L}[\delta(t)] = \int_{-\infty}^{\infty} \delta(t)e^{-st}dt = 1 \text{ for all } s.$$

The other Laplace transforms are

$$e^{-t}u(t) \leftrightarrow \frac{1}{s+1}, \ \text{Re}\{s\} > -1$$

$$e^{2t}u(t) \leftrightarrow \frac{1}{s-2}, \ \text{Re}\{s\} > 2$$

$$\therefore \quad X(s) = 1 - \frac{4/3}{s+1} + \frac{1/3}{s-2}$$

$$X(s) = \frac{(s-1)^2}{(s+1)(s-2)}, \quad \text{ROC: } \text{Re}\{s\} > 2.$$

Example 7.25

Find the Laplace transform of the signal

$$x(t) = \begin{cases} e^{at} & 0 < t < T \\ 0 & \text{otherwise} \end{cases}.$$

Solution
Given

$$X(s) = \int_0^T e^{-at} e^{-st} dt = \frac{1}{(s+a)} e^{-(a+s)t} \Big|_0^T$$

$$X(s) = \frac{1}{(s+a)} \left[1 - e^{-(s+a)T}\right], \quad \text{ROC}: \ \text{Re}\{s\} > -a.$$

Example 7.26

Find the Laplace transform of the signal $x(t) = e^{-|a|t}$.

Solution

Given $x(t) = e^{-|a|t}$. It can be divided into the sum of a right-sided and left-sided signals, that is,

$$x(t) = e^{-at} u(t) + e^{+at} u(-t).$$

We know that

$$e^{-at} u(t) \leftrightarrow \frac{1}{s+a}, \quad \text{Re}\{s\} > -a.$$

$$e^{at} u(-t) \leftrightarrow \frac{-1}{s-a}, \quad \text{Re}\{s\} > +a.$$

If $a \leq 0$, there is no common region of convergence and Laplace transform does not exist.
If $a > 0$, the Laplace transform of $x(t)$ is

$$e^{-|a|t} \leftrightarrow \frac{1}{s+a} - \frac{1}{s-a} = \frac{-2a}{s^2 - a^2}, \quad -a < \text{Re}\{s\} < +a.$$

Example 7.27

Find the Laplace transform for $x(t) = e^{-2t} u(t) + e^{-3t} u(t)$.

Solution

Given $x(t) = e^{-2t} u(t) + e^{-3t} u(t)$. From the properties of Laplace transform we know that

$$e^{-2t} u(t) \leftrightarrow \frac{1}{s+2}, \quad \text{Re}\{s\} > -2$$

$$e^{-3t} u(t) \leftrightarrow \frac{1}{s+3}, \quad \text{Re}\{s\} > -3$$

$$\therefore \quad X(s) = \mathcal{L}[e^{-2t} u(t) + e^{-3t} u(t)]$$

$$= \frac{1}{s+2} + \frac{1}{s+3} = \frac{2s+5}{s^2+5s+6}.$$

By equating the denominator polynomial to zero, the poles are at $s = -2$ and $s = -3$. By equating the numerator polynomial to zero, the zero is at $s = -5/2$.

Since the signals are left-sided signals, the region of convergence is ROC: Re$\{s\} > -2$. The pole–zero plot with ROC is shown in Fig. 7.14.

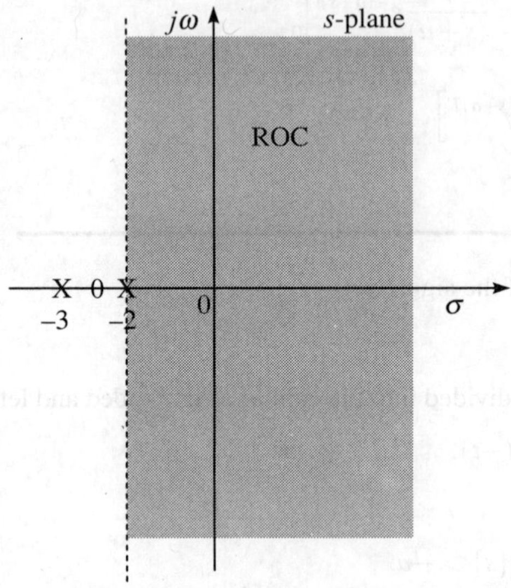

Fig. 7.14 *Pole–zero plot and ROC for* $x(t) = e^{-2t}u(t) + e^{-3t}u(t)$

Example 7.28

Find the Laplace transform of signal. $f(t) = e^{-5t}[u(t) - u(t-5)]$ and its ROC.

Solution

Given $f(t) = e^{-5t}[u(t) - u(t-5)]$.

$$f(t) = e^{-5t}u(t) - e^{-5t}u(t-5) = e^{-5t}u(t) - e^{-25}e^{-5(t-5)}u(t-5).$$

From the properties of Laplace transform, we know that

$$e^{-5t}u(t) \leftrightarrow \frac{1}{s+5}, \quad \text{Re}\{s\} > -5$$

$$e^{-5(t-5)}u(t-5) \leftrightarrow \frac{e^{-5s}}{s+5}, \quad \text{Re}\{s\} > -5$$

$$X(s) = \mathcal{L}[e^{-5t}u(t) - e^{-25}e^{-5(t-5)}u(t-5)]$$

$$= \frac{1}{s+5} - \frac{e^{-25}e^{-5s}}{s+5} = \frac{1 - e^{-5(s-5)}}{s+5}, \quad \text{ROC}: \ \text{Re}\{s\} > -5.$$

By equating the denominator polynomial to zero, the pole is at $s = -5$. Since the signals are left-sided signals, the region of convergence is ROC: Re $\{s\} > -5$.

Example 7.29

Find the Laplace transform of $x(t) = e^{-2t}(4\cos 5t - 3\sin 5t)u(t)$.

Solution

Given $x(t) = e^{-2t}(4\cos 5t - 3\sin 5t)u(t)$. From the properties of Laplace transform we know that

$$e^{-2t}\cos(5t)u(t) \leftrightarrow \frac{s+2}{(s+2)^2+25}$$

$$e^{-2t}\sin(5t)u(t) \leftrightarrow \frac{5}{(s+2)^2+25}.$$

$$\therefore \quad X(s) = \mathcal{L}[e^{-2t}(4\cos 5t - 3\sin 5t)u(t)] = \frac{4(s+2)}{(s+2)^2+25} - \frac{15}{(s+2)^2+25}$$

$$= \frac{4s-7}{(s+2)^2+25}.$$

Example 7.30

Find the Laplace transform of $x(t) = e^{-t}\cos(\omega_0 t + \varphi)u(t)$.

Solution

Given $x(t) = e^{-t}\cos(\omega_0 t + \varphi)u(t)$. Expanding,

$$x(t) = e^{-t}(\cos(\omega_0 t)\cos\varphi - \sin(\omega_0 t)\sin\varphi)u(t).$$

From the properties of Laplace transform we know that

$$e^{-t}\cos(\omega_0 t)u(t) \leftrightarrow \frac{s+1}{(s+1)^2+\omega_0^2}$$

$$e^{-t}\sin(\omega_0 t)u(t) \leftrightarrow \frac{\omega_0}{(s+1)^2+\omega_0^2}.$$

$$\therefore \quad X(s) = \mathcal{L}\left[e^{-t}\cos(\omega_0 t + \varphi)u(t)\right] = \frac{\cos(\varphi)(s+1)}{(s+1)^2+\omega_0^2} - \frac{\sin(\varphi)\omega_0}{(s+1)^2+\omega_0^2}.$$

$$= \frac{\cos(\varphi)(s+1) - \sin(\varphi)\omega_0}{(s+1)^2+\omega_0^2}.$$

Example 7.31

Find the Laplace transform of $x(t) = \sinh(at)u(t)$ and $x(t) = e^{-2t}\cos(3t)u(t)$.

Solution

(a) Given $x(t) = \sinh(at)u(t)$.

We know that $\sinh a(t) = \dfrac{e^{at} - e^{-at}}{2}$.

From the basic properties, the Laplace transforms are

$$e^{-at}u(t) \leftrightarrow \frac{1}{s+a}, \quad \text{Re}\{s\} > -a,$$

$$e^{at}u(t) \leftrightarrow \frac{1}{s-a}, \quad \text{Re}\{s\} > +a.$$

$$\mathcal{L}[\sinh(at)u(t)] = \frac{1}{2}\mathcal{L}[e^{at}u(t)] - \frac{1}{2}\mathcal{L}[e^{-at}u(t)].$$

$$\therefore \quad X(s) = \frac{1}{2}\left[\frac{1}{s-a} - \frac{1}{s+a}\right] = \frac{a}{s^2-a^2}, \quad \text{Re}\{s\} > a$$

(b) Given $x(t) = e^{-2t}\cos(3t)u(t)$.

We know that

$$\cos(3t)u(t) = \frac{e^{3jt}+e^{-3jt}}{2}.$$

The Laplace transforms are

$$\mathcal{L}[\cos(3t)u(t)] = \frac{1}{2}L[e^{3jt}u(t)] + \frac{1}{2}\mathcal{L}[e^{-3jt}u(t)]$$

$$= \frac{1}{2}\left[\frac{1}{s-3j} + \frac{1}{s+3j}\right] = \frac{s}{s^2+9}.$$

Now using the properties of Laplace transforms,

$$\therefore \quad X(s) = \mathcal{L}[e^{-2t}\cos(3t)u(t)] = \left.\frac{s}{s^2+9}\right|_{s=s+2} = \frac{s+2}{(s+2)^2+9}.$$

Example 7.32

Find the Laplace transform of $x(t) = \dfrac{1-e^t}{t}u(t)$.

Solution

Given $x(t) = \dfrac{1-e^t}{t}u(t)$. The Laplace transform of $1-e^t$ is

$$(1-e^t)u(t) \leftrightarrow \frac{1}{s} - \frac{1}{s-1}.$$

From the integration property,

$$\mathcal{L}\left[\frac{1-e^t}{t}u(t)\right] = \int_s^\infty \left(\frac{1}{s} - \frac{1}{s-1}\right) ds.$$

$$\therefore \quad X(s) = [\log s - \log(s-1)]_s^\infty = \log\left(\frac{s-1}{s}\right).$$

Example 7.33

Find the Laplace transform of $f(t) = \dfrac{\sin(at)u(t)}{t}$.

Solution

Given $f(t) = \dfrac{\sin(at)u(t)}{t}$. We know that

$$\mathcal{L}[\sin(at)u(t)] = \frac{a}{s^2+a^2}.$$

Using integration property,

$$F(s) = \mathcal{L}\left[\frac{\sin(at)u(t)}{t}\right] = \int_s^{\infty}\left(\frac{a}{s^2+a^2}\right)ds$$

$$F(s) = a\left(\frac{1}{a}\tan^{-1}\left(\frac{s}{a}\right)\right)_s^{\infty}$$

$$= \frac{\pi}{2} - \tan^{-1}\left(\frac{s}{a}\right) = \cot^{-1}\left(\frac{s}{a}\right).$$

Example 7.34

Find the Laplace transform of $f(t) = \sin^2 t$.

Solution

We know that $f(t) = \sin^2 t = \dfrac{1-\cos 2t}{2}$.

Also, $\mathcal{L}[\sin^2 t] = \mathcal{L}\left[\dfrac{1-\cos 2t}{2}\right]$.

$$\text{Thus, } \mathcal{L}\left[\frac{(1-\cos 2t)u(t)}{2}\right] = \mathcal{L}\left[\frac{u(t)}{2} - \frac{\cos(2t)u(t)}{2}\right]$$

$$= \frac{1}{2s} - \frac{s}{2(s^2+4)} = \frac{2}{s(s^2+4)}.$$

Example 7.35

Find the Laplace transform of $f(t) = t\sin\omega t$ using differentiation property.

Solution

Given $f(t) = t\sin\omega t$. We know that the Laplace transform

$$\sin\omega t \leftrightarrow \frac{\omega}{s^2+\omega^2}.$$

Using differentiation property,

$$t\sin\omega t \leftrightarrow \frac{d}{ds}\left(\frac{\omega}{s^2+\omega^2}\right).$$

$$\therefore \quad \mathcal{L}[t\sin\omega t] = F(s) = \frac{2\omega s}{(s^2+\omega^2)^2}.$$

Example 7.36

If $\mathcal{L}[e^{2t}] = \dfrac{1}{s-2}$, find the Laplace transform of $f(t) = t^2e^{2t}$.

Solution

Given $\mathcal{L}[e^{2t}] = \dfrac{1}{s-2}$.

Using differentiation property,

$$te^{2t} \leftrightarrow \frac{-d}{ds}\left(\frac{1}{s-2}\right).$$

$$\mathcal{L}[te^{2t}] = -\frac{d}{ds}\left(\frac{1}{s-2}\right) = \frac{1}{(s-2)^2}$$

Also, $\mathcal{L}[t^2e^{2t}] = -\dfrac{d^2}{ds^2}\left(\dfrac{1}{s-2}\right) = \dfrac{2}{(s-2)^3}$.

Example 7.37

If $\mathcal{L}[\sin t] = \dfrac{1}{s^2+1}$, find the Laplace transform of $\dfrac{\sin t}{t}$.

Solution

Given $\mathcal{L}[\sin t] = \dfrac{1}{s^2+1}$. Using integration property,

$$\mathcal{L}\left\{\frac{\sin t}{t}\right\} = \int_s^\infty \frac{ds}{s^2+1} = -\tan^{-1}\left(\frac{1}{s}\right)\Big|_s^\infty = \tan^{-1}\left(\frac{1}{s}\right).$$

Example 7.38

(a) Determine the Laplace transform of $\dfrac{\sin^2 t}{t}$. (b) In addition, evaluate the integral $\displaystyle\int_0^\infty \frac{e^{-t}\sin^2 t}{t}dt$.

Solution

(a) The Laplace Transform of $\sin^2 t$ can be evaluated by

$$\mathcal{L}\left[\sin^2 t\right] = \mathcal{L}\left[\frac{1-\cos 2t}{2}\right] = \frac{1}{2}\left(\frac{1}{s} - \frac{s}{s^2+4}\right) = \frac{2}{s(s^2+4)}.$$

Now, using integration property,

$$\mathcal{L}\left[\frac{\sin^2 t}{t}\right] = \int_s^\infty \frac{2}{s(s^2+4)}ds$$

$$= \frac{1}{2}\int_s^\infty \left(\frac{1}{s} - \frac{s}{s^2+4}\right)ds$$

$$= \frac{1}{2}\left[\ln(s) - \frac{1}{2}\ln(s^2+4)\right]_s^\infty$$

$$= \left[\frac{1}{4}\ln(\frac{s^2}{s^2+4})\right]_s^\infty = \frac{1}{4}\ln\left(\frac{s^2+4}{s^2}\right). \quad \left(\text{since } \lim_{s\to\infty}\left(\ln\frac{s^2}{s^2+4}\right) = \ln(1) = 0\right).$$

(b) Now, the integral $\int_0^\infty e^{-t}\left(\frac{\sin^2 t}{t}\right) dt$ can be viewed as

$$\mathcal{L}\left[\frac{\sin^2 t}{t}\right] = \int_0^\infty e^{-st}\left(\frac{\sin^2 t}{t}\right) dt \text{ as } s = 1.$$

Thus, $$\int_0^\infty e^{-t}\left(\frac{\sin^2 t}{t}\right) dt = \frac{1}{4}\left(\ln\left(\frac{s^2+4}{s^2}\right)\right)_{s=1} = \frac{1}{4}\ln 5.$$

Example 7.39

If $\mathcal{L}[\cos 2t] = \frac{s}{s^2+4}$, find the Laplace transform of $e^{-t}\cos 2t$.

Solution

Given $\mathcal{L}[\cos 2t] = \frac{s}{s^2+4}$. Using shifting property,

$$\mathcal{L}[e^{-t}\cos 2t] = \frac{s+1}{(s+1)^2+4} = \frac{s+1}{s^2+2s+5}.$$

Example 7.40

Find the Laplace transform of (a) $f(t) = t^3$ and (b) $g(t) = \begin{cases} (t-2)^3 & t > 2 \\ 0 & t < 2 \end{cases}$.

Solution

Given (a) $f(t) = t^3$. The Laplace transform is

$$\mathcal{L}[t^3] = \frac{3!}{s^4} = \frac{6}{s^4}.$$

(b) Given $g(t) = (t-2)^3 \quad \text{for } t > 2$.

$$g(t) = (t-2)^3 u(t-2).$$

Using shifting property,

$$\mathcal{L}[g(t)] = \frac{6}{s^4}e^{2s}.$$

Example 7.41

Calculate the Laplace transform of $f(t) = \begin{cases} e^t & 0 \le t \le 2\pi \\ e^t + \cos t & t > 2\pi \end{cases}$.

Solution

Given the function

$$f(t) = \begin{cases} e^t & 0 \le t \le 2\pi \\ e^t + \cos t & t > 2\pi \end{cases}$$

the first term e^t exists for $t > 0$ and the second term $\cos t$ exists for $t > 2\pi$.

Therefore the function $f(t)$ can be written as

$$f(t) = e^t u(t) + cos(t - 2\pi)u(t - 2\pi).$$

Taking Laplace transform,

$$\mathcal{L}[f(t)] = \mathcal{L}[[e^t u(t)] + \mathcal{L}[[u(t - 2\pi)\cos(t - 2\pi)] = \frac{1}{s-1} + \frac{se^{-2\pi s}}{1+s^2}.$$

Example 7.42

If the Laplace transform of $f(t) = \dfrac{\sin t}{t}$ is $F(s) = \tan^{-1}\left(\dfrac{1}{s}\right)$, find the Laplace transform of $\dfrac{\sin at}{t}$.

Solution

Given $f(t) = \dfrac{\sin t}{t}$, and $F(s) = \tan^{-1}\left(\dfrac{1}{s}\right)$. We know, using shifting property, that

$$\mathcal{L}\left[\frac{\sin at}{at}\right] = \frac{1}{a}F\left(\frac{s}{a}\right) = \frac{1}{a}\tan^{1}\left(\frac{a}{s}\right).$$

$$\therefore \quad \mathcal{L}\left[\frac{\sin at}{t}\right] = a\mathcal{L}\left[\frac{\sin at}{at}\right] = \tan^{1}\left(\frac{a}{s}\right).$$

Example 7.43

Find the inverse Laplace transform of $X(s) = \dfrac{s}{s^2 - a^2}$.

Solution

Given $X(s) = \dfrac{s}{s^2 - a^2}$. Using partial fractions,

$$\frac{s}{s^2 - a^2} = \frac{1}{2}\left[\frac{1}{s-a} + \frac{1}{s+a}\right].$$

The inverse Laplace transform is

$$\mathcal{L}^{-1}\left[\frac{s}{s^2 - a^2}\right] = \frac{1}{2}\mathcal{L}^{-1}\left[\frac{1}{s-a}\right] + \frac{1}{2}\mathcal{L}^{-1}\left[\frac{1}{s+a}\right]$$

$$= \frac{1}{2}e^{at} - \frac{1}{2}e^{-at} = \cosh at.$$

Example 7.44

If $\mathcal{L}^{-1}\left[\dfrac{1}{s^2+4}\right] = \dfrac{1}{2}\sin 2t$, find the inverse Laplace transform of $\dfrac{1}{s(s^2+4)}$.

Solution

Given $\mathcal{L}^{-1}\left[\dfrac{1}{s^2+4}\right] = \dfrac{1}{2}\sin 2t$.

Let $F(s) = \dfrac{1}{s^2+4}$.

Using integration property, we know that $\frac{F(s)}{s} \leftrightarrow \int_0^t f(t)dt$. That is,

$$\mathcal{L}^{-1}\left[\frac{1}{s(s^2+4)}\right] = \int_0^t \frac{1}{2}\sin 2\tau d\tau = \frac{1-\cos 2t}{4}.$$

Example 7.45

If $\mathcal{L}^{-1}\left[\frac{1}{s^2+1}\right] = \sin t$, find the inverse Laplace transform of $\frac{1}{s^3(s^2+1)}$.

Solution

Given $\mathcal{L}^{-1}\left[\frac{1}{s^2+1}\right] = \sin t$. Using integration property, we know that

$$\frac{F(s)}{s} \leftrightarrow \int_0^{t'} f(t)dt,$$

and $\frac{F(s)}{s^3} \leftrightarrow \int_0^t \int_0^t \int_0^t f(t)dt.$

That is,

$$\mathcal{L}^{1}\left[\frac{1}{s^3(s^2+1)}\right] = \int_0^t \int_0^t \int_0^t (\sin t)dtdtdt = \int_0^t \int_0^t (1-\cos t)dtdt = \int_0^t (t-\sin t)dt$$

$$= \left[\frac{t^2}{2} + \cos t\right]_0^t = \frac{t^2}{2} + \cos t - 1.$$

Example 7.46

Evaluate inverse Laplace transform of $X(s) = \tan^{-1}\left(\frac{1}{s}\right)$.

Solution

Given $X(s) = \tan^{-1}\left(\frac{1}{s}\right)$, it is easy to evaluate the inversion of the derivative of $\tan^{-1}\left(\frac{1}{s}\right)$.

We know that $\frac{d\tan^{-1}(s)}{ds} = \frac{1}{s^2+1}$.

Thus, $\frac{d(\tan^{-1}(1/s))}{ds} = \frac{-(1/s)^2}{(1/s)^2+1} = -\frac{1}{s^2+1}$.

The inverse Laplace transform is

$$\mathcal{L}^{-1}\left[\frac{d(\tan^{-1}(1/s))}{ds}\right] = \mathcal{L}^{-1}\left[\frac{dX(s)}{ds}\right] = -\sin t.$$

Using the differentiation property,

$$= -tf(t) = -\sin t$$

$$\therefore \quad f(t) = \frac{\sin t}{t}.$$

Example 7.47

Evaluate $\mathcal{L}^{-1}\left[\ln\left(1+\frac{1}{s}\right)\right]$.

Solution

Given $\mathcal{L}^{-1}\left[\ln\left(1+\frac{1}{s}\right)\right]$.

Let $F(s)=\ln\left(1+\frac{1}{s}\right)$. Taking the differentiation,

$$\frac{dF(s)}{ds}=\frac{d}{ds}\left(\ln\left(1+\frac{1}{s}\right)\right)=\frac{-1}{s(s+1)}=-\frac{1}{s}+\frac{1}{s+1}.$$

$$\mathcal{L}^{-1}\left[\frac{dF(s)}{ds}\right]=(e^{-t}-1)u(t).$$

We know that $\frac{dF(s)}{ds}\leftrightarrow -tf(t)$. Hence,

$$-1+e^{-t}=-tf(t).$$

$$\therefore\quad f(t)=\frac{1-e^{-t}}{t}.$$

Example 7.48

Find the inverse Laplace transform of $F(s)=\frac{s}{(s^2+1)^2}$.

Solution

Given $F(s)=\frac{s}{(s^2+1)^2}$. We know that the Laplace transforms of $\sin t$ is

$$\mathcal{L}[\sin t]=\frac{1}{s^2+1}.$$

Using the differentiation property,

$$\mathcal{L}[t\sin t]=\frac{-d}{ds}\left(\frac{1}{s^2+1}\right)=\frac{2s}{(s^2+1)^2}$$

$$f(t)=\mathcal{L}^{-1}\left[\frac{s}{(s^2+1)^2}\right]=\frac{t\sin t}{2}.$$

Example 7.49

Find the inverse Laplace transform of $X(s)=\frac{1-e^{-\pi s/2}}{1+s^2}$.

Solution

Given $X(s) = \dfrac{1-e^{-\pi s/2}}{1+s^2}$. Expanding,

$$\frac{1-e^{-\pi s/2}}{1+s^2} = \frac{1}{s^2+1} - \frac{e^{-\pi s/2}}{s^2+1}.$$

From basic properties,

$$\mathcal{L}\left[\frac{1-e^{-\pi s/2}}{1+s^2}\right] = \sin(t)u(t) - \sin(t-\pi/2)u(t-\pi/2).$$

$$= \sin(t)u(t) + \cos(t)u(t-\pi/2).$$

Example 7.50

Find the inverse Laplace transform of $X(s) = \dfrac{6s-4}{s^2-4s+20}$.

Solution

Given $X(s) = \dfrac{6s-4}{s^2-4s+20}$. Using partial fractions,

$$\frac{6s-4}{s^2-4s+20} = \frac{6(s-2)+8}{(s-2)^2+16} = \frac{6(s-2)}{(s-2)^2+16} + \frac{8}{(s-2)^2+16}.$$

From the basic properties, the relevant Laplace transforms are

$$e^{2t}\cos(4t)u(t) \leftrightarrow \frac{(s-2)}{(s-2)^2+16},$$

$$e^{2t}\sin(4t)u(t) \leftrightarrow \frac{4}{(s-2)^2+16}.$$

$$x(t) = \mathcal{L}^{-1}\left[\frac{6s-4}{s^2-4s+20}\right] = (6e^{2t}\cos 4t + 2e^{2t}\sin 4t)u(t)$$

$$= 2e^{2t}(3\cos 4t + \sin 4t)u(t).$$

Example 7.51

Find the inverse Laplace transform of $X(s) = \dfrac{2s-3}{s^2-2s+10}$.

Solution

Given $X(s) = \dfrac{2s-3}{s^2-2s+10}$.

Expanding,

$$\frac{2s-3}{s^2-2s+10} = \frac{2(s-1)-1}{(s-1)^2+9} = \frac{2(s-1)}{(s-1)^2+9} - \frac{1}{(s-1)^2+9}.$$

From the basic properties, the Laplace transforms are

$$e^t \cos(3t)u(t) \leftrightarrow \frac{(s-1)}{(s-1)^2+9},$$

$$e^t \sin(3t)u(t) \leftrightarrow \frac{3}{(s-1)^2+9}.$$

$$x(t) = \mathcal{L}^{-1}\left[\frac{2s-3}{s^2-2s+10}\right] = \left(2e^t\cos 3t - \frac{1}{3}e^t \sin 3t\right)u(t)$$

$$= e^t\left(2\cos 3t - \frac{1}{3}\sin 3t\right)u(t).$$

Example 7.52

Find the inverse Laplace transform of $X(s) = \dfrac{e^{-s}(s-2)}{s^2-4s+3}$.

Solution

Given $X(s) = \dfrac{e^{-s}(s-2)}{s^2-4s+3}$. Using partial fractions,

$$\frac{s-2}{s^2-4s+3} = \frac{s-2}{(s-1)(s-3)} = \frac{A}{(s-1)} + \frac{B}{(s-3)}.$$

$$A = \left.\frac{s-2}{(s-3)}\right|_{s=1} = \frac{1}{2}; \quad B = \left.\frac{s-2}{(s-1)}\right|_{s=3} = \frac{1}{2}.$$

Thus $X(s) = \dfrac{e^{-s}(s-2)}{s^2-4s+3} = \dfrac{1}{2}\left(\dfrac{e^{-s}}{s-1} + \dfrac{e^{-s}}{s-3}\right)$.

$\therefore$ From basic properties,

$$x(t) = \frac{1}{2}\left(e^{-t}u(t-1) + e^{-3t}u(t-1)\right)$$

$$= \frac{1}{2}\left(e^{-t} + e^{-3t}\right)u(t-1).$$

Example 7.53

If $x(t) = e^{-t}u(t)$, find the inverse Laplace transform of $e^{-3s}X(2s)$.

Solution

Given $x(t) = e^{-t}u(t)$,

$$X(s) = \frac{1}{s+1}.$$

Using scaling property,

$$X(at) \leftrightarrow \frac{1}{|a|}X\left(\frac{s}{a}\right).$$

Given $a = 2$,

$$\therefore \quad X(2s) \leftrightarrow \frac{1}{2}x(t/2).$$

Using shifting property,

$$x(t-T) \leftrightarrow e^{Ts}X(s).$$

Given $T = -3$,

$$x(t-3) \leftrightarrow e^{3s}X(s).$$

The inverse Laplace transform of $e^{-3s}X(2s)$ is

$$\mathcal{L}^{-1}\left[e^{-3s}X(2s)\right] = \frac{1}{2}x\left(\frac{(t-3)}{2}\right) = \frac{1}{2}e^{-(t-3)/2}u\left(\frac{t-3}{2}\right).$$

Example 7.54

Find the inverse Laplace transform of $F(s) = \dfrac{1+e^{-2s}}{3s^2+2s}$.

Solution

Given $F(s) = \dfrac{1+e^{-2s}}{3s^2+2s} = \dfrac{1+e^{-2s}}{s(3s+2)}$. Using partial fraction expansion,

$$\frac{1}{s(3s+2)} = \frac{A}{s} + \frac{B}{3s+2}.$$

$$A = \left.\frac{1}{(3s+2)}\right|_{s=0} = \frac{1}{2}; \quad B = \left.\frac{1}{s}\right|_{s=-2/3} = \frac{-3}{2}.$$

Thus, $F(s) = (1+e^{-2s})\left(\dfrac{1}{2s} - \dfrac{3}{2(3s+2)}\right)$

$$= \frac{1}{2s} - \frac{1}{2(s+2/3)} + \frac{e^{-2s}}{2s} - \frac{e^{-2s}}{2(s+2/3)}.$$

Then $f(t) = \mathcal{L}^{-1}[F(s)] = \dfrac{1}{2}\left(1-e^{-2t/3}\right)u(t) + \dfrac{1}{2}(1-e^{-2(t-2)/3})u(t-2)$.

Example 7.55

Evaluate $\mathcal{L}^{-1}\left[\ln\left(\dfrac{s+a}{s+b}\right)\right]$.

Solution

Given $\mathcal{L}^{-1}\left[\ln\left(\dfrac{s+a}{s+b}\right)\right]$.

Let $F(s) = \ln\left(\dfrac{s+a}{s+b}\right)$. Taking differentiation,

$$\frac{dF(s)}{ds} = \frac{d}{ds}\left(\ln\left(\frac{s+a}{s+b}\right)\right) = \frac{1}{(s+a)} - \frac{1}{(s+b)}.$$

$$\mathcal{L}^{-1}\left[\frac{dF(s)}{ds}\right] = \mathcal{L}^{-1}\left[\frac{1}{(s+a)} - \frac{1}{(s+b)}\right].$$

We know that $\frac{dF(s)}{ds} \leftrightarrow -tf(t)$.

$$\therefore \quad f(t) = \frac{e^{-bt} - e^{-at}}{t}.$$

Example 7.56

Find the Laplace transform of a periodic triangular waveform $f(t)$ shown in Fig. 7.15.

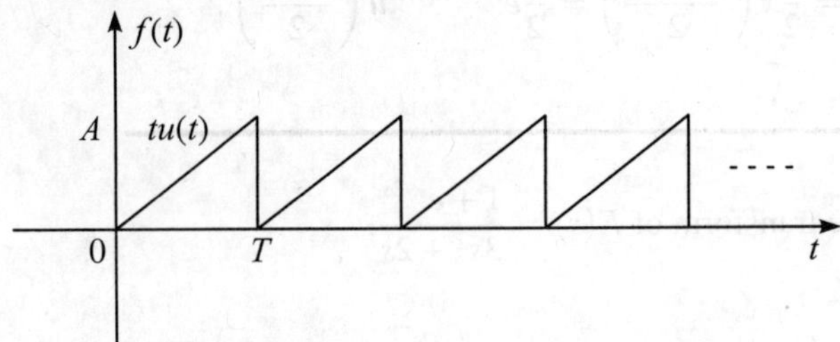

Fig. 7.15 *Periodic triangular waveform*

Solution

Given $f(t)$ is a periodic triangular waveform. The basic waveform for one cycle shown in Fig. 7.16 is expressed as

$$f_1(t) = \frac{A}{T}[tu(t) - (t-T)u(t-T)] - Au(t-T).$$

Now, the Laplace transform is

$$\mathcal{L}[f_1(t)] = \frac{A}{T}[\mathcal{L}[tu(t)] - \mathcal{L}[(t-T)u(t-T)] - \mathcal{L}[Tu(t-T)]]$$

$$= \frac{A}{T}\left[\frac{1}{s^2} - \frac{1}{s^2}e^{-sT} - \frac{T}{s}e^{-sT}\right]$$

$$= \frac{A}{s^2T}\left[1 - e^{-sT} - Tse^{-sT}\right].$$

Then the Laplace transform of the periodic triangular waveform is

$$F(s) = \left(\frac{1}{1-e^{-sT}}\right)F_1(s)$$

$$= \left(\frac{1}{1-e^{-sT}}\right)\frac{A}{s^2T}\left[1 - e^{-sT} - Tse^{-sT}\right]$$

$$= \frac{A}{s^2T}\left(\frac{1 - e^{-sT} - Tse^{-sT}}{1-e^{-sT}}\right).$$

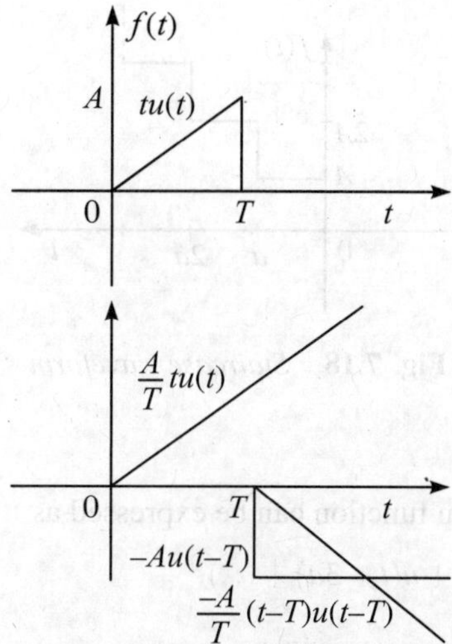

Fig. 7.16 *Synthesis of a periodic triangular waveform*

Example 7.57

Find the Laplace transform of a rectangular pulse shown in Fig. 7.17.

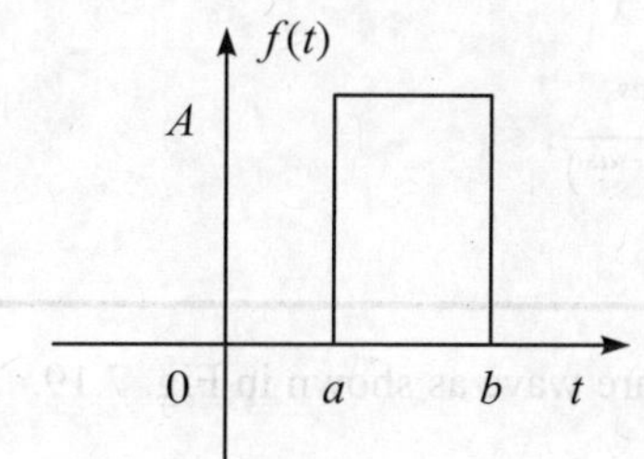

Fig. 7.17 *Rectangular pulse*

Solution

The rectangular pulse can be expressed as

$$f(t) = u(t-a) - u(t-b).$$

$$\mathcal{L}[f(t)] = \mathcal{L}[u(t-a)] - \mathcal{L}[u(t-b)]$$

$$= \frac{e^{-as}}{s} - \frac{e^{-bs}}{s} = \frac{1}{s}(e^{-as} - e^{-bs}).$$

Example 7.58

Find the Laplace transform of the staircase waveform shown in Fig. 7.18.

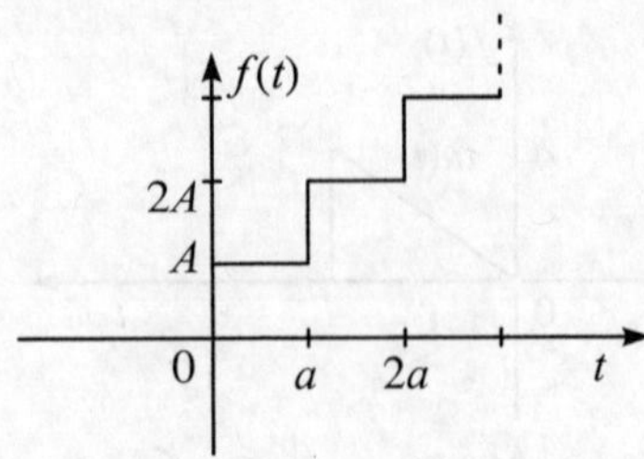

Fig. 7.18 *Staircase waveform*

Solution

Using waveform synthesis, the given function can be expressed as

$$f(t) = A(u(t-a) + u(t-2a) + u(t-3a) + \ldots).$$

$$\mathcal{L}[f(t)] = \mathcal{L}[u(t-a)] + \mathcal{L}[u(t-2a)] + \mathcal{L}[u(t-3a)] + \ldots$$

$$= \frac{A}{s}(e^{-as} + e^{-2as} + e^{-3as} + \ldots).$$

We know that

$$1 + x + x^2 + \cdots = \sum_{n=0}^{\infty} x^n = \frac{1}{1-x}, \quad |x| < 1.$$

Then, for $s > 0$, $\mathcal{L}[f(t)] = \dfrac{Ae^{-as}}{s(1-e^{-as})}$.

Example 7.59

Find the Laplace transform of a square wave as shown in Fig. 7.19. (Dec 2011)

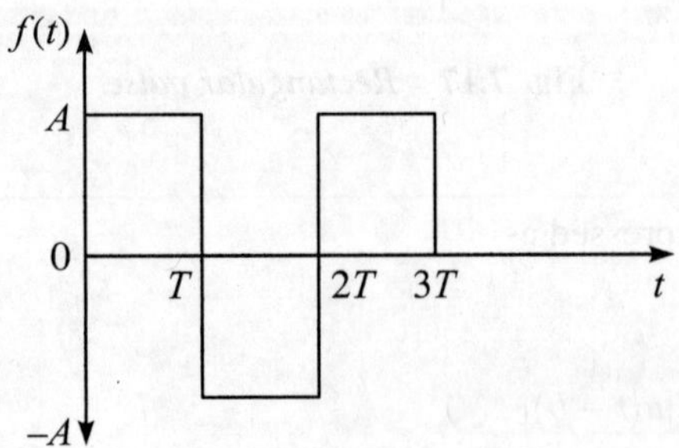

Fig. 7.19 *Square wave*

Solution

Using waveform synthesis, the given function can be expressed as

$$f(t) = A\,(u(t) - 2u(t-T) + 2u(t-2T) - 2u(t-3T) + \ldots).$$

$$\mathcal{L}[f(t)] = \frac{A}{s}\,(1 - 2e^{-Ts} + 2e^{-2Ts} - 2e^{-3Ts} + \ldots)$$

$$= \frac{A}{s}\left(2\left(1 - e^{-Ts} + e^{-2Ts} - e^{-3Ts} + \ldots\right) - 1\right)$$

$$= \frac{A}{s}\left[\frac{2}{1+e^{-Ts}} - 1\right] = \frac{A}{s}\left[\frac{1-e^{-Ts}}{1+e^{-Ts}}\right] = \frac{A}{s}\left[\frac{e^{Ts/2} - e^{-Ts/2}}{e^{Ts/2}+e^{-Ts/2}}\right] = \frac{A}{s}\tanh\left(\frac{Ts}{2}\right).$$

Example 7.60

A system is described by a differential equation $\frac{d^2y(t)}{dt^2} + 5\frac{dy(t)}{dt} + 4y(t) = x(t)$ with the initial conditions $y(0) = -2$, $y'(0) = -1$. Find the transfer function and the output response of the system if the input is $x(t) = e^{-2t}u(t)$.

Solution

Given $\frac{d^2y(t)}{dt^2} + 5\frac{dy(t)}{dt} + 4y(t) = x(t)$. For transfer function, assume that all initial conditions are zero. Applying Laplace transforms on both sides,

$$s^2Y(s) + 5Y(s) + 4Y(s) = X(s).$$

$$(s^2 + 5s + 4)Y(s) = X(s).$$

Therefore the transfer function

$$H(s) = \frac{Y(s)}{X(s)} = \frac{1}{s^2 + 5s + 4}.$$

For output response the initial conditions are $y(0) = -2$, $y'(0) = -1$ and the input is $x(t) = e^{-2t}u(t)$. Then the Laplace transform for differential equation is,

$$s^2Y(s) - sy(0^-) - y'(0^-) + 5sY(s) - 5y(0^-) + 4Y(s) = \frac{1}{s+2}$$

$$s^2Y(s) - s(-2) - (-1) + 5sY(s) - 5(-2) + 4Y(s) = \frac{1}{s+2}$$

$$s^2Y(s) + 2s + 1 + 5sY(s) + 10 + 4Y(s) = \frac{1}{s+2}$$

$$(s^2 + 5s + 4)Y(s) = -2s - 11 + \frac{1}{s+2}.$$

$$Y(s) = \frac{-2s-11}{(s^2+5s+4)} + \frac{1}{(s+2)(s^2+5s+4)} = \frac{-(2s^2+15s+21)}{(s+2)(s+1)(s+4)}.$$

Using partial fraction expansion,

$$\frac{-(2s^2+15s+21)}{(s+2)(s+1)(s+4)} = \frac{A}{(s+1)} + \frac{B}{(s+2)} + \frac{C}{(s+4)}.$$

$$A = \left.\frac{-(2s^2+15s+21)}{(s+2)(s+4)}\right|_{s=-1} = \frac{-8}{3}; \quad B = \left.\frac{-(2s^2+15s+21)}{(s+1)(s+4)}\right|_{s=-2} = \frac{-1}{2};$$

$$C = \left.\frac{-(2s^2+15s+21)}{(s+2)(s+1)}\right|_{s=-4} = \frac{7}{6}.$$

$$\therefore \quad Y(s) = \frac{-(2s^2+15s+21)}{(s+2)(s+1)(s+4)} = \frac{-8}{3(s+1)} - \frac{1}{2(s+2)} + \frac{7}{6(s+4)}.$$

Taking the inverse Laplace transform,

$$y(t) = \frac{-8}{3}e^{-t}u(t) - \frac{1}{2}e^{-2t}u(t) + \frac{7}{6}e^{-4t}u(t).$$

Summary

1. The unilateral Laplace transform of a function $f(t)$, defined for all real numbers $t \geq 0$, is $F(s) = \int_{\sigma^-}^{\infty} f(t)e^{-st}dt$, where $s = \sigma + j\omega$ is a complex frequency. The notation for the Laplace transform is $F(s) = \mathcal{L}[f(t)]$.
2. The inverse Laplace transform of a function $F(s)$ is $f(t) = \dfrac{1}{2\pi j}\displaystyle\int_{\sigma-j\infty}^{\sigma+j\infty} F(s)e^{st}\, ds f(t) = \mathcal{L}^{-1}[F(s)]$.
3. The *bilateral Laplace transform* or two-sided Laplace transform is obtained by extending the limits of integration to the entire real axis, i.e., $-\infty \leq t \leq \infty$. Bilateral Laplace transform pair is $F(s) = \int_{-\infty}^{\infty} f(t)e^{-st}dt,\ f(t) = \dfrac{1}{2\pi j}\displaystyle\int_{\sigma-j\infty}^{\sigma+j\infty} F(s)e^{st}ds.$
4. The Fourier transform of $f(t)$ converges if $f(t)$ is absolutely integrable. That is, $\int_{-\infty}^{\infty} |f(t)e^{-\sigma t}|\, dt < \infty$.
5. The range values of the complex variable $s = \sigma + j\omega$ for which the Laplace transform converges is called the region of convergence (ROC). ROC is the shaded area on the pole–zero plot of the s-plane, where $X(s)$ exists for all the values of s in the shaded area. The poles of $X(s)$ lie outside the ROC since $X(s)$ does not converge at the poles; the zeros, on the other hand, may lie inside or outside the ROC.
6. The properties of the ROC are:
   (a) The ROC does not contain any poles.
   (b) If $x(t)$ is a finite duration function, then the ROC is the entire s-plane except at $s = 0$ and $s = \infty$.
   (c) If $x(t)$ is a right-sided signal, that is, $(0 < t < \infty)$, then the ROC is $\mathrm{Re}\{s\} > \sigma_{\max}$, where $\sigma_{\max}$ is the maximum value of all the poles of $X(s)$.
   (d) If $x(t)$ is a left-sided signal, that is, $(-\infty < t < 0)$, then the ROC is $\mathrm{Re}\{s\} < \sigma_{\min}$, where $\sigma_{\min}$ is the maximum value of all the poles of $X(s)$.
   (e) If $x(t)$ is a two-sided signal, that is, $(-\infty < t < \infty)$, then the ROC is of the form $\sigma_1 < \mathrm{Re}\{s\} < \sigma_2$, where σ_1 and σ_2 are the real parts of the two poles of $X(s)$.
7. The Fourier transform of a signal is simply equal to its Laplace transform evaluated on the $j\omega$ axis, provided that it is included in the region of convergence. That is,

   $$\mathcal{L}[f(t)] = \mathcal{F}[f(t)e^{-\sigma t}].$$
8. Laplace transforms for causal periodic signals can be derived easily from the basic causal waveforms comprising exponential, ramp, step or delta functions.

9. The Laplace transforms of some common signals are as given below.

Signals of time function $x(t)$	La place transform $X(s)$	Region of convergence (ROC)
$\delta(t)$	1	All s
$u(t)$	$\frac{1}{s}$	$\text{Re}\{s\} > 0$
$-u(-t)$	$\frac{1}{s}$	$\text{Re}\{s\} < 0$
$\frac{t^{n-1}}{(n-1)!}u(t)$	$\frac{1}{s^n}$	$\text{Re}\{s\} > 0$
$\frac{t^{n-1}}{(n-1)!}u(-t)$	$\frac{1}{s^n}$	$\text{Re}\{s\} < 0$
$e^{-at}u(t)$	$\frac{1}{s+a}$	$\text{Re}\{s\} > -a$
$-e^{-at}u(-t)$	$\frac{1}{s+a}$	$\text{Re}\{s\} < -a$
$te^{at}u(t)$	$\frac{1}{(s+a)^2}$	$\text{Re}\{s\} > -\text{Re}\{a\}$
$-te^{at}u(-t)$	$\frac{1}{(s+a)^2}$	$\text{Re}\{s\} < -\text{Re}\{a\}$
$\frac{t^{n-1}}{(n-1)!}e^{-at}u(t)$	$\frac{1}{(s+a)^n}$	$\text{Re}\{s\} > -a$
$-\frac{t^{n-1}}{(n-1)!}e^{-at}u(-t)$	$\frac{1}{(s+a)^n}$	$\text{Re}\{s\} < -a$
$\delta(t-T)$	e^{-st}	All s
$[\cos\omega_0 t]u(t)$	$\frac{s}{s^2+\omega_0^2}$	$\text{Re}\{s\} > 0$
$[\sin\omega_0 t]u(t)$	$\frac{s}{s^2+\omega_0^2}$	$\text{Re}\{s\} > 0$
$[e^{-at}\cos\omega_0 t]u(t)$	$\frac{s+a}{(s+a)^2+\omega_0^2}$	$\text{Re}\{s\} > -\text{Re}\{a\}$
$[e^{-at}\sin\omega_0 t]u(t)$	$\frac{\omega_0}{(s+a)^2+\omega_0^2}$	$\text{Re}\{s\} < -\text{Re}\{a\}$

10. The summary of Laplace transform properties are given below.

Property/Theorem	Time domain signals $x(t)$	Laplace transforms $X(s)$	Region of convergence (ROC) R
Linearity	$a_1x_1(t)+a_2x_2(t)$	$a_1X_1(s)+a_2X_2(s)$	At least $R_1 \cap R_2$
Time shifting	$x(t-t_o)$	$e^{-st_0}X(s)$	R
Shifting in the s-domain	$e^{s t_0}x(t)$	$X(s-s_0)$	Shifted version of R (i.e., s is in the ROC if $s-s_0$ is in R.)

Time scaling	$x(at)$	$\frac{1}{\lvert a\rvert}X(\frac{s}{a})$	Scaled ROC (i.e., s is in the ROC if $\frac{s}{a}$ is in R)
Conjugation	$x^*(t)$	$X^*(S^*)$	R
Time convolution	$x_1(t) * x_2(t)$	$X_1(s)X_2(s)$	At least $R_1 \cap R_2$
Frequency convolution	$x_1(t)x_2(t)$	$X_1(s) * X_2(s)$	At least $R_1 \cap R_2$
Differentiation in the time domain	$\frac{d}{dx}x(t)$	$sX(s)$	At least R
Differentiation in the s-domain	$-tx(t)$	$\frac{d}{ds}X(s)$	R
Integration in the time domain	$\int_{-\infty}^{t} x(\tau)d(\tau)$	$\frac{1}{s}X(s)$	At least $R \cap \{\mathrm{Re}\{s\} > 0\}$
Initial value theorem	$x(0^+) = \lim_{t\to 0} x(t)$	$\lim_{s\to\infty} sX(s)$	–
Final value theorem	$x(\infty) = \lim_{t\to\infty} x(t)$	$\lim_{s\to 0} sX(s)$	–

11. A periodic waveform can be synthesised using the basic waveforms as follows. Let $X(s)$ be the Laplace transform of a causal periodic waveform $x(t)$ of period T. Then, $X(s) = \left(\frac{1}{1-e^{-sT}}\right) X_1(s)$, where $X_1(s)$ is the Laplace transform of a basic waveform $x_1(t)$ for the first period of $x(t)$.
12. The inverse Laplace transform involves complex integration in the complex plane, which can be solved either by contour integral methods, or simply by using partial fractions and then using a table of basic Laplace transform pairs. The region of convergence of each individual fraction must be found to select the correct time-domain signal.
13. The transfer function of a system is $H(s) = \frac{Y(s)}{X(s)}$ when all initial conditions are zero or $H(s) = \frac{Y(s)}{X(s)} = \frac{(b_M s^M + b_{M-1}s^{M-1} + \cdots + b_0)}{(a_N s^N + a_{N-1}s^{N-1} + \cdots + a_0)}$.
14. The impulse response of the system is $h(t) = \mathcal{L}^{-1}[H(s)] = \mathcal{L}^{-1}\left[\frac{Y(s)}{X(s)}\right]$.
15. The output response of the system is $y(t) = \mathcal{L}^{-1}[Y(s)] = \mathcal{L}^{-1}[H(s)X(s)]$.
16. The frequency response is $H(\omega) = \frac{Y(s)}{X(s)}\Big|_{s=j\omega} = \frac{Y(j\omega)}{X(j\omega)}$.
17. $H(\omega) = |H(\omega)||\phi(\omega)$, where the magnitude response is $|H(\omega)| = \frac{|Y(j\omega)|}{|X(j\omega)|}$ and the phase response is $\angle\phi(\omega) = \tan^{-1}\left(\frac{Y(j\omega)}{X(j\omega)}\right)$.
18. An LTI system is stable if and only if the ROC of its transfer function contains the $j\omega$-axis.
19. A causal system with a proper rational transfer function $H(s)$ is stable if and only if all of its poles are in the left-half of the s-plane; that is, all the poles have negative real parts.

Review Questions

1. Define Laplace transform and its inverse. (Dec 2011)
2. Define the region of convergence and state its properties. (Dec 2011)
3. What is the condition to be satisfied for the existence of the Laplace transform?
4. When is a function $x(t)$ said to be Laplace transformable?
5. Define Laplace transform of signal $x(t)$ and its region of convergence.
6. Find the Laplace transform of the following signals:

   (i) Impulse function, (ii) unit step function, (iii) unit ramp function and (iv) $A\sin(\omega_o t)u(t)$.
7. Distinguish between Fourier transform and Laplace transform.
8. Define Laplace transform of signal $x(t)$ and its region of convergence.
9. State the properties of the ROC of the Laplace transform.
10. State any four properties of the Laplace transform. (Dec 2011)
11. Bring out the relationship between Laplace and Fourier transforms. (Dec 2011)
12. Explain the significance of pole–zero plot in the Laplace domain. How will it help to verify the stability of a system? (June 2014)
13. State and prove time differentiation and s-domain differentiation properties of the Laplace transform.
14. State and prove the initial and final value theorems of the Laplace transform.
15. State and prove the differentiation theorem for the Laplace transforms.
16. Write any two properties of the Laplace transform.
17. List the advantages and limitations of the Laplace transform.
18. State and prove the initial value and final value theorems with respect to the Laplace transform. (Dec 2011)

Problems

1. Determine the Laplace transform for the following signals:

   (a) $e^{-t}u(t-1)$ (b) $(\sin t + \cos t)u(t)$

   (c) $t^2u(t-1)$ (d) $(\delta(t)+3\delta(t-1))$
2. Find the Laplace transform of the following signals and their ROCs:

   (a) $f(t) = e^{-t}u(t) + e^{2t}u(t)$ (b) $f(t) = e^{-t}u(t) - e^{2t}u(-t)$

   (c) $f(t) = -e^{-t}u(-t) - e^{2t}u(-t)$ (d) $f(t) = -e^{-t}u(-t) + e^{2t}u(t)$
3. Find the Laplace transform of the signal $f(t) = e^{-3t}u(t) + e^{-4t}u(t)$ and its ROC.
4. Find the Laplace transform of $[e^{-2t}+e^{3t}]u(t)$ and its region of convergence.
5. Find the Laplace transform of $x(t) = -te^{-2t}u(t)$ and its region of convergence.
6. Find the Laplace transform of $x(t) = e^{2t}u(t) + e^{-2t}u(-t)$ and its region of convergence.
7. Find the Laplace transform of $x(t) = e^{-2t}u(-t) + e^{-3t}u(-t)$ and its region of convergence.
8. Find Laplace transform of $x(t) = 3e^{-2t}u(t) + 4e^{3t}u(-t)$.

9. Prove that the signals $x(t) = e^{-at}u(t)$ and $x(t) = e^{-at}u(-t)$ have the same $X(s)$ and differ only in ROC. (Nov 2015)
10. Find the Laplace transforms and sketches the ROC of: (May 2015)
    (i) $x(t) = u(t-5)$ and (ii) $x(t) = e^{j\omega t}u(t)$.
11. Find the Laplace transform of the signal $x(t) = e^{-b|t|}$ and plot the region of convergence (ROC). (June 2014)
12. Find the Laplace transform of the following: (Dec 2015)
    (i) $te^{-at}u(t)$ and (ii) $\cos\omega_0 t u(t)$.
13. Find the Laplace transform of the signal $x(t) = e^{-at}u(t) + e^{-bt}u(-t)$ and plot ROC. (Dec 2015)
14. Find the Laplace transform of the following signals using its properties.
    (a) $x(t) = u(t+4) + \delta(t-4)$ (b) $x(t) = te^{-t}u(t) + t^2e^{-3t}u(t)$
    (c) $x(t) = te^{-2t}\sin(2t)u(t)$ (d) $x(t) = \dfrac{\cos 3t}{t}u(t)$
15. Given the Laplace transform $X(s) = \dfrac{5s}{s^2-25}$, find the Laplace transform of $x(4t)$.
16. Determine the initial value for the following Laplace transforms:
    (a) $X(s) = \dfrac{4(s+2)}{(s^2+4s+8)}$ (b) $X(s) = \dfrac{5}{(s^2+s-5)}$
17. Determine the final value for the following Laplace transforms:
    (a) $X(s) = \dfrac{4(s-7)}{s(s+14)}$ (b) $X(s) = \dfrac{s^2+2s+9}{3s^2+s+2}$
18. Determine the initial value and final value of signal $x(t)$ whose Laplace transform is $X(s) = \dfrac{2s+5}{s(s+3)}$. (Dec 2015)
19. Find the final value of the function $F(s)$ given by $\dfrac{(s-1)}{s(s^2-1)}$. (Dec 2011)
20. Determine the time signals corresponding to the following Laplace transforms:
    (a) $X(s) = \dfrac{s}{(s+2)^2+1}$ (b) $X(s) = \dfrac{10s}{(2s^2-8)}$
    (c) $X(s) = \dfrac{s^2}{(s-1)^4}$ (d) $X(s) = \dfrac{2(s^3+1)}{s(s+1)(s+2)}$
21. Find the inverse Laplace transform of the following functions:
    (a) $X(s) = \dfrac{1+e^{-2S}}{s(s+5)}$ (b) $X(s) = \ln\left(1+\dfrac{16}{s^2}\right)$
    (c) $X(s) = \ln\left(\dfrac{s+3}{s+2}\right)$ (d) $X(s) = \dfrac{s^2+6s+7}{s^2+3s+2}$, ROC : $\text{Re}\{s\} > -1$
22. Using the convolution theorem of Laplace transform, find $y(t) = x_1(t) * x_2(t)$ for the following functions:
    (a) $x_1(t) = 2e^{-2t}u(t)$, $x_2(t) = u(t-4)$ (b) $x_1(t) = tu(t)$, $x_2(t) = u(t-4)$
    (c) $x_1(t) = tu(t)$, $x_2(t) = tu(t)$ (d) $x_1(t) = e^{-at}u(t)$, $x_2(t) = e^{-bt}u(t)$

23. Find the inverse of the following bilateral Laplace transform, $X(s) = \dfrac{3}{(s+2)(s+5)}$, when

(a) ROC : $-5 < \mathrm{Re}\{s\} < -2$ (b) ROC : $\mathrm{Re}\{s\} > -2$

(c) ROC : $\mathrm{Re}\{s\} < -5$ (d) ROC : $-5 > \mathrm{Re}\{s\} > -2$

24. Find the inverse Laplace transform of $F(S) = \dfrac{3S+6}{(S+4)^2(S+1)}$.

25. Find the inverse Laplace transform of the following function:

$$F(S) = e^{-2S}\left(\frac{2S+5}{S^2+5S+6}\right).$$

26. Find the inverse Laplace transform of $X(s) = \dfrac{5s+4}{s^2+2s+1}$, $\mathrm{Re}(s) < -1$. (Nov-2015)

27. Find the inverse Laplace transform of $X(s) = \dfrac{(-5s-7)}{(s+1)(s-1)(s+2)}$. (May 2015)

28. Find the inverse Laplace transform of $\dfrac{(s-1)}{(s)(s+1)}$. (Dec 2011)

29. Find the Laplace transform of a full-wave rectifier with frequency ω_0.

30. Find the Laplace transform of $f(t) = \sin a(t-b)u(t-b)$.

31. Find the Laplace transform of the following signals: (a) $f(t) = 10\sin(100\pi t)u(t)$ and, (b) $f(t) = \cos 20(t-2)u(t-2)$.

32. Find the Laplace transform of the periodic rectangular wave shown in Fig. P7.32. (Dec 2011)

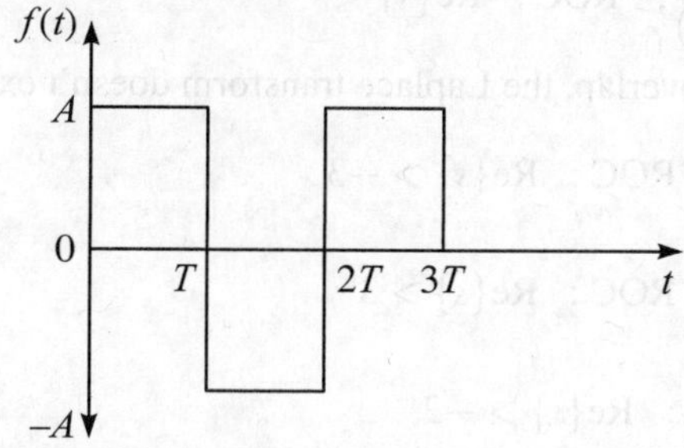

Fig. P7.32 *Periodic rectangular waveform*

33. Determine the Laplace transform of (Dec 2011, May 2013)

(i) $f(t) = e^{-at}\sin\omega t$ and (ii) $f(t) = e^{-at}\cos h\omega t$.

34. Find the Laplace transform of the waveform shown in Fig. P7.34. (May 2013)

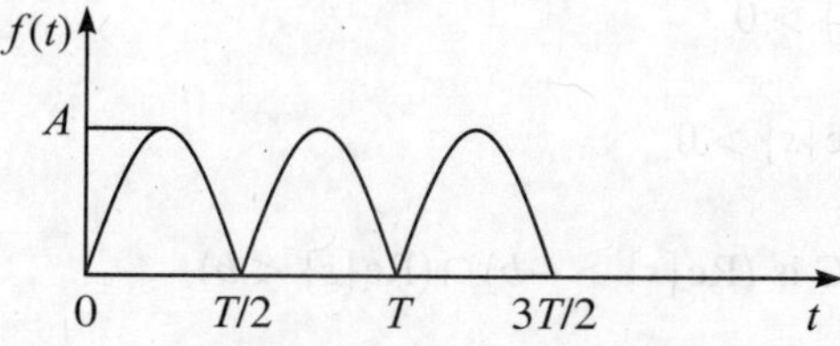

Fig. P7.34 *Rectified sine wave*

35. Find the Laplace transform of the following signal and its ROC:
$x(t) = e^{-5t}[u(t) - u(t-5)]$. (Jan 2015)
36. Find the Laplace transform of the signal shown in Fig. P7.36. (Jan 2015)

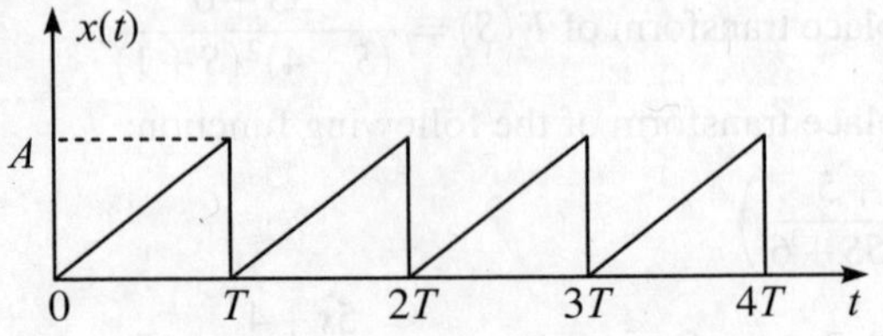

Fig. P7.36 *Triangular waveform*

Answers

1. (a) $\frac{e^{-(S+1)}}{S+1}$ (b) $\frac{s+1}{s^2+1}$ (c) $\frac{e^{-s}}{s^3}$ (d) $1+3e^{-s}$
2. (a) $F(s) = \frac{2s-3}{(s+1)(s-2)}$; ROC: $\text{Re}\{s\} > 2$
(b) $F(s) = \frac{2s-3}{(s+1)(s-2)}$; ROC: $-1 < \text{Re}\{s\} < 2$
(c) $F(s) = \frac{2s-3}{(s+1)(s-2)}$; ROC: $\text{Re}\{s\} < -1$
(d) Since ROC doesn't overlap, the Laplace transform doesn't exists.
3. $F(s) = \frac{2s+7}{(s+3)(s+4)}$; ROC: $\text{Re}\{s\} > -3$
4. $F(s) = \frac{2s-1}{(s-3)(s+2)}$; ROC: $\text{Re}\{s\} > 3$
5. $X(s) = \frac{-1}{(s+2)^2}$; ROC: $\text{Re}\{s\} > -2$
6. No ROC. The Laplace transform does not exist.
7. $X(s) = \frac{-(2s+5)}{(s+3)(s+2)}$; ROC: $\text{Re}\{s\} < -3$
8. $X(s) = \frac{-s-17}{(s+2)(s+3)}$
10. (i) $X(s) = \frac{e^{-5s}}{s}$, $\text{Re}\{s\} > 0$
(ii) $X(s) = \frac{1}{s-j\omega}$, $\text{Re}\{s\} > 0$
11. $X(s) = \frac{-2b}{s^2-b^2}$; ROC is $(\text{Re}\{s\} > -b) \cap (\text{Re}\{s\} < b)$
12. (i) $\frac{1}{(s+a)^2}$; ROC: $\text{Re}\{s\} > -a$ (ii) $\frac{s}{s^2+\omega_0^2}$

13. $X(s) = \dfrac{a+b}{(s+a)(s+b)}$

ROC: $(\text{Re}\{s\} > -a) \cap (\text{Re}\{s\} < -b)$

Note: If $-a > -b$, the ROC does not exist. Hence the Laplace transform for the given signal does not exist.

14. (a) $X(s) = e^{-4s} + \dfrac{e^{4s}}{s}$; ROC: $\text{Re}\{s\} > 0$

(b) $X(s) = \dfrac{1}{(s+1)^2} + \dfrac{2}{(s+3)^3}$; ROC: $\text{Re}\{s\} > -1$

(c) $X(s) = \dfrac{4(s+2)}{(s^2+4s+8)^2}$; ROC: $\text{Re}\{s\} > -2$

(d) No convergence. The Laplace transform does not exist.

15. $X(s) = \dfrac{25s}{s^2 - 25^2}$.

16. (a) 4 (b) 0.

17. (a) -2 (b) 0.

18. $x(0) = 2,\ x(\infty) = \dfrac{5}{3}$.

19. 1.

20. (a) $x(t) = (e^{-2t}\cos t - 2e^{-2t}\sin t)u(t)$.

(b) $x(t) = 5\cosh(2t)u(t)$.

(c) $x(t) = te^t\left(1 + t + \dfrac{t^2}{6}\right)u(t)$.

(d) $x(t) = 2\delta(t) + u(t) - 7e^{-2t}u(t)$.

21. (a) $x(t) = (e^{-5t} - 1)u(t) + (e^{-5(t-2)} - 1)u(t-2)$.

(b) $x(t) = \dfrac{2(1-\cos 5t)}{t}u(t)$

(c) $x(t) = \dfrac{e^{-2t} - e^{-3t}}{t}u(t)$

(d) $x(t) = \delta(t) + 2e^{-t}u(t) + e^{-2t}u(t)$

22. (a) $y(t) = u(t-4) - e^{-2(t-4)}u(t-4)$ (b) $y(t) = \dfrac{(t-4)^2}{2}u(t-4)$

(c) $y(t) = \dfrac{t^3u(t)}{6}$ (d) $y(t) = \dfrac{1}{a-b}(e^{-at} - e^{-bt})u(t)$

23. (a) $x(t) = -e^{-5t}u(t) - e^{-2t}u(-t)$ (b) $x(t) = -e^{-5t}u(t) + e^{-2t}u(t)$

(c) $x(t) = e^{-5t}u(-t) - e^{-2t}u(-t)$ (d) No ROC overlap. Laplace transform does not exists.

24. $x(t) = (2te^{4t} - \dfrac{1}{3}e^{4t} + \dfrac{1}{3}e^{-t})u(t)$

25. $x(t) = e^{-2t}u(t) + e^{-3t}u(t)$

26. $x(t) = e^{-t}u(-t)(-5+t)$

27. $x(t) = e^{-t}u(t) - 2e^{t}u(t) + 3e^{-2t}u(t)$

28. $x(t) = -u(t) + 2e^{-t}u(t)$

29. $F(s) = \dfrac{\omega_0}{(s^2+\omega_0^2)}\coth\left(\dfrac{s\pi}{2\omega_0}\right)$; ROC : $\text{Re}\{s\} > 0$

30. $F(s) = \dfrac{ae^{-bs}}{s^2+a^2}$

31. (a) $F(s) = \dfrac{1000\pi}{s^2-100^2}$; $F(s) = \dfrac{se^{-2s}}{s^2-100^2}$

32. $F(s) = \dfrac{A}{s}\left[\dfrac{1-e^{-sT}}{1+e^{-sT}}\right] = \dfrac{A}{s}\tanh\left(\dfrac{sT}{2}\right)$

33. (i) $F(s) = \dfrac{\omega}{(s+a)^2+\omega^2}$ (ii) $F(s) = \dfrac{(s+a)}{(s+a)^2-\omega^2}$

34. $F(s) = \dfrac{A\omega_0}{s^2+\omega^2}\left(\dfrac{1+e^{-sT/2}}{1-e^{-sT/2}}\right) = \dfrac{A\omega_0}{s^2+\omega^2}\coth\left(\dfrac{sT}{4}\right)$

35. $X(s) = \dfrac{1}{s+5}\left(1-e^{-5(s+5)}\right)$

36. $X(s) = \dfrac{A}{Ts^2}\left(1-e^{-4sT}\right) - \dfrac{A}{s}e^{-sT}\left(1+e^{-sT}+e^{-2sT}+e^{-3sT}\right)$

Multiple-choice Questions

1. The Laplace transform of a unit ramp function starting at $t = a$ is

   (a) $\dfrac{1}{(s+a)^2}$ (b) $\dfrac{e^{-as}}{(s+a)^2}$

   (c) $\dfrac{e^{-as}}{s^2}$ (d) $\dfrac{a}{s^2}$

2. If $\mathcal{L}[f(t)] = \dfrac{2(s+1)}{s^2+2s+5}$, then $f(0^+)$ and $f(\infty)$ are given by

   (a) 0, 2 respectively (b) 2, 0 respectively

   (c) 0, 1 respectively (d) 2/5, 0 respectively

3. The final value theorem is used to find the

   (a) steady state value of the system output
   (b) initial value of the system output
   (c) transient behaviour of the system output
   (d) none of these

4. The Laplace transform of $e^{\alpha t}\cos(\alpha t)$ is equal to

   (a) $\dfrac{(s-\alpha)}{(s-\alpha)^2+\alpha^2}$ (b) $\dfrac{(s+\alpha)}{(s+\alpha)^2+\alpha^2}$

   (c) $\dfrac{1}{(s-\alpha)^2}$ (d) none of the above

5. If $\mathcal{L}[f(t)] = \dfrac{\omega}{(s^2+\omega^2)}$, then the value of $\lim_{t\to\infty} f(t)$

(a) cannot be determined (b) is zero

(c) is unity (d) is infinite

6. If $\mathcal{L}[f(t)] = F(s)$, then $\mathcal{L}[f(t-T)]$ is equal to

(a) $e^{sT}F(s)$ (b) $e^{-sT}F(s)$ (c) $\dfrac{F(s)}{1+e^{-sT}}$ (d) $\dfrac{F(s)}{1-e^{sT}}$

7. Given that $\mathcal{L}[f(t)] = \dfrac{s+2}{s^2+1}$, $\mathcal{L}[g(t)] = \dfrac{s^2+1}{(s+3)(s+2)}$, if $h(t) = \int_0^1 f(\tau)g(t-\tau)d\tau$, then $\mathcal{L}[h(t)]$ is

(a) $\dfrac{s^2+1}{(s+3)}$ (b) $\dfrac{1}{(s+3)}$

(c) $\dfrac{s^2+1}{(s+3)(s+2)} + \dfrac{s+2}{s^2+1}$ (d) none of the above

8. The Laplace transform of $i(t)$ is given by $I(s) = \dfrac{2}{s(1+s)}$. As $t\to\infty$, the value of $i(t)$ tends to

(a) 0 (b) 1 (c) 2 (d) ∞

9. If the Laplace transform of a signal $y(t)$ is $Y(s) = \dfrac{1}{s(s-1)}$. Then $y(t)$ is

(a) -1 (b) 0 (c) 1 (d) unbounded

10. Laplace transforms of the functions $tu(t)$ and $u(t)\sin(t)$ are respectively:

(a) $\dfrac{1}{s^2}, \dfrac{1}{s^2+1}$ (b) $\dfrac{1}{s^2}, \dfrac{s}{s^2+1}$

(c) $s, \dfrac{1}{s^2+1}$ (d) $s, \dfrac{s}{s^2+1}$

11. The Laplace transform of a function $f(t)u(t)$, where $f(t)$ is periodic with period T, is $A(s)$ times the Laplace transform of its first period. Then

(a) $A(s) = s$ (b) $A(s) = 1/(1-\exp(-Ts))$

(c) $A(s) = 1/(1+\exp(-Ts))$ (d) $A(s) = \exp(Ts)$

12. If $F(s) = \mathcal{L}[f(t)] = \dfrac{K}{(s+1)(s^2+4)}$, then $\lim_{t\to\infty} f(t)$ is given by

(a) $K/4$ (b) zero (c) infinite (d) undefined

13. The inverse Laplace transform of the function $\dfrac{(s+5)}{(s+1)(s+3)}$ is

(a) $2e^{-t}-e^{-3t}$ (b) $2e^{-t}+e^{-3t}$ (c) $e^{-t}-2e^{-3t}$ (d) $e^{-t}+2e^{-3t}$

14. The Laplace transform of a continuous-time signal $x(t)$ is $X(s) = \dfrac{5-s}{s^2-s-2}$. If the Fourier transform of this signal exists, then $x(t)$ is

(a) $e^{2t}u(t)-2e^{t}u(t)$ (b) $-e^{2t}u(-t)+2e^{-t}u(t)$

(c) $-e^{2t}u(-t)-2e^{-t}u(t)$ (d) $e^{2t}u(-t)-2e^{-t}u(t)$

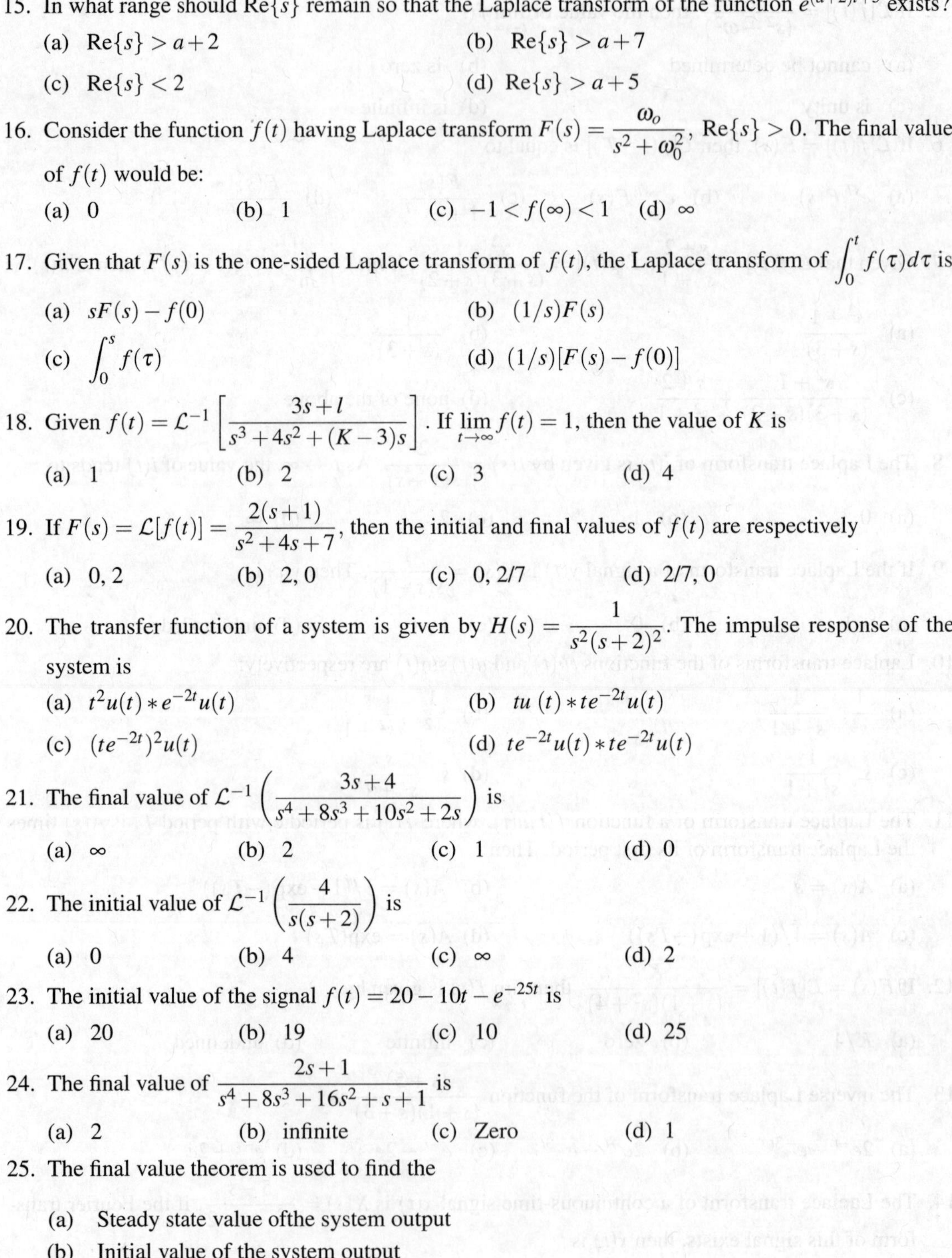

15. In what range should $\text{Re}\{s\}$ remain so that the Laplace transform of the function $e^{(a+2)t+5}$ exists?

(a) $\text{Re}\{s\} > a+2$ (b) $\text{Re}\{s\} > a+7$

(c) $\text{Re}\{s\} < 2$ (d) $\text{Re}\{s\} > a+5$

16. Consider the function $f(t)$ having Laplace transform $F(s) = \dfrac{\omega_o}{s^2+\omega_0^2}$, $\text{Re}\{s\} > 0$. The final value of $f(t)$ would be:

(a) 0 (b) 1 (c) $-1 < f(\infty) < 1$ (d) ∞

17. Given that $F(s)$ is the one-sided Laplace transform of $f(t)$, the Laplace transform of $\int_0^t f(\tau)d\tau$ is

(a) $sF(s) - f(0)$ (b) $(1/s)F(s)$

(c) $\int_0^s f(\tau)$ (d) $(1/s)[F(s) - f(0)]$

18. Given $f(t) = \mathcal{L}^{-1}\left[\dfrac{3s+l}{s^3+4s^2+(K-3)s}\right]$. If $\lim_{t\to\infty} f(t) = 1$, then the value of K is

(a) 1 (b) 2 (c) 3 (d) 4

19. If $F(s) = \mathcal{L}[f(t)] = \dfrac{2(s+1)}{s^2+4s+7}$, then the initial and final values of $f(t)$ are respectively

(a) 0, 2 (b) 2, 0 (c) 0, 2/7 (d) 2/7, 0

20. The transfer function of a system is given by $H(s) = \dfrac{1}{s^2(s+2)^2}$. The impulse response of the system is

(a) $t^2u(t) * e^{-2t}u(t)$ (b) $tu\,(t) * te^{-2t}u(t)$

(c) $(te^{-2t})^2u(t)$ (d) $te^{-2t}u(t) * te^{-2t}u(t)$

21. The final value of $\mathcal{L}^{-1}\left(\dfrac{3s+4}{s^4+8s^3+10s^2+2s}\right)$ is

(a) ∞ (b) 2 (c) 1 (d) 0

22. The initial value of $\mathcal{L}^{-1}\left(\dfrac{4}{s(s+2)}\right)$ is

(a) 0 (b) 4 (c) ∞ (d) 2

23. The initial value of the signal $f(t) = 20 - 10t - e^{-25t}$ is

(a) 20 (b) 19 (c) 10 (d) 25

24. The final value of $\dfrac{2s+1}{s^4+8s^3+16s^2+s+1}$ is

(a) 2 (b) infinite (c) Zero (d) 1

25. The final value theorem is used to find the

(a) Steady state value ofthe system output

(b) Initial value of the system output

(c) Transient behaviour of the system output

(d) None of these

26. The Laplace transform of a square wave with peak amplitude A and period T is

(a) $\dfrac{1+e^{-sT}}{1-e^{-sT}}$ (b) $\dfrac{A}{s}\left[\dfrac{1-e^{-sT}}{1+e^{-sT}}\right]$

(c) $\dfrac{A}{s}\left[\dfrac{1+e^{-sT}}{1-e^{-sT}}\right]$ (d) $\dfrac{1-e^{-sT}}{1+e^{-sT}}$

27. The inverse Laplace transform of the function $F(s) = 2\log\left(\dfrac{s+2}{s}\right)$ is

(a) $\dfrac{2-e^{-2t}}{t}$ (b) $\dfrac{e^{-2t}}{t}$

(c) $\dfrac{2}{t}$ (d) $\dfrac{2+e^{-2t}}{t}$

28. The inverse Laplace transform of the function $F(s) = \dfrac{6}{s^4}$ is

(a) 3 (b) t^2 (c) t^3 (d) $3t$

29. The Laplace transform of the first derivative of a function $f(t)$ is

(a) $\dfrac{F(s)}{s}$ (b) $sF(s) - f(0)$ (c) $F(s) - f(0)$ (d) $f'(0)$

30. If $\delta(t)$ denotes a unit impulse function, then the Laplace transform of $\dfrac{d^2\delta(t)}{dt^2}$ is

(a) 1 (b) s^2 (c) s (d) s^{-2}

Answers

1. (c)	2. (b)	3. (a)	4. (a)	5. (a)	6. (b)	7. (b)	8. (c)	9. (d)	10. (a)
11. (b)	12. (d)	13. (a)	14. (c)	15. (a)	16. (c)	17. (b)	18. (d)	19. (b)	20. (b)
21. (b)	22. (a)	23. (b)	24. (d)	25. (a)	26. (b)	27. (a)	28. (c)	29. (b)	30. (b)

8

The Z-Transform

Learning Objectives

The main objective of this chapter is to understand thoroughly the basic concepts of Z-transforms. We will learn how discrete-time signals and systems are analysed using the Z-transform. By reading this chapter, students will be able to relate to the following topics:

- Discrete-time signal representations
- Concept of Z-transform
- Distinction between Laplace, Fourier and Z-transform
- Region of convergence in Z-transform
- Properties of Z-transform
- Inverse Z-transform
- Applications of Z-transform in linear systems

8.1 Introduction

In the previous chapter we looked at the Laplace transform as an extension of the complex Fourier transform for continuous-time signals and systems. There are many classes of signals and systems for which the Fourier transform does not converge but Laplace transform does. The Laplace transform can perform analysis of both stable and unstable systems.

In a similar way, the basic mathematical tool for analysing and synthesising discrete-time signals and systems is the *Z*-transform. The *Z*-transform can be derived directly by applying Laplace transform on discrete-time signals; it is the discrete-time counterpart of the Laplace transform.

We know that discrete-time LTI systems are represented by *n*th order difference equations similar to differential equations for continuous time LTI systems. The *Z*-transform converts the *n*th order difference equations with initial conditions into simple linear algebraic equations in the *z*-domain. The system initial conditions are applied along with the transformation. By solving these algebraic equations, the required output in the *z*-domain is obtained. The output in discrete-time can be obtained by using the inverse *Z*-transform.

We will study the concept of *Z*-transform, its convergence and properties in this chapter.

8.2 Fundamental Difference between Continuous-and Discrete-time Signals

As stated in the first chapter, signals are functions of time that represent the physical or electrical quantities such as a room temperature, atmospheric pressure, the speed of a vehicle, a voltage or a statistical data. The continuous-time signal $f(t)$ is defined continuously for all values of time t. A sound signal in the atmosphere, or the function $f(t) = e^{-t} + \sin t$, $-\infty < t < \infty$, are examples of continuous time signals. Discrete-time signals are functions of discrete-time variable; they are defined only at specific instants of time interval and is represented by a sequence of numbers like $\{f[nT], n = 0, \pm 1, \pm 2, \ldots\}$, where n is an integer number representing discrete-time. Or, by considering $T = 1$,

$$f[n] = \{\ldots.f[-2], f[-1], \underset{\uparrow}{f[0]}, f[1], f[2], \ldots.\}. \tag{8.1}$$

The arrow indicates the $n = 0$ term.

As an example, the room temperature measured every hour in a day or a function $f[n] = [1, 2, 4, 6, 4, 2]$ are discrete-time signals. Figure 8.1 shows the waveforms of a continuous-time signal and a discrete-time signal.

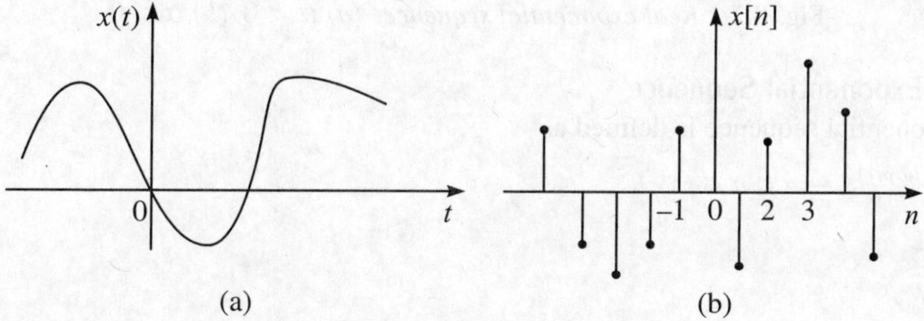

Fig. 8.1 *(a) Continuous-time signal and (b) discrete-time signal*

A continuous-time signal is an analog signal. A discrete-time signal or sampled signal can be derived from a continuous -time signal by sampling the continuous-time signal at equal sampling intervals of T_s seconds.

The continuous-time signal can be obtained by using a lowpass reconstruction filter as explained in Chapter 6. In analog signal processing, continuous-time signals are processed using passive circuit elements such as resistors, capacitors, inductors, and active components such as operational amplifiers and other nonlinear devices, whereas in digital signal processing, discrete-time signals are converted into digital signals for processing using digital computers. Analog-to-digital converters (A/D), and digital-to-analog converters (D/A) are used to convert continuous-time signals to digital signal and vice-versa.

8.3 Discrete-time Signal Representation Using Complex Exponential Components

A discrete exponential signal or sequence can be categorised as:

1. Discrete Complex Exponential Sequence

Recall from Chapter 1 that the discrete-time complex exponential sequence is defined as

$$x[n] = ce^{\alpha n}, \quad -\infty < n < \infty, \tag{8.2}$$

where c and α are complex numbers. Depending upon the values of these parameters, the complex exponential can take on several different characteristics.

2. Real Exponential Sequence

A real exponential *sequence* is defined as $x[n] = ce^{\alpha n}$, where both c and α are real parameters as explained in section 1.3. If α is positive, then as n increases $x[n]$ is a growing exponential; if α is negative, then as n increases $x[n]$ is a decaying exponential; and for $\alpha = 0$, $x[n] = c$ is a constant. For real values of c and α, the waveform is shown in Fig. 8.2.

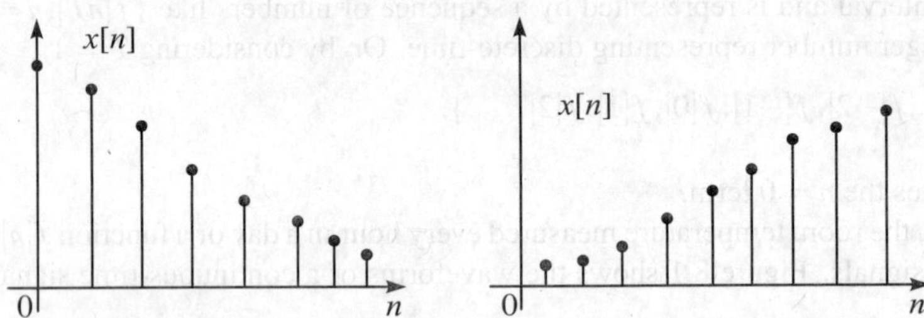

Fig. 8.2 *Real exponential sequence: (a) $\alpha < 0$, (b) $\alpha > 0$*

3. Imaginary Exponential Sequence

Imaginary exponential sequence is defined as

$$x\,[n] = e^{j\omega_0 n}, \qquad -\infty < n < \infty$$

$$\text{where } \omega_0 = \frac{2\pi}{N} \tag{8.3}$$

is the fundamental frequency. An important property of this signal is that it is periodic with period N since $x[n] = x[n+N]$ for all n. The signals $e^{j\omega_0 n}$ and $e^{-j\omega_0 n}$ both are periodic with the same fundamental period.

if $\omega_0 = 0$, then $x[n] = 1$, which is constant for any value of N.

8.4 Discrete-time Signal Representation Using Sinusoidal Components

A sinusoidal sequence can be defined as

$$x[n] = A\cos(\omega_0 n + \phi), \tag{8.4}$$

where A is the amplitude, $\omega_0 = 2\pi f_0 = \dfrac{2\pi}{N}$ the angular frequency in radians per second, and ϕ the phase shift in radians.

8.4.1 Sinusoidal Signals in Terms of Complex Exponential

A sinusoidal sequence can be written in terms of periodic complex exponentials with the same fundamental period by using Euler's relation,

$$A\cos(\omega_0 n + \phi) = \frac{A}{2}e^{j\phi}e^{j\omega_0 n} + \frac{A}{2}e^{-j\phi}e^{-j\omega_0 n}. \tag{8.5}$$

Note that the two exponentials in Eq. (8.5) have complex amplitudes. Alternatively, a sinusoidal sequence can be expressed in term of the complex exponential sequences as

$$A\cos(\omega_0 n+\phi) = A\mathrm{Re}\left\{e^{j(\omega_0 n+\phi)}\right\}$$

$$A\sin(\omega_0 n+\phi) = A\,\mathrm{Im}\{e^{j(\omega_0 n+\phi)}\} \tag{8.6}$$

where Re{.} denotes the real part and Im{.} denotes the imaginary part of the complex signal.

For example, if $A=1$ and $\phi=0$, the expressions $\mathrm{Re}\left\{e^{j\omega_0 n}\right\}=\cos(\omega_0 n)$ and $\mathrm{Im}\left\{e^{j\omega_0 n}\right\}=\sin(\omega_0 n)$ are shown in Fig. 8.3. If $\omega_0=0$, then $x[n]=A\cos\phi$ is a constant for any value of T.

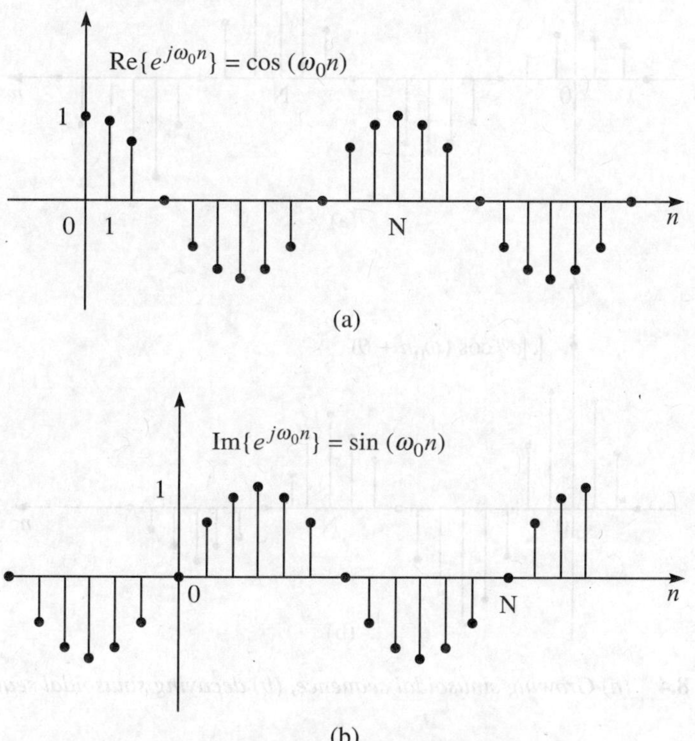

Fig. 8.3 *(a) The real part and (b) imaginary part of a complex exponential sequence*

Note: The sinusoidal sequence is also periodic with fundamental frequency ω_0, or time period N. Since $x[n]$ is periodic, $x[n]=x[n+N]$ for all n.

Similarly, the complex exponential sequence can be expressed in terms of sinusoidal sequences as

$$x[n] = ce^{j(\omega_0 n+\phi)} = c\left[\cos(\omega_0 n+\phi)+j\sin(\omega_0 n+\phi)\right]. \tag{8.7}$$

8.4.2 Damped Sinusoidal Signals

Consider a complex exponential $x[n]=ce^{\alpha n}$, where c and α are complex variables. If c is expressed in polar form and α in rectangular form, that is, $c=|c|e^{j\theta}$ and $\alpha=r+j\omega_0$, then, $ce^{at}=|c|e^{j\theta}e^{(r+j\omega_0)n}=|c|e^{nr}e^{j(\omega_0 n+\theta)}$.

Using Euler's relation,

$$ce^{an} = |c|e^{nr}\cos[\omega_0 n + \theta] + j|c|e^{nr}\sin[\omega_0 n + \theta]. \tag{8.8}$$

For $r = 0$, the real and imaginary parts of a complex exponential are sinusoidal. For $r > 0$ they correspond to sinusoidal signals multiplied by a growing exponential, and for $r < 0$ they correspond to sinusoidal signals multiplied by a decaying exponential. These two cases are shown in Fig. 8.4.

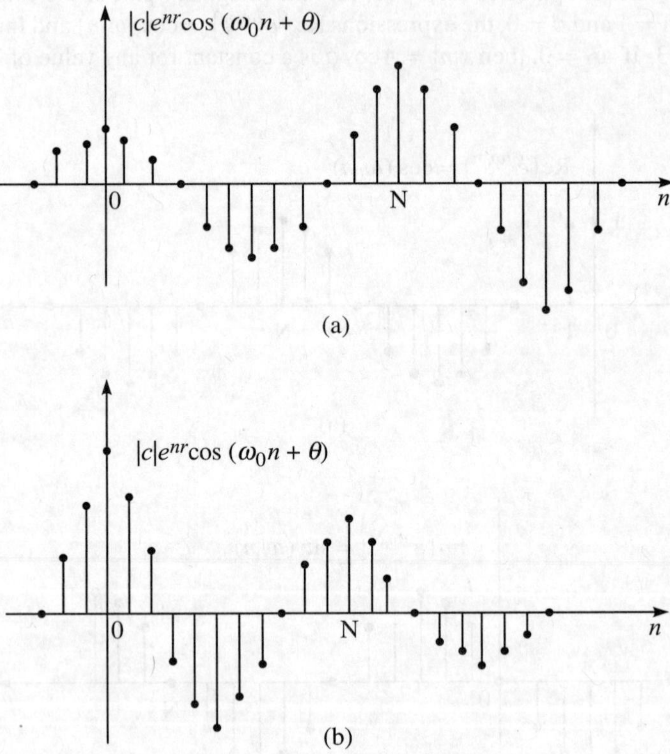

Fig. 8.4 *(a) Growing sinusoidal sequence, (b) decaying sinusoidal sequence*

8.5 Periodicity of Discrete-time Signal Using Complex Exponential Signal

An important property of the complex exponential or sinusoidal sequence is that it is periodic with single fundamental frequency ω_0 or time period N.

8.5.1 Periodicity of a Discrete-time Sequence

A discrete-time signal is periodic if it repeats itself at every N number of samples, i.e., a discrete-time signal $x[n]$ is said to be periodic if it satisfies the condition

$$x[n] = x[n+N],$$

where N is a positive integer. This condition is true if $N = \dfrac{2\pi k}{\omega_0}$ is an integer for any integer k or,

$$\frac{N}{k} = \frac{2\pi}{\omega_0} = \text{rational number.} \tag{8.9}$$

The smallest integer N is called fundamental period. $\omega_0 = \dfrac{2\pi}{N}$ rad/s is the fundamental angular frequency.

Sum of Periodic Sequences

If a sequence $x[n] = x_1[n] + x_2[n]$ is a sum of two periodic sequences with time periods N_1, N_2 and if $x[n]$ is periodic with period N, then the ratio of periods is a rational number.

$$\text{That is, } mN_1 = kN_2 = N \text{ or } \frac{N_1}{N_2} = \frac{k}{m} = \text{rational number.} \tag{8.10}$$

$$\text{Then the fundamental period is } N = \text{LCM}[N_1, N_2]. \tag{8.11}$$

The condition can be extended for a sum of n number of periodic sequences.

8.5.2 Periodicity of a Complex Exponential

Let the complex exponential signal be $x[n] = ce^{j\omega_0 n}$.

If it is periodic with period N, that is, $x[n+N] = x[n]$, then $N = \dfrac{2\pi k}{\omega_0}$ is an integer for any integer number k, or $\dfrac{N}{k} = \dfrac{2\pi}{\omega_0} = $ rational number.

Proof: Given $x[n] = ce^{j\omega_0 n} \quad -\infty < n < \infty$.

For any integer number $N, x[n+N] = ce^{j\omega_0(n+N)}$ or $x[n+N] = ce^{j\omega_0 n}e^{j\omega_0 N}$.

$$\text{If } \omega_0 > 0, \qquad e^{j\omega_0 N} = e^{j2\pi k} = 1 \text{ for any integer } k. \tag{8.12}$$

Then $x[n+N] = x[n]$ satisfies the periodic property $x[n+N] = x[n]$.

Therefore, from the Eq. (8.12),

$$\omega_0 N = 2\pi k \text{ or } \frac{N}{k} = \frac{2\pi}{\omega_0} = \text{ rational number.} \qquad \text{(Proved)}$$

Thus the fundamental period $N = \dfrac{2\pi}{\omega_0}k$ is the smallest positive value of N for which the periodicity holds.

Note: All complex exponential or sinusoidal sequences are not periodic. For example, if $\omega_0 = \sqrt{2}$, the sinusoidal sequence is not periodic since $\dfrac{N}{k} = \dfrac{2\pi}{\omega_0} = \dfrac{2\pi}{\sqrt{2}} = \sqrt{2}\pi$ is not a rational number.

Example 8.1

Determine whether or not each of the following discrete-time signals is periodic. If the signal is periodic, determine its fundamental period.

(a) $x[n] = e^{jn3\pi/8}$; (b) $x[n] = \sin(\sqrt{2}n + 30)$; (c) $x[n] = \cos\left(\dfrac{n}{2}\right)\cos\left(\dfrac{n\pi}{6}\right)$.

Solution

(a) Given $x[n] = e^{jn3\pi/8}$. The fundamental frequency is $\omega_0 = 3\pi/8$ rad/s and the period is

$N = \frac{2\pi k}{\omega_0} = \frac{2\pi k}{3\pi/8} = \frac{16k}{8}$.

For $k = 8$, the minimum period is $N = 16$. The signal is periodic.

(b) Given $x[n] = \sin(\sqrt{2}n + 30)$. The fundamental frequency is $\omega_0 = \sqrt{2}$ rad/s.

The period $N = 2\pi k/\omega_0 = \frac{2\pi k}{\sqrt{2}} = \sqrt{2}\pi k$ is not a rational number for all k. Therefore the signal is non-periodic.

(c) Given $x[n] = \cos\left(\frac{n\pi}{2}\right)\cos\left(\frac{n\pi}{6}\right) = \frac{1}{2}\left[\cos\left(\frac{2\pi}{3}n\right) + \cos\left(\frac{\pi}{3}n\right)\right]$.

The signal frequencies are $\omega_1 = 2\pi 1/3$ and $\omega_2 = \pi/3$.

Period $N_1 = 2\pi k/\omega_0 = \frac{2\pi k}{2\pi/3} = 3k$ is a rational number for all k.

Period $N_2 = 2\pi k/\omega_0 = \frac{2\pi k}{\pi/3} = 6k$ is a rational number for all k.

Fundamental period N = LCM $[N_1, N_2] = 6$. Therefore the signal is periodic.

8.6 Z-Transform of a Discrete Sequence

The Z-transform of a discrete-time sequence $x[n]$ is defined as

$$X(z) = \mathcal{Z}[x[n]] = \sum_{n=-\infty}^{\infty} x[n]z^{-n} \tag{8.13}$$

where $z = re^{j\omega}$ is a complex variable defined on a complex z-plane in the region of convergence. The z-plane is a polar plot with radius r and ω the angular frequency. Another notation for the Z-transform is $\mathcal{Z}[x[n]]$, where Z[.] is the Z-transform operator.

The inverse Z-transform is

$$x[n] = \mathcal{Z}^{-1}[X(z)] = \frac{1}{2\pi j}\oint_c X(z)z^{n-1}dz, \tag{8.14}$$

where the integration takes the path along a circle of radius $|z| = r$ in the anticlockwise direction and $Z^{-1}[.]$ is the inverse Z-transform operator. It is very difficult to evaluate the integration directly. Instead the inverse Z-transform can be evaluated using the standard table of Z-transform pairs.

The Z-transform pair of $x[n]$ and $X(z)$ is denoted by

$$x[n] \leftrightarrow X(z). \tag{8.15}$$

In general, if the input sequence extends from $-\infty \leq n \leq \infty$, the Z-transform of Eq.(8.13) is called the two-sided Z-transform or the bilateral Z-transform. In practice most of the causal signals start only at $n = 0$, that is, they are non-zero only for $n \geq 0$. Then the Z-transform is called the one-sided Z-transform or unilateral Z-transform and is defined as

$$X(z) = \sum_{n=0}^{\infty} x[n]z^{-n}. \tag{8.16}$$

Example 8.2

Determine the Z-transform of the sequences (a) $x[n] = [1, 2, 3, 0, 4]$, and (b) $x[n] = \{3, 2, 1, \underset{\uparrow}{0}, 1, 2, 3\}$.

Solution

(a) Given $x[n] = [1, 2, 3, 0, 4]$. Taking the Z-transform,

$$X(z) = \sum_{n=-\infty}^{\infty} x[n]z^{-n} = \sum_{n=0}^{4} x[n]z^{-n}$$

$$X(z) = 1 + 2z^{-1} + 3z^{-2} + 4z^{-4}.$$

(b) $x[n] = \{3, 2, 1, \underset{\uparrow}{0}, 1, 2, 3\}$

Taking the Z-transform,

$$X(z) = \sum_{n=-\infty}^{\infty} x[n]z^{-n} = \sum_{n=-3}^{3} x[n]z^{-n}$$

$$X(z) = 3z^3 + 2z^2 + z + z^{-1} + 2z^{-2} + 3z^{-3}.$$

8.7 Distinction between Laplace, Fourier and Z-Transforms

The Z-transform can be derived from the Laplace transform applied to the sampled signal. Let $x(t)$ be a continuous-time signal sampled by an ideal impulse train $\delta_T(t)$. The sampled signal is

$$x_s(t) = x(t)\delta_T(t) = \sum_{n=\mathrm{r}}^{\infty} x(t)\delta(t - nT). \tag{8.17}$$

Since the $\delta(t - nT)$ is non-zero only at sampling instants $t = nT$, $x(t)$ can be replaced with $x[nT]$.

$$x_s(t) = \sum_{n=-\infty}^{\infty} x[nT]\delta(t - nT).$$

Taking Laplace transform on both sides yields

$$x_s(s) = \mathcal{L}[x_s(t)] = \int_{-\infty}^{\infty} \sum_{n=\infty}^{\infty} x[nT]\delta(t - nT)e^{-st}\,dt.$$

Interchanging integration and summation,

$$X_s(s) = \sum_{n=\infty}^{\infty} x[nT] \int_{-\infty}^{\infty} \delta(t - nT)e^{-st}\,dt.$$

Using the shifting property,

$$X_s(s) = \sum_{n=\infty}^{\infty} x(nT)e^{-snT}.$$

Finally, the complex variable z is defined as $z = e^{sT}$. Then

$$\mathcal{L}[x_s(t)] = X(z) = \sum_{n=\infty}^{\infty} x[nT]z^{-n}. \tag{8.18}$$

Thus the Laplace transform of a sampled signal $x(nT)$ is equal to the Z-transform, where $z = e^{sT}$. For convenience, assume the sample time $T = 1$ second.

$$X(z) = \sum_{n=\infty}^{\infty} x[n]z^{-n}.$$

For causal signals where $x(t) = 0$ for $t \leq 0$, the sampled signal is defined in $0 \leq n \leq \infty$. The one- sided Z-transform is

$$X(z) = \sum_{n=0}^{\infty} x[n]z^{-n}.$$

It is noted that the $x[n]$ denotes the sample value and z^{-n} denotes the sample occurring at the nth sample period after the reference time. We know the Laplace operator $s = \sigma + j\omega$. Then the Z-transform operator is

$$z = e^{(\sigma+j\omega)T} = e^{\sigma T} e^{j\omega T} = re^{j\theta}.$$

The magnitude is $r = |z| = e^{\sigma T}$ and the angle is $\theta = \angle z = \omega T$. (8.19)

For $\sigma = 0$, or $s = j\omega$, then $z = e^{(\sigma+j\omega)} = e^{j\omega}$, and the Z-transform becomes

$$X(\omega) = \sum_{n=0}^{\infty} x[n]e^{-j\omega n}. \quad (8.20)$$

This represents the Fourier transform or frequency response of $x[n]$.

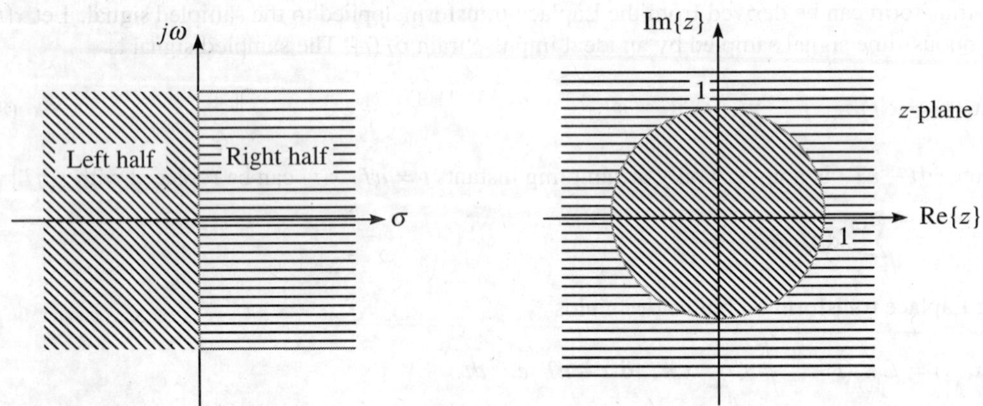

Fig. 8.5 *Mapping of the s-plane and the z-plane*

Mapping of s-plane and z-plane: The s-plane is a rectangular plot with σ along horizontal axis and $j\omega$ along vertical axis, whereas the z-plane is polar plot with radius $r = |z| = e^{\sigma T}$ and angle $\theta = \angle z = \omega T$. Therefore the mapping of the s-plane and z-plane is as follows:

1. The right half of the s-plane where $\sigma > 0$ corresponds to $|z| > 1$, while the left half of the s-plane where $\sigma < 0$ corresponds to $|z| < 1$.
2. The line $\sigma = 0$, or $s = j\omega$ on the s-plane corresponds to $|z| = 1$, a unit circle on the z-plane.

The mapping of the s-plane into the z-plane is shown in Fig. 8.5. The $s = j\omega$ axis is mapped to $|z| = 1$, a unit circle; the left half of the s-plane is mapped to the interior of the unit circle; and the right half of the s-plane is mapped to the outside of the unit circle in the z-plane. The planes provide the pole–zero plot. The poles in the left half of the s-plane correspond to poles inside the $|z| = 1$ circle. The poles in the right half of the s-plane correspond to poles outside the $|z| = 1$ circle. The poles on the $s = j\omega$ line of the s-plane correspond to poles on the $|z| = 1$ circle, since $s = 0$ corresponds to $z = 1$.

8.8 Region of Convergence (ROC) in Z-Transform

For a given sequence $x[n]$, the Z-transform exists if the infinite sum of series $x[n]z^{-n}$ converges. The range of values of the complex variable z for which the Z-transform converges is called the region of convergence (ROC). The region of convergence of $X(z)$ can be represented graphically with a pole-zero plot on the z-plane, where $z = re^{j\omega T}$, r is a radius and ωT is an angle in polar form.

Poles and Zeros of $X(z)$

The general polynomial form of $X(z)$ is given by

$$X(z) = \frac{N(z)}{D(z)} = \frac{b_0 + b_1 z^{-1} + b_2 z^{-2} + \cdots + b_m z^{-m}}{1 + a_1 z^{-1} + a_2 z^{-2} + \cdots + a_n z^{-n}}, \tag{8.21}$$

where a and b are real coefficients, m and n are positive integers. The roots of the numerator polynomial $N(z)$ are called the zeros of $X(z)$ because $X(z) = 0$ for these values of z. Similarly, the roots of the denominator polynomial $D(z)$ are called the poles of $X(z)$ because $X(z)$ is infinite for these values of z.

$X(z)$ can be represented graphically in polar form by locating poles and zeros in the z-plane as shown in Fig. 8.6. Generally, the symbol cross mark ("X") is used to indicate single pole location and the symbol circle mark ("0") to indicate zero location. Multiple pole is indicated with multiple cross symbols.

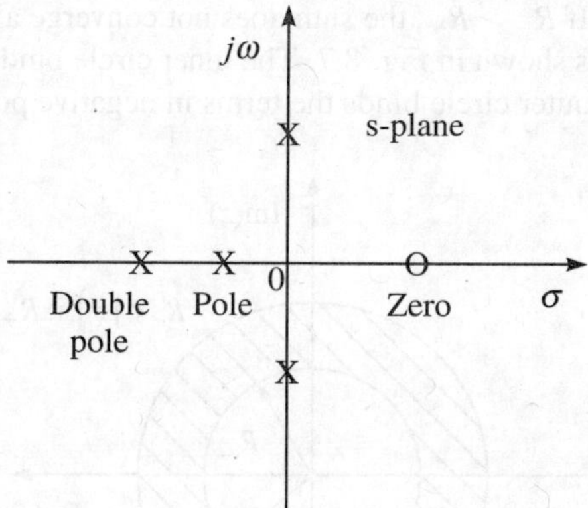

Fig. 8.6 *Pole–zero plot in the s-plane*

The Region of Convergence

Consider the Z-transform of a sequence $x[n]$ where

$$X(z) = \mathcal{Z}[x[n]] = \sum_{n=\infty}^{\infty} x[n]z^{-n}.$$

ROC is the range of complex values of z for which $\sum|x(n)z^{-n}|$ has a finite value. If we represent z in polar form as $z = re^{j\theta}$, we have

$$\sum_{n=\infty}^{\infty}|x[n]z^{-n}| = \sum_{n=\infty}^{\infty}|x[n](re^{j\theta})^{-n}| = \sum_{n=\infty}^{-1}|x[n]r^{-n}e^{-jn\theta}| + \sum_{n=0}^{\infty}|x[n]r^{-n}e^{-jn\theta}|$$

$$= \sum_{m=1}^{\infty}|x[-m]|r^{m} + \sum_{n=0}^{\infty}|x[n]|r^{-n}. \tag{8.22}$$

For the infinite sum $\sum_{n=-\infty}^{\infty}|x(n)z^{-n}|$ to be finite, each of the two sums in Eq. (8.22) must be finite. To make the sums finite, let the discrete-time sequence $x[n]$ be written as

$|x[n]| \leq MR_{+}^{n}$ $n < 0$ represents a left-sided sequence and
$|x[n]| \leq MR_{-}^{n}$ $n \geq 0$ represents a right-sided sequence,

where M, R_{+}, and R_{-} are positive real numbers. Then, substitute in Eq.(8.22)

$$\sum_{n=\infty}^{\infty}|x(n)z^{-n}| \leq M\left\{\sum_{m=1}^{\infty}\left(\frac{r}{R_{+}}\right)^{m} + \sum_{n=0}^{\infty}\left(\frac{R_{-}}{r}\right)^{n}\right\}$$

$$\sum_{n=-\infty}^{\infty}|x(n)z^{-n}| \leq M\left\{\sum_{m=1}^{\infty}R_{+}^{-m}r^{m} + \sum_{n=0}^{\infty}R_{-}^{n}r^{-n}\right\}. \tag{8.23}$$

The sums in Eq. (8.23) are finite if and only if $\frac{r}{R_{+}} < 1$ in the first term and $\frac{R_{-}}{r} < 1$ in the second term.

That is, the sequence $x[n]$ converges absolutely for all z in the region $\frac{r}{R_{+}} < 1$ and $\frac{R_{-}}{r} < 1$, or $R_{-} \leq r \leq R_{+}$; $R_{-} \leq |z| \leq R_{+}$. If $R_{-} \geq R_{+}$, the sum does not converge absolutely for any value of z.

The region of convergence is shown in Fig. 8.7. The inner circle binds the terms in positive powers of z away from the origin. The outer circle binds the terms in negative powers of z away from large $|z|$ values.

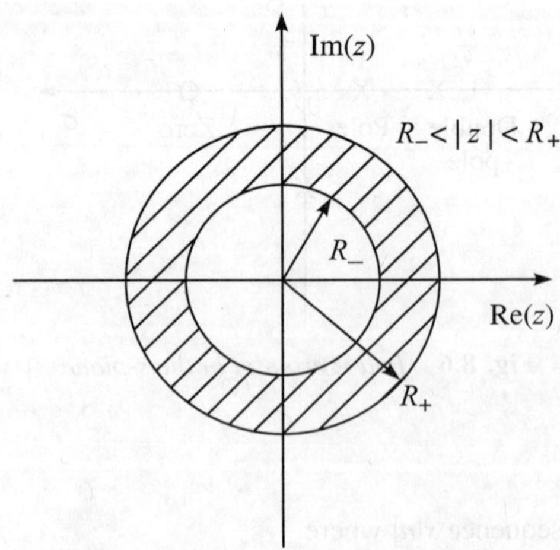

Fig. 8.7 *Region of convergence*

8.9 Constraints on ROC for Various Classes of Signals

Assume that $X(z)$ is a rational function of z. The region of convergence of $X(z)$ depends on the discrete-time nature of $x[n]$. Also, $X(z)$ depends only on magnitude $|z|$ but not on angle. Therefore if a particular value of $|z|$ is in the ROC, then all values of z lie in the same circle of radius $|z|$. Thus the ROC is in the form of a circular ring in the z-plane centered about the origin. The inner circle may extend to the origin and/or the outer circle may extend to infinity.

The sequence $x[n]$ may be a finite length or infinite length and it can be classified into six categories.

1. Right-sided sequence with finite duration; that is, $x[n]$ exists in a finite interval $N_1 \leq n \leq N_2$, where N_1 and N_2 are finite positive numbers.
2. Left-sided sequence with finite duration; that is, $x[n]$ exists in a finite interval $N_1 \leq n \leq N_2$, where N_1 and N_2 are finite negative numbers.
3. Two-sided sequence with finite duration; that is, $x[n]$ exists in a finite interval $N_1 \leq n \leq N_2$, where N_1 is a finite negative number and N_2 is a finite positive number. It is also called bilateral sequence with finite duration.
4. Right-sided sequence with infinite duration; that is, $x[n]$ exists in a interval $N \leq n \leq \infty$, where N is a finite positive number or zero. It is a causal sequence.
5. Left-sided sequence with infinite duration; that is, $x[n]$ exists in a interval $-\infty \leq n \leq N$, where N is a finite negative number or zero. It is non-causal sequence.
6. Two-sided sequence with infinite duration; that is, $x[n]$ is an infinite duration sequence existing on both right side as well as left side. It is also called bilateral sequence. It is non-causal sequence.

8.9.1 Properties of the ROC

Property 1: The ROC does not contain any poles since at all pole values $X(z)$ becomes infinite and does not converge.

Property 2: If $x[n]$ is a finite sequence and $X(z)$ converges for some value of z, then the ROC is the entire z-plane except possibly $z = 0$ and/or $z = \infty$.

(a) Right-sided sequence with finite duration; ROC is the entire z-plane except at $z = 0$.

(b) Left-sided sequence with finite duration: ROC is the entire z-plane except at $z = \infty$.

(c) Two-sided sequence with finite duration: ROC is the entire z-plane except at $z = 0$ and $z = \infty$.

Property 3: If $x[n]$ is a right-sided sequence, that is, $x[n]$ exists in a interval $N \leq n \leq \infty$ where N is a finite number and $X(z)$ converges for some value of z, then the ROC is in the form $|z| > r_{\max}$ or $\infty > |z| > r_{\max}$, where $r_{\max}$ is the largest value of any of the poles of $X(z)$. The ROC is the exterior of the circle $|z| = r_{\max}$ in the z-plane with the possible exception of $z = \infty$ (see Fig. 8.8(a)).

Property 4: If $x[n]$ is a left-sided sequence, that is, $x[n]$ exists in a interval $-\infty \leq n \leq N$ where N is a finite number and $X(z)$ converges for some value of z, then the ROC is of the form $|z| < r_{\min}$ or $0 < |z| < r_{\min}$, where $r_{\min}$ is the smallest magnitude of any of the poles of X(z). Thus, the ROC is the interior of the circle $|z| = r_{\min}$ in the z-plane with the possible exception of z = 0 (see Fig. 8.8(b)).

Property 5: If $x[n]$ is a two-sided sequence of infinite length, that is, $x[n]$ is an infinite-duration sequence existing on both right side as well as left side and $X(z)$ converges for some value of z, then the ROC is of the form $r_1 < |z| < r_2$, where r_1 and r_2 are the magnitudes of the two poles of $X(z)$.

Thus, the ROC is an annular ring in the z-plane between the circles $|z| = r_1$ and $|z| = r_2$ not containing any poles (see Fig. 8.8(c)).

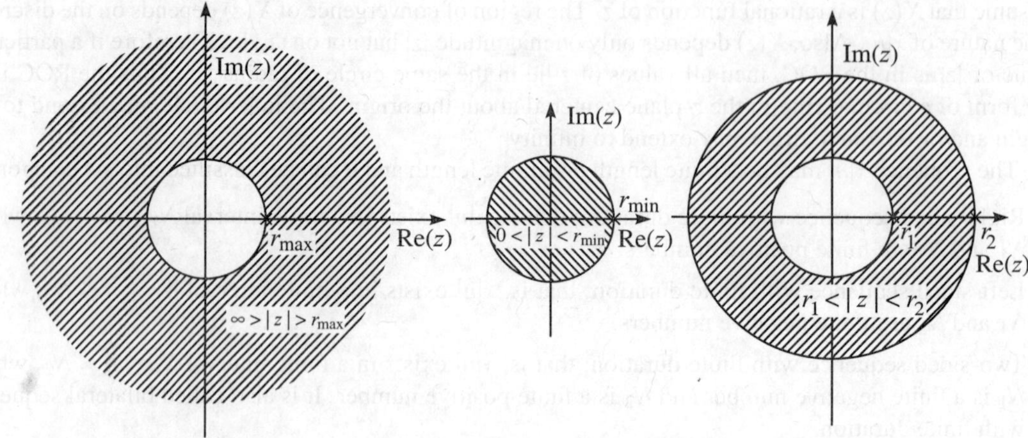

Fig. 8.8 *Region of convergence for (a) right-sided sequence, (b) left-sided sequence and (c) two-sided sequence*

8.9.2 Z-Transform for Some Basic Sequences

1. Unit Impulse Sequence

We know that the unit impulse sequence is defined as $\delta[n] = \begin{cases} 1 & n = 0 \\ 0 & \text{otherwise} \end{cases}$. Its Z-transform is

$$\mathcal{Z}[\delta[n]] = \sum_{n=r}^{\infty} \delta[n] z^{-n} = 1 \quad 0 < |z| \leq \infty. \tag{8.24}$$

The ROC is entire z-plane except $z = \infty$. The Z-transform pair is $\delta[n] \leftrightarrow 1$.

Also, for delayed unit impulse sequence $\delta[n-N]$ the z-transform is z^{-N} and for $\delta[n+N]$ the z-transform is

$$Z[\delta[n+N]] = \sum_{n=-\infty}^{\infty} \delta[-n+N] z^{-n+N} z^{-N} = z^{N}. \tag{8.25}$$

For example, suppose that

$$x[n] = \delta[n+2] + 2\delta[n+1] + 3\delta[n] + 2\delta[n-1] + \delta[n-2].$$

Then the Z-transform is $X(z) = z^2 + 2z^1 + 3 + 2z^{-1} + z^{-2}$, which converges absolutely for all z except $z = 0$ and $z = \infty$.

2. Unit Step Sequence

We know that the unit step sequence $u[n] = \begin{cases} 1 & n \geq 0 \\ 0 & \text{otherwise.} \end{cases}$ Its Z-transform is

$$\mathcal{Z}[u[n]] = \sum_{n=0}^{\infty} u[n] z^{-n} = \frac{z}{z-1}; \quad |z| > 1. \tag{8.26}$$

The ROC is $|z| > 1$ and the Z-transform pair is $u[n] \leftrightarrow \dfrac{z}{z-1}$.

3. Unit Ramp Sequence

We know that the unit ramp sequence $r[n] = \begin{cases} n & n \geq 0 \\ 0 & \text{otherwise} \end{cases}$.

$r[n] = nu[n]$.

The Z-transform is

$$\begin{aligned}
\mathcal{Z}[r[n]] &= \sum_{n=0}^{\infty} nu[n]z^{-n} = \sum_{n=0}^{\infty} nz^{-n} = z^{-1} + 2z^{-2} + 3z^{-3} + \ldots. \\
&= z^{-1}(1+z^{-1}+z^{-2}+\ldots.) + z^{-2}(1+z^{-1}+z^{-2}+\ldots.) + z^{-3}(1+z^{-1}+z^{-2}+\ldots.) + \ldots \\
&= (z^{-1}+z^{-2}+z^{-3}+\ldots.)(1+z^{-1}+z^{-2}+\ldots.) \\
&= z^{-1}(1+z^{-1}+z^{-2}+\ldots.)(1+z^{-1}+z^{-2}+\ldots.) = z^{-1}\left(\frac{1}{1-z^{-1}}\right)\left(\frac{1}{1-z^{-1}}\right) \\
&= z^{-1}\left(\frac{1}{1-z^{-1}}\right)^2 = \frac{z}{(z-1)^2} \text{ for } |z| > 1.
\end{aligned} \tag{8.27}$$

The ROC is $|z| > 1$ and the Z-transform pair is $r[n] \leftrightarrow \dfrac{z}{(z-1)^2}$.

4. Unilateral Exponential Sequence

We know that the unilateral exponential sequence is $x[n] = a^n u[n]$. From the basic definition of Z-transform,

$$\mathcal{Z}[x[n]] = \sum_{n=\infty}^{\infty} x[n]z^{-n}.$$

Replace $x[n]$ by $a^n u[n]$.

$$X(z) = \mathcal{Z}[a^n u[n]] = \sum_{n=-\infty}^{\infty} a^n u[n]z^{-n} = \sum_{n=0}^{\infty} a^n z^{-n} = \sum_{n=0}^{\infty} (az^{-1})^n.$$

The series expansion is

$$X(z) = 1 + (az^{-1}) + (az^{-1})^2 + (az^{-1})^3 + \ldots + (az^{-1})^n + \ldots.$$

Sum of infinite series yields

$$X(z) = \frac{1}{1-az^{-1}}, \quad \text{if } |az^{-1}| < 1.$$

$$\therefore \quad \mathcal{Z}[a^n u[n]] = \frac{z}{z-a} \qquad |z| > |a|. \tag{8.28}$$

The ROC is $|z| > a$ and the Z-transform pair is $a^n u[n] \leftrightarrow \dfrac{Z}{z-a}$.

The pole is at $z = a$ and zero is at $z = 0$. Hence the ROC of a right-sided sequence (causal sequence) is outside the circle of outer most pole at $|z| = |a|$. The pole-zero plot with ROC is shown in Fig. 8.9.

For example, consider a right-sided sequence $x[n] = \begin{cases} (-\frac{1}{2})^n & n \geq 0 \\ 0 & n < 0 \end{cases}$, or $x[n] = (-1/2)^n u[n]$.

The Z-transform is

$$X(z) = \frac{z}{z+\frac{1}{2}}$$

and $X(z)$ converges absolutely for all $|z| > \frac{1}{2}$. The ROC is the exterior of the circle $|z| = \frac{1}{2}$.

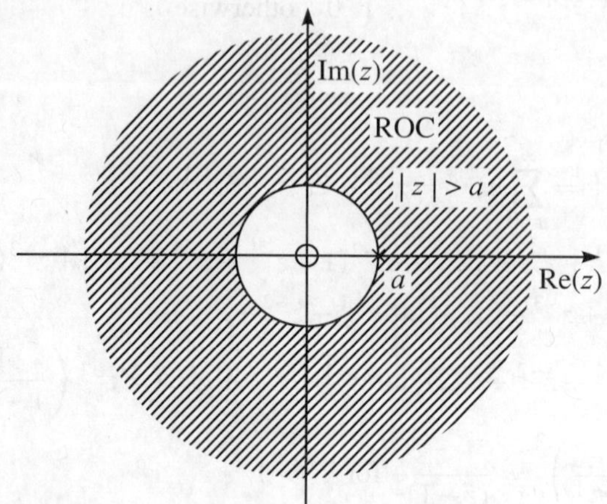

Fig. 8.9 *The pole–zero plot with ROC on z-plane of a right-sided sequence*

5. Rectangular Sequence

We know that the rectangular sequence for the period N is

$$G_N[n] = u[n] - u[n-N].$$

$$u[n] - u[n-N] = \delta[n] + \delta[n-1] + \delta[n-2] + \ldots\ldots + \delta[n-N].$$

The Z-transform is

$$\mathcal{Z}[G_N[n]] = \sum_{n=0}^{N-1} z^{-n} = 1 + z^{-1} + z^{-2} + z^{-3} + \ldots\ldots + z^{-N+1}.$$

The sum of finite series yields

$$\mathcal{Z}[G_N[n]] = \frac{1 - z^{-N}}{1 - z^{-1}}. \tag{8.29}$$

The ROC is $0 < |z| \leq \infty$. Therefore, for finite sequence, the ROC is the entire z-plane.

6. Unilateral Cosine and Sine Sequences

The unilateral cosine sequence is $x[n] = \cos(\omega_o n)u[n]$. Expanding, we get

$$\cos(\omega_0 n) = \frac{1}{2}e^{jn\omega_0} + \frac{1}{2}e^{-jn\omega_0}.$$

The Z-transform is

$$\mathcal{Z}[\cos(\omega_o n)u[n]] = \sum_{n=0}^{\infty}\left[\frac{1}{2}e^{jn\omega_0} + \frac{1}{2}e^{-jn\omega_0}\right]z^{-n}$$

$$= \frac{1}{2}\sum_{n=0}^{\infty}(e^{j\omega_0}z^{-1})^n + \frac{1}{2}\sum_{n=0}^{\infty}(e^{-j\omega_0}z^{-1})^n$$

$$= \frac{1}{2}\left(\frac{z}{z-e^{j\omega_0}} + \frac{z}{z-e^{-j\omega_0}}\right) = \frac{z}{2}\left(\frac{2z-e^{-j\omega_0}-e^{j\omega_0}}{z^2-z(e^{j\omega_0}+e^{-j\omega_0})+1}\right)$$

$$= \frac{z(z-\cos\omega_0)}{z^2-2z\cos\omega_0+1}. \tag{8.30}$$

The ROC is $|z|>1$ and the Z-transform pair is $\cos(\omega_0 n)u[n] \leftrightarrow \dfrac{z(z-\cos\omega_0)}{z^2-2z\cos\omega_0+1}$.

The unilateral sine sequence is $x[n]=\sin(\omega_0 n)u[n]$. Expanding, we get

$$\sin(\omega_0 n) = \frac{1}{2j}e^{jn\omega_0} - \frac{1}{2j}e^{-jn\omega_0}.$$

The Z-transform is

$$\mathcal{Z}[\sin(\omega_o n)u[n]] = \sum_{n=0}^{\infty}\left[\frac{1}{2j}e^{jn\omega_0} - \frac{1}{2j}e^{-jn\omega_0}\right]z^{-n}$$

$$= \frac{1}{2j}\sum_{n=0}^{\infty}(e^{j\omega_0}z^{-1})^n - \frac{1}{2j}\sum_{n=0}^{\infty}(e^{-j\omega_0}z^{-1})^n$$

$$= \frac{1}{2j}\left[\frac{Z}{z-e^{j\omega_0}} - \frac{Z}{z-e^{-j\omega_0}}\right] = \frac{Z}{2j}\left(\frac{e^{j\omega_0}-e^{-j\omega_0}}{z^2-z(e^{j\omega_0}+e^{-j\omega_0})+1}\right)$$

$$= \frac{z\sin\omega_0}{z^2-2z\cos\omega+1}. \tag{8.31}$$

The ROC is $|z|>1$ and the Z-transform pair is

$$\sin(\omega_0 n)u[n] \leftrightarrow \frac{z\sin\omega_0}{z^2-2z\cos\omega_0+1}.$$

Example 8.3

Find the Z-transform of an anti-causal sequence $x[n]=-a^n u[-n-1]$ and its ROC.

Solution

Given $x[n]=-a^n u[-n-1]$, a left-sided signal. From the basic definition of Z-transform,

$$X(z) = \sum_{n=-\infty}^{\infty} x[n]z^{-n}.$$

Then, $X(z) = \sum_{n=-\infty}^{\infty}[-a^n u[-n-1]]z^{-n}$.

Since $u[-n-1] = \begin{cases} 1 & n \le -1 \\ 0 & \text{otherwise} \end{cases}$.

$$X(z) = -\sum_{n=\infty}^{-1} a^n z^{-n} = -\sum_{n=-\infty}^{-1}(az^{-1})^n = -\sum_{n=1}^{\infty}(az^{-1})^{-n} = -\sum_{n=1}^{\infty}(a^{-1}z)^n.$$

The series expansion is

$$X(z) = -\left[(a^{-1}z)+(a^{-1}z)^2+(a^{-1}z)^3+\ldots\ldots\right]$$
$$= -(a^{-1}z)\left[1+(a^{-1}z)+(a^{-1}z)^2+\ldots\ldots\right].$$

The sum of this infinite series yields

$$X(z) = -\left(\frac{z}{a}\right)\left(\frac{1}{1-\frac{z}{a}}\right), \text{ if } \left|\frac{z}{a}\right|<1$$
$$= -\left(\frac{z}{a}\right)\left(\frac{a}{a-z}\right).$$

$$\therefore \quad \mathcal{Z}[-a^n u(-n-1)] = \frac{z}{z-a}, \quad |z|<|a|.$$

The pole is at $z=a$ and zero is at $z=0$. Hence the ROC of the left-sided sequence (anti-causal sequence) is inside the circle of the inner most pole at $|z|=|a|$. The pole-zero plot with ROC is shown in Fig. 8.10.

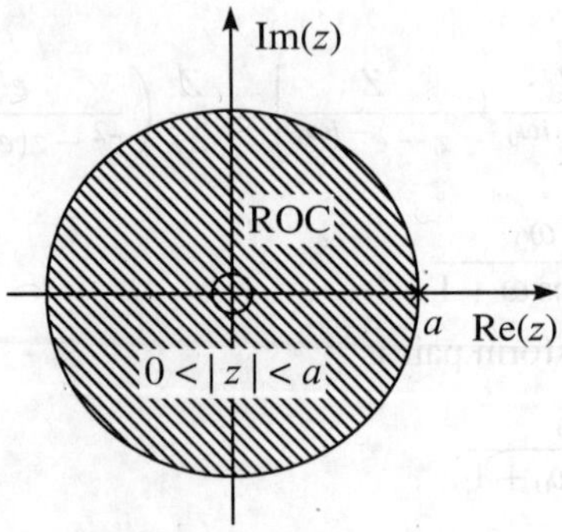

Fig. 8.10 *The pole–zero plot with ROC on z-plane of a left-sided sequence*

Note: The Z-transform of $a^n u[n]$ is identical to the Z-transform of $-a^n u[-n-1]$ only the ROC is different. To determine the time sequences we need both Z-transform and ROC.

Example 8.4

Find the Z-transform of the two sided sequence $x[n] = a^n u[n] - b^n u[-n-1]$. Plot the region of convergence and the zeros and poles for $b>a$.

Solution

The given $x[n]$ can be represented as $x[n] = x_1[n] + x_2[n]$, where

$$x_1[n] = a^n u[n]$$

$$x_2[n] = -b^n u[-n-1].$$

The Z-transform of $x_1[n]$ and $x_2[n]$ are

$$X_1(z) = \mathcal{Z}[x_1[n]] = \frac{z}{z-a} \quad |z|>a$$

$$X_2(z) = \mathcal{Z}[x_2[n]] = \frac{z}{z-b} \quad |z| < b$$

so that

$$X(z) = X_1(z) + X_2(z) = \frac{z}{z-a} + \frac{z}{z-b} \quad a < |z| < b$$

$$= \frac{2z\left(z - \dfrac{a+b}{2}\right)}{(z-a)}, \quad a < |z| < b.$$

The poles are at $z = a$ and $z = b$ and zeros are at $z = 0$ and $z = \dfrac{a+b}{2}$. Hence the ROC of both- sided sequence (non-causal sequence) is a ring located between two poles at $z = a$ and $z = b$. The pole- zero plot with ROC is shown in Fig. 8.11.

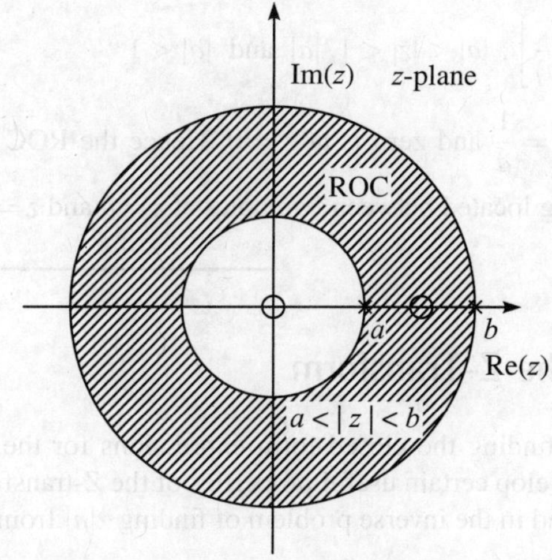

Fig. 8.11 *Pole–zero plot and ROC on z-plane of a two sided sequence*

Note: If $b \leq a$, the ROC does not exist.

Example 8.5

Find the Z-transform of a non-causal sequence $x[n] = a^{|n|}$.

Solution

Given $x[n] = a^{|n|}$, from the basic definition of Z-transform,

$$\mathcal{Z}[x[n]] = \sum_{n=\infty}^{\infty} x[n]z^{-n}.$$

$$\text{Then } X(z) = \sum_{n=-\infty}^{\infty} a^{|n|}z^{-n} = \sum_{n=-\infty}^{-1} a^{-n}z^{-n} + \sum_{n=0}^{\infty} a^n z^{-n}$$

$$= \sum_{n=-\infty}^{-1} (az)^{-n} + \sum_{n=0}^{\infty} (az^{-1})^n = \sum_{n=1}^{\infty} (az)^n + \sum_{n=0}^{\infty} (az^{-1})^n.$$

The series expansion is

$$X(z) = \left[az + (az)^2 + (az)^3 + \ldots\ldots\right] + \left[1 + (az^{-1})^1 + (az^{-1})^2 + (az^{-1})^3 \ldots\ldots\right]$$

$$= az\left[1 + az + (az)^2 + (az)^3 + \ldots\right] + \left[1 + (az^{-1})^1 + (az^{-1})^2 + (az^{-1})^3 \ldots\ldots\right].$$

The sum will converge if $|a| < 1$. The sum of this infinite series yields

$$X(z) = az\left[\frac{1}{1-az}\right] + \left[\frac{1}{1-\frac{a}{z}}\right], \text{ for } |az| < 1 \text{ and } |\frac{a}{z}| < 1$$

$$= \left[\frac{-z}{z-1/a}\right] + \left[\frac{z}{z-a}\right].$$

Thus,

$$X(z) = \left[\frac{z(a-1/a)}{(z-1/a)(z-a)}\right], \ |a| < |z| < 1/|a| \text{ and } |a| < 1.$$

The poles are at $z = a$ and $z = \dfrac{1}{a}$ and zero is at $z = 0$. Hence the ROC of the both-sided sequence (non-causal sequence) is a ring located between two poles at $z = a$ and $z = \dfrac{1}{a}$.

8.10 Properties of the Z-Transform

There are many methods of finding the closed-form expressions for the Z-transform of a given sequence. In this section we develop certain useful properties of the Z-transform that can be used both in determining $X(z)$ from $x[n]$ and in the inverse problem of finding $x[n]$ from the given $X(z)$.

1. Linearity (Superposition)

Let signals $x_1[n]$ and $x_2[n] \ldots x_m[n]$ have Z-transform $X_1(z)$, $X_2(z) \ldots\ldots X_m(z)$ respectively. The linearity property states that if

$$x_1[n] \leftrightarrow X_1(z) \text{ with ROC } R_1$$

$$x_2[n] \leftrightarrow X_2(z) \text{ with ROC } R_2$$

$$x_m[n] \leftrightarrow X_m(z) \text{ with ROC } R_m$$

then the linear combination of the signals satisfy

$$\mathcal{Z}[a_1x_1[n] + a_2x_2[n] + \ldots + a_mx_m[n]] = a_1X_1(z) + a_2X_2(z) + \ldots + a_mX_m(z)$$

$$a_1x_1[n] + a_2x_2[n] + \ldots + a_mx_m[n] \leftrightarrow a_1X_1(z) + a_2X_2(z) + \ldots + a_mX_m(z),$$

where $a_1, a_2, \ldots a_m$ are arbitrary constants.

Proof: Consider signals $x_1[n]$ and $x_2[n] \ldots x_m[n]$ having Z-transforms, $X_1(z)$, $X_2(z) \ldots\ldots X_m(z)$ respectively. Then,

$$\mathcal{Z}[a_1x_1[n]+a_2x_2[n]+\ldots+a_mx_m[n]] = \sum_{n=-\infty}^{\infty} (a_1x_1[n]+a_2x_2[n]+\ldots+a_mx_m[n])z^{-n}$$

$$= a_1 \sum_{n=-\infty}^{\infty} x_1[n]z^{-n} + a_2 \sum_{n=-\infty}^{\infty} x_2[n]z^{-n} + a_3 \sum_{n=-\infty}^{\infty} x_3[n]z^{-n} + \ldots + a_m \sum_{n=-\infty}^{\infty} x_m[n]z^{-n}$$

$$= a_1X_1(z)+a_2X_2(z)+\ldots+a_mX_m(z).$$

The combined ROC is the common area or intersection of all ROCs, that is, the ROC is $R_1 \cap R_2 \cap R_3 \ldots\ldots \cap R_m$. The Z-transform pair is

$$a_1x_1[n]+a_2x_2[n]+\ldots+a_mx_m[n] \leftrightarrow a_1X_1(z)+a_2X_2(z)+\ldots+a_mX_m(z).$$

Example 8.5

Find the Z-transform of $x[n]$ a signal that is the sum of three sequences

$$x[n] = (\frac{1}{4})^n u[n] + \left(\frac{1}{2}\right)^n u[n] + \delta[n-1].$$

Solution

The Z-transform of signal $x[n]$ is given by

$$X(z) = \sum_{n=-\infty}^{\infty} x[n]z^{-n}.$$

For the signal $x_1[n] = \left(\frac{1}{4}\right)^n$,

$$X_1(z) = \frac{1}{1-\frac{1}{4}z^{-1}} = \frac{z}{z-\frac{1}{4}};\ \text{ROC is } z > \frac{1}{4}.$$

For signal $x_2[n] = \left(\frac{1}{2}\right)^n u[n]$,

$$X_2(z) = \frac{1}{1-\frac{1}{2}z^{-1}} = \frac{z}{z-\frac{1}{2}};\ \text{ROC is } z > \frac{1}{2}.$$

And for the signal $x_3[n] = \delta[n-1]$,

$X_3(z) = z^{-1}$, ROC is all z, except $z = 0$.

Using linearity property, the sum of all signals is

$$X(z) = \frac{z}{z-\frac{1}{4}} + \frac{z}{z-\frac{1}{2}} + z^{-1}.$$

The ROC is $z > \frac{1}{4} \cap z > \frac{1}{2} \cap$ all z. Therefore the ROC is $z > \frac{1}{2}$.

The equation, therefore, has two poles at $z = 1/4$ and $z = 1/2$. Thus the region of convergence is $|z| > \frac{1}{2}$. The pole-zero plot and ROC for the Z-transform of the combined signal are shown in Fig. 8.12.

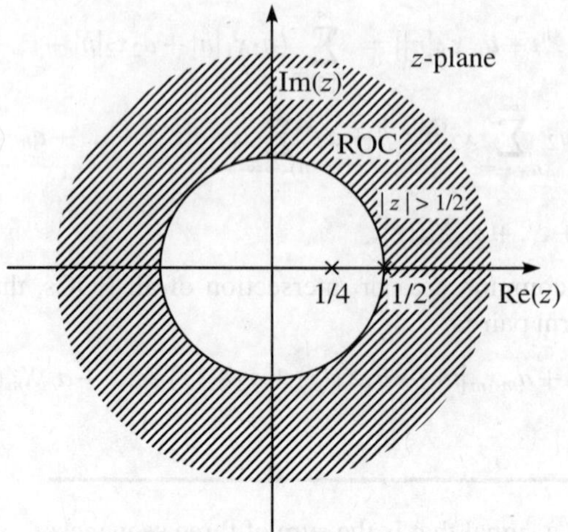

Fig. 8.12 *ROC for the given sequence*

2. Time Shifting Property

Bilateral sequence with zero initial conditions: The time shifting property states that if $x[n] \leftrightarrow X(z)$ with ROC R and zero as initial conditions, then

$$\mathcal{Z}[x[n \pm m]] = Z^{\pm m}X(z), \text{ with ROC } R. \tag{8.33}$$

Proof: Given $x[n] \leftrightarrow X(z)$ with ROC R, the Z-transform is

$$\mathcal{Z}[x[n-m]] = \sum_{n=-\infty}^{\infty} x[n-m]z^{-n}.$$

Let $k = n - m$.

$$X(z) = \sum_{k=-\infty}^{\infty} x(k)z^{-(k+m)} = z^{-m} \sum_{k=-\infty}^{\infty} x(k)z^{-k} = z^{-m}X(z).$$

Similarly, $\mathcal{Z}[x[n+m]] = z^m X(z)$.

Therefore $\mathcal{Z}[x[n \pm m]] = Z^{\pm m}X(z)$.

Unilateral lateral sequence with initial conditions: Consider a unilateral sequence $x[n]$ with initial conditions. The time shifting property states that if $x[n] \leftrightarrow X(z)$ with ROC R, then by shifting to the left,

$$\mathcal{Z}[x[n+m]u[n]] = z^m X(z) - z^m \sum_{k=0}^{m-1} x[k]z^{-k}, \quad \text{ROC} : R \tag{8.34}$$

and by shifting to the right

$$\mathcal{Z}[x[n-m]u[n]] = z^{-m}X(z) + z^{-m}\sum_{k=1}^{m} x[-k]z^k, \quad \text{ROC} : R. \tag{8.35}$$

Proof: Given $x[n] \leftrightarrow X(z)$ with ROC : R.

(a) Then the Z-transform of $x[n+m]u[n]$ is

$$\mathcal{Z}[x[n+m]u[n]] = \sum_{n=0}^{\infty} x[n+m]z^{-n} = z^m \sum_{n=0}^{\infty} x[n+m]z^{-(n+m)}.$$

Letting $k = n+m$,

$$= z^m \left(\sum_{m}^{\infty} x(k)z^{-k} \right) = z^m \left(\sum_{k=0}^{\infty} x[k]z^{-k} - \sum_{k=0}^{m-1} x[k]z^{-k} \right)$$

$$\mathcal{Z}[x[n+m]u[n]] = z^m X(z) - z^m \sum_{k=0}^{m-1} x[k]z^{-k}, \quad \text{where, } m \text{ is an integer.}$$

(b) The Z-transform of $x[n-m]u[n]$ is

$$\mathcal{Z}[x[n-m]u[n]] = \sum_{n=0}^{\infty} x[n-m]z^{-n} = z^{-m} \sum_{n=0}^{\infty} x[n-m]z^{-(n-m)}.$$

Let $k = n-m$. Then

$$\mathcal{Z}[x[n-m]u[n]] = z^{-m} \left(\sum_{-m}^{\infty} x[k]z^{-k} \right), \quad \text{where } m \text{ is an integer.}$$

$$= z^{-m} \left(\sum_{k=0}^{\infty} x[k]z^{-k} + \sum_{k=-m}^{-1} x[k]z^{-k} \right) = z^{-m} \left(X(z) - \sum_{k=1}^{m} x[-k]z^{k} \right)$$

$$\mathcal{Z}[x[n-m]u[n]] = z^{-m}X(z) - z^{-m} \sum_{k=1}^{m} x[-k]z^{k}, \quad \text{where } m \text{ is an integer.}$$

For example, if $m = 2$, we have

$Z[x[n+2]u[n]] = z^2X(z) - z^2x[0] - zx[1]$ and

$Z[x[n-2]u[n]] = z^{-2}X(z) - z^{-1}x[-1] - x[-2]$.

where $x[0], x[1], x[-1]$ and $x[-2]$ are the initial conditions.

Note: For unilateral sequence with zero initial conditions, the Z-transform is $Z[x[n \pm m]] = Z^{\pm m}X(z)$.

Example 8.7

Find the Z-transform of $x[n] = (0.5)^{n-2}u[n-2]$.

Solution

We know that the Z-transform of $(0.5)^n u[n]$ is

$$X_1(z) = \mathcal{Z}[(0.5)^n u[n]] = \frac{z}{z-0.5}.$$

Using time shifting property, n is delayed by $n-2$.

$$X(z) = \mathcal{Z}[(0.5)^{n-2}u[n-2]] = z^{-2}X_1(z) = z^{-2}\frac{z}{z-0.5} = \frac{z^{-1}}{z-0.5}.$$

The ROC is $|z| > 1/2$.

Example 8.8

If x $[n] = (0.5)^n u[n]$ with initial conditions $x[-1] = 2$, $x[-2] = 4$, calculate $\mathcal{Z}[x[n-2]u[n]]$.

Solution

Given $x[n] = (0.5)^n u[n]$. Using scaling property,

$$X(z) = \frac{z}{z-1/2}. \quad \text{ROC is } |z| > 1/2.$$

Using time shifting property for causal signal,

$$\mathcal{Z}[x[n-2]u[n]] = z^{-2}X(z) + z^{-1}x[-1] + x[-2]$$

$$= z^{-2}\left(\frac{z}{z-1/2}\right) + x[-1]z^{-1} + x[-2].$$

The initial conditions are $x[-1] = 2$, $x[-2] = 4$. Therefore,

$$\mathcal{Z}[x[n-2]u[n]] = z^{-2}\left(\frac{z}{z-1/2}\right) + 2z^{-1} + 4$$

$$= z^{-2}\frac{z}{z-1/2} + \frac{(4+2z^{-1})(z-1/2)}{z-1/2}$$

$$\mathcal{Z}[x[n-2]u[n]] = \frac{z^{-1} + (4z - z^{-1})}{z-1/2} = \frac{4z}{z-1/2}.$$

3. Time Reversal Property

If $x[n] \leftrightarrow X(z)$ with ROC $= R$, then $\mathcal{Z}[x[-n]] = X\left(\frac{1}{z}\right)$ with ROC $= \frac{1}{R}$. (8.36)

Proof: Given $x[n] \leftrightarrow X(z)$ with ROC$= R$. The Z-transform of $x[-n]$ is

$$\mathcal{Z}[x[-n]] = \sum_{n=-\infty}^{\infty} x[-n]^{-n}.$$

Letting $k = -n$,

$$= \sum_{k=\infty}^{\infty} x[k]z^k = \sum_{k=-\infty}^{\infty} x[k](z^{-1})^{-k}$$

$$\mathcal{Z}[x[-n]] = X(z^{-1}).$$

Since the sequence is in negative axis, the ROC is $1/R$.

4. Time Expansion Property

The time expansion of discrete-time sequences can be accomplished by inserting a number of zeros between successive values of a discrete-time sequence $x[n]$. These types of sequences have an important role in the analysis of discrete-time signals and system. Consider a sequence and its Z-transform $X(z)$.

Then the sequence for a change in time instants, $x_k[n]$ is defined as

$$x_k[n] = \begin{cases} x\left[\frac{n}{k}\right] & \text{if } n \text{ is a multiple of } k\,, \\ 0 & \text{if } n \text{ is not a multiple of } k \end{cases}$$

The time extension property states that if $x[n] \leftrightarrow X(z)$ with ROC$=R$, then the Z-transform of $x_k[n]$ is

$$\mathcal{Z}[x_k[n]] = X(z^k). \tag{8.37}$$

The ROC depends on the location of poles on the z-plane.

Proof: Given $x[n] \leftrightarrow X(z)$. The Z-transform of $x_k[n]$

$$\mathcal{Z}[x_k[n]] = Z\left[x\left[\frac{n}{k}\right]\right] = \sum_{n=-\infty}^{\infty} x\left[\frac{n}{k}\right] z^{-n}.$$

Letting $\frac{n}{k} = m$,

$$\mathcal{Z}[x_k[n]] = \sum_{m=-\infty}^{\infty} x[m] z^{-km} = \sum_{m=-\infty}^{\infty} x[m](z^k)^{-m} = X(z^k).$$

Note 1: If the $x[n]$ is even sequence, that is, $x[n] = x[-n]$, then using time expansion property with $k = -1$, the Z-transform is $\mathcal{Z}[x_{-1}[n]] = X\left(\frac{1}{z}\right)$. Therefore, for even sequence, the Z-transform satisfies

$$X(z) = X\left(\frac{1}{z}\right). \tag{8.38}$$

Note 2: Similarly, if $x[n]$ is odd sequence, that is, $x[n] = -x[-n]$, then the Z-transform is $-\mathcal{Z}[x_{-1}[n]] = -X\left(\frac{1}{z}\right)$. Therefore, for odd sequence the Z-transform satisfies

$$X(z) = -X\left(\frac{1}{z}\right). \tag{8.39}$$

5. Scaling Property

If $x[n] \leftrightarrow X(z)$ with ROC $= R$, then the scaling property states that the Z-transform of $a^n x[n]$ is

$$\mathcal{Z}[a^n x[n]] = X\left(\frac{z}{a}\right) \text{ with ROC} = aR. \tag{8.40}$$

Proof: Given $x[n] \leftrightarrow X(z)$ with ROC $= R$. The Z-transform of $a^n x[n]$ is

$$\mathcal{Z}[a^n x[n]] = \sum_{n=-\infty}^{\infty} a^n x[n] z^{-n} = \sum_{n=-\infty}^{\infty} x[n]\left(\frac{z}{a}\right)^{-n} = X\left(\frac{z}{a}\right).$$

Similarly,

$$\mathcal{Z}[a^{-n} x[n]] = X(az) \text{ with ROC} = \frac{R}{a}.$$

Example 8.9

Find the Z-transform of $x[n] = (0.5e^{j\omega T})^n u[n]$.

Solution

Given $x[n] = (0.5e^{j\omega T})^n u[n]$. We know that the Z-transform of $u[n] \leftrightarrow \frac{z}{z-1}$, ROC is $z > 1$.

Using scaling property,

$$\mathcal{Z}[a^n x[n]] = X\left(\frac{z}{a}\right) \text{ with ROC} = aR.$$

$$\therefore \quad \mathcal{Z}\left[(0.5e^{j\omega T})^n u[n]\right] = u\left(\frac{z}{0.5e^{j\omega T}}\right); \text{ ROC} = \frac{z}{0.5e^{j\omega T}} > 1.$$

$$\text{Then, } X(z) = \frac{\dfrac{z}{0.5e^{j\omega T}}}{\dfrac{z}{0.5e^{j\omega T}} - 1} = \frac{z}{z - 0.5e^{j\omega T}}.$$

ROC is $|z| > 0.5e^{j\omega T}$ or *ROC* is $|z| > 0.5$.

---

6. Time Multiplication or Differentiation in the Z-Domain

If $x[n] \leftrightarrow X(z)$ with ROC $= R$, then time multiplication property states that the Z-transform is

$$\mathcal{Z}[nx(n)] = -z\frac{dX(z)}{dz} \text{ and ROC} = R.$$

Proof: Given $x[n] \leftrightarrow X(z)$ with ROC$= R$. The Z-transform of $x[n]$ is

$$X(z) = \mathcal{Z}[x[n]] = \sum_{n=-\infty}^{\infty} x[n]z^{-n}.$$

Taking differentiation with respect to z,

$$\frac{dX(z)}{dz} = \frac{d}{dz}\left[\sum_{n=-\infty}^{\infty} x[n]z^{-n}\right]$$

$$= \sum_{n=-\infty}^{\infty} \frac{d}{dz}(x[n]z^{-n}) = \sum_{n=-\infty}^{\infty} (-n)x[n]z^{-n-1}$$

$$\sum_{n=-\infty}^{\infty} (nx[n])z^{-n} = -z\frac{dX(z)}{dz}.$$

$$Z[nx[n]] = -z\frac{dX(z)}{dz} \text{ and ROC is } R.$$

Example 8.10

Given $x[n] = nu[n]$, find the Z-transform of the signal $(n-1)x[n] + 3^n x[n]$.

Solution

Given $x[n] = nu[n]$ and the signal $(n-1)x[n] + 3^n x[n]$.

Since $Z[u[n]] = \frac{z}{z-1}$, using time multiplication property,

$$X(z) = Z[nu[n]] = -z\frac{d}{dz}\left(\frac{z}{z-1}\right)$$

or $X(z) = \dfrac{z}{(z-1)^2}$.

The first term is $(n-1)x[n] = nx[n] - x[n] = n^2u[n] - nu[n]$.

Its Z-transform is

$$\mathcal{Z}[(n-1)x[n]] = \mathcal{Z}[n^2u[n]] - \mathcal{Z}[nu[n]]$$

$$= -z\frac{d}{dz}\left(\frac{z}{(z-1)^2}\right) - \frac{z}{(z-1)^2}$$

$$= \frac{z(z+1)}{(z-1)^3} - \frac{z}{(z-1)^2} = \frac{z^2+z-(z^2-z)}{(z-1)^3} = \frac{2z}{(z-1)^3}.$$

The Z-transform of the second term is, using scaling property,

$$\mathcal{Z}[3^n x[n]] = X\left[\frac{z}{3}\right] = \frac{(z/3)}{(z/3-1)^2} = \frac{3z}{(z-3)^2}.$$

The Z-transform of the combined signal is

$$\mathcal{Z}[(n-1)x[n] + 3^n x[n]] = \frac{2z}{(z-1)^3} + \frac{3z}{(z-3)^2} \text{ and ROC is } |z| > 3.$$

7. Complex Conjugation Property

If $x[n] \leftrightarrow X(z)$ with ROC $= R$, then $x^*[n] \leftrightarrow X^*(z^*)$ with ROC $= R$. (8.42)

Proof: Given $x[n] \leftrightarrow X(z)$ with ROC $= R$. The Z-transform of $x^*[n]$ is

$$\mathcal{Z}[x^*[n]] = \sum_{n=-\infty}^{\infty} x^*[n]z^{-n}$$

$$= \sum_{n=-\infty}^{\infty} x^*[n]z^{-n} = \left(\sum_{n=-\infty}^{\infty} x[n](z^*)^{-n}\right)^*$$

$$= \left(\sum_{n=-\infty}^{\infty} x[n](z^*)^{-n}\right)^* = (X(z^*))^* = X^*(z^*).$$

$x^*[n] \leftrightarrow X^*(z^*)$ with ROC $= R$.

8. Convolution in Time Domain

If two sequences are given by $x[n] \leftrightarrow X(z)$ with ROC R_1 and $y[n] \leftrightarrow Y(z)$ with ROC R_2, then the Z-transform of $x[n] * y[n]$ is

$$\mathcal{Z}[x[n] * y[n]] = X(z)Y(z) \text{ with ROC } R_1 \cap R_2,$$

where the convolution of $x[n]$ and $y[n]$ is

$$x[n] * y[n] = \sum_{m=-\infty}^{\infty} x[m]y[n-m]. \quad (8.43)$$

Proof: Given two sequences $x[n] \leftrightarrow X(z)$ and $y[n] \leftrightarrow Y(z)$. Then the convolution of the two signals is

$$x[n] * y[n] = \sum_{m=-\infty}^{\infty} x[m]y[n-m].$$

The Z-transform is

$$\mathcal{Z}[x[n] * y[n]] = \sum_{n=-\infty}^{\infty} \left(\sum_{m=-\infty}^{\infty} x[m]y[n-m] \right) z^{-n}$$

$$= \sum_{m=-\infty}^{\infty} x[m] \sum_{n=-\infty}^{\infty} y[n-m]z^{-n} = \sum_{m=r}^{\infty} x[m]z^{-m} \sum_{n=-\infty}^{\infty} y[n-m]z^{-(n-m)}$$

$$= X(z)Y(z).$$

The ROC is the common region of the two ROCs, i.e., ROC $R_1 \cap R_2$.

Thus the Z-transform of convolution of two sequences gives the multiplication in the z domain.

Example 8.11

Convolve the following two signals, $x_1[n] = [1,1,1]$ and $x_2[n] = [0.5,1,0.5]$.

Solution

$$x_1[n] = [1,1,1],\ x_2[n] = [0.5,1,0.5].$$

$$x_1[n] = \delta[n] + \delta[n+1] + \delta[n+2]$$

$$x_2[n] = 0.5\delta[n] + \delta[n+1] + 0.5\delta[n+2]$$

Taking Z-transform,

$$X_1(z) = 1 + z^{-1} + z^2$$

$$X_2(z) = 0.5 + z^{-1} + 0.5z^2.$$

The Z-transform of convolution of $x[n] = x_1[n] * x_2[n]$ is

$$X(z) = X_1(z)X_2(z) = (1 + z^{-1} + z^{-2})(0.5 + z^{-1} + 0.5z^{-2})$$

$$= 0.5 + (1+0.5)z^{-1} + (0.5+0.5+1)z^{-2} + (1+0.5)\, z^{-3} + 0.5z^{-4}.$$

$$X(z) = 0.5 + 1.5z^{-1} + 2z^{-2} + 1.5z^{-3} + 0.5z^{-4}.$$

Taking the inverse Z-transform, by power series,

$$x[n] = 0.5\delta[n] + 1.5\delta[n+1] + 2\delta[n+2] + 1.5\delta[n+3] + 0.5\delta[n+4].$$

9. Convolution in z-Domain or Multiplication of Two Sequences

Consider two complex sequences with Z-transforms given by

$$x[n] \leftrightarrow X(z) \text{ with } R_1 \text{ and } y[n] \leftrightarrow Y(z) \text{ with ROC } R_2.$$

Then the multiplication of the two sequences gives

$$x[x]y[n] \leftrightarrow \frac{1}{2\pi j}\oint_c X(v)Y\left(\frac{z}{v}\right)v^{-1}dv \quad \text{with ROC } R_1 \cap R_2. \tag{8.44}$$

Proof: Given two complex sequences $x[n] \leftrightarrow X(z)$ with ROC R_1 and $y[n] \leftrightarrow Y(z)$ with ROC R_2. We know that the inverse Z-transform is

$$x[n] = \frac{1}{2\pi j}\oint_c X(v)v^{n-1}dv$$

$$x[n]y[n] = \frac{1}{2\pi j}\oint_c X(v)v^{n-1}dvy[n].$$

Taking Z-transform,

$$= \frac{1}{2\pi j}\sum_{n=-\infty}^{\infty}\oint_c X(v)v^{n-1}dvy[n]z^{-n} = \frac{1}{2\pi j}\oint_c X(v)\left(\sum_{n=-\infty}^{\infty} y[n]v^n\right)v^{-1}z^n dv$$

$$= \frac{1}{2\pi j}\oint_c X(v)\left(\sum_{n=-\infty}^{\infty} y[n]\left(\frac{z}{v}\right)^{-n}\right)v^{-1}dv = \frac{1}{2\pi j}\oint_c X(v)Y\left(\frac{z}{v}\right)v^{-1}dv.$$

$$x[n]y[n] \leftrightarrow \frac{1}{2\pi j}\oint X(v)Y\left(\frac{z}{v}\right)v^{-1}dv. \qquad \text{(Proved)}$$

10. The Initial Value Theorem

The initial value theorem states that, if $x[n]$ is a causal sequence i.e., $x[n] = 0$ for $n < 0$ and If $x[n] \leftrightarrow X(z)$ with ROC$= R$, then

$$x[0] = \lim_{n\to 0} x[n] = \lim_{z\to\infty} X(z).$$

Proof: Given $x[n] \leftrightarrow X(z)$ with ROC $= R$. For causal sequences the Z-transform is

$$X(z) = \sum_{n=0}^{\infty} x[n]z^{-n} = x[0] + x[1]z^{-1} + x[2]z^{-2} + \cdots$$

As $z \to \infty$, $z^{-n} \to 0$. Taking limits on both sides,

$$\lim_{z\to\infty} X(z) = \lim_{z\to\infty}\left[x[0] + x[1]z^{-1} + x[2]z^{-2} + \cdots\right] = x[0] + x[1](0) + x[2](0) + \ldots. = x[0].$$

Therefore, $x[0] = \lim_{n\to 0} x[n] = \lim_{z\to\infty} X(z)$.

One consequence of the initial-value theorem is that for a causal sequence if $x[0]$ is finite. then $\lim_{z\to\infty} X(z)$ is finite. Consequently, with $X(z)$ expressed as a ratio of polynomials in z, the order of the numerator polynomial cannot be greater than the order of the denominator polynomial ; or, equivalently, the number of zeros of $X(z)$ cannot be greater then the numbers of finite poles.

Example 8.12

Find the initial value of the signal having Z-transform $X(z) = \dfrac{2+z^{-1}}{1-z^{-1}+18z^{-2}}$.

Solution

Given $X(z) = \dfrac{2+z^{-1}}{1-z^{-1}+1.8z^{-2}}$. From the initial value theorem we know that

$$x[0] = \lim_{n\to 0} x[n] = \lim_{z\to\infty} X(z).$$

$$= \lim_{z\to\infty} X(z) = \lim_{z\to\infty} \frac{2+z^{-1}}{1-z^{-1}+1.8z^{-2}}$$

$$= \frac{2+0}{1-0+0} = 2.$$

The initial value is 2.

---

11. Final Value Theorem

The final value theorem states that if x $[n]$ is a causal sequence, i.e., if $x[n] = 0$ for $n < 0$ and if $x[n] \leftrightarrow X(z)$ with ROC $= R$ then,

$$x[\infty] = \lim_{n\to\infty} x[n] = \lim_{z\to 1}(z-1)X(z). \tag{8.46}$$

Proof: Given $x[n] \leftrightarrow X(z)$ with ROC $= R$. For causal sequences the Z-transform of $x[n]$ is

$$X(z) = \sum_{n=0}^{\infty} x[n]z^{-n}.$$

To find out the final value of the sequence, consider the Z-transform of $x[n+1]$. Using shifting property,

$$\mathcal{Z}[x[n+1]] = zX(z) - x[0]$$

$$\mathcal{Z}[x[n+1] - x[n]] = zX(z) - x[0] - X(z)$$

$$= (z-1)X(z) - x[0].$$

$$\sum_{n=0}^{\infty} [x[n+1] - x[n]]z^{-n} = (z-1)X(z) - x[0].$$

$$(z-1)X(z) = \sum_{n=0}^{\infty} [x[n+1] - x[n]]z^{-n} + x[0].$$

Taking limits on both sides,

$$\lim_{z\to 1}(z-1)X(z) = x[0] + \lim_{z\to 1} \sum_{n=0}^{\infty} [x[n+1] - x[n]]z^{-n}$$

$$= x[0] + [x[1] - x[0]] + [x[2] - x[1]] + [x[3] - x[2]] + \ldots [x[\infty] - x[\infty - 1]] = x[\infty]$$

That is, $\lim_{z\to 1}(z-1)X(z) = x(\infty)$.

Example 8.13

Find the final value of the signal having Z-transform $X(z) = \dfrac{2z^{-1}}{1-1.8z^{-1}+0.8z^{-2}}$.

Solution

Given $X(z) = \dfrac{2z^{-1}}{1-1.8z^{-1}+0.8z^{-2}}$. From the final value theorem we know that $x(\infty) = \lim_{z\to 1}(z-1)X(z)$. Then,

$$(z-1)X(z) = \frac{2z(z-1)}{z^2-1.8z+0.8} = \frac{2z(z-1)}{(z-1)(z-0.8)} = \frac{2z}{(z-0.8)}.$$

Now, $x(\infty) = \lim_{z\to 1}(z-1)X(z) = \lim_{z\to 1}\dfrac{2z}{(z-0.8)} = 10.$

Therefore, the final value is 10.

12. Correlation of two sequences

If two sequences are given as $x[n] \leftrightarrow X(z)$ with ROC R_1 and $y[n] \leftrightarrow Y(z)$ with ROC R_2, then the Z-transform of $x[n] \otimes y[n]$ is $\mathcal{Z}[x[n] \otimes y[n]] = X(z)Y(z^{-1})$ with ROC $R_1 \cap R_2$.

The correlation is

$$x[n] \otimes y[n] = Z^{-1}[X(z)Y(z^{-1})], \tag{8.47}$$

where the correlation of $x[n]$ and $y[n]$ is

$$x[n] \otimes y[n] = \sum_{m=-\infty}^{\infty} x[m]y[m-n].$$

Proof: Given two sequences $x[n] \leftrightarrow X(z)$ and $y[n] \leftrightarrow Y(z)$, the correlation of two sequences is

$$x[n] \otimes y[n] = \sum_{m=-\infty}^{\infty} x[m]y[m-n].$$

The *Z*-transform is

$$\begin{aligned}
\mathcal{Z}[x[n] \otimes y[n]] &= \sum_{n=-\infty}^{\infty}\left(\sum_{m=-\infty}^{\infty} x[m]y[m-n]\right) z^{-n} \\
&= \sum_{m=-\infty}^{\infty} x[m] \sum_{n=-\infty}^{\infty} y[m-n]z^{-n} = \sum_{m=-\infty}^{\infty} x[m]z^{-m} \sum_{n=-\infty}^{\infty} y[m-n]z^{-(n-m)} \\
&= \sum_{m=-\infty}^{\infty} x[m]z^{-m} \sum_{n=-\infty}^{\infty} y[m-n]\left(\frac{1}{z}\right)^{-(m-n)} \\
&= X(z)Y\left(\frac{1}{z}\right) = X(z)Y(z^{-1}).
\end{aligned}$$

The ROC is the common region of the two ROCs, i.e., ROC $R_1 \cap R_2$.

Example 8.14

Using *Z*-transform find the correlation of two sequences $x[n] = u(n) - u(n-4)$, and $y[n] = u(n-1)$.

Solution

Given the of two sequences $x[n] = u(n) - u(n-4)$, and $y[n] = u(n-1)$.

$$x[n] = u(n) - u(n-4) = \delta(n) + \delta(n-1) + \delta(n-2) + \delta(n-3).$$

The Z-transform is $X(z) = 1 + z^{-1} + z^{-2} + z^{-3}$.

$$Y(z) = \frac{1}{z-1} \text{ and } Y(z^{-1}) = \frac{1}{z^{-1}-1} = \frac{z}{1-z}.$$

The Z-transform of correlation of two sequences is

$$\mathcal{Z}[x[n] \otimes y[n]] = X(z)Y(z^{-1})$$

$$= (1 + z^{-1} + z^{-2} + z^{-3})\frac{z}{1-z} = \frac{z}{1-z} + \frac{1}{1-z} + \frac{z^{-1}}{1-z} + \frac{z^{-2}}{1-z}.$$

Applying inverse Z-transform,

$$x[n] \otimes y[n] = -u(n) - u(n-1) - u(n-2) - u(n-3).$$

13. Parseval's Theorem

Consider two complex sequences with Z-transform given by $x[n] \leftrightarrow X(z)$ with ROC R_1 and $y[n] \leftrightarrow Y(z)$ with ROC R_2. Then the Parseval's theorem states that

$$\sum_{n=-\infty}^{\infty} x[n]y^*[n] = \frac{1}{2\pi} \oint X(z)Y^*\left(\frac{1}{z^*}\right) z^{-1}dz, \quad \text{ROC } R_1 \cap R_2. \tag{8.48}$$

Proof: Given two complex sequences $x[n] \leftrightarrow X(z)$ with ROC R_1 and $y[n] \leftrightarrow Y(z)$ with ROC R_2. We know that the inverse Z-transform is

$$x[n] = \frac{1}{2\pi j} \oint_c X(z)z^{n-1}dz.$$

Now taking the summation,

$$\sum_{n=-\infty}^{\infty} x[n]y^*[n] = \sum_{n=-\infty}^{\infty} \frac{1}{2\pi j} \oint_c X(z)z^{n-1}dzy^*[n].$$

Interchanging summation and integration,

$$= \frac{1}{2\pi j} \oint_c X(z) \sum_{n=-\infty}^{\infty} y^*[n]z^{n-1}dz = \frac{1}{2\pi j} \oint_c X(z) \left(\sum_{n=-\infty}^{\infty} y^*[n]z^n\right) z^{-1}dz$$

$$= \frac{1}{2\pi j} \oint_c X(z) \left(\sum_{n=-\infty}^{\infty} y^*[n] \left(\frac{1}{z}\right)^n\right) z^{-1}dz = \frac{1}{2\pi j} \oint_c X(z)Y^*\left(\frac{1}{z^*}\right) z^{-1}dz.$$

Since from complex conjugation property, $\sum_{n=-\infty}^{\infty} y^*[n] \left(\frac{1}{z}\right)^{-n} = Y^*\left(\frac{1}{z^*}\right)$,

$$\therefore \quad \sum_{n=-\infty}^{\infty} x[n]y^*[n] = \frac{1}{2\pi j} \oint_c X(z)Y^*\left(\frac{1}{z^*}\right) z^{-1}dz. \qquad \text{(Proved)}$$

8.11 Inverse Z-Transform

Inverse Z-transform is a process of recovering the discrete-time sequence $x[n]$ from the Z-transform of $X(z)$. Direct integration in the transform requires solving of complex integrals, which is a very difficult task. Indirectly, the inverse of Z-transform can be evaluated by using the basic Z-transform pairs with region of convergence.

The following methods are considered:

1. Power series expansion (long division method)
2. Partial fraction expansion
3. Residue methods

1. Power Series Expansion (Long Division Method)

If $x[n] \leftrightarrow X(z)$ with ROC $= R$, $X(z)$ can be represented in polynomial form as

$$X(z) = \frac{N(z)}{D(z)} = \frac{b_m z^m + b_{m-1} z^{m-1} + \cdots + b_0}{a_n z^n + a_{n-1} z^{n-1} + \cdots + a_0}. \tag{8.49}$$

From definition, $X(z)$ can be expressed as a power series in terms of z or z^{-1}.

$$\therefore \quad X(z) = \sum_{n=0}^{\infty} x[n] z^{-n} = x[0] + x[1]z^{-1} + x[2]z^{-2} + \cdots \tag{8.50}$$

By long division, the power series can be generated by dividing the numerator of $X(z)$ by the denominator; the value of $x[n]$ are the coefficients of the power series. This method is applicable only for unilateral sequences, either left-sided or right-sided. If the ROC is given for a left-sided sequence, the power series of $X(z)$ is in z and if ROC is given for right-sided sequence, the power series of $X(z)$ is in z^{-1}. For example, let $X(z) = \dfrac{a}{z-a}$ for $|z| < a$. Since the given ROC is for a left-sided sequence, we can express $X(z)$ in terms of z by using long division method.

$$\begin{array}{r|l} & -1 - \frac{z}{a} - \frac{z^2}{a^2} - \cdots \\ -a+z & a \\ & \underline{a - z} \\ & z \\ & \underline{z - \frac{z^2}{a}} \\ & \frac{z^2}{a} \end{array}$$

and so on.

$$X(z) = -1 - \frac{Z}{a} - \frac{z^2}{a^2} - \cdots$$

Taking inverse Z-transform by using Z-transform pairs,

$$x[n] = -\delta[n] - \frac{1}{a}\delta[n+1] - \frac{1}{a^2}\delta[n+2] - \frac{1}{a^3}\delta[n+3] - \cdots.$$

$$x[n] = -[\delta[n] + a^{-1}\delta[n+1] + a^{-2}\delta[n+2] + \ldots\ldots a^{-m}\delta[n+m] + \ldots..]$$

$$x[n] = -a^n u(-n).$$

We can also represent it in terms of coefficients as

$$x[n] = \left\{ \cdots - \frac{1}{a^3}, -\frac{1}{a^2}, -\frac{1}{a}, \underset{\underset{n=0}{\uparrow}}{-1} \right\}$$

The series $X(z)$ converges absolutely for $|z| < a$.

For a right-sided sequence, let $X(z) = \frac{a}{z-a}$ for $|z| > a$. Since the given ROC is for a right-sided sequence, we can express $X(z)$ in terms of z^{-1} by using long division method.

The long division is

$$\begin{array}{r|l} & \frac{a}{z} + \frac{a^2}{z^2} + \frac{a^3}{z^3} \cdots \\ -a+z & a \\ & a - \frac{a^2}{z} \\ \hline & \frac{a^2}{z} \\ & \frac{a^2}{z} - \frac{a^3}{z^2} \\ \hline & \frac{a^3}{z^2} \cdots \end{array}$$

Thus $X(z) = \frac{a}{z} + \frac{a^2}{z^2} + \frac{a^3}{z^3} + \ldots$.

Taking inverse Z-transform using Z-transform pairs,

$$x[n] = a\delta[n-1] + a^2\delta[n-2] + a^3\delta[n-3] + \ldots\ldots + a^m\delta[n-m] + \ldots . x[n] = a^n u(n-1).$$

We can also represent it in terms of coefficients as

$$x[n] = [0, a, a^2, a^3 \ldots].$$

Note: In long division, if $X(z)$ converges for ROC $|z| < R$, the series is obtained by beginning the division with the lowest order of z. If X (z)converges for ROC $|z| > R$, the series is obtained by beginning the division with highest power of z.

Example 8.15

Find the inverse Z-transform using power series expansion $X(z) = \log\left(\frac{z}{z-2}\right) \quad |z| > 2$.

Solution

Given $X(z) = \log\left(\frac{z}{z-2}\right) \quad |z| > 2$

$$X(z) = -\log(1 - 2z^{-1}).$$

Since the given ROC is for a right-sided sequence, using power series expansion, we know that

$$\log(1-a) = -\sum_{n=1}^{\infty} \frac{1}{n}(a)^n; \quad |a| < 1.$$

$$X(z) = -\log(1-2z^{-1}) = \sum_{n=1}^{\infty} \frac{1}{n}(2z^{-1})^n$$

$$= \sum_{n=1}^{\infty} \frac{1}{n}(2z^{-1})^n = \sum_{n=1}^{\infty} \frac{2^n}{n} z^{-n}.$$

Taking inverse Z-transform, from definition

$$x(n) = \frac{2^n}{n}u(n-1).$$

---

2. Partial Fraction Expansion

Let $x[n] \leftrightarrow X(z)$ with ROC $= R$. Assume that $X(z)$ is a proper rational function. $X(z)$ can be represented in polynomial form as

$$X(z) = \frac{N(z)}{D(z)} = \frac{b_m z^m + b_{m-1}z^{m-1} + \cdots + b_0}{a_n z^n + a_{n-1}z^{n-1} + \cdots + a_0} \quad \text{with } n > m. \tag{8.51}$$

In some cases $X(z)$ may be expressed in terms of z^{-1} rather than z. In such cases, apply partial fraction expansion directly on $X(z)$ or convert $X(z)$ in terms of z by factoring out the highest order of z^{-1} from the polynomials. To get the basic z transform pairs in the expansions, the best practice is to perform partial fraction expansion on $\frac{X(z)}{z}$ instead of $X(z)$.

The partial fraction expansion is obtained by factoring the denominator polynomial into a product form given by

$$\frac{X(z)}{z} = \frac{K(b_m z^m + b_{m-1}z^{m-1} + \cdots + b_0)}{(z-p_1)(z-p_2)(z-p_2)\cdots(z-p_n)} \quad \text{with } n > m. \tag{8.52}$$

where p_k, $k = 1,2,3\ldots.$ are poles and K is a multiplying factor.

Case 1 - Poles are distinct: Assume that the poles of $X(z)$ are distinct, that is, none is repeated. The partial fraction expansion of $\frac{X(z)}{z}$ is in the form

$$\frac{X(z)}{z} = \sum_{i=0}^{n} \frac{A_i}{z-p_i} \tag{8.53}$$

where A_k, $k = 1,2,3\ldots.$ the coefficients can be obtained as

$$A_k = \left((z-p_k)\frac{X(z)}{z}\right)_{z=p_k} \qquad k = 0,1,\cdots,n. \tag{8.54}$$

Case 2 - Poles are repeated m times: If $X(z)$ has repeated pole m times, then the partial fraction expansion of $\frac{X(z)}{z}$ is in the form

$$\frac{X(z)}{z} = \sum_{k=0}^{m} \frac{B_k}{(z-p)^k}. \tag{8.55}$$

For $k < m$,

$$B_k = \frac{1}{(m-k)!}\left[\frac{d^{m-k}}{dz^{m-k}}\left(\frac{(z-p_1)^m X(z)}{z}\right)\right]_{z=p_k} \tag{8.56}$$

For $k = m$,

$$B_m = \left.\frac{(z-p)^m X(z)}{z}\right|_{z=p_k}.$$

Case 3 - Complex poles: If $X(z)$ has complex poles, then the partial fraction expansion of $\frac{X(z)}{z}$ is in the form

$$\frac{X(z)}{z} = \frac{C}{z-p} + \frac{C^*}{z-p*},$$

where p^* and p are complex conjugate poles and C^* is the complex conjugate of C.

Case 4 - Improper rational function: Assume that $X(z)$ is not a proper rational function. $X(z)$ can be represented in polynomial form as

$$X(z) = \frac{b_m z^m + b_{m-1} z^{m-1} + \cdots + b_0}{a_n z^n + a_{n-1} z^{n-1} + \cdots + a_0} \quad \text{with} \quad n \leq m.$$

To convert $X(z)$ as proper rational function, using long division $X(z)$ can be represented in polynomial form as

$$X(z) = Q(z) + \frac{N_1(z)}{D(z)} = Q(z) + X_1(z) \tag{8.59}$$

where $Q(z) = 1$ for $n = m$, and $Q(z)$ is a polynomial in z of the order $m-n$. Then apply the partial fraction expansion on the rational function $\frac{X_1(z)}{z}$.

Note 1: If $X(z)$ has combination of distinct, repeated and complex poles, then the partial fraction expansion is algebraic sum of the individual cases.

Note 2: The inverse Z-transform of $X(z)$ can be obtained by multiplying the partial fraction factors with z. Each term depends on the ROC of $X(z)$. The sequence may be either a left-sided sequence or a right-sided sequence.

3. Residue Method or Contour Integral

The inverse Z-transform of $X(z)$ can be obtained directly by using inverse formula. We know that

$$X(z) = \sum_{n=-\infty}^{\infty} x(n) z^{-n}. \tag{8.60}$$

Proof: The inverse Z-transform of $X(z)$ can be obtained directly by using inverse formula. We know that $X(z) = \sum_{n=-\infty}^{\infty} x(n) z^{-n}$.

To calculate $x[n]$, consider a closed contour C in the counterclockwise direction in the region of convergence. Multiply by z^{m-1} on both sides and integrate along the contour C.

$$\oint_c z^{m-1} X(z) dz = \oint \left[\sum_{n=-\infty}^{\infty} x(n) z^{-n} \right] z^{m-1} dz.$$

Changing the order of the integration, we have

$$\oint_c X(z) z^{m-1} dz = \sum_{n=-\infty}^{\infty} x[n] \oint_c z^{-m-n-1} dz.$$

According to Cauchy theorem,

$$\oint_c z^{k-1} dz = \begin{cases} 2\pi j & k=0 \\ 0 & k \neq 0 \end{cases}$$

Substitute for $\oint_c z^{m-n-1}dz = \begin{cases} 2\pi j & m=n \\ 0 & m \neq n \end{cases}$.

Then at $m = n$,

$$\oint_c X(z)z^{n-1}dz = 2\pi jx[n]$$

or

$$x[n] = \frac{1}{2\pi j}\oint_c X(z)z^{n-1}dz. \quad \text{(Proved)}$$

The above equation is the expression for inverse Z-transform of $X(z)$. To solve the contour integral, from the residue theorem,

$$\frac{1}{2\pi}\oint X(z)z^{n-1} = \sum_m \text{Residues of } [X(z)z^{n-1}]_{z=p_m}.$$

Thus

$$x[n] = \sum_m \text{Residues of } [X(z)z^{n-1}]_{z=p_m} \quad (8.61)$$

where $z = p_m$ are the poles of the $X(z)z^{n-1}$ inside contour C.

$$x[n] = \sum_m \left[(z-z_m)X(z)z^{n-1}\right]_{z=p_m}.$$

If a pole of the $X(z)z^{n-1}$ is outside of contour C, then $x[n] = 0$ for that pole.

If z has repeated poles of r times, then

$$\text{Residue of } [X(z)z^{n-1}]_{z=p_m} = \frac{1}{(r-1)!}\frac{d^{r-1}}{dz^{r-1}}\left[(z-z_m)^r X(z)z^{n-1}\right]_{z=p_m}. \quad (8.62)$$

Example 8.16

Find the inverse Z-transform of $X(z) = \dfrac{2z(z-0.25)}{(z-0.2)(z-0.3)}$.

Solution

Given $X(z) = \dfrac{2z(z-0.25)}{(z-0.2)(z-0.3)}$. By dividing with z and taking partial fraction expansion,

$$\frac{X(z)}{z} = \frac{2(z-0.25)}{(z-0.2)(z-0.3)} = \frac{A}{z-0.2} + \frac{B}{z-0.3}.$$

$$A = \left.\frac{2(z-0.25)}{(z-0.3)}\right|_{z=0.2} = \frac{2(0.2-0.25)}{(0.2-0.3)} = 1$$

$$B = \left.\frac{2(z-0.25)}{(z-0.2)}\right|_{z=0.3} = \frac{2(0.3-0.25)}{(0.3-0.2)} = 1$$

$$X(z) = \frac{z}{z-0.2} + \frac{z}{z-0.3}.$$

The inverse Z-transform is

$$x[n] = [(0.2)^n + (0.3)^n]u[n].$$

8.12 Applications of z-Transform on Linear Systems

8.12.1 The Transfer Function of LTI Systems

The transfer function of a system completely describes the behaviour of the system with any type of input. Consider an LTI discrete-time system with impulse response $h[n]$ as shown in Fig. 8.13. Let $x[n], y[n]$ and $h[n]$ have Z-transform $X(z), Y(z)$ and $H(z)$ respectively.

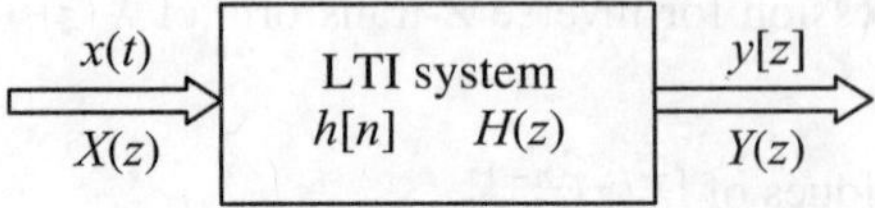

Fig. 8.13 *Discrete-time LTI system*

The transfer function of a system is defined as the ratio of the Z-transform of the output signal to the Z-transform of the input signal when all initial conditions are zero. Mathematically,

$$H(z) = \frac{Y(z)}{X(z)}.$$

We know that If $x[n]$ is the input signal, then the output signal can be obtained from the convolution given by $y[n] = x[n] * h[n]$.

The Z-transform of the output signal is

$$\mathcal{Z}[y[n]] = Y(z) = X(z)H(z) \text{ or, } H(z) = \frac{Y(z)}{X(z)}. \tag{8.63}$$

The Z-transform converts the convolution of time sequences in to multiplication of Z-transform.

Inverse Z-transform of the transfer function gives the impulse response of the system. That is,

$$h[n] = \mathcal{Z}^{-1}[H(z)] = \mathcal{Z}^{-1}\left[\frac{Y(z)}{X(z)}\right]. \tag{8.64}$$

Also, the inverse Z-transform of the output function gives the output signal. That is,

$$y[n] = \mathcal{Z}^{-1}[Y(z)] = \mathcal{Z}^{-1}[H(z)X(z)]. \tag{8.65}$$

8.12.2 Systems Described by Linear Difference Equations

Consider an LTI system in discrete time described by an Nth order linear difference equation with constant coefficients:

$$\sum_{k=0}^{N} a_k y[n-k] = \sum_{k=0}^{M} b_k x[n-k],$$

where a_k, $k = 1,2,3\ldots N$ and b_k, $k = 1,2,3\ldots M$ are real coefficients, N = order of the system and $N > M$. Performing Z-transform for the unilateral input and output sequences and zero initial conditions, then

$$a_N z^{-N} Y(z) + a_{N-1} z^{-N-1} Y(z) + \ldots + a_0 Y(z) = b_M z^{-M} X(z) + b_{M-1} z^{-M-1} X(z) + \ldots + b_0 X(z)$$

$$(a_N z^{-N} + a_{N-1} z^{-N-1} + \ldots + a_0) Y(s) = (b_M z^{-M} + b_{M-1} z^{-M-1} + \ldots + b_0) X(z)$$

$$\text{or,} \quad Y(z) = \frac{(b_M z^M + b_{M-1} z^{-M-1} + \ldots + b_0)}{(a_N z^N + a_{N-1} z^{-N-1} + \ldots + a_0)} X(z)$$

$$\text{or,} \quad H(z) = \frac{Y(z)}{X(z)} = \frac{(b_M z^M + b_{M-1} z^{-M-1} + \ldots + b_0)}{(a_N z^N + a_{N-1} z^{-N-1} + \ldots + a_0)} = \frac{\sum_{k=0}^{M} b_k z^{-k}}{\sum_{k=0}^{N} a_k z^{-k}}. \tag{8.67}$$

Note 1: To find out system transfer function, all initial conditions of the input signal should be zero. Transfer function is independent of input signal.

Note 2: Transfer function is a rational function in z.

Note 3: Frequency response can be obtained by setting $H(\omega) = H(z)|_{z=e^{j\omega}}$.

Example 8.17

Find the transfer function of a system represented by

$$y[n] - 0.5y[n-1] + 0.25y[n-2] = x[n] + 0.5x[n-1].$$

Solution

Given $y[n] - 0.5y[n-1] + 0.25y[n-2] = x[n] + 0.5x[n-1]$. Taking the Z-transform of each term. Assume all initial conditions are zero,

$$Y(z) - 0.5z^{-1}Y(z) + 0.25z^{-2}Y(z) = X(z) + 0.5z^{-1}X(z)$$

$$(1 - 0.5z^{-1} + 0.25z^{-2})Y(z) = (1 + 0.5z^{-1})X(z).$$

The transfer function is $H(z) = \dfrac{Y(z)}{X(z)} = \dfrac{1 + 0.5z^{-1}}{1 - 0.5z^{-1} + 0.25z^{-2}}$.

$$H[z] = \frac{z(z+0.5)}{z^2 - 0.5z + 0.25}.$$

8.12.3 Stability and Causality

Recall from Chapter 1 that if a system is stable, for all bounded inputs the output is bounded. The necessary and sufficient condition for a causal LTI discrete-time system to be BIBO-stable is: if $|x[n]| \leq M$ for all n, where M is a finite number, then the output is $|y[n]| < \infty$ for all n, and the impulse response of the system is

$$\sum_{k=-\infty}^{\infty} |h[k]| < \infty \text{ for all } k. \tag{8.68}$$

Proof: Given a discrete-time LTI system. Let the input be $|x[n]| \leq M$ for all n, where M is a finite number, and if $h[n]$ is the impulse response of the system, the output $y[n]$ is

$$y[n] = x[n] * h[n] = \sum_{k=-\infty}^{\infty} x[n-k]h[k].$$

$$|y(t)| = \left| \sum_{k=-\infty}^{\infty} h[k]x[n-k] \right| \leq \sum_{k=-\infty}^{\infty} |h[k]||x[n-k]|$$

$$|y(t)| \leq M \sum_{k=-\infty}^{\infty} |h[k]|. \tag{8.69}$$

Then for bounded output, the condition is

$$\sum_{k=-\infty}^{\infty} |h[k]| < \infty.$$

The magnitude of the Z-transform is

$$|H(z)| = \left| \sum_{n=-\infty}^{\infty} h[n]z^{-n} \right|$$

$$|H(z)| \leq \sum_{n=-\infty}^{\infty} h(n)||z^{-n}|.$$

For stable system, the condition is

$$|H(z)| \leq \sum_{n=-\infty}^{\infty} |h(n)| < \infty. \tag{8.70}$$

Causality

We know that if a system is causal, then $h[n] = 0$ for $n < 0$. That is, the impulse response exists for $0 < n < \infty$; the system is called right-sided causal system. Otherwise, the system is called left-sided anti-causal system.

The necessary and sufficient condition for a causal LTI discrete-time system to be BIBO-stable is $|H(z)| \leq \sum_{n=0}^{\infty} |h(n)| < \infty$; ROC is $|z| > 1$. That is, the impulse response $h[n]$ should be absolutely summable and the discrete transformation of the impulse response exists. The ROC of $H(z)$ must include the unit circle and all poles should lie inside the unit circle in the z-plane.

From the pole-zero plot it is observed that for a right-sided causal system, a pole inside the unit circle in the z-plane represents an exponentially decaying function in the impulse response, while a pole outside the unit circle represents an exponentially increasing function. Similarly, for a lefl-sided anti causal system, a pole inside the unit circle in the z-plane represents an exponentially increasing function in the impulse response, while a pole outside the unit circle represents an exponentially decaying function. A pole on the unit circle represents a complex sinusoidal term in the impulse function.

We know that the impulse response of a stable system should contain only exponentially decaying functions. Exponentially increasing functions and sinusoidal functions leads to instability. The conditions for stability and causality are:

Causol and stable system: The ROC of $H(z)$ must include the unit circle and all poles are inside the unit circle in the z-plane.

Anti-causal and stable system: The ROC of $H(z)$ must include the unit circle and all poles are outside the unit circle in the z-plane.

Example 8.18

Determine which of the following causal systems are stable.

(i) $H(z) = \dfrac{(z-0.1)(z-2)}{z(z-0.5)(z+0.25)}$ (ii) $H(z) = \dfrac{z(z-2)}{z^2+2.5z-1}$

Solution

(i) Given $H(z) = \dfrac{(z-1)(z-2)}{z(z-0.5)(z+0.25)}$. The system has poles at $z=0, z=0.25$, and $z=0.5$. Since the system is causal the ROC is $|z|>0.5$. Here the ROC includes unit circle and all poles lie within the unit circle. Therefore the system is stable.

(ii) $H(z) = \dfrac{z(z-2)}{z^2+2.5z-1} = \dfrac{z(z-2)}{(z+2.85)(z-0.35)}$

The system has poles at $z=-2.85$, and $z=0.35$. Since the system is causal, the ROC is $|z|>2.85$. Here the ROC does not include the unit circle and one pole is outside the unit circle. Therefore the system is unstable.

8.12.4 Steady State Frequency Response of Linear Discrete-time System

The steady state frequency response of a linear discrete-time system can be obtained from the Z- transform by substituting $z=e^{j\omega}$ in the system transfer function. Therefore, the steady state frequency response is

$$H(e^{j\omega}) = \left.\frac{Y(z)}{X(z)}\right|_{z=e^{j\omega}} = \frac{Y(e^{j\omega})}{X(e^{j\omega})}. \tag{8.71}$$

$H(e^{j\omega})$ is a continuous complex function with frequency ω and periodic with period of sampling frequency $\omega_s = \dfrac{2\pi}{T}$. For $T=1$, $\omega_s = 2\pi$.

$$H(e^{j\omega}) = |H(e^{j\omega})| \underline{|\phi(e^{j\omega})} \tag{8.72}$$

where the magnitude response is

$$|H(e^{j\omega})| = \frac{|Y(e^{j\omega})|}{|X(e^{j\omega})|}. \tag{8.73}$$

And the phase response is

$$\underline{|\phi(e^{j\omega})} = \tan^{-1}\left(\frac{Y(e^{j\omega})}{X(e^{j\omega})}\right).$$

DC response: The dc response of the system is the steady state frequency response at $\omega=0$, or $z=e^{j0}=1$. Therefore the dc response is

$$H(1) = \left.\frac{Y(z)}{X(z)}\right|_{z=1}.$$

Example 8.19

Obtain the steady state frequency response for the given difference equations of the system $y[n]-ay[n-1]=kx[n]$, where a and k are constants.

Solution

Given $y[n] - ay[n-1] = kx[n]$; a and k are constants. Taking Z-transform on both sides,

$$Y(z) - az^{-1}Y(z) = kX(z)$$

$$Y(z)(1 - az^{-1}) = kX(z).$$

The transfer function is

$$H(z) = \frac{Y(z)}{X(z)} = \frac{k}{1 - az^{-1}}$$

and the steady state frequency response is

$$H(e^{j\omega}) = \frac{Y(z)}{X(z)}\bigg|_{z=e^{j\omega}} = \frac{k}{1 - ae^{-j\omega}} = \frac{k}{1 - a(\cos\omega - j\sin\omega)} = \frac{k}{1 - a\cos\omega + ja\sin\omega}.$$

The magnitude response is $|H(e^{j\omega})| = \dfrac{k}{\sqrt{(1 - a\cos\omega)^2 + (a\sin\omega)^2}}$.

Phase response is $\underline{|\phi(e^{j\omega})} = -\tan^{-1}\left(\dfrac{a\sin\omega}{1 - a\cos\omega}\right)$.

The dc response is $H(1) = \dfrac{k}{1-a}$.

Example 8.20

Obtain the dc response for the given system $H(z) = \dfrac{5z(z-2)(z-3)}{z^2 + 4z - 2}$.

Solution

Given $H(z) = \dfrac{5z(z-2)(z-3)}{z^2 + 4z - 2}$, the dc response is

$$H(1) = \frac{5(1-2)(1-3)}{1+4-2} = 3.33.$$

Additional Problems

Example 8.21

Find the Z-transform of $x[n] = a^{n-1}u(n-1)$.

Solution

Given $x[n] = a^{n-1}u(n-1)$. We know that the Z-transform of $a^n u[n]$ is

$$X_1(z) = \mathcal{Z}[a^n u[n]] = \frac{z}{z-a}.$$

Using time shifting property, n is delayed by $n-1$:

$$X(z) = \mathcal{Z}[a^{n-1}u[n-1]] = z^{-1}X_1(z) = z^{-1}\frac{z}{z-a} = \frac{1}{z-a}.$$

Example 8.22

If $x[n] = (0.5)^n u[n]$, find $X_1(z) = \mathcal{Z}[x[n+2]u[n]]$ with initial conditions $x[1] = 1/2$ and $x[0] = 1$.

Solution

Given $x[n] = (0.5)^n u[n]$, then $X(z) = \dfrac{z}{z-0.5}$.

Now, $X_1(z) = \mathcal{Z}[x[n+2]u[n]]$

$$X_1(z) = z^2 X(z) - z^2 x[0] - zx[1].$$

Given the initial conditions, $x[0] = 1$ and $x[1] = 1/2$,

$$X_1[z] = z^2\left(\frac{z}{z-0.5}\right) - z^2 - z\left(\frac{1}{2}\right) = \frac{0.25z}{z-0.5}.$$

Example 8.23

Find the Z-transform of $x[n]$, a signal that is the sum of two real exponentials

$$x[n] = 7\left(\frac{1}{3}\right)^n u[n] - 6\left(\frac{1}{2}\right)^n u[n] \text{ and plot the ROC.}$$

Solution Given the signal $x[n] = 7\left(\frac{1}{3}\right)^n u[n] - 6\left(\frac{1}{2}\right)^n u[n]$. It is a right-sided causal sequence.

The Z-transform is $X(z) = \sum\limits_{n=-\infty}^{\infty} x[n]z^{-n}$.

$$X[z] = \sum_{n=-\infty}^{\infty}\left(7\left(\frac{1}{3}\right)^n u[n] - 6\left(\frac{1}{2}\right)^n u[n]\right)z^{-n}$$

$$= 7\sum_{n=-\infty}^{\infty}\left(\frac{1}{3}z^{-1}\right)^n - 6\sum_{n=-\infty}^{\infty}\left(\frac{1}{2}z^{-1}\right)^n = \frac{7}{1-\frac{1}{3}z^{-1}} - \frac{6}{1-\frac{1}{2}z^{-1}}$$

$$= \frac{7z}{z-\frac{1}{3}} - \frac{6z}{z-\frac{1}{2}} = \frac{z(z-3/2)}{(z-1/3)(z-1/2)}.$$

The equation for causal sequence has two poles., $z = 1/3$ and $z = 1/2$, and two zeros at $z = 0$ and $z = 3/2$. Thus the region of convergence is $|z| > \dfrac{1}{2}$. The pole-zero plot and ROC for the Z-transform of the combined signal are shown in Fig. 8.14.

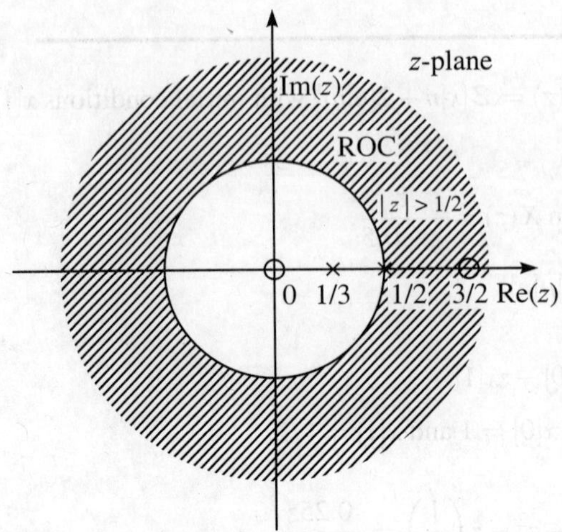

Fig. 8.14 *Pole–zero plot and ROC for the given sequence*

Example 8.24

Find the Z-transform of a two-sided sequence of infinite length $x[n] = \left(\frac{1}{2}\right)^{|n|}$ for $-\infty < n < \infty$.

Solution

Given sequence $x[n] = \begin{cases} \left(\frac{1}{2}\right)^{-n} & n < 0 \\ \left(\frac{1}{2}\right)^{n} & n \geq 0 \end{cases}$ or $x[n] = \left(\frac{1}{2}\right)^{-n} u[-n-1] + \left(\frac{1}{2}\right)^{n} u[n]$.

The Z-transform is

$$X(z) = \sum_{n=-\infty}^{\infty} \left(\frac{1}{2}\right)^{|n|} z^{-n} = \sum_{n=-\infty}^{-1} \left(\frac{1}{2}\right)^{-n} z^{-n} + \sum_{n=0}^{\infty} \left(\frac{1}{2}\right)^{n} z^{-n}$$

$$= \sum_{n=-\infty}^{-1} \left(\frac{z}{2}\right)^{-n} + \sum_{n=0}^{\infty} \left(\frac{1}{2z}\right)^{n} = \sum_{n=1}^{\infty} \left(\frac{z}{2}\right)^{n} + \sum_{n=0}^{\infty} \left(\frac{1}{2z}\right)^{n}$$

$$= \frac{\frac{z}{2}}{1-\frac{z}{2}} + \frac{1}{1-\frac{1}{2z}} \frac{z}{2} = \frac{z}{2-z} + \frac{z}{z-1/2} \quad \text{for } |z| < 2 \text{ and } |z| > \frac{1}{2}.$$

Since the first sum converges absolutely for $\left|\frac{z}{2}\right| < 1$ and the second for $\left|\frac{1}{2z}\right| < 1$ or $|z| < 2$ and $|z| > \frac{1}{2}$,

$$X(z) = \frac{3z/2}{(z-2)(z-1/2)} \quad \text{for ROC } \frac{1}{2} < |z| < 2.$$

Therefore, $X(z) = \dfrac{3z/2}{(z-2)(z-1/2)}$ with ROC $\frac{1}{2} < |z| < 2$.

Example 8.25

Find the two-sided Z-transform of the signal $x[n] = \begin{cases} \left(\frac{1}{3}\right)^n & n \geq 0 \\ (-2)^n & n \leq -1 \end{cases}$. (Dec 2011)

Solution

Given sequence $x[n] = \begin{cases} \left(\frac{1}{3}\right)^n & n \geq 0 \\ (-2)^n & n \leq -1, \end{cases}$ from the basic definition of Z-transform,

$$X(z) = \sum_{n=0}^{\infty} \left(\frac{1}{3}\right)^n z^{-n} + \sum_{n=-\infty}^{-1} (-2)^n z^{-n}$$

$$= \sum_{n=-\infty}^{-1} \left(\frac{-2}{z}\right)^n + \sum_{n=0}^{\infty} \left(\frac{1}{3z}\right)^n == \sum_{n=1}^{\infty} \left(\frac{-z}{2}\right)^n + \sum_{n=0}^{\infty} \left(\frac{1}{3z}\right)^n$$

$$= \frac{-\frac{z}{2}}{1+\frac{z}{2}} + \frac{1}{1-\frac{1}{3z}} = \frac{7z}{3(z+2)\left(z-\frac{1}{3}\right)}.$$

The equation has two poles, $z = -2$, $z = 1/3$, and one zero at $z = 0$. Here the first sum converges absolutely for $|z| < 2$ and, the second for $|z| > \frac{1}{3}$. ROC is of the form $\frac{1}{3} < |z| < 2$.

Thus the region of convergence is $\frac{1}{3} < |z| < 2$. The pole–zero plot and ROC for the Z-transform of the combined signal are shown in Fig. 8.15.

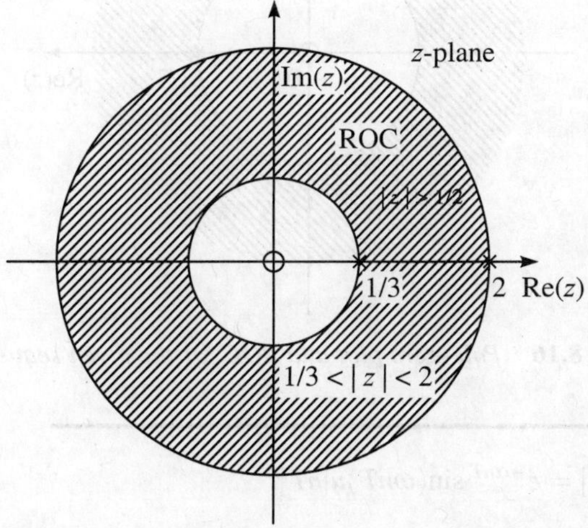

Fig. 8.15 *Pole–zero plot and ROC for the given sequence*

Example 8.26

Find the Z-transform of $x[n] = \left(\frac{1}{3}\right)^n \sin\left(\frac{\pi}{4}n\right) u[n]$.

Solution Given a right sided sequence

$$x[n] = \left(\frac{1}{3}\right)^n \sin\left(\frac{\pi}{4}n\right) u[n]$$

$$= \left(\frac{1}{3}\right)^n \frac{(e^{j\pi n/4} - e^{-j\pi n/4})}{2j} u[n] = \frac{1}{2j}\left(\frac{1}{3}e^{j\pi/4}\right)^n u[n] - \frac{1}{2j}\left(\frac{1}{3}e^{-j\pi/4}\right)^n u[n].$$

Using scaling property, the Z-transform is

$$X(z) = \frac{1}{2j}\left(\frac{z}{z-\frac{1}{3}e^{j\pi/4}} - \frac{z}{z-\frac{1}{3}e^{-j\pi/4}}\right) = \frac{\frac{1}{3}z\sin(\pi/4)}{\left(z^2 - \frac{1}{3}z\cos(\pi/4)+1\right)} = \frac{\frac{1}{3\sqrt{2}}z}{\left(z^2 - \frac{1}{3\sqrt{2}}z+1\right)}.$$

For convergence of $X[z]$, both sums must converge, which requires that $\left|\frac{1}{3}e^{j\frac{\pi}{4}}z^{-1}\right| < 1$ and $\left|\frac{1}{3}e^{-j\frac{\pi}{4}}z^{-1}\right| < 1$, or equivalently, ROC is $|z| > \frac{1}{3}$. The pole–zero plot with ROC is shown in Fig. 8.16.

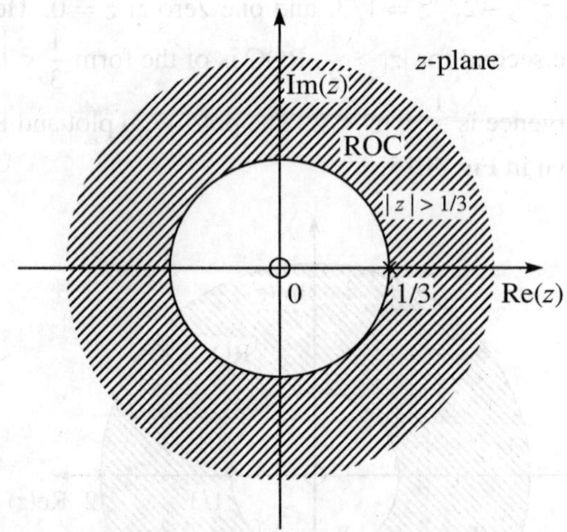

Fig. 8.16 *Pole–zero plot and ROC for the given sequence*

Example 8.27

Find the Z-transform of $x[n] = e^{-\alpha nT}\sin(\omega nT)u[nT]$.

Solution

Given $x[n] = e^{-\alpha nT}\sin(\omega nT)u[nT] = (e^{-\alpha T})^n \frac{(e^{j\omega T} - e^{-j\omega T})}{2j} u[nT]$

$$= \frac{1}{2j}[e^{-\alpha T}e^{j\omega T}]^n u[nT] - \frac{1}{2j}[e^{-\alpha T}e^{-j\omega T}]^n u[nT].$$

The Z-transform of this signal, using scaling property, is

$$X(z) = \frac{1}{2j}\left[\frac{z}{z - e^{-\alpha T}e^{j\omega T}} - \frac{z}{z - e^{-\alpha T}e^{-j\omega T}}\right] \text{ for } |z| > |e^{-\alpha T}|$$

$$= \frac{ze^{-\alpha T}\sin(\omega T)}{(z^2 - 2ze^{-\alpha T}\cos(\omega T) + e - 2\alpha T)}.$$

Example 8.28

If $x[n] = 2^n u[n]$, find the Z-transform of $nx[n] + 3^n x[n] + x[n-2] + x[n+1]$.

Solution

Given $x[n] = 2^n u[n]$, then $X(z) = \dfrac{z}{z-2}$.

Given the signal $nx[n] + 3^n x[n] + x[n-2] + x[n+1]$, the first term is solved by using time multiplication property:

$$\mathcal{Z}[nx[n]] = -z\frac{d}{dz}\left(\frac{z}{z-2}\right) = -z\left(\frac{(z-2) - z(1)}{(z-2)^2}\right) = \frac{2z}{(z-2)^2};$$

the second term by using scaling property:

$$\mathcal{Z}[3^n x[n]] = X(z/3) = \frac{(z/3)}{(z/3) - 2} = \frac{z}{z-6}.$$

The third term by the right shifting property is

$$\mathcal{Z}[x[n-2]] = z^{-2}X(z) = z^{-2}\frac{z}{z-2} = \frac{z^{-1}}{z-2};$$

and the last term by the left shifting property for causal signal $x[0] = 1$:

$$\mathcal{Z}[x[n+1]] = zX(z) - zx[0] = z\left(\frac{z}{z-2}\right) - (z)(1) = \frac{z^2 - z(z-2)}{z-2} = \frac{2z}{z-2}.$$

The Z-transform of the given sequence is

$$nx[n] + 3^n x[n] + x(n-2) + x[n+1] = \frac{2z}{(z-2)^2} + \frac{Z}{z-6} + \frac{z^{-1}}{z-2} + \frac{2z}{z-2}.$$

Example 8.29

Find the Z-transform of $x[n] = 7\left(\dfrac{1}{3}\right)^n \cos\left(\dfrac{\pi}{4} + \dfrac{2n\pi}{6}\right)u[n]$.

Solution

Given $x[n] = 7\left(\dfrac{1}{3}\right)^n \cos\left(\dfrac{\pi}{4} + \dfrac{2n\pi}{6}\right)u[n]$. Let

$$x_1[n] = \cos\left(\frac{\pi}{4} + \frac{2n\pi}{6}\right)u[n] = \left(\cos\left(\frac{\pi}{4}\right)\cos\left(\frac{2n\pi}{6}\right) - \sin\left(\frac{\pi}{4}\right)\sin\left(\frac{2n\pi}{6}\right)\right)u[n]$$

$$x_1[n] = 0.707\left(\cos\left(\frac{2n\pi}{6}\right) - \sin\left(\frac{2n\pi}{6}\right)\right)u[n].$$

Using Z-transform pair,

$$\cos\left(\frac{2\pi n}{6}\right)u[n] \leftrightarrow \frac{z\left(z-\cos\left(\frac{2\pi}{6}\right)\right)}{z^2-2z\cos\left(\frac{2\pi}{6}\right)+1} = \frac{z(z-0.5)}{z^2-2(z)(0.5)+1} = \frac{z(z-0.5)}{z^2-z+1}$$

and $\sin\left(\frac{2\pi n}{6}\right)u[n] \leftrightarrow \frac{z\sin(\pi/3)}{z^2-2z\cos(\pi/3)+1} = \frac{0.866z}{z^2-z+1}$.

$$\text{Therefore, } X_1(z) = 0.707\left(\frac{z(z-0.5)}{z^2-z+1} - \frac{0.866z}{z^2-z+1}\right) = \frac{0.707z^2-0.966z}{z^2-z+1}.$$

Now, for the given signal $x[n] = 7\left(\frac{1}{3}\right)^n \cos\left(\frac{\pi}{4}+\frac{2n\pi}{6}\right)u[n]$, the Z-transform, using scaling property is

$$X(z) = 7\left(\frac{0.707z^2-0.966z}{z^2-z+1}\right)\Bigg|_{z=3z} = \frac{44.54z^2-20.28z}{9z^2-3z+1}.$$

Example 8.30

Find the Z-transform of: (a) $x_1[n] = \sin(45°n)(\delta[n]+\delta[n-1])$; (b) $x_2[n] = (0.3)^n\sin(30°n)u[n]$; (c) $x_3[n] = e^{-n-2}\sin\left(\frac{\pi}{2}n\right)u[n]$; and (d) $x_4[n] = \sin\left(\frac{n\pi}{8}-\frac{\pi}{4}\right)u[n-2]$.

Solution

(a) Given $x_1[n] = \sin(45°n)(\delta[n]+\delta[n-1])$,

$$x_1[n] = \sin(45°n)(\delta[n]+\delta[n-1]) = \sin(0)\delta[n]+\sin(45°)\delta[n-1]$$

$$X_1[z] = 0+\frac{1}{\sqrt{2}}z^{-1} = 0.707z^{-1}.$$

(b) Given $x_2[n] = (0.3)^n\sin(30°n)u[n]$. Using Z-transform pair,

$$a^n\sin\omega_0 n)u[n] \leftrightarrow \frac{za\sin\omega_0)}{z^2-2za\cos\omega_0)+a^2}$$

$$X_2(z) = \frac{(0.3)z\sin 30}{z^2-2(.3)z\cos 30+(0.3)^2}$$

$$X_2(z) = \frac{(0.3)(0.5)z}{z^2-2(.3)(0.866)z+(0.3)^2} = \frac{z(0.15)}{z^2-0.52z+0.09}.$$

(c) Given $x_3[n] = e^{-n-2}\sin\left(\frac{\pi}{2}n\right)u[n] = e^{-2}(e^{-1})^n\sin\left(\frac{\pi}{2}n\right)u[n]$.

Using Z-transform pair $a^n\sin(\omega_0 n)u[n] \leftrightarrow \frac{za\sin(\omega_0)}{z^2-2za\cos(\omega_0)+a^2}$,

$$X_3(z) = e^{-2}\frac{(e^{-1})z\sin 90}{z^2-2(e^{-1})z\cos 90+(e^{-1})^2} = = \frac{e^{-3}z}{z^2+e^{-2}}.$$

(d) Given $x_4[n] = \sin\left(\frac{n\pi}{8} - \frac{\pi}{4}\right) u[n-2], = \sin\left(\frac{(n-2)\pi}{8}\right) u[n-2]$.

Now, let $x[n] = \sin\left(\frac{n\pi}{8}\right) u[n]$.

From the Z-transform pair, $X(z) = \dfrac{z \sin\left(\frac{\pi}{8}\right)}{z^2 - 2z\cos\left(\frac{\pi}{8}\right) + 1}$.

Using time shifting property, $X_4(z) = z^2 X(z)$.

The Z-transform of $\sin\left(\frac{(n-2)\pi}{8}\right) u[n-2]$ is

$$X_4(z) = \frac{z^{-1} \sin\left(\frac{\pi}{8}\right)}{z^2 - 2z\cos\left(\frac{\pi}{8}\right) + 1}, \quad |z| > 1.$$

Example 8.31

Using Z-transform, find the convolution of $x[n] = [1, 2, -1, 0, 3]$, and $y[n] = [1, 2, -1]$.

Solution

Given $x[n] = [1, 2, -1, 0, 3]$, and $y[n] = [1, 2, -1]$. The convolution ofthe Z-transform is $\mathcal{Z}[x[n] * y[n]] = X(z)Y(z)$.

Now, $x[n] = \delta[n] + 2\delta[n-1] - \delta[n-2] + 3\delta[n-4]$.

Then, $X(z) = 1 + 2z^{-1} - z^{-2} + 3z^{-4}$.

$y[n] = \delta[n] + 2\delta[n-1] - \delta[n-2]$.

Then, $Y(z) = 1 + 2z^{-1} - z^{-2}$.

$$\therefore \quad \mathcal{Z}[x[n] * y[n]] = X(z)Y(z) = (1 + 2z^{-1} - z^{-2} + 3z^{-4})(1 + 2z^{-1} - z^{-2})$$

$$= 1 + 4z^{-1} + 2z^{-2} - 4z^{-3} + 4z^{-4} + 6z^{-5} - 3z^{-6}.$$

Then, taking the inverse Z-transform,

$$x[n] * y[n] = 1 + 4\delta[n-1] + 2\delta[n-2] - 4\delta[n-3] + 4\delta[n-4] + 6\delta[n-5] - 3\delta[n-6]$$

$$= [1, 4, 2, -4, 4, 6, -3].$$

Example 8.32

What is the inverse Z-transform of $X(z) = 1 + 2z^{-2} + 3z^{-4}$?

Solution

Given $X(z) = 1 + 2z^{-2} + 3z^{-4}$. From the definition of series expansion, $x[0] = 1$, $x[2] = 2$, $x[4] = 3$.

$x[n] = \delta[n] + 2\delta[n-2] + 3\delta[n-4]$.

Example 8.33

Find the inverse Z-transform of $X(z) = \dfrac{1}{2z-1}$.

Solution

Given $X(z) = \dfrac{1}{2z-1} = \dfrac{0.5}{z-0.5}$.

Using partial fraction expansion,

$$\frac{X(z)}{z} = \frac{0.5}{z(z-0.5)} = \frac{A}{z} + \frac{B}{z-0.5}.$$

$$A = \left.\frac{0.5}{z-0.5}\right|_{z=0} = -1,\ B = \left.\frac{0.5}{z}\right|_{z=05} = 1.$$

Then $X(z) = \dfrac{z}{z-0.5} - 1$.

Using Z-transform pair,

$$x[n] = (0.5)^n u[n] - \delta[n] = \delta[n] + (0.5)^n u[n-1] - \delta[n]$$

$$= (0.5)^n u[n-1].$$

Another method using shifting property:

Given $X(z) = \dfrac{0.5}{z-0.5} = 0.5z^{-1}\left(\dfrac{z}{z-0.5}\right)$. We know that $\dfrac{z}{z-0.5} \leftrightarrow (0.5)^n u(n)$. Using shifting property,

$$x[n] = 0.5(0.5)^{n-1} u(n-1) = (0.5)^n u(n-1).$$

Example 8.34

Find the inverse Z-transform of $X(z) = \dfrac{1}{(z-1)(z-0.5)}$.

Solution

Given $X(z) = \dfrac{1}{(z-1)(z-0.5)}$. Then

$$\frac{X(z)}{z} = \left[\frac{1}{z(z-1)(z-0.5)}\right].$$

Taking the partial fraction expansion,

$$\frac{1}{z(z-1)(z-0.5)} = \frac{A}{z-1} + \frac{B}{z-0.5} + \frac{C}{z}.$$

$$A = \left.\frac{1}{z(z-0.5)}\right|_{z=1} = 2;\quad B = \left.\frac{1}{z(z-1)}\right|_{z=05} = -4;\quad C = \left.\frac{1}{(z-0.5)(z-1)}\right|_{z=0} = 2.$$

$$\frac{F(z)}{z} = \frac{2}{z-1} - \frac{4}{z-0.5} + \frac{2}{z}.$$

$$X(z) = \left(\frac{2z}{z-1}\right) - \left(\frac{4z}{z-0.5}\right) + 2.$$

Then by applying the Z-transform,

$$x[n] = (2 - 4(0.5)^{n-1})u[n] + \delta[n].$$

Example 8.35

Find the inverse Z-transform of $F(z) = \dfrac{z^2+2z-1}{z^2+3z+2}$.

Solution

$F(z) = \dfrac{z^2+2z-1}{z^2+3z+2}$. Since the order of z is equal in the numerator and in the denominator, we can write

$$F(z) = \frac{z^2+2z-1}{z^2+3z+2} = 1 - \frac{z+3}{z^2+3z+2}.$$

Now, let $F_0(z) = \dfrac{z+3}{z^2+3z+2}$.

Using partial fraction expansion,

$$F_0(z) = \frac{z+3}{(z+1)(z+2)} = \frac{A}{z+1} + \frac{B}{z+2}.$$

$$A = \left.\frac{z+3}{(z+2)}\right|_{z=-1} = 2, \quad B = \left.\frac{z+3}{(z+1)}\right|_{z=-2} = -1.$$

$$F(z) = 1 - \frac{2}{z+1} + \frac{1}{z+2} = 1 - 2z^{-1}\left(\frac{z}{z+1}\right) + z^{-1}\left(\frac{z}{z+2}\right).$$

Applying Z-transform pairs,

$$f[n] = \delta[n] - 2(-1)^{n-1}u[n-1] + (-2)^{n-1}u[n-1].$$

Example 8.36

Find the inverse Z-transform of $F(z) = \dfrac{z(z+1)}{(z-1)(z-0.25)^2}$.

Solution

The above equation can be expressed using partial fraction expansion as

$$\frac{F(z)}{z} = \frac{(z+1)}{(z-1)(z-0.25)^2} = \frac{A}{z-1} + \frac{B}{(z-0.25)^2} + \frac{C}{(z-0.25)}.$$

$$A = \left.\frac{(z+1)}{(z-0.25)^2}\right|_{z=1} = \frac{2}{(0.75)^2} = 3.56; \quad B = \left.\frac{(z+1)}{(z-1)}\right|_{z=0.25} = \frac{-1.25}{0.75} = -1.67.$$

To get C, take derivative with respect to z.

$$C = \frac{d}{dz}\left[\frac{(z+1)}{(z-1)}\right]\Bigg|_{z=0.25} = \frac{1(z-1)-1(z+1)}{(z-1)^2}\Bigg|_{z=0.25} = -3.56.$$

$$F(z) = \frac{(z+1)}{(z-1)(z-0.25)^2} = \frac{3.56z}{z-1} + \frac{-1.67z}{(z-0.25)^2} + \frac{-3.56z}{(z-0.25)}$$

$$= \frac{3.56z}{z-1} - 6.68\frac{0.25z}{(z-0.25)^2} - \frac{3.56z}{(z-0.25)}.$$

Applying inverse Z-transform,

$$f[n] = [3.56 - 6.68 \times n(0.25)^n - 3.56(0.25)^n]u[n].$$

Example 8.37

Find the inverse Z-transform of $X(z) = \dfrac{z}{z(z-1)(z-2)^2}$ $|z| > 2$.

Solution

Given $X(z) = \dfrac{z}{z(z-1)(z-2)^2}$.

$$\frac{X(z)}{z} = \frac{1}{z(z-1)(z-2)^2}.$$

Taking the partial fraction expansion,

$$\frac{1}{z(z-1)(z-2)^2} = \frac{\mathrm{A}}{z} + \frac{B}{z-1} + \frac{C}{(z-2)^2} + \frac{D}{(z-2)}.$$

$$\mathrm{A} = \frac{1}{(z-1)(z-2)^2}\Bigg|_{z=0} = -0.25;\ B = \frac{1}{z(z-2)^2}\Bigg|_{z=1} = 1;\ C = \frac{1}{z(z-1)}\Bigg|_{z=2} = 0.5.$$

To get D, the first derivative is

$$D = \frac{d}{dz}\left(\frac{1}{z(z-1)}\right)\Bigg|_{z=2} = \left[\frac{-(2z-1)}{z^2(z-1)^2}\right]_{z=2} = \frac{-3}{4}.$$

$$\frac{F(z)}{z} = \frac{1}{z-1} - \frac{0.75}{z-2} + \frac{0.5}{(z-2)^2} - \frac{0.25}{z}.$$

$$F(z) = \frac{z}{z-1} - \frac{0.75z}{z-2} + \frac{0.5z}{(z-2)^2} - 0.25.$$

Then by applying inverse Z-transform,

$$x[n] = (1 - 0.75(2)^n + 0.5n(2)^n)u[n] + 0.25\delta[n].$$

Example 8.38

Using power series expansion method, determine the inverse Z-transform of $X(z) = \dfrac{z}{2z^2-3z+1}$ $|z|>2$.

Solution

Given $X(z) = \dfrac{z}{2z^2-3z+1}$. Since ROC $|z|>2$, the sequence is a right-sided sequence and the $X(z)$ should be in terms of z^{-1}. Using long division method,

$$\begin{array}{r l}
 & \frac{1}{2}z^{-1}+\frac{3}{4}z^{-2}+\frac{5}{8}z^{-3}+\frac{9}{16}z^{-4}+\ldots \\ \hline
2z^2-3z+1 \,| & z \\
 & z-\frac{3}{2}+\frac{1}{2}z^{-1} \\ \hline
 & \frac{3}{2}-\frac{1}{2}z^{-1} \\
 & \frac{3}{2}-\frac{9}{4}z^{-1}+\frac{3}{4}z^{-2} \\ \hline
 & \frac{5}{4}z^{-1}-\frac{3}{4}z^{-2} \\
 & \frac{5}{4}z^{-1}-\frac{15}{8}z^{-2}+\frac{5}{8}z^{-3} \\ \hline
 & \frac{9}{8}z^{-2}-\frac{5}{8}z^{-3} \\
 & \ldots\ldots\ldots\ldots\ldots
\end{array}$$

$$X(z) = \frac{1}{2}z^{-1}+\frac{3}{4}z^{-2}+\frac{5}{8}z^{-3}+\frac{9}{16}z^{-4}+\ldots.$$

Then, by applying inverse Z-transform,

$$x[n] = \frac{1}{2}\delta[n-1]+\frac{3}{4}\delta[n-2]+\frac{5}{8}\delta[n-3]+\frac{9}{16}\delta[n-4]+\ldots.$$

Example 8.39

Using power series expansion method, determine the inverse Z-transform of $X(z) = \dfrac{z+1}{z^2-3z+2}$ if (a) $x[n]$ is causal and (b) $x[n]$ is non-causal.

Solution

Given $X(z) = \dfrac{z}{2z^2-3z+2}$.

(a) Since $x[n]$ is causal, the sequence is a right-sided sequence. $X(z)$ should be in terms of z^{-1}
Using long division method,

$$
\begin{array}{r|l}
 & z^{-1}+4z^{-2}+10z^{-3}+22z^{-4}+\ldots\ldots \\
2-3z+z^2 & z+1 \\
 & z-3+2z^{-1} \\
 & \quad 4-2z^{-1} \\
 & \quad 4-12z^{-1}+8z^{-2} \\
 & \quad\quad 10z^{-1}-8z^{-2} \\
 & \quad\quad 10z^{-1}-30z^{-2}+20z^{-3} \\
 & \quad\quad\quad 22z^{-2}-20z^{-3} \\
 & \quad\quad\quad \ldots\ldots\ldots\ldots
\end{array}
$$

$$X(z) = z^{-1}+4z^{-2}+10z^{-3}+22z^{-4}+\ldots.$$

Applying inverse Z-transform pair,

$$x[n] = \delta[n-1]+4\delta[n-2]+10\delta[n-3]+22\delta[n-4]+\ldots.$$

$$x[n] = \{\underset{\uparrow}{0},1,4,10,22,\ldots.\}$$

(b) Since $x[n]$ is non-causal, the sequence is a left-sided sequence. $X(z)$ should be in terms of z. Using long division method,

$$
\begin{array}{r|l}
 & \frac{1}{2}+\frac{5}{4}z+\frac{13}{8}z^2+\frac{29}{16}z^3+\ldots\ldots \\
2-3z+z^2 & 1+z \\
 & 1-\frac{3}{2}z+\frac{1}{2}z^2 \\
 & \quad \frac{5}{2}z-\frac{1}{2}z^2 \\
 & \quad \frac{5}{2}z-\frac{15}{4}z^2+\frac{5}{4}z^3 \\
 & \quad\quad \frac{13}{4}z^2-\frac{5}{4}z^3 \\
 & \quad\quad \frac{13}{4}z^2-\frac{39}{8}z^3+\frac{13}{8}z^4 \\
 & \quad\quad\quad \frac{29}{8}z^3-\frac{13}{8}z^4 \\
 & \quad\quad\quad \ldots\ldots\ldots\ldots
\end{array}
$$

$$X(z) = \frac{1}{2}+\frac{5}{4}z+\frac{13}{8}z^2+\frac{29}{16}z^3+\ldots.$$

Applying inverse Z-transform pair,

$$x[n] = \frac{1}{2}\delta[n]+\frac{5}{4}\delta[n+1]+\frac{13}{8}\delta[n+2]+\frac{29}{16}\delta[n+3]+\ldots.$$

$$x[n] = \left\{ \ldots\ldots\ldots, \frac{29}{16}, \frac{13}{8}, \frac{5}{4}, \underset{\uparrow}{\frac{1}{2},} \right\}.$$

Example 8.40

Determine the inverse Z-transform of $X(z) = \dfrac{z+2}{2z^2-7z+3}$ by the partial fraction expansion method ifthe ROCs are (i) $|z|>3$; (ii) $|z|<\dfrac{1}{2}$ and (iii) $\dfrac{1}{2}<|z|<3$.

Solution

Given $X(z) = \dfrac{z+2}{2z^2-7z+3} = \dfrac{0.5(z+2)}{(z-3)(z-0.5)}$. Then

$$\frac{F(z)}{z} = \frac{0.5(z+2)}{z(z-3)(z-0.5)}.$$

Taking the partial fraction expansion,

$$\frac{0.5(z+2)}{z(z-3)(z-0.5)} = \frac{A}{z-0.5} + \frac{B}{z-3} + \frac{C}{z}$$

$$A = \left.\frac{0.5(z+2)}{z(z-3)}\right|_{z=0.5} = -1; \quad B = \left.\frac{0.5(z+2)}{z(z-0.5)}\right|_{z=3} = \frac{1}{3}; \quad C = \left.\frac{0.5(z+2)}{(z-3)(z-0.5)}\right|_{z=0} = \frac{2}{3}.$$

$$\frac{F(z)}{z} = \frac{1/3}{z-3} - \frac{1}{z-0.5} + \frac{\frac{2}{3}}{z}$$

$$F(z) = \left(\frac{z/3}{z-3}\right) - \left(\frac{z}{z-0.5}\right) + \frac{2}{3}.$$

(i) If the ROCs is $|z|>3$, all terms are right-sided.

Therefore,

$$f[n] = \left(\frac{1}{3}(3)^n - (0.5)^n\right) u[n] + \frac{2}{3}\delta[n].$$

(ii) If the ROCs is $|z|<\dfrac{1}{2}$, all terms are left-sided.

$$f[n] = -\left(\frac{1}{3}(3)^n - (0.5)^n\right) u[-n-1] + \frac{2}{3}\delta[n].$$

(iii) If the ROCs is $\dfrac{1}{2}<|z|<3$, the first term is left-sided.

$$f[n] = -\left(\frac{1}{3}(3)^n\right) u[-n-1] - (0.5)^n u(n) + \frac{2}{3}\delta[n].$$

Example 8.41

Solve for the impulse response $y[n]-0.5y[n-1]=\delta[n]$.

Solution

Given the difference equation $y[n]-0.5y[n-1]=\delta[n]$.

Taking the Z-transform of each term (assume all initial conditions are zero),

$$Y(z)-0.5z^{-1}Y(z)=1$$

$$Y(z)(1-0.5z^{-1})=1$$

$$Y[z]=\frac{1}{1-0.5z^{-1}}=\frac{z}{z-0.5}.$$

Taking inverse Z-transform, the impulse response is $y[n]=(0.5)^n u[n]$.

Example 8.42

Solve for the impulse response $y[n+1]-0.5y[n]=\delta[n+1]$.

Solution

Given $y[n+1]-0.5y[n]=\delta[n+1]$. Taking the Z-transform on both sides and neglecting initial conditions,

$$zY(z)-0.5Y(z)=z.$$

$$Y(z)(z-0.5)=z.$$

$$Y(z)=\frac{z}{z-0.5}.$$

Taking inverse Z-transform, $y[n]=(0.5)^n u[n]$.

Example 8.43

Solve $y[n]-0.3y[n-1]=u[n]$ and $y[-1]=0$.

Solution

Given $y[n]-0.3y[n-1]=u[n]$ and $y[-1]=0$. Applying Z-transform on both sides,

$$Y(z)-0.3(z^{-1}Y(z)-y[1])=\frac{z}{z-1}$$

$$Y[z]=\frac{z}{(1-0.3z^{-1})(z-1)}$$

$$Y[z]=\frac{z^2}{(z-0.3)(z-1)}.$$

Applying partial fraction expansion for

$$\frac{Y(z)}{z}=\frac{z}{(z-0.3)(z-1)}=\frac{A}{(z-0.3)}+\frac{B}{(z-1)},$$

$$A = \left.\frac{z}{(z-1)}\right|_{z=0.3} = \frac{-0.3}{0.7} = -0.429; \; B = \left.\frac{z}{(z-0.3)}\right|_{z=1} = \frac{1}{0.7} = 1.429.$$

Then, $Y(z) = \frac{-0.429z}{(z-0.3)} + \frac{1.429z}{(z-1)}$.

Applying inverse Z-transform,

$$y[n] = [1.429 - 0.429(0.3)^n]u[n].$$

Example 8.44

The transfer function of an LTI system is given by $H(z) = \frac{3-4z^{-1}}{1-3.5z^{-1}+1.5z^{-2}}$. Determine the impulse response for (i) stable system, (ii) causal system.

Solution

Given $H(z) = \frac{3-4z^{-1}}{1-3.5z^{-1}+1.5z^{-2}}$.

$$\frac{H(z)}{z} = \frac{3z-4}{z^2-3.5z+1.5}.$$

Using partial fractions expansion,

$$\frac{H(z)}{z} = \frac{3z-4}{(z-3)(z-0.5)} = \frac{A}{(z-3)} + \frac{B}{(z-0.5)}.$$

$$A = \left.\frac{3z-4}{(z-0.5)}\right|_{z=3} = 2, \quad B = \left.\frac{3z-4}{(z-3)}\right|_{z=0.5} = 1.$$

$$\frac{H(z)}{z} = \frac{2}{(z-3)} + \frac{1}{(z-0.5)} \quad H(z) = \frac{2z}{(z-3)} + \frac{z}{(z-0.5)}.$$

The poles are $z = 0$ and $z = 0.5$.

(i) For stable system the ROC must include the unit circle and the pole inside the unit circle. $z = 0.5$ represents a right-sided (causal) sequence and the pole outside the unit circle. $z = 3$ represents a lefl-sided (anti causal) sequence. The ROC is $0.5 < |z| < 3$.

The impulse response is $h[n] = (0.5)^n u[n] - 2(3)^n u[-n-1]$.

(ii) For causal system the ROC is $|z| > 3$. The ROC does not include the unit circle and hence the system is unstable. The impulse response is $h[n] = (0.5)^n u[n] + 2(3)^n u[n]$.

Example 8.45

Determine the stability of an LTI system given by $y[n] - 0.7y[n-1] + 0.12y[n-2] = x[n-1] + x[n-2]$.

Solution

Given $y[n] - 0.7y[n-1] + 0.12y[n-2] = x[n-1] + x[n-2]$. Taking Z-transform on both sides,

$$Y(z)(1 - 0.7z^{-1} + 0.12z^{-2}) = (z^{-1} + z^{-2})X(z).$$

$$H(z) = \frac{Y(z)}{X(z)} = \frac{(z^{-1}+z^{-2})}{(1-0.7z^{-1}+0.12z^{-2})} = \frac{(z+1)}{(z^2-0.7z+0.12)} = \frac{(z+1)}{(z-0.3)(z-0.4)}.$$

The poles are $z = 0.3$ and $z = 0.4$. The ROC is $|z| > 0.4$.

The system is causal and stable since the ROC includes the unit circle and the poles are inside the unit circle.

Example 8.46

Obtain the step response of causal system $y[n] - 0.85y[n-1] = 0.15x[n]$. Assume all initial conditions are zero.

Solution

Given $y[n] - 0.85y[n-1] = 0.15x[n]$. For step response, $x[n] = u[n]$.

$$y[n] - 0.85y[n-1] = 0.15u[n].$$

Take the Z-transform,

$$Y(z)\left(1-0.85z^{-1}\right) = \frac{0.15z}{(z-1)}$$

$$Y(z) = \frac{0.15z^2}{(z-1)(z-0.85)}.$$

Using partial fractions expansion,

$$\frac{Y(z)}{z} = \frac{0.15z}{(z-1)(z-0.85)} = \frac{A}{(z-1)} + \frac{B}{(z-0.85)}.$$

$$A = \left.\frac{0.15z}{(z-0.85)}\right|_{z=1} = 1; \quad B = \left.\frac{0.15z}{(z-1)}\right|_{z=0.85} = -0.85.$$

$$Y(z) = \frac{z}{(z-1)} - \frac{0.85z}{(z-0.85)}.$$

The output response is $y[n] = u[n] - (0.85)^{n+1}u[n]$.

Summary

1. The magnitude of a discrete-time signal is defined only at specific instants of time interval and it is represented with a sequence of numbers: $f[n] = \{\ldots.f[-2], f[-1], \underset{\uparrow}{f[0]}, f[1], f[2], \ldots.\}$.
2. A discrete-time signal can be derived by sampling a continuous signal at equal sampling intervals of T_s seconds. A continuous-time signal can be derived from a discrete-time signal using a lowpass reconstruction filter.
3. $x(n) = ce^{an}$ is a real exponential sequence, where both c and a are real.
4. The complex exponential sequence is defined as $x(n) = ce^{j(\omega_0 n+\phi)}$.

5. A discrete-time sinusoidal function is defined as $x[n] = A\cos(\omega_0 n + \phi)$, where $\omega_0 = 2\pi f_0 = \dfrac{2\pi}{N}$.

6. A discrete-time signal $x[n]$ is said to be periodic signal if $x[n] = x[n+N]$, where $N = \dfrac{2\pi k}{\omega_0}$ is an integer for any integer k or $\dfrac{N}{k} = \dfrac{2\pi}{\omega_0}$ = rational number. The smallest integer N is called fundamental period. $\omega_0 = \dfrac{2\pi}{N}$ rad/s is the fundamental angular frequency.

7. If a sequence $x[n] = x_1[n] + x_2[n]$ is a sum of two periodic sequences with time periods N_1, N_2 and if $x[n]$ is periodic with period N, then the ratio of periods is a rational number. That is, $mN_1 = kN_2 = N$ or $\dfrac{N_1}{N_2} = \dfrac{k}{m}$ = rational number.

8. The two-sided Z-transform or bilateral Z-transform for a discrete-time sequence $x[n]$ is defined as $X(z) = \mathcal{Z}[x[n]] = \sum_{n=-\infty}^{\infty} x[n]z^{-n}$.

9. The one-sided Z-transform or unilateral Z-transform and is defined as $X(z) = \sum_{n=0}^{\infty} x[n]z^{-n}$.

10. The inverse Z-transform is $x[n] = Z^{-1}[X(z)] = \dfrac{1}{2\pi j}\oint X(z)z^{n-1}dz$.

11. The notation for Z-transform pair is $x[n] \leftrightarrow X(z)$.

12. The Laplace transform of a sampled signal $x(nT)$ is equal to the Z-transform: $\mathcal{L}[x_s(t)] = X(z) = \sum_{n=\infty}^{\infty} x[nT]z^{-n}$. For convenience, always assume $T = 1$. where $z = e^{sT}$ and $s = \sigma + j\omega$.

13. For $\sigma = 0$, or $s = j\omega$, by substituting $z = e^{(\sigma+j\omega)} = e^{j\omega}$, the Z-transform becomes $X(\omega) = \sum_{n=0}^{\infty} x[n]e^{-j\omega n}$ and represents the Fourier transform or frequency response of $x[n]$ and it is given as $X(\omega) = \sum_{n=0}^{\infty} x[n]e^{-j\omega n}$.

14. The range of values of the complex variable z for which the Z-transform converges is called the region of convergence (ROC). The region of convergence of $X(z)$ can be represented graphically with a pole–zero plot on the z-plane, where $z = re^{j\omega T}$, r is the radius and ωT is the angle in polar form.

15. The ROC is in the form of a circular ring in the z-plane centered about the origin.

16. Classification of sequences:

   (i) Right-sided sequence with finite duration, that is, $x[n]$ exists in a finite interval $N_1 \le n \le N_2$, where N_1 and N_2 are finite positive numbers.

   (ii) Left-sided sequence with finite duration, that is, $x[n]$ exists in a finite interval $N_1 \le n \le N_2$, where N_1 and N_2 are finite negative numbers.

   (iii) Two-sided sequence with finite duration that is, $x[n]$ exists in a finite interval $N_1 \le n \le N_2$, where N_1 is a finite negative number and N_2 is a finite positive number. It is also called bilateral sequence with finite duration.

   (iv) Right-sided sequence with infinite duration, that is, $x[n]$ exists in a interval $N_1 \le n \le \infty$, where N is a finite positive number or zero. It is a causal sequence.

(v) Left-sided sequence with infinite duration, that is, $x[n]$ exists in a interval $-\infty \leq n \leq N$, where N is a finite negative number or zero. It is non-causal sequence.

(vi) Two-sided sequence with infinite duration, that is, $x[n]$ is an infinite duration sequence existing on both right side as well as left side. It is also called bilateral sequence. It is non-causal sequence.

17. Properties of the ROC

Property 1: The ROC does not contain any poles since at all pole of $X(z)$ becomes infinite and does not converge.

Property 2: If $x[n]$ is a finite sequence and $X(z)$ converges for some value of z, then the ROC is the entire z-plane except possibly $z = 0$ and/or $z = \infty$.

(a) Right-sided sequence with finite duration, ROC is the entire z-plane except at $z = 0$.
(b) Left-sided sequence with finite duration, ROC is the entire z-plane except at $z = \infty$.
(c) Two-sided sequence with finite duration, ROC is the entire z-plane except at $z = 0$ and $z = \infty$.

Property 3: If $x[n]$ is a right-sided sequence with infinite duration, the ROC is in the form $|z| > r_{\max}$ or $\infty > |z| > r_{\max}$, where $r_{\max}$ is the largest value of any of the poles of $X(z)$.

Property 4: If $x[n]$ is a left-sided sequence with infinite duration, the ROC is of the form $|z| < r_{\min}$ or $0 < |z| < r_{\min}$, where $r_{\min}$ is the smallest magnitude of any of the poles of $X(z)$.

Property 5: If $x[n]$ is a two-sided sequence with infinite duration, the ROC is of the form $r_1 < |z| < r_2$, where r_1 and r_2 are the magnitudes of the two poles of $X(z)$.

18. The Z-transform of some common signals are given in the table below.

Signals of time function $x(t)$	Z-transform, $X(s)$	Region of convergence, ROC
$\delta[n]$	1	Complete z-plane
$u[n]$	$\frac{z}{z-1}$	$\|z\| > 1$
$-u(-n-1)$	$\frac{z}{z-1}$	$\|z\| < 1$
$r[n] = nu[n]$	$\frac{z}{(z-1)^2}$	$\|z\| > 1$
$a^n u[n]$	$\frac{z}{z-a}$	$\|z\| > \|a\|$
$e^{\alpha n}u[n]$	$\frac{z}{z-e^{\alpha}}$	$\|z\| > \|e^{\alpha}\|$
$-a^n u(-n-1)$	$\frac{z}{z-a}$	$\|z\| < \|a\|$
$\delta[n-m]$	z^{-m}	Complete z-plane
$\cos(\omega_o n)u[n]$	$\frac{z(z-\cos\omega_0)}{z^2 - 2z\cos\omega_0 + 1}$	$\|z\| > 1$
$\sin(\omega_o n)u[n]$	$\frac{z\sin\omega_0}{z^2 - 2z\cos\omega_0 + 1}$	$\|z\| > 1$

$[a^n \cos \omega_o n]\, u[n]$	$\dfrac{z(z-a\cos\omega_0)}{z^2-2za\cos\omega_0+a^2}$	$\lvert z\rvert > a$
$[a^n \sin \omega_o n]\, u[n]$	$\dfrac{za\sin\omega_0}{z^2-2za\cos\omega_0+a^2}$	$\lvert z\rvert > a$

19. Let signals $x_1[n]$ and $x_2[n]$ have Z-transforms $X_1(z)$ and $X_2(z)$ respectively, Also, $x_1[n] \leftrightarrow X_1(z)$ with ROC R_1 and $x_2[n] \leftrightarrow X_2(z)$ with ROC. The properties of Z-transform are given in the table next.

Property	Time domain signal $x[n]$	Z-transform, $X(z)$	Region of convergence, ROC R
Linearity	$a_1x_1[n]+a_2x_2[n]$	$a_1X_1(z)+a_2X_2(z)$	$R_1 \cap R_2$
Time Shifting	$x[n \pm m]$	$Z^{\pm m}X(z)$	R
	$x[n+m]u[n]$	$x^mX(s)-z^m\sum_{k=0}^{m-1} x[k]z^{-k}$	R
	$x[n-m]u[n]$	$z^{-m}X(z)+z^{-m}\sum_{k=1}^{m} x[-k]z^{k}$	R
Time reversal	$x[-n]$	$X\left(\frac{1}{z}\right)$	$\frac{1}{R}$
Time expansion	$x_k[n] = \begin{cases} x[\frac{n}{k}] & n = mk \\ 0 & n \neq mk \end{cases}$ where m is an integer	$X(z^k)$	R
Symmetry	Even sequence, $x[n]=x[-n]$	$X(z)=X\left(\frac{1}{z}\right)$	R
	Odd sequence, $x(n)=-x(-n)$.	$X(z)=-X\left(\frac{1}{z}\right)$	R
Time scaling	$a^nx[n]$	$X\left(\frac{z}{a}\right)$	aR
	$a^{-n}x[n]$	$X(az)$	$\frac{R}{a}$
Conjugation	$x^*[n]$	$X^*(z^*)$	R
Real part of the sequence	Re $\{x[n]\} = \frac{1}{2}(x[n]+x^*[n])$	$=\frac{1}{2}(X(z)+X^*(z^*))$	R
Imaginary part of the sequence	Im $\{x[n]\} = \frac{1}{2j}(x[n]-x^*[n])$	$=\frac{1}{2j}(X(z)-X^*(z^*))$	R

Time convolution	$x[n] * y[n] = \sum_{m=-\infty}^{\infty} x[m]y[n-m]$	$X(z)Y(z)$	$R_1 \cap R_2$
Frequency convolution	$x[n]y[n]$	$\frac{1}{2\pi j}\oint X(v)Y\left(\frac{z}{v}\right)v^{-1}dv$	At least $R_1 \cap R_2$
Correlation of two sequences	$x[n] \otimes y[n] = \sum_{m=-\infty}^{\infty} x[m]y[m-n]$	$X(z)Y(z^{-1})$	$R_1 \cap R_2$
Time multiplication or differentiation in the z domain	$nx(n)$	$-z\frac{dX(z)}{dz}$	R
Intial value theorem	$x[0] = \lim_{n\to\infty} x[n]$	$\lim_{z\to\infty} X(z)$	–
Final value theorem	$x[\infty] = \lim_{n\to\infty} x[n]$	$\lim_{z\to 1}(z-1)X(z)$	–
Parseval's theorem	$\sum_{n=-\infty}^{\infty} x[n]y^*[n]$	$\frac{1}{2\pi j}\oint X(z)Y^*\left(\frac{1}{z^*}\right)z^{-1}dz$	$R_1 \cap R_2$

20. The inverse Z-transform can be obtained by direct integration of complex variables using residue methods, which is very difficult to evaluate. Indirectly, the inverse Z-transform can be evaluated using the basic Z-transform pairs with region of convergence. The methods are: (1) power series expansion (long division method), and (2) partial fraction expansion.
21. The transfer function of a system is $H(z) = \frac{Y(z)}{X(z)}$ with all initial conditions zero. Or,
$$H(z) = \frac{Y(z)}{X(z)} = \frac{(b_M z^{-M} + b_{M-1}z^{-M-1} + \cdots + b_0)}{(a_N z^{-N} + a_{N-1}z^{-N-1} + \cdots + a_0)}.$$
22. The impulse response of the system is $h[n] = z^{-1}[H(z)] = z^{-1}\left[\frac{Y(z)}{X(z)}\right]$.
23. The output response ofthe system is $y[n] = z^{-1}[Y(z)] = z^{-1}[H(z)X(z)]$.
24. The necessary and sufficient condition for a causal LTI discrete-time system to be BIBO-stable is $|H(z)| \leq \sum_{n=0}^{\infty} |h(n)| < \infty$; ROC is $|z| > 1$.

Review Questions

1. What are the fundamental differences between continuous-and discrete-time signals? (May 2015)
2. Distinguish between Fourier transform, Laplace transform and Z-Transform. (June 2014)
3. Compare Laplace transform and Fourier transform in detail. (May 2015)
4. Explain briefly ROC and its important properties with regards to Z-transform. (May 2015)
5. Explain the properties of the region of convergence of $X(z)$. (May 2012)

6. Distinguish between one-sided and two-sided Z-transforms and thier region of convergence. (May 2012)
7. Define the region of convergence and state its properties with respect to Z-transform. (Dec 2015)
8. Define region of convergence and state its properties with regard to Z-transform. (May 2013)
9. Explain the concept of Z-transform of a discrete-time sequence. (May 2012)
10. State and prove time shifting and time convolution properties of Z-transform. (May 2012)
11. Using scaling property determine the Z-transform of $a^n cos\omega n$ and find its ROC. (Dec 2016)
12. State and prove time-reversal, time-shifting and scaling properties with regard to Z-transform (Dec 2016)
13. State and prove differentiation property of Z-transform. (Dec 2014)
14. State and prove initial and final value theorems of Z-transform. (Dec 2014)
15. State and prove integration and differentiation properties of Z-transform. (Nov 2015)
16. State and prove correlation between two sequences. (Dec 2016)
17. State and prove Parseval's theorem. (Dec 2011)
18. Explain the concept of inverse Z-transform. (Dec 2011)
19. What are the methods by which the inverse Z-transform can be found? (May 2012)
20. Find the transfer function of a causal LTI system using Z-transform. (May 2015)

Problems

1. Determine the Z-transform and ROC for the following time signals.

   (a) $3\delta[n-2]+\delta[n-5]$; (b) $n\delta[n-3]$; (c) $((3)^{n+1}-(2)^{n+1})u[n]$.

2. Determine the Z-transform and ROC for the following time signals:

   (a) $\left(\frac{1}{2}\right)^n (u[n]-u[n-10])$; (b) $\{3,2,1,\underset{\uparrow}{0},1,2,\ 3\}$ and (c) $e^{an}u[-n-1]$.

3. Determine the Z-transform and ROC for the following time signals.:

   (a) $(2)\ u[-n-3]$; (b) $\left(\frac{1}{4}\right)^n u[n+5]$ and (c) $n[n-1]u[n-1]$.

4. Determine the Z-transform and ROC for the following time signals.

   (a) $x[n]=\sin\left(\frac{n\pi}{2}-\frac{3\pi}{2}\right)u[n-3]$; (b) $x[n]=\cos\left(\frac{n\pi}{8}-\frac{\pi}{4}\right)u[n-2]$

5. Find the Z-transform for bilateral sequence $x[n]=\left(\frac{1}{2}\right)^n u(-n)$ and its ROC, plotting the poles and zeros in the z-plane.

6. Find the Z-transform of the bilateral sequence given by $x(n)=\begin{cases}1 & 0\le n\le 10\\ 0 & \text{elsewhere.}\end{cases}$

7. Prove that the sequences $x_1[n]=a^n u(n)$ and $x_2[n]=-a^n u(-n-1)$ have the same $X(z)$ and differ only in ROCs. Plot their ROCs.

8. Find the convolution ofthe two sequences $x[n] = \left(\frac{1}{2}\right)^n u[n]$ and $h[n] = \left(\frac{1}{3}\right)^n u[n]$ using Z- transform.

9. Find the Z-transform of the convolution of the two sequences $x[n] = n\left(\frac{1}{3}\right)^n u[n]$ and $h[n] = u[n] - u[n-3]$.

10. Determine the Z-transform of the following signals: (a) $x[n] = \{1,2,0,2\}$; (b) $x[n] = \{1,2,-1,2,9\}$, (c) $x[n] = u(n) - u(n-5)$. (Dec 2011)

11. Find the two sided Z-transform of the signal $x[n] = \begin{cases} (0.33)^n & n \geq 0 \\ (-2)^n & n \leq -1 \end{cases}$. (Dec 2011)

12. Using differentiation property, find the Z-transform of $x[n] = n^2 u(n)$. (Dec 2011)

13. Obtain the Z-transform of $x[n] = -a^n u(-n-1)$ and find its ROC. (Dec 2011)

14. Given the Z-transform, $X(z) = z^{-1} + bz^{-4} - 2z^{-7}, |z| > 0$, determine the time sequence $x[n]$.

15. Determine the inverse of the Z-transform for the $F(z)$ given below.

(a) $\dfrac{3}{z-2}$, $|z| > 2$; (b) $\dfrac{2+3z^{-1}}{(z+1)\left(1+\frac{1}{4}z^{-1} - \frac{1}{8}z^{-2}\right)}$, $|z| > 1$; (c) $\dfrac{z - \frac{1}{2Z}}{1-2}$, $|z| > 2$

16. Determine the inverse of the Z-transform for

(a) $\dfrac{Z}{3z^2 - 4z + 1}$, $|z| > 1$; (b) $\dfrac{2z-7}{z^2 - 5z + 6}$, $|z| < 2$; (c) $\dfrac{z^2 - az}{(z-a)^3}$, $|z| > |a|$.

17. Determine the inverse of the Z-transform for function $X(z) = \ln\left(\dfrac{1}{1-0.3z}\right)$, $|z| < 0.3$.

18. Find the inverse of the Z-transform of $X(z) = \dfrac{1}{2^{10}}\left(\dfrac{2^{10} - z^{-10}}{1 - \frac{1}{2}z^{-1}}\right)$, $|z| > \dfrac{1}{2}$.

19. Find out the left-sided sequence having Z-transform $X(z) = \dfrac{1}{(1 - \frac{1}{2} - z^{-1}), (1 - z^{-1})}$

20. Find the inverse Z-transform using partial fraction method for

$X(z) = \dfrac{z(z-2)}{z^2 + 2.5z - 1}, |z| > 2.85.$

21. Using power series expansion, find the time sequence $x[n]$ for the given $X(z) = e^Z |z| < \infty$.

22. Using power series expansion, find the time sequence $x[n]$ of $X(z) = \dfrac{1 - 0.5z^{-1}}{1 + 0.5z^{-1}}$ for (i) $|z| > 0.5$ and (ii) $|z| < 0.5$.

23. Using the power series expansion, find the time sequence $x[n]$ for

$X(z) = \dfrac{z^3 - 10z^2 - 4z + 4}{2z^2 - 2z - 4} |z| < 1.$

24. Find inverse Z-transform of $X(z) = \dfrac{1}{\left(1 - \frac{1}{3}z^{-1}\right)\left(1 - \frac{1}{6}z^{-1}\right)}$; ROC $|z| > \dfrac{1}{3}$. (Dec 2015)

25. Find the inverse Z-transform of $X(z) = \frac{(z^2+z)}{(z-1)(z-3)}$, ROC $z > 3$ using (a) partial fraction method, and (b) residue method. (Dec 2015)

26. Find the inverse Z-transform of:

(a) $X(z) = \frac{1}{(1-0.5z^{-1}+0.5z^{-2})}$ for ROC $|z| > 1$ and (b) $\frac{1}{(z^2-1.2z+0.2)}$ (Dec 2015)

27. Find the inverse Z-transform of $X(z) = \frac{z}{(z+2)(z-3)}$ when (a) ROC : $|z| < 2$ and (b) ROC : $2 < |z| < 3$. (Jan 2015)

28. Find the inverse Z-transform of $X(z) = \frac{z}{(z+2)(z-3)}, |z| > 2$. (Jan 2015)

29. Find the inverse Z-transform of $X(z) = \frac{(2+z^{-1})}{(1-0.5z^{-1})}$ with ROC $|z| > \frac{1}{2}$ using power series expansion. (May 2015)

30. Determine the inverse Z-transform of $X(z) = \frac{z}{(3z^2-4z+1)}$ if the region of convergence is (a) $z > 1$; (b) $z < \frac{1}{3}$; (c) $\frac{1}{3} < z < 1$. (Dec 2011)

31. Find the transfer function of a causal LTI system if the input signal is $x[n] = \left(\frac{-1}{2}\right)^n u[n]$ and the output sequence is $y[n] = 3(-1)^n u[n] + \left(\frac{1}{3}\right)^n u[n]$.

32. Solve the following difference equation. Assume all initial conditions are zero.

$$y[n] - 0.5y[n-1] + 0.38y[n-2] = x[n] + 0.5x[n-1].$$

33. Solve the following difference equation: $18y[n] - 9y[n-1] + y[n-2] = 0$. The initial conditions are $y[-2] = 9$, $y[-1] = 1$.

34. Given the Z-transform $X(z) = \frac{-3z-1}{2-5z^{-1}+2z^{-2}}$ and determine the time sequence for given three of ROC:

1. $|z| > 2$ 2. $|z| < 0.5$ 3. $0.5 < |z| < 2$.

35. Determine the Z-transform, locations of poles and zeros and sketch the ROC of signal $x[n] = -u(-n-1) + \left(\frac{1}{2}\right)^n u(n)$. (May 2015)

36. Represent the pole-zero pattem of $x[n] = \begin{cases} a^n, & 0 \le n \le N-1,\ a > 0 \\ 0, & \text{otherwise} \end{cases}$. (June 2014)

37. Determine the impulse and unit step response of the systems described by the difference equation $y(n) = 0.6y(n-1) - 0.08y(n-2) + x(n)$.

(Dec 2011, May 2013)

38. A system has an impulse response $h(n) = \{1,2,3\}$ and output response $y(n) = \{1,1,2,-1,3\}$. Determine the input sequence $x[n]$. (Dec 2011)

Answers

1. (a) $3z^{-2}+z^{-5}$ ROC: All z except at $z=0$

   (b) $3z^{-3}$ ROC: All z except at $z=0$

   (c) $\dfrac{3z}{z-3}-\dfrac{2z}{z-2},\ |z|>3$

2. (a) $\dfrac{2z}{(2z-1)}(1-(2z)^{-10}),\ |z|>\dfrac{1}{2}$;

   (b) $3z^2+2z^2+z^1+z^{-1}+2z^{-2}+3z^{-2}$; (c) $\dfrac{-z}{z-e^a}\ |z|>e^a$

3. (a) $\dfrac{z^3}{4(2-z)},\ |z|<2$; (b) $\dfrac{z^6}{z-\frac{1}{4}},\ |z|>\dfrac{1}{4}$; (c) $\dfrac{z(z+1)}{(z-1)^3}-\dfrac{z}{(z-1)^2},\ |z|>1$

4. (a) $\dfrac{z^{-2}}{(z^2+1)},\ |z|>1$; (b) $\dfrac{z^{-1}\left(z-\cos\left(\frac{\pi}{8}\right)\right)}{z^2-2z\cos\left(\frac{\pi}{8}\right)+1},\ |z|>1$

5. $\dfrac{1}{1-2z},\ |z|<0.5$

6. $\dfrac{1-z^{-11}}{1-z^{-1}},\ |z|>1$

8. $y[n]=3\left(\dfrac{1}{2}\right)^n u[n]-2\left(\dfrac{1}{3}\right)^n u[n]$

9. $Y(z)=\dfrac{1}{2}\dfrac{z(1+z^{-1}+z^{-2})}{\left(z-\frac{1}{3}\right)^2}$

10. (a) $X(z)=1+2z^{-1}+2z^{-3}$; (b) $X(z)=1+2z^{-1}-z^{-2}+2z^{-3}+9z^{-4}$;

    (c) $X(z)=1+z^{-1}+z^{-2}+z^{-3}+z^{-4}$

11. $X(z)=\dfrac{3z}{z-1},\ |z|>\dfrac{1}{3},\quad \dfrac{-z}{z+2},\ |z|<2$

12. $X(z)=\dfrac{z^3-z}{(z-1)^4}$ with ROC $|z|>0$.

13. $X(z)=\dfrac{z}{z-a}$ with ROC $|z|<a$.

14. $x[n]=\delta[n-1]+3\delta[n-4]-2\delta[n-7]$

15. (a) $x[n]=3(2)^{n-1}u[n-1]$

    (b) $x(n)=\left(-\dfrac{8}{5}(-1)^n+\dfrac{8}{3}\left(-\dfrac{1}{2}\right)^n+\dfrac{14}{15}\left(\dfrac{1}{4}\right)^n\right)u(n)$

    (c) $x[n]=-\left((2)^{n+1}+\left(\dfrac{1}{2}\right)^{n+1}\right)u[n]$

16. (a) $x[n] = \frac{1}{2}\left(u[n] - \left(\frac{1}{3}\right)^n u[n]\right), \ |z| > 1$

(b) $x[n] = -\frac{7}{6}\delta[n] - \frac{3}{2}(2)^n u[-n-1] + \frac{1}{3}(3)^n u[-n-1]$

17. $\frac{-(0.3)^n}{n} u[n-1]$

18. $x[n] = \begin{cases} \left(\frac{1}{2}\right)^n & 0 \le n \le 9 \\ 0 & \text{otherwise} \end{cases}$

19. $x[n] = -2u[-n-1] + \left(\frac{1}{2}\right)^n u[-n-1]$

20. $x[n] = (1.5156(-2.85)^n - 0.5156(0.35)^n)u[n]$

22. (i) $\left\{\underset{\uparrow}{1}, -1, \frac{1}{2}, -\frac{1}{4}, \frac{1}{8}, \ldots\right\}$, (ii) $\left\{\ldots 32, -16, 8, -4, 2, \underset{\uparrow}{0}\right\}$

23. $\left\{\cdots \frac{5}{4}, -\frac{3}{2}, -1, \underset{\uparrow}{0}\right\}$

24. $x[n] = 2\left(\frac{1}{3}\right)^n u[n] - \left(\frac{1}{6}\right)^n u[n]$

25. $x[n] = -u[n] + 2(3)^n u[n]$

26. (a) $x[n] = -(0.5)^n u[n] + 2u[n]$; (b) $x[n] = \frac{5}{4}\left[(-1)^{n-1}u[n-1] - (0.2)^{n-1}u[n-1]\right]$

27. (a) $x[n] = \frac{1}{5}\left[(-2)^n - (3)^n\right]u[-n-1]$; (b) $x[n] = \frac{-1}{5}(-2)^n u[-n] - \frac{1}{5}(3)^n u[-n-1]$

28. $x[n] = \frac{-1}{4}\delta[n] + u[n] - \frac{3}{4}(2)^n u[n] + \frac{1}{2}n(2)^n u[n]$

29. $x[n] = 4\left(\frac{1}{2}\right)^n u[n] - 2\delta[n]$

30. (a) For ROC $z > 1$, $x[n] = \frac{1}{2}u[n] - \frac{1}{2}\left(\frac{1}{3}\right)^n u[n]$

(b) For ROC $z < \frac{1}{3}$, $x[n] = -\frac{1}{2}u[-n-1] + \frac{1}{2}\left(\frac{1}{3}\right)^n u[-n-1]$

(c) For ROC $\frac{1}{3} < z < 1$, $x[n] = -\frac{1}{2}u[-n-1] - \frac{1}{2}\left(\frac{1}{3}\right)^n u[n]$

31. $\frac{3}{2}(-1)^n u(n) + \frac{5}{2}\left(\frac{1}{3}\right)^n u(n)$

32. $\frac{1 + 0.5z^{-1}}{(1 - 0.5z^{-1} + 0.38z^{-2})}$

33. $y[n] = 2\left(\frac{1}{6}\right)^{n+1} u[n] - \left(\frac{1}{3}\right)^{n+1} u[n]$

34. i) $\frac{-14}{3}(2)^n u(n) + \frac{5}{12}\left(\frac{1}{2}\right)^n u(n)$

ii) $\frac{14}{3}(2)^n u(-n-1) + \frac{5}{12}\left(\frac{1}{2}\right)^n u(-n-1)$

iii) $\frac{14}{3}(2)^n u(-n-1) + \frac{5}{12}\left(\frac{1}{2}\right)^n u(n)$

35. $X(z) = \frac{z(2z-3/2)}{(z-1)(z-1/2)}$ with ROC $\frac{1}{2} < |z| < 1$

36. $X(z) = \frac{1-(az^{-1})^N}{1-az^{-1}},\ a > 0$

37. Impulse response: $y[n] = 2(0.4)^n u[n] - (0.2)^n u[n]$.

Step response: $y[n] = \frac{25}{12}u[n] - \frac{4}{3}(0.4)^n u[n] + \frac{1}{4}(0.2)^n u[n]$.

38. $x[n] = \{1, -1, 1\}$

Multiple-choice Questions

1. For what value of $|z|$ the Fourier and Z-transform are equal?.

(a) 1 (b) 0 (c) Infinity (d) -1

2. Which of the following Z-transform statement is correct?

(i) Z-transform of $\delta[n]$ is 1 (ii) Z-transform $\delta[n-1]$ is z^{-1}

(iii) Z -transform $\delta[n+1]$ is z (iv) Z-transform $6\ [n-1] + 6[n+1]$ is $z^{-1} + z$

(a) Only i (b) i, ii, iii (c) i, iv (d) i, ii, iii. iv

3. Which of the following ROC statement is correct?

(i) ROC of Z–transform of $\delta[n]$ is entire z-plane.

(ii) ROC of the Z-transform of $\delta[n-1]$ is the entire z-plane except $z = 0$.

(iii) ROC of Z-transform of $\delta[n+1]$ is entire z-plane except $z = \infty$.

(iv) Z-transform of $\delta[n-1] + \delta[n+1]$ is entire z-plane except $z = 0$ and $z = \infty$.

(a) Only i (b) i, ii, iii (c) i,iv (d) i, ii, iii, iv

4. What is the Z-transform of $-a^n u[-n-1]$?

(a) $\frac{z}{z-a}$, ROC : $|z| > a$ (b) $\frac{z}{z+a}$, ROC : $|z| > a$

(c) $\frac{z}{z-a}$, ROC : $|z| < a$ (d) $\frac{z}{z+a}$, ROC : $|z| < a$

5. What is the Z-transform of $\cos[n\theta]u[n]$?

(a) $\frac{1-z^{-1}\cos(\theta)}{1-2z^{-1}\cos(\theta)+z^{-2}}$ (b) $\frac{1+z^{-1}\cos(\theta)}{1-2z^{-1}\cos(\theta)+z^{-2}}$

(c) $\frac{1+z^{-1}\cos(\theta)}{1+2z^{-1}\cos(\theta)+z^{-2}}$ (d) $\frac{1-z^{-1}\cos(\theta)}{1-2z^{-1}\cos(\theta)+z^{-2}}$

6. What is the Z-transform of $\sin[n\theta]u[n]$?

(a) $\dfrac{z^{-1}\sin(\theta)}{1-2z^{-1}\cos(\theta)+z^{-2}}$ (b) $\dfrac{z^{-1}\cos(\theta)}{1-2z^{-1}\cos(\theta)+z^{-2}}$

(c) $\dfrac{z^{-1}\cos(\theta)}{1-2z^{-1}\cos(\theta)+z^{-2}}$ (d) $\dfrac{z^{-1}\sin(\theta)}{1-2z^{-1}\cos(\theta)-z^{-2}}$

7. What is the ROC of Z-transform of $\left(\frac{1}{2}\right)^n u[n]+\left(\frac{1}{3}\right)^n u[n]+\left(\frac{1}{4}\right)^n u[n]$?

(a) $|z|>\frac{1}{2}$ (b) $|z|>\frac{1}{3}$ (c) $|z|>\frac{1}{4}$ (d) $|z|>1$

8. What is the ROC of Z-transform of $\left(\frac{1}{2}\right)^n u[-n-1]+\left(\frac{1}{3}\right)^n u[-n-1]+\left(\frac{1}{4}\right)^n u[-n-1]$?

(a) $|z|<\frac{1}{2}$ (b) $|z|<\frac{1}{3}$ (c) $|z|<\frac{1}{4}$ (d) $|z|<1$

9. Which ofthe following statement is correct?

(i) ROC does not contain any pole.

(ii) If $x[n]$ is a right-sided sequence, then ROC is outside the circle of outermost pole.

(iii) If $x[n]$ is left-sided, then ROC is inside the circle of innermost pole.

(iv) If $x[n]$ is both-sided, then ROC is finite duration ring between two poles.

(a) i, ii & iii (b) ii, iii & iv (c) i, iii & iv (d) all

10. If Z-transform of $x[n]=X(z)$, then the Z-transform of $x[-n]$ is

(a) $X(-z)$ (b) $X(z^{-1})$ (c) $X(1-z)$ (d) $X(1+z)$

11. If the Z-transform of $x[n]=X(z)$, then the Z-transform of $x^*[n]$ is

(a) $X^*(z)$ (b) $X(z)$ (c) $X^*(z^*)$ (d) $X(z^*)$

12. If the Z-transform of $x[n]=X(z)$, then the Z-transform of $nx[n]$ is

(a) $z\frac{d}{dz}X(z)$ (b) $-z\frac{d}{dz}X(z)$ (c) $\frac{d}{dz}X(z)$ (d) $-\frac{d}{dz}X(z)$

13. If the Z-transform ofx $[n]=X(z)$, then the Z-transform of $a^n x[n]$ is

(a) $X(az)$ (b) $X(a+z)$ (c) $X(a/z)$ (d) $X(z/a)$

14. If Z-transform ofx $[n]=X(z)$, then the Z-transform of $x[n-a]$ is

(a) $z^{-a}X(z)$ (b) $z^{a}X(z)$ (c) $X(z-a)$ (d) $X(z+a)$

15. Which ofthe following is initial value theorem?

(a) $x[0]=\lim_{n\to 0}[x[n]]=\lim_{z\to\infty}[zX(z)]$

(b) $x[0]=\lim_{n\to 0}[x[n]]=\lim_{z\to\infty}[X(z)]$

(c) $x[0]=\lim_{n\to 0}[x[n]]=\lim_{z\to\infty}[z+X(z)]$

(d) $x[0]=\lim_{n\to 0}[x[n]]=\lim_{z\to\infty}[z-X(z)]$

16. Which ofthe following is final value theorem?

(a) $x[\infty] = \lim_{n\to\infty}[x[n]] = \lim_{z\to 1}[zX(z)]$

(b) $x[\infty] = \lim_{n\to\infty}[x[n]] = \lim_{z\to 0}[X(z)]$

(c) $x[\infty] = \lim_{n\to\infty}[x[n]] = \lim_{z\to 1}\left[\left(\frac{z-1}{z}\right)X(z)\right]$

(d) $x[\infty] = \lim_{n\to\infty}[x[n]] = \lim_{z\to 1}\left[\left(\frac{z+1}{z}\right)X(z)\right]$

17. Which of the following method is used to obtain inverse Z-transform?

(a) Power series method (Long division method)

(b) Partial fraction method

(c) Residue method (Inverse integration method)

(d) All

18. What is the Z-transform of convoluted sequence $a^n u[n] \otimes b^n u[n]$?

(a) $\frac{z^2}{(z-a)(z-b)}$ (b) $\frac{z}{z-a} + \frac{z}{z-b}$

(c) $\frac{z}{(z-a)(z-b)}$ (d) $\frac{z^2}{(z+a)(z+b)}$

19. What is the Z-transform of convoluted sequence $u[n] \otimes nu[n]$?

(a) $\frac{z^2}{z-1}$ (b) $\frac{z^2}{(z-1)^2}$ (c) $\frac{z^2}{(z-1)^3}$ (d) $\frac{z^2}{(z-1)^4}$

20. If $\mathcal{Z}[x[n]] = X(z) = \frac{(z+1)(3z+1)}{(2z-1)(4z-1)}$, then the initial value of $x[n]$ is

(a) 0 (b) 8/3 (c) 3/8 (d) 1/2

21. If $\mathcal{Z}[x[n]] = X(z) = \frac{(z+1)(3z+1)}{(2z-1)(4z-1)}$, then the final value is value of $x[n]$ is

(a) 0 (b) 8/3 (c) 3/8 (d) 1/2

22. If $X(z) = \frac{z}{1+z^2}$, then $x[n]$ is

(a) $\sin\left(\frac{n\pi}{2}\right)u[n]$ (b) $\sin\left(\frac{n\pi}{2}\right)$ (c) $\cos\left(\frac{n\pi}{2}\right)u[n]$ (d) $\cos\left(\frac{n\pi}{2}\right)$

23. if $X(z) = \frac{z^2}{1+z^2}$, then $x[n]$ is

(a) $\sin\left(\frac{n\pi}{2}\right)u[n]$ (b) $\sin\left(\frac{n\pi}{2}\right)$ (c) $\cos\left(\frac{n\pi}{2}\right)u[n]$ (d) $\cos\left(\frac{n\pi}{2}\right)$

24. If $x[n] = \{1, 0, -1, 0, 1, 0, -1, \ldots\ldots\ldots\}$, then $X(z)$ is

(a) $\frac{z^2}{1+z^2}$ (b) $\frac{z}{1+z^2}$ (c) $\frac{z^2}{1+z}$ (d) $\frac{z}{1+z}$

25. If $x[n] = \{1,0,1,0,1,0,1,\ \}$, then $X(z)$ is

(a) $\frac{z^2}{z^2-1}$ (b) $\frac{z^2}{z^2+1}$ (c) $\frac{z}{z^2-1}$ (d) $\frac{z}{1+z}$

26. What is the causality of a system if $H(z) = \frac{z(z+1)}{(z-0.5)(z-0.25)}$ and $|z| > \frac{1}{2}$?

(a) Non-causal (b) Causal

(c) Depends on $z = \frac{1}{4}$ (d) None of the above

27. What is the causality of a system if $H(z) = \frac{z(z+1)}{(z-0.5)(z-0.25)}$ and $|z| < \frac{1}{4}$?

(a) Non-causal (b) Causal

(c) Depends on $z = \frac{1}{2}$ (d) None of these

28. What is necessary and sufficient condition that a causal system has a rational system function $H(z)$?

(a) The degree of numerator cannot be greater than the degree of the denominator.

(b) ROC of $H(z)$ should be outside the circle of outermost pole.

(c) Both (a) and (b)

(d) None of these

29. What is the condition for a system to be stable?

(i) Output should be bounded if input is bounded

(ii) Impulse response is absolutely summable

(iii) ROC of $H(z)$ includes the unit circle

(iv) All of the poles of causal system should be inside the unit circle.

Choose the correct answer:

(a) i, ii, iii, iv (b) only iv (c) iii, iv (d) ii, iii, iv

30. For what value of a the system having impulse response $h(n) = a^n u(n)$ is stable?

(a) $|a| < 1$ (b) $a > 1$ (c) $a < 1$ (d) $a < 0$

31. For a given discrete LTI system having difference equation $y(n) - 4y(n-1) = x(n) - 3x(n-1)$, the transfer function is

(a) $H(z) = \frac{z-3}{z-4}$ (b) $H(z) = \frac{z-4}{z-3}$ (c) $H(z) = \frac{z+3}{z+4}$ (d) $H(z) = \frac{z+4}{z+3}$

32. For a given discrete LTI system having difference equation $y(n) - 4y(n-1) = x(n)$, the impulse response is

(a) $h(n) = 0.25^n u(n)$ (b) $h(n) = 4^n u(n)$

(c) $h(n) = -4^n u(n)$ (d) None of these

33. For a given discrete LTI system having difference equation $y(n)-4y(n-1)=x(n)-x(n-1)$, the unit step response is
(a) $h(n)=0.25^n u(n)$ (b) $h(n)=4^n u(n)$
(c) $h(n)=-4^n u(n)$ (d) None of these
34. For a given discrete LTI system having difference equation $y(n)-16y(n-4)=x(n)$, the poles are
(a) 2, 4, $2i$, $4i$ (b) 2, 4, -2, -4 (c) ± 2, $\pm 2i$ (d) None of these
35. The number of zeros located at $z=0$ of a given discrete LTI system having difference equation $y(n)-16y(n-4)=x(n)$ are:
(a) 0 (b) 1 (c) 2 (d) 4
36. For what value of a is the given system having difference equation $y(n)-ay(n-1)+0.1y(n-2)=x(n)$ stable?
(a) $-1.1<a<1.1$ (b) $a>1.1$
(c) For all values of a (d) $a<-1.1$

Answers

1. (a)	2. (d)	3. (d)	4. (c)	5. (a)	6. (a)	7. (a)	8. (c)	9. (a)	10. (b)
11. (c)	12. (b)	13. (d)	14. (a)	15. (b)	16. (c)	17. (d)	18. (a)	19. (c)	20. (c)
21. (a)	22. (a)	23. (c)	24. (a)	25. (a)	26. (b)	27. (a)	28. (c)	29. (a)	30. (a)
31. (a)	32. (b)	33. (b)	34. (c)	35. (d)	36. (a)				

9

Random Processes — Temporal Characteristics

Learning Objectives

The main objective of this chapter is to understand thoroughly the basic concepts of random signals and processes. We will learn how random signals are represented using stochastic processes. An understanding of the basic concepts on probability theory and random variables are a prerequisite for this chapter. By reading this chapter, students will be able to relate to the following topics:

- Random process and classification of random processes
- Distribution and density functions
- Stationarity and statistical independence
- Stationary processes, wide-sense and strict-sense stationarity
- Time averages, ergodicity and ergodic processes
- Correlation functions and their properties
- Covariance function and its properties
- Gaussian and Poisson random processes
- Linear system with random processes
- Mean and mean-squared value of system response
- Correlation functions of input and output of a linear system

9.1 Introduction

Broadly, signals are divided into two categories - deterministic and random signals. Unlike deterministic signals, random signals cannot be predicted by mathematical equations.

Any signal which is random in its nature is called a random signal or a stochastic signal. An example is the thermal noise generated by the random motion of electrons in a conductor. A random signal may also have any one of its parameters in random. Random signals may be either periodic or aperiodic, analog or digital, depending on the application. For example, a periodic sinusoidal signal with a random amplitude or phase is a random signal.

A random process is a family of random signals or functions defined with random variables on a sample space S and as a function of time T. The analysis of random processes involve probabilistic concepts in addition to the deterministic signal analysis that has been dealt with so far in this book. This chapter introduces the concepts of random processes, also called stochastic processes.

Stochastic processes deal with randomly varying time waveforms such as message signals, speech, music, background noise, variations of weather conditions, variations of signals at the receiver end, etc. They are described statistically, since a complete knowledge about their origin is not known. Statistical measures are used to characterise ones degree of knowledge about the randomness of a process. Probability distribution and probability density functions give the complete statistical characteristics of random signals. In fact, most information-bearing signals possess a certain degree of unpredictability and hence fall in this category.

Since random processes have infinite energy, transform tools such as Fourier transform, Laplace transform, Z-transform, etc., are not applicable. Hence, finite energy functions like correlation and covariance functions in the time domain, for which transform tools exist, are used to analyse random processes.

Readers are advised to refer to any textbook on probability theory and random variables to refresh their understanding of the basic concepts.

9.2 The Random Process Concept

A random process is a family of functions $\{X \leq x(t,\ s), s \in S \text{ and } t \in T\}$ defined with random variables $\{X \leq x\}$ on a sample space S and as a function of time T. Thus a random process is a function of two variables—sample space and time. For example, let us consider an experiment which measures the number of calls received during peak hours. The calls received (sample space) is a random process as the number of calls received at different time intervals may be different. Thus the number of calls is a function of both sample space and time.

A set of random variables can be derived from a random process at a fixed time t_K. So, at $t = t_K$, $X(t_K, s) = X_K(s)$ is a random variable over the sample space S at time t_K. Similarly, a sample function of the process can be obtained at a particular sample point $s_i \in S$. So at $S = S_i$, the process $X(t,\ s_i) = X_i(t)$ is a sample function or a realisation of the process.

The family of sample (or time) functions over the sample space S is called an ensemble. Figure 9.1 illustrates a continuous random process $X(t,\ S)$ with a family of time functions $\{x_1(t),\ x_2(t),\ x_3(t), \ldots, x_n(t)\}$.

From Fig. 9.1 we can obtain random variables from the random process at fixed time values. At time t_1, X_1 is a random variable with set of outcomes $x_1,\ x_2,\ x_3 \ldots x_n$.

$$\therefore\quad X_i = X(t_1) = \{x_1,\ x_2, x_3 \ldots x_n\}.$$

Similarly, the random variables defined at different time values are

$$X_k = X(t_k),\ k = 1, 2, \ldots, n. \tag{9.1}$$

Therefore, $X_1,\ X_2,\ X_3, \ldots, X_n$ are random variables defined at time values $t_1,\ t_2,\ t_3 \ldots t_n$ respectively.

If $f_X(x)$ is a probability density function, then the expected value of X_k is called the statistical average or ensemble average. That is, $E[X_k] = \int_{-\infty}^{\infty} X_k f_{X_k}(x)dx.$ (9.2)

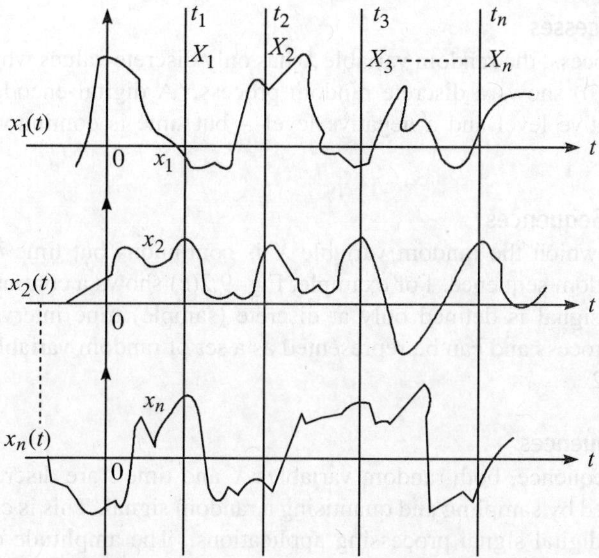

Fig. 9.1 *A continuous random process*

A random process can also be represented as a time function over the entire time. That is,

$$X(t,s_i) = X_i(t) \tag{9.3}$$

Here, the time functions $x_1(t), x_2(t), \ldots\ldots, x_n(t)$ are random processes. The expected value of a time function is called the time average of that function.

It can be expressed as

$$A[x_i(t)] = \frac{1}{T}\int_0^T x(t)dt. \tag{9.4}$$

$A[\cdot]$ represents the time average.

If both sample space s and time t are fixed, then $X(t_k,\ s_i)$ is simply a real number. A physical example of a continuous stochastic process is the fluctuations of voltage and phase at the receiving antenna. The received signal is a sample function.

$$X_i(t) = A(s_i)\cos[\omega t + (\phi(s_i)], \tag{9.5}$$

where $A(s_i)$ is a random amplitude, $\phi(s_i)$ is a random phase and $s_i \in S$ is a sample space. The random variables at time $t = t_K$ are

$$A(t_K, s_i) \text{ and } \{\phi(t_K, s_i),\ i = 1, 2, \cdots n.$$

9.3 Classification of Random Processes

Random processes are mainly classified into the following four types based on time t and amplitude of the random variable X:

Continuous Random Processes

A random process is said to be continuous if both the random variable X and time are continuous over the entire time. For example, Fig. 9.2(a) shows a continuous random process. The fluctuations of a noise voltage in any network is a continuous random process.

Discrete Random Processes

In a discrete random process, the random variable X has only discrete values while time t is continuous. For example, Fig. 9.2(b) shows a discrete random process. A digital encoded signal has only two discrete values—a positive level and a negative level— but time is continuous. So, it is a discrete random process.

Continuous Random Sequences

A random process for which the random variable X is continuous but time t has discrete values is called a continuous random sequence. For example, Fig. 9.2(c) shows a continuous random sequence. A continuous random signal is defined only at discrete (sample) time intervals. It is also called a discrete time random process and can be represented as a set of random variables $\{X\ (t_K)\}$ for sample times $t_K,\ k = 0,\ \pm 1, \pm 2 \ldots$.

Discrete Random Sequences

In a discrete random sequence, both random variables X and time t are discrete. A discrete random sequence can be obtained by sampling and quantising a random signal. This is called the digital process and is mostly used in digital signal processing applications. The amplitude of the sequence can be quantised into two levels or multi-levels as shown in Fig. 9.2 (d) and (e).

9.4 Deterministic and Non-deterministic Processes

In general, a random process may be deterministic or non-deterministic.

A process is called a deterministic random process if the future values of any sample function can be predicted from its past values. For example, $X(t) = -A\sin(\omega_0 t + \theta)$, where the parameters A, ω_0 or θ may be random variables, is a deterministic random variable because the future values of the sample function can be predicted from its known shape. If the future values of a sample function cannot be predicted from the observed past values, the process is a called non-deterministic process. For example, the random process shown in Fig. 9.2(a) is a non-deterministic process.

9.5 Distribution and Density Functions of Random Processes

A random variable obtained from a random process at a particular instant of time t will have statistical properties related to its probability density function.

If two random variables are obtained from the random process at two different times, they will have all the statistical properties related to their joint probability density function. Similarly, N random variables defined at N time instants will have the statistical properties related to their N-dimensional joint probability density function.

9.5.1 Joint Distribution Functions of a Random Process

Consider a random process $X(t)$. For a single random variable at time t_1, $X_1 = X(t_1)$, the cumulative distribution function is defined as

$$F_X(x_1;t_1) = P\{(X(t_1) \le x_1\}, \tag{9.6}$$

where x_1 is any real number. The function $F_X(x_1;t_1)$ is known as the first order distribution function of $X(t)$.

For two random variables at time instants t_1 and t_2, $X(t_1) = X_1$ and $X(t_2) = X_2$, the joint distribution is called the second order joint distribution function of the random process $X(t)$. It is given by

$$F_1(x_1, x_2; t_1, t_2) = P\{X(t_1) \leq x_1, X(t_2) \leq x_2\}. \tag{9.7}$$

In general, for N random variables at N time instants, $X(t_i) = X_i$, $i = 1, 2, \cdots N$, the Nth order joint distribution function of $X(t)$ is defined as

$$F_X(x_1, x_2, \ldots x_N; t_1, t_2 \ldots t_N) = P\{X(t_1) \leq x_1, X(t_2) \leq x_2, \ldots\ldots, X(t_N) \leq x_N\}. \tag{9.8}$$

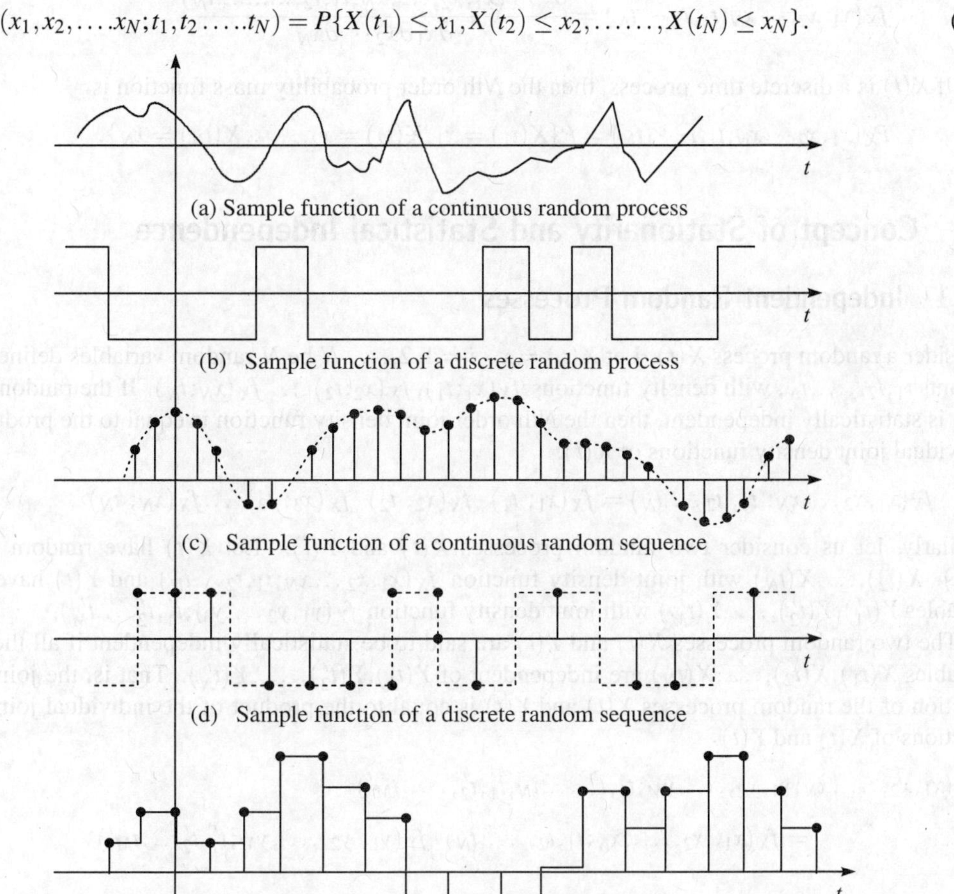

Fig. 9.2 *Classification of random processes*

9.5.2 Joint Density Functions of a Random Process

Joint density functions of a random process can be obtained from the derivatives of the distribution function.

(1) First order density function

$$f_X(x_1; t_1) = \frac{dF_X(x_1; t_1)}{dx_1} \tag{9.9}$$

(2) Second order joint density function

$$f_X(x_1,x_2;t_1,t_2) = \frac{\partial^2 F_X(x_1,x_2;t_1,t_2)}{\partial x_1 \partial x_2} \tag{9.10}$$

(3) Nth order joint density function

$$f_X(x_1,x_2,\cdots x_N;t_1,t_2\cdots t_N) = \frac{\partial^N F_X(x_1,x_2,\ldots,x_N;t_1,t_2,\ldots\ldots,t_N)}{\partial x_1 \partial x_2 \cdots \partial x_N} \tag{9.11}$$

(4) If $X(t)$ is a discrete time process, then the Nth order probability mass function is

$$P_X(x_1,x_2\cdots x_N;t_1,t_2\cdots t_N) = P\{X(t_1)=x_1,X(t_2)=x_2,\ldots.,X(t_N)=x_N\}. \tag{9.12}$$

9.6 Concept of Stationarity and Statistical Independence

9.6.1 Independent Random Processes

Consider a random process $X(t)$. Let $X(t_i)=x_i,\ i=1,2,\ldots.,N$ be N random variables defined at time instants $t_1,t_2,\ldots.,t_N$, with density functions $f_X(x_1;t_1), f_X(x_2;t_2)\ldots.\ f_X(x_N;t_N)$. If the random process $X(t)$ is statistically independent, then the Nth order joint density function is equal to the product of the individual joint density functions of $X(t)$.

$$f_X(x_1,x_2\ldots x_N;\ t_1,\ t_2\ldots t_N) = f_X(x_1;\ t_1)\cdot f_X(x_2;\ t_2)\cdot f_X(x_3;\ t_3)\ldots.f_X(x_N;\ t_N) \tag{9.13}$$

Similarly, let us consider two random processes $X(t)$ and $Y(t)$. Let $X(t)$ have random variables $X(t_1),\ X(t_2),\ldots,X(t_N)$ with joint density function $f_X(x_1,x_2\ldots x_N;t_1,t_2\ldots t_N)$ and $Y(t)$ have random variables $Y(t_1'),Y(t_2'),\ldots.Y(t_M')$ with joint density function $f_Y(y_1,y_2\ldots y_M;t_1',t_2'\ldots t_M')$.

The two random processes $X(t)$ and $Y(t)$ are said to be statistically independent if all the random variables $X(t_1),X(t_2),\ldots.,X(t_N)$ are independent of $Y(t_1'),Y(t_2'),\ldots.,Y(t_N')$. That is, the joint density function of the random processes $X(t)$ and $Y(t)$ is equal to the product of the individual joint density functions of $X(t)$ and $Y(t)$.

$$\begin{aligned} &f_{X,Y}(x_1,x_2,\ldots.,x_N;y_1,y_2,\ldots,y_M;t_1,t,\ldots.,t_N,t_1',t_2',\ldots.,t_M{}') \\ &\qquad = f_X(x_1,\ x_2,\ldots.,x_N;t_1,t_2,\ldots.,t_N)\cdot f_Y(y_1,\ y_2,\ldots.,y_M;t_1',t_2',\ldots t_M{}') \end{aligned} \tag{9.14}$$

9.6.2 Statistical Properties of Random Processes

Mean

The mean value of a random process $X(t)$ is equal to the expected value of the random process $X(t)$. It is denoted as

$$\begin{aligned} \overline{X(t)} &= E[X(t)] \\ &= \int_{-\infty}^{\infty} x f_X(x;t)dx, \end{aligned} \tag{9.15}$$

where $f_X(x;t)$ is the probability density function of the random process $X(t)$. The mean value $E[X(t)]$ is also called the ensemble average of $X(t)$.

Autocorrelation

Consider a random process $X(t)$. Let X_1 and X_2 be two random variables defined at times t_1 and t_2 respectively with joint density function $f_X(x_1,x_2;t_1,t_2)$. The correlation of X_1 and X_2, $E[X_1X_2] = E[X(t_1)X(t_2)]$ is called the autocorrelation function of the random process $X(t)$. It is denoted as

$$R_{XX}(t_1,t_2) = E[X(t_1)X(t_2)] = E[X_1X_2]$$

or $$R_{XX}(t_1,t_2) = \int_{-\infty}^{\infty}\int_{-\infty}^{\infty} x_1x_2 f_X(x_1,x_2;t_1,t_2)dx_1dx_2. \quad (9.16)$$

Cross-correlation

Consider two random processes $X(t)$ and $Y(t)$ defined with random variables X and Y at time instants t_1 and t_2 respectively. The joint density function is $f_{X,Y}(x,y;\ t_1,t_2)$. Then the correlation of X and Y, $E[XY] = E[X(t_1)Y(t_2)]$ is called the cross-correlation function of the random processes $X(t)$ and $Y(t)$. It is denoted as

$$R_{XY}(t_1,t_2) = E[X(t_1)Y(t_2)] = E[XY]$$

or $$R_{XY}(t_1,t_2) = \int_{-\infty}^{\infty}\int_{-\infty}^{\infty} xy f_{X,Y}(x,\ y;t_1,t_2)dx\,dy. \quad (9.17)$$

9.7 Stationary Processes — First Order Stationary Processes

A random process is said to be stationary if all its statistical properties such as mean, moments, variances, etc., do not change with time. The stationarity which depends on the density functions of the random variables of the process has different levels or orders.

First Order Stationary Processes

A random process is said to be stationary to order one or first order stationary if its first order density function does not change with time or shift in time value. If $X(t)$ is a first order stationary process, then

$$f_X(x_1;t_1) = f_X(x_1;t_1+\Delta t) \text{ for any } t_1, \quad (9.18)$$

where Δt is a shift in time value.

From (9.18) we can say that $f_X(x_1,t_1)$ is independent of t_1, and so the mean value of the process is constant. Therefore, the condition for a process to be a first order stationary random process is that its mean value must be constant at any time instant, i.e.,

$$E[X(t)] = \overline{X} = \text{constant}. \quad (9.19)$$

Proof:

Consider a random process $X(t)$ with random variables $X_1 = X(t_1)$ and $X_2 = X(t_2)$ at time instants t_1 and $t_2 = t_1 + \Delta t$ respectively. The mean values are

$$E[X_1] = E[X(t_1)] = \int_{-\infty}^{\infty} x_1 f_X(x_1;t_1)dx_1$$

and $$E[X_2] = E[X(t_2)] = \int_{-\infty}^{\infty} x_1 f_X(x_1;t_2)dx_1.$$

Now substitute $t_2 = t_1 + \Delta t$,

$$E[x(t_1+\Delta t)] = \int_{-\infty}^{\infty} x_1 f_X(x_1;t_1+\Delta t)dx_1.$$

For a stationary random process, we know that

$$f_X(x_1;t_1+\Delta t) = f_X(x_1;t_1).$$

$$\therefore \quad E[X(t_1+\Delta t)] = \int_{-\infty}^{\infty} x_1 f_X(x_1;t_1)dx$$

$$E[X(t_1+\Delta t)] = E[X(t_1)] = \text{a constant.} \qquad \text{(Proved)}$$

9.8 Second Order Stationary Processes

A random process is said to be stationary to order two or a second order stationarity process if its second order joint density function does not change with time or shift in time value. In other words,

$$f_X(x_1,x_2;t_1,t_2) = f_X(x_1,x_2;t_1+\Delta t,t_2+\Delta t) \quad \text{for all} \quad t_1,\ t_2 \ \text{and} \ \Delta t. \tag{9.20}$$

It is a function of time differences $(t_2,\ t_1)$ and not of absolute time t. Note that a second order stationary process is also a first order stationary process.

Thus the *stationary condition* for a process to be a second order random process is that its autocorrelation function should depend only on time the differences and not on the absolute time.

That is, if X_1 and X_2 are the two random variables of a random process $X(t)$ defined at time values t_1 and t_2 respectively, then the autocorrelation function is

$$R_{XX}(t_1,t_2) = E[X(t_1)X(t_2)].$$

If $\tau = t_2 - t_1$, then

$$R_{XX}(t_1,t_1+\tau) = E[X(t_1)X(t_1+\tau)] = R_{XX}(\tau). \tag{9.21}$$

$R_{XX}(\tau)$ should be independent of time t.

9.8.1 Wide Sense Stationary Processes (WSS)

If a random process $X(t)$ is a second order stationary process, then it is called a wide sense stationary process (WSS) or a weak sense stationary process X. However the converse is not true. The conditions for a wide sense stationary process are:;

$$(1)\ E[X(t)] = \overline{X} = \text{ constant} \tag{9.22}$$

$$(2)\ E[X(t)X(t+\tau)] = R_{XX}(\tau) \text{ is independent of absolute time } t. \tag{9.23}$$

9.8.2 Jointly Wide Sense Stationary Processes

Consider two random processes $X(t)$ and $Y(t)$. If they are jointly wide sense stationary (WSS), then the cross-correlation function of $X(t)$ and $Y(t)$ is a function of the time difference $\tau = t_2 - t_1$ only and not the absolute time. That is, the cross-correlation function

$$R_{XY}(t_1,t_2) = E[X(t_1),Y(t_2)]. \tag{9.24}$$

If the time difference $\tau = t_2 - t_1$ and $t = t_1$, then

$$R_{XY}(t,t+\tau) = E[X(t),Y(t+\tau)] = R_{XY}(\tau). \tag{9.25}$$

Therefore, the conditions for a process to be a joint wide sense stationary process are

(1) $E[X(t)] = \overline{X} =$ constant

(2) $E[Y(t)] = \overline{Y} =$ constant

(3) $E[X(t)Y(t+\tau)] = R_{XY}(\tau)$ is independent of time t. (9.26)

9.9 *N*th Order and Strict-Sense Stationarity

A random process $X(t)$ is said to be strict sense stationary if its Nth order joint density function does not change with time or shift in time value. That is,

$$f_X(x_1, x, \ldots, x_N; t_1 t_2, \ldots, t_N) = f_X(x_1, x_2, \ldots, x_N; t_1 + \Delta t, \ldots, t_2 + \Delta t, \ldots, t_N + \Delta t) \quad (9.27)$$

for all $t_1, t_2, \ldots, t_N$ and Δt.

A process that is stationary to all orders $n = 1, 2, \ldots, N$, is called a strict sense stationary process.

A strict sense stationary process is also called an Nth order stationary process or stationary to order N. Note that a strict sense stationary (SSS) process is also a WSS process but the converse is not true.

If the stationary conditions of a random process $X(t)$ does not hold for all N but holds for $N \le K$, then the process $X(t)$ is stationary to order K or a Kth order stationary process.

Example 9.1

A random process $Y(t)$ is given by $Y(t) = X(t)\cos(\omega t + \theta)$, where $X(t)$ is a wide sense stationary random process, ω is a constant and θ is a random phase independent of $X(t)$, uniformly distributed on $(\pi, -\pi)$. Find (a) $E[Y(t)]$ and (b) $R_{YY}(\tau)$.

Solution

Given $Y(t) = X(t)\cos(\omega t + \theta)$.

Since the phase θ is uniformly distributed on $(\pi, -\pi)$, the density function is

$$f_\theta(\theta) = \begin{cases} \frac{1}{2\pi} & -\pi < \theta < \pi \\ 0 & \text{otherwise.} \end{cases}$$

(a) Mean value of $Y(t)$:

$$E[Y(t)] = E[X(t)\cos(\omega t + \theta)].$$

Since $X(t)$ and θ are independent

$$E[Y(t)] = E[X(t)]E[\cos(\omega t + \theta)].$$

Now,

$$E[\cos(\omega t + \theta)] = \int_{-\infty}^{\infty} \cos(\omega t + \theta) f_\theta(\theta) d\theta$$

$$= \int_{-\pi}^{\pi} \cos(\omega t + \theta) \times \frac{1}{2\pi} d\theta$$

$$= \frac{1}{2\pi}[\sin(\omega t + \theta)]_{-\pi}^{\pi}$$

$$= \frac{1}{2\pi}[\sin(\omega t+\pi) - \sin(\omega t - \pi)] = 0.$$

$$\therefore \quad E[Y(t)] = E[X(t)](0) = 0.$$

(b) Autocorrelation of $Y(t)$:

$$R_{YY}(\tau) = E[Y(t)Y(t+\tau)]$$

$$= E[X(t)\cos(\omega t+\theta)X(t+\tau)\cos(\omega(t+\tau)+\theta)]$$

$$= E[X(t)X(t+\tau)\cos(\omega t+\theta)\cos(\omega t+\omega\tau+\theta)]$$

$$= E[X(t)X(t+\tau)]E[\cos(\omega t+\theta)\cos(\omega t+\omega\tau+\theta))]$$

$$= R_{XX}(\tau)E\left[\frac{1}{2}(\cos(2\omega t+\omega\tau+2\theta)+\cos(\omega\tau))\right]$$

$$= \frac{1}{2}R_{XX}(\tau)[E[\cos(2\omega t+\omega\tau+2\theta)]+E(\cos\omega\tau)]$$

$$= \frac{R_{XX}(\tau)}{2}[0+\cos(\omega\tau)] = \frac{1}{2}R_{XX}(\tau)\cos\omega\tau.$$

$$\therefore \quad R_{YY}(\tau) = \frac{1}{2}R_{XX}(\tau)\cos\omega\tau.$$

9.10 Time Averages and Ergodicity

9.10.1 Time Average Function

Consider a random process $X(t)$. Let $x(t)$ be a sample function which exists for all time at a fixed value in the given sample space S. The average value of $x(t)$ taken over all times is called the time average of $x(t)$. It is also called mean value of $x(t)$. It can be expressed as,

$$\bar{x} = A[x(t)] = \lim_{T\to\infty}\frac{1}{2T}\int_{-T}^{T} x(t)dt. \tag{9.28}$$

9.10.2 Time Autocorrelation Function

Consider a random process $X(t)$. The time average of the product $X(t)$ and $X(t+\tau)$ is called time autocorrelation function of $X(t)$ and is denoted by,

$$\mathbb{R}_{XX}(\tau) = A[X(t)X(t+\tau)] \tag{9.29}$$

or $$\mathbb{R}_{XX}(\tau) = \lim_{T\to\infty}\frac{1}{2T}\int_{-T}^{T} x(t)x(r+\tau)dt \tag{9.30}$$

where τ is a time difference or shift in time t.

Time Mean Square Function

If $\tau = 0$, the time average of $X^2(t)$ is called time mean square value of $X(t)$. It is denoted by

$$A[X^2(t)] = \lim_{T\to\infty} \frac{1}{2T}\int_{-T}^{T} x^2(t)dt. \tag{9.31}$$

9.10.3 Time Cross-correlation Function

Let $X(t)$ and $Y(t)$ be two random processes with sample functions $x(t)$ and $y(t)$ respectively. The time average of the product of $x(t)$ and $y(t+\tau)$ is called time cross-correlation function of $X(t)$ and $Y(t)$. It is denoted as

$$\mathbb{R}_{XY}(\tau) = \lim_{T\to\infty} \frac{1}{2T}\int_{-T}^{T} x(t)y(t+\tau)dt. \tag{9.32}$$

Note: The ensemble averages and time averages are not always the same.

It is observed that the time averages of all sample functions of random processes are random variables, therefore the expected value of time averages gives the ensemble averages, that is,

(1) $E[A[x(t)]] = E[X(t)] = \overline{X}$ (9.33)

$E[\overline{x}] = \overline{X}.$

(2) $E[A(X(t)X(t+\tau))] = E[X(t)X(t+\tau)] = R_{XX}(\tau)$ (9.34)

$E[\mathbb{R}_{XX}(\tau)] = R_{XX}(\tau).$

(3) $E[\mathbb{R}_X(\tau)] = R_{XY}(\tau).$ (9.35)

If all time averages are constant, that is, $\overline{x} =$ constant and $\mathbb{R}_{XX}(\tau) =$ constant,

then $E[\overline{x}] = \overline{x} = \overline{X}$ (9.36)

$$E[\mathbb{R}_{XX}(\tau)] = \mathbb{R}_{XX}(\tau) = R_{XX}(\tau) \tag{9.37}$$

and $E[\mathbb{R}_{XY}(\tau)] = \mathbb{R}_{XY}(\tau) = R_{XY}(\tau).$ (9.38)

9.11 Ergodic Theorem and Ergodic Processes

The ergodic theorem states that for any random process $X(t)$, all the time averages of the sample functions of $X(t)$ are equal to the corresponding statistical or ensemble averages of $X(t)$, i.e.,

$$\overline{x} = \overline{X} \text{ and } \mathbb{R}_{XX}(\tau) = R_{XX}(\tau). \tag{9.39}$$

Ergodic Processes

Random processes that satisfy the ergodic theorem are called ergodic processes. The analysis of ergodicity is extremely complex. In most physical applications it is assumed that all stationary processes are ergodic processes.

9.11.1 Mean Ergodic Processes

A random process $X(t)$ is said to be mean ergodic or ergodic in the mean if the time average of any sample function $x(t)$ is equal to its statistical average $\overline{X}$, which is constant, and the probability of all other sample functions is equal to one. That is,

$$E[X(t)] = \overline{X} = A[x(t)] = \overline{x} \tag{9.40}$$

with probability one for all $x(t)$.

9.11.2 Jointly Ergodic Processes

Let $X(t)$ and $Y(t)$ be two random processes with sample function $x(t)$ and $y(t)$ respectively. The two random processes are said to be jointly ergodic if they are individually ergodic and their time cross-correlation functions are equal to their respective ensemble cross-correlation functions. That is,

$$(1)\ \overline{x} = \overline{X},\ \overline{y} = \overline{Y}, \tag{9.41}$$

$$(2)\ \mathbb{R}_{XX}(\tau) = R_{XX}(\tau),\ \mathbb{R}_{XY}(\tau) = R_{XY}(\tau) \text{ and } \mathbb{R}_{YY}(\tau) = R_{YY}(\tau). \tag{9.42}$$

9.12 Autocorrelation Function and Its Properties

9.12.1 Autocorrelation Ergodic Processes

A stationary random process $X(t)$ is said to be autocorrelation ergodic or ergodic in the autocorrelation if and only if the time autocorrelation function of any sample function $x(t)$ is equal to the statistical autocorrelation function of $X(t)$, that is,

$$A[X(t)X(t+\tau)] = E[X(t)X(t+\tau)]$$

$$\mathbb{R}_{xx}(\tau) = R_{XX}(\tau). \tag{9.43}$$

9.12.2 Properties of Autocorrelation Functions

Consider that a random process $X(t)$ is atleast wide sense stationary, that is, the autocorrelation function $R_{XX}(t, t+\tau)$ is a function of only the time difference $\tau = t_2 - t_1$. Then the following are the properties of the autocorrelation function of $X(t)$.

1. The mean square value of $X(t)$ is

   $$E[X^2(t)] = R_{XX}(0). \tag{9.44}$$

   It is equal to the power of the process $X(t)$.

   Proof:

   We know that for $X(t)$, $R_{XX}(\tau) = E[X(t)X(t+\tau)]$.

   If $\tau = 0$, $R_{XX}(0) = E[X(t)X(t)] = E[X^2(t)]$.

2. Autocorrelation function is maximum at the origin, i.e.,

   $$|R_{XX}(\tau)| \leq R_{XX}(0). \tag{9.45}$$

Proof:

Consider two random variables $X(t_1)$ and $X(t_2)$ of $X(t)$, defined at time values t_1 and t_2 respectively. Consider a positive quantity

$$[X(t_1) \pm X(t_2)]^2 \geq 0.$$

Take the expected value

$$E[X(t_1) \pm X(t_2)]^2 \geq 0$$

$$E[X^2(t_1) + X^2(t_2) \pm 2X(t_1)X(t_2)] \geq 0$$

$$E[X^2(t_1)] + E[X^2(t_2)] \pm 2E[X(t_1)X(t_2)] \geq 0.$$

From (9.44), $R_{XX}(0) + R_{XX}(0) \pm 2R_{XY}(t_1, t_2) \geq 0$.

Given $X(t)$ is WSS and $\tau = t_2 - t_1$,

$$\therefore \quad 2R_{XX}(0) \pm 2R_{XX}(\tau) \geq 0$$

$$R_{XX}(0) \pm R_{XX}(\tau) \geq 0$$

$$R_{XX}(0) \geq R_{XX}(\tau)$$

$$\text{or} \quad R_{XX}(\tau) \leq R_{XX}(0).$$

That is, the autocorrelation function at the origin has maximum value. So, $R_{XX}(\tau)$ is bounded by its value at the origin.

3. $R_{XX}(\tau)$ is an even function of τ, i.e.,

$$R_{XX}(-\tau) = R_{XX}(\tau) \tag{9.46}$$

Proof:

We know that $\quad R_{XX}(\tau) = E[X(t)X(t+\tau)]$.

Let $\tau = -\tau$.

Then $\quad R_{XX}(-\tau) = E[X(t)X(t-\tau)]$.

Also, let $u = t - \tau$.

$$\therefore \quad R_{XX}(-\tau) = E[X(u+\tau)X(u)] = E[X(u)X(u+\tau)]$$

$$\text{or} \quad R_{XX}(-\tau) = R_{XX}(\tau).$$

4. If a random process $X(t)$ has a non-zero mean value, $E[X(t)] \neq 0$ and ergodic with no periodic components, then

$$\lim_{|\tau| \to \infty} R_{XX}(\tau) = \overline{X}^2. \tag{9.47}$$

Proof:

Consider a random process $X(t)$, with random variables $X(t_1)$ and $X(t_2)$. Given the mean value is $E[X(t)] = \overline{X} \neq 0$, the autocorrelation function is independent. That is,

$$R_{XX}(\tau) = E[X(t)X(t+\tau)] = E[X(t_1)X(t_2)].$$

Since the process has no periodic components as $|\tau| \to \infty$, the random variables become independent. That is,

$$\lim_{|\tau|\to\infty} R_{XX}(\tau) = E[X(t_1)] \cdot E[X(t_2)].$$

Since $X(t)$ is ergodic,

$$E[X(t)] = E[X(t_2)] = \overline{X}.$$

$$\therefore \quad \lim_{|\tau|\to\infty} R_{XX}(\tau) = \overline{X}^2. \tag{9.48}$$

5. If $X(t)$ is periodic, then its autocorrelation function is also periodic.

   Proof:

   Consider a random process $X(t)$ which is periodic with period T_0.

   Then $X(t) = X(t \pm T_0)$ or $X(t+\tau) = X(t+\tau \pm T_0)$.

   Now we have,

   $$R_{XX}(\tau) = E[X(t)X(t+\tau)].$$

   Then $R_{XX}(\tau \pm T_0) = E[X(t)X(t+\tau \pm T_0)]$.

   Given $X(t)$ is WSS, $R_{XX}(\tau \pm T_0) = E[X(t)X(t+\tau)]$

   $$R_{XX}(\tau \pm T_0) = R_{XX}(\tau).$$

   Therefore, $R_{YX}(\tau)$ is periodic.

6. If $X(t)$ is ergodic, has zero mean, and no periodic component, then

   $$\lim_{|\tau|\to\infty} R_{XX}(\tau) = 0. \tag{9.49}$$

   Proof:

   From eq. (9.47), it is proved that

   $$\lim_{|\tau|\to\infty} R_{XX}(\tau) = \overline{X}^2$$

   where $\overline{X}$ is a mean value of $X(t)$. If $\overline{X} = 0$, then

   $$\lim_{|\tau|\to\infty} R_{XX}(\tau) = 0.$$

7. The autocorrelation function of a random process $R_{XX}(\tau)$ cannot have any arbitrary shape.

   Proof:

   The autocorrelation function $R_{XX}(\tau)$ is an even function of τ, and has maximum value at the origin. $R_{XX}(\tau)$ is also related to the power density spectrum through the Fourier transform. The form of the spectrum is not arbitrary. Hence the autocorrelation function also cannot have an arbitrary shape.

8. If a random process $X(t)$ with zero mean has the dc component A as $Y(t) = A + X(t)$, then

   $$R_{YY}(\tau) = A^2 + R_{XX}(\tau). \tag{9.50}$$

Proof:

Given a random process $Y(t) = A + X(t)$,

$$R_{YY}(\tau) = E[Y(t)Y(t+\tau)]$$
$$= E[(A + X(t))(A + X(t+\tau)]$$
$$= E[A^2 + AX(t) + AX(t+\tau) + X(t)X(t+\tau)]$$
$$= E[A^2] + AE[X(t)] + AE[X(t+\tau)] + E[X(t)X(t+\tau)]$$
$$= A^2 + 0 + 0 + R_{XX}(\tau)$$

$$R_{YY}(\tau) = A^2 + R_{XX}(\tau). \tag{9.51}$$

9. If the random process $Z(t)$ is a sum of two random processes $X(t)$ and $Y(t)$, that is,

$$Z(t) = X(t) + Y(t),$$

then $$R_{ZZ}(\tau) = R_{XX}(\tau) + R_{XY}(\tau) + R_{YX}(\tau) + R_{YY}(\tau) \tag{9.52}$$

Proof:

Given $Z(t) = X(t) + Y(t)$.

Then $$R_{ZZ}(\tau) = E[Z(t)Z(t+\tau)]$$
$$= E[(X(t) + Y(t))[X(t+\tau) + Y(t+\tau)]$$
$$R_{ZZ}(\tau) = E[(X(t)X(t+\tau) + X(t)Y(t+\tau) + Y(t)X(t+\tau) + Y(t)Y(t+\tau)]$$
$$= E[(X(t)X(t+\tau)] + E\mathrm{X(t)}(t+\tau)] + E[Y(t)X(t+\tau)] + E[Y(t)Y(t+\tau)]$$

$$\therefore \quad R_{ZZ}(\tau) = R_{XX}(\tau) + R_{XY}(\tau) + R_{YX}(\tau) + R_{YY}(\tau).$$

9.13 Cross-correlation Function and Its Properties

9.13.1 Cross-correlation Ergodic Processes

Two stationary random processes $X(t)$ and $Y(t)$ are said be cross-correlation ergodic or ergodic in cross-correlation if and only if its time cross-correlation function of sample functions $x(t)$ and $y(t)$ is equal to the statistical cross-correlation function of $X(t)$ and $Y(t)$, that is,

$$A[X(t)Y(t+\tau)] = E[X(t)Y(t+\tau)]$$

or $$\mathbb{R}_{XY}(\tau) = R_{XY}(\tau). \tag{9.53}$$

9.13.2 Properties of Cross-correlation Functions

Consider that two random processes $X(t)$ and $Y(t)$ are atleast jointly wide sense stationary. That is, the cross-correlation function $R_{XY}(t,\ t+\tau)$ is a function of only the time difference $\tau = t_2 - t_1$. Then following are the properties of a cross-correlation function.

1. $R_{XY}(\tau) = R_{YX}(-\tau)$ is a symmetry property.

   Proof:

   The cross-correlation function for $X(t)$ and $Y(t)$ is

   $$R_{XY}(\tau) = E[X(t)Y(t+\tau)].$$

   Also $R_{YX}(\tau) = E[Y(t)X(t+\tau)]$.

   Let $\tau = -\tau$. Then $R_{YX}(-\tau) = E[Y(t)X(t-\tau)]$. Put $u = t-\tau$.

   $$\text{Then } R_{YX}(-\tau) = E[Y(u+\tau)X(u)]$$
   $$= E[X(u)Y(u+\tau)]$$
   $$\therefore \quad R_{YX}(-\tau) = R_{XY}(\tau) \tag{9.54}$$

2. If $R_{XX}(\tau)$ and $R_{YY}(\tau)$ are the autocorrelation functions of $X(t)$ and $Y(t)$ respectively, then the cross-correlation satisfies the inequality:

   $$|R_{YY}(\tau)| \leq \sqrt{R_{XX}(0)R_{YY}(0)}. \tag{9.55}$$

   Proof:

   Consider two random processes $X(t)$ and $Y(t)$ with autocorrelation functions $R_{XX}(\tau)$ and $R_{YY}(\tau)$. Also consider the inequality

   $$E\left[\frac{X(t)}{\sqrt{R_{XX}(0)}} \pm \frac{Y(t+\tau)}{\sqrt{R_{YY}(0)}}\right]^2 \geq 0$$

   $$E\left[\frac{X^2(t)}{\sqrt{R_{XX}(0)}} + \frac{Y^2(t+\tau)}{R_{YY}(0)} \pm \frac{2X(t)Y(t+\tau)}{\sqrt{R_{XX}(0)R_{YY}(0)}}\right] \geq 0$$

   $$\text{or} \quad \frac{E[X^2(t)]}{R_{XX}(0)} + \frac{E[Y^2(t+\tau)]}{R_{YY}(0)} \pm \frac{2E[X(t)Y(t+\tau)]}{\sqrt{R_{XX}(0)R_{YY}(0)}} \geq 0.$$

   We know that $E[X^2(t)] = R_{XX}(0)$.

   $$E[Y^2(t+\tau)] = R_{YY}(0) \text{ and } E[X(t)Y(t+\tau)] = R_{XY}(\tau). \tag{9.56}$$

   $$\therefore \quad \frac{R_{XY}(0)}{R_{XX}(0)} + \frac{R_{YY}(0)}{R_{YY}(0)} \pm 2\frac{R_{XY}(\tau)}{\sqrt{R_{XX}(0)R_{YY}(0)}} \geq 0$$

   $$2 \pm \frac{2R_{XY}(\tau)}{\sqrt{R_{XX}(0)R_{YY}(0)}} \geq 0$$

   $$\frac{\pm R_{XY}(\tau)}{\sqrt{R_{XX}(0)R_{YY}(0)}} \leq 1$$

   $$\text{or} \quad |R_{XY}(\tau)| \leq \sqrt{R_{XX}(0)R_{YY}(0)}. \tag{9.57}$$

Hence the absolute value of the cross-correlation function is always less than or equal to the geometric mean of the autocorrelation functions of $X(t)$ and $Y(t)$ at $\tau = 0$.

3. If $X(t)$ and $Y(t)$ are two random processes with autocorrelation functions $R_{XX}(\tau)$ and $R_{YY}(\tau)$, then the cross-correlation function satisfies the inequality

$$|R_{XY}(\tau)| \leq \frac{1}{2}[R_{XX}(0) + R_{YY}(0)]. \tag{9.58}$$

Proof:

We know that the geometric mean of any two positive numbers cannot exceed their arithmetic mean, that is, if $R_{XX}(\tau)$ and $R_{YY}(\tau)$ are two positive quantities, then at $\tau = 0$.

$$\sqrt{R_{XX}(0)R_{YY}(0)} \leq \frac{1}{2}[R_{XX}(0) + R_{YY}(0)].$$

From (9.54), we know that

$$|R_{XY}(\tau)| \leq \sqrt{R_{XX}(0)R_{YY}(0)}.$$

Therefore, $|R_{XY}(\tau)| \leq \frac{1}{2}[R_{XX}(0) + R_{YY}(0)]$.

4. If two random processes $X(t)$ and $Y(t)$ are statistically independent and are at least wide sense stationary, then

$$R_{XY}(\tau) = \overline{X}\,\overline{Y}. \tag{9.59}$$

Proof:

Let two random processes $X(t)$ and $Y(t)$ be jointly WSS. We know that

$$R_{XY}(\tau) = E[X(t)Y(t+\tau)].$$

Since $X(t)$ and $Y(t)$ are independent,

$$R_{XY}(\tau) = E[X(t)]E[Y(t+\tau)] = \overline{X}\,\overline{Y}.$$

Note: If the two random processes are orthogonal, then $R_{XY}(t) = 0$.

5. If the two random processes $X(t)$ and $Y(t)$ have zero mean and are jointly WSS, then

$$\lim_{\tau \to \infty} R_{XY}(\tau) = 0. \tag{9.60}$$

Proof:

For two random processes, $X(t)$ and $Y(t)$ which are jointly WSS, we have

$$R_{XY}(\tau) = E[X(t)Y(t+\tau)].$$

Take the limits

$$\lim_{\tau \to \infty} R_{XY}(\tau) = \lim_{\tau \to \infty} E[X(t)Y(t+\tau)].$$

$\tau \to \infty$, the random processes $X(t)$ and $Y(t)$ can be considered as independent processes.

$$\therefore \quad \lim_{\tau \to \infty} R_{XY}(\tau) = E[X(t)]E[Y(t+\tau)] = \overline{X} \cdot \overline{Y}.$$

Given that $\overline{X} = \overline{Y} = 0$.

Then $\lim_{\tau \to \infty} R_{XY}(\tau) = 0$.

Similarly, $\lim_{\tau \to \infty} R_{YX}(\tau) = 0$.

9.14 Covariance and Its Properties

9.14.1 Autocovariance Function

The autocovariance function is a measure of interdependence between two random variables. The same concept of random variables can be extended to random processes. Consider random processes $X(t)$, $X(t+\tau)$ at two time intervals t and $t+\tau$. The autocovariance function is defined and can be expressed as

$$C_{XY}(t,t+\tau) = E[(X(t) - E[X(t)])(X(t+\tau) - E[X(t+\tau)])]$$

or $$C_{XX}(t,t+\tau) = R_{XX}(t,t+\tau) - E[X(t)]E[X(t+\tau)]. \quad (9.61)$$

Note 1: If $X(t)$ is at least a wide sense stationary random process, then

$$C_{XX}(\tau) = R_{XX}(\tau) - \overline{X}^2$$

The autocovariance function is not a function of absolute time t.

Note 2: At $\tau = 0$

$$C_{XX}(0) = R_{XX}(0) - \overline{X}^2 = E[X^2] - \overline{X}^2 = \sigma_X^2. \quad (9.62)$$

Therefore, at $\tau = 0$, the autocovariance function becomes the variance of the random process, i.e., $\sigma_X^2 = C_{XX}(0)$.

Note 3: The autocorrelation coefficient of the random process $X(t)$ is defined as

$$\rho_{XX}(t,t+\tau) = \frac{C_{XX}(t,t+\tau)}{\sqrt{C_{XX}(t,t)C_{XX}(t+\tau,t+\tau)}}. \quad (9.63)$$

If $\tau = 0$,

$$\rho_{XX}(0) = \frac{C_{XX}(0)}{C_{XX}(0)} = 1.$$

Also $|\rho_{XX}(t,t+\tau)| \leq 1$.

9.14.2 Cross-covariance Function

If two random processes $X(t)$ and $Y(t)$ have random variables at $X(t)$ and $Y(t+\tau)$, then the cross-covariance function is defined as

$$C_{XY}(t,t+\tau) = E[(X(t) - E[X(t)])(Y(t+\tau) - E[Y(t+\tau)])]$$

or $$C_{XY}(t,t+\tau) = R_{XY}(t,t+\tau) - E[X(t)]E[Y(t+\tau)]. \quad (9.64)$$

Note 1: If the random processes are at least jointly wide sense stationary, then

$$C_{XY}(\tau) = R_{XY}(\tau) - \overline{XY}. \tag{9.65}$$

The cross-covariance function is not a function of absolute time t.

Note 2: If two random processes $X(t)$ and $Y(t)$ are uncorrelated, then

$$C_{XY}(t, t+\tau) = 0$$

or $$R_{XY}(t, t+\tau) = E[X(t)]E[Y(t+\tau)]. \tag{9.66}$$

This is the condition for two random processes to be statistically independent. Therefore, independent random processes are uncorrelated. But the converse is not always true.

Note 3: The cross-correlation coefficient of the random processes $X(t)$ and $Y(t)$ is defined as

$$\rho_{XY}(t, t+\tau) = \frac{C_{XY}(t, t+\tau)}{\sqrt{C_{XX}(t,t)C_{YY}(t+\tau, t+\tau)}}. \tag{9.67}$$

If $\tau = 0$,

$$\rho_{XY}(0) = \frac{C_{XY}(0)}{\sqrt{C_{XY}(0)C_{YY}(0)}} = \frac{C_{XY}(0)}{\sigma_X \sigma_Y}.$$

Example 9.2

If a random process $X(t) = A\cos\omega t + B\sin\omega t$ is given, where A and B are uncorrelated, zero mean random variables having variance σ^2, find (a) autocorrelation and (b) show that $X(t)$ is wide sense stationary (WSS). (May 2011, Nov 2010)

Solution

Given a random process

$$X(t) = A\cos\omega t + B\sin\omega t$$

$$E[AB] = E[A]E[B] = 0$$

and $\sigma_A^2 = E[A^2] = \sigma_B^2 = E[B^2] = \sigma^2$.

(a) The autocorrelation function:

$$\begin{aligned} R_{XX}(\tau) &= E[X(t)X(t+\tau)] \\ &= E[(A\cos\omega t + B\sin\omega t)(A\cos\omega(t+\tau) + B\sin\omega(t+\tau))] \\ &= E[A^2\cos\omega t\cos\omega(t+\tau) + AB\cos\omega t\sin\omega(t+\tau) \\ &\quad + AB\sin\omega t\cos\omega(t+\tau) + B^2\sin\omega t\sin\omega(t+\tau)]. \end{aligned}$$

$$\begin{aligned} \therefore \quad R_{XX}(\tau) &= E[A^2]E[\cos\omega t\cos\omega(t+\tau)] \\ &\quad + E[AB]E[\cos\omega t\cos\omega(t+\tau) + \sin\omega t\sin\omega(t+\tau)] \\ &\quad + E[B^2]E[\sin\omega t\sin\omega(t+\tau)]. \end{aligned}$$

$$\therefore \quad R_{XX}(\tau) = \sigma^2(E[\cos\omega t \cos\omega(t+\tau) + \sin\omega t \sin\omega(t+\tau)])$$
$$= \sigma^2(E[\cos(\omega t + \omega\tau - \omega t]) = \sigma^2 E[\cos\omega\tau] = \sigma^2\cos\omega\tau$$

(b) The mean value :

$$E[X(t)] = E[A\cos\omega t + B\sin\omega t]$$
$$= E[A\cos\omega t] + E[B\sin\omega t] = E[A]\cos\omega t + E[B]\sin\omega t = 0.$$

Therefore, The mean value $E[X(t)] = 0 =$ constant and the autocorrelation function $R_{XX}(\tau) = \sigma^2\cos\omega\tau$ is independent of time t. Hence the random process is WSS.

Example 9.3

A random process is given as $X(t) = At$, where A is a uniformly distributed random variable on $(0,2)$. Find whether $X(t)$ is WSS or not.

Solution

The condition for WSS is $E[X(t)] = \overline{X} =$ constant and $R_{XX}(\tau)$ is independent of time t.

Now given $X(t) = At$

$$\text{and } f_A(A) = \begin{cases} \frac{1}{2} & 0 \le A < 2 \\ 0 & \text{otherwise.} \end{cases}$$

$$E[X(t)] = \int_{-\infty}^{\infty} x(t) f_A(A)dA$$
$$= \int_0^2 \frac{1}{2}At\,dA = \frac{t}{2}\left[\frac{A^2}{2}\right]_0^2 = \frac{t}{2}\times\frac{4}{2} = t.$$
$$\therefore \quad E[X(t)] = t$$

and the autocorrelation function is

$$R_{XX}(\tau) = E[X(t)X(t+\tau)]$$
$$= E[At\,A(t+\tau)] = t(t+\tau)E[A^2]$$
$$= t(t+\tau)\int_0^2 A^2 f_A(A)dA$$
$$= t(t+\tau)\int_0^2 \frac{A^2}{2}dA = \frac{t(t+\tau)}{2}\times\left.\frac{A^3}{3}\right|_0^2$$
$$R_{XX}(\tau) = \frac{4t(t+\tau)}{3}.$$

Here mean value is not constant and $R_{XX}(\tau)$ is a function of t. Hence $X(t)$ is not a WSS.

9.15 Gaussian Random Processes

Consider a continuous random process $X(t)$. Let N random variables $X_1 = X(t_1)$, $X_2 = X(t_2), \ldots,$ $X_N = X(t_N)$ be defined at time instants $t_1, t_2, t_3, \ldots, t_N$ respectively.

If these random variables are jointly Gaussian for any $N = 1, 2, \ldots$. and at any time instants $t_1, t_2, \ldots, t_N$, then the random process $X(t)$ is called a Gaussian random process. The joint density function for a Gaussian random variable is given as

$$f_X(x_1, x_2 \ldots x_N; t_1, t_2 \ldots t_N) = \frac{\exp\left\{\frac{-1}{2}[x - \overline{X}]^t [C_{XX}^{-1}[x - \overline{X}]\right\}}{\sqrt{(2\pi)^N |[C_{XX}]|}}, \tag{9.68}$$

where $\overline{X} = E[X_i] = E[x(t)]$.

$[C_{XX}]$ = covariance matrix and its elements are

$$C_{ik} = C_{XX_k} = E[(X_i - \overline{X}_i)(X_k - \overline{X}_k)]$$

$$= E[(x(t_i) - \overline{X}(t_i))(x(t_k) - \overline{X}(t_k))]$$

$$C_{ik} = C_{XX}(t_i, t_k).$$

C_{ik} is the autocovariance of $X(t_i)$ and $X(t_k)$. Also, by expanding the above equation we get

$$C_{XX}(t_i, t_k) = R_{XX}(t_i, t_k) - E[X(t_i)]E[X(t_k)], \tag{9.69}$$

where $R_{XX}(t_i, t_k)$ is the autocorrelation function of X.

If the process is wide sense stationary, then it should satisfy the following conditions:

1. Mean value will be constant, i.e.,

$$\overline{X}_i = E[X(t_i)] = \overline{X}, \text{a constant.} \tag{9.70}$$

2. The mean autocorrelation and autocovariance functions will depend only on time differences and not on absolute time.

$$C_{XX}(t_i, t_k) = C_{XX}(t_k - t_i)$$

$$R_{XX}(t_i, t_k) = R_{XX}(t_k - t_i). \tag{9.71}$$

Note:

1. The wide sense stationary (WSS) Gaussian process is also a strictly stationary process.
2. If two random processes $X(t)$ and $Y(t)$ are said to be jointly Gaussian, the random processes, $X(t_1), X(t_2), \ldots, X(t_N)$, $Y(t'_N), Y(t'_2), \ldots, Y(t'_M)$ defined at times $t_1, t_2, \ldots, t_N$ and t'_1, $t'_2, \ldots, t'_M$, respectively are jointly Gaussian for any time $t_1, t_2, \ldots, t_N; t'_1$, $t'_2, \ldots, t'_M$.

9.16 Poisson Random Processes

The Poisson process $X(t)$ is a discrete random process which represents the number of times that some event has occurred as a function of time. $X(t)$ is integer valued, non-decreasing sample functions, such as check-in registers, arrival of a customer, arrival of vehicles at a particular point, etc.

In these functions, a single event occurs at a random time. Counting the number of occurrences with time is a Poisson process. It is, therefore, also called a counting process. Figure 9.3 shows the sample function of a Poisson counting process.

The conditions for a Poisson process $X(t)$ are

1. $X(0) = 0$.

2. Only one event occurs in any instant time, i.e., in an infinitesimal time interval.
3. The number of events that occur in any given time interval is independent of the number events in any other non-overlapping time intervals. i.e., $X(t)$ has independent increments.

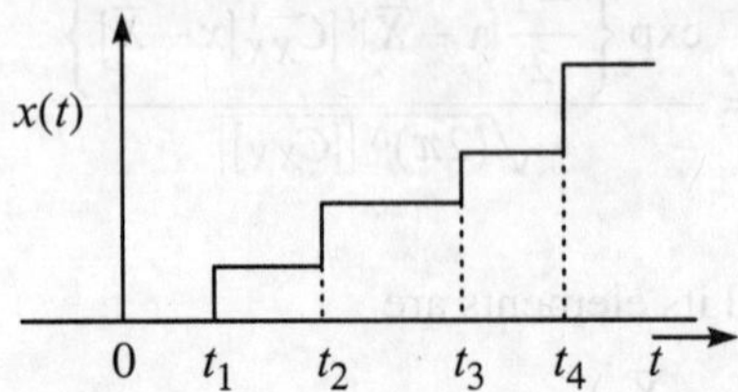

Fig. 9.3 *A Poisson counting process*

Probability Density Function

If the number of occurrences of an event in any finite interval of time is described by a Poisson distribution with the average rate of occurrences λ, then the probability of exactly k occurrences over a time interval (0, t) is

$$P[X(t)=k]=\frac{(\lambda t)^k e^{-\lambda t}}{k!},\ k=0,1,2\ldots. \tag{9.72}$$

and the probability density function is

$$f_X(x)=\sum_{k=0}^{\infty}\frac{(\lambda t)^k e^{-\lambda/t}}{k!}\delta(x-k). \tag{9.73}$$

Mean Value

The mean value of a Poisson density function is

$$E[X(t)]=\int_{-\infty}^{\infty}xf_X(x)dx=\int_{-\infty}^{\infty}x\sum_{k=0}^{\infty}\frac{(\lambda t)^k e^{-\lambda t}}{k!}\delta(x-k)dx.$$

Since $\int_{-\infty}^{\infty}x\delta(x-k)dx=k$,

$$E[X(t)]=\sum_{k=0}^{\infty}\frac{k(\lambda t)^k e^{-\lambda t}}{k!}. \tag{9.74}$$

By using infinite series expansion we get

$$=e^{-\lambda t}\left(\lambda t+\frac{2(\lambda t)^2}{2!}+\frac{3(\lambda t)^3}{3!}+\frac{4(\lambda t)^4}{4!}+\ldots\right)$$

$$=e^{-\lambda t}\left(\lambda t(1+\frac{(\lambda t)}{1!}+\frac{(\lambda t)^2}{2!}+\frac{(\lambda t)^3}{3!}+\ldots\right)=e^{-\lambda t}\lambda t e^{\lambda t}=\lambda t.$$

Also, we know from the Poisson density function of a random variable that the mean value is λt.

So, $E[X(t)]=\lambda t$.

$$\therefore\quad E[X(t)]=\sum_{k=0}^{\infty}\frac{k(\lambda t)^k e^{-\lambda t}}{k!}=\lambda t \tag{9.75}$$

Second Moment
The second moment is

$$E[X^2(t)] = \int_{-\infty}^{\infty} x^2 f_X(x)dx$$

$$\int_{-\infty}^{\infty} x^2 \sum_{k=0}^{\infty} \frac{(\lambda t)^k e^{-\lambda t}}{k!} \delta(x-k)dx. \tag{9.76}$$

Since $\int_{-\infty}^{\infty} x^2 \delta(x-k)dx = k^2$

$$E[X^2(t)] = \sum_{k=0}^{\infty} \frac{k^2(\lambda t)^k e^{-\lambda t}}{k!}. \tag{9.77}$$

By using infinite series expansion we get

$$= e^{-\lambda t} \sum_{k=0}^{\infty} \frac{(k+k(k-1))(\lambda t)^k}{k!} = e^{-\lambda t} \sum_{k=0}^{\infty} \frac{k(\lambda t)^k}{k!} + e^{-\lambda t} \sum_{k=0}^{\infty} \frac{k(k-1)(\lambda t)^k}{k!}$$

$$= e^{-\lambda t} \left(\lambda t + \frac{2(\lambda t)^2}{2!} + \frac{3(\lambda t)^3}{3!} + \frac{4(\lambda t)^4}{4!} + \ldots \right)$$

$$+ e^{-\lambda t} \left(\frac{2(\lambda t)^2}{2!} + \frac{(3\times 2)(\lambda t)^3}{3!} + \frac{(4\times 3)(\lambda t)^4}{4!} + \ldots \right)$$

$$= e^{-\lambda t} \lambda t \left(1 + \frac{(\lambda t)}{1!} + \frac{(\lambda t)^2}{2!} + \frac{(\lambda t)^3}{3!} + \ldots \right)$$

$$+ e^{-\lambda t} (\lambda t)^2 \left(1 + \frac{(\lambda t)}{1!} + \frac{(\lambda t)^2}{2!} + \frac{(\lambda t)^3}{3!} + \ldots \right)$$

$$= e^{-\lambda t} \lambda t e^{\lambda t} + e^{-\lambda t} (\lambda t)^2 e^{\lambda t} = \lambda t + (\lambda t)^2.$$

Also, we know that the second moment is

$$E[X^2] = \sigma_X^2 + \overline{X}^2$$

$$= \lambda t + (\lambda t)^2 = \lambda t + \lambda^2 t^2.$$

$$\therefore \quad E[X^2(t)] = \sum_{k=0}^{\infty} \frac{k^2(\lambda t)^k e^{-\lambda t}}{k!} = \lambda t + \lambda^2 t^2. \tag{9.78}$$

9.17 Mean Value of System Response

Although the concepts of random processes can be applied to various kinds of systems, here we shall concentrate on linear systems since many real time systems can be easily approximated to linear time invariant systems. This section explains the methods of describing the output response of a linear time invariant system (LTI) when a continuous random process is applied at the input. Here the concepts of statistical properties, i.e., mean, mean square and correlation functions of input and output response are explained.

Consider a continuous LTI system with impulse response $h(t)$. Assume that the system is always causal and stable. When a continuous time random process $X(t)$ with ensemble members $x(t)$ is applied on this system, the output response is also a continuous time random process $Y(t)$ with ensemble members $y(t)$. If the random processes X and Y are discrete time signals, then the linear system is called a discrete time system.

9.17.1 System Response

Let a random process $X(t)$ be applied to a continuous linear time invariant system whose inipulse response is $h(t)$ as shown in Fig. 9.6. Then the output response $y(t)$ is also a random process. It can be expressed by the convolutional integral,

$$Y(t) = h(t) * X(t).$$

That is, the output response is $Y(t) = \int_{-\infty}^{\infty} h(\tau)X(t-\tau)d\tau.$ (9.79)

The system transfer function is

$$H(\omega) = \frac{Y(\omega)}{X(\omega)}.$$

$X(t)$ → $h(t)$ $H(\omega)$ → $Y(t)$

$X(\omega)$ $Y(\omega)$

Fig. 9.6 *Linear time invariant ystem for a random process*

Example 9.4

Find the transfer function of the network shown in Fig. 9.4.

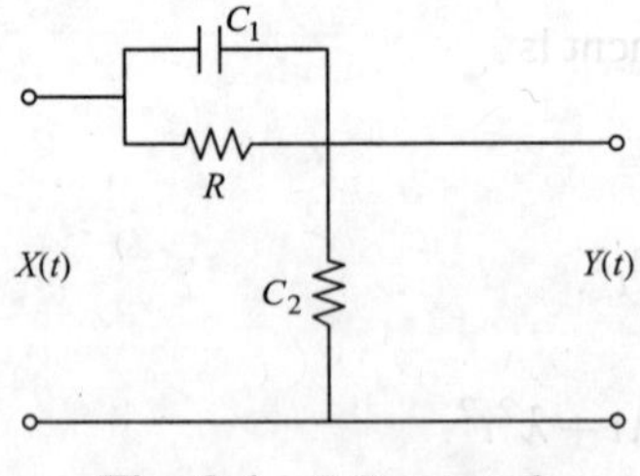

Fig. 9.4 *RC network*

Solution

Let $X(\omega)$ and $Y(\omega)$ be the input and output functions of the given network. The network is drawn with reactance values as shown in Fig. 9.5.

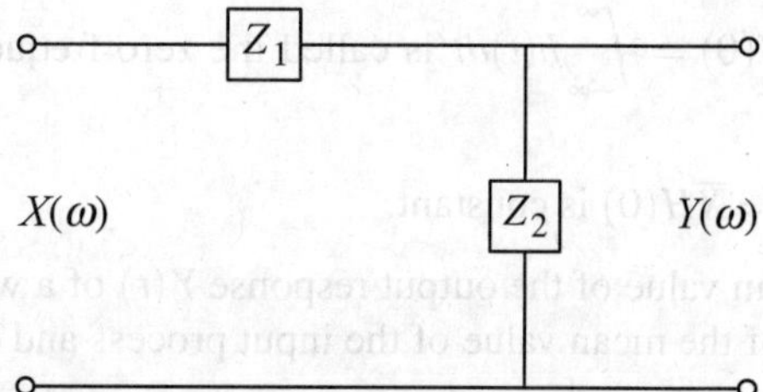

Fig. 9.5 *(a) Network with reactances* **Fig. 9.5** *(b) Network with impedances*

The impedances are

$$Z_1 = \frac{R \times \dfrac{1}{j\omega C_1}}{R + \dfrac{1}{j\omega C_1}} = \frac{R}{1 + j\omega C_1 R}$$

$$Z_2 = \frac{1}{j\omega C_2}.$$

From Fig. 9.5(b), $Y(\omega) = \frac{Z_2}{Z_1+Z_2} X(\omega)$. Therefore, If $H(\omega)$ is the transfer function, then

$$H(\omega) = \frac{Y(\omega)}{X(\omega)}.$$

$$\therefore \quad H(\omega) = \frac{Z_2}{Z_1 + Z_2} = \frac{\dfrac{1}{j\omega C_2}}{\dfrac{R}{1 + j\omega C_1 R_1} + \dfrac{1}{j\omega C_2}} = \frac{1 + j\omega C_1 R}{j\omega R C_2 + 1 + j\omega C_1 R}$$

$$H(\omega) = \frac{1 + j\omega C_1 R}{1 + j\omega R(C_1 + C_2)}.$$

9.17.2 Mean Value of Output Response

Consider that the random process $X(t)$ is a wide sense stationary process.

Mean value of output response $= E[Y(t)]$

Then, $E[Y(t)] = E[h(t) * X(t)]$

$$= E\left[\int_{-\infty}^{\infty} h(\tau) X(t-\tau) d\tau\right] = \int_{-\infty}^{\infty} h(\tau) E[X(t-\tau)] d\tau.$$

But $E[X(t-\tau)] = \overline{X} =$ constant, since $X(t)$ is WSS.

$$\text{Then} \quad E[Y(t)] = \overline{Y} = \overline{X} \int_{-\infty}^{\infty} h(t) dt. \tag{9.80}$$

Also, if $H(\omega)$ is the Fourier transform of $h(t)$,

then, $H(\omega) = \int_{-\infty}^{\infty} h(t) e^{-j\omega t} dt.$

At $\omega = 0$, $H(0) = \int_{-\infty}^{\infty} h(t)dt$ is called the zero frequency response of the system. Substituting it in Eq. (9.80),

$$E[Y(t)] = \overline{Y} = \overline{X}H(0) \text{ is constant.} \tag{9.81}$$

Thus the mean value of the output response $Y(t)$ of a wide sense stationary random process is equal to the product of the mean value of the input process and the zero frequency response of the system.

9.18 Mean-square Value of Output Response

The mean square value of the output response is

$$\begin{aligned}
E[Y^2(t)] &= E[(h(t)^*X(t))^2] \\
&= E[(h(t)^*X(t))(h(t)^*X(t))] \\
&= E\left[\int_{-\infty}^{\infty} h(\tau_1)X(t-\tau_1)d\tau_1 \int_{-\infty}^{\infty} h(\tau_2)X(t-\tau_2)d\tau_2\right] \\
&= E\left[\int_{-\infty}^{\infty}\int_{-\infty}^{\infty} X(t-\tau_1)X(t-\tau_2)h(\tau_1)h(\tau_2)d\tau_1 d\tau_2\right]
\end{aligned}$$

$$E[Y^2(t)] = \left[\int_{-\infty}^{\infty}\int_{-\infty}^{\infty} E[X(t-\tau_1)X(t-\tau_2)]h(\tau_1)h(\tau_2)d\tau_1 d\tau_2\right], \tag{9.82}$$

where τ_1 and τ_2 are the shift in time variables.

If the input $X(t)$ is a wide sense stationary random process, then

$$E[X(t-\tau_1)X(t-\tau_2)] = R_{XX}(\tau_1-\tau_2).$$

Equation (9.82) becomes

$$E[Y^2(t)] = [\int_{-\infty}^{\infty}\int_{-\infty}^{\infty} R_{XX}(\tau_1-\tau_2)h(\tau_1)h(\tau_2)d\tau_1 d\tau_2]. \tag{9.83}$$

The above expression is independent of time t. It represents the output power.

9.19 Autocorrelation Function of Response

The autocorrelation of $Y(t)$ is

$$\begin{aligned}
R_{YY}(t_1,t_2) &= E[Y(t_1)Y(t_2)] \\
&= E[(h(t_1)^*X(t_1)(h(t_2)^*X(t_2))] \qquad (9.84) \\
&= E\left[\int_{-\infty}^{\infty} h(\tau_1)X(t_1-\tau_1)d\tau_1 \int_{-\infty}^{\infty} h(\tau_2)X(t_2-\tau_2)d\tau_2\right] \\
&= E\left[\int_{-\infty}^{\infty}\int_{-\infty}^{\infty} X(t_1-\tau_1)X(t_2-\tau_2)h(\tau_1)h(\tau_2)d\tau_1 d\tau_2\right] \\
&= \int_{-\infty}^{\infty}\int_{-\infty}^{\infty} E[X(t_1-\tau_1)X(t_2-\tau_2)]h(\tau_1)h(\tau_2)d\tau_1 d\tau_2. \qquad (9.85)
\end{aligned}$$

We know that $E[X(t_1-\tau_1)X(t_2-\tau_2)] = R_{XX}(t_2-t_1+\tau_1-\tau_2)$.

If the input $X(t)$ is a wide sense stationary random process and the time difference

$$\tau = t_2 - t_1 \text{ and } t = t_1,$$

then $E[X(t-\tau_1)X(t+\tau-\tau_2)] = R_{XX}(\tau+\tau_1-\tau_2)$.

Equation (9.85) becomes

$$R_{YY}(t,t+\tau) = R_{YY}(\tau).$$

$$\text{Therefore, } R_{YY}(\tau) = E[Y(t)Y(t+\tau)] = \left[\int_{-\infty}^{\infty}\int_{-\infty}^{\infty} R_{xx}(\tau+\tau_1-\tau_2)h(\tau_1)h(\tau_2)d\tau_1 d\tau_2\right]. \tag{9.86}$$

If $R_{XX}(\tau)$ is the autocorrelation function of $X(t)$, then

$$R_{YY}(\tau) = R_{XX}(\tau) * h(\tau) * h(-\tau). \tag{9.87}$$

It is observed that the output autocorrelation function is a function of only τ. Hence the output random process $Y(t)$ is also wide sense stationary process.

Taking Fourier transform, $R_{YY}(\omega) = R_{XX}(\omega)|H(\omega)|^2$.

9.20 Cross-Correlation Functions of Input and Output of a Linear System

If the input $X(t)$ is a wide sense stationary random process, then the cross-correlation function of input $X(t)$ and output $Y(t)$ is

$$R_{XY}(t,t+\tau) = E[X(t)Y(t+\tau)]$$

$$R_{XY}(\tau) = E\left[X(t)\int_{-\infty}^{\infty} h(\tau_1)X(t+\tau-\tau_1)d\tau_1\right] = \int_{-\infty}^{\infty} E[X(t)X(t+\tau-\tau_1)]h(\tau_1)d\tau_1$$

$$R_{XY}(\tau) = \int_{-\infty}^{\infty} R_{XX}(\tau-\tau_1)h(\tau_1)d\tau_1, \tag{9.88}$$

which is the convolution of $R_{XX}(\tau)$ and $h(\tau)$.

$$\therefore \quad R_{XY}(\tau) = R_{XX}(\tau) * h(\tau). \tag{9.89}$$

Similarly, we can show that

$$R_{YX}(\tau) = \int_{-\infty}^{\infty} R_{XX}(\tau-\tau_1)h(-\tau_1)d\tau_1.$$

$$\text{Therefore, } R_{YX}(\tau) = R_{XX}(\tau) * h(-\tau). \tag{9.90}$$

This shows that $X(t)$ and $Y(t)$ are jointly WSS. If we substitute Eq. (9.89) and Eq. (9.90) in Eq. (9.90), we get

$$R_{YY}(\tau) = R_{XY}(\tau) * h(-\tau) \tag{9.91}$$

$$\text{and } R_{YY}(\tau) = R_{YX}(\tau) * h(\tau). \tag{9.92}$$

These expressions show the relationship between the autocorrelation functions and the cross correlation functions.

Taking Fourier transform, $R_{YY}(\omega) = R_{XY}(\omega)H^*(\omega)$ and $R_{YY}(\omega) = R_{YX}(\omega)H(\omega)$.

Example 9.5

Two random processes are defined as

$$X(t) = A\cos(\omega t + \theta) \quad \text{and} \quad Y(t) = B\sin(\omega t + \theta)$$

where θ is a uniform random variable on $(0, 2\pi)$, and A, B and ω are constants. Find (a) the cross-correlation function $R_{XY}(t, t+\tau)$ and (b) verify whether $X(t)$ and $Y(t)$ are jointly WSS or not.

Solution

Given two random variables

$$X(t) = A\cos(\omega t + \theta) \text{ and } Y(t) = B\sin(\omega t + \theta)$$

$$\text{and } f_\theta(\theta) = \begin{cases} \frac{1}{2\pi} & 0 < \theta < 2\pi \\ 0 & \text{otherwise.} \end{cases}$$

(a) Now the cross-correlation function is

$$\begin{aligned} R_{XY}(t, t+\tau) &= E[X(t)Y(t+\tau)] \\ &= E[A\cos(\omega t + \theta)B\sin(\omega(t+\tau) + \theta)] \\ &= \frac{AB}{2}E[\sin(2\omega t + \omega\tau + 2\theta) + \sin\omega\tau] \end{aligned}$$

$$R_{XY}(t, t+\tau) = \frac{AB}{2}E[\sin(2\omega t + \omega\tau + 2\theta)] + \frac{AB}{2}E[\sin\omega\tau].$$

Now take

$$\begin{aligned} E[\sin(2\omega t + \omega\tau + 2\theta)] &= \int_{-\infty}^{\infty} \sin(2\omega t + \omega\tau + 2\theta) f_\theta(\theta)\, d\theta \\ &= \frac{1}{2\pi}\int_0^{2\pi} \sin(2\omega t + \omega\tau + 2\theta)\, d\theta \\ &= \frac{1}{2\pi}\left[\frac{\cos(2\omega t + \omega\tau + 2\theta)}{2}\right]_0^{2\pi} \\ &= \frac{1}{4\pi}[\cos(2\omega t + \omega\tau) - \cos(2\omega\tau + \omega\tau)] \\ &= 0. \end{aligned}$$

So, $R_{XY}(t, t+\tau) = 0 + \sin\omega t$.

$$\therefore \quad R_{XY}(\tau) = \frac{AB}{2}\sin\omega\tau.$$

(b) The condition for two processes to be jointly WSS is

$$E[X(t)] = \overline{X} = \text{constant}; \qquad E[Y(t)] = \overline{Y} = \text{constant}$$

and $R_{XY}(\tau)$ is independent of time.

Now, $E[X(t)] = E[A\cos(wt+\theta)]$

$$= \int_{-\infty}^{\infty} A\cos(\omega t+\theta) f_{\Theta}(\theta) d\theta$$

$$= \int_{0}^{2\pi} \frac{A}{2\pi} \cos(\omega t+\theta) d\theta$$

$$= \frac{A}{2\pi}[-\sin(\omega t+\theta)]_0^{2\pi} = 0$$

and $E[Y(t)] = E[B\sin(\omega t+\theta)]$

$$= \int_{0}^{2\pi} \frac{B}{2\pi} \sin(\omega t+\theta) d\theta$$

$$= \frac{B}{2\pi}[\cos(\omega t+\theta)]_0^{2\pi} = 0.$$

Therefore, $E[X(t)] = E[Y(t)] = 0 =$ constant

and $R_{XY}(\tau) = \frac{AB}{2}\sin\omega\tau$ is independent of time.

Hence $X(t)$ and $Y(t)$ are jointly WSS.

Example 9.6

A random process $X(t)$ is applied to a network with impulse response $h(t) = e^{-bt}u(t)$, where $b > 0$ is constant. The cross-correlation of $X(t)$ with the output $Y(t)$ is known to have the form $R_{XY}(\tau) = u(\tau)\tau e^{-b\tau}$. Find the autocorrelation of $Y(t)$. (Feb 2006, May 2009)

Solution

Given $h(t) = e^{-bt}u(t)$ and $R_{XY}(\tau) = \tau e^{-b\tau}u(\tau)$.

The autocorrelation of $Y(t)$ is

$$R_{YY}(\tau) = R_{XY}(\tau) * h(-\tau).$$

Taking Fourier transform,

$$R_{YY}(\omega) = R_{XY}(\omega)H^*(\omega).$$

$$\text{Now } H(\omega) = F[e^{-bt}u(t)] = \frac{1}{b+j\omega}.$$

$$H^*(\omega) = \frac{1}{b-j\omega}.$$

$$R_{XY}(\omega) = F[\tau e^{-b\tau}u(\tau)] = \frac{1}{(b+j\omega)^2},$$

$$\therefore \quad R_{YY}(\omega) = \frac{1}{(b+j\omega)^2} \times \frac{1}{b-j\omega}$$

and $R_{YY}(\tau) = F^{-1}\left[\frac{1}{(b+j\omega)(b^2+\omega^2)}\right]$.

Using partial fractions,

$$\frac{1}{(b+j\omega)^2(b-j\omega)} = \frac{1}{4b^2}\left[\frac{1}{b+j\omega} + \frac{1}{b-j\omega}\right] + \frac{1}{2b}\left[\frac{1}{(b+j\omega)^2}\right].$$

$$\therefore \quad R_{YY}(\tau) = \frac{1}{4b^2}\left[e^{-b\tau}u(\tau) + e^{b\tau}u(\tau)\right] + \frac{1}{2b}\tau e^{-b\tau}u(\tau)$$

$$R_{YY}(\tau) = \left(\frac{1}{2b^2}\cosh b\tau + \frac{1}{2b}\tau e^{-b\tau}\right)u(\tau).$$

Additional Problems

Example 9.7

If $Y_1(t) = X_1\cos\omega t + X_2\sin\omega t$ and $Y_2(t) = X_1\sin\omega t + X_2\cos\omega t$, where X_1 and X_2 are zero mean independent random variables with unity variance, show that the random processes $Y_1(t)$ and $Y_2(t)$ are individually WSS but not jointly WSS.

Solution

The conditions of a random process $X(t)$ to be WSS are

(i) Mean value is constant

(ii) Autocorrelation is independent of time t.

Given $Y_1(t) = X_1\cos\omega t + X_2\sin\omega t$,

$E[X_1] = E[X_2] = 0$, and $\sigma_{X_1}^2 = \sigma_X^2 = 1$.

(1) Mean value of $Y_1(t)$ is

$$E[Y_1(t)] = E[X_1\cos\omega t + X_2\sin\omega t] = E[X_1]\cos\omega t + E[X_2]\sin\omega t$$

$$E[Y_1(t)] = 0 = \text{ constant.}$$

Also, mean value of $Y_2(t)$ is

$$E[Y_2(t)] = E[X_1\sin\omega t + X_2\cos\omega t] = E[X_1]\sin\omega t + E[X_2]\cos\omega t.$$

$$E[Y_2(t)] = 0 = \text{ constant}$$

(2) The autocorrelation of $Y_1(t)$ is

$$\begin{aligned}
R_{Y_1Y_1}(\tau) &= E[(X_1\cos\omega t + X_2\sin\omega t)(X_1\cos\omega(t+\tau) + X_2\sin\omega(t+\tau)]\\
&= E[X_1^2\cos\omega t\cos\omega(t+\tau) + X_1X_2\cos\omega t\sin\omega(t+\tau)\\
&\quad + X_1X_2\sin\omega t\cos\omega(t+\tau) + X_2^2\sin\omega t\sin\omega(t+\tau)]\\
&= E[X_1^2]\cos\omega t\cos\omega(t+\tau) + E[X_1X_2]\cos\omega t\sin\omega(t+\tau)\\
&\quad + E[X_1X_2]\sin\omega t\cos\omega(t+\tau) + E[X_2^2]\sin\omega t\sin\omega(t+\tau).
\end{aligned}$$

Since X_1 and X_2 are independent,

$$E[X_1X_2] = 0 \quad \text{and} \quad E[X_1^2] = \sigma_{X_1}^2 + \overline{X}^2 = 1$$

$$E[X_2^2] = \sigma_{X_2}^2 + \overline{X}_2^2 = 1.$$

$$\therefore \quad R_{Y_1Y_1}(\tau) = \cos\omega t \cos\omega(t+\tau) + \sin\omega t \sin(\omega t + \tau)$$

$$= \cos(\omega t + \omega\tau - \omega t) = \cos\omega\tau.$$

It is independent of t. The autocorrelation function of $Y_2(t)$ is

$$R_{Y_2Y_2}(\tau) = E[(X_1 \sin\omega t + X_2 \cos\omega t)(X_1 \sin\omega(t+\tau) + X_2 \cos\omega(t+\tau))]$$

$$= E[X_1^2]\sin\omega t \sin\omega(t+\tau) + E[X_1X_2](\sin\omega t \cos\omega(t+\tau)$$

$$+ \cos\omega t \sin\omega(t+\tau)) + E[X_2^2]\cos\omega t \cos\omega(t+\tau).$$

Since $[X_1^2] = E[X_2^2] = 1$ and $E[X_1X_2] = 0$,

$$R_{Y_2Y_2}(\tau) = \cos\omega t \cos\omega(t+\tau) + \sin\omega t \sin\omega(t+\tau)$$

$$= \cos(\omega t + \omega\tau - \omega t) = \cos\omega\tau.$$

$R_{Y_2Y_2}(\tau) = \cos\omega\tau$ is also independent of time.

Therefore the random process $Y_1(t)$ and $Y_2(t)$ are individually WSS. Now the cross-correlation function of $Y_1(t)$ and $Y_2(t)$ is

$$R_{Y_1Y_2}(\tau) = E[Y_1(t)Y_2(t+\tau)]$$

$$= E[(X_1 \cos\omega t + X_2 \sin\omega t)(X_1 \sin\omega(t+\tau) + X_2 \cos\omega(t+\tau))]$$

$$= [X_1^2]\cos\omega t \sin\omega(t+\tau) + E[X_1X_2]\cos\omega t \cdot \cos\omega(t+\tau)$$

$$+ E[X_1X_2]\sin\omega t \sin\omega(t+\tau) + E[X_2^2]\sin\omega t \cos\omega(t+\tau).$$

since $E[X_1^2] = E[X_2^2] = 1$ and $E[X_1X_2] = 0$.

$$\text{Now,} \quad R_{Y_1Y_2}(\tau) = \cos\omega t \sin\omega(t+\tau) + \sin\omega t \cos\omega(t+\tau)$$

$$= \sin(\omega t + \omega\tau + \omega t) = \sin(2\omega t + \omega\tau) = \sin\omega(2t+\tau).$$

The cross-correlation is a function of time t. Therefore, $Y_1(t)$ and $Y_2(t)$ are not jointly WSS.

Example 9.8

For a given random process $X(t)$, the mean value is $\overline{X} = 6$ and autocorrelation is

$$R_{XX}(\tau) = 36 + 25e^{-|\tau|}.$$

Find (a) the average power of the process $X(t)$ and (b) variance of $X(t)$.

Solution

Given $R_{XX}(\tau) = 36 + 25e^{-|\tau|}$.

(a) The average power is

$$E[X^2(t)] = R_{XX}(0)$$

$$\therefore \quad R_{XX}(0) = 36 + 25e^{-(0)} = 36 + 25 = 61$$

$\therefore$ The average power is 61 watts.

(b) Variance of $X(t)$ is

$$\sigma_X^2 = E[X^2] - \overline{X}^2 = R_{XX}(0) - \overline{X}^2 = 61 - 6^2 = 61 - 36 = 25.$$

Example 9.9

The autocorrelation function of a stationary random process $X(t)$ is given by

$$R_{XX}(\tau) = 36 + \frac{16}{1+8\tau^2}.$$

Find the mean, mean square and variance of the process.

Solution

Given the autocorrelation function $R_{XX}(\tau) = 36 + \dfrac{16}{1+8\tau^2}$.

(a) Since the process is stationary, we know that

$$\overline{X}^2 = \lim_{\tau\to\infty} R_{XX}(\tau) = \lim_{\tau\to\infty}\left[36 + \frac{16}{1+8\tau^2}\right].$$

$$\overline{X}^2 = 36 + 0 = 36$$

$\therefore$ The mean value is $\overline{X} = 6$.

(b) We know that the mean square value is

$$E[X^2] = R_{XX}(0) = 36 + \frac{16}{1+8(0)} = 36 + 16 = 52.$$

(c) Variance of X is given by

$$\sigma_X^2 = E[X^2] - \overline{X}^2 = 52 - 36 = 16.$$

Example 9.10

Given a WSS Gaussian random process $X(t_i)$, $t_i = 1,2,3$, the autocorrelation function is $R_{XX}(\tau) = 16e^{-2|\tau|}$ with mean $\overline{X} = 3$. Find the covariance matrix.

Solution

Given $R_{XX}(\tau) = 16e^{-2|\tau|}$, $\overline{X} = 3$ and the random process is $X(t_i)$, $t_i = 1,2,3$. Since it is wide sense stationary,

$$\tau = t_k - t_i, \; t_i = t_k = 1,2,3.$$

$$\therefore \quad R_{XX}(\tau) = R_{XX}(t_k - t_i) = 16e^{-2|t_k - t_i|}.$$

The autocorrelation matrix is

$$[R_X] = \begin{bmatrix} 16 & 16e^{-2} & 16e^{-4} \\ 16e^{-2} & 16 & 16e^{-2} \\ 16e^{-4} & 16e^{-2} & 16 \end{bmatrix} = \begin{bmatrix} 16 & 2.16 & 0.29 \\ 2.16 & 16 & 2.16 \\ 0.21 & 2.16 & 16 \end{bmatrix}.$$

The covariance of $X(t)$ is

$$C_{XX}(t_k - t_i) = R_{XX}(\tau) - \overline{X}^2.$$

$\therefore$ The covariance matrix is

$$[C_X] = \begin{bmatrix} 16-9 & 16e^{-2}-9 & 16e^{-4}-9 \\ 16e^{-2}-9 & 16-9 & 16e^{-2}-9 \\ 16e^{-4}-9 & 16e^{-2}-9 & 16-9 \end{bmatrix} = \begin{bmatrix} 7 & -6.83 & -8.71 \\ -6.83 & 7 & -6.83 \\ -8.71 & -2.83 & 7 \end{bmatrix}$$

Example 9.11

A random process is described by $X(t) = A$, where A is a continuous random variable uniformly distributed on $(0,1)$. Show that $X(t)$ is stationary process. (Feb 2007, Jan 2015)

Solution

Given $X(t) = A$, where A is a continuous random variable with uniform distribution on $(0,1)$

$$\therefore \quad f_A(A) = \begin{cases} 1 & 0 \le A \le 1 \\ 0 & \text{otherwise} \end{cases}$$

(i) Mean value of $X(t)$ is

$$E[X(t)] = \int_0^1 X(t) f_A(A) dA = \int_0^1 A dA = \frac{1}{2} \text{ (constant)}.$$

(ii) The autocorrelation of $X(t)$ is

$$R_{XX}(\tau) = E[X(t)X(t+\tau)]$$

$$= \int_0^1 A\,A f_A(A) dA = \int_0^1 A^2 dA = \left.\frac{A^3}{3}\right|_0^1 = \frac{1}{3}.$$

$R_{XX}(\tau)$ is independent of time.

$\therefore$ The given function $X(t)$ is a stationary process.

Example 9.12

Consider two random processes, $X(t) = A\cos(\omega_1 t + \theta)$ and $Y(t) = B\cos(\omega_2 t + \phi)$, where A, B, ω_1 and ω_2 are constants, while θ and ϕ are statistically independent random variables uniformly distributed on $(0, 2\pi)$.

(i) Show that $X(t)$ and $Y(t)$ are jointly WSS.

(ii) If $\theta = \phi$, show that $X(t)$ and $Y(t)$ are not jointly WSS unless $\omega_1 = \omega_2$. (Feb 2008, Dec 2013)

Solution

Given the random processes are

$$X(t) = A\cos(\omega_1 t + \theta) \text{ and } Y(t) = B\cos(\omega_2 t + \phi).$$

Also, θ and ϕ are uniformly distributed over $(0, 2\pi)$.

Then $f_\Theta(\theta) = \begin{cases} \frac{1}{2\pi} & 0 \le \theta \le 2\pi \\ 0 & \text{otherwise} \end{cases}$

and $f_\Phi(\phi) = \begin{cases} \frac{1}{2\pi} & 0 \le \phi \le 2\pi \\ 0 & \text{otherwise.} \end{cases}$

(i) For the processes to be jointly WSS,

$$E[X(t)] = E[A\cos(\omega_1 t + \theta)]$$

$$= A\int_0^\infty \cos(\omega_1 t + \theta) f_\Theta(\theta)d\theta = \frac{A}{2\pi}\int_0^{2\pi} \cos(\omega_1 t + \theta)d\theta$$

$$= \frac{A}{2\pi}(\sin(\omega_1 t + \theta)|_0^{2\pi}) = \frac{A}{2\pi}(\sin(\omega_1 t) - \sin(\omega_1 t)) = 0.$$

$\therefore$ $E[X(t)] = 0$ and the autocorrelation of $X(t)$ is

$$E[X(t)X(t+\tau)] = A^2\int_0^\infty \cos(\omega_1 t + \theta)\cos(\omega_1(t+\tau) + \theta))f_\Theta(\theta)d\theta$$

$$= \frac{A^2}{2\pi}\int_0^{2\pi} \frac{1}{2}[\cos\omega\tau + \cos(2\omega_1 t + \omega_1\tau + 2\theta)]d\theta$$

$$= \frac{A^2}{4\pi}(\cos\omega_1\tau)[2\pi + 0].$$

$$R_{XX}(\tau) = \frac{A^2}{2}\cos\omega_1\tau.$$

Similarly,

$$E[Y(t)] = 0 \text{ and } R_{YY}(\tau) = \frac{B^2}{2}\cos\omega_2\tau.$$

Now the cross-correlation function between $X(t)$ and $Y(t)$ is

$$R_{XY}(\tau) = E[X(t)Y(t+\tau)]$$

$$= \int_0^\infty\int_0^\infty X(t)Y(t+\tau)f_{\Theta,\phi}(\theta,\phi)d\theta d\phi.$$

Since θ and ϕ are statistically independent,

$$f_{\Theta,\phi}(\theta,\phi) = f_\Theta(\theta)f_\Phi(\phi).$$

$$\therefore\quad R_{XY}(\tau) = \int_0^\infty X(t)f_\Theta(\theta)d\theta\int_0^\infty Y(t+\tau)f_\Phi(\phi)d\phi$$

$$= \frac{AB}{4\pi^2}\int_0^{2\pi}\cos(\omega_1 t+\theta)d\theta\int_0^{2\pi}\cos(\omega_2(t+\tau)+\phi)d\phi$$

$$= \frac{AB}{16\pi^2}(0) = 0.$$

$$R_{XY}(\tau) = 0.$$

Therefore, the mean values of $X(t)$ and $Y(t)$ are constant. The autocorrelation of both $X(t)$ and $Y(t)$ are independent of t. The cross-correlation is also independent of t. Hence the random processes are jointly WSS.

(ii) If $\theta = \phi$, the expression for cross-correlation between $X(t)$ and $Y(t)$ is

$$R_{XY}(\tau) = E[X(t)Y(t+\tau)]$$

$$= E[A\cos(\omega_1 t+\theta)B\cos(\omega_2(t+\tau)+\theta)]$$

$$= \frac{AB}{2}E[\cos((\omega_2-\omega_1)t+\omega_2\tau)+\cos((\omega_2+\omega_1)t+\omega_2\tau+2\theta)]$$

$$= \frac{AB}{4\pi}\int_0^{2\pi}\cos((\omega_2-\omega_1)t+\omega_2\tau)d\theta+\frac{AB}{4\pi}\int_0^{2\pi}\cos((\omega_2+\omega_1)t+\omega_2\tau+2\theta)d\theta.$$

The second term is equal to zero.

$$\therefore\quad R_{XY}(\tau) = \frac{AB}{4\pi}\int_0^{2\pi}\cos((\omega_2-\omega_1)t+\omega_2\tau)d\theta$$

$$R_{XY}(\tau) = \frac{AB}{2}\cos((\omega_2-\omega_1)t+\omega_2\tau).$$

The cross-correlation function is a function of both t and τ. Therefore the given $X(t)$ and $Y(t)$ are not jointly WSS.

But if $\omega_2 = \omega_1$, then $R_{XY}(\tau) = \frac{AB}{2}\cos\omega_2\tau$.

The cross-correlation function is independent of time t. Therefore the given processes are jointly WSS at $\theta = \phi$ and $\omega_1 = \omega_2$.

Example 9.13

A random process $Y(t) = X(t) - X(t+\tau)$ is defined in terms of a process $X(t)$ that is at least WSS.

(a) Show that the mean value of $Y(t)$ is zero even if $X(t)$ has a non-zero mean value.

(b) Show that $\sigma_Y^2 = 2[R_{XX}(0) - R_{XX}(\tau)]$.

(c) If $Y(t) = X(t) + X(t+\tau)$, find $E[Y(t)]$ and σ_Y^2. (Nov 2007, Sept 2014)

Solution

Given the random process $Y(t) = X(t) - X(t+\tau)$ where $X(t)$ is a WSS random process.

(a) The mean value of $Y(t)$ is

$$E[Y(t)] = E[X(t) - X(t+\tau)] = E[X(t)] - E[X(t+\tau)].$$

For a stationary process, we know that $E[X(t)] = E[X(t+\tau)]$.

Given $E[X(t)] \neq 0$,

then $E[Y(t)] = 0$.

(b) The variance of $Y(t)$ is

$$\sigma_Y^2 = E[Y^2(t)] - E[Y(t)]^2 = E[Y^2(t)] - 0 = E[Y^2(t)].$$

$$\sigma_Y^2 = E[[X(t) - X(t+\tau)]^2]$$

$$= E[X^2(t)] + E[X^2(t+\tau)] - 2E[X(t)]E[X(t+\tau)]$$

$$\sigma_Y^2 = E[X^2(t)] + E[X^2(t)] - 2E[X(t)]E[X(t+\tau)]$$

$$\sigma_Y^2 = R_{XX}(0) + R_{XX}(0) - 2R_{XX}(\tau)$$

$$\sigma_Y^2 = 2[R_{XX}(0) - R_{XY}(\tau)].$$

(a) If the random process is given by $Y(t) = X(t) + X(t+\tau)$, then the mean value is

$$E[Y(t)] = E[X(t) + X(t+\tau)]$$

$$= E[X(t)] + E[X(t+\tau)] = 2E[X(t)].$$

$\therefore \quad \overline{Y(t)} = 2\overline{X(t)}$.

The variance of $Y(t)$ is

$$\sigma_Y^2 = E[Y^2(t)] - E[Y(t)]^2$$

$$\sigma_Y^2 = E[(X(t) + X(t+\tau)]^2] - (2\overline{X(t)})^2.$$

$$\therefore \quad \sigma_Y^2 = E[X^2(t)] + E[X^2(t+\tau)] + 2E[X(t)]E[X(t+\tau)] - 4\overline{X(t)}^2$$

$$\sigma_Y^2 = R_{XX}(0) + R_{XX}(0) + 2R_{XX}(\tau) - 4\overline{X(t)}^2$$

$$\sigma_Y^2 = 2[R_{XX}(0) + R_{XX}(\tau)] - 4\overline{X(t)}^2.$$

Example 9.14

Determine if the random process $X(t) = A$, where A is a random variable with mean $\overline{A}$ and variance σ_A^2, is mean ergodic. **(Feb 2007, Jan 2015)**

Solution

Given $X(t) = A$, and $\overline{X(t)} = A$.

If $X(t)$ is mean ergodic, then $E[X(t)] = A[X(t)]$.

The time average of $X(t)$ is

$$A[X(t)] = \lim_{T\to\infty} \frac{1}{2T}\int_{-T}^{T} X(t)dt = \lim_{T\to\infty} \frac{1}{2T}\int_{-T}^{T} A\,dt = A.$$

$\therefore \quad \overline{X(t)} = A[X(t)] = A$.

Therefore the random process is ergodic in mean.

Example 9.15

Statistically independent zero mean random processes $X(t)$ and $Y(t)$ have autocorrelation functions $R_{XX}(\tau) = e^{-|\tau|}$ and $R_{YY}(\tau) = \cos(2\pi\tau)$ respectively.

(a) Find the autocorrelation function of the sum $W_1(t) = X(t) + Y(t)$.

(b) Find the autocorrelation function of the difference $W_2(t) = X(t) - Y(t)$.

(c) Find the cross-correlation function of $W_1(t)$ and $W_2(t)$. (Nov 2007, Sept 2014)

Solution

Given $X(t)$ and $Y(t)$ are two statistically independent zero mean random processes. The autocorrelation functions are

$$R_{XX}(\tau) = e^{-|\tau|}$$

$$R_{YY}(\tau) = \cos(2\pi\tau).$$

(a) Given $W_1(t) = X(t) + Y(t)$, the autocorrelation function is

$$R_{W_1W_1}(\tau) = E[W_1(t)W_1(t+\tau)]$$

$$R_{W_1W_1}(\tau) = E[(X(t)+Y(t))(X(t+\tau)+Y(t+\tau))]$$

$$= E[X(t)X(t+\tau) + X(t)Y(t+\tau) + X(t+\tau)Y(\tau) + Y(t)Y(t+\tau)]$$

$$\therefore \quad R_{W_1W_1}(\tau) = E[X(t)X(t+\tau)] + E[X(t)Y(t+\tau)] + E[Y(t)X(t+\tau)] + E[Y(t)Y(t+\tau)]$$

$$R_{W_1W_1}(\tau) = R_{XX}(\tau) + R_{XY}(\tau) + R_{YX}(\tau) + R_{YY}(\tau).$$

Since $X(t)$ and $Y(t)$ are statistically independent,

$$R_{XY}(\tau) = R_{YX}(\tau) = 0.$$

$$\therefore \quad R_{W_1W_1}(\tau) = R_{XX}(\tau) + R_{YY}(\tau) = e^{-|\tau|} + \cos 2\pi\tau.$$

(b) Given $W_2(t) = X(t) - Y(t)$, the autocorrelation function of $W_2(t)$ is

$$R_{W_2W_2}(\tau) = E[W_2(t)W_2(t+\tau)]$$

$$= E[(X(t)-Y(t))(X(t+\tau)-Y(t+\tau))]$$

$$= E[X(t)X(t+\tau)] - E[X(t)Y(t+\tau)] - E[Y(t)X(t+\tau)] + E[Y(t)Y(t+\tau)]$$

$$R_{W_2W_2}(\tau) = R_{XX}(\tau) - R_{XY}(\tau) - R_{YX}(\tau) + R_{YY}(\tau)$$

$$= R_{XX}(\tau) + R_{YY}(\tau) = e^{-|\tau|} + \cos(2\pi\tau).$$

(c) The cross-correlation of $W_1(t)$ and $W_2(t)$ is

$$R_{W_1W_2}(\tau) = E[W_1(t)W_2(t+\tau)]$$

$$= E[(X(t)+Y(t))(X(t+\tau)-Y(t+\tau))]$$

$$= R_{XX}(\tau) - R_{XY}(\tau) + R_{YX}(\tau) - R_{YY}(\tau)$$

$$= R_{XX}(\tau) - R_{YY}(\tau).$$

$$\therefore \quad R_{W_1W_2}(\tau) = e^{-|\tau|} - \cos(2\pi\tau).$$

Example 9.16

A random process is described by $X(t) = A^2 \cos^2(\omega_c t + \theta)$, where A and ω_c are constants and θ is a random variable uniformly distributed between $\pm\pi$. Is $X(t)$ wide sense stationary?

(May 2010, Sept 2014)

Solution

Given $X(t) = A^2 \cos^2(\omega_c t + \theta)$, θ is uniformly distributed between $\pm\pi$.

The density function is

$$f_\Theta(\theta) = \begin{cases} \frac{1}{2\pi} & -\pi \le \theta \le \pi \\ 0 & \text{otherwise.} \end{cases}$$

The mean value of $X(t)$ is

$$E[X(t)] = \int_{-\infty}^{\infty} x(t) f_\Theta(\theta) d\theta = \int_{-\pi}^{\pi} A^2 \cos^2(\omega_c t + \theta) \times \frac{1}{2\pi} d\theta$$

$$= \frac{A^2}{2\pi}\left[\int_{-\pi}^{\pi} \cos^2(\omega_c^2 t + \theta) d\theta\right] = \frac{A^2}{4\pi}\int_{-\pi}^{\pi} [1 + \cos 2(\omega_c t + \theta)] d\theta$$

$$= \frac{A^2}{4\pi}\left[2\pi + \frac{\sin 2(\omega_c t + \theta)}{2}\Bigg|_{-\pi}^{\pi}\right]$$

$$= \frac{A^2}{4\pi}[2\pi + 0] = \frac{A^2}{2}.$$

$$\therefore \quad E[X(t)] = \frac{A^2}{2} \quad \text{constant.}$$

The autocorrelation function is

$$R_{XX}(\tau) = E[X(t)X(t+\tau)]$$

$$= E[A^2 \cos^2(\omega_c t + \theta) A^2 \cos^2(\omega_c(t+\tau) + \theta)]$$

$$= A^4 E[\cos^2(\omega_c t + \theta) \cdot \cos^2(\omega_c t + \omega_c \tau + \theta)]$$

$$= \frac{A^4}{2\pi}\int_{-\pi}^{\pi} \cos^2(\omega_c t + \theta) \cos^2(\omega_c t + \omega_c \tau + \theta) d\theta$$

$$= \frac{A^4}{2\pi}\left[\int_{-\infty}^{\infty}\left(\frac{1 + \cos(2\omega_c t + 2\theta)}{2}\right)\left(\frac{1 + \cos(2\omega_c t + 2\omega_c \tau + 2\theta)}{2}\right) d\theta\right]$$

$$= \frac{A^4}{8\pi}\Bigg[\int_{-\pi}^{\pi} \{(1 + \cos(2\omega_c t + 2\theta)) + \cos(2\omega_c t + 2\omega_c \tau + 2\theta)$$

$$+ \cos(2\omega_c t + 2\theta)\cos(2\omega_c t + 2\omega_c \tau + 2\theta)\} d\theta\Bigg]$$

$$= \frac{A^4}{8\pi}\left[\int_{-\pi}^{\pi} d\theta + \int_{-\pi}^{\pi} \cos(2\omega_c t + 2\theta)d\theta + \int_{-\pi}^{\pi} \cos(2\omega_c t + 2\omega_c \tau + 2\theta)d\theta\right.$$

$$\left. + \frac{1}{2}\int_{-\pi}^{\pi} \cos(2\omega_c \tau)d\theta + \frac{1}{2}\int_{-\pi}^{\pi} \cos(4\omega_c \tau + 2\omega_c \tau + 4\theta)d\theta\right]$$

$$= \frac{A^4}{8\pi}\left[2\pi + \frac{\sin(2\omega_c + 2\theta)}{2} - \pi\pi + \frac{\sin(2\omega_c t + 2\omega_c \tau + 2\theta)}{2}\right]_{-\pi}^{\pi}$$

$$\left. + \int_{-\pi}^{\pi}\left[\frac{\cos 2\omega_c \tau}{2}\right]d\theta + \int_{-\pi}^{\pi}\left(\frac{\cos(4\omega_c t + 2\omega_c \tau + 4\theta)}{4}\right)d\theta\right]$$

$$= \frac{A^4}{8\pi}[2\pi + 0 + 0 + \pi\cos 2\omega_c \tau + 0]$$

$$= \frac{A^4}{8\pi}[2\pi + \pi\cos 2\omega_c \tau] = \frac{A^4}{8}[2 + \cos 2\omega_c \tau].$$

$$\therefore \quad R_{XX}(\tau) = \frac{A^4}{8}(2 + \cos 2\omega_c \tau).$$

Since the mean value of $X(t)$ is constant and autocorrelation of $X(t)$ is independent of time t, the random process $X(t)$ is wide sense stationary.

Example 9.17

Prove that the random process $X(t) = A\cos(\omega_c t + \theta)$ is wide sense stationary if it is assumed that ω_c is a constant and θ is a uniformly distributed variable in the interval $(0, 2\pi)$. (Jan 2015)

Solution

Given $X(t) = A\cos(\omega_c t + \theta)$, where θ is uniformly distributed between$(0, 2\pi)$.

The density function is

$$f_\Theta(\theta) = \begin{cases} \frac{1}{2\pi} & 0 < \theta < 2\pi \\ 0 & \text{elsewhere.} \end{cases}$$

For a process to be wide sense stationary, mean value of $X(t)$ is

$$E[X(t)] = \int_{-\infty}^{\infty} x(t) f_\Theta(\theta)d\theta\infty$$

$$= \int_0^{\pi} A\cos(\omega_c t + \theta)\frac{1}{2\pi}d\theta$$

$$= \frac{A}{2\pi}\int_0^{2\pi} \cos(\omega_c t + \theta)d\theta = \frac{A}{2\pi}[\sin(\omega_c t + \theta)]_0^{2\pi} = 0$$

and autocorrelation of $X(t)$ is

$$R_{XX}(t,t+\tau) = E[X(t)X(t+\tau)]$$

$$= \int_0^{2\pi} A\cos(\omega_c t+\theta)\, A\cos(\omega_c(t+\tau)+\theta)\frac{1}{2\pi}d\theta$$

$$= \frac{A^2}{2\pi}\int_0^{2\pi} \cos(\omega_c t+\theta)\cos(\omega_c(t+\tau)+\theta)d\theta$$

$$= \frac{A^2}{4\pi}\left[\int_0^{2\pi}\cos\omega_c\tau d\theta + \int_0^{2\pi}\cos(2\omega_c t+2\theta+\tau)d\theta\right]$$

$$= \frac{A^2}{4\pi}(\cos\omega_c\tau)(2\pi) = \frac{A^2}{2}\cos\omega_c\tau.$$

Since the mean value is constant and autocorrelation is independent of time, $X(t)$ is wise sense stationary.

Example 9.18

Let $X(t)$ be a stationary continuous random process that is differentiable. Denote its time derivative by $\dot{X}(t)$.

(a) Show that $E[\dot{X}(t)] = 0$.

(b) Find $R_{\dot{X}\dot{X}}(\tau)$ in terms of $R_{XX}(\tau)$. (Feb 2007, Feb 2008, Sept 2014)

Solution

(a) Given $X(t)$ is a stationary random process.

The time derivative of $X(t)$ is $\dot{X}(t)$. From derivative principles,

$$\dot{X}(t) = \lim_{\Delta\to\infty}\frac{X(t+\Delta)-X(t)}{\Delta}.$$

The mean values is

$$E[\dot{X}(t)] = \lim_{\Delta\to\infty}\frac{E[X(t+\Delta)-X(t)]}{\Delta}$$

$$= \lim_{\Delta\to\infty}\frac{E[X(t+\Delta)]-E[X(t)]}{\Delta}.$$

We know that if $X(t)$ is stationary, then

$$E[X(t+\tau)] = E[X(t)] = E[X(t+\Delta)].$$

$$\therefore \quad E[\dot{X}(t)] = 0.$$

(b) We know that $R_{XX}(\tau) = E[X(t)X(t+\tau)]$, with the derivative of $X(t)$ as

$$R_{\dot{X}\dot{X}}(\tau) = E[\dot{X}(t)\dot{X}(t+\tau)].$$

Consider $X(t)$ has N random variables with joint density function. Then

$$R_{\dot{X}\dot{X}}(\tau) = \int_{-\infty}^{\infty}\int_{-\infty}^{\infty}\cdots\int_{-\infty}^{\infty} \dot{x}(t)\dot{x}(t+\tau) f_X(x_1,x_2,\cdots,x_N;t_1,t_2\cdots t_N)dx_1dx_2\cdots dx_N$$

$$= \int_{-\infty}^{\infty}\int_{-\infty}^{\infty}\cdots\int_{-\infty}^{\infty} \frac{\partial x(t)}{dt}\cdot\frac{\partial x(t+\tau)}{dt} f_X(x_1,x_2,\cdots x_N;t_1,t_2\cdots t_N)dx_1dx_2\cdots dx_N$$

Interchanging integration and differentiation,

$$R_{\dot{X}\dot{X}}(\tau) = \frac{\partial^2}{\partial t^2}\left[\int_{-\infty}^{\infty}\int_{-\infty}^{\infty}\cdots\int_{-\infty}^{\infty} x(t)x(t+\tau) f_X(x_1,x_2,\cdots,x_N;t_1,t_2,\cdots t_N)dx_1dx_2\cdots dx_N\right].$$

$$\therefore \quad R_{\dot{X}\dot{X}}(\tau) = \frac{\partial^2}{\partial t^2}[R_{XX}(\tau)].$$

Example 9.19

Given $\overline{X} = 6$ and $R_{XX}(t,t+\tau) = 36+25\exp(-\tau)$ for a random process $X(t)$. Indicate which of the following statements are true based on what is known with certainty: $X(t)$

(a) is first order stationary.

(b) has total average power of 61 W.

(c) is ergodic.

(d) is wide sense stationary.

(e) has a periodic component.

(f) has an AC power of 36 W. (Feb 2007, May 2009, May 2013)

Solution

Given the random process $X(t)$ with:
mean value $\overline{X} = 6$
autocorrelation $R_{XX}(t,t+\tau) = 36+25\exp(-\tau)$.

(a) First order stationary:

Since the mean value $\overline{X(t)} = 6$, a constant, the first order density function does not change with time t.

$\therefore$ $X(t)$ is first order stationary.

(b) We know that the average power is

$$P_{avg} = R_{XX}(0) = 36+25\exp(-0) = 36+25 = 61 \text{ W}.$$

The average power is 61 W.

(c) If the process is ergodic, the time averages are equal to ensemble averages.

Here, the time average is

$$A[X(t)] = A[\overline{X(t)}] = \lim_{T\to\infty}\frac{1}{2T}\int_{-T}^{T} 6dt = 6,$$

$$\therefore \quad A[X(t)] = \overline{X(t)}.$$

Also, the time autocorrelation function is

$$A[R_{XX}(t,t+\tau)] = \lim_{T\to\infty}\frac{1}{2T}\int_{-T}^{T}(36+25e^{-|\tau|})dt$$

$$= 36+25e^{-|\tau|} = R_{XX}(t,t,+\tau).$$

Hence the process is ergodic.

(d) Wide sense stationary (WSS):

The conditions for the process to be WSS is

$$E[X] = 6 = \text{ constant}$$

and $R_{XX}(t,t+\tau) = R_{XX}(\tau) = 36+25e^{-|\tau|}$

is independent of absolute time t. Therefore the process is WSS.

(e) The process has no periodic components:

Since

$$\lim_{\tau\to\infty} R_{XX}(t,\ t_1+\tau) = E[X]^2 = \lim_{\tau\to\infty} 36+25e^{-|\tau|} = 36.$$

$$E[X]^2 = 36.$$

(f) The ac power of the process is the variance of $X(t)$.

$$\therefore \quad \sigma_X^2 = E[X(t)^2] - E[X(t)]^2$$

$$E[X(t)^2] = R_{XX}(t,t+\tau)|_{\tau=0} = R_{XX}(0) = 61.$$

$$\therefore \quad \sigma_X^2 = 61-6^2 = 25.$$

The ac power is 25 watts.

Example 9.20

A random process is defined by $X(t) = X_0 + Vt$, where X_0 and V, are statistically independent random variables uniformly distributed on intervals $[X_{01}, X_{02}]$ and $[V_1, V_2]$ respectively. Find

(a) mean,

(b) autocorrelation and

(c) autocovariance function.

(d) Is $X(t)$ stationary in any sense? If so, state the type. (Feb 2008, Dec 2013)

Solution

The given random process is

$$X(t) = X_0 + Vt,$$

where X_0 and V are statistically independent.

$$f_{X_0}(x_0) = \begin{cases} \dfrac{1}{X_{02}-X_{01}} & X_{01} < X_0 < X_{02} \\ 0 & \text{elsewhere} \end{cases}$$

$$f_V(v) = \begin{cases} \dfrac{1}{V_2 - V_1} & V_1 < V < V_2 \\ 0 & \text{elsewhere} \end{cases}$$

(a) The mean value of $X(t)$ is

$$E[X(t)] = E[X_0 + Vt] = E[X_0] + E[Vt]$$

$$= \int_{-\infty}^{\infty} x_0 f_{X_0}(x_0)dx_0 + \int_{-\infty}^{\infty} Vt f_\nabla(V)dV$$

$$= \frac{1}{X_{02} - X_{01}} \int_{X_{01}}^{X_{02}} x_0 dx_0 + \frac{t}{V_2 - V_1} \int_{V_1}^{V_2} V dV$$

$$= \frac{X_{02}^2 - X_{01}^2}{2(X_{02} - X_{01})} + \frac{t(V_2^2 - V_1^2)}{2(V_2 - V_1)}$$

$$E[X(t)] = \frac{1}{2}[X_{02} + X_{01} + tV_2 + tV_1].$$

$$R_{XX}\ (t, t+\tau) = E[X(t)X(t+\tau)]$$

$$= E[(X_0 + Vt)(X_0 + V(t+\tau))]$$

$$= E[(X_0 + Vt)(X_0 + Vt + V\tau)]$$

$$= E[X_0^2 + X_0Vt + X_0V\tau + X_0Vt + V^2t^2 + V^2t\tau]$$

$$R_{XX}\ (t, t+\tau) = E[X_0^2] + (2t+\tau)E[X_0V] + (t^2 + t\tau)E[V^2].$$

Now, $E[X_0^2] = \int_{X_{01}}^{X_{02}} x_0^2 f_{X_0}(x_0)dx_0 = \dfrac{1}{X_{02} - X_{01}} \int_{X_{01}}^{X_{02}} x_0^2 dx_0$

$$E[X_0^2] = \frac{1}{(X_{02} - X_{01})}\left[\frac{X_{02}^3 - X_{01}^3}{3}\right] = \frac{X_{02}^2 + X_{01}^2 + X_{01}X_{02}}{3}.$$

$$E[V^2] = \int_{V_2}^{V_1} V^2 f_V(v)dV = \frac{1}{V_2 - V_1}\int_{V_2}^{V_1} V^2 dV$$

$$= \frac{1}{V_2 - V_1}\left[\frac{V_2^3 - V_1^3}{3}\right] = \frac{V_2^2 + V_1^2 + V_1V_2}{3}.$$

Since X_0 and V are independent

$$E[X_0V] = E[X_0]E[V]$$

$$= \frac{1}{X_{02} - X_{01}} \int_{X_{01}}^{X_{02}} x_0 dx_0 \frac{1}{V_2 - V_1} \int_{V_2}^{V_1} V dv$$

$$E[X_0V] = \frac{X_{02}^2 - X_{01}^2}{2(X_{02} - X_{01})} \frac{V_2^2 - V_1^2}{2(V_2 - V_1)} = \frac{(X_{02} + X_{01})(V_2 + V_1)}{4}.$$

$$\therefore\ R_{XX}(t, t+\tau) = \frac{X_{02}^2 + X_{01}^2 + X_{01}X_{02}}{3} + \frac{(2t+\tau)(X_{02} + X_{01})(V_2 + V_1)}{4} + \frac{t(t+\tau)(V_2^2 + V_1^2 + V_1V_2)}{3}$$

(c) The autocovariance function of $X(t)$ is

$$C_{XX}\ (t, t+\tau) = R_{XX}(t, t+\tau) - E[X(t)]E[X(t+\tau)]$$

Now, $E[X(t+\tau)] = \frac{1}{2}[(X_{02} + X_{01}) + (t+\tau)(V_2 + V_1)]$

and $E[X]E[X(t+\tau)] = \frac{1}{2}[X_{02} + X_{01} + t(V_2 + V_1)]\frac{1}{2}[X_{02} + X_{01} + (t+\tau)(V_2 + V_1)]$

$$= \frac{1}{4}\left[(X_{02} + X_{01})^2 + (2t+\tau)(X_{02} + X_{01})(V_2 + V_1) + t(t+\tau)(V_2 + V_1)^2\right].$$

$$\therefore\ C_{XX}(t, t+\tau) = \frac{X_{02}^2 + X_{01}^2 + X_{02}X_{01}}{3} + \frac{(2t+\tau)}{4}(X_{02} + X_{01})(V_2 + V_1)$$

$$+ \frac{t(t+\tau)}{3}(V_2^2 + V_1^2 + V_1V_2) - \frac{1}{4}(X_{02} + X_{01})^2$$

$$- \frac{(2t+\tau)}{4}(X_{02} + X_{01})(V_2 + V_1) - \frac{t(t+\tau)}{4}(V_2 + V_1)^2$$

$$= \frac{X_{02}^2 + X_{01}^2 + X_{02}X_{01}}{3} + \frac{t(t+\tau)}{3}(V_2^2 + V_1^2 + V_1V_2)$$

$$- \frac{X_{02}^2 + X_{01}^2 + 2X_{02}X_{01}t(t+\tau)}{44}(V_2^2 + V_1^2 + 2V_1V_1)$$

$$C_{XX}\ (t, t+\tau) = \frac{X_{02}^2 + X_{01}^2 - 2X_{02}X_{01}}{12} + \frac{t(t+\tau)}{12}(V_2^2 + V_1^2 - 2V_1V_2).$$

(d) 1. The mean of $X(t)$ is not constant.
2. The autocorrelation function depends on time t.
3. The autocovariance function depends on time t.

Hence the given random process is not stationary.

Example 9.21

Let $X(t)$ be a WSS process with autocorrelation function $R_{XX}(\tau) = e^{-a|\tau|}$, where $a > 0$ is a constant. Assume $X(t)$ amplitude modulates a carrier $\cos\omega_0 t + \theta$ as shown in Fig. 9.7, where ω_0 is a constant and θ is a random variable uniformly distributed on $(-\pi, \pi)$. If $X(t)$ and θ are statically independent, determine the autocorrelation function of $Y(t)$. (Feb 2008, March 2014)

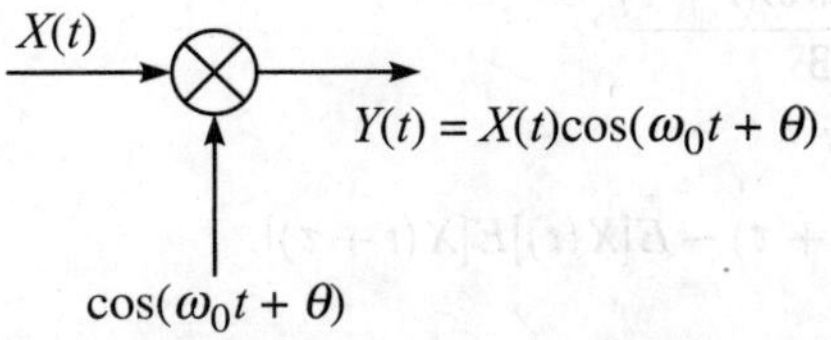

Fig. 9.7 *Modulation system*

Solution

Given a WSS random process $X(t)$ has autocorrelation $R_{XX}(\tau) = e^{-a|\tau|}a > 0$, where θ is a uniformly distributed random variable on $(-\pi, \pi)$.

$$\therefore \quad f_\theta(\theta) = \begin{cases} \dfrac{1}{2\pi} & -\pi < \theta < \pi \\ 0 & \text{elsewhere} \end{cases}$$

and $X(t)$ and θ are statistically independent. From Fig. 9.7, $Y(t) = X(t)\cos(\omega_0 t + \theta)$. The autocorrelation of $Y(t)$ is

$$\begin{aligned} R_{YY}(t, t+\tau) &= E[Y(t)Y(t+\tau)] \\ &= E[X(t)\cos(\omega_0 t + \theta)X(t+\tau)\cos(\omega_0(t+\tau)+\theta)] \\ &= E[X(t)X(t+\tau)]E[\cos(\omega_0 t + \theta)\cos(\omega_0(t+\tau)+\theta)] \\ &= R_{XX}(\tau)\frac{1}{2}E[\cos\omega_0\tau + \cos(2\omega_0 t + 2\theta + \omega_0\tau)] \\ &= \frac{R_{XX}(\tau)}{2}\left[\int_{-\pi}^{\pi}\frac{1}{2\pi}\cos\omega_0\tau d\theta + \int_{-\pi}^{\pi}\frac{1}{2\pi}\cos(2\omega_{0f} + 2\theta + \omega_0\tau)d\theta\right] \\ &= \frac{R_{XX}(\tau)}{2}[\cos\omega_0\tau + 0] \end{aligned}$$

$$R_{YY}(\tau) = \frac{1}{2}R_{XX}(\tau)\cos\omega_0\tau$$

Example 9.22

Consider a random process $X(t) = A\cos\omega t$, where ω is a constant and A is a random variable uniformly distributed over $(0, 1)$. Find the autocorrelation and autocovariance of $X(t)$. (May 2010)

Solution

Given a random process $X(t) = A\cos\omega t$

$$f_A(A) = \begin{cases} 1 & 0 < A < 1 \\ 0 & \text{otherwise} \end{cases}$$

$$= \int_0^1 A\cos\omega t\, A\, \cos\omega(t+\tau)dA = \int_0^1 A^2\cos\omega t\cos\omega(t+\tau)dA$$

$$= \cos\omega t\, \cos\omega(t+\tau)\frac{A^3}{3}\bigg|_0^1$$

$$\therefore \quad R_{XX}(t,\, t+\tau) = \frac{\cos\omega t \cos\omega(t+\tau)}{3}.$$

(b) Autocovariance of $X(t)$ is

$$C_{XX}(t,\, t+\tau) = R_{XX}(t,\, t+\tau) - E[X(t)]E[X(t+\tau)].$$

Now,

$$E[X(t)] = \int_0^1 x(t) f_A(A) dA = \cos\omega t \int_0^1 A\, dA = \frac{\cos\omega t}{2}$$

and $E[X(t+\tau)] = \dfrac{\cos\omega(t+\tau)}{2}$.

$$\therefore \quad C_{XX}(t, t+\tau) = \frac{\cos\omega t \cos\omega(t+\tau)}{3} - \frac{\cos\omega t \cos\omega(t+\tau)}{4}$$

$$= \frac{\cos\omega t \cos\omega(t+\tau)}{12}.$$

Example 9.23

A stationary random process $X(t)$ with mean 2 has the autocorrelation function $R_{XX}(\tau) = 4 + \exp(-|\tau|/10)$. Find the mean and variance of $Y = \int_0^1 X(t)dt$. (May 2013)

Solution

Given $E[X(t)] = 2$, $R_{XX}(\tau) = 4 + e^{-|\tau|/10}$ and $Y(t) = \int_0^1 X(t)dt$.

(a) Mean value of $Y(t)$ is

$$E[Y(t)] = E\left[\int_0^1 X(t)dt\right] = \int_0^1 E[X(t)]dt$$

$$E[Y(t)] = \int_0^1 2dt = 2.$$

(b) Variance of $Y(t)$ is

$$\sigma_Y^2 = E[Y^2(t)] - E[Y(t)]^2.$$

Now, for $E[Y^2(t)]$ we know that if $Y(t) = \int_0^T X(t)dt$, then

$$E[Y^2(t)] = \int_{-T}^{T} (1-|\tau|) R_{XX}(\tau) d\tau$$

$$= \int_{-1}^{1} (1-|\tau|)(4 + e^{-1\tau|/10}) d\tau$$

$$E[Y^2(t)] = \int_{-1}^{0} (1+\tau)(4 + e^{\tau/10}) d\tau + \int_0^1 (1-\tau)(4 + e^{-\tau/10}) d\tau$$

$$= \int_{-1}^{0} 4d\tau + \int_{-1}^{0} 4\tau d\tau + \int_{-1}^{0} e^{\tau/10} d\tau + \int_{-1}^{0} \tau e^{\tau/10} d\tau$$

$$+ \int_{0}^{1} 4d\tau - \int_{0}^{1} 4\tau d\tau + \int_{0}^{1} e^{-\tau/10} - \int_{0}^{1} \tau e^{-\tau/10} d\tau$$

$$= 4\tau\Big|_{-1}^{0} + \frac{4\tau^2}{2}\Big|_{-1}^{0} + 10\tau e^{\tau/10}\Big|_{-1}^{0} - 100e^{\tau/10}\Big|_{-1}^{0}$$

$$+ 4\tau\Big|_{-1}^{0} - \frac{4\tau^2}{2}\Big|_{0}^{1} - 10\tau e^{-\tau/10}\Big|_{0}^{1} + 100e^{-\tau/10}\Big|_{0}^{1}$$

$$= 4 - 2 + 10(1 - e^{-0.1}) + 10e^{-0.1} - 100(1 - e^{-0.1})$$

$$+ 4 - 2 - 10(e^{-0.1} - 1) + 10e^{-0.1} + 100(e^{-0.1} - 1)$$

$$= 200e^{-0.1} - 200 + 20 + 4$$

$$E[Y^2(t)] = 200e^{-0.1} - 176 = 4.967.$$

$$\therefore \quad \sigma_Y^2 = E[Y^2(t)] - E[Y(t)]^2.$$

$$\sigma_Y^2 = 4.967 - 2^2.$$

Variance of $Y(t)$ is $= 0.967$.

Example 9.24

A stationary process has an autocorrelation function given by

$$R(\tau) = \frac{2.5\tau^2 + 36}{625\tau^2 + 4}.$$

Find the mean value, mean square value and variance of the process. (May 2011)

Solution

Given $R(\tau) = \dfrac{25\tau^2 + 36}{6.25\tau^2 + 4}$

(i) The mean square value is

$$E[X^2] = R(\tau)|_{\tau=0} = \frac{36}{4} = 9.$$

(ii) Mean value is

$$\overline{X}^2 = \lim_{\tau \to \infty} R(\tau)$$

$$\overline{X}^2 = \frac{25 + \frac{36}{\tau^2}}{6.25 + \frac{4}{\tau^2}} = \frac{25}{6.25} = 4$$

$$\overline{X} = 2$$

(iii) Variance

$$\sigma_X^2 = E[X^2] - \overline{X}^2 = 9 - 4 = 5$$

Example 9.25

Assume that an Ergodic random process $x(t)$ has an autocorrelation function

$$R_{XY}(\tau) = 18 + \frac{2}{6+\tau^2}(1+4\cos(2\tau))$$

(i) Find $|\overline{X}|$.

(ii) Does this process have a periodic component?

(iii) What is the average power in $x(t)$? (May 2011)

Solution

$$R_{XY}(\tau) = 18 + \frac{2}{6+\tau^2}(1+4\cos(2\tau)).$$

(i) Mean value:

$$\overline{X}^2 = \lim_{\tau\to\infty} R_{XX}(\tau)$$

$$|\overline{X}^2| = \lim_{\tau\to\infty} |R_{XX}(\tau)|$$

$$|R_{XX}(\tau)| = 18 + \frac{2}{6+\tau^2}(1+4) = 18 + \frac{10}{6+\tau^2}$$

$$|\overline{X}|^2 = \lim_{\tau\to\infty}\left(18 + \frac{10}{6+\tau^2}\right) = 18.$$

$$|\overline{X}| = \sqrt{18} = 4.2426.$$

(ii) For finding if there is a periodic component, writing

$$\lim_{\tau\to\infty} R_{XX}(\tau) = \lim_{\tau\to\infty}\left((18 + \frac{2}{6+\tau^2}\right)(1+4\cos(2\tau)) = \infty$$

shows the process has a periodic component.

(iii) Average power of $X(t)$ is

$$E[X^2] = R_{XX}(0) = 18 + \frac{2}{6}(1+4\cos(0)) = 18 + \frac{10}{6} = 19.667.$$

Example 9.26

A stationary ergodic random process has the autocorrelation function with the periodic components as $R_{XX}(\tau) = 25 + \frac{4}{1+6\tau^2}$. Find the mean and variance of $X(t)$. (May 2011)

Solution

Given $R_{XX}(\tau) = 25 + \dfrac{4}{1+6\tau^2}$, the mean value is obtained from

The mean value:

$$\overline{X}^2 = \lim_{\tau \to \infty} R_{XX}(\tau) = 25$$

$$\therefore \quad \overline{X} = 5.$$

Mean square value is

$$E[X^2] = R_{XX}(0) = 25 + 4 = 29$$

$$\therefore \quad \text{Variance } \sigma_X^2 = E[X^2] - \overline{X}^2 = 29 - 25 = 4.$$

Example 9.27

A Gaussian random process is known to be WSS with a mean of $\overline{X} = 4$ and autocorrelation $R_{XX}(\tau) = 25e^{-3|\tau|/2}$. Find the covariance function necessary to specify the joint density of the random process and give the covariance matrix for $x_i = x(t_1)$, where $i = 1, 2, \ldots, N$. (Sept 2014)

Solution

Given a WSS Gaussian random process $X(t)$ with $\overline{X}(t) = 4$ and $R_{XX}(\tau) = 25e^{-3|\tau|/2}$. N random variables are given by $X(t_i)$, $i = 1, 2, \ldots N$ with $t_k - t_i = (k-i) = \tau$, i and $k = 1, 2, 3, \ldots, N$.

The autocorrelation function for N random variables is

$$R_{XX}(t_i, t_k) = R_{XX}(t_k - t_i) = R_{XX}(\tau)$$

$$R_{XX}(t_i, t_k) = 25e^{-3|t_k - t_i|/2}.$$

The covariance function is defined as

$$C_{XX}\ (t_i,\ t_k) = R_{XX}(t_i, t_k) - E[X(t_i)]E[X(t_k)].$$

But, given $E[X(t_i)] = E[X(t_k)] = E[x(t)] = 4$.

$$C_{XX}\ (t_i,\ t_k) = 25e^{-3|t_k - t_i|/2}k - 4 \times 4 = 25e^{-3|t_k - t_i|/2}k - 16$$

$\therefore$ The covariance matrix is

$$[C_X] = \begin{bmatrix} 25-16 & 25e^{-3/2}-16 & 25e^{-3}-16 & \ldots & 25e^{-3(|1-N|)/2}-16 \\ 25e^{-3/2}-16 & 25-16 & 25e^{-3/2}-16 & \ldots & 25e^{-3(|2-N|)/2}-16 \\ 25e^{-3}-16 & 25e^{-3/2}-16 & 25-16 & \ldots & 25e^{-3(|3-N|)/2}-16 \\ \ldots & \ldots & \ldots\ \ldots & & \ldots \\ 25e^{-3(|N-1|)/2}-16 & 25e^{-3(|N-2|)/2}-16 & 25e^{-3(|N-3|)/2}-16 & \ldots & 25-16 \end{bmatrix}.$$

For example, if $N = 2$,

$$[C_X] = \begin{bmatrix} 9 & -10.42 \\ -10.42 & 9 \end{bmatrix}.$$

Example 9.28

Telephone calls are initiated through an exchange mean average rate of 75 per minute and are described by a Poisson process. Find the probability that more than 3 calls are initiated in any 5 second period.

(Feb. 2007, Dec. 2013)

Solution

Given $\lambda = 75$ per minute $= \dfrac{75}{60}$ per second, time period $t = 5$ second.

$$\therefore \quad \lambda t = \frac{75}{60} \times 5 = 6.25.$$

Number of calls initiated $N = 3$.

Since the process is a Poisson's process, we know that the probability of the number of calls initiated being $\leq N$ is

$$P[X(t) \leq N] = \sum_{t=0}^{N} \frac{(\lambda t)^k e^{-\lambda t}}{k!}, \; k = 0, 1, 2, 3, \ldots.$$

Thus, the probability that the number of calls initiated is more than 3 is

$$P[X(t) > 3] = 1 - P[X(t) \leq 3]$$

$$= 1 - \sum_{k=0}^{3} \frac{(\lambda t)^k e^{-\lambda t}}{k!} = 1 - e^{-6.25} \sum_{k=0}^{3} \frac{(6.25)^k}{k!}$$

$$= 1 - \left[(6.25)^0 e^{-6.25} + 6.25 e^{-6.25} + \frac{(6.25)^2}{2!} e^{-6.25} + \frac{(6.25)^3}{3!} e^{-6.25} \right]$$

$$= 1 - 0.1302.$$

$$P[X(t) > 3] = 0.869.$$

Example 9.29

Find the transfer function $H(\omega)$ for the network shown in Fig. 9.8. If $X(t) = e^{j\omega t}$, show that $Y(t) = X(t)H(\omega)$.

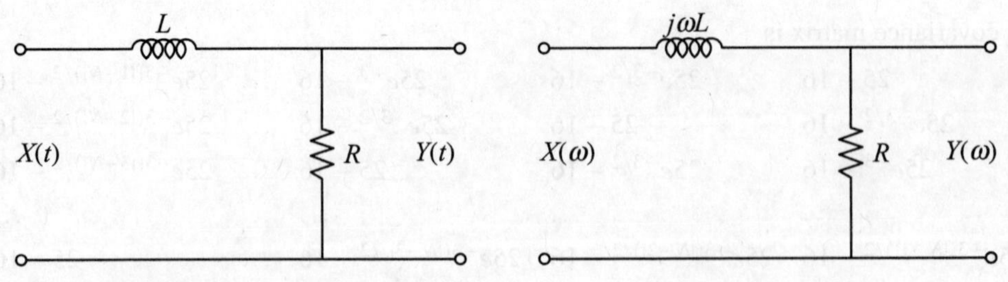

Fig. 9.8 *RL circuit* **Fig. 9.9** *RL circuit*

Solution

The given Fig. 9.8 can be drawn in the frequency domain as shown in Fig. 9.9. The output response is

$$Y(\omega) = \frac{RX(\omega)}{R + j\omega L} = \frac{x(\omega)}{1 + j\omega(L/R)}.$$

$$\therefore \quad \text{The transfer function } H(\omega) = \frac{Y(\omega)}{X(\omega)} = \frac{1}{1 + j\omega(L/R)}.$$

In the time domain, the output is the convolution of $X(t)$ and $h(t)$.

$$Y(t) = h(t) * X(t). \tag{9.93}$$

$$\therefore \quad Y(t) = \int_{-\infty}^{\infty} h(\tau) e^{J\omega(t-\tau)} d\tau = e^{+J\omega t} \left[\int_{-\infty}^{\infty} h(\tau) e^{J\omega\tau} d\tau \right]$$

$$Y(t) = e^{+j\omega t} H(\omega).$$

Therefore, the output is $Y(t) = X(t)H(\omega)$.

Example 9.30

Find the transfer function of the network shown in Fig. 9.10. (Dec 2013)

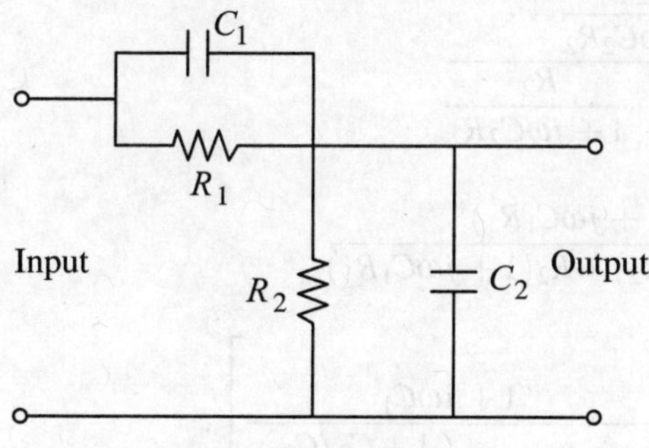

Fig. 9.10 *RC network*

Solution

Let $X(\omega)$ and $Y(\omega)$ be the input and output functions. The given network is drawn with reactance values as shown in Fig. 9.11.

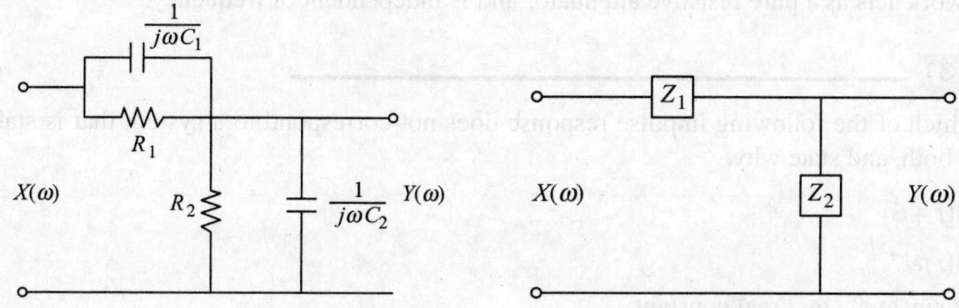

Fig. 9.11 (a) *RC network with reactance* **Fig. 9.11 (b)** *With impedance*

The impedances are

$$Z_1 = R_1 \frac{\dfrac{1}{j\omega C}}{R_1 + \dfrac{1}{j\omega C}} = \frac{R_1}{1 + j\omega R_1 C_2}$$

and $$Z_2 = \frac{R_2 \times \dfrac{1}{j\omega C_2}}{R_2 + \dfrac{1}{j\omega C_2}} = \frac{R_2}{1 + j\omega R_2 C_2}.$$

Now from Fig. 9.14(b),

$$Y(\omega) = \frac{Z_2}{Z_1 + Z_2} X(\omega).$$

∴ The transfer function is

$$H(\omega) = \frac{Y(\omega)}{X(\omega)} = \frac{Z_2}{Z_1 + Z_2}.$$

Substituting the impedance values,

$$H(\omega) = \frac{\dfrac{R_2}{1 + j\omega C_2 R_2}}{\dfrac{R_1}{1 + j\omega C_1 R_1} + \dfrac{R_2}{1 + j\omega C_2 R_2}}$$

$$= \frac{R_2(1 + j\omega C_1 R_1)}{R_1(1 + j\omega C_2 R_2) + R_2(1 + j\omega C_1 R_1)}$$

$$H(\omega) = \left(\frac{R_2}{R_1 + R_2}\right) \left[\frac{1 + j\omega C_1}{1 + j\omega R_1 C_1 \left(\dfrac{1 + C_2/C_1}{1 + R_1/R_2}\right)}\right].$$

Note: If $C_2/C_1 = R_1/R_2$ or $C_1 R_1 = C_2 R_2$, the transfer function becomes

$$H(\omega) = \frac{R_2}{R_1 + R_2}.$$

Then the network acts as a pure resistive attenuator and is independent of frequency.

Example 9.31

Determine which of the following impulse response does not correspond to a system that is stable or realisable, or both, and state why.

(i) $h(t) = u(t+3)$

(ii) $h(t) = u(t)e^{-t^2}$

(ii) $h(t) = e^t \sin(\omega_0 t)$, ω_0 : real constant

(iv) $h(t) = u(t)e^{-3t}$, ω_0 : real constant (Feb2007, March 2014)

Solution

(i) Given $h(t) = u(t+3)$. The system is unstable and unrealisable because:

(1) For stable system, we know that $y(t) = \int_{-\infty}^{\infty} h(\tau)x(t-\tau)d\tau < \infty$.

Let $x(t) = A$, finite constant. Then, the output

$$y(f) = \int_{-\infty} \infty Au(\tau+3)d\tau = A\int_{-\infty}^{\infty} u(\tau+3)d\tau = \infty.$$

(2) Also, for the system to be causal, $h(t) = 0$ for $t < 0$.

But $h(-1) = u(2) \neq 0$.

The system is non-causal and hence unstable and unrealisable.

(ii) Given $h(t) = u(t)e^{-t^2}$.

The system is stable and realisable because:

(1) For a stable system,

$$y(t) = \int_{-\infty}^{\infty} Au(\tau)e^{-\tau^2}d\tau = A\int_{-\infty}^{\infty} u(\tau)e^{-\tau^2}d\tau$$

$$= A\int_{-\infty}^{\infty} e^{-\tau^2}d\tau = A\sqrt{\pi}.$$

Therefore it is bounded Bounded output.

(2) For the system to be causal, $h(t) = 0, \quad t < 0$

Hence $u(t) = 0$, $t < 0$.

System is causal and therefore it is stable and realisable.

(iii) Given $h(t) = e^t \sin\omega_0 t)$, ω_0 : real constant.

The system is unstable and unrealisable because:

(1) For a stable system,

$$y(t) = \int_{-\infty}^{\infty} Ae^{\tau}\sin(\omega_0\tau)d\tau = \infty.$$

Since it is an exponential function, it grows without bound.

(2) For a causal system,

$$h(t) = 0 \quad \text{for} \quad t < 0.$$

Here $e^t \sin(\omega_0 t) \neq 0$ for $t < 0$. System is non-causal.

Therefore the system is unstable and unrealisable.

(iv) Given $h(t) = u(t)e^{-3t}$, ω_0 : real constant.

The system is stable and realisable because:

(1) For a stable system

$$y(t) = \int_{-\infty}^{\infty} Au(\tau)e^{-3\tau}d\tau = A\int_{-\infty}^{\infty} e^{-3\tau}d\tau = +A/3.$$

Therefore, bounded output.

(2) For a causal system

$$h(t) = 0 \text{ for } t < 0$$

Here $e^{-3\tau}u(t) = 0$ for $t < 0$; the system is causal.
Therefore, the system is stable and realisable.

Example 9.32

A signal $x(t) = u(t)e^{-\alpha t}$ is applied to a network having an impulse response $h(t) = Wu(t)e^{-Wt}$. Here, α and W are real positive constants and $u(t)$ is a unit step function. Find (a) $Y(t)$ and (b) the output spectrum. (Nov 2008, Sept 2014)

Solution

Given $x(t) = u(t)e^{-\alpha t}$; $h(t) = Wu(t)e^{-Wt}$.

(a) We know that the system response is y(t)=x(t)*h(t).

$$\therefore \quad y(t) = \int_{-\infty}^{\infty} h(\tau)x(t-\tau)d\tau$$

$$y(t) = \int_{-\infty}^{\infty} Wu(\tau)e^{-W\tau}u(t-\tau)e^{-\alpha(t-\tau)}d\tau.$$

Since $u(\tau)u(t-\tau) = \begin{cases} 1 & 0 < \tau \leq t \\ 0 & \text{otherwise.} \end{cases}$

$$\therefore \quad y(t) = \int_0^t We^{-\alpha t}e^{-(W-\alpha)\tau}d\tau$$

$$= We^{-\alpha t}\left[\frac{e^{-\tau(W-\alpha)}}{-(W-\alpha)}\right]_0^t = \frac{We^{-\alpha t}}{W-\alpha}[1-e^{\alpha t}e^{-Wt}]$$

$$\therefore \quad y(t) = \frac{W}{W-\alpha}[e^{-\alpha t} - e^{-Wt}] \quad \text{for } t > 0$$

$$\text{or} \quad y(t) = \frac{W}{W-\alpha}[e^{-\alpha t} - e^{-Wt}]u(t).$$

(b) The input spectrum is

$$X(\omega) = \mathcal{F}[x(t)]$$

$$X(\omega) = \mathcal{F}[e^{-\alpha t}u(t)] = \frac{1}{\alpha + j\omega}$$

$$\text{and} \quad H(\omega) = \mathcal{F}[h(t)]$$

$$= \mathcal{F}[We^{-Wt}u(t)] = \frac{W}{W+j\omega}.$$

The output spectrum is

$$Y(\omega) = X(\omega)H(\omega) = \frac{W}{(\alpha+j\omega)(W+j\omega)}.$$

We can also find the output spectrum using Fourier transform

$$Y(\omega) = \mathcal{F}[y(t)]$$

$$= \frac{W}{W-\alpha}\left[F\left[e^{-\alpha t}\right] - F\left[e^{-Wt}\right]\right]$$

$$= \frac{W}{W-\alpha}\left[\frac{1}{\alpha+j\omega} - \frac{1}{W+j\omega}\right]$$

$$= \frac{W}{W-\alpha}\left[\frac{W+j\omega-\alpha-j\omega}{(\alpha+j\omega)(W+j\omega)}\right]$$

$$= \frac{W}{(\alpha+j\omega)(W+j\omega)}.$$

Example 9.33

Two systems have transfer function $H_1(\omega)$ and $H_2(\omega)$.

(a) Show that the transfer function $H(\omega)$ of the cascade of the two, which means that the output of the first feeds the input of the second system, is $H(\omega) = H_1(\omega)H_2(\omega)$.

(b) For a cascade of N systems with transfer functions $H_n(\omega)$, $n = 1,2,3\ldots..N$, show that

$$H(\omega) = \prod_{n=1}^{N} H_n(\omega).$$

(May 2009, March 2014)

Solution

(a) Consider the cascade system as shown in Fig. 9.12(a).

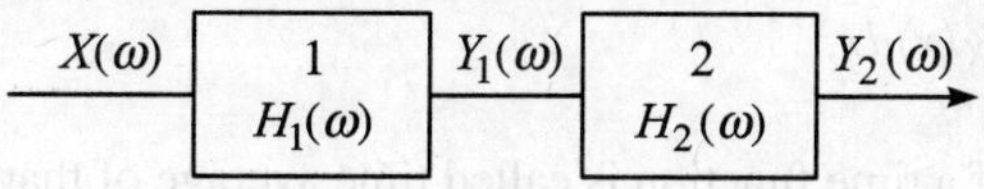

$X(\omega)$ → [1, $H_1(\omega)$] → $Y_1(\omega)$ → [2, $H_2(\omega)$] → $Y_2(\omega)$

Fig. 9.12 (a) *Cascade of two systems*

We know that $Y_1(t) = X(\omega)H_1(\omega)$

and $\quad Y_2(\omega) = H_2(\omega)Y_1(\omega)$

or $\quad Y_2(\omega) = H_2(\omega)H_1(\omega)X(\omega).$

If $H(\omega)$ is transfer function of the system, then

$$H(\omega) = \frac{Y_2(\omega)}{X(\omega)}.$$

So $H(\omega) = H_1(\omega)H_2(\omega)$.

(b) If the third system shown in Fig. 9.12(b) is cascaded to the first two systems, then

$$Y_3(\omega) = Y_2(\omega)H_3(\omega)$$

$$= H_3(\omega)H_2(\omega)H_1(\omega)X(\omega)$$

or $H(\omega) = \dfrac{Y_3(\omega)}{X(\omega)} = H_1(\omega)H_2(\omega)H_3(\omega)$.

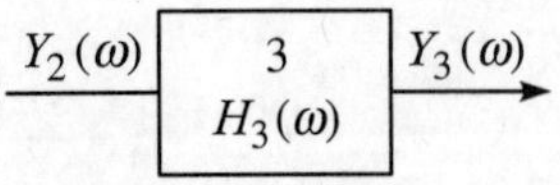

Fig. 9.12 (b) *Third system*

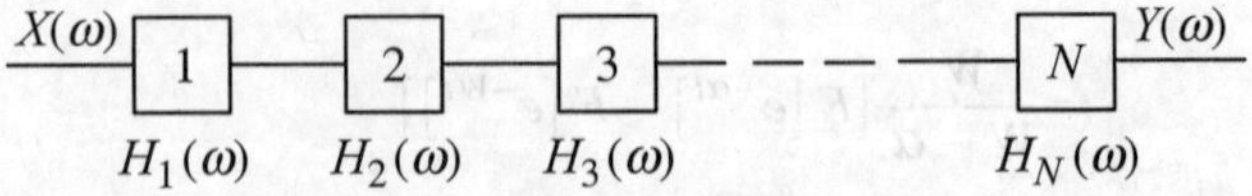

Fig. 9.12 (c) *Cascade of N system*

Similarly, for a cascade of N systems as shown in Fig. 9.12(c), the output function is

$$Y(\omega) = H_1(\omega)H_2(\omega)H_3(\omega)\ldots.H_N(\omega)X(\omega).$$

If $H(\omega)$ is the overall transfer function, then

$$Y(\omega) = H(\omega)X(\omega).$$

$$\therefore \quad H(\omega) = H_1(\omega)H_2(\omega)H_3(\omega)\ldots..H_N(\omega)$$

or $H(\omega) = \prod_{n=1}^{N} H_n(\omega)$. (Proved).

Summary

1. A random process is a family of functions $\{X \leq x(t,s), s \in S \text{ and } t \in T\}$ defined with random variables $\{X \leq x\}$ on a sample space S and with a function of time T.
2. If $f_X(x)$ is a probability density function, then the expected value of X is called the statistical average or ensemble average. That is,
$$E[X] = \int_{-\infty}^{\infty} X f_X(x)dx.$$
3. The expected value of a time function is called time average of that function. It is expressed as
$$A[x_i(t)] = \frac{1}{T}\int_0^T x(t)dt.$$
4. A process is called deterministic random processes if its future values of any sample function can be predicted from its past values. If future values of any sample function cannot be predicted from the observed past values, the process is called non-deterministic process.
5. A random process is said to be continuous if both the random variable X and time are continuous over the entire time.
6. In a discrete random process, a random variable X has only discrete values, while time t is continuous.
7. A random process for which the random variable X is continuous and the time t has discrete values is called a continuous random sequence.
8. In discrete random sequence, both random variables X and time t are discrete.
9. A random variable can be obtained from a random process at a particular instant of time t. The random variable has all the statistical properties related to its probability density function.

10. For a single random variable at time, $t_1, X_1 = X(t_1)$, the cumulative distribution function is a first-order distribution function of $X(t)$ defined as $F_X(x_1;t_1) = P\{(X(t_1) \le x_1\}$, where x_1 is any real number.

11. For two random variables at time instants t_1 and $t_2, X(t_1) = X_1$ and $X(t_2) = X_2$, the joint distribution is called the second-order joint distribution function of the random process $X(t)$ and it is given by $F_X(x_1,x_2;t_1,t_2) = P\{X(t_1) \le x_1, X(t_2) \le x_2\}$.

12. For N random variables at N time instants, $X(t_i) = X_i, i = 1,2,\ldots N$, the Nth-order joint distribution function of is defined as

$$F_X(x_1,x_2,\ldots x_N;t_1,t_2\ldots t_N) = P\{X(t_1) \le x_1, X(t_2) \le x_2, \ldots X(t_N) \le x_N\}.$$

13. Fist-order density function is $f_X(x_1;t_1) = \dfrac{dF_X(x_1;t_1)}{dx_1}$.

14. Second-order joint density function is $f_X(x_1,x_2;t_1,t_2) = \dfrac{\partial^2 F_X(x_1,x_2;t_1,t_2)}{\partial x_1 \partial x_2}$.

15. Nth-order joint density function is $f_X(x_1,x_2\ldots . x_N;t_1,t_2\ldots . t_N) = \dfrac{\partial^N F_X(x_1,x_2\ldots x_N.;t_1,t_2\ldots\ldots t_N)}{\partial x_1 \partial x_2 \ldots \partial x_N}$.

16. If $X(t)$ is a discrete-time process, then the Nth order probability mass function is

$$P_X(x_1,x_2\ldots . x_N;t_1,t_2\ldots . t_N) = P\{X(t_1) = x_1, X(t_2) = x_2, \ldots . X(t_N) = x_N\}.$$

17. If $X(t)$ is statistically independent random process, then the Nth order joint density function is equal to the product of the individual joint function of $X(t)$.

$$f_X(x_1,x_2\ldots x_N;t_1,t_2\ldots t_N) = f_X(x_1;t_1)\cdot f_X(x_2;t_2)\cdot f_X(x_3;t_3)\ldots f_X(x_N;t_N)$$

18. The two random processes $X(t)$ and $Y(t)$ are said to be statistically independent if all the random variables $X(t_1), X(t_2)\ldots X(t_N)$ are independent of $Y(t_1'), Y(t_2')\ldots . Y(t_N')$. That is, the joint density function of the random processes $X(t)$ and $Y(t)$ is equal to the product of the individual joint density function of $X(t)$ and $Y(t)$.

$$f_{XY}(x_1,x_2\ldots x_N;y_1,y_2\ldots y_M;t_1,t_2\ldots t_N,t_1',t_2'\ldots t_M') = f_X(x_1,x_2,\ldots x_N;t_1,t_2\ldots t_N)\cdot$$
$$f_Y(y_1,y_2\ldots y_M;t_1',t_2'\ldots t_M')$$

19. The mean value of a random process $X(t)$ is $\overline{X(t)} = E[X(t)] = \int_{\Gamma}^{\infty} x f_X(x;t)dx$.

20. The autocorrelation function of the random process $X(t)$ is

$$R_{XX}(t_1,t_2) = E[X(t_1)X(t_2)] = E[X_1X_2]$$

or $$R_{XX}(t_1,t_2) = \int_{-\infty}^{\infty}\int_{-\infty}^{\infty} x_1x_2 f_X(x_1,x_2;t_1,t_2)dr_1dr_2.$$

21. The cross-correlation function of the random processes $X(t)$ and $Y(t)$ is

$$R_{XY}(t_1,t_2) = E[X(t_1)Y(t_2)] = E[XY] \text{ or } R_{XY}(t_1,t_2) = \int_{-\infty}^{\infty}\int_{-\infty}^{\infty} xy f_{XY}(x,y:t_1,t_2)dxdy.$$

22. For a first-order stationary process, $f_X(x_1;t_1) = f_X(x_1;t_1+\Delta t)$ for any t_1, the condition for first-order stationary random process is

$$\overline{X(t)} = E[X(t)] = \int_{-\infty}^{\infty} x f_X(x;t)dx = \text{ constant.}$$

23. A random process is said to be stationary to order two or a second-order stationarity process if its second order joint density function does not change with time or shift in the time value. In other words,

$$f_X(x_1,x_2;t_1,t_2) f_X(x_1,x_2;t_1,t_2) = f_X(x_1,x_2;t_1+\Delta t, t_2+\Delta t) \text{ for all } t_1,\ t_2 \text{ and } \Delta t.$$

24. The stationary condition for a second-order random process is its autocorrelation function and it depends only on the time differences and not on absolute time. That is,

$$R_{XX}(t_1,t_1+\tau) = E[X(t_1)X(t_1+\tau)] = R_{XX}(\tau).$$

25. If a random process $X(t)$ is a second-order stationary process, then it is called wide-sense stationary process (WSS) or weak sense stationary process X. However, the converse is not true.
26. The conditions for wide-sense stationary process are:

$$E[X(t)] = \overline{X} = \text{constant and } E[X(t)X(t+\tau)] = R_{XX}(\tau) \text{ is independent of absolute time } t.$$

27. The conditions for joint wide-sense stationary process are:

$$E[X(t)] = \overline{X} = \text{constant}\ ;\ E[Y(t)] = \overline{Y} = \text{constant and}$$

$$E[X(t)Y(t+\tau)] = R_{XY}(\tau) \text{ is independent of time } t.$$

28. A random process $X(t)$ is said to be strict-sense stationary if its Nth order joint density function does not change with time or shift in the time value. That is,

$$f_X(x_1,x_2\ldots x_N;t_1t_2\ldots t_N) = f_X(x_1,x_2\ldots x_N;t_1+\Delta t,\ldots t_2+\Delta t,\ldots t_N+\Delta t) \text{ for all}$$

$$t_1,t_2\ldots t_N \text{ and } \Delta t.$$

29. If a random process is stationary to all orders $n = 1,2,\ldots N$, it is called strict-sense stationary process.
30. The strict-sense stationary process is also called Nth-order stationary process or stationary to order-N. Note that strict-sense stationary (SSS) process is also WSS process, but the converse is not true.
31. If the stationary conditions of a random process $X(t)$ does not hold for all N but holds for $N \leq K$, then the process $X(t)$ is stationary to order K or a Kth order stationary process.
32. The average value of $x(t)$ taken over all time is called the time average of $x(t)$. It is also called mean value of $x(t)$. It can be expressed as,

$$\overline{x} = A[x(t)] = \lim_{T\to\infty} \frac{1}{2T}\int_{-T}^{T} x(t)dt.$$

33. The time autocorrelation function of $X(t)$ is $\mathbb{R}_{XX}(\tau)\ A[X^2(t)] = \lim_{T\to\infty} \frac{1}{2T}\int_{-T}^{T} x^2(t+\tau)dt.$
34. The time mean square value of $X(t)$ is $A[X^2] = \lim_{T\to\infty} \frac{1}{2T}\int_{-T}^{T} x^2(t)dt.$
35. The time cross correlation function of $X(t)$ and $Y(t)$ is

$$\mathbb{R}_{XY}(\tau) = A[X(t)Y(t+\tau)] = \lim_{T\to\infty} \frac{1}{2T} x(t)y(t+\tau)dt.$$

36. The ensemble averages and time averages are always not the same.

37. The expected value of time averages gives ensemble averages. That is,

$$E[[x(t)]] = E[X(t)] = \overline{X}\ , E[A(X(t)X(t+\tau))] = E[X(t)X(t+\tau)] = R_{XX}(\tau)$$

and $E[-\mathbb{R}_{XY}(\tau)] = R_{XY}(\tau)$.

38. Ergodic theorem states that for any random process $X(t)$ all the time averages of sample functions of $X(t)$ are equal to the corresponding statistical or ensemble averages of $X(t)$. That is, $\overline{x} = \overline{X}$ and $\mathbb{R}_{XX}(\tau) = R_{XX}(\tau)$.

39. The random processes that satisfy the ergodic theorem are called ergodic processes. Two random processes are said to be jointly ergodic if they are individually ergodic and also, the time cross-correlation function is equal to the ensemble cross-correlation function of $X(t)$. That is, $\overline{x} = \overline{X}$, $\overline{y} = \overline{Y}$, $\mathbb{R}_{XX}(\tau) = R_{XX}(\tau)$, $\mathbb{R}_{XY}(\tau) = R_{XY}(\tau)$ and $\mathbb{R}_{YY}(\tau) = R_{YY}(\tau)$.

40. If a random process is mean-ergodic or ergodic in the mean, then $E[X(t)] = \overline{X} = A[x(t)] = \overline{x}$ with probability one for all $x(t)$.

41. If a stationary random process is autocorrelation ergodic or ergodic in the autocorrelation, then $A[X(t)X(t+\tau)] = E[X(t)X(t+\tau)]$ or $\mathbb{R}_{XX}(\tau) = R_{XX}(\tau)$.

42. Two stationary random processes are said to be cross-correlation ergodic or ergodic in cross correlation if $A[X(t)Y(t+\tau)] = E[X(t)Y(t+\tau)]$ or $\mathbb{R}_{XY}(\tau) = R_{XY}(\tau)$.

43. *Properties of autocorrelation function*
   (i) $E[X^2(t)] = R_{XX}(0)$
   (ii) $|R_{XX}(\tau)| \le R_{XY}(0)$
   (iii) $R_{XX}(-\tau) = R_{XX}(\tau)]$
   (iv) If a random process $X(t)$ has non-zero mean value, $E[X(t)] \neq 0$, and is ergodic with no periodic components, then $\lim\limits_{|\tau|\to\infty} R_{XX}(\tau) = \overline{X}^2$.
   (v) If $X(t)$ is periodic, then its autocorrelation function is also periodic.
   (vi) If $X(t)$ is ergodic, of zero mean and has no periodic component, then $\lim\limits_{|\tau|\to\infty} R_{XX}(\tau) = 0$.
   (vii) The autocorrelation function of a random process $R_{XX}(\tau)$ cannot have any arbitrary shape.
   (viii) If $Y(t) = A + X(t)$, then $R_{YY}(\tau) = A^2 + R_{XX}(\tau)$.
   (ix) If $Z(t) = X(t) + Y(t)$, then $R_{ZZ}(\tau) = R_{XX}(\tau) + R_{XY}(\tau) + R_{YX}(\tau) + R_{YY}(\tau)$.

44. *Properties of cross correlation function*
   (i) $R_{XY}(\tau) = R_{YX}(-\tau)$
   (ii) $|R_{XY}(\tau)| \le \sqrt{R_{XX}(0)R_{YY}(0)}$
   (iii) $|R_{XY}(\tau)| \le \frac{1}{2}[R_{XX}(0) + R_{YY}(0)]$
   (iv) If two random processes $X(t)$ and $Y(t)$ are statistically independent and are at least wide sense stationary, then $R_{XY}(\tau) = \overline{XY}$.
   (v) If the two random processes $X(t)$ and $Y(t)$ have zero mean and are jointly WSS, then $\lim\limits_{\tau\to\infty} R_{XY}(\tau) = 0$.

45. The auto-covariance function is a measure of interdependence between two random variables of the random processes $X(t)$. It is given by

$$C_{XX}(t, t+\tau) = E[(X(t) - E[X(t)])(X(t+\tau) - E[X(t+\tau)])] \quad \text{or}$$

$$C_{XX}(t, t+\tau) = R_{XX}(t, t+\tau) - E[X(t)]E[X(t+\tau)].$$

46. If $X(t)$ is at least a wide sense stationary random process, then $C_{XX}(\tau) = R_{XX}(\tau) - \overline{X}^2$.

47. At $\tau = 0$, the auto-covariance function becomes the variance of the random process, that is,

$$C_{XX}(0) = R_{XX}(0) - \overline{X}^2 = E[X^2] - \overline{X}^2 = \sigma_X^2.$$

48. The autocorrelation coefficient of the random process $X(t)$ is defined as

$$R_{XX}(t, t+\tau) = \frac{C_{XX}(t, t+\tau)}{\sqrt{C_{XX}(t,t)C_{XX}(t+\tau, t+\tau)}} \text{ and } R_{XX}(0) = \frac{C_{XY}(0)}{C_{XY}(0)} = 1, \ |R_M(t, t+\tau)| \leq 1.$$

49. If two random processes $X(t)$ and $Y(t)$ have random variables at $X(t)$ and $Y(t+\tau)$, then the cross covariance function is defined as

$$C_{XY}(t,\ t+\tau) = E[(X(t) - E[X(t)])(Y(t+\tau) - E[Y(t+\tau)])] \quad \text{or}$$

$$C_{XY}(t, t+\tau) = R_{XY}(t, t+\tau) - E[X(t)]E[Y(t+\tau)].$$

50. If the random processes are at least jointly wide-sense stationary, then $C_{XY}(\tau) = R_{XY}(\tau) - \overline{XY}$.

51. If the two random processes $X(t)$ and $Y(t)$ are uncorrelated or are statistically independent, then

$$C_{XY}(t, t+\tau) = 0 \text{ or } R_{XY}(t, t+\tau) = E[X(t)]E[Y(t+\tau)].$$

52. The cross-correlation coefficient of the random processes is defined as

$$R_{XY}(t, t+\tau) = \frac{C_{XY}(t, t+\tau)}{\sqrt{C_{XX}(t,t)C_{1Y}(t, t+\tau)}} \text{ and } R_{XY}(0) = \frac{C_{XY}(0)}{\sqrt{C_{XY}(0)C_{YY}(0)}} = \frac{C_{XY}(0)}{\sigma_X \sigma_Y}.$$

53. If the random variables are jointly Gaussian for any $N = 1, 2 \ldots$ and at any time instants $t_1, t_2 \ldots t_N$, then the random processes are called Gaussian random processes.

54. The joint density function for Gaussian random variable is given as

$$f_X(x_1, x_2 \ldots x_N; t_1, t_2 \ldots J_N) = \frac{\exp\left\{\frac{-1}{2}[x - \overline{X}]^{-1}[C_{XX}]^{-1}[x - \overline{X}]\right\}}{\sqrt{(2\pi)^N |[C_{XX}]|}}.$$

55. The wide-sense stationary (WSS) Gaussian process is also strictly stationary process.

56. If two random processes $X(t)$ and $Y(t)$ are said to be jointly Gaussian if the random processes $X(t_1), X(t_2) \ldots X(t_N)$, $Y(t_1'), Y(t_2') \ldots Y(t_M')$ defined at times $t_1, t_2 \ldots t_N$ and $t_1', t_2' \ldots t_M'$ respectively are jointly Gaussian for any time $t_1, t_2 \ldots t_N$; $t_1', t_2' \ldots t_M'$.

57. The Poisson process $X(t)$ is a discrete random process which represents the number of times that some event has occurred as a function of time.

58. The conditions for Poisson process $X(t)$ are

    (i) $X(0) = 0$.
    (ii) The event occurs only once at any instant of time, i.e., in an infinitesimal time interval.
    (iii) The number that occurs in any given time interval is independent of the number in any other non-overlapping time interval, i.e., $X(t)$ has independent increments.

59. Poisson distribution function is

$$P[X(t) = k] = \frac{(\lambda t)^k e^{-\lambda t}}{k!}, \ K = 0, 1, 2 \ldots.$$

and the probability density function is

$$f_X(x) = \sum_{k=0}^{\infty} \frac{(\lambda t)^k e^{-\lambda /}}{k!} \delta(x-k),$$

where λ is the average rate of occurrences and k is the number of occurrences over a time interval $(0, t)$.

60. The mean value of Poisson process $X(t)$ is

$$E[X(t)] = \int_{-\infty}^{\infty} x f_X(x) dx = \sum_{k=0}^{\infty} \frac{k(\lambda t)^k e^{-\lambda t}}{k!}.$$

61. The second moment of Poisson process $X(t)$ is

$$E[X^2(t)] = \sum_{k=0}^{\infty} \frac{k^2(\lambda t)^k e^{-\lambda t}}{k!} = \lambda t + \lambda^2 t^2.$$

62. The output response of an LTI system is

$$Y(t) = h(t) * X(t), \qquad Y(t) = \int_{-\infty}^{\infty} h(\tau) X(t-\tau) d\tau.$$

63. Mean value of output response $= E[Y(t)]$

$$E[Y(t)] = \overline{Y} = \overline{X} H(0) \text{ is constant.}$$

64. Mean square value of the output response is $E[Y^2(t)] = E[(h(t) * X(t))^2]$

$$E[Y^2(t)] = \left[\int_{-\infty}^{\infty}\int_{-\infty}^{\infty} R_{XX}(\tau_1 - \tau_2) h(\tau_1) h(\tau_2) d\tau_1 d\tau_2\right].$$

65. The autocorrelation of $Y(t)$ is

$$R_{YY}(t_1, t_2) = E[Y(t_1) Y(t_2)]$$

$$R_{YY}(\tau) = R_{XX}(\tau) * h(\tau) * h(-\tau)$$

$$R_{XY}(\tau) = R_{XX}(\tau) * h(\tau)$$

$$R_{YX}(\tau) = R_{XX}(\tau) * h(-\tau)$$

$$R_{YY}(\tau) = R_{XY}(\tau) * h(-\tau)$$

and $R_{YY}(\tau) = R_{YX}(\tau) * h(\tau)$

Review Questions

1. Explain the concept of a random process. (Nov 2010, May 2013)
2. Explain how random processes are classified with the help of a neat sketch. (Nov 2006, Dec 2013)
3. Briefly explain the distribution and density functions in the context of stationary and independent random processes. (May 2014)

4. Distinguish between stationary and non-stationary random processes. (May 2009, Nov 2010, Sept 2014)
5. State the condition for a wide sense stationary random process. (May 2009, Nov 2010, Jan 2015)
6. Explain in brief the concepts of time average and ergodicity.
7. Write about the following ergodic processes: (i) Mean ergodic process, (ii) Correlation ergodic process.
8. State the properties of the autocorrelation function. (Nov 2010, May 2014)
9. State the properties of the cross-correlation function. (Feb 2008, May 2011, Jan 2015)
10. Explain stationarity and ergodic random processes. (May 2011, Jan 2015)
11. Define the following functions.
    (i) Autocorrelation function
    (ii) Cross-correlation function (Feb 2008, Jan 2015)
12. Two systems have transfer functions $H_1(\omega)$ and $H_2(\omega)$. Show that the transfer function $H(\omega)$ of the cascade of the two, $H(\omega) = H_1(\omega)H_2(\omega)$. (Nov 2007, Sept 2014)
13. For a cascade of N-systems with transfer functions $H_n(\omega)$, $n = 1,2,\ldots,N$, show that $H(\omega) = \Pi H_n(\omega)$. (Nov 2007, Sept 2014)
14. How is the autocorrelation function of white noise represented? What is its significance? (May 2009, March 2014)
15. Derive the relation between input and output autocorrelation function of a random process of an LTI system. (May 2009, May 2010, Jan 2015)

Problems

1. A random process is defined by $Y(t) = X(t)\cos(\omega_0 t + \theta)$, where $X(t)$ is a WSS random process that amplitude modulates a carrier of constant angular frequency ω_0 with a random phase θ independent of $X(t)$ and uniformly distributed on $(-\pi, \pi)$.
   (a) Find $E[Y(t)]$.
   (b) Find the autocorrelation function of $Y(t)$.
   (c) Is $Y(t)$ wide sense stationary?
2. A random process is defined as $X(t) = a\cos(\omega t + \theta)$, where ω and θ are constants and a is a random variable uniformly distributed over $(-A, A)$. Check $X(t)$ for stationarity.
3. A random process is given as $X(t) = at + b$, where b is a constant, and a is a random variable uniformly distributed over $(-2, 2)$. Check $X(t)$ for stationarity.
4. Determine whether or not the stochastic process $X(t) = A\sin t + B\cos t$ is ergodic. A and B are normally distributed independent random variables with zero means and equal standard deviation.
   (a) Find $E[Y(t)]$.
   (b) Find the autocorrelation function of $Y(t)$.
   (c) Is $Y(t)$ wide sense stationary?

5. For a stationary ergodic random process with no periodic components, the autocorrelation function is given by $R_{XX}(\tau) = 36 + \frac{5}{1+7\tau^2}$. Find the mean and variance of the process $X(t)$.
6. If $X(t)$ is a random process with mean 3 and autocorrelation of $9 + 4e^{-0.2\tau}$, find the mean and variances of the random variable $X(t)$.
7. A random process is given as $X(t) = K$, where K is a random variable uniformly distributed over $(-1, 1)$. Find (a) $E[X(t)]$, (b) $R_{XX}(t_1, t_2)$ and (c) check $X(t)$ for stationarity.
8. Two random processes $U(t)$ and $V(t)$ are defined as $U(t) = X(t) + Y(t)$ and $V(t) = 2X(t) + 3Y(t)$, where $X(t)$ and $Y(t)$ are two independent random processes. Find $R_U(\tau)$, $R_V(\tau)$, $R_{UV}(\tau)$ and $R_{VU}(\tau)$ in terms of $R_X(\tau)$ and $R_Y(\tau)$.
9. A random process is given by $X(t) = A\cos(\omega_0 t + \theta)$, where A and θ are independent random variables and θ has a uniform probability density function over $(-\pi, \pi)$. Show that $E[X(t)] = 0$ and $E[X^2(t)] = \frac{A^2}{2}$.
10. Show that for the autocorrelation function of the process $X(t) = A\cos(100t + \theta)$, where A is a normal random variable with zero mean and unity variance, θ is uniformly distributed in the interval $(-\pi, \pi)$. Assume that A and θ are independent random variables.
11. A random process $X(t)$ is defined as $X(t) = \begin{cases} A & 0 \le t \le 1 \\ 0 & \text{otherwise,} \end{cases}$ where A is a random variable that is uniformly distributed from $-\theta$ to θ. Prove that the autocorrelation function of $X(t)$ is $\theta^2/3$.
12. For the random process $X(t) = A\cos\omega t$, where ω is a constant and A is a uniform random variable over $(0, 1)$, find the mean square value of $X(t)$.
13. A stationary process has an autocorrelation function given by $R(\tau) = \frac{25\tau^2 + 36}{6.25\tau^2 + 4}$. Find the mean value, mean square value and variance of the process.
14. The function time $Z(t) = X_1\cos\omega_0 t - X_2\sin\omega_0 t$ is a random process. If X_1 and X_2 are independent Gaussian random variables, each with zero mean and variance σ^2, find $E[Z]$, $E[Z^2]$ and var (Z).
15. Consider a random process $X(t) = 10\cos(100t + \phi)$, where ϕ is a uniformly distributed random variable in the interval $(-\pi, \pi)$. Show that the process is ergodic in the mean and correlation sense.
16. A voltage $V(t)$ is a Gaussian ergodic random process with zero mean and variance of 25 volts^2 measured by a dc meter, a true rms meter, and a meter which first squares $V(t)$ and then reads its dc component. Find the output of each meter.
17. A particular commercial system for controlling a petroleum distillation plant has failures (resulting in plant down time) that occur at the average rate of two per 30 days. Assume that the number of failures is a Poisson process. Find the probability that one failure will occur during the first 30 days and no other failures will occur for the next 30 days.
18. A complex random process $Z(t) = X(t) + jY(t)$ is defined by jointly stationary real processes $X(t)$ and $Y(t)$. Show that $E[|Z(t)|^2] = R_{XX}(0) + R_{YY}(0)$.
19. Let $X_1(t), X_2(t).Y_1(t)$ and $Y_2(t)$ be real random processes and define

$$Z_1(t) = X_1(t) + jY_1(t),\ Z_2(t) = X_2(t) + jY_2(t).$$

Find the expressions for the cross-correlation function of $Z_1(t)$ and $Z_2(t)$ if:

(a) All the real processes are correlated.

(b) They are uncorrelated.

(c) They are uncorrelated with zero mean.

20. It is desired to generate a random signal $X(t)$ with autocorrelation function, $R_X(\tau) = 5\eta e^{-5|\tau|}$, by passing white noise $n(t)$ with power spectral density $S_n(f) = \eta/2$ watts/Hz through an LTI system. Obtain an expression for the transfer function $H(f)$ of the LTI system.

21. A WSS random process $X(t)$ is applied to the input of an LTI system whose impulse response is $5te^{-2t}u(t)$. The mean of $X(t)$ is 3. Find the mean of the output of the system.

Answers

1. (a) 0, $\frac{1}{2}R_{XK}$ (b) $(\tau)\cos\omega_0\tau$, (c) Yes

4. Not ergodic process

5. $\pm 6, 5$

6. 3,13

13 2,9,5

14. 0, σ^2, σ^2

16. DC meter reads zero volts, true rms meter reads 5 volts and the third meter reads 25 volts^2.

17. 0.0366

19. (a) $R_{X_1X_2}(\tau) - R_{Y_1Y_2}(\tau) - jR_{X_1Y_2}(\tau) - jR_{Y_1X_2}(\tau)$ (b)$Z_1^*(t)Z_2(t+\tau)$ (c) 0

20. $\dfrac{100}{25+\omega^2}$

21. $\dfrac{15}{4}$

Multiple-choice Questions

1. $X(t_1) = X_1$ and $X(t_2) = X_2$. The correlation between X_1 and X_2 is $R(t_1, t_2) =$

(a) $\int_{-\infty}^{\infty} f(x_1, x_2; t_1, t_2)dx_1$ (b) $\int_{-\infty}^{\infty} f(x_1, x_2, t_1.t_2)dx_2$

(c) $\int_{-\infty}^{\infty}\int_{-\infty}^{\infty} x_1, x_2 f(x_1, x_2; t_1, t_2)dt_1dt_2$ (d) $\int_{-\infty}^{\infty}\int_{-\infty}^{\infty} x_1,\ x_2 f(x_1, x_2; t_1, t_2)dx_1dx_2$

2. The autocovariance cov $X(t_1,\ t_2)$ of a random process $X(t)$ is

(a) $R_{XX}(t_1, t_2) - E[X(t_1)]$ (b) $R_{XX}(t_1, t_2) + E[X(t_2)]$

(c) $R_{XX}(t_1, t_2) - E[X(t_2)]E[X(t_2)]$ (d) $R_{XX}(t_1, t_2) + E[X(t_1)] + E[X(t_1)]$

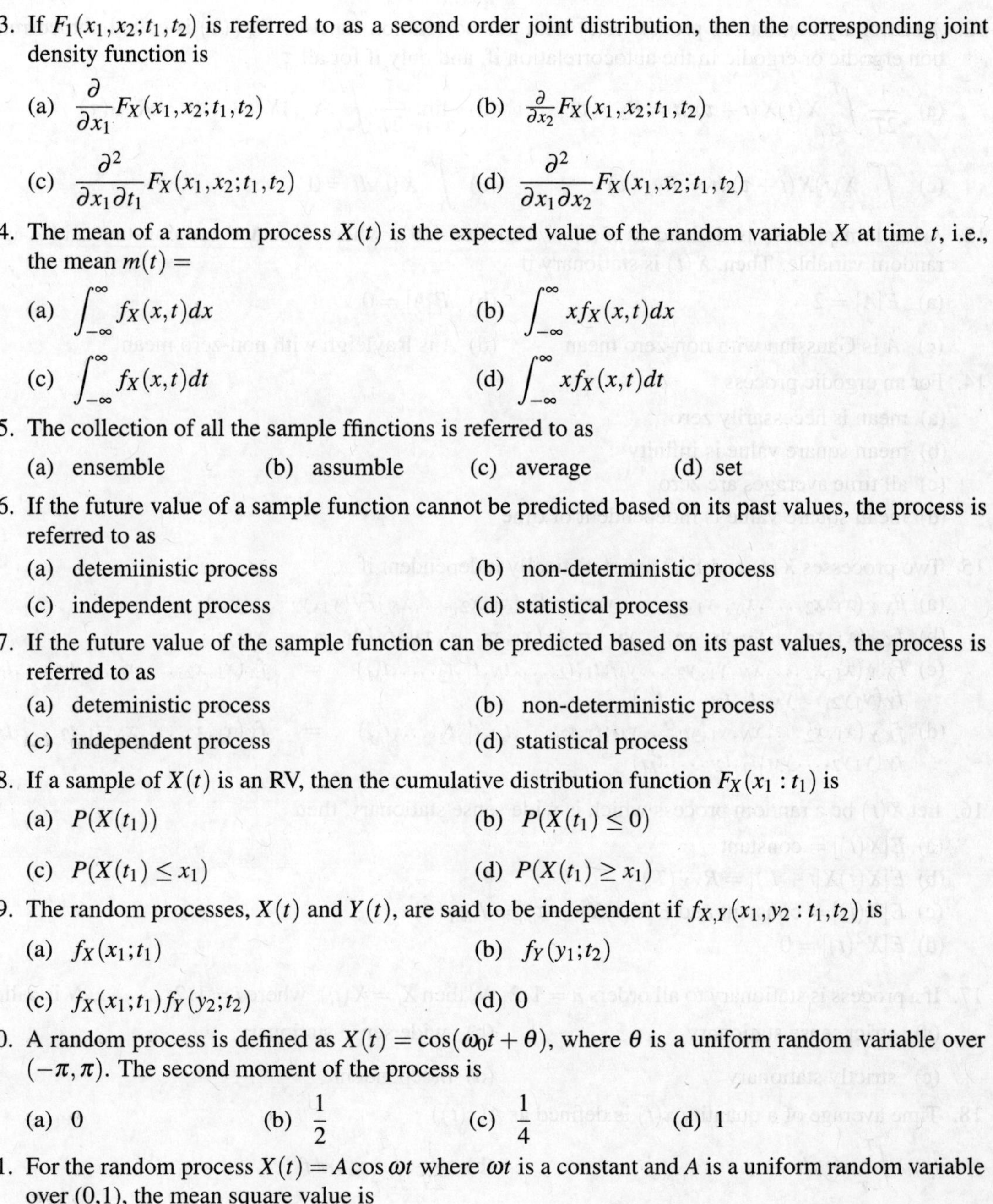

3. If $F_1(x_1,x_2;t_1,t_2)$ is referred to as a second order joint distribution, then the corresponding joint density function is

(a) $\frac{\partial}{\partial x_1}F_X(x_1,x_2;t_1,t_2)$ (b) $\frac{\partial}{\partial x_2}F_X(x_1,x_2;t_1,t_2)$

(c) $\frac{\partial^2}{\partial x_1 \partial t_1}F_X(x_1,x_2;t_1,t_2)$ (d) $\frac{\partial^2}{\partial x_1 \partial x_2}F_X(x_1,x_2;t_1,t_2)$

4. The mean of a random process $X(t)$ is the expected value of the random variable X at time t, i.e., the mean $m(t) =$

(a) $\int_{-\infty}^{\infty} f_X(x,t)dx$ (b) $\int_{-\infty}^{\infty} xf_X(x,t)dx$

(c) $\int_{-\infty}^{\infty} f_X(x,t)dt$ (d) $\int_{-\infty}^{\infty} xf_X(x,t)dt$

5. The collection of all the sample ffinctions is referred to as

(a) ensemble (b) assumble (c) average (d) set

6. If the future value of a sample function cannot be predicted based on its past values, the process is referred to as

(a) detemiinistic process (b) non-deterministic process

(c) independent process (d) statistical process

7. If the future value of the sample function can be predicted based on its past values, the process is referred to as

(a) deteministic process (b) non-deterministic process

(c) independent process (d) statistical process

8. If a sample of $X(t)$ is an RV, then the cumulative distribution function $F_X(x_1 : t_1)$ is

(a) $P(X(t_1))$ (b) $P(X(t_1) \leq 0)$

(c) $P(X(t_1) \leq x_1)$ (d) $P(X(t_1) \geq x_1)$

9. The random processes, $X(t)$ and $Y(t)$, are said to be independent if $f_{X,Y}(x_1,y_2 : t_1,t_2)$ is

(a) $f_X(x_1;t_1)$ (b) $f_Y(y_1;t_2)$

(c) $f_X(x_1;t_1)f_Y(y_2;t_2)$ (d) 0

10. A random process is defined as $X(t) = \cos(\omega_0 t + \theta)$, where θ is a uniform random variable over $(-\pi,\pi)$. The second moment of the process is

(a) 0 (b) $\frac{1}{2}$ (c) $\frac{1}{4}$ (d) 1

11. For the random process $X(t) = A\cos\omega t$ where ωt is a constant and A is a uniform random variable over (0,1), the mean square value is

(a) $\frac{1}{3}$ (b) $\frac{1}{3}\cos\omega t$ (c) $\frac{1}{3}\cos^2\omega t$ (d) $\frac{1}{9}$

12. A stationary continuous process $X(t)$ with autocorrelation function $R_{XX}(\tau)$ is called autocorrelation ergodic or ergodic in the autocorrelation if, and only if for all τ

(a) $\frac{1}{2T}\int_{-T}^{T} X(t)X(t+\tau)dt = R_{XX}(\tau)$ (b) $\lim_{T\to\infty}\frac{1}{2T}\int_{-T}^{T} X(t)X(t+\tau)dt = R_{XX}(\tau)$

(c) $\int_{-\infty}^{\infty} X(t)X(t+\tau)dt = R_{YX}(\tau)$ (d) $\int_{\Gamma}^{\infty} X(t)dt = 0$

13. A random process is defined as $X(t) = A\cos(\omega t+\theta)$, where ω and θ are constant and A is a random variable. Then, $X(t)$ is stationary if

(a) $E[A] = 2$ (b) $E[A] = 0$

(c) A is Gaussian with non-zero mean (d) A is Rayleigh with non-zero mean

14. For an ergodic process
   (a) mean is necessarily zero
   (b) mean square value is infinity
   (c) all time averages are zero
   (d) mean square value is independent of time

15. Two processes $X(t)$ and $Y(t)$ are statistically independent if
   (a) $F_{X,Y}(x_1,x_2,\ldots,x_N,\ y_1,y_2,\ldots,y_M) = F_X(x_1,x_2,\ldots,x_N)F_Y(y_1,y_2\ldots,y_M)$
   (b) $f_{XY}(x_1,x_2,\ldots x_N,y_1,\ y_2,\ldots y_M) = f_X(x_1,x_2,\ldots,x_N)f_Y(y_1,y_2,\ldots,y_M)$
   (c) $F_{X,Y}(x_1,x_2\ldots,x_N,y_1,y_2,\ldots y_M;t_1,t_2,\ldots,t_N,t'_1,t'_2,\ldots,t'_M) = f_X(x_1,x_2,\ldots,x_N;t_1,t_2,\ldots,t_N)$ $f_Y(y_1y_2,\ldots y_M;t'_1,t_{2'}\ldots.t'_M)$
   (d) $f_{X,Y}(x_1,x_2\ldots,x_N,y_1,y_2,\ldots y_M;t_1,t_2,\ldots,t_N,t'_1,t'_2,\ldots,t'_M) = f_X(x_1,x_2,\ldots,x_N;t_1,t_2,\ldots,t_N)$ $f_Y(y_1y_2,\ldots y_M;t'_1,t_{2'}\ldots.t'_M)$

16. Let $X(t)$ be a random process which is wide sense stationary, then
   (a) $E[X(t)] =$ constant
   (b) $E[X(t)X[t+T)] = R_{XX}(T)$
   (c) $E[X(t)] =$ constant and $E[X(t)X[t+\tau)] = R_{XX}(\tau)$
   (d) $E[X^2(t)] = 0$

17. If a process is stationary to all orders $n = 1,2,\ N$, then $X_i = X(t_i)$, where $i = 1,2,\ldots\ldots.,N$ is called

(a) strict sense stationary (b) wide sense stationary

(c) strictly stationary (d) independent

18. Time average of a quantity $x(t)$ is defined as $A(x(t))$

(a) $\int_{-T}^{T} x(t)dt$ (b) $\frac{1}{2T}\int_{-T}^{T} x(t)df$

(c) $\lim_{T\to\infty}\frac{1}{2T}\int_{-T}^{T} x(t)dt$ (d) $\lim_{T\to\infty}\int_{-T}^{T} x(t)dt$

19. Consider a random process $X(t)$ defined as $X(t) = A\cos\omega t + B\sin\omega t$, where ω is a constant and A and B are random variables. Which of the following is a condition for the process to be stationary?

(a) $E(A) = 0,\ E(B) = 0$ (b) $E(AB) \neq 0$

(c) $E(A) \neq 0.\ E(B) \neq 0$ (d) A and B should be independent

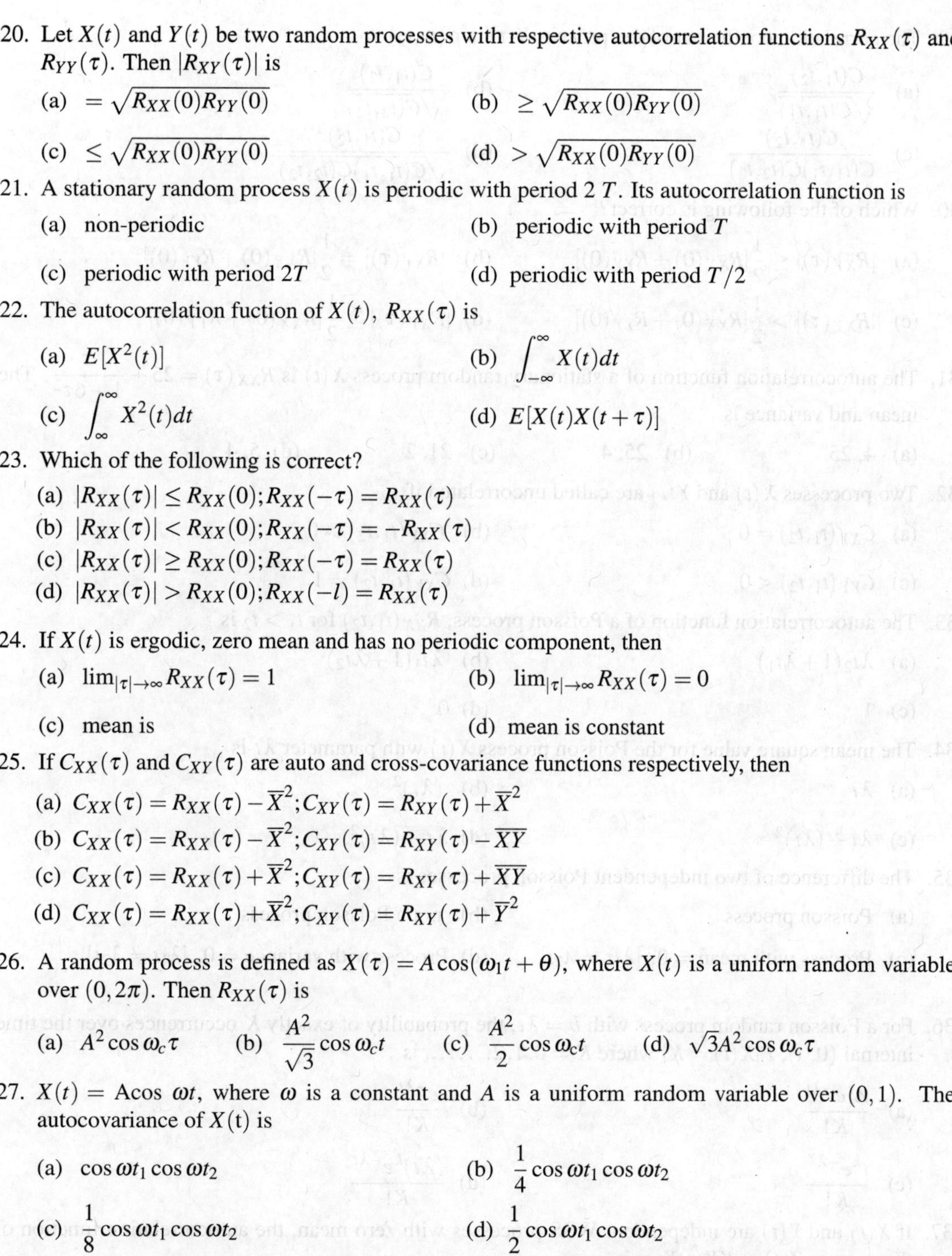

20. Let $X(t)$ and $Y(t)$ be two random processes with respective autocorrelation functions $R_{XX}(\tau)$ and $R_{YY}(\tau)$. Then $|R_{XY}(\tau)|$ is
(a) $= \sqrt{R_{XX}(0)R_{YY}(0)}$ (b) $\geq \sqrt{R_{XX}(0)R_{YY}(0)}$
(c) $\leq \sqrt{R_{XX}(0)R_{YY}(0)}$ (d) $> \sqrt{R_{XX}(0)R_{YY}(0)}$

21. A stationary random process $X(t)$ is periodic with period 2 T. Its autocorrelation function is
(a) non-periodic (b) periodic with period T
(c) periodic with period $2T$ (d) periodic with period $T/2$

22. The autocorrelation fuction of $X(t)$, $R_{XX}(\tau)$ is
(a) $E[X^2(t)]$ (b) $\int_{-\infty}^{\infty} X(t)dt$
(c) $\int_{\infty}^{\infty} X^2(t)dt$ (d) $E[X(t)X(t+\tau)]$

23. Which of the following is correct?
(a) $|R_{XX}(\tau)| \leq R_{XX}(0); R_{XX}(-\tau) = R_{XX}(\tau)$
(b) $|R_{XX}(\tau)| < R_{XX}(0); R_{XX}(-\tau) = -R_{XX}(\tau)$
(c) $|R_{XX}(\tau)| \geq R_{XX}(0); R_{XX}(-\tau) = R_{XX}(\tau)$
(d) $|R_{XX}(\tau)| > R_{XX}(0); R_{XX}(-l) = R_{XX}(\tau)$

24. If $X(t)$ is ergodic, zero mean and has no periodic component, then
(a) $\lim_{|\tau|\to\infty} R_{XX}(\tau) = 1$ (b) $\lim_{|\tau|\to\infty} R_{XX}(\tau) = 0$
(c) mean is (d) mean is constant

25. If $C_{XX}(\tau)$ and $C_{XY}(\tau)$ are auto and cross-covariance functions respectively, then
(a) $C_{XX}(\tau) = R_{XX}(\tau) - \overline{X}^2; C_{XY}(\tau) = R_{XY}(\tau) + \overline{X}^2$
(b) $C_{XX}(\tau) = R_{XX}(\tau) - \overline{X}^2; C_{XY}(\tau) = R_{XY}(\tau) - \overline{XY}$
(c) $C_{XX}(\tau) = R_{XX}(\tau) + \overline{X}^2; C_{XY}(\tau) = R_{XY}(\tau) + \overline{XY}$
(d) $C_{XX}(\tau) = R_{XX}(\tau) + \overline{X}^2; C_{XY}(\tau) = R_{XY}(\tau) + \overline{Y}^2$

26. A random process is defined as $X(\tau) = A\cos(\omega_1 t + \theta)$, where $X(t)$ is a uniforn random variable over $(0, 2\pi)$. Then $R_{XX}(\tau)$ is
(a) $A^2 \cos\omega_c\tau$ (b) $\frac{A^2}{\sqrt{3}}\cos\omega_c t$ (c) $\frac{A^2}{2}\cos\omega_c t$ (d) $\sqrt{3}A^2\cos\omega_c\tau$

27. $X(t) =$ Acos ωt, where ω is a constant and A is a uniform random variable over $(0,1)$. The autocovariance of X(t) is
(a) $\cos\omega t_1 \cos\omega t_2$ (b) $\frac{1}{4}\cos\omega t_1 \cos\omega t_2$
(c) $\frac{1}{8}\cos\omega t_1 \cos\omega t_2$ (d) $\frac{1}{2}\cos\omega t_1 \cos\omega t_2$

28. The autocovariance $C(t_1, t_2)$ of a process $X(t)$ is (assume that $E[X(t)] = \eta(t)$)
(a) $C(t_1,\ t_2) = R(t_1,\ t_2)$ (b) $C(t_1,\ t_2) = R(t_1,\ t_2) - |\eta(t)|^2$
(c) $C(t_1,\ t_2) = R_2(t_1,\ t_2) + |\eta(t)|^2$ (d) $C(t_1,\ t_2) = R_2(t_1,\ t_2) - |\eta(t)|^2$

29. The correlation coefficient of the process $X(t)$ is $\rho(t_1,t_2) =$

(a) $\frac{C(t_1,t_2)}{\sqrt{C(t_1,t_1)}}$ (b) $\frac{C(t_1,t_2)}{\sqrt{C(t_2,t_2)}}$

(c) $\frac{C(t_1,t_2)}{C(t_1,t_1)C(t_2,t_2)}$ (d) $\frac{C(t_1,t_2)}{\sqrt{C(t_1,t_1)C(t_2,t_2)}}$

30. Which of the following is correct?

(a) $|R_{XY}(\tau)| \leq \frac{1}{2}[R_{XX}(0)+R_{YY}(0)]$ (b) $|R_{XY}(\tau)| = \frac{1}{2}[R_{XX}(0)+R_{YY}(0)]$

(c) $|R_{XY}(\tau)| > \frac{1}{2}[R_{XX}(0)+R_{YY}(0)]$ (d) $|R_{XY}(\tau)| \geq \frac{1}{2}[R_{XX}(0)+R_{YY}(0)]$

31. The autocorrelation function of a stationary random process $X(t)$ is $R_{XX}(\tau) = 25 + \frac{\tau}{1+6\tau^2}$. The mean and variance is

(a) 4, 25 (b) 25, 4 (c) 21, 2 (d) 5, 4

32. Two processes $X(t)$ and $Y(t)$ are called uncorrelated if

(a) $C_{XY}(t_1,t_2) = 0$ (b) $C_{XY}(t_1,t_2) > 0$

(c) $C_{YY}(t_1,t_2) < 0$ (d) $C_{XY}(t_1,t_2) = 1$

33. The autocorrelation function of a Poisson process, $R_{XX}(t_1,t_2)$ for $t_1 > t_2$ is

(a) $\lambda t_2(1+\lambda t_1)$ (b) $\lambda t_1(1+\lambda t_2)$

(c) 1 (d) 0

34. The mean square value for the Poisson process $X(t)$ with parameter λt is

(a) λt (b) $(\lambda t)^2$

(c) $\lambda t + (\lambda t)^2$ (d) $\lambda t \cdot (\lambda t)^2$

35. The difference of two independent Poisson processes is

(a) Poisson process (b) not a Poisson process

(c) Process with mean $= 0$, $[\lambda_1 t \neq \lambda_2 t]$ (d) Process with variance $= 0$, $[\lambda_1 t \neq \lambda_2 t]$

36. For a Poisson random process with $b = \lambda t$, the probability of exactly K occurrences over the time internal (0, t), $P[X(t) = K]$ where $K = 0, 1, 2, \ldots\ldots$, is

(a) $\frac{\lambda^t e^{-\lambda t}}{K!}$ (b) $\frac{e^{-\lambda t}}{K!}$

(c) $\frac{t^k e^{-\lambda \gamma}}{K!}$ (d) $\frac{(\lambda t)^k e^{-\lambda t}}{K!}$

37. If $X(t)$ and $Y(t)$ are independent WSS processes with zero mean, the autocorrelation function of the process $Z(t) = KX(t)Y(t)$ is

(a) $R_{XX}(\tau)R_{YY}(\tau)$ (b) $KR_{XX}(\tau)R_{YY}(\tau)$

(c) $K^2 R_{XX}(\tau)R_{YY}(\tau)$ (d) $\frac{R_{XX}(\tau)R_{YY}(\tau)}{K}$

38. The mean square value of a random process whose autocorrelation function $\frac{\eta}{4RC}e^{|-\tau|/RC}$ is

(a) $\frac{\eta}{4}$ (b) $\frac{\eta}{4RC}$ (c) $\frac{\eta\tau}{4RC}$ (d) $\frac{\eta|\tau|}{4RC}$

39. A random process $X(t)$ has $R_{XX}(\tau) = A^2 + Be^{-|\tau|}$, where A and B are positive constants. The mean value of the response of a system having an impulse response $h(t) = e^{-kt}$ for $t > 0$, where K is a real positive constant, for which $X(t)$ is its input is

(a) $\frac{A}{K}$ (b) AK (c) AK^2 (d) A

40. $X(t)$ is a WSS process with zero mean. It is the input of an LTI system with $H(\omega) = \frac{1}{j\omega+2}$. If $R_{XX}(\tau) = e^{-2|\tau|}$, the area enclosed by the autocorrelation function of output process is

(a) 1 (b) $\frac{1}{2}$ (c) $\frac{1}{16}$ (d) $\frac{1}{4}$

41. If $R_{XX}(\tau) = 3\delta(\tau)$ and $H(\omega) = \frac{1}{6+j\omega}$, then the mean square value of output is

(a) 3 (b) 4 (c) $\frac{1}{3}$ (d) $\frac{1}{4}$

42. The cross-correlation between $X(t)$ and $Y(t)$ is $R_{XY}(\tau) =$

(a) $h(\tau)^* R_{XX}(\tau)$ (b) $h(-\tau)^* R_{XX}(\tau)$

(c) $h(-\tau)^* R_{XY}(\tau)$ (d) $h(\tau)^* R_{YX}(\tau)$

43. A random process $X(t)$ of mean 3 is applied to a delay element. The mean of the output process is

(a) 2 (b) 3 (c) 1.5 (d) 9

Answers

1. (d)	2. (c)	3. (d)	4. (b)	5. (a)	6. (b)	7. (a)	8. (c)	9. (c)	10. (b)
11. (c)	12. (b)	13. (b)	14. (d)	15. (d)	16. (c)	17. (a)	18. (c)	19. (a)	20. (c)
21. (c)	22. (d)	23. (a)	24. (b)	25. (b)	26. (c)	27. (d)	28. (b)	29. (d)	30. (a)
31. (a)	32. (a)	33. (a)	34. (c)	35. (b)	36. (d)	37. (c)	38. (b)	39. (a)	40. (d)
41. (d)	42. (a)	43. (b)							

10

Random Processes — Spectral Characteristics

Learning Objectives

The main objective of this chapter is to understand thoroughly the concepts of power density spectrum, cross-power spectrum and their properties. We will learn how the spectrum is processed through linear system. By reading this chapter, students will be able to relate to the following topics:

- Power density spectrum and its properties
- Relationship between power spectrum and autocorrelation function
- Cross-power spectrum and its properties
- Relationship between cross-power spectrum and cross-correlation function
- Spectral characteristics of system response
- Power density spectrum of response
- Cross-power spectral density of input and output of a linear system

10.1 Power Spectrum

In the previous chapter the characteristics of random processes regarding correlation and covariance functions defined in the time domain were studied. This chapter explores the important concept of characterising random processes in the frequency domain. These characteristics are called spectral properties. The concepts depend heavily on the theory of Fourier transforms.

10.1.1 Power Density Spectrum

Consider a random process $X(t)$. The amplitude of the random process, when it varies randomly with time, does not satisfy Dirichlet's conditions. Therefore, it is not possible to apply Fourier transform directly on the random process for a frequency domain analysis. Hence the autocorrelation function of a WSS random process is used to study spectral characteristics such as power density spectrum or power spectral density (psd).

Definition 1

The power spectrum density (power spectral density function) of a wide sense stationary random process $X(t)$ is defined as the Fourier transform of the autocorrelation function $R_{XX}(\tau)$ of $X(t)$. It can be expressed as

$$S_{XX}(\omega) = \int_{-\infty}^{\infty} R_{XX}(\tau)e^{-j\omega\tau}d\tau. \tag{10.1}$$

We can obtain the autocorrelation function from the power spectral density by taking the the inverse Fourier transform, i.e.,

$$R_{XX}(\tau) = \frac{1}{2\pi}\int_{-\infty}^{\infty} S_{XX}(\omega)e^{j\omega\tau}d\omega. \tag{10.2}$$

Therefore, the power density spectrum $S_{XX}(\omega)$ and the autocorrelation function $R_{XX}(\tau)$ are Fourier transform pairs.

Note: If $X(t)$ is a real process, then $R_{XX}(\tau)$ and $S_{XX}(\omega)$ are real and even functions.

Definition 2

The power spectral density can also be defined as,

$$S_{XX}(\omega) = \lim_{T\to\infty} \frac{E\left[|X_T(\omega)|^2\right]}{2T}, \tag{10.3}$$

where $X_T(\omega)$ is a Fourier transform of $X(t)$ in the interval $[-T,T]$.

10.1.2 Average Power of the Random Process

The average power P_{XX} of a WSS random process $X(t)$ is defined as the time average of its second moment or autocorrelation function at $\tau = 0$.

Mathematically,

$$P_{XX} = A\left\{E\left[X^2(t)\right]\right\} \tag{10.4}$$

$$P_{XX} = \lim_{T\to\infty} \frac{1}{2T}\int_{-T}^{T} E\left[X^2(t)\right]dt$$

or $$P_{XX} = R_{XX}(\tau)|_{\tau=0}. \tag{10.5}$$

We know that from the power density spectrum,

$$R_{XX}(\tau) = \frac{1}{2\pi}\int_{-\infty}^{\infty} S_{XX}(\omega)e^{j\omega\tau}d\omega.$$

At $\tau = 0$

$$P_{XX} = R_{XX}(0) = \frac{1}{2\pi}\int_{-\infty}^{\infty} S_{XX}(\omega)d\omega.$$

Therefore, the average power of $X(t)$ is

$$P_{XX} = \frac{1}{2\pi}\int_{-\infty}^{\infty} S_{XX}(\omega)d\omega. \tag{10.6}$$

Proof: Consider a WSS random process $X(t)$.
To obtain the Fourier transform, let $X_T(t)$ be defined in the interval $[-T,T]$, that is,

$$X_T(t) = X(t) \quad -T < t < T.$$

The Fourier transform is

$$X_T(\omega) = \int_{-\infty}^{\infty} X_T(t)e^{-j\omega t}dt = \int_{-T}^{T} X(t)e^{-j\omega t}dt.$$

Now, using Parseval's theorem, the energy contained in $X_T(t)$ is

$$\text{Energy} = \int_{-T}^{T} X_T^2(t)dt = \frac{1}{2\pi}\int_{-\infty}^{\infty} |X_T(\omega)|^2 d\omega. \tag{10.7}$$

The average power over the interval $[-T,T]$ is

$$P(T) = \frac{1}{2T}[\text{energy}]$$

$$P(T) = \frac{1}{2T}\int_{-T}^{T} X_T^2(t)dt = \frac{1}{2\pi}\int_{-T}^{T} \frac{|X_T(\omega)|^2}{2T}d\omega. \tag{10.8}$$

The average power of the random process $X(t)$ in the entire time interval $-\infty < T \leq \infty$ is equal to the mean value of $P(T)$ as $T \to \infty$.

$$\therefore \quad P_{XX} = \lim_{T\to\infty} \frac{1}{2T}\int_{-T}^{T} E[X^2(t)]dt = \frac{1}{2\pi}\int_{-\infty}^{\infty} \lim_{T\to\infty} \frac{E\left[|X_T(\omega)|^2\right]d\omega}{2T}. \tag{10.9}$$

Now,

$$P_{XX} = \lim_{T\to\infty} \frac{1}{2T}\int_{-T}^{T} E[X^2(t)]dt$$

$$P_{XX} = A\{E[X^2(t)]\} \quad \text{(Proved)}. \tag{10.10}$$

Thus the average power P_{XX} is the time average of its second moment.

Also, $$P_{XX} = \frac{1}{2\pi}\int_{-\infty}^{\infty} \lim_{T\to\infty} \frac{E\left[|X_T(\omega)|^2\right]}{2T}d\omega. \tag{10.11}$$

If power density spectrum is defined as

$$S_{XX}(\omega) = \lim_{T\to\infty} \frac{E[|X_T(\omega)|^2]}{2T}, \tag{10.12}$$

then $$P_{XX} = \frac{1}{2\pi}\int_{-\infty}^{\infty} S_{XX}(\omega)d\omega \tag{10.13}$$

$$P_{XX} = R_{XX}(0) \quad \text{(Proved)}.$$

The average power P_{XX} is the autocorrelation of $X(t)$ at the origin.

10.2 Properties of the Power Density Spectrum

The properties of the power density spectrum $S_{XX}(\omega)$ for a wide sense stationary random process $X(t)$ are given below.

(1) $S_{XX}(\omega) \geq 0$

Proof: From the definition, the expected value of a non-negative function $E\left[|X_T(\omega)|^2\right]$ is always non-negative.

Hence $S_{XX}(\omega) \geq 0.$

(2) The power spectral density at zero frequency is equal to the area under the curve of the autocorrelation $R_{XX}(\tau)$.

$$\text{That is, } S_{XX}(0) = \int_{-\infty}^{\infty} R_{XX}(\tau)d\tau \tag{10.14}$$

Proof: From the definition, we know that

$$S_{XX}(\omega) = \int_{-\infty}^{\infty} R_{XX}(\tau)e^{-jw\tau}d\tau.$$

At $\omega = 0$,

$$S_{XX}(0) = \int_{-\infty}^{\infty} R_{XX}(\tau)d\tau.$$

(3) The power density spectrum of a real process $X(t)$ is an even function, i.e.,

$$S_{XX}(-\omega) = S_{XX}(\omega). \tag{10.15}$$

Proof: Consider a WSS real process $X(t)$.

$$\text{Then, } S_{XX}(\omega) = \int_{-\infty}^{\infty} R_{XX}(\tau)e^{-j\omega\tau}d\tau.$$

$$\text{Also, } S_{XX}(-\omega) = \int_{-\infty}^{\infty} R_{XX}(\tau)e^{j\omega\tau}d\tau.$$

Substituting $\tau = -\tau$,

$$S_{XX}(-\omega) = \int_{-\infty}^{\infty} R_{XX}(-\tau)e^{-j\omega\tau}d\tau.$$

Since $X(t)$ is real, from the properties of autocorrelation of Eq. (9.46) we know that,

$$R_{XX}(-\tau) = R_{XX}(\tau).$$

$$S_{XX}(-\omega) = \int_{-\infty}^{\infty} R_{XX}(\tau)e^{-j\omega\tau}d\tau = S_{XX}(\omega).$$

$$\therefore \quad S_{XX}(\omega) \text{ is an even function.} \tag{10.16}$$

(4) $S_{XX}(\omega)$ is always a real function.

Proof: From Eq. (10.3) we know that

$$S_{XX}(\omega) = \lim_{T\to\infty} \frac{E\left[|X_T(\omega)|^2\right]}{2T}.$$

Since the function $|X_T(\omega)|^2$ is a real function, $S_X(\omega)$ is always real.

(5) If $S_{XX}(\omega)$ is a psd of the WSS random process $X(t)$, then

$$\frac{1}{2\pi}\int_{-\infty}^{\infty} S_{XX}(\omega)d\omega = A\left\{E\left[X^2(t)\right]\right\} = R_{XX}(0)$$

(or) the time average of the mean square value of a WSS random process equals the area under the curve of the power spectral density.

Proof: From Eq. (9.34) we know that

$$R_{XX}(\tau) = A\{E[X(t+\tau)X(t)]\} = \frac{1}{2\pi}\int_{-\infty}^{\infty} S_{XX}(\omega)e^{j\omega\tau}d\omega. \tag{10.17}$$

Now at $\tau = 0$, $\quad R_{XX}(0) = A\{E[X^2(t)]\} = \dfrac{1}{2\pi}\displaystyle\int_{-\infty}^{\infty} S_{XX}(\omega)d\omega$

$=$ Area under the curve of $S_{XX}(\omega)$.

(6) If $X(t)$ is a WSS random process with psd $S_{XX}(\omega)$, then the psd of the derivative of $X(t)$ is equal to ω^2 times the psd $S_{XX}(\omega)$. That is,

$$S_{\dot{X}\dot{X}}(\omega) = \omega^2 S_{XY}(\omega). \tag{10.18}$$

Proof: From Eq. (10.3), we know that

$$S_{XX}(\omega) = \lim_{T\to\infty}\frac{E\left[|X_T(\omega)|^2\right]}{2T}$$

and $\quad X_T(\omega) = \displaystyle\int_{-T}^{T} X(t)e^{-j\omega t}dt.$

$$\dot{X}_T(\omega) = \frac{dX_T(\omega)}{dt} = \frac{d}{dt}\int_{-T}^{T} X(t)e^{-j\omega t}dt$$

$$= \int_{-T}^{T} X(t)\frac{d}{dt}e^{-j\omega t}dt = \int_{-T}^{T} X(t)(-j\omega)e^{-j\omega t}dt$$

$$= (-j\omega)\int_{-T}^{T} X(t)e^{-j\omega t}dt.$$

$\therefore \quad \dot{X}_T(\omega) = (-j\omega)X_T(\omega).$

$$S_{\dot{X}\dot{X}}(\omega) = \lim_{T\to\infty}\frac{E\left[\left|\dot{X}_T(\omega)\right|^2\right]}{2T}$$

$$= \lim_{T\to\infty}\frac{E\left[|(-j\omega)X_T(\omega)|^2\right]}{2T} = \lim_{T\to\infty}\frac{\omega^2 E\left[|X_T(\omega)|^2\right]}{2T}$$

$$S_{\dot{X}\dot{X}}(\omega) = \omega^2\lim_{T\to\infty}\frac{E\left[|X_T(\omega)|^2\right]}{2T} = \omega^2 S_{XX}(\omega) \qquad \text{(Proved).}$$

10.3 Relationship between Power Spectrum and Autocorrelation Function

The power density spectrum and the time average of the autocorrelation function form a Fourier transform pair.

That is, $$S_{XX}(\omega) = \int_{-\infty}^{\infty} R_{XX}(\tau)e^{-j\omega\tau}d\tau \quad (10.19)$$

and $$R_{XX}(\tau) = \frac{1}{2\pi}\int_{-\infty}^{\infty} S_{XX}(\omega)e^{j\omega\tau}d\tau. \quad (10.20)$$

Proof: Consider a WSS random process $X(t)$. We know that the time average of the autocorrelation function is

$$A[R_{XX}(t,t+\tau)] = A\{E[X(t)X(t+\tau)]\}.$$

For a WSS random process,

$$A[R_{XX}(t,t+\tau)] = R_{XX}(\tau).$$

Now, the power spectral density is

$$S_{XX}(\omega) = \lim_{T\to\infty} \frac{E\left[|X_T(\omega)|^2\right]}{2T}.$$

But we know that $|X_T(\omega)|^2 = X_T(\omega)X_T^*(\omega)$.

Let $$X_T(\omega) = \int_{-T}^{T} X(t_1)e^{-j\omega t_1}dt_1$$

and $$X_T^*(\omega) = \int_{-T}^{T} X(t)e^{j\omega t}dt.$$

$$\therefore \quad S_{XX}(\omega) = \lim_{T\to\infty} E\left[\frac{1}{2T}\int_{-T}^{T} X(t)e^{j\omega t}dt \int_{-T}^{T} X(t_1)e^{-j\omega t_1}dt_1\right]$$

$$S_{XX}(\omega) = \lim_{T\to\infty} \frac{1}{2T}\int_{-T}^{T}\int_{-T}^{T} E[X(t_1)X(t)]e^{-j\omega(t_1-t)}dtdt_1.$$

We know that $E[X(t_1)X(t)] = R_{XX}(t,t_1)$

$$\therefore \quad S_{XX}(\omega) = \lim_{T\to\infty} \frac{1}{2T}\int_{-T}^{T}\int_{-T}^{T} R_{XX}(t,t_1)e^{-j\omega(t_1-t)}dt_1dt. \quad (10.21)$$

By taking inverse Fourier transform,

$$\frac{1}{2\pi}\int_{-\infty}^{\infty} S_{XX}(\omega)e^{j\omega\tau}d\omega$$

$$= \lim_{T\to\infty} \frac{1}{2T}\int_{T}^{T}\int_{-T}^{T} R_{XX}\,(t,t_1)\left(\frac{1}{2\pi}\right)\int_{-\infty}^{\infty} e^{j\omega\tau}e^{-j\omega(t_1-t)}d\omega dtdt_1$$

$$= \lim_{T\to\infty} \frac{1}{2T}\int_{-T}^{T}\int_{-T}^{T} R_{XX}\,(t,t_1)\left(\frac{1}{2\pi}\right)\int_{-\infty}^{\infty} e^{-j\omega(t_1-t-\tau)}d\omega dtdt_1$$

$$= \lim_{T\to\infty} \frac{1}{2T} \int_{-T}^{T}\int_{-T}^{T} R_{XX}(t,t_1)\delta(t_1 - t - \tau)dtdt_1.$$

Since $\int_{-\infty}^{\infty} e^{-j\omega(t_1-t-\tau)}d\omega = 2\pi\delta(t_1 - t - \tau)$,

$$\therefore \frac{1}{2\pi}\int_{-\infty}^{\infty} S_{XX}(\omega)e^{j\omega\tau}d\omega = \lim_{T\to\infty}\frac{1}{2T}\int_{-T}^{T} R_{XX}(t,t_1)\int_{-T}^{T}\delta(t_1 - t - \tau)dt_1 dt. \quad (10.22)$$

Since $\int_{-T}^{T}\delta(t_1 - t - \tau)dt_1 = 1$, let $t_1 = t + \tau$.

So, $\frac{1}{2\pi}\int_{-\infty}^{\infty} S_{XX}(\omega)e^{j\omega\tau}d\tau = \lim_{T\to\infty}\frac{1}{2T}\int_{-T}^{T} R_{XX}(t,t+\tau)dt$

$$= A[R_{XX}(t,t+\tau)].$$

$$\therefore \quad A[R_{XX}\,(t,t+\tau)] = \frac{1}{2\pi}\int_{-\infty}^{\infty} S_{XX}(\omega)e^{j\omega\tau}d\tau. \quad (10.23)$$

Taking Fourier transform,

$$S_{XX}(\omega) = \int_{-\infty}^{\infty} A[R_{XX}(t,t+\tau)]e^{-j\omega\tau}d\tau.$$

For a WSS, $R_{XX}(t,\ t+\tau) = R_{XX}(\tau)$.

$\therefore$ $R_{XX}(\tau)$ and $S_{XX}(\omega)$ are a Fourier transform pair.

Note 1: For a WSS random process $X(t)$, the Fourier transform pair

$$S_{XX}(\omega) = \int_{-\infty}^{\infty} R_{XX}(\tau)e^{-j\omega\tau}d\tau \text{ and } R_{XX}(\tau) = \frac{1}{2\pi}\int_{-\infty}^{\infty} S_{XX}(\omega)e^{j\omega\tau}d\omega$$

is called Wiener–Khintchine relations.

Note 2: The knowledge of the power spectral density of a process $X(t)$ gives the complete form of autocorrelation function when $X(t)$ is at least wide sense stationary. For a non-stationary random process, we can get only the time average of the autocorrelation function.

Example 10.1

Determine which of the following functions are valid power density spectrums and why.

(a) $\frac{\cos 8(\omega)}{2+\omega^4}$ (b) $e^{-(\omega-1)^2}$ (c) $\frac{\omega^2}{\omega^6+3\omega^2+3}$

Solution

(a) Given $S_{XX}(\omega) = \frac{\cos 8(\omega)}{2+\omega^4}$.

From the properties of psd,

(i) The given function $S_{XX}(\omega)$ is real and positive;

(ii) $S_{XX}(-\omega) = \dfrac{\cos 8(-\omega)}{2+(-\omega)^4} = \dfrac{\cos 8\omega}{2+\omega^4} = S_{XX}(\omega).$

The function is even. Hence the given function is a valid psd.

(b) Given $S_{XX}(\omega) = e^{-(\omega-1)^2}$. From the properties of psd,

$$S_{XX}(-\omega) = e^{-(\omega-1)^2} = e^{-(-\omega-1)^2} = e^{-(\omega+1)^2}.$$

$$\therefore \quad S_{XX}(-\omega) \neq S_{XX}(\omega).$$

Therefore, the given function is not a valid psd.

(c) Given $S_{XX}(\omega) = \dfrac{\omega^2}{\omega^6+3\omega^2+3}$. From the properties of psd,

(i) The given function is real and positive.

(ii) $$S_{XX}(-\omega) = \frac{(-\omega)^2}{(-\omega)^6+3(-\omega)^2+3} = \frac{\omega^2}{\omega^6+3\omega^2+3}.$$

$\therefore \quad S_{XX}(-\omega) = S_{XX}(\omega)$ is an even function. Therefore, the given function is a valid psd.

Bandwidth of the Power Density Spectrum

Baseband Process

Assume that $X(t)$ is a lowpass process with its spectral components clustered at the origin, $\omega = 0$. As the frequency increases, the spectrum magnitude decreases as shown in Fig. 10.1. The normalisation of a power spectrum is a measure of its spread. It is called *rms bandwidth.*

The rms bandwidth is obtained from

$$W_{\text{rms}}^2 = \frac{\int_{-\infty}^{\infty} \omega^2 S_{XX}(\omega)d\omega}{\int_{-\infty}^{\infty} S_{XX}(\omega)d\omega}. \tag{10.24}$$

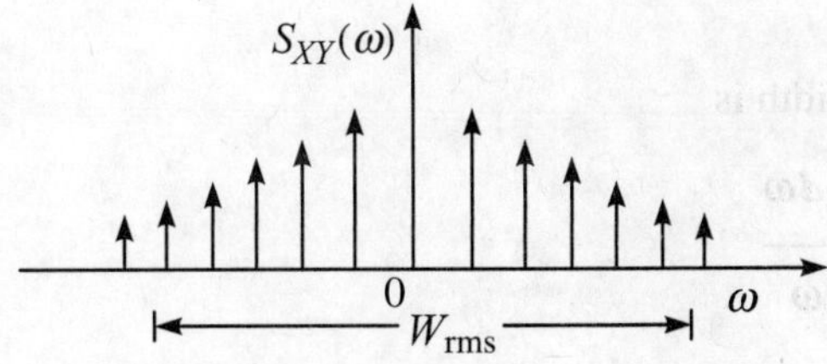

Fig. 10.1 *Baseband process*

Bandpass Process

In a bandpass process, the spectral components are clustered near some frequencies $\overline{\omega}_0$ and $-\overline{\omega}_0$ as shown in Fig. 10.2. The mean frequency $\overline{\omega}_0$ can be expressed as

$$\overline{\omega}_0 = \frac{\int_0^{\infty} \omega^2 S_{XX}(\omega) d\omega}{\int_0^{\infty} S_{XX}(\omega) d\omega} \tag{10.25}$$

and the rms bandwidth is

$$W_{\text{rms}}^2 = \frac{4\int_0^{\infty} (\omega - \overline{\omega}_0)^2 S_{XX}(\omega) d\omega}{\int_0^{\infty} S_{XX}(\omega) d\omega} \tag{10.26}$$

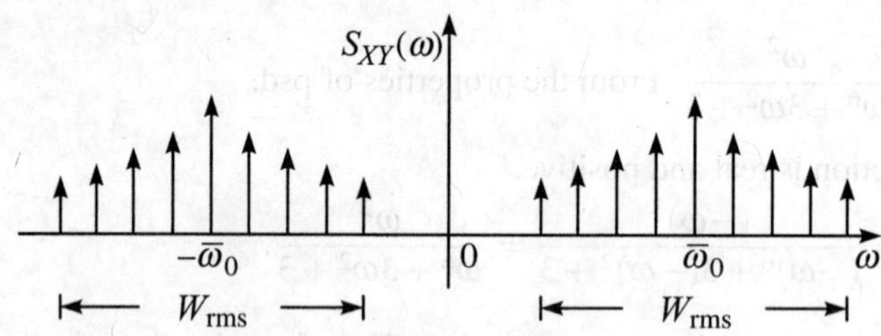

Fig. 10.2 *Spectrum of a bandpass process*

Example 10.2

The power density spectrum of a baseband random process $X(t)$ is

$$S_{XX}(\omega) = \frac{2}{\left[1 + \left(\frac{\omega}{2}\right)^2\right]^2}$$

Find the rms bandwidth.

Solution

Given power density spectrum

$$S_{XX}(\omega) = \frac{2}{\left[1 + \left(\frac{\omega}{2}\right)^2\right]^2}$$

We know that the rms bandwidth is

$$W_{rms}^2 = \frac{\int_{-\infty}^{\infty} \omega^2 S_{XX}(\omega) d\omega}{\int_{-\infty}^{\infty} S_{XX}(\omega) d\omega}$$

Now,

$$\int_{-\infty}^{\infty} \omega^2 S_{XX}(\omega) d\omega = \int_{-\infty}^{\infty} \frac{2\omega^2}{\left(1 + \frac{\omega^2}{4}\right)^2} d\omega$$

$$= \int_{-\infty}^{\infty} \frac{32\omega^2}{(4 + \omega^2)^2} d\omega$$

$$= 32\int_{-\infty}^{\infty} \frac{\omega^2}{(4+\omega^2)^2} d\omega$$

$$= 32\left[\frac{-\omega}{2(4+\omega^2)} + \frac{1}{2\times 2}\tan^{-1}\left(\frac{\omega}{2}\right)\right]_{-\infty}^{\infty}$$

$$= 32\left[0+\frac{\pi}{4}\right] = 8\pi$$

and

$$\int_{-\infty}^{\infty} S_{XX}(\omega)d\omega = \int_{-\infty}^{\infty} \frac{2}{\left(1+\frac{\omega^2}{4}\right)^2} d\omega$$

$$= \int_{-\infty}^{\infty} \frac{32}{(4+\omega^2)^2} d\omega$$

$$= 32\left[\frac{\omega}{8(4+\omega^2)} + \frac{1}{2\times 8}\tan^{-1}\left(\frac{\omega}{2}\right)\right]_{-\infty}^{\infty}$$

$$= 32\left[0+\frac{\pi}{16}\right] = 2\pi$$

$$\therefore \quad W_{\text{rms}}^2 = \frac{8\pi}{2\pi} = 4$$

$$W_{\text{rms}} = \sqrt{4} = 2 \text{ rad/s}$$

10.4 Cross Power Density Spectrum

Definition 1

Consider two real random processes $X(t)$ and $Y(t)$. If $X(t)$ and $Y(t)$ are jointly wide sense stationary random processes, then the cross power density spectrum is defined as the Fourier transform of the cross correlation function of $X(t)$ and $Y(t)$. It is expressed as

$$S_{XY}(\omega) = \int_{-\infty}^{\infty} R_{XY}(\tau)e^{-j\omega\tau}d\tau \tag{10.27}$$

$$\text{and} \quad S_{YX}(\omega) = \int_{-\infty}^{\infty} R_{YX}(\tau)e^{-j\omega\tau}d\tau. \tag{10.28}$$

By taking inverse Fourier transform, we can obtain the cross correlation functions:

$$R_{XY}(\tau) = \frac{1}{2\pi}\int_{-\infty}^{\infty} S_{XY}(\omega)e^{j\omega\tau}d\omega \tag{10.29}$$

$$\text{and} \quad R_{YX}(\tau) = \frac{1}{2\pi}\int_{-\infty}^{\infty} S_{YX}(\omega)e^{j\omega\tau}d\omega. \tag{10.30}$$

Therefore, the cross psd and cross correlation functions are a Fourier transform pair.

Definition 2

If $X_T(\omega)$ and $Y_T(\omega)$ are the Fourier transforms of $X(t)$ and $Y(t)$ respectively in the interval $[-T,T]$, then the cross power density spectrum is defined as

$$S_{XY}(\omega) = \lim_{T\to\infty} \frac{E[X_T^*(\omega)Y_T(\omega)]}{2T} \tag{10.31}$$

and $$S_{YX}(\omega) = \lim_{T\to\infty} \frac{E[Y_T^*(\omega)X_T(\omega)]}{2T}. \tag{10.32}$$

10.4.1 Average Cross Power

The average cross power P_{XY} of the WSS random processes $X(t)$ and $Y(t)$ is defined as the cross correlation function at $\tau = 0$. That is,

$$P_{XY} = \lim_{T\to\infty} \frac{1}{2T}\int_{-T}^{T} R_{XY}(t,t)dt$$

or $$P_{XY} = R_{XY}(\tau)|_{\tau=0} = R_{XY}(0). \tag{10.33}$$

Also, $$P_{XY} = \frac{1}{2\pi}\int_{-\infty}^{\infty} S_{XY}(\omega)d\omega$$

and $$P_{YX} = \frac{1}{2\pi}\int_{-\infty}^{\infty} S_{YX}(\omega)d\omega. \tag{10.34}$$

Proof: Consider two real random processes $X(t)$ and $Y(t)$. Let $X_T(t)$ and $Y_T(t)$ be defined in the interval $[-T,T]$ of $X(t)$ and $Y(t)$ respectively. That is,

$$X_T(t) = X(t),\ -T < t \le T$$

and $$Y_T(t) = Y(t),\ -T < t \le T.$$

Using Fourier transforms,

$$X_T(\omega) = \int_{-\infty}^{\infty} X_T(t)e^{-j\omega t}dt = \int_{-T}^{T} X(t)e^{-j\omega t}dt$$

and $$Y_T(\omega) = \int_{-\infty}^{\infty} Y_T(t)e^{-j\omega t}dt = \int_{-T}^{T} Y(t)e^{-j\omega t}dt.$$

Now, the cross power in the two processes $X(t)$ and $Y(t)$ within the interval $-T < t \le T$ is defined as,

$$P_{XY}(T) = \frac{1}{2T}\int_{-\infty}^{\infty} X_T(t)Y_T(t)dt = \frac{1}{2T}\int_{-T}^{T} X(t)Y(t)dt.$$

Applying Parseval's theorem,

$$P_{XY}(T) = \frac{1}{2T}\int_{-T}^{T} X(t)Y(t)dt = \frac{1}{2\pi}\int_{-\infty}^{\infty} \frac{X_T^*(\omega)Y_T(\omega)}{2T}d\omega. \tag{10.35}$$

The average cross power is

$$\overline{P}_{XY} = E[P_{XY}(T)].$$

The total average cross power is

$$P_{XY} = \lim_{T\to\infty} E[P_{XY}(T)] = \lim_{T\to\infty} \frac{1}{2T}\int_{-T}^{T} E[X(t)Y(t)]dt$$

$$P_{XY} = \lim_{T\to\infty} \frac{1}{2T}\int_{-T}^{T} R_{XY}(t,t)dt.$$

$$\therefore \quad P_{XY} = \lim_{T\to\infty} \frac{1}{2T}\int_{-T}^{T} R_{XY}(t,t)dt = \frac{1}{2\pi}\int_{-\infty}^{\infty} \lim_{T\to\infty} \frac{E[X_T^*(\omega)Y_T(\omega)]}{2T}d\omega. \tag{10.36}$$

Now from Eq. (10.31), the cross power spectral density is

$$S_{XY}(\omega) = \lim_{T\to\infty} \frac{E[X_T^*(\omega)Y_T(\omega)]}{2T}. \tag{10.37}$$

Then, $P_{XY} = \frac{1}{2\pi}\int_{-\infty}^{\infty} S_{XY}(\omega)d\omega.$

$$\therefore \quad P_{XY} = R_{XY}(0) \quad \text{(Proved)}. \tag{10.38}$$

Similarly, $P_{YX} = R_{YX}(0)$.

Since $S_{XY}(\omega) = S_{YX}^*(\omega)$, (10.39)

$$\therefore \quad P_{XY} = P_{YX}^*. \tag{10.40}$$

10.5 Properties of Cross Power Density Spectrum

The properties of the cross power spectrum for real random processes $X(t)$ and $Y(t)$ are given below.

1. $S_{XY}(\omega) = S_{YX}(-\omega) = S_{YX}^*(\omega)$. (10.42)

   Proof: Consider the cross correlation function $R_{XY}(\tau)$. The cross power density spectrum is

   $$S_{XY}(\omega) = \int_{-\infty}^{\infty} R_{XY}(\tau)e^{-j\omega\tau}d\tau.$$

   Let $\tau = -\tau$.

   Then, $S_{XY}(\omega) = \int_{-\infty}^{\infty} R_{XY}(-\tau)e^{j\omega\tau}d\tau.$

   Since $R_{XY}(-\tau) = R_{YX}(\tau)$,

   $$S_{XY}(\omega) = \int_{-\infty}^{\infty} R_{YX}(\tau)e^{-j(-\omega)\tau}d\tau.$$

   $$\therefore \quad S_{XY}(\omega) = S_{YX}(-\omega) = S_{YX}^*(\omega) \quad \text{(Proved)}.$$

2. The real part of $S_{XY}(\omega)$ and the real part of $S_{YX}(\omega)$ are even functions of ω, i.e., $\mathrm{Re}[S_{XY}(\omega)]$ and $\mathrm{Re}[S_{YX}(\omega)]$ are even functions.

   Proof: Consider $S_{XY}(\omega) = \int_{-\infty}^{\infty} R_{XY}(\tau)e^{-j\omega\tau}d\tau.$

Since we know $e^{-j\omega\tau} = \cos\omega\tau - j\sin\omega\tau$,

$$\text{Re}[S_{XY}(\omega)] = \int_{-\infty}^{\infty} R_{XY}(\tau)\cos\omega\tau d\tau.$$

Since $\cos\omega\tau$ is an even function, i.e., $\cos\omega\tau = \cos(-\omega\tau)$,

$$\text{Re}[S_{XY}(\omega)] = \int_{-\infty}^{\infty} R_{XY}(\tau)\cos\omega\tau d\tau = \int_{-\infty}^{\infty} R_{XY}(\tau)\cos(-\omega\tau)d\omega. \quad (10.43)$$

$\text{Re}[S_{XY}(\omega)] = \text{Re}[S_{XY}(-\omega)]$ is an even function.

Similarly, $\text{Re}[S_{YX}(\omega)]$ is also an even function. (10.44)

3. The imaginary part of $S_{XY}(\omega)$ and the imaginary part of $S_{YX}(\omega)$ are odd functions of ω, i.e., $\text{Im}[S_{XY}(\omega)]$ and $\text{Im}[S_{YX}(\omega)]$ are odd functions. (10.45)

Proof: Consider

$$s_{XY}(\omega) = \int_{-\infty}^{\infty} R_{XY}(\tau)e^{-j\omega\tau}d_T$$

$$S_{XY}(\omega) = \int_{-\infty}^{\infty} R_{XY}(\tau)(\cos\omega\tau - j\sin\omega\tau)d\tau.$$

$$\text{Im}[S_{XY}(\omega)] = \int_{-\infty}^{\infty} R_{XY}(\tau)(-\sin\omega\tau)d_T = -\int_{-\infty}^{\infty} R_{XY}(\tau)(\sin\omega\tau)d\tau.$$

Now,

$$\text{Im}[S_{XY}(-\omega)] = -\int_{-\infty}^{\infty} R_{XY}(\tau)\sin(-\omega\tau)d\tau = -\int_{-\infty}^{\infty} R_{XY}(\tau)(-\sin\omega\tau)d\tau$$

$$= \int_{-\infty}^{\infty} R_{XY}(\tau)\sin\omega\tau d\tau.$$

$$\text{Im}[S_{XY}(-\omega)] = -\text{Im}[S_{XY}(\omega)].$$

$\therefore$ $\text{Im}[S_{XY}(\omega)]$ is an odd function of ω.

Similarly, $\text{Im}[S_{YX}(\omega)]$ is also an odd function.

4. $S_{XY}(\omega) = 0$ and $S_{YX}(\omega) = 0$ if $X(t)$ and $Y(t)$ are orthogonal. (10.46)

Proof: From the properties of cross correlation function, we know that the random processes $X(t)$ and $Y(t)$ are said to be orthogonal if and only if their cross correlation function is zero,

i.e., $R_{XY}(\tau) = R_{YX}(\tau) = 0$.

We know that, $S_{XY}(\omega) = \int_{-\infty}^{\infty} R_{XY}(\tau)e^{-j\omega\tau}d\tau$.

$\therefore$ $S_{XY}(\omega) = 0$.

Also, $S_{YX}(\omega) = 0$.

5. If $X(t)$ and $Y(t)$ are uncorrelated and have constant mean values $\overline{X}$ and $\overline{Y}$, then

$$S_{XY}(\omega) = 2\pi\overline{XY}\delta\omega. \quad (10.47)$$

Proof: We know that

$$S_{XY}(\omega) = \int_{-\infty}^{\infty} R_{XY}(\tau)e^{-j\omega\tau}d\tau$$

$$= \int_{-\infty}^{\infty} E[X(t)Y(t+\tau)]e^{-j\omega\tau}d\tau.$$

Since $X(t)$ and $Y(t)$ are uncorrelated, we know that

$$E[X(t)Y(t+\tau)] = E[X(t)]E[Y(t+\tau)].$$

$$\therefore \quad S_{XY}(\omega) = \int_{-\infty}^{\infty} E[X(t)]E[Y(t+\tau)]e^{-j\omega\tau}d\tau$$

$$= \int_{-\infty}^{\infty} \overline{X}\overline{Y}e^{-j\omega\tau}d\tau = \overline{X}\overline{Y}\int_{-\infty}^{\infty} e^{-j\omega\tau}d\tau$$

$$S_{XY}(\omega) = \overline{X}\overline{Y}2\pi\delta(\omega).$$ (Proved).

10.6 Relationship between Cross-power Spectrum and Cross-correlation Function

The time average of the cross-correlation function and cross-power spectral density are Fourier transform pairs.

$$S_{XY}(\omega) = F\{A[R_{XY}(t,\, t+\tau)]\} \text{ and } S_{YX}(\omega) = F\{A[R_{YX}(t,\, t+\tau)]\}. \tag{10.48}$$

Also, if $X(t)$ and $Y(t)$ are jointly WSS random processes, then

$$S_{XY}(\omega) = F[R_{XY}(\tau)] \text{ and } S_{YX}(\omega) = F[R_{YX}(\tau)].$$

Proof: Consider $X(t)$ and $Y(t)$ are real random processes. Let $X_T(t)$ and $Y_T(t)$ be defined in the interval $[-T, T]$, then the Fourier transforms are

$$X_T(\omega) = \int_{-T}^{T} X(t)e^{-j\omega t}dt,$$

$$Y_T(\omega) = \int_{-T}^{T} Y(t_1)e^{-j\omega t_1}dt_1.$$

Then $$X_T(\omega)Y_T(\omega) = \int_{-T}^{T} X(t)e^{-j\omega t}dt \int_{-T}^{T} Y(t_1)e^{-j\omega t}dt_1. \tag{10.49}$$

From Eq. (10.32) we know that,

$$S_{XY}(\omega) = \lim_{T\to\infty} \frac{E[X_T^*(\omega)Y_T(\omega)]}{2T}$$

$$= \lim_{T\to\infty} \frac{1}{2T}E\left[\int_{-T}^{T} X(t)e^{j\omega t}dt \int_{-T}^{T} Y(t_1)e^{-j\omega t_1}dt_1\right]$$

$$= \lim_{T\to\infty} \frac{1}{2T}\int_{-T}^{T}\int_{-T}^{T} E[X(t)Y(t_1)]e^{j\omega t}e^{-j\omega t_1}dt dt_1$$

$$S_{XY}(\omega) = \lim_{T\to\infty} \frac{1}{2T}\int_{-T}^{T}\int_{-T}^{T} R_{XY}(t,t_1)e^{-j\omega(t_1-t)}dt dt_1. \tag{10.50}$$

Let $t_1 = t+\tau$.

$$\therefore \quad S_{XY}(\omega) = A\left[\int_{-T}^{T} R_{XY}(t,t+\tau)e^{-j\omega\tau}d\tau\right]$$

$$= \int_{-T}^{T} A\left[R_{XY}(t,t+\tau)\right]e^{-j\omega\tau}d\tau. \tag{10.51}$$

$\therefore \quad S_{XY}(\omega) =$ Fourier transform of $A\left[R_{XY}(t,t+\tau)\right]$.

Now take $S_{XY}(\omega) = \int_{-T}^{T} A\left[R_{XY}(t,t+\tau)\right]e^{-J\omega\tau}d\tau.$

Take the inverse Fourier transform:

$$\frac{1}{2\pi}\int_{-\infty}^{\infty} S_{XY}(\omega)e^{j\omega\tau_1}d\omega = \frac{1}{2\pi}\int_{-\infty}^{\infty}\int_{-T}^{T} A\left[R_{XY}(t,t+\tau)\right]e^{-j\omega\tau}e^{+j\omega\tau_1}d\tau d\omega$$

$$= \int_{-T}^{T} A\left[R_{XY}(t,t+\tau)\right]\frac{1}{2\pi}e^{j\omega(\tau_1-\tau)}d\tau d\omega$$

$$= \int_{-T}^{T} A\left[R_{XY}(t,t+\tau)\right]\delta(\tau_1-\tau)d\tau$$

$$= A\left[R_{XY}(t,t+\tau)\right]\int_{-T}^{T}\delta(\tau_1-\tau)d\tau$$

$$\frac{1}{2\pi}\int_{-\infty}^{\infty} S_{XY}(\omega)e^{j\omega\tau_1}d\omega = A\left[R_{XY}(t,t+\tau)\right].$$

$\therefore \quad A\left[R_{XY}(t,t+\tau)\right] =$ Inverse Fourier transform of $S_{XY}(\omega)$

or $\quad A[R_{XY}(t,t+\tau)] = F^{-1}[S_{XY}(\omega)]$.

Similarly, $\quad A[R_{YX}(t,t+\tau)] = F^{-1}[S_{YX}(\omega)]$. (10.52)

For jointly wide sense stationary random processes $X(t)$ and $Y(t)$, we know that

$$A[R_{XY}(t,t+\tau)] = R_{XY}(\tau) \text{ and } A[R_{YX}(t,t+\tau)] = R_{YX}(\tau). \tag{10.53}$$

$$\therefore \quad R_{XY}(\tau) = F^{-1}[S_{XY}(\omega)]$$

$$R_{YX}(\tau) = F^{-1}[S_{YX}(\omega)]$$

or $\quad S_{XY}(\omega) = F[R_{XY}(\tau)]$

and $\quad S_{YX}(\omega) = F[R_{YX}(\tau)]$. (Proved).

Example 10.3

The cross-psd is given as

$$S_{XY}(\omega) = \frac{1}{(a+j\omega)} \quad a > 0,\ a \text{ is a constant}$$

Find the cross-correlation function and average power.

Solution

Given $S_{XY}(\omega) = \dfrac{1}{(a+j\omega)}$,
the cross correlation function is

$$R_{XY}(\tau) = \text{ Inverse Fourier transform of } S_{XY}(\omega)$$

$$= F^{-1}\left[\frac{1}{(a+j\omega)}\right].$$

We know from the Fourier transform that

$$\frac{1}{(a+j\omega)} \leftrightarrow e^{-at}u(t).$$

$$\therefore \quad R_{XY}(\tau) = e^{-a\tau}u(\tau).$$

The average power is $R_{XY}(0) = 1$ watt.

10.7 Spectral Characteristics of a System Response

Consider that the random process $X(t)$ is a wide sense stationary process with the autocorrelation function $R_{XX}(\tau)$ applied through an LTI system. It is noted that the output response $Y(t)$ is also WSS and the processes $X(t)$ and $Y(t)$ are jointly WSS. We can obtain the power spectral characteristics of the output process $Y(t)$ by taking the Fourier transform of the correlation functions.

10.7.1 Power Density Spectrum of Response

Consider that a random process $X(t)$ is applied on an LTI system having a transfer function $H(\omega)$. The output response is $Y(t)$. If the power spectrum of the input process is $S_{XX}(\omega)$, then the power spectrum of the output response is given by

$$S_{YY}(\omega) = |H(\omega)|^2 S_{XX}(\omega). \tag{10.54}$$

Proof: Let $R_{YY}(\tau)$ be the autocorrelation of the output response $Y(t)$. Then the power spectrum of the response is the Fourier transform of $R_{YY}(\tau)$.

$$\therefore \quad S_{YY}(\omega) = F[R_{YY}(\tau)] = \int_{-\infty}^{\infty} R_{YY}(\tau)e^{-j\omega\tau}d\tau. \tag{10.55}$$

We know that $R_{YY}(\tau) = \displaystyle\int_{-\infty}^{\infty}\int_{-\infty}^{\infty} R_{XX}(\tau+\tau_1-\tau_2)h(\tau_1)h(\tau_2)d\tau_1 d\tau_2.$

Then $S_{YY}(\omega)=\int_{-\infty}^{\infty}\int_{-\infty}^{\infty}\int_{-\infty}^{\infty}R_{XX}(\tau+\tau_1-\tau_2)h(\tau_1)h(\tau_2)d\tau_1 d\tau_2 e^{-j\omega\tau}d\tau$

$$=\int_{-\infty}^{\infty}h(\tau_1)\int_{-\infty}^{\infty}h(\tau_2)\int_{-\infty}^{\infty}R_{XX}\int_{-\infty}^{\infty}R_{XX}(\tau+\tau_1-\tau_2)e^{-j\omega\tau}d\tau d\tau_2 d\tau_1$$

$$=\int_{-\infty}^{\infty}h(\tau_1)e^{j\omega\tau_1}\int_{-\infty}^{\infty}h(\tau_2)e^{-j\omega\tau_2}\int_{-\infty}^{\infty}R_{XX}(\tau+\tau_1-\tau_2)e^{-j\omega\tau}e^{-j\omega\tau_1}e^{j\omega\tau_2}d\tau d\tau_1 d\tau_2$$

Let $\tau+\tau_1-\tau_2=t,\ d\tau=dt$

$$S_{YY}(\omega)=\int_{-\infty}^{\infty}h(\tau_1)e^{j\omega\tau_1}d\tau_1\int_{-\infty}^{\infty}h(\tau_2)e^{-j\omega\tau_2}d\tau_2\int_{-\infty}^{\infty}R_{XX}(t)e^{-j\omega t}dt.$$

We know that,

$$S_{YY}(\omega)=H^*(\omega)H(\omega)S_{XX}(\omega)=H(-\omega)H(\omega)S_{XX}(\omega).$$

$\therefore\quad S_{YY}(\omega)=|H(\omega)|^2S_{XX}(\omega).$ (Proved).

Similarly, we can prove that the cross power spectral density function is

$$S_{XY}(\omega)=S_{XX}(\omega)H(\omega) \tag{10.56}$$

and $S_{YX}(\omega)=S_{XX}(\omega)H(-\omega).$ (10.57)

Likewise,

$$S_{YY}(\omega)=S_{XY}(\omega)H(-\omega) \tag{10.58}$$

and $S_{XY}(\omega)=S_{YX}(\omega)H(\omega).$ (10.59)

Example 10.4

A random process $X(t)$ whose mean value is 2 and autocorrelation function is $R_{XX}(\tau)=4e^{-2|\tau|}$ is applied to a system whose transfer function is $\frac{1}{(2+j\omega)}$. Find the mean value, autocorrelation, power density spectrum and average power of the output signal $Y(t)$.

Solution

Given $R_{XX}(\tau)=4e^{-2|\tau|}$, $\overline{X}=2$ and $H(\omega)=\frac{1}{2+j\omega}$.

We know that $S_{XX}(\omega))=F[R_{XX}(\tau)]$.

$$\therefore\ S_{XX}(\omega)=F[4e^{-2|\tau|}]=4\frac{4}{4+\omega^2}=\frac{16}{\omega^2+4}$$

The mean value of $Y(t)$ is $E[Y(t)]=E[X]H(0)$.

$$\overline{Y}=\overline{X}H(0).$$

$$\overline{Y}=2\cdot\frac{1}{2}=1$$

The power density spectrum of $Y(t)$ is

$$S_{YY}(\omega) = |H(\omega)|^2 S_{XX}(\omega)$$

$$= \frac{1}{\omega^2+4} \times \frac{16}{\omega^2+4} = \frac{16}{(\omega^2+4)^2}.$$

The autocorrelation of $Y(t)$ is

$$R_{YY}(\tau) = F^{-1}[S_{YY}(\omega)]$$

$$= F^{-1}\left[\frac{16}{(\omega^2+4)^2}\right] = \frac{16}{4}\tau e^{-2|\tau|}.$$

Average power of output, $P_{YY} = R_{YY}(0) = 0$.

Average power of input, $P_{XX} = R_{XX}(0) = 4$.

10.7.2 Spectrum Bandwidth

The power spectral density is mostly concentrated at a certain frequencies and decreases at other frequencies. The bandwidth of the spectrum is the range of frequencies having significant values. It is defined as "the measure of spread of the spectral density" and is also called rms bandwidth or normalised bandwidth. It is given by

$$W_{\text{rms}}^2 = \frac{\int_{-\infty}^{\infty} \omega^2 S_{XX}(\omega)d\omega}{\int_{-\infty}^{\infty} S_{XX}(\omega)d\omega}. \tag{10.60}$$

Additional Problems

Example 10.5

Consider the random process

$$X(t) = A\cos(\omega t + \theta)$$

where A and ω are real constants and θ is a random variable uniformly distributed over $[0, 2\pi]$. Find the average power P_{XX}.

Solution

We know that $P_{XX} = A\{E[X^2(t)]\}$.

$$\text{Now } X(t) = \text{A}\cos\,(\omega t + \theta) \tag{10.61}$$

$$\text{and } f_\theta(\theta) = \begin{cases} \dfrac{1}{2\pi} & 0 < \theta \le 2\pi \\ 0 & \text{otherwise.} \end{cases}$$

$$\therefore \quad E[X^2(t)] = \int_{-\infty}^{\infty} X^2(t) f_\theta(\theta)d\theta$$

$$= \int_0^{2\pi} A^2 \cos^2(\omega t + \theta) \frac{1}{2\pi} d\theta$$

$$= \frac{A^2}{2\pi} \left[\int_0^{2\pi} \frac{1 + \cos(2\omega t + 2\theta)}{2} d\theta \right]$$

$$= \frac{A^2}{4\pi} \left[\int_0^{2\pi} d\theta + \int_0^{2\pi} \cos(2\omega t + 2\theta) d\theta \right]$$

$$= \frac{A^2}{4\pi} \left[2\pi + \left. \frac{\sin(2\omega t + 2\theta)}{2} \right|_0^{2\pi} \right]$$

$$= \frac{A^2}{4\pi} \left[2\pi - \frac{\sin(2\omega t)}{2} \right]$$

$$E[X^2(t)] = \frac{A^2}{2} - \frac{A^2}{8\pi} \sin 2\omega t.$$

The time average power is

$$P_{XX} = A\left[E[X^2(t)]\right] = \lim_{T\to\infty} \frac{1}{2T} \int_{-T}^{T} \left[\frac{A^2}{2} - \frac{A^2}{8\pi} \sin(2\omega t) \right] dt$$

$$= \frac{1}{2T} \frac{A^2}{2} (2T) - 0$$

$$P_{XX} = \frac{A^2}{2}.$$

Example 10.6

The psd of $X(t)$ is given by

$$S_{XX}(\omega) = \begin{cases} 1 + \omega^2 & \text{for } |\omega| < 1 \\ 0 & \text{otherwise.} \end{cases}$$

Find the autocorrelation function. (May 2011)

Solution

The autocorrelation function is

$$R_{XX}(\tau) = \frac{1}{2\pi} \int_{-\infty}^{\infty} S_{XX}(\omega) e^{j\omega\tau} d\omega = \frac{1}{2\pi} \int_{-1}^{1} (1 + \omega^2) e^{j\omega\tau} d\omega.$$

$$R_{XX}(\tau) = \frac{1}{2\pi} \left[\int_{-1}^{1} e^{j\omega\tau} d\omega + \int_{-1}^{1} \omega^2 e^{j\omega\tau} d\omega \right]$$

Now take,

$$\int_{-1}^{1} \omega^2 e^{j\omega\tau} d\omega = \left. \frac{e^{j\omega\tau}}{j\tau} \omega^2 \right|_{-1}^{1} - \int_{-1}^{1} \frac{e^{j\omega\tau}}{j\tau} (2\omega) d\omega$$

$$= \frac{e^{j\tau} - e^{-j\tau}}{j\tau} - \left[\frac{2e^{j\omega\tau}}{(j\tau)^2}\omega \Bigg|_{-1}^{1} - \int_{-1}^{1} \frac{2e^{j\omega\tau}}{(j\tau)^2} d\omega \right]$$

$$= \frac{e^{j\tau} - e^{-j\tau}}{j\tau} - \frac{2(e^{j\tau} + e^{-j\tau})}{-\tau^2} + \frac{2}{-\tau^2} \cdot \frac{e^{j\omega\tau}}{j\tau} \Bigg|_{-1}^{1}$$

$$\therefore \quad \int_{-1}^{1} \omega^2 e^{j\omega\tau} d\omega = \frac{e^{j\tau} - e^{-j\tau}}{j\tau} + \frac{2(e^{j\tau} + e^{-j\tau})}{\tau^2} - \frac{2(e^{j\tau} - e^{-j\tau})}{j\tau^3}.$$

Also, $\displaystyle\int_{-1}^{1} e^{j\omega\tau} d\omega = \frac{e^{j\omega\tau}}{j\tau}\Bigg|_{-1}^{1} = \frac{e^{j\tau} - e^{-j\tau}}{j\tau}.$

$$\therefore \quad R_{XX}(\tau) = \frac{1}{2\pi} \left[\frac{e^{j\tau} - e^{j\tau}}{j\tau} + \frac{e^{j\tau} - e^{-j\tau}}{j\tau} + \frac{2}{\tau^2}(e^{j\tau} + e^{-j\tau}) - \frac{2}{j\tau^3}(e^{j\tau} - e^{-j\tau}) \right]$$

$$= \frac{1}{2\pi} \left[\frac{2\sin\tau}{\tau} + \frac{2\sin\tau}{\tau} + \frac{4\cos\tau}{\tau^2} - \frac{4}{\tau^3}\sin\tau \right]$$

$$= \frac{1}{\pi} \left[\frac{2\sin\tau}{\tau} + \frac{2\cos\tau}{\tau^2} - \frac{2}{\tau^3}\sin\tau \right]$$

$$R_{XX}(\tau) = \frac{2}{\pi\tau^3}\left[\tau^2 \sin\tau + \tau\cos\tau - \sin\tau\right].$$

Example 10.7

The autocorrelation of a WSS random process $X(t)$ is given by $R_{XX}(\tau) = A\cos(\omega_0\tau)$, where A and ω_0 are constants. Find psd.

Solution

We know that the power spectral density

$$S_{XX}(\omega) = \text{Fourier transform of } R_{XX}(\tau).$$

$$S_{XX}(\omega) = \int_{-\infty}^{\infty} R_{XX}(\tau) e^{-j\omega\tau} d\tau = \int_{-\infty}^{\infty} A\cos\omega_0\tau e^{-j\omega\tau} d\tau$$

$$S_{XX}(\omega) = \int_{-\infty}^{\infty} A \frac{e^{j\omega_0\tau} + e^{-j\omega_0\tau}}{2} e^{-j\omega\tau} d\tau$$

$$= \frac{A}{2}\left[\int_{-\infty}^{\infty} (e^{j(\omega-\omega_0)\tau} d\tau + e^{-j(\omega-\omega_0)\tau}) d\tau\right]$$

$$= \frac{A}{2}\left[\int_{-\infty}^{\infty} e^{j(\omega-\omega_0)\tau} d\tau + \int_{-\infty}^{\infty} e^{-j(\omega-\omega_0)\tau} d\tau\right]$$

$$= \frac{A}{2}\left[2\pi\delta(\omega-\omega_0) + 2\pi\delta(\omega+\omega_0)\right]$$

$$S_{XX}(\omega) = A\pi[\delta(\omega-\omega_0)+\delta(\omega+\omega_0)].$$

The power density spectrum is shown in Fig. 10.3.

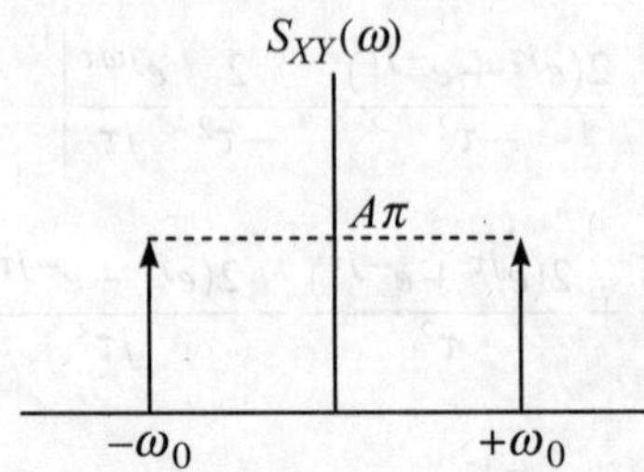

Fig. 10.3 *Power density spectrum of $A\cos\omega_0\tau$*

Example 10.8

Find the psd of a WSS random process $X(t)$ whose autocorrelation function is $R_{XX}(\tau) = ae^{-b|\tau|}$.

Solution

We know that the power spectral density

$$S_{XX}(\omega) = \text{Fourier transform of } R_{XX}(\tau).$$

$$S_{XX}(\omega) = \int_{-\infty}^{\infty} R_{XX}(\tau)e^{-j\omega\tau}d\tau = \int_{-\infty}^{\infty} a\,e^{-b|\tau|}e^{-j\omega\tau}d\tau$$

$$= a\int_{-\infty}^{0} e^{b\tau}e^{-j\omega\tau} + a\int_{0}^{\infty} e^{-b\tau}e^{-j\omega\tau}d\tau$$

$$= a\int_{-\infty}^{0} e^{\tau(b-j\omega)}d\tau + a\int_{0}^{\infty} e^{-\tau(b+j\omega)}d\tau$$

$$= a\frac{e^{(b-j\omega)\tau}}{b-j\omega}\Bigg|_{-\infty}^{0} + a\frac{e^{-(b+j\omega)\tau}}{-(b+j\omega)}\Bigg|_{0}^{\infty}$$

$$= \frac{a}{b-j\omega}[1-0] - \frac{a}{b+j\omega}[0-1]$$

$$= \frac{a}{b-j\omega} + \frac{a}{b+j\omega} = \frac{a(b+j\omega+b-j\omega)}{b^2+\omega^2}.$$

$$S_{XY}(\omega) = \frac{2ab}{b^2+\omega^2}.$$

Example 10.9

A random process has autocorrelation function

$$R_{XX}(\tau) = \begin{cases} 1-|\tau|, & |\tau| \le 1 \\ 0 & \text{otherwise.} \end{cases}$$

Find the psd and draw plots. (Nov 2010)

Solution

The autocorrelation function is given as $R_{XX}(\tau) = \begin{cases} 1-|\tau| & |\tau| \le 1 \\ 0 & \text{otherwise.} \end{cases}$

The power spectral density is

$$S_{XX}(\omega) = \int_{-\infty}^{\infty} R_{XX}(\tau)e^{-j\omega\tau}d\tau$$

$$S_{XX}(\omega) = \int_{-1}^{1}(1-|\tau|)e^{-j\omega\tau}d\tau$$

$$S_{XX}(\omega) = \int_{-1}^{0}(1+\tau)e^{-j\omega\tau}d\tau + \int_{0}^{1}(1-\tau)e^{-j\omega\tau}d\tau$$

$$= \int_{-1}^{0} e^{-j\omega\tau}d\tau + \int_{-1}^{0} \tau e^{-j\omega\tau}d\tau + \int_{0}^{1} e^{-j\omega\tau}d\tau - \int_{0}^{1} \tau e^{-j\omega\tau}d\tau.$$

We know that $\int e^{-j\omega\tau}d\tau = \dfrac{e^{-j\omega\tau}}{-j\omega}$

and, $\int \tau e^{-j\omega\tau}d\tau = \dfrac{e^{-j\omega\tau}}{-j\omega}\tau - \int \dfrac{e^{-j\omega\tau}}{(-j\omega)}d\tau$

$$= -\frac{e^{-j\omega\tau}\cdot\tau}{j\omega} - \frac{e^{-j\omega\tau}}{(-j\omega)^2}$$

or $\int \tau e^{-j\omega\tau}d\tau = \dfrac{-\tau e^{-j\omega\tau}}{j\omega} + \dfrac{e^{-j\omega\tau}}{\omega^2}.$

Now $S_{XX}(\omega) = \left.\dfrac{-e^{-j\omega\tau}}{j\omega}\right|_{-1}^{0} + \left.\dfrac{-e^{-j\omega\tau}}{j\omega}\right|_{0}^{1} + \left.\left(\dfrac{-\tau e^{-j\omega\tau}}{j\omega} + \dfrac{e^{-j\omega\tau}}{\omega^2}\right)\right|_{-1}^{0} - \left.\left(\dfrac{-\tau e^{-j\omega\tau}}{j\omega} + \dfrac{e^{-j\omega\tau}}{\omega^2}\right)\right|_{0}^{1}$

$$S_{XX}(\omega) = \frac{-1}{j\omega}(1-e^{j\omega}) - \frac{1}{j\omega}(e^{-j\omega}-1) - \left(\frac{+e^{j\omega}}{j\omega} - \frac{1}{\omega^2} + \frac{e^{+j\omega}}{\omega^2}\right) - \left(\frac{-e^{-j\omega}}{j\omega} + \frac{e^{-j\omega}}{\omega^2} - \frac{1}{\omega^2}\right)$$

$$S_{XX}(\omega) = \frac{-1}{j\omega} + \frac{e^{j\omega}}{j\omega} - \frac{e^{-j\omega}}{j\omega} + \frac{1}{j\omega} - \frac{e^{j\omega}}{j\omega} + \frac{1}{\omega^2} - \frac{e^{j\omega}}{\omega^2} + \frac{e^{-j\omega}}{j\omega} - \frac{e^{-j\omega}}{\omega^2} + \frac{1}{\omega^2}$$

$$= \frac{2}{\omega^2} - \frac{e^{j\omega}}{\omega^2} - \frac{e^{-j\omega}}{\omega^2} = \frac{-1}{\omega^2}(e^{j\omega} + e^{-j\omega} - 2)$$

$$= \frac{1}{\omega^2}\left[\frac{(e^{j\omega/2})^2 + (e^{j\omega/2})^2 - 2e^{j\omega/2}e^{-j\omega/2}}{j^2}\right]$$

$$= \frac{(e^{j\omega/2} - e^{-j\omega/2})^2}{j^2\omega^2} = \frac{(e^{j\omega/2} - e^{-j\omega/2})^2}{(2j)^2(\omega/2)^2}.$$

$$\therefore\ S_{XX}(\omega) = \frac{\sin^2(\omega/2)}{(\omega/2)^2} = \left(\frac{\sin(\omega/2)}{(\omega/2)}\right)^2$$

$$S_{XX}(\omega) = \mathrm{Sa}^2(\omega/2).$$

Plots of autocorrelation function $R_{XX}(\tau)$ and the power density spectrum are shown in Fig. 10.4.

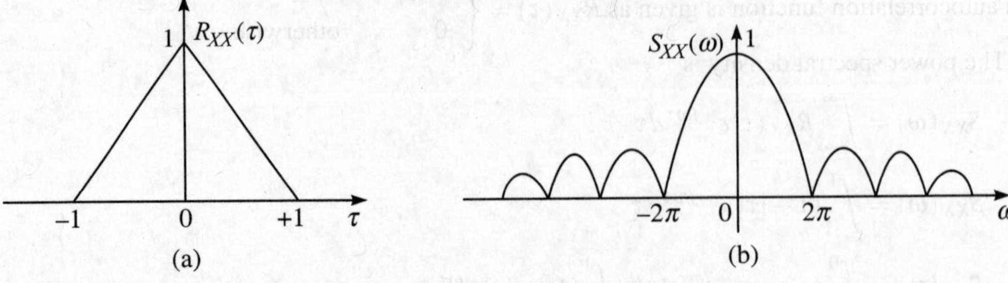

Fig. 10.4 (a) *Autocorrelation function* **Fig. 10.4 (b)** *Power spectral density*

Example 10.10

The cross power spectrum of $X(t)$ and $Y(t)$ is defined as

$$S_{XY}(\omega) = \begin{cases} K + \frac{j\omega}{W} & -W < \omega < W \\ 0 & \text{otherwise} \end{cases}$$

where $W > 0$ and K are constants. Find the cross correlation function.

Solution

We know that the cross correlation function is

$$R_{XY}(\tau) = \frac{1}{2\pi}\int_{-\infty}^{\infty} S_{XY}(\omega)e^{+j\omega\tau}d\omega$$

$$= \frac{1}{2\pi}\int_{-W}^{W}\left(K + \frac{j}{W}\omega\right)e^{j\omega\tau}d\omega$$

$$= \frac{1}{2\pi}\left[\int_{-W}^{W} Ke^{j\omega\tau}d\omega + \frac{j}{W}\int_{-W}^{W}\omega e^{j\omega\tau}d\omega\right]$$

$$R_{XY}(\tau) = \frac{K}{2\pi}\int_{-W}^{W} e^{j\omega\tau}d\omega + \frac{j}{2\pi W}\int_{-W}^{W}\omega e^{j\omega\tau}d\omega.$$

Now, the integration

$$\int_{-W}^{W}\omega e^{j\omega\tau}d\omega = \frac{e^{j\omega\tau}}{j\tau}(\omega) - \int_{-W}^{W}\frac{e^{j\omega\tau}}{j\tau}d\omega$$

$$= \left[\frac{\omega e^{j\omega\tau}}{j\tau} - \frac{e^{j\omega\tau}}{(j\tau)^2}\right]_{-W}^{W}$$

$$= e^{j\omega\tau}\left[\frac{W}{j\tau} + \frac{1}{\tau^2}\right] - e^{-j\omega\tau}\left[\frac{-W}{j\tau} + \frac{1}{\tau^2}\right]$$

$$= \frac{We^{jW\tau}}{j\tau} + \frac{e^{jW\tau}}{\tau^2} + \frac{We^{-jW\tau}}{j\tau} - \frac{e^{-jW\tau}}{\tau^2}$$

and

$$\int_{-W}^{W} e^{j\omega\tau} d\tau = \left.\frac{e^{j\omega\tau}}{j\tau}\right|_{-W}^{W} = \frac{e^{j\omega\tau} - e^{-j\omega\tau}}{j\tau}.$$

$$\therefore \quad R_{XY}(\tau) = \frac{K}{2\pi}\left[\frac{e^{jW\tau} - e^{-jW\tau}}{j\tau}\right] + \frac{j}{2\pi W}\left[\frac{W}{j\tau}(e^{-jW\tau} + e^{jW\tau}) + \frac{1}{\tau^2}(e^{jw\tau} - e^{-jW\tau})\right]$$

$$= \frac{K}{2\pi\tau}[2\sin W\tau] + \frac{1}{2\pi\tau}[2\cos W\tau] - \frac{1}{2\pi W\tau^2}[2\sin W\tau]$$

$$R_{XY}(\tau) = \left(\frac{K}{\pi\tau} - \frac{1}{\pi W\tau^2}\right)\sin W\tau + \frac{\cos W\tau}{\pi\tau}.$$

Example 10.11

Find the autocorrelation function and power spectral density of the random process $X(t) = \text{A}\cos(\omega_0 t + \theta)$, where θ is a random variable over the ensemble and is uniformly distributed over the range $(0, 2\pi)$. (May 2013)

Solution

Given the random process $X(t) = A\cos(\omega_0 t + \theta)$ and $f_\theta(\theta) = \dfrac{1}{2\pi} \quad 0 \le \theta \le 2\pi$.

(i) The autocorrelation function is

$$R_{XX}(\tau) = E[X(t)X(t+\tau)]$$

$$= E[A\cos(\omega_0 + \theta)A\cos(\omega_0(t+\tau) + \theta)]$$

$$= A^2 E[\cos(\omega_0 t + \theta)\cos(\omega_0 t + \omega_0\tau + \theta)]$$

$$= \frac{A^2}{2}E[\cos(2\omega_0 t + \omega_0\tau + 2\theta) + \cos\omega_0\tau]$$

$$= \frac{A^2}{2}\{E[\cos\omega_0\tau] + E[\cos(2\omega_0 t + \omega_0\tau + 2\theta)]\}$$

$$= \frac{A^2}{2}\int_0^{2\pi}\frac{\cos\omega_0\tau}{2\pi}d\theta + \frac{A^2}{2}\int_0^{2\pi}\frac{1}{2\pi}[\cos(2\omega_0 t + \omega_0\tau + 2\theta)]d\theta$$

$$= \frac{A^2}{2}\cos\omega_0\tau + \frac{A^2}{4\pi}\left[\frac{\sin(2\omega_0 t + \omega_0\tau + 2\theta)}{2}\right]_0^{2\pi}$$

$$= \frac{A^2}{2}\cos\omega_0\tau + 0.$$

$$\therefore \quad R_{XX}(\tau) = \frac{A^2}{2}\cos\omega_0\tau.$$

(ii) The power spectral density is

$$S_{XX}(\omega) = \text{ Fourier transform of } R_{XX}(\tau)$$

$$= \frac{A^2}{2}\int_{-\infty}^{\infty} \cos\omega_0\tau e^{-j\omega\tau}d\tau = \frac{A^2}{2}\int_{-\infty}^{\infty}(\frac{e^{j\omega_0\tau}+e^{-j\omega_0\tau}}{2})e^{-j\omega\tau}d\tau$$

$$= \frac{A^2}{4}\left[\int_{-\infty}^{\infty} e^{-j(\omega-\omega_0)\tau}d\tau + \int_{-\infty}^{\infty} e^{-j(\omega+\omega_0)\tau}d\tau\right]$$

$$= \frac{A^2}{4}2\pi[\delta(\omega-\omega_0)+\delta(\omega+\omega_0)]$$

$$S_{XX}(\omega) = \frac{A^2\pi}{2}[\delta(\omega-\omega_0)+\delta(\omega+\omega_0)].$$

Example 10.12

The autocorrelation function of a WSS random process is $R_{XX}(\tau) = a\exp(-(\tau/b)^2)$. Find the power spectral density and normalised average power of the signal. (May 2011, Sept 2014)

Solution

Given the autocorrelation $R_{XX}(\tau) = a\exp(\frac{-\tau^2}{b^2})$,
the power spectral density is

$$S_{XX}(\omega) = \int_{-\infty}^{\infty} R_{XX}(\tau)e^{-j\omega\tau}d\tau$$

$$= \int_{-\infty}^{\infty} a\exp\left(\frac{-\tau^2}{b^2}\right)e^{-j\omega\tau}d\tau = a\int_{-\infty}^{\infty}\exp\left(\frac{-\tau^2}{b^2} - j\omega\tau\right)d\tau.$$

Converting the exponential term to get a perfect square:

$$S_{XX}(\omega) = a\int_{-\infty}^{\infty}\exp\left(-\left(\frac{\tau}{b}\right)^2 - \frac{2j\omega\tau}{b}\frac{b}{2} + \left(\frac{b}{2}j\omega\right)^2 - \left(\frac{b}{2}j\omega\right)^2\right)d\tau$$

$$= a\int_{-\infty}^{\infty}\exp\left(-\left(\frac{\tau}{b}+\frac{j\omega b}{2}\right)^2\right)\exp\left(\frac{j\omega b}{2}\right)^2 d\tau$$

Let $\left(\frac{\tau}{b}+\frac{j\omega b}{2}\right) = \frac{Z}{\sqrt{2}}$.

$$d\tau = \frac{b}{\sqrt{2}}dz.$$

$$S_{XX}(\omega) = a\int_{-\infty}^{\infty} e^{\frac{-z^2}{2}}e^{\frac{-\omega^2b^2}{4}}\frac{bdz}{4} = \frac{abe^{\frac{-\omega^2b^2}{4}}}{\sqrt{2}}\int_{-\infty}^{\infty} e^{\frac{-z^2}{2}}dz.$$

$$S_{XX}(\omega) = ab\sqrt{\pi}e^{\frac{-\omega^2b^2}{4}}\frac{1}{\sqrt{2\pi}}\int_{-\infty}^{\infty} e^{\frac{-z^2}{2}}dz.$$

We know that $\frac{1}{\sqrt{2\pi}}\int_{-\infty}^{\infty} e^{\frac{-z^2}{2}}dz = 1.$

$$\therefore \quad S_{XX}(\omega) = ab\sqrt{\pi}e^{\frac{-\omega^2 b^2}{4}}.$$

(ii) The average power of the signal is $P_{XX} = R_{XX}(0)$.

Given $R_{XX}(\tau) = a\exp\left(\frac{-\tau^2}{b^2}\right)$,

at $\tau = 0$, $R_{XX}(0) = a\exp(0) = a$ watts.
Or, we can also find from $S_{XX}(\omega)$,

$$\text{Average power } P_{XX} = \frac{1}{2\pi}\int_{-\infty}^{\infty} S_{XX}(\omega)d\omega$$

$$P_{XX} = \frac{1}{2\pi}\int_{-\infty}^{\infty} ab\sqrt{\pi}e^{\frac{-\omega^2 b^2}{4}}\,d\omega.$$

$$\text{Let} \quad \frac{\omega^2 b^2}{4} = \frac{z^2}{2} \text{ or } \frac{\omega b}{2} = \frac{z}{\sqrt{2}}.$$

$$\omega = \frac{z\sqrt{2}}{b}; \quad d\omega = \frac{\sqrt{2}}{b}dz.$$

$$\therefore \quad P_{XX} = \frac{ab}{2\sqrt{\pi}}\int_{-\infty}^{\infty} e^{-z^2/2}\frac{\sqrt{2}}{b}dz = a\frac{1}{\sqrt{2\pi}}\int_{-\infty}^{\infty} e^{-z^2/2}dz$$

$$P_{XX} = a(1) = a \text{ watts}.$$

Example 10.13

Two independent stationary random processes $X(t)$ and $Y(t)$ have power spectrum densities $S_{XY}(\omega) = \dfrac{16}{\omega^2+16}$ and $S_{XY}(\omega) = \dfrac{\omega^2}{\omega^2+16}$ respectively with zero means. Let another random process $U(t) = X(t)+Y(t)$. Then find (i) psd of $U(t)$, (ii) $S_{XY}(\omega)$ and (iii) $S_{XU}(\omega)$. (Sept 2014)

Solution

Given $X(t)$ and $Y(t)$ are independent random processes.

$$\overline{X} = \overline{Y} = 0.$$

$$S_{XX}(\omega) = \frac{16}{\omega^2+16}, \quad S_{YY}(\omega) = \frac{\omega^2}{\omega^2+16} \quad \text{and} \quad U(t) = X(t)+Y(t).$$

(i) Power density of $U(t)$ is

$$S_{UU}(\omega) = S_{XX}(\omega) + S_{YY}(\omega).$$

$$\therefore \quad S_{UU}(\omega) = \frac{16}{\omega^2+16} + \frac{\omega^2}{\omega^2+16} = \frac{16+\omega^2}{16+\omega^2} = 1.$$

(ii) Since $X(t)$ and $Y(t)$ are independent and uncorrelated,

$$S_{XY}(\omega) = 2\pi\overline{X}\overline{Y}\delta(\omega) = 0.$$

(iii) $S_{XU}(\omega)$ = Fourier transform of $R_{XU}(\tau)$.

Now,

Now, $R_{XU}(\tau) = E[X(t)U(t+\tau)]$

$$= E[X(t)(X(t+\tau)+Y(t+\tau))]$$

$$= E[X(t)X(t+\tau)+X(t)Y(t+\tau)]$$

$$= E[X(t)X(t+\tau)+E[X(t)Y(t+\tau)]$$

$$= R_{XX}(\tau)+R_{XY}(\tau).$$

$\therefore\ S_{XU}(\omega)$ = Fourier transform of $[R_{XX}(\tau)+R_{XY}(\tau)]$

$$= S_{XX}(\omega)+S_{XY}(\omega)$$

$$= \frac{16}{\omega^2+16}+0 = \frac{16}{\omega^2+16}.$$

Example 10.14

Find the cross correlation function for the psd $S_{XY}(\omega) = \frac{1}{25+\omega^2}$. (Dec 2013)

Solution

The given psd is $S_{XY}(\omega) = \frac{1}{25+\omega^2}$.

The cross correlation function is

$$R_{XY}(\tau) = \mathcal{F}^{-1}[S_{XY}(\omega)]$$

$$= \mathcal{F}^{-1}\left(\frac{1}{25+\omega^2}\right).$$

We know from the Fourier transform that

$$\frac{2a}{a^2+\omega^2} \leftrightarrow e^{-a|t|},\ a>0,\ a \text{ constant}$$

or $$\frac{1}{a^2+\omega^2} \leftrightarrow \frac{1}{2a}e^{-a|t|}.$$

$$\therefore\ \frac{1}{25+\omega^2} \leftrightarrow \frac{1}{2\times 5}e^{-5|t|}.$$

$$\therefore\ R_{XY}(\tau) = \frac{1}{10}e^{-5|\tau|}.$$

Example 10.15

Find the cross power spectral density if (a) $R_{XY}(\tau) = \frac{A^2}{2}\sin(\omega_0\tau)$ and (b) $R_{XY}(\tau) = \frac{A^2}{2}\cos(\omega_0\tau)$. (Dec 2013)

Solution

(a) Given $R_{XY}(\tau) = \dfrac{A^2}{2}\sin(\omega_0\tau)$.

The power spectral density

$$S_{XY}(\omega) = \mathcal{F}\left[\frac{A^2}{2}\sin\omega_0\tau\right] = \frac{jA^2\pi}{2}[\delta(\omega+\omega_0) - \delta(\omega-\omega_0)]$$

(b) $R_{XX}(\tau) = \dfrac{A^2}{2}\cos(\omega_0\tau)$.

$$S_{XY}(\omega) = F\left[\frac{A^2}{2}\cos(\omega_0\tau)\right].$$

$$S_{XY}(\omega) = \frac{A^2\pi}{2}[\delta(\omega+\omega_0) + \delta(\omega-\omega_0)].$$

Both power density spectrums are shown in Fig. 10.5.

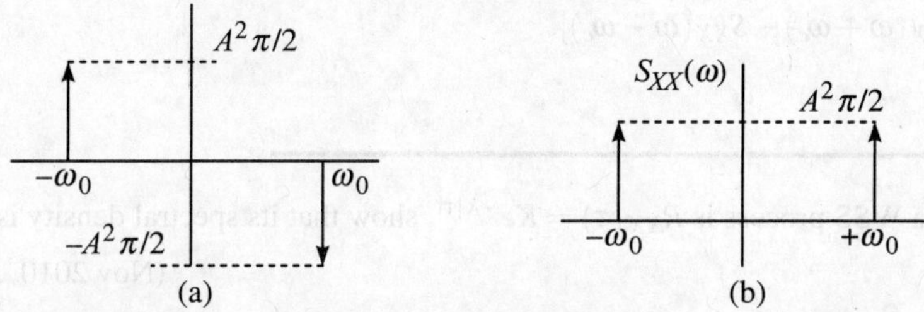

Fig. 10.5 (a) *psd for* $\dfrac{A^2}{2}\sin\omega_0\tau$ **Fig. 10.5 (b)** *psd for* $\dfrac{A^2}{2}\cos s\omega_0\tau$

Example 10.16

$X(t)$ is a stationary random process with spectral density $S_{XX}(\omega)$. $Y(t)$ is another independent random process, $Y(t) = A\cos(\omega_c t + \theta)$, where θ is a random variable uniformly distributed over $(-\pi, \pi)$. Find the spectral density function of $Z(t) = X(t)Y(t)$. (Sept 2014)

Solution

Given two independent random processes $X(t)$ and $Y(t) = A\cos(\omega_c t + \theta)$.

The power spectral density of $X(t)$ is $S_{XX}(\omega)$. θ is uniformly distributed over $(-\pi - \pi)$.

$$\therefore \quad f_\theta(\theta) = \begin{cases} \frac{1}{2\pi} & -\pi \le \theta \le \pi \\ 0 & \text{otherwise.} \end{cases}$$

The autocorrelation of $Y(t)$ is

$$R_{YY}(\tau) = E[Y(t)Y(t+\tau)]$$

$$= \int_{-\infty}^{\infty} y(t)y(t+\tau)f_\theta(\theta)d\theta$$

$$= \frac{1}{2\pi}\int_{-\pi}^{\pi} A^2\cos(\omega_c t+\theta)\cos(\omega_c(t+\tau)+\theta)d\theta$$

$$= \frac{A^2}{4\pi}\left[\int_{-\pi}^{\pi}\cos(\omega_c\tau)d\theta + 0\right] = \frac{A^2}{2}\cos\omega_c\tau.$$

Now, $Z(t) = X(t)Y(t)$.

$$R_{ZZ}(\tau) = E[X(t)Y(t)X(t+\tau)Y(t+\tau)].$$

Since $X(t)$ and $Y(t)$ are independent,

$$R_{ZZ}(\tau) = E[X(t)X(t+\tau)]E[Y(t)Y(t+\tau)]$$

$$R_{ZZ}(\tau) = R_{XX}(\tau)\frac{A^2}{2}\cos\omega_c\tau.$$

The power density spectrum is

$$S_{ZZ}(\omega) = F[R_{ZZ}(\tau)] = \frac{A^2}{2} \times F[R_{XX}(\tau)\cos\omega_c\tau]$$

$$S_{ZZ}(\omega) = \frac{A^2}{2}\pi[S_{XX}(\omega+\omega_c) + S_{XX}(\omega-\omega_c)].$$

Example 10.17

If the autocorrelation of a WSS process is $R_{XX}(\tau) = Ke^{-K|\tau|}$, show that its spectral density is given by $S(\omega) = \dfrac{2}{1+(\omega/K)^2}$. (Nov 2010, Jan 2015)

Solution

Given the random process is WSS. The autocorrelation function is $R_{XX}(\tau) = Ke^{-K|\tau|}$. We know that the power spectral density is

$$S_{XX}(\omega) = F[R_{XX}(\tau)]$$

$$= \int_{-\infty}^{\infty} Ke^{-K|\tau|}e^{-j\omega\tau}d\tau$$

$$= K\left[\int_{-\infty}^{0} e^{K\tau}e^{-j\omega\tau}d\tau + \int_{0}^{\infty} e^{-K\tau}e^{-j\omega\tau}d\tau\right]$$

$$= K\left[\left.\frac{e^{(K-j\omega)\tau}}{K-j\omega}\right|_{-\infty}^{0} + \left.\frac{e^{(-K-j\omega)\tau}}{-(K+j\omega)}\right|_{0}^{\infty}\right]$$

$$= K\left[\frac{1}{K-j\omega} + \frac{1}{K+j\omega}\right] = \frac{2K^2}{K^2+\omega^2}$$

$$\therefore \quad S_{XX}(\omega) = \frac{2}{1+(\omega/K)^2}.$$

Example 10.18

Consider a WSS random process $X(t)$ with power spectral density $S_{XX}(\omega)$. Another random process is given by $Y(t) = X(t+T) + X(t-T)$, where T is a constant. Find the power spectrum of $Y(t)$.

(Sept 2014)

Solution

Given a WSS random process $X(t)$ with psd $S_{XX}(\omega)$ and $Y(t) = X(t+T) + X(t-T)$.

Let the autocorrelation function of $X(t)$ be $R_{XX}(\tau)$. We know that the autocorrelation function is

$$\begin{aligned}
R_{YY}(\tau) &= E[Y(t)Y(t+\tau)] \\
&= E[(X(t+T)+X(t-T))(X(t+T+\tau)+X(t-T+\tau))] \\
&= E[X(t+T)X(t+T+\tau)+X(t+T)X(t-T+\tau) \\
&\quad +X(t-T)X(t+T+\tau)+X(t-T)X(t-T+\tau)] \\
&= [X(t+T)X(t+T+\tau)]+E[X(t+T)X(t-T+\tau)] \\
&\quad +E[X(t-T)X(t+T+\tau)]+E[X(t-T)X(t-T+\tau)].
\end{aligned}$$

Let $t+T = t_1$ and $t-T = t_2$. Then

$$\begin{aligned}
R_{YY}(\tau) = E[X(t_1)X(t_1+\tau)]+E[X(t_1)X(t_1+\tau-2T)] \\
+E[X(t_2)X(t_2+\tau+2T)]+E[X(t_2)X(t_2+\tau)].
\end{aligned}$$

$$\therefore \quad R_{YY}(\tau) = R_{XX}(\tau)+R_{XX}(\tau-2T)+R_{XX}(\tau+2T)+R_{XX}(\tau)$$

or $\quad R_{YY}(\tau) = 2R_{XX}(\tau)+R_{XX}(\tau-2T)+R_{XX}(\tau+2T).$

Now the power density spectrum is

$$\begin{aligned}
S_{YY}(\omega) &= F[R_{YY}(\tau)] \\
&= \int_{-\infty}^{\infty} [2R_{XX}(\tau)+R_{XX}(\tau-2T)+R_{XX}(\tau+2T)]e^{-j\omega\tau}d\tau \\
&= 2\int_{-\infty}^{\infty} R_{XY}(\tau)e^{-j\omega\tau}d\tau + \int_{-\infty}^{\infty} R_{XX}(\tau-2T)e^{-j\omega\tau}d\tau \\
&\quad + \int_{-\infty}^{\infty} R_{XX}(\tau+2T)e^{-j\omega\tau}d\tau \\
&= 2S_{XX}(\omega) + \int_{-\infty}^{\infty} R_{XX}(\tau-2T)e^{-j\omega(\tau-2T)}e^{-j\omega 2T}d\tau \\
&\quad + \int_{-\infty}^{\infty} R_{XX}(\tau+2T)e^{-j\omega(\tau+2T)}e^{+j\omega 2T}d\tau \\
&= 2S_{XX}(\omega)+e^{-j\omega 2T}S_{XX}(\omega)+e^{j\omega 2T}S_{XX}(\omega) \\
&= (2+e^{j\omega 2T}+e^{-j\omega 2T})S_{XX}(\omega) \\
&= (2+2\cos 2\omega T)S_{XX}(\omega)
\end{aligned}$$

$$S_{YY}(\omega) = 4\cos^2(\omega\tau)S_{XX}(\omega).$$

Example 10.19

Consider the random process $X(t) = 4\cos(\omega_0 t + \theta)$, where A and ω_0 are real constants and θ is a random variable uniformly distributed on the interval $(0, \pi/2)$. Find the average power of $X(t)$.

(Jan 2015)

Solution

Given the random process $X(t) = 4\cos(\omega_0 t + \theta)$ and θ is uniformly distributed over $(0, \pi/2)$.

$$\therefore \quad f_\Theta(\theta) = \begin{cases} \dfrac{2}{\pi} & 0 < \theta < \pi/2 \\ 0 & \text{otherwise.} \end{cases}$$

The power is $\quad P_{XX} = E[X^2(t)]$.

$$P_{XX} = \int_{-\infty}^{\infty} X^2(t) f_\Theta(\theta) d\theta$$

$$= \int_0^{\pi/2} A^2 \cos(\omega_0 t + \theta) \frac{2}{\pi} d\theta$$

$$= \frac{2A^2}{\pi} \left[\int_0^{\pi/2} \frac{1 + \cos 2(\omega_0 t + \theta)}{2} d\theta \right]$$

$$= \frac{A^2}{\pi} \left[\int_0^{\pi/2} d\theta + \int_0^{\pi/2} \cos 2(\omega_0 t + \theta) d\theta \right]$$

$$= \frac{A^2}{\pi} \left[\frac{\pi}{2} + \frac{\sin 2(\omega_0 t + \theta)}{2} \Bigg|_0^{\pi/2} \right]$$

$$= \frac{A^2}{2\pi} [\pi + (\sin(2\omega_0 t + \pi) - \sin 2\omega_0 t)]$$

$$= \frac{A^2}{2\pi} [\pi - 2\sin 2\omega_0 t]$$

$$P_{XX} = \frac{A^2}{2} - \frac{A^2}{\pi} \sin 2\omega_0 t.$$

Since the power is a function of time, the time average power is

$$P_{XX} = A[P_{XX}] = \lim_{T\to\infty} \frac{1}{2T} \int_{-T}^{T} P_{XX} dt$$

$$P_{XX} = \lim_{T\to\infty} \frac{1}{2T} \int_{-T}^{T} \left(\frac{A^2}{2} - \frac{A^2}{\pi} \sin 2\omega t \right) dt$$

$$P_{XX} = \lim_{T\to\infty} \frac{1}{2T} \left[\frac{A^2}{2} \times 2T - 0 \right]$$

$$P_{XX} = \frac{A^2}{2}.$$

Example 10.20

The power spectral density of a stationary random process is given by

$$S_{XX}(\omega) = \begin{cases} A & -K < \omega < K \\ 0 & \text{otherwise.} \end{cases}$$

Find the aucorrelation function. (Feb 2008, Sept 2014)

Solution

We know that the autocorrelation function is

$$R_{XX}(\tau) = F^{-1}[S_{XX}(\omega)]$$

$$= \frac{1}{2\pi}\int_{-\infty}^{\infty} S_{XX}(\omega)e^{j\omega\tau}d\omega = \frac{1}{2\pi}\int_{-K}^{K} A.e^{j\omega\tau}d\omega$$

$$= \frac{A}{2\pi}\left[\frac{e^{j\omega\tau}}{j\tau}\right]_{-K}^{K}$$

$$R_{XX}(\tau) = \frac{A}{j2\pi\tau}\left[e^{jK\tau} - e^{-jK\tau}\right] = \frac{A}{\pi\tau}\left[\frac{e^{jK\tau} - e^{-jK\tau}}{2j}\right].$$

$$R_{XX}(\tau) = \frac{A}{\pi\tau}\sin(K\tau) = \frac{AK\sin(K\tau)}{\pi K\tau}$$

$$\therefore \quad R_{XX}(\tau) = \frac{AK}{\pi}\mathrm{Sa}(K\tau).$$

Example 10.21

A stationary random process $X(t)$ has autocorrelation function $R_{XX}(\tau) = 10 + 5\cos(2\tau) + 10e^{-2|\tau|}$. Find the dc and ac average powers of $X(t)$. (Nov 2003, Dec 2013)

Solution

Given $R_{XX}(\tau) = 10 + 5\cos(2\tau) + 10e^{-2|\tau|}$.

(a) The average power of the dc component (without τ) is

$$P_{\text{dc}} = 10 \text{ watts.}$$

(b) The average power of the ac component (with τ at $\tau = 0$) is

$$P_{\text{ac}} = 5 + 10 = 15 \text{ watts.}$$

P_{ac} is also called variance of the process $X(t)$.

(c) The average power of the process $X(t)$ is

$$P_{XX} = R_{XX}(0) = 10 + 5 + 10 = 25 \text{ watts.}$$

Example 10.22

Assume that the ergodic random process $X(t)$ has an autocorrelation function

$$R_{XX}(\tau) = 18 + \frac{2}{6+\tau^2}[1+4\cos(12\tau)]$$

What is the average power of $X(t)$? (May 2010, May 2011)

Solution

Given $X(t)$ is an ergodic process.

$$R_{XX}(\tau) = 18 + \frac{2}{6+\tau^2}[1+4\cos(12\tau)].$$

The average power in $X(t)$ is

$$P_{XX} = R_{XX}(\tau)|_{\tau=0}$$

$$= 18 + \frac{2}{6+0}[1+4\cos 0] = 18 + \frac{1}{3}[1+4] = 18 + \frac{5}{3}$$

$\therefore \quad P_{XX} = 19.67$ watts

Example 10.23

The cross spectral density of two random processes $X(t)$ and $Y(t)$ is given as

$$S_{XY}(\omega) = \frac{1}{-\omega^2 + j4\omega + 4}.$$

Find the cross correlation function. (May 2013)

Solution

Given the cross spectral density

$$S_{XY}(\omega) = \frac{1}{-\omega^2 + j4\omega + 4} = \frac{1}{(2+j\omega)^2}.$$

The cross-correlation function is

$$R_{XY}(\tau) = \mathcal{F}^{-1}[S_{XY}(\omega)] = F^{-1}\left[\frac{1}{(2+j\omega)^2}\right]$$

$$= \mathcal{F}^{-1}\left[\frac{1}{(j\omega+2)^2}\right] = \tau e^{-2\tau}u(\tau). \qquad \text{(From Appendix B)}$$

$\therefore \quad R_{XY}(\tau) = \tau e^{-2\tau}$ for $\tau > 0$.

Example 10.24

The power density spectrum of a WSS process is

$$S_{XX}(\omega) = 4\pi\delta(\omega) + 3\pi\delta(\omega - 5\pi) + 3\pi\delta(\omega + 5\pi) + 2\pi\delta(\omega - 4) + 2\pi\delta(\omega + 4)$$

(a) What are the frequencies in $X(t)$?

(b) Find mean variance and average power of $X(t)$. (Dec 2013)

Solution

Given the random process is WSS with psd,

$$S_{XX}(\omega) = 4\pi\delta(\omega) + 3\pi\delta(\omega - 5\pi) + 3\pi\delta(\omega + 5\pi) + 2\pi\delta(\omega - 4) + 2\pi\delta(\omega + 4).$$

Using inverse Fourier transformation, the autocorrelation function is

$$R_{XX}(\tau) = 2 + \frac{3}{2}e^{-5\pi\tau} + \frac{3}{2}e^{5\pi\tau} + e^{-4\tau} + e^{4\tau}.$$

Since $2\pi\delta(\omega - \omega_0) \leftrightarrow e^{-j\omega_0\tau}$,

$$R_{XX}(\tau) = 2 + \frac{3}{2}(e^{-j5\pi\tau} + e^{j5\pi\tau}) + (e^{-j4\tau} + e^{j4\tau})$$

$$R_{XX}(\tau) = 2 + 3\cos 5\pi\tau + 2\cos 4\tau.$$

(a) Therefore the frequencies in $X(t)$ are $\omega_1 = 5\pi$ and $\omega_2 = 4$.

$$\therefore \quad f_1 = \frac{5\pi}{2\pi} = 2.5 \text{ Hz} \quad \text{and} \quad f_2 = \frac{4}{2\pi} = \frac{2}{\pi} \text{ Hz}.$$

(b) The mean value of $X(t)$ is the dc component of $R_{XX}(\tau)$.

$$\therefore \quad E[X(t)] = 2.$$

The average power or the mean square value of $X(t)$ is $R_{XX}(0)$.

$$\therefore \quad R_{XX}(0) = 2 + 3 + 2 = 7.$$

The variance of $X(t)$ is

$$\sigma_X^2 = R_{XX}(0) - E[X(t)]^2 = 7 - 2^2 = 3.$$

Example 10.25

Given a random process $X(t) = A\cos\omega_0 t$, where ω_0 is a constant and A is uniformly distributed with mean 5 and variance 2. Find the average power of $X(t)$. (Jan 2015)

Solution

Given $X(t) = A\cos\omega_0 t$, $E[A] = 5$, $\sigma_A^2 = 2$.

Then, $E[A^2] = \sigma_A^2 + E[A]^2 = 2 + 5^2 = 27$.

The autocorrelation of $X(t)$ is

$$\begin{aligned} R_{XX}(t, t+\tau) &= E[X(t)X(t+\tau)] \\ &= E[A\cos(\omega_0 t)A\cos\omega_0(t+\tau)] \\ &= E[A^2\cos(\omega_0 t)\cos\omega_0(t+\tau)] \\ &= E[A^2]\cos(\omega_0 t)\cos\omega_0(t+\tau). \end{aligned}$$

$$\therefore \quad R_{XX}(t, t+\tau) = 27\cos(\omega_0 t)\cos\omega_0(t+\tau).$$

We know that the mean square value

$$E[X^2(t)] = \lim_{\tau\to 0} R_{XX}(t, t+\tau)$$
$$E[X^2(t)] = 27\cos^2\omega_0 t.$$

The average power is

$$P_{XX} = A\left[\overline{X^2(t)}\right]$$

$$P_{XX} = \lim_{T\to\infty}\frac{1}{2T}\int_{-T}^{T} E[X^2(t)]dt$$

$$= \lim_{T\to\infty}\frac{1}{2T}\int_{-T}^{T} 27\cos^2\omega_0 t\,dt$$

$$= \lim_{T\to\infty}\frac{27}{2T}\int_{-T}^{T}\left[\frac{1+\cos 2\omega_0 t}{2}\right]dt$$

$$= \frac{27}{4T}[2T-0] = \frac{27}{2} = 13.5 \text{ watts.}$$

Example 10.26

Find the average power of the random process $X(t)$ which has the power spectral density

$$S_{XX}(\omega) = \begin{cases} 1-\dfrac{\omega}{4\pi} & |\omega| < 4\pi \\ 0 & \text{otherwise.}\end{cases}$$ (Sept 2014)

Solution

Given psd of $X(t)$ is

$$S_{XX}(\omega) = \begin{cases} 1-\dfrac{\omega}{4\pi} & |\omega| < 4\pi \\ 0 & \text{otherwise.}\end{cases}$$

We know that the average power

$$P_{XX} = E[X^2(t)] = \frac{1}{2\pi}\int_{-\infty}^{\infty} S_{XX}(\omega)d\omega$$

$$= \frac{1}{2\pi}\int_{-4\pi}^{4\pi}\left(1-\frac{\omega}{4\pi}\right)d\omega$$

$$= \frac{1}{2\pi}\left[\omega - \frac{\omega^2}{8\pi}\right]_{-4\pi}^{4\pi}$$

$$= \frac{1}{2\pi}\left\{4\pi-(-4\pi) - \frac{1}{8\pi}\left[(4\pi)^2-(-4\pi)^2\right]\right\}$$

$$= \frac{1}{2\pi}[8\pi] = 4 \text{ watts.}$$

Example 10.27

A random process has the power density spectrum $\xi_{XX}(\omega) = \frac{6\omega^2}{1+\omega^2}$. Find the average power in the process. (May 2011)

Solution

Given $\xi_{XX}(\omega) = \dfrac{6\omega^2}{1+\omega^2}$

Average power of $X(t)$ is

$$P_{xX} = E[X^2(t)] = \frac{1}{2\pi}\int_{-\infty}^{\infty} \xi_m(\omega)d\omega$$

$$= \frac{1}{2\pi}\int_{-\infty}^{\infty} \frac{6\omega^2}{1+\omega^4}d\omega$$

$$= \frac{6}{2\pi}\left[\int_{-\infty}^{\infty} \frac{\omega^2}{1+\omega^4}d\omega\right] \qquad \text{(From Appendix A)}$$

$$P_{XX} = \frac{6}{2\pi}\left[\frac{\pi}{2\sqrt{2}}\right] = \frac{3}{2\sqrt{2}} = 1.06 \text{ watts.}$$

Example 10.28

Find the average power of the WSS random process $X(t)$ which has the power spectral density

$$S_{XX}(\omega) = \frac{\omega^2-17}{(\omega^2+49)(\omega^2+16)} \qquad \text{(Dec 2013).}$$

Solution

We know that the average power of $X(t)$ is

$$P_{XX} = E[X^2(t)] = \frac{1}{2\pi}\int_{-\infty}^{\infty} S_{XX}(\omega)d\omega$$

$$= \frac{1}{2\pi}\int_{-\infty}^{\infty} \frac{\omega^2-17}{(\omega^2+49)(\omega^2-16)}d\omega$$

$$= \frac{1}{2\pi}\int_{-\infty}^{\infty} \left[\frac{2}{\omega^2+49} - \frac{1}{\omega^2+16}\right] d\omega$$

$$= \frac{1}{2\pi}\int_{-\infty}^{\infty} \frac{2}{\omega^2+49}d\omega - \frac{1}{2\pi}\int_{-\infty}^{\infty} \frac{1}{\omega^2+16}d\omega$$

$$= \frac{1}{2\pi}\left[\frac{2}{7}\tan^{-1}(\omega/7) - \frac{1}{4}\tan^{-1}(\omega/4)\right]_{-\infty}^{\infty} \qquad \text{(From Appendix C)}$$

$$= \frac{1}{2\pi}\left[\frac{2\pi}{7} - \frac{\pi}{4}\right] = \frac{1}{2}\left[\frac{2}{7} - \frac{1}{4}\right].$$

$$P_{XX} = \frac{1}{56} \text{ watts}$$

Example 10.29

A WSS random process $X(t)$ has psd

$$S_{XX}(\omega) = \frac{\omega^2}{\omega^4 + 10\omega^2 + 9}. \qquad \text{(May 2013)}$$

Find the autocorrelation and mean square value of the process.

Solution

Given WSS random process with psd $S_{XX}(\omega) = \dfrac{\omega^2}{\omega^4 + 10\omega^2 + 9}$.

The autocorrelation function

$$R_{XX}(\tau) = \mathcal{F}^{-1}[S_{XX}(\omega)]$$

$$= \mathcal{F}^{-1}\left[\frac{\omega^2}{\omega^4 + 10\omega^2 + 9}\right]$$

$$= \mathcal{F}^{-1}\left[\frac{\omega^2}{(\omega^2+1)(\omega^2+9)}\right]$$

$$= \mathcal{F}^{-1}\left[\frac{1}{8}\left(\frac{9}{\omega^2+9} - \frac{1}{\omega^2+1}\right)\right].$$

We know that $\dfrac{2a}{\omega^2 + a^2} \leftrightarrow e^{-a|\tau|}$. (From Appendix A)

$$\therefore \quad S_{XX}(\omega) = F^{-1}\left[\frac{1}{8}\left(\frac{9}{6} \times \frac{6}{\omega^2+3^2} - \frac{1}{2} \times \frac{2}{\omega^2+1}\right)\right].$$

$$\therefore \quad R_{XX}(\tau) = \frac{3}{16}e^{-3|\tau|} - \frac{1}{16}e^{-|\tau|}.$$

The mean square value of the process $X(t)$ is

$$E[X^2(t)] = R_{XX}(0) = \frac{3}{16} - \frac{1}{16} = \frac{1}{8} \text{ watts.}$$

Example 10.30

The spectral density of a WSS random process $X(t)$ is given by

$$S_{xx}(\omega) = \frac{\omega^2}{\omega^4 + 13\omega^2 + 36} \qquad \text{(May 2013)}$$

Find the autocorrelation and average power of the process.

Solution

Given WSS random process with psd

$$S_{XX}(\omega) = \frac{\omega^2}{\omega^4 + 13\omega^2 + 36}.$$

The autocorrelation function is

$$R_{XX}(\tau) = \mathcal{F}^{-1}[S_{XX}(\omega)]$$

$$= \mathcal{F}^{-1}\left[\frac{\omega^2}{\omega^4+13\omega^2+36}\right] = \mathcal{F}^{-1}\left[\frac{\omega^2}{(\omega^2+4)(\omega^2+9)}\right]$$

$$= \mathcal{F}^{-1}\left[\frac{-4/5}{(\omega^2+4)} + \frac{9/5}{(\omega^2+9)}\right]$$

$$= \mathcal{F}^{-1}\left[\frac{-4}{5}\times\frac{4}{4(\omega^2+4)} + \frac{9}{5}\times\frac{6}{6(\omega^2+9)}\right].$$

We know that $\frac{2a}{\omega^2+a^2} \leftrightarrow e^{-a|\tau|}$.

$$R_{XX}(\tau) = F^{-1}\left[\frac{-1}{5}\times\frac{4}{\omega^2+4} + \frac{9}{30}\times\frac{6}{\omega^2+9}\right] = \frac{-1}{5}e^{-4|\tau|} + \frac{9}{30}e^{-3|\tau|}.$$

The average power is

$$P_{XX} = R_{XX}(\tau)|_{\tau=0} = \frac{-1}{5} + \frac{9}{30} = 0.1 \text{ watts.}$$

Example 10.31

The power density spectrum of a random process is given by $S_{XX}(\omega) = \frac{16}{\omega^2+16}$. Find out whether it is a valid spectral density or not. If it is transmitted through a system as shown in Fig. 10.6, find the output power density spectrum.

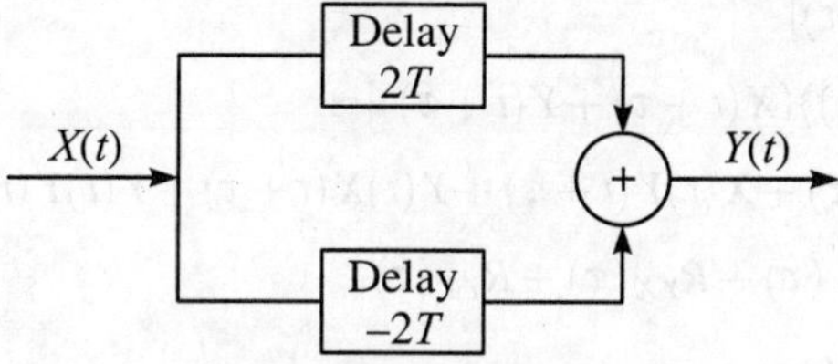

Fig. 10.6 *A delay system*

Solution

Given $S_{XX}(\omega) = \frac{16}{\omega^2+16}$.

For a valid spectral density, $S_{XX}(\omega)$ should satisfy

1. $S_{XX}(\omega) > 0$ for any value of ω
2. $S_{XX}(\omega) = S_{XX}(-\omega)$.

It is observed that $S_{XX}(\omega)$ satisfies both conditions. Therefore, $S_{XX}(\omega)$ is a valid spectral density. From Fig. 10.31, the output is

$$Y(t) = X(t+2T) + X(t-2T).$$

The Fourier transform is

$$Y(\omega) = e^{j2T\omega}X(\omega) + e^{-j2T\omega}X(\omega) = X(\omega)(e^{j2T\omega} + e^{-j2T\omega}).$$

$$\therefore \quad H(\omega) = \frac{Y(\omega)}{X(\omega)} = e^{j2T\omega} + e^{-j2T\omega} = 2\cos(2\omega T).$$

We know that

$$S_{YY}(\omega) = |H(\omega)|^2 S_{XX}(\omega)$$

$$S_{YY}(\omega) = 4\cos^2(2\omega T)\frac{16}{\omega^2+16} = \frac{64\cos^2(2\omega T)}{\omega^2+16}.$$

Example 10.32

The stationary random processes $X(t)$ and $Y(t)$ have spectral density functions $S_{XX}(\omega) = \frac{16}{\omega^2+16}$, and $S_{YY}(\omega) = \frac{\omega^2}{\omega^2+16}$ respectively. Let another stationary random process be $U(t) = X(t) + Y(t)$. Find the spectral densities $S_{UU}(\omega), S_{XU}(\omega)$, and $S_{YU}(\omega)$. Assume that $X(t)$ and $Y(t)$ are uncorrelated with zero mean values. (Jan 2015)

Solution

Given that $S_{XX}(\omega) = \dfrac{16}{\omega^2+16}$, $S_{YY}(\omega) = \dfrac{\omega^2}{\omega^2+16}$, $U(t) = X(t) + Y(t)$.

$S_{UU}(\omega) = F[R_{UU}(\tau)]$

The autocorrelation function is

$$\begin{aligned} R_{UU}(\tau) &= E[U(t)U(t+\tau)] \\ &= E[(X(t)+Y(t))(X(t+\tau)+Y(t+\tau))] \\ &= E[X(t)X(t+\tau) + X(t)Y(t+\tau) + Y(t)X(t+\tau) + Y(t)Y(t+\tau)] \\ &= R_{XX}(\tau) + R_{XY}(\tau) + R_{YX}(\tau) + R_{YY}(\tau). \end{aligned}$$

$\therefore$ The power spectral density is

$$\begin{aligned} S_{UU}(\omega) &= F[R_{XX}(\tau)] + F[R_{XY}(\tau)] + F[R_{YX}(\tau)] + F[R_{YY}(\tau)] \\ &= S_{XX}(\omega) + S_{XY}(\omega) + S_{YX}(\omega) + S_{YY}(\omega). \end{aligned}$$

Since $X(t)$ and $Y(t)$ are uncorrelated and have zero mean values, then,

$$S_{XY}(\omega) = S_{YX}(\omega) = \delta(\omega)\overline{XY} = 0.$$

$$\therefore \quad S_{UU}(\omega) = \frac{16}{\omega^2+16} + 0 + 0 + \frac{\omega^2}{\omega^2+16} = \frac{\omega^2+16}{\omega^2+16} = 1$$

and $\quad S_{XU}(\omega) = F[R_{XU}(\omega)].$

$$\begin{aligned} \text{Now,} \quad R_{XU}(\omega) &= E[X(t)U(t+\tau)] \\ &= E[X(t)(X(t+\tau) + Y(t+\tau))]. \end{aligned}$$

$$R_{XU}(\omega) = R_{XX}(\tau) + R_{XY}(\tau).$$

$$\therefore \quad S_{XU}(\omega) = S_{XX}(\omega) + S_{XY}(\omega)$$

or $$S_{XU}(\omega) = \frac{16}{\omega^2 + 16}.$$

Similarly,

$$S_{YU}(\omega) = S_{YX}(\omega) + S_{YY}(\omega) = \frac{\omega^2}{\omega^2 + 16}.$$

Example 10.33

A random process $Y(t)$ is obtained from the system as shown in Fig. 10.7. Find the output spectral density in terms of input spectral density. A_0 and ω_0 are constants, $X(t)$ is the input random process and its power spectral density is $S_{XX}(\omega)$. (Sept 2014)

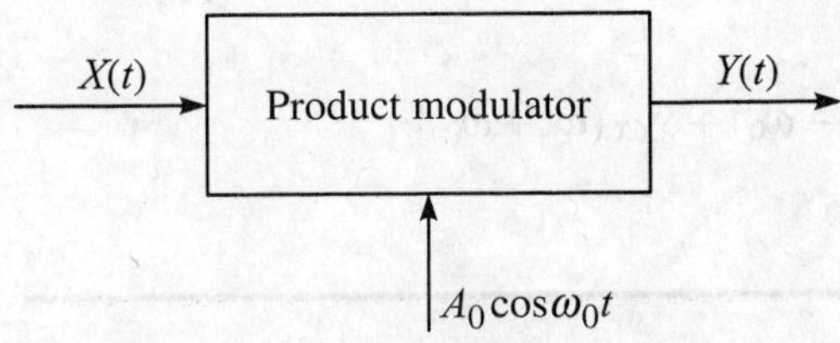

Fig. 10.7 *Product modulator*

Solution

From Fig. 10.7, $Y(t)$ is given by

$$Y(t) = X(t)A_0 \cos \omega_0 t.$$

$$\begin{aligned} R_{YY}(t, t+\tau) &= E[Y(t)Y(t+\tau)] \\ &= E[X(t)A_0 \cos(\omega_0 t) X(t+\tau) A_0 \cos \omega_0 (t+\tau)] \\ &= A_0^2 E[X(t)X(t+\tau)] \cos(\omega_0 t) \cos \omega_0 (t+\tau) \\ &= A_0^2 R_{XX}(\tau) \cos(\omega_0 t) \cos \omega_0 (t+\tau) \end{aligned}$$

$$R_{YY}(t, t+\tau) = A_0^2 R_{XX}(\tau) \frac{1}{2} (\cos \omega_0 \tau + \cos(2\omega_0 t + \omega_0 \tau)).$$

Since R_{YY} depends on t, the output is not a wide sense stationary random process. The output spectral density is

$$S_{YY}(\omega) = \mathcal{F}\{A[R_{YY}(t, t+\tau)]\}.$$

We know that,

$$A[R_{YY}\ (t, t\ +\tau)] = \lim_{T \to \infty} \frac{1}{2T} \int_{-T}^{T} R_{YY}(t, t+\tau) dt.$$

$$R_{YY}(\tau) = \lim_{T\to\infty} \frac{1}{2T}\int_{-T}^{T} \frac{A_0^2}{2} R_{XX}(\tau)(\cos\omega_0\tau + \cos(2\omega_0 t + \omega_0\tau))dt$$

$$= \lim_{T\to\infty} \frac{1}{2T}\frac{A_0^2}{2} R_{XX}(\tau)\left[t\cos\omega_0\tau + \frac{\sin(2\omega_0 t + \omega_0\tau)}{2\omega_0}\right]_{T}^{T}$$

$$= \lim_{T\to\infty} \frac{1}{2T}\frac{A_0^2}{2} R_{XX}(\tau)[2T\cos\omega_0\tau + 0]$$

$$= \frac{A_0^2}{2} R_{XX}(\tau)\cos\omega_0\tau.$$

The power spectral density is

$$S_{YY}(\omega) = \mathcal{F}[R_{XX}(\tau)]$$

$$= \mathcal{F}\left[\frac{A_0^2}{2} R_{XX}(\tau)\cos\omega_0\tau\right]$$

$$= \frac{A_0^2}{2}\mathcal{F}[R_{XX}(\tau)\cos\omega_0\tau] \qquad \text{(From Appendix B)}$$

$$= \frac{A_0^2}{4}[S_{XX}(\omega - \omega_0) + S_{XX}(\omega + \omega_0)].$$

Example 10.34

If the input auto correlation function is $R_{XX}(\tau) = Ae^{-\alpha|\tau|}$, where A and α are constants, find the output spectral density and draw the output power spectrum for

$$S_{YY}(\omega) = \frac{A_0^2}{4}[S_{XX}(\omega - \omega_0) + S_{XX}(\omega + \omega_0)]. \qquad \text{(March 2004)}$$

Solution

Given $R_{XX}(\tau) = Ae^{-\alpha|\tau|}$. The power spectral density is $S_{XX}(\omega) = F[R_{XX}(\tau)]$.

$$\therefore\ S_{XX}(\omega) = \int_{-\infty}^{\infty} A\, e^{-\alpha|\tau|} e^{-j\omega\tau} d\tau$$

$$= \int_{-\infty}^{0} Ae^{\alpha\tau}e^{-j\omega\tau}d\tau + \int_{0}^{\infty} Ae^{-\alpha\tau}e^{-j\omega\tau}d\tau$$

$$= A\left[\int_{-\infty}^{0} e^{(\alpha - j\omega)\tau}d\tau + \int_{0}^{\infty} e^{-(\alpha + j\omega)\tau}d\tau\right]$$

$$= A\left[\frac{1}{\alpha - j\omega} + \frac{1}{\alpha + j\omega}\right]$$

$$S_{XX}(\omega) = \frac{A2\alpha}{\omega^2 + \alpha^2}.$$

Now $S_{XX}(\omega-\omega_0)=\dfrac{2A\alpha}{(\omega-\omega_0)^2+\alpha^2}$

and $S_{XX}(\omega+\omega_0)=\dfrac{2A\alpha}{(\omega+\omega_0)^2+\alpha^2}.$

Given the output power density, $S_{YY}(\omega)=\dfrac{A_0^2}{4}[S_{XX}(\omega-\omega_0)+S_{XX}(\omega+\omega_0)]$,

$$\therefore\quad S_{YY}(\omega)=A\left(\frac{A_0^2}{4}\right)\left[\frac{2\alpha}{(\omega-\omega_0)^2+\alpha^2}+\frac{2\alpha}{(\omega+\omega_0)+\alpha^2}\right].$$

Figure. 10.8 shows the input autocorrelation and its density spectrum and Fig. 10.9 shows the output power density spectrum.

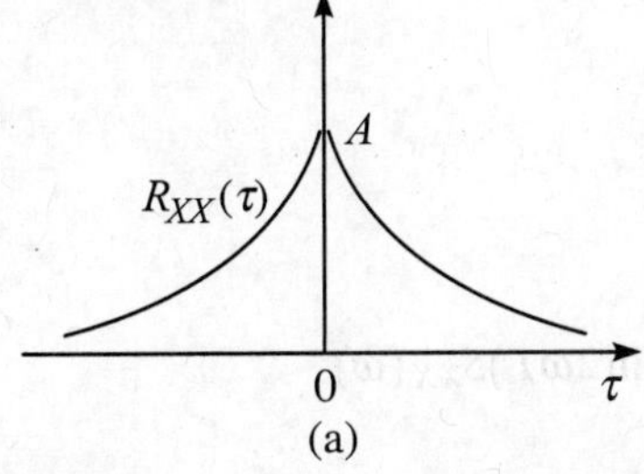

Fig. 10.8 (a) *Input correlation function*

Fig. 10.8 (b) *Input power spectral density*

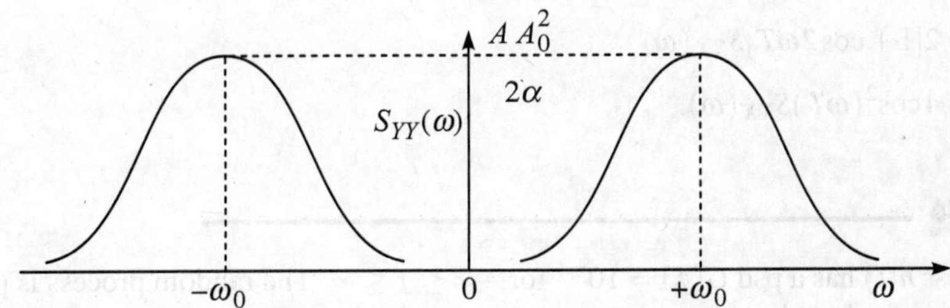

Fig. 10.9 *Output power density spectrum*

Example 10.35

A WSS random process $X(t)$ with psd $S_{XX}(\omega)$ is applied at the input of a delay system shown in Fig. 10.10. Find the psd of $Y(t)$. (March 2014)

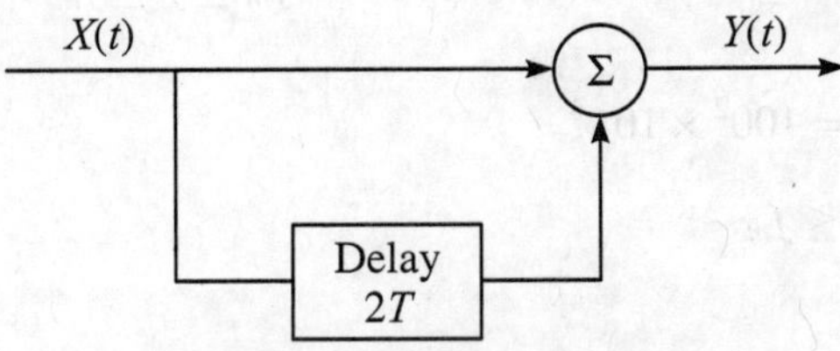

Fig. 10.10 *Delay system*

Solution

Given the psd of $X(t)$ is $S_{XX}(\omega)$. From Fig. 10.10,

$$Y(t) = X(t) + X(t-2T).$$

Taking Fourier transform

$$Y(\omega) = X(\omega) + X(\omega)\, e^{-j\omega 2T}$$

$$Y(\omega) = X(\omega)[1 + e^{-j2\omega T}].$$

The system transfer function is

$$H(\omega) = \frac{Y(\omega)}{X(\omega)} = 1 + e^{-j2\omega T}.$$

Output psd is $S_{YY}(\omega) = |H(\omega)|^2 S_{XX}(\omega)$

$$= |1 + e^{-j2\omega T}|^2 S_{XX}(\omega)$$

$$\begin{aligned} S_{YY}(\omega) &= |1 + e^{-j2\omega T}|^2 S_{XX}(\omega) \\ &= (1 + \cos 2\omega T - j\sin 2\omega T)(1 + \cos 2\omega T + j\sin 2\omega T) S_{XX}(\omega) \\ &= [(1 + \cos 2\omega T)^2 + \sin^2 2\omega T] S_{XX}(\omega) \\ &= [1 + \cos^2 2\omega T + 2\cos 2\omega T + \sin^2 2\omega T] S_{XX}(\omega) \\ &= 2[1 + \cos 2\omega T] S_{XX}(\omega) \end{aligned}$$

$$S_{YY}(\omega) = 4\cos^2(\omega T) S_{XX}(\omega).$$

Example 10.36

A random process $n(t)$ has a psd $G(f) = 10^{-4}$ for $-\infty \le f \le \infty$. The random process is passed through an LPF whose transfer function is

$$H(f) = \begin{cases} 100 & -f_m \le f \le f_m \\ 0 & \text{otherwise} \end{cases}.$$

Find the psd of the waveform at the output of the filter. (May 2011)

Solution

Given $G(f) = 10^{-4}$ $\quad -\infty \le f \le \infty$ and $H(f) = 100 \quad -f_m \le f \le f_m$

The output psd is

$$G_0(f) = |H(f)|^2 G(f) = 100^2 \times 10^{-4}$$

$$G_0(f) = 1 \quad -f_m \le f \le f_m.$$

Example 10.37

Consider a linear system as shown in Fig. 10.11.

$$x(t) \rightarrow \boxed{\frac{1}{6+j\omega}} \rightarrow y(t)$$

Fig. 10.11 *Linear system*

$x(t)$ is the input and $y(t)$ is the output of the system. The autocorrelation of $x(t)$ is $R_{XX}(\tau) = 5\delta(\tau)$. Find the psd, autocorrelation function and mean square value of the output $y(t)$. (Nov 2010)

Solution

Given $H(\omega) = \dfrac{1}{6+5\omega}$, $R_{XX}(\tau) = 5\delta(\tau)$

(i) psd of the output is $S_{YY}(\omega) = S_{XX}(\omega)\, |H(\omega)\,|^2$.
psd of the input $S_{XX}(\omega) = F[R_{XX}(\tau)] = 5$.

$$\therefore \quad S_{YY}(\omega) = 5 \times \frac{1}{6^2+\omega^2} = \frac{5}{\omega^2+36}.$$

(ii) Output autocorrelation function:

$$R_{YY}(\tau) = \mathcal{F}^{-1}[S_{YY}(\omega)] = \mathcal{F}^{-1}\left[\frac{5}{\omega^2+36}\right].$$

Since $\dfrac{2a}{\omega^2+a^2} \leftrightarrow e^{-a|\tau|}$,

$$R_{YY}(\tau) = \mathcal{F}^{-1}\left[\frac{5(12)}{\omega^2+6^2}\left(\frac{1}{12}\right)\right]$$

$$= \frac{5}{12}\mathcal{F}^{-1}\left[\frac{12}{\omega^2+6^2}\right] = \frac{5}{12}e^{-6|\tau|}.$$

(iii) Mean square value:

$$E[X^2] = R_{YY}(0) = \frac{5}{12}e^{-(0)} = \frac{5}{12}\text{ watts} = 0.4167 \text{ watts}.$$

Example 10.38

A received SSB signal has a power spectrum which extends over the frequency range from $f_C = 1$ MHz to $f_C + f_m = 1.003$ MHz. The signal is accompanied by noise with uniform spectral density 10^{-9} watts/Hz.

(a) The noise $n(t)$ is expressed as

$$n(t) = n_I(t)\cos 2\pi f_C t - n_s(t)\sin 2\pi f_C t.$$

Find the power spectral densities of the quadrature components in the spectral range specified as $f_C \le f \le f_C + f_m$.

(b) The signal plus its accompanying noise is multiplied by a local carrier $\cos(2\pi f_C t)$. Plot the psd of the noise at the output of the multiplier. (Dec 2013)

Solution

Given SSB has frequencies $f_c = 1$ MHz to $f_c + f_m = 1.003$ MHz.

$\therefore\ f_m = 3$ kHz.

Noise psd $S_N(f) = 10^{-9}$ watts/Hz.

(a) Given the noise $n(t) = n_I(t)\cos 2\pi f_c t - n_s(t)\sin 2\pi f_c t$.

The power densities of the in-phase and quadrature phase components are:

$$S_{N_I}(f) = S_{N_s}(f) = S_N(f_c - f_m) + S_N(f_c + f_m)$$

$$= 10^{-9} + 10^{-9} = 2 \times 10^{-9} \text{ watts/Hz.}$$

(b) When the signal is multiplied by a local carrier $\cos(2\pi f_c t)$, its power will be reduced 4 times. The output power will have uniform power spectral density of $(10^{-9}/4)$ watts/Hz, with the following frequency components: (i) $f_c + f_c = 2f_c$ (ii) $f_c - f_c = 0$ (iii) $f_c + f_c + f_m = 2f_c + f_m$ (iv) $f_c - f_c + f_m = f_m$. The plot of the output psd is shown in Fig. 10.12.

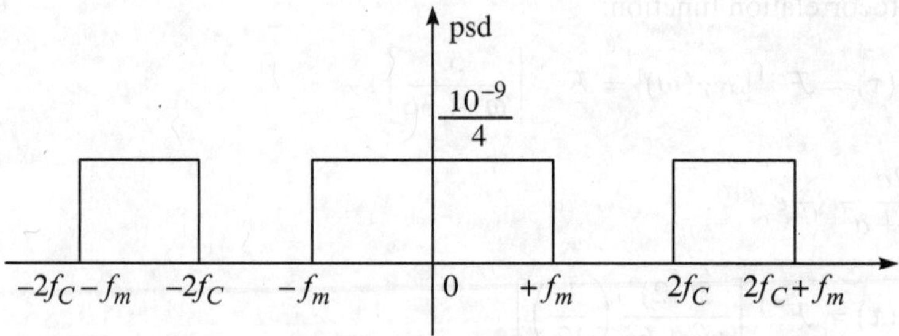

Fig. 10.12 *Output psd*

Example 10.39

A stationary random process $X(t)$ having an autocorrelation function $R_{XX}(\tau) = 2e^{-4|\tau|}$ is applied to the network shown in Fig. 10.13. Find (i) $S_{XX}(\omega)$, (ii) $|H(\omega)|^2$ and (iii) $S_{YY}(\omega)$. (March 2014)

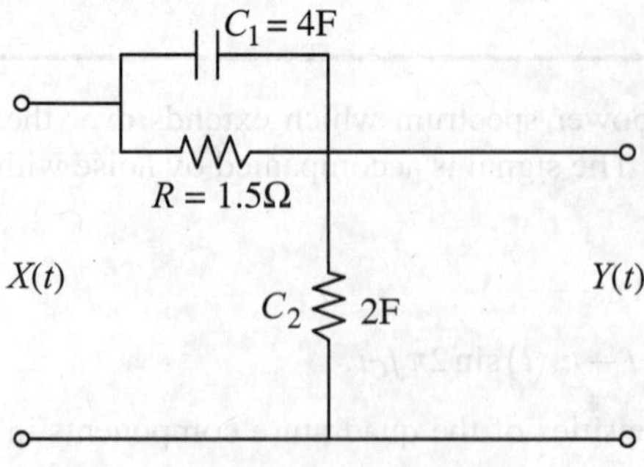

Fig. 10.13 *RC network*

Solution

Given $R_{XX}(\tau) = 2e^{-4|\tau|}$.

(i) $S_{XX}(\omega) = F[R_{XX}(\tau)] = \int_{-\infty}^{\infty} 2e^{-4|\tau|}e^{-j\omega\tau}d\tau$

$$= 2\int_{-\infty}^{0} e^{4\tau}e^{-j\omega\tau}d\tau + 2\int_{0}^{\infty} e^{4\tau}e^{-j\omega\tau}d\tau$$

$$= 2\int_{-\infty}^{0} e^{-\tau(j\omega-4)}d\tau + 2\int_{0}^{\infty} e^{-\tau(j\omega+4)}d\tau$$

$$= 2\left[\frac{e^{-\tau(j\omega-4)}}{-(j\omega-4)}\right]_{-\infty}^{0} + 2\left[\frac{e^{-\tau(j\omega+4)}}{-(j\omega+4)}\right]_{0}^{\infty}$$

$$= 2\left[\frac{-1}{(j\omega-4)} + \frac{1}{j\omega+4}\right] = \frac{2\times 8}{\omega^2+4^2}$$

$$= \frac{16}{\omega^2+16}.$$

(ii) The circuit with reactance values is shown in Fig. 10.14.

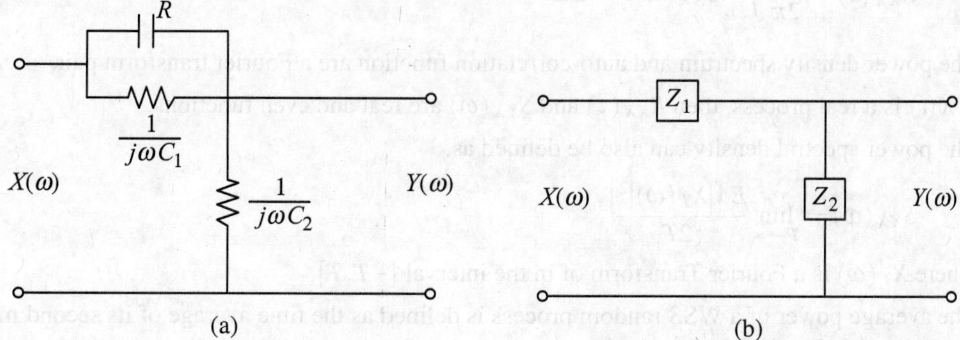

Fig. 10.14 (a) *RC network with reactance* **Fig. 10.14 (b)** *RC network with impedance*

The impedances are

$$Z_1 = \frac{R\frac{1}{j\omega C_1}}{R+\frac{1}{j\omega C_1}} = \frac{R}{1+j\omega C_1 R} \text{ and } Z_2 = \frac{1}{j\omega C_2}.$$

The transfer function

$$H(\omega) = \frac{Y(\omega)}{X(\omega)} = \frac{Z_2}{Z_1+Z_2} = \frac{\frac{1}{j\omega_2 C_2}}{\frac{1}{j\omega C_2}+\frac{R}{j\omega C_1 R}}$$

$$= \frac{1+j\omega C_1 R}{1+j\omega C_1 R + j\omega C_2 R} = \frac{1+j\omega C_1 R}{1+j\omega R(C_1+C_2)}.$$

Now, given $R = 1.5\ \Omega$, $C_1 = 4$ F, $C_2 = 2$ F.

$$H(\omega) = \frac{1+j\omega 6}{1+j\omega 9}.$$

$$\therefore \quad |H(\omega)|^2 = H(\omega)H^*(\omega) = \frac{(1+j\omega 6)(1-j\omega 6)}{(1+j\omega 9)(1-j\omega 9)} = \frac{1+36\omega^2}{1+81\omega^2}.$$

(iii) The output power density spectrum is

$$S_{YY}(\omega) = |H(\omega)|^2 S_{XX}(\omega) = \frac{(1+36\omega^2)}{(1+81\omega^2)} \times \frac{16}{\omega^2+16}$$

$$= \frac{16(1+36\omega^2)}{(16+\omega^2)(1+81\omega^2)}.$$

Summary

1. The power spectrum density of a wide sense stationary random process is defined as the Fourier transform of the auto correlation function.

$$S_{XX}(\omega) = \int_{-\infty}^{\infty} R_{XX}(\tau)e^{-j\omega\tau}d\tau.$$

2. Inverse Fourier transform of power density spectrum gives autocorrelation function

$$R_{XX}(\tau) = \frac{1}{2\pi}\int_{-\infty}^{\infty} S_{XX}(\omega)e^{j\omega\tau}d\omega.$$

3. The power density spectrum and auto-correlation function are a Fourier transform pair.
4. If $X(t)$ is a real process, then $R_{XX}(\tau)$ and $S_{XX}(\omega)$ are real and even functions.
5. The power spectral density can also be defined as,

$$S_{XX}(\omega) = \lim_{T\to\infty} \frac{E\left[|X_T(\omega)|^2\right]}{2T}$$

where $X_T(\omega)$ is a Fourier Transform of in the interval $[-T,T]$.

6. The average power of a WSS random process is defined as the time average of its second moment or autocorrelation function at $\tau = 0$.

$$P_{XX} = A\left\{E\left[X^2(t)\right]\right\} \text{ or } P_{XX} = R_{XX}(\tau)|_{\tau=0} \text{ or} P_{XX} = \frac{1}{2\pi}\int_{-\infty}^{\infty} S_{XX}(\omega)d\omega$$

7. *The properties of the power density spectrum for a wide sense stationary random process $X(t)$*

   (i) $S_{XX}(\omega) \geq 0$

   (ii) $S_{XX}(0) = \int_{-\infty}^{\infty} R_{XX}(\tau)d\tau$

   (iii) $S_{XX}(-\omega) = S_{XX}(\omega)$, $X(t)$ is real

   (iv) $S_{XX}(\omega)$ is always a real function

   (v) $\frac{1}{2\pi}\int_{-\infty}^{\infty} S_{XX}(\omega)d\omega = A\left\{E\left[X^2(t)\right]\right\} = R_{XX}(0)$

   (vi) $S_{\dot{X}\dot{X}}(\omega) = \omega^2 S_{XX}(\omega)$

   (vii) $S_{XX}(\omega) = \int_{-\infty}^{\infty} R_{XX}(\tau)e^{-j\omega\tau}d\tau$

(viii) $R_{XX}(\tau) = \frac{1}{2\pi}\int_{-\infty}^{\infty} S_{XX}(\omega)e^{j\omega\tau}d\tau$

8. For *WSS* random process $X(t)$, the Fourier Transform pairs are called Wiener-Khintchine relations.

9. For bandpass process, the rms bandwidth of power spectrum is $W_{\text{rms}}^2 = \frac{\int_{-\infty}^{\infty}\omega^2 S_{XX}(\omega)d\omega}{\int_{-\infty}^{\infty} S_{XX}(\omega)d\omega}$.

10. For bandpass process, the rms bandwidth of power spectrum is

$$W_{\text{rms}}^2 = \frac{4\int_0^{\infty}(\omega-\overline{\omega}_0)^2 S_{XX}(\omega)d\omega}{\int_0^{\infty} S_{XX}(\omega)d\omega} \quad \text{where} \quad \overline{\omega}_0 = \frac{\int_0^{\infty}\omega^2 S_{XX}(\omega)d\omega}{\int_0^{\infty} S_{XX}(\omega)d\omega}.$$

11. The cross power density spectrum is defined as the Fourier transform of the cross-correlation function $X(t)$ and $Y(t)$. It is expressed as $S_{XY}(\omega) = \int_{-\infty}^{\infty} R_{XY}(\tau)e^{-j\omega}d\tau$ and $S_{YX}(\omega) = \int_{-\infty}^{\infty} R_{YX}(\omega)\, e^{-j\omega\tau}d\tau$.

12. Inverse Fourier transform of cross power density spectrum gives cross correlation function.

$$R_{XY}(\tau) = \frac{1}{2\pi}\int_{-\infty}^{\infty} S_{XY}(\omega)e^{j\omega\tau}d \text{ and } R_{YX}(\tau) = \frac{1}{2\pi}\int_{-\infty}^{\infty} S_{YX}(\omega)e^{j\omega\tau}d\omega$$

13. Also, $S_{XY}(\omega) = \lim_{T\to\infty}\frac{E[X_T^*(\omega)Y_T(\omega)]}{2T}$ and $S_{YX}(\omega) = \lim_{T\to\infty}\frac{E[Y_T^*(\omega)X_T(\omega)]}{2T}$

14. Average power $P_{YX} = \frac{1}{2\pi}\int_{-\infty}^{\infty} S_{YX}(\omega)d\omega$ or $P_{YX} = R_{YX}(0)$ and $P_{XY} = P_{YX}^*$

15. *The properties of the cross power density spectrum for a wide sense stationary random process $X(t)$ and $Y(t)$*
    (i) $S_{XY}(\omega) = S_{YX}(-\omega) = S_{YX}^*(\omega)$
    (ii) $\text{Re}[S_{XY}(\omega)]$ and $\text{Re}[S_{YX}(\omega)]$ are even functions.
    (iii) $\text{Im}[S_{XY}(\omega)]$ and $\text{Im}[S_{XY}(\omega)]$ are odd functions.
    (iv) $S_{XY}(\omega) = 0$ and $S_{YX}(\omega) = 0$ if $X(t)$ and $Y(t)$ are orthogonal.
    (v) $S_{XY}(\omega) = \mathcal{F}\{A[R_{XY}(t,t+\tau)]\}$
    (vi) $S_{YX}(\omega) = \mathcal{F}\{A[R_{YX}(t,t+\tau)]\}$
    (vii) If $X(t)$ and $Y(t)$ are jointly WSS random processes, then $S_{XY}(\omega) = \mathcal{F}[R_{XY}(\tau)]$ and $S_{YX}(\omega) = \mathcal{F}[R_{YX}(\tau)]$

16. If a random process $X(t)$ is applied on an LTI system having a transfer function $H(\omega)$ and the output response is $Y(t)$, the output power spectrum is $S_{YY}(\omega) = |H(\omega)|^2 S_{XX}(\omega)$, where $S_{XX}(\omega)$ is the input power spectrum.

17. The cross power spectrums are $S_{XY}(\omega) = S_{XX}(\omega)H(\omega)$, $S_{YX}(\omega) = S_{XX}(\omega)H(-\omega)$, $S_{YY}(\omega) = S_{XY}(\omega)H(-\omega)$ and $S_{XY}(\omega) = S_{YX}(\omega)H(\omega)$.

18. RMS bandwidth or normalised bandwidth is given by $W_{\text{rms}}^2 = \frac{\int_{-\infty}^{\infty}\omega^2 S_{XX}(\omega)d\omega}{\int_{-\infty}^{\infty} S_{XX}(\omega)d\omega}$.

Review Questions

1. State and prove any four properties of the power spectral density of a random process. (Feb 2007, Feb 2008, Jan 2015)
2. State and prove the Wiener-Khintchin relations.

   or

   Power spectrum and autocorrelation functions are a Fourier transform pair. Prove this statement. (Feb 2007, Nov 2007, May 2011, Jan 2015)
3. The psd of a random process is given by

$$\delta_{XY}(\omega) = \begin{cases} \pi, & |\omega| < 1 \\ 0, & \text{elsewhere} \end{cases}$$

   Find its autocorrelation function. (Feb 2008, Dec 2013)
4. If the psd of $X(t)$ is $S_{XX}(\omega)$, find the psd of $\dfrac{dx(t)}{dt}$. (Nov 2007, Feb 2008, Dec 2013)
5. Prove that $S_{XX}(\omega) = S_{XX}(-\omega)$. (Nov 2007, Feb 2008, Sept 2014)
6. If $R(\tau) = ae^{-b|\tau|}$, find the spectral density function, where a and b are constants. (Nov 2007, Feb 2008, Sept 2014)
7. Write a short note on the cross power density spectrum. (May 2009, May 2011, Jan 2015)
8. State and prove any three properties of the cross power density spectrum. (May 2009, May 2011, Sept 2014)
9. Derive the relation between cross power spectrum and cross correlation function. (Nov 2010, Sept 2014)
10. Prove that the power spectrum and autocorrelation function of the random process form a Fourier transform pair.
11. Derive the relation between autocorrelation function and power spectral density.
12. Derive the relation between the PSD of the input and the output random process of an LTI system.
13. Explain Spectral characteristics of a linear system response.

Problems

1. Determine the autocorrelation function and power spectral density of the random process, $X(t) = A\cos(\omega_0 t + \theta)$, where θ is a random variable over the ensemble and is uniformly distributed over the range $(0, 2\pi)$.
2. Find the psd of a random process $X(t)$ if $E[X(t)] = 1$ and $R_{XX}(\tau) = 1 + e^{-a|\tau|}$.
3. Find the psd and mean square value of the random process $X(t) = \text{A} \cos(\omega_c t + \theta)$, where θ is a uniform random variable over $(0, 2\pi)$.
4. A wide sense stationary noise process $N(t)$ has an autocorrelation function $R_{NN}(\tau) = ke^{-3|\tau|}$, where k is a constant. Find its power spectral density.
5. Consider a random process $X(t) = \cos(\omega t + \theta)$, where ω is a real constant and θ is a uniform random variable in $(0, \pi/2)$. Find the average power in the process.

6. Find the power spectrum for the random process $X(t)$ whose autocorrelation function is given as

$$R_{XX}(\tau) = \begin{cases} A\,(1+\tau/T) & -T<\tau<0 \\ A(1-\tau/T) & 0<\tau<T \\ 0 & \text{elsewhere.} \end{cases}$$

7. Show that the power spectrum of the autocorrelation function $R_{XX}(\tau) = e^{-a|\tau|}[1+a\,|\tau|]$ is equal to $\dfrac{4a^2}{(a^2+\omega^2)^2}$.

8. Find the psd of the random process $X(t) = 50\cos(100\pi t+\theta)$, where θ is a random variable uniformly distributed over $[0,2\pi]$.

9. Find the autocorrelation of a random process whose psd is $\dfrac{16}{4+\omega^2}$.

10. Determine the autocorrelation function and the power of a random process $X(t)$ with psd as shown in Fig. P10.1.

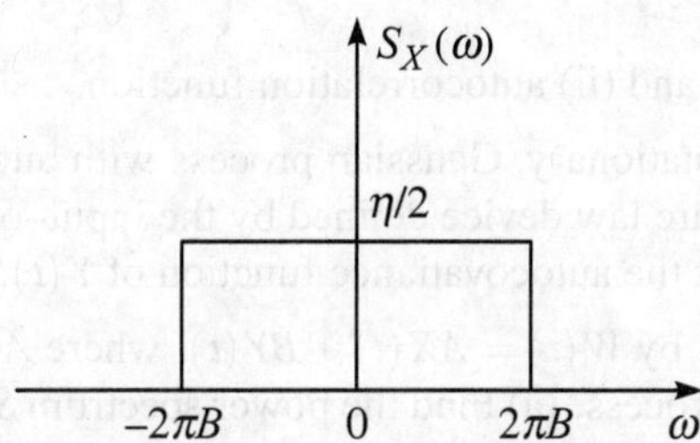

Fig. P10.1 *Power spectral density*

11. The power spectral density of a stationary random process is given by

$$S_{XY}(\omega) = \begin{cases} A, & -K<\omega<K \\ 0, & \text{otherwise.} \end{cases}$$

Determine the autocorrelation function and mean square value.

12. Find the autocorrelation function and psd of $m(t)\cos(\omega_C t+\theta)$, where $m(t)$ is a wide sense stationary random process and θ is a uniform random variable over $(0,2\pi)$ and independent of $m(t)$.

13. Determine the cross-correlation function corresponding to the cross power density spectrum

$$S_{XY}(\omega) = \frac{8}{(a+j\omega)^3} \quad \text{where } a>0 \text{ is a constant.}$$

14. Consider a cross-power spectrum given by

$$S_{XY}(\omega) = \begin{cases} a+jb\omega/K & -K<\omega<K \\ 0, & \text{elsewhere} \end{cases}$$

where $K>0$, a, and b are real constants. Find the cross-correlation function.

15. The cross-correlation of jointly WSS stationary processes $X(t)$ and $Y(t)$ is assumed to be $R_{XY}(\tau) = Bu(\tau)\exp(-W\tau)$, where $B>0$ and $W>0$ are constants.

(a) Find $R_{YX}(\tau)$. (b) Find $S_{XY}(\omega)$ and $S_{YX}(\omega)$.

16. A random process $n(t)$ has a psd $G(f) = 10^{-4}$ for $-\infty \le f \le \infty$. The random process is passed through an LPF whose transfer function is

$$H(f) = \begin{cases} 100 & -f_M \le f \le f_M \\ 0 & \text{otherwise.} \end{cases}$$

Find the psd of the waveform at the output of the filter.

17. Consider a linear system having transfer function $H(\omega) = \dfrac{1}{6+j\omega}$. $X(t)$ is the input and $Y(t)$ is the output of the system. The autocorrelation of $X(t)$ is $R_{XX}(\tau) = 3.\delta(\tau)$. Find the psd, autocorrelation function and mean square value of the output $Y(t)$.

18. A random process $Y(t)$ has the power spectral density $S_{YY}(\omega) = \dfrac{9}{\omega^2+64}$. Find (i) the average power of the process and (ii) the atucorrelation function.

19. A random process $X(t)$ has PSD given by $S_{XX}(\omega)\begin{cases} 4-\dfrac{\omega^2}{9} & |\omega| \le 6 \\ 0 & \text{elsewhere.} \end{cases}$

Find (i) the average power and (ii) autocorrelation function.

20. Let $X(t)$ be a zero mean, stationary, Gaussian process with autocorrelation function $R_X(\tau)$. This process is applied to a square law device defined by the input–output relation $Y(t) = X^2(t)$, where $Y(t)$ is the output. Find out the autocovatiance function of $Y(t)$.

21. A random process is given by $W(t) = AX(t) + BY(t)$, where A and B are real constants and $X(t)$ and $Y(t)$ are jointly WSS process. (a) Find the power spectrum $S_{WW}(\omega)$ of $W(t)$. (b) Find $S_{WW}(\omega)$ if $X(t)$ and $Y(t)$ are uncorrelated. (c) Find the cross power spectrum of $S_{WW}(\omega)$ and $S_{YW}(\omega)$.

22. Given a cross power spectrum, $S_{XY}(\omega) = \dfrac{8}{(a+j\omega)^3}$, where $a > 0$ is a constant, find the cross correlation function.

23. Find the power spectral density of a WSS random process $X(t)$, where the autocorrelation function is $R_{XX}(\tau) = a\, e^{-b|\tau|}$.

24. The stationary random process $X(t)$ has a mean value $E[X(t)] = 4$ and autocorrelation function $R_{XX}(\tau) = 16 + 2e^{-|\tau|}$. Find (i) the mean value and (ii) the variance of the random variable Y, where $Y = \int_0^2 X(t)dt$.

25. A WSS process $X(t)$ with a mean value 5 and power spectrum $S_{XX}(\omega) = 50\pi\,\delta(\omega) + \dfrac{3}{1+\left[\frac{\omega}{2}\right]^2}$ is applied to a network with impulse response $h(t) = 4e^{-4|t|}$.

   (i) Find $H(\omega)$ of the network.
   (ii) FInd the power spectrum of the response $y(t)$.

26. A stationary random process has an autocorrelation function of $R_X(\tau) = 16 - e^{-5|\tau|}\cos 20\pi\tau + 8\cos 10\pi\tau$. Find the variance of this process.

27. For a random process $x(t)$, assume that $R_{XX}(\tau) = \rho e^{-\tau^2/2a^2}$, where $\rho > 0$ and $a < 0$ are constants. Find the power density spectrum of $x(t)$.

Answers

1. $R_{XX}(t) = \frac{A^2}{2}\cos\omega_0\tau,\ S_{XY}(\omega) = \frac{\pi A^2}{2}[\delta(\omega+\omega_0)+\delta(\omega-\omega_0)]$
2. $\delta(\omega) + \frac{2a}{\omega^2+a^2}$
3. $psd = \frac{\pi A^2}{2}[\delta(\omega+\omega_C)+\delta(\omega-\omega_C)]$, Mean square value $= \frac{A^2}{2}$
4. $S_{nn}(\omega) = \frac{6K}{9+\omega^2}$
5. $\omega/2$
6. $AT\left(\frac{\sin\left(\frac{\omega T}{2}\right)}{\omega T/2}\right)^2$
8. $625[\delta(\omega-100\pi)+\delta(\omega+100\pi)]$
9. $4e^{-2|\tau|}$
10. $R_X(\tau) = \eta BSa(2\pi B\tau)$, Power $= \eta B$
11. $\frac{Ak}{\pi}$
12. Autocorrelation function $= \frac{1}{2}R_m(\tau)\cos\omega_C\tau$

    psd $= \frac{1}{4}[S_m(\omega+\omega_C)+S_m(\omega-\omega_C)]$, where $S_m(\omega)$ is the psd of $m(t)$
13. $4\tau^2e^{-a\tau}u(t)$
14. $R_{XY}(\tau) = \frac{1}{\pi K\tau^2}[(aK\tau-b)\sin(K\tau)+bK\tau\cos(K\tau)]$
15. $Be^{-\omega\tau}u(t),\ \frac{B}{W+j\omega},\ \frac{B}{W-j\omega}$
16. 1 W/ Hz
17. $\frac{3}{36+\omega_2},\ \frac{3}{12}e^{-|\tau|},\ \frac{1}{4}$
18. (i) $R_{XX} = \frac{9}{16}e^{-8|t|}$; (ii) $P_XX = \frac{9}{16}$ watts
19. (i) $R_{XX} = \frac{35}{9\pi\tau}\sin 6t - \frac{2}{9\pi\tau^2}\cos 6t + \frac{2}{9\pi\tau^3}\sin 6t$; (ii) $P_{XX} = \frac{16}{\pi}$ watts
20. $C_{YY}(\tau) = R_X(\tau) - R_X(0)$
21. (a) $S_{WW}(\omega) = A^2S_{XX}(\omega)+ABS_{XY}(\omega)+ABS_{YX}(\omega)+B^2S_{YY}(\omega)$

    (b) $S_{WW}(\omega) = A^2S_{XX}(\omega)+B^2S_{YY}(\omega)$

    (c) $S_{XW}(\omega) = AS_{XX}(\omega)+BS_{XY}(\omega)$; $S_{YW}(\omega) = AS_{YX}(\omega)+BS_{YY}(\omega)$;
22. $R_{XY}(\tau) = 4\tau^2e^{-a\tau}u(\tau)$
23. $S_{XX}(\omega) = \frac{2ab}{\omega^2+b^2}$
24. (i) 8; (ii) $\sigma_Y^2 = 8e^{-2} - 64$

25. (i) $H(\omega) = \dfrac{2}{1+\omega^2/16}$; (ii) $S_{YY}(\omega) = \dfrac{12}{(1+\omega^2/16)^2(1+\omega^2/4)} + \dfrac{200\pi\delta(\omega)}{(1+\omega^2/16)^2}$

26. $\sigma_X^2 = 7$

27. $S_{XX}(\omega) = \rho a\sqrt{2\pi}e^{-\omega^2 a^2/2}$

Multiple-choice Questions

1. The output of a filter is given by $Y(t) = X(t+T) - X(t-T)$, where $X(t)$ is a WSS process with power spectrum $S_{XX}(\omega)$ and T is a constant. The power spectrum of $Y(t)$ is

   (a) $\dfrac{\sin^2(\omega T)S_{XX}(\omega)}{9}$ (b) $\sin^2(\omega T)S_{XX}(\omega)$

   (c) $\dfrac{\sin^2(\omega T)S_{XX}(\omega)}{4}$ (d) $4\sin^2(\omega T)S_{XX}(\omega)$

2. The psd of a random process whose autocorrelation function is $a.e^{-b|\tau|}$ is

   (a) $\dfrac{a}{a^2+\omega^2}$ (b) $\dfrac{2ab}{a^2+\omega^2}$

   (c) $\dfrac{2ab}{b(a^2+\omega^2)}$ (d) $\dfrac{2ab}{a^2+\omega^2}$

3. The psd of a random process $X(t) = A\cos(Bt+Y)$, where Y is a uniform random variable over $(0, 2\pi)$ is

   (a) $\dfrac{\pi A^2}{2}$ (b) $\dfrac{\pi A^2}{2}[\delta(\omega+B) - \delta(\omega-B)]$

   (c) $\dfrac{\pi A^2}{2}[\delta(\omega+B) + \delta(\omega-B)]$ (d) $\dfrac{\pi A^2}{2}[\delta(\omega-B) - \delta(\omega+B)]$

4. The autocorrelation function of a random process whose psd $\dfrac{4}{1+\omega^2/4}$ is

   (a) $e^{-2|\tau|}$ (b) $2e^{-2|\tau|}$ (c) $3e^{-2|\tau|}$ (d) $4e^{-2|\tau|}$

5. The power density spectrum or power spectral density of $X_T(t) = X(t)$ for $-T < t < T$; $X_T(t) = 0$ elsewhere is $S_{XX}(\omega) =$

   (a) $\dfrac{1}{2\pi}\displaystyle\int_{-\infty}^{\infty} \dfrac{|X_T(\omega)|^2}{2T}$ (b) $\displaystyle\int_{-\infty}^{\infty} \dfrac{|X_T(\omega)|^2}{2T} d\omega$

   (c) $\lim\limits_{T\to\infty} \dfrac{E\{|X_T(\omega)|^2\}}{2T}$ (d) $\dfrac{1}{2\pi}\lim\limits_{T\to\infty} \dfrac{E\{|X_T(\omega)|^2\}}{2T}$

6. The average power of the random process having psd $S_{XX}(\omega)$ is $P_{XX} =$

   (a) $\displaystyle\int_{-\infty}^{\infty} S_{XX}(\omega)d\omega$ (b) $\dfrac{1}{2\pi}\displaystyle\int_{-\infty}^{\infty} S_{XX}(\omega)d\omega$

   (c) $2\pi\displaystyle\int_{-\infty}^{\infty} S_{XX}(\omega)d\omega$ (d) zero

7. The time average of the autocorrelation function and the power spectral density form a pair of

   (a) Fourier transform (b) Laplace transform

   (c) Z-transform (d) convolution

8. The rms band width in terms of psd is

(a) $\dfrac{\int_{-\infty}^{\infty} \omega^2 S_{XX}(\omega)d\omega}{\int_{-\infty}^{\infty} S_{XX}(\omega)d\omega}$ (b) $\dfrac{\int_{-\infty}^{\infty} S_{XX}(\omega)d\omega}{\int_{-\infty}^{\infty} \omega^2 S_{XX}(\omega)d\omega}$

(c) $\dfrac{\int_{-\infty}^{\infty} S_{XX}(\omega)d\omega}{2\pi\int_{-\infty}^{\infty} \omega^2 S_{XX}(\omega)d\omega}$ (d) $\dfrac{2\pi\int_{-\infty}^{\infty} S_{XX}(\omega)d\omega}{\int_{-\infty}^{\infty} S_{XX}(\omega)d\omega}$

9. The power spectral density of WSS is always

(a) negative (b) non negative

(c) finite (d) can be negative or positive

10. The average power of the periodic random process whose autocorrelation function is $R_{XX}(\tau) = \exp\left[-\dfrac{\tau^2}{2\sigma^2}\right]$ is

(a) 0 (b) 1 (c) 2 (d) 3

11. If $S_{XX}(\omega) = \dfrac{16}{\omega^2+16}$, $S_{YY}(\omega) = \dfrac{\omega^2}{\omega^2+16}$ and $X(t)$ and $Y(t)$ are of zero mean, $U(t) = X(t) + Y(t)$., $S_{XU}(\omega)$ is

(a) $\dfrac{\omega^2}{\omega^2+16}$ (b) $\dfrac{\omega}{\omega^2+16}$

(c) $\dfrac{4\omega}{\omega^2+16}$ (d) $\dfrac{16}{\omega^2+16}$

12. If $Y(t) = X(t) - X(t-a)$ is a random process and $X(t)$ is a WSS process with $a > 0$, a constant, the power spectral density of $Y(t)$ in terms of the corresponding quantities of $X(t)$ is

(a) $S_{XX}(\omega)\sin^2\left(\dfrac{\omega a}{2}\right)$ (b) $2S_{XX}(\omega)\sin^2\left(\dfrac{\omega a}{2}\right)$

(c) $3S_{XX}(\omega)\sin^2\left(\dfrac{\omega a}{2}\right)$ (d) $4S_{XX}(\omega)\sin^2\left(\dfrac{\omega a}{2}\right)$

13. A random process is given by $Z(t) = AX(t) + BY(t)$. where A and B are real constants. $X(t)$ and $Y(t)$ are jointly WSS processes. The power spectrum $S_{ZZ}(\omega)$ is

(a) $ABS_{XY}(\omega) + ABS_{YX}(\omega)$
(b) $A^2 + B^2 + ABS_{XY}(\omega) + ABS_{YX}(\omega)$
(c) $A^2S_{XX}(\omega) + ABS_{XY}(\omega) + ABS_{YX}(\omega) + B^2S_{YY}(\omega)$
(d) 0

14. A random process is given by $Z(t) = AX(t) + BY(t)$, where A and B are real constants, $X(t)$ and $Y(t)$ are jointly WSS processes. If $X(t)$ and $Y(t)$ are uncorrelated, then $S_{ZZ}(\omega)$ is

(a) $A^2 + B^2$ (b) $A^2S_{XX}(\omega) + B^2S_{YY}(\omega)$

(c) $ABS_{XY}(\omega) + ABS_{YX}(\omega)$ (d) 0

15. For a WSS process, psd at zero frequency gives
    (a) the area under the graph of a power spectral density
    (b) area under the graph of the autocorrelation of the process
    (c) mean of the process
    (d) variance of the process

16. The mean square value of a WSS process equals
    (a) the area under the graph of the psd
    (b) the area under the graph of the autocorrelation of process
    (c) zero
    (d) mean of the process

17. $x(t) = A\cos(\omega_0 t + \theta)$, where A and ω_0 are constants and θ is a random variable uniformly distributed over $(0, \pi)$. The average power of $X(t)$ is
    (a) θ (b) $\frac{A^2}{2}$ (c) $\frac{A^2}{4}$ (d) $\frac{A^2}{8}$

18. A WSS process, $X(t)$, has an autocorrelation function $R_{XX}(\tau) = e^{-3|\tau|}$. Then psd is
    (a) $\frac{6}{9+\omega^2}$ (b) $\frac{9}{6+\omega^2}$ (c) $\frac{3}{9+\omega^2}$ (d) $\frac{9}{3+\omega^2}$

19. The real and imaginary parts of $S_{YX}(\omega)$ is an ____ and ____ function of ω respectively.
    (a) odd, odd (b) odd, even (c) even, odd (d) even, even

20. If the cross correlation function of two random processes is $R_{XY}(t, t+\tau) = \frac{AB}{2}[\sin\omega_0\tau + \cos\omega_0(2t+\tau)]$, the cross power spectrum is $S_{XY}(\omega) =$
    (a) $\frac{j\pi AB}{2}[\delta(\omega-\omega_0) - \delta(\omega+\omega_0)]$ (b) $\frac{\pi AB}{2}[\delta(\omega-\omega_0) - \delta(\omega+\omega_0)]$
    (c) $\frac{AB}{2}[\delta(\omega-\omega_0) - \delta(\omega+\omega_0)]$ (d) $-j\frac{\pi AB}{2}[\delta(\omega-\omega_0) - \delta(\omega+\omega_0)]$

21. If $X(t)$ and $Y(t)$ are uncorrelated and of constant means, $E[X]$ and $E[Y]$ respectively, then $S_{XY}(\omega)$ is
    (a) $E[X]E[Y]$ (b) $2E[X]E[Y]\delta(\omega)$
    (c) $2\pi E[X]E[Y]\delta(\omega)$ (d) $\frac{E[X]E[Y]\delta(\omega)}{2}$

22. The cross spectral density of two random processes, $X(t)$ and $Y(t)$. $S_{XY}(\omega)$ is
    (a) $\lim_{T\to\infty}\frac{E[X_T^*(\omega)Y_T(\omega)]}{2T}$ (b) $\frac{E[X_T^*(\omega)Y_T(\omega)]}{2T}$
    (c) $\lim_{T\to\infty}\frac{E[X_T^*(\omega)Y_T(\omega)]}{2\pi}$ (d) $\frac{E[X_T^*(\omega)Y_T(\omega)]}{2\pi}$

23. The cross spectral density $S_{YX}(\omega) =$
    (a) $S_{XY}(\omega)$ (b) $S_{XX}(-\omega)$
    (c) $S_{YX}(-\omega)$ (d) $-S_{YX}(\omega)$

24. If $X(t)$ and $Y(t)$ are orthogonal, then

(a) $S_{XY}(\omega) = 0$ (b) $S_{XY}(\omega) = 1$

(c) $S_{XY}(\omega) > 1$ (d) $S_{XY}(\omega) < 1$

25. The cross power P_{XY} is given by

(a) $\int_{-\infty}^{\infty} S_{XY}(\omega)d\omega$ (b) $2\pi \int_{-\infty}^{\infty} S_{XY}(\omega)d\omega$

(c) $\frac{1}{2\pi}\int_{-\infty}^{\infty} S_{YX}(\omega)d\omega$ (d) $\frac{1}{2\pi}\int_{-\infty}^{\infty} S_{XY}(\omega)d\omega$

26. The average power P_{XY} is

(a) P_{YX}^* (b) $-P_{YX}$ (c) $-P_{YX}^*$ (d) P_{YX}

27. The cross spectral density of two random processes $X(t)$ and $Y(t)$ is $S_{XY}(\omega) = a + \frac{jb\omega}{K}$. The area enclosed by their autocorrelation function is

(a) $\frac{a}{K}$ (b) a (c) Ka (d) $a + \frac{1}{K}$

28. If $S_{XY}(\omega) = \frac{8}{(\alpha + j\omega)^3}$, then $R_{XY}(\tau)$ is

(a) $u(\tau)\tau^2 e^{-\alpha t}$ (b) $4u(\tau)\tau^2 e^{-\alpha\tau}$

(c) $8u(\tau)\tau^2 e^{-\alpha\tau}$ (d) $12u(\tau)\tau^2 e^{-\alpha\tau}$

29. A random process is given by $Z(t) = AX(t) + BY(t)$. where A and B are real constants. $X(t)$ and $Y(t)$ are jointly WSS processes. The cross power spectrum $S_{XZ}(\omega)$ is

(a) $AS_{YX}(\omega) + BS_{YY}(\omega)$ (b) $AS_{XX}(\omega) + BS_{YY}(\omega)$

(c) $AS_{XX}(\omega) + BS_{XY}(\omega)$ (d) $AS_{YY}(\omega) + BS_{XY}(\omega)$

30. For the above problem, $S_{YZ}(\omega)$

(a) $AS_{YX}(\omega) + BS_{YY}(\omega)$ (b) $AS_{XX}(\omega) + BS_{YY}(\omega)$

(c) $AS_{XX}(\omega) + BS_{XY}(\omega)$ (d) $AS_{YY}(\omega) + BS_{XX}(\omega)$

31. The autocorrelation function of a WSS random process $X(t)$ is $R_{XX}(\tau) = 4 + 2e^{-2|\tau|}$ The area enclosed by the psd curve of $X(t)$ is

(a) 6 (b) 2 (c) 8 (d) 4

32. If the autocorrelation function is $R(\tau) = e^{-\alpha\tau|}[1 + \alpha|\tau|]$, then the power spectrum of a random process is

(a) $\frac{4}{(\alpha^2 + \omega^2)^2}$ (b) $\frac{4\alpha^3}{(\alpha^2 + \omega^2)^2}$

(c) $\frac{\alpha^3}{(\alpha^2 + \omega^2)^2}$ (d) $\frac{\alpha^3}{4(\alpha^2 + \omega^2)^2}$

33. If $X(t)$ is an ergodic process with autocorrelation function of the form

$$R_{XX}(\tau) = \begin{cases} 1 - |\tau| & \text{for } |\tau| \leq 1 \\ 0 & \text{for } |\tau| \geq 1 \end{cases}$$

the spectral density $S_{XX}(\omega)$ is

(a) $\frac{\sin \omega/2}{\omega/2}$ (b) $\frac{\sin^2 \omega/2}{\omega/2}$

(c) $\frac{\sin^2 \omega/2}{(\omega/2)^2}$ (d) $\left[\frac{\sin^2 \omega/2}{\omega/2}\right]$

34. If $S_{XX}(\omega) = \frac{\eta}{2}$ for $-2\pi B \leq \omega \leq 2\pi B$, then $R_{XX}(\tau)$ is

(a) $\eta B \operatorname{sinc}(2\pi B\tau)$ (b) $B \operatorname{sinc}(2\pi B\tau)$

(c) $\frac{\operatorname{sinc}(2\pi B\tau)}{\eta B}$ (d) $\frac{\operatorname{sinc}(2\pi B\tau)}{B}$

35. The autocorrelation function of a WSS process is $R(\tau) = Ke^{-K|\tau|}$ Then its spectral density is $S(\omega) =$

(a) $\frac{1}{1+(\omega/K)^2}$ (b) $\frac{2}{1+(\omega/K)^2}$

(c) $1+(\omega/K)^2$ (d) $\frac{1+(\omega/K)^2}{2}$

36. The autocorrelation function of a process with psd of $\frac{4}{1+0.25\omega^2}$ is

(a) $4\exp(-|\tau|)$ (b) $\exp(-|\tau|)$

(c) $\exp(-2|\tau|)$ (d) 1

37. If $Z(t) = X(t)+Y(t)$, where $X(t)$ and $Y(t)$ are two WSS processes and $X(t)$ and $Y(t)$ are uncorrelated with zero mean, then $R_{ZZ}(T) =$

(a) $R_{XY}(T)+R_{YX}(T)$ (b) $R_{XX}(T)+R_{YY}(T)$

(c) $R_{XY}(T)-R_{YX}(T)$ (d) $R_{XX}(T)-R_{YY}(T)$

38. For the above problem, $S_{ZZ}(\omega)$ is

(a) $S_{XX}(\omega)-S_{YY}(\omega)$ (b) $S_{XY}(\omega)-S_{YX}(\omega)$

(c) $S_{YX}(\omega)+S_{YY}(\omega)$ (d) $S_{XY}(\omega)+S_{YX}(\omega)$

39. White noise with a two sided psd $\frac{\eta}{2}$ is passed through an RC lowpass network with time constant $T = RC$ and thereafter through an ideal amplifier with voltage gain 10. The expression for mean square value of output noise is

(a) $\frac{\eta}{RC}$ (b) $\frac{\eta}{25RC}$

(c) $\frac{25\eta}{RC}$ (d) $\frac{RC}{\eta}$

40. A random process with psd K watts/Hz for $-\infty < f < \infty$ is passed through a system with transfer function $H(f) = 2$ for $-B \leq f \leq B$ and zero elsewhere. The output power of the system is

(a) BK (b) $2BK$

(c) $\frac{BK}{2}$ (d) B^2K

41. If the input power spectral density of a system is $S_{XX}(\omega) = N_0$ and its output psd is $S_{YY}(\omega) = N_0\left[\dfrac{\omega^2+9}{\omega^2+16}\right]$, then the transfer function of the system is

(a) $\dfrac{\omega - j3}{\omega - j4}$ (b) $\dfrac{\omega + j3}{\omega + j4}$

(c) $\dfrac{\omega + j4}{\omega + j3}$ (d) $\dfrac{\omega - j4}{\omega - j3}$

42. Let $X(t)$ be a WSS process with autocorrelation function $R_{XX}(\tau)$ and power spectrum $S_{XX}(\omega)$. If $X'(t) = \dfrac{d}{dt}X(t)$, then $R_{X'X'}(\tau)$ is

(a) $\dfrac{d}{d\tau}R_{XX}(\tau)$ (b) $\dfrac{d^2}{d\tau^2}R_{XX}(\tau)$

(c) $\dfrac{-d^2}{d\tau^2}R_{XX}(\tau)$ (d) $-\dfrac{d}{d\tau}R_{XX}(\tau)$

43. If $S_{XX}(\omega)$ is the power spectrum of the input process $X(t)$ and $|H(\omega)|^2$ is the power transfer function of the system, then the average power P_{YY} is

(a) $\int_{-\infty}^{\infty} S_{XX}(\omega)|H(\omega)|^2 d\omega$ (b) $\dfrac{1}{2\pi}\int_{-\infty}^{\infty} S_{XX}(\omega)|H(\omega)|^2 d\omega$

(c) $\int_{-\infty}^{\infty} \dfrac{S_{XX}(\omega)}{|H(\omega)|^2} d\omega$ (d) $\dfrac{1}{2\pi}\int_{-\infty}^{\infty} \dfrac{S_{XX}(\omega)}{|H(\omega)|^2} d\omega$

44. The output power density of $Y(t)$ can be obtained by, $S_{YY}(\omega) =$

(a) $S_{XX}(\omega)|H(\omega)|$ (b) $\dfrac{S_{XX}(\omega)}{|H(\omega)|^2}$

(c) $|H(\omega)|S_{XX}(\omega)$ (d) $|H(\omega)|^2 S_{XX}(\omega)$

45. The cross power density of $X(t)$ and $Y(t)$ can be obtained by $S_{YX}(\omega) =$

(a) $H(\omega)S_{XX}(\omega)$ (b) $H^*(\omega)S_{XX}(\omega)$

(c) $\dfrac{H(\omega)}{S_{XX}(\omega)}$ (d) $\dfrac{H^*(\omega)}{S_{XX}(\omega)}$

Answers

1. (d)	2. (b)	3. (c)	4. (d)	5. (c)	6. (b)	7. (a)	8. (a)	9. (b)	10. (b)
11. (d)	12. (d)	13. (c)	14. (b)	15. (b)	16. (a)	17. (b)	18. (a)	19. (c)	20. (a)
21. (c)	22. (a)	23. (b)	24. (a)	25. (d)	26. (a)	27. (a)	28. (b)	29. (c)	30. (a)
31. (a)	32. (b)	33. (d)	34. (a)	35. (b)	36. (c)	37. (b)	38. (b)	39. (c)	40. (b)
41. (b)	42. (b)	43. (b)	44. (d)	45. (b)					

Simulation with MATLAB Programming

This appendix provides a brief introduction to MATLAB software and the use of its computer simulation capability to solve problems in this book. Also, several MATLAB built-in functions and user-defined functions related to the subject are introduced. MATLAB is a product of the MathWorks™ Inc. and an acronym for MATrix LABoratory. It is a high-performance language enriched with built-in mathematical tools, editing and debugging tools with data structures that supports object-oriented programming. It is an excellent tool for teaching and research and is considered a standard in most research laboratories, universities, technical institutes and industries worldwide.

The desktop for MATLAB, a special window with graphical interface called the MATLAB desktop, contains the following:

- Command Window
- Command History
- Workspace
- Current Directory
- Help Browser
- Start Button

The prompt (>>) in the Command Window appears for writing and editing programs.

For example, $x = 1 + 2 * 7$

$x = 15$

Some of the general-purpose commands are:

help - Displays all commands with topics
demo - Displays the MATLAB demos menu and runs demonstrations
who - Lists current variables
whos - Lists current variables, long form
clc - Clears the screen
clear - Clears variables and functions from memory
quit - Quits MATLAB session
exit - Exits from MATLAB

what - Lists MATLAB-specific files in the directory
pwd - Displays current working directory
format - changes output format

For computation, we can use either a command window or an editor file. Small programs can be directly typed in the command prompt (≫). Editor/Debugger file is used for executing saved programs and user defined functions.

The following steps are involved in writing and executing/ debugging programs:

1. From the File menu on the toolbar, create *new m-file*. In the *Editor Window* type the program or open a previously saved file. The program should be saved with the file name extension *.m*, like abc.m.
2. To execute a program, type the file name *without* the *.m* extension at the ≫ prompt in the command window.
3. To get help with MATLAB topics, use the MATLAB User's Guide and the references provided.

Note: Use the commands `clear all; close all; clc;` to clear variables and screen.

Example A1

Generate a continuous-time sinusoidal signal of given frequency, duration and phase angle. That is, $y_1(t) = a \times \cos(2\pi ft)$ and $y_2(t) = a \times \cos(2\pi ft - \theta)$.

Program

```
% Generation of a continuous-time sinusoidal signal of
% given frequency, duration and phase angle.
clear all;
close all; a
clc;
disp('----- First Signal ------')
f = input('Enter frequency (in Hz): ');
duration = input('Enter time duration of the signal
(in milli seconds): ');
duration = duration*10^(-3);
a = 4;
time_period = 1/f;
num_cycles = duration/time_period;
t1 = -num_cycles*time_period/2;
t2 = num_cycles*time_period/2;
num_samples = 1000;
t = linspace(t1, t2, num_samples);
y1 = a*cos(2*pi*f*t);
disp('------ SECOND SIGNAL ------')
phase_shift = input('Enter amount of phase shift (in radiance)
% relative to first signal : ');
y2 = a*cos(2*pi*f*t-phase_shift);
subplot(2,1,1),plot(t,y1), title(' First Signal')
subplot(2,1,2),plot(t,y2), title(' Second Signal')
```

Output

—— First Signal ——
Enter frequency (in Hz): 200
Enter time duration of the signal (in milli seconds): 20
—— Second Signal ——
Enter amount of phase shift (in radiance) relative to first signal : 20

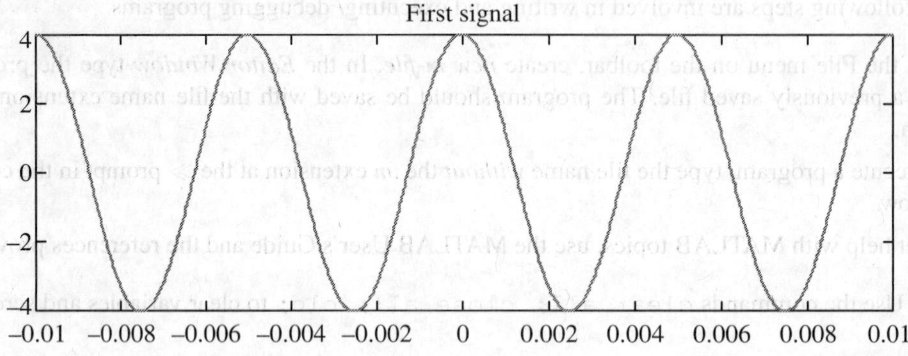

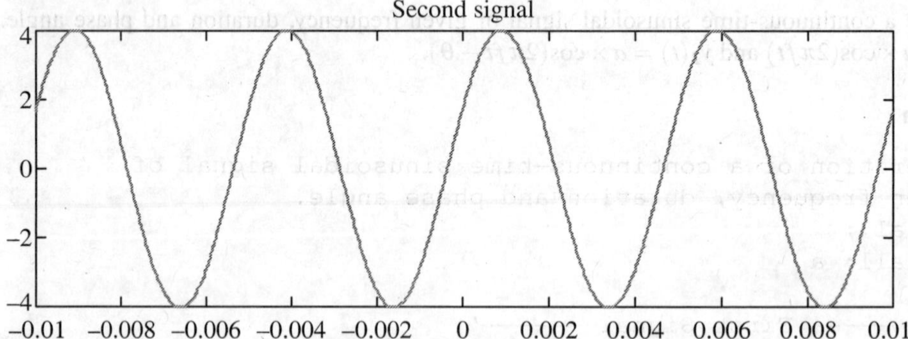

Example A2

Generate (a) a rectangular pulse having a width of 20 ms; (b) a triangular pulse having a width of 0.4 s; (c) a sinc function, and; (d) a signum function.

Program

```
% Generate 2 seconds of a rectangular pulse with a sample rate of
% 10 Khz and a width of 20 ms.

fs = 10000; % Number of samples per second
t = −1:1/fs:1; % Time vector
x1 = rectpuls(t,20e−3); % Generating rectangular pulse
subplot(2,2,1), plot(t,x1), axis([−0.1 0.1 −0.2 1.2]) % Plotting
the pulse
xlabel('Time (sec)');ylabel('Amplitude'); title('Rectangular
Function')
```

```
% Create a triangular pulse with width 0.4.

fs = 10000;          % Sampling frequency (samples/sec)
t = -1:1/fs:1;       % Time Vector
w = .4;              % Triangle Width
x2 = tripuls(t,w);   % Sampled aperiodic triangle
subplot(2,2,2), plot(t,x2); axis([-1.0 1.0 -0.2 1.2]) % Plotting
the pulse
xlabel('Time (sec)');ylabel('Amplitude'); title('Triangular
Function')

% Example of a sinc function for a linearly spaced vector:
t = linspace(-5,5);
x3 = sinc(t);% Generating sinc pulse
subplot(2,2,3), plot(t,x3);
xlabel('Time (sec)');ylabel('Amplitude'); title('Sinc Function')

% Example of a sign function for a linearly spaced vector:
t = linspace(-5,5);
x4 = sign(t);% Generating signum pulse
subplot(2,2,4), plot(t,x4);axis([-5.0 5.0 -1.2 1.2]) % Plotting
the pulse
xlabel('Time (sec)');ylabel('Amplitude'); title('Signum Function')
```

Output

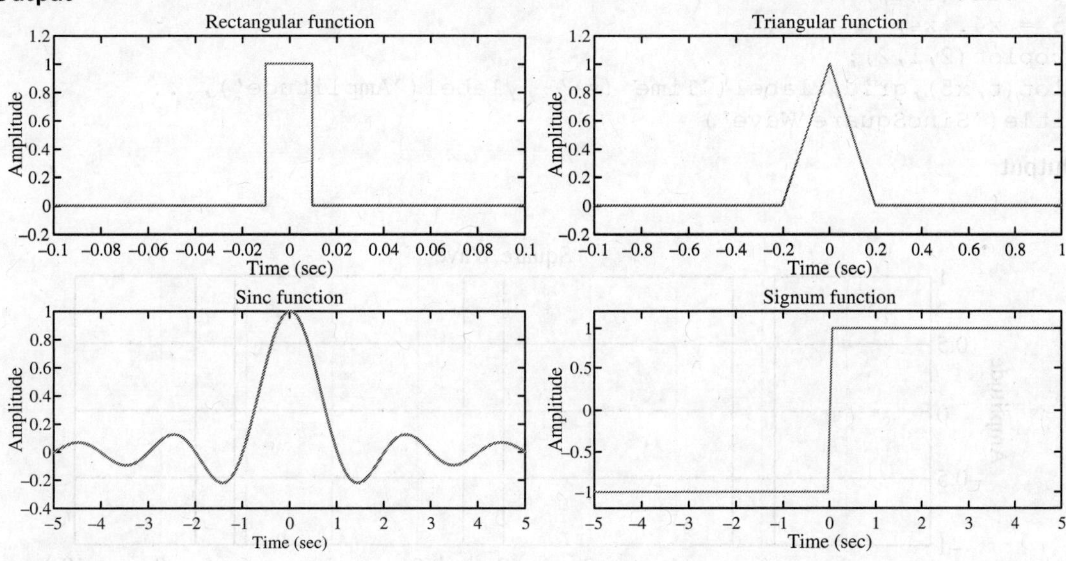

Example A3

Generate the following periodic waveforms: (a) rectangular pulse waveform; (b) triangular waveform; (c) sawtooth waveform; and (d) sinc square waveform.

Program

```
%Generation of periodic Waveforms
t = -10:0.001:10; %Generation of Rectangular Pulse Waveform
duty =15;
figure(1);
subplot(2,1,1);
x1 =square(t,duty);
plot(t,x1),grid,xlabel('Time (s)'),ylabel('Amplitude'),...
title('Square Wave');
%Generation of Triangular Waveform
subplot(2,1,2);
Td = -10:1:10;
x2 =pulstran(t,Td,'tripuls');
plot(t,x2),grid,xlabel('Time (s)'),ylabel('Amplitude'),...
title('Triangle Wave');
%Generation of SawTooth Waveform
Wd = 1;
x3 =sawtooth(t,Wd);
figure(2);
subplot(2,1,1);
plot(t,x3),grid,xlabel('Time (s)'),ylabel('Amplitude'),...
title('Triangle Wave');
%Generation of SincSquare Waveform
x4 =sinc(t/2);
x5 = x4.*x4;
subplot(2,1,2);
plot(t,x5),grid,xlabel('Time (s)'),ylabel('Amplitude'),...
title('SincSquare Wave')
```

Output

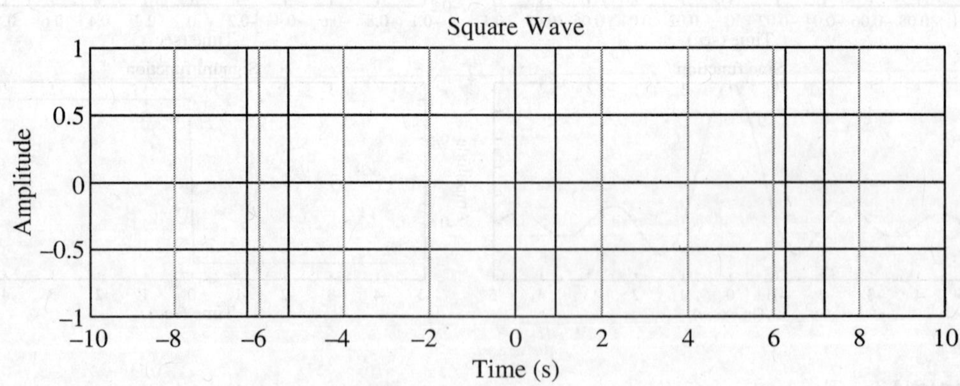

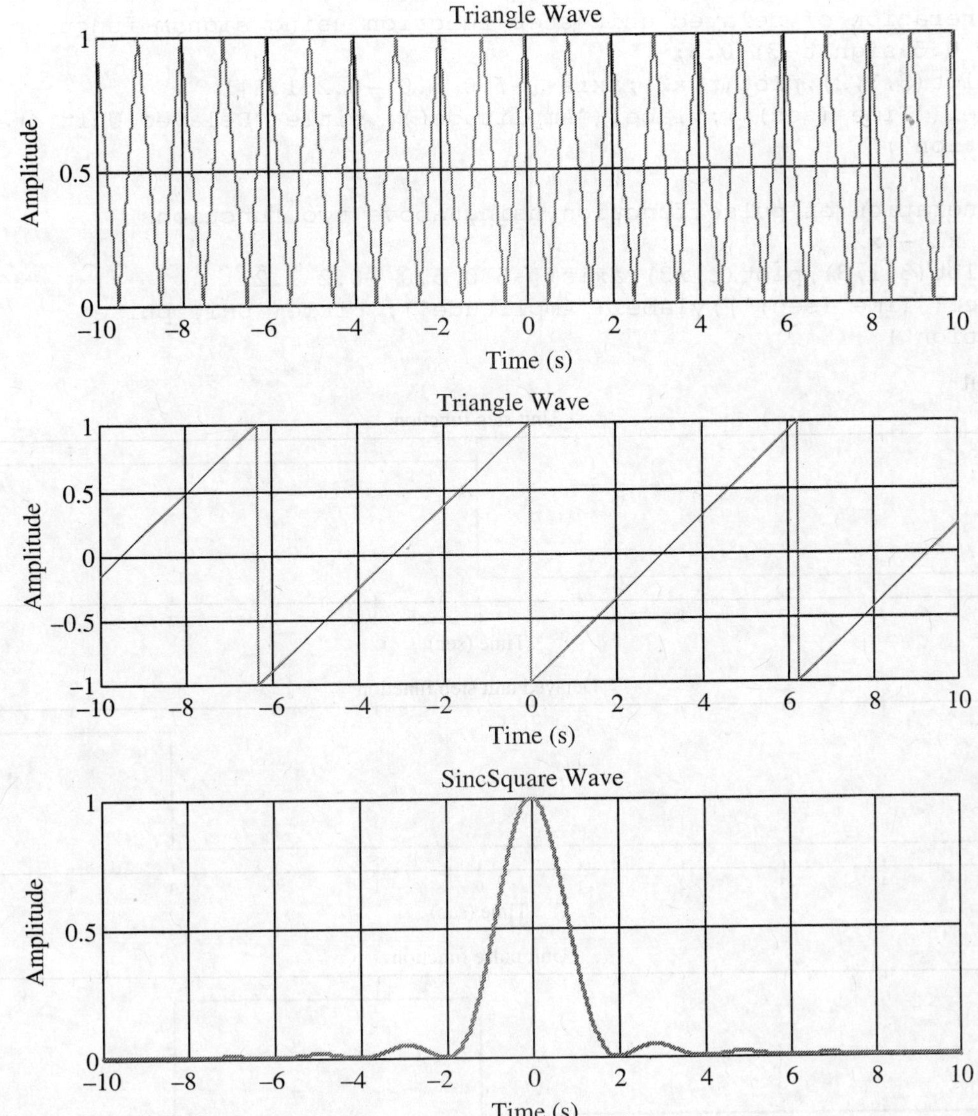

Example A4

Generate the following using signum functions: (a) a unit step function; (b) a delayed unit step function and ; (c) a pulse function.

Program

```
% Generation of unit step function using signum function
t = linspace(-5,5,1000);
x1 = 0.5*sign(t)+0.5;
subplot(3,1,1),plot(t,x1);axis([-5.0 5.0 -0.2 1.2])
xlabel('Time (sec)');ylabel('Amplitude'); title('Unit step Function')
```

```
% Generation of delayed unit step function using signum function
x2 = 0.5*sign(t-3)+0.5;
subplot(3,1,2),plot(t,x2);axis([-5.0 5.0 -0.2 1.2])
xlabel('Time (sec)');ylabel('Amplitude'); title('Delayed Unit step
Function')

% Generation of pulse function using above two functions
x3 = x1 - x2;
subplot(3,1,3),plot(t,x3);axis([-5.0 5.0 -0.2 1.2])
xlabel('Time (sec)');ylabel('Amplitude'); title('Unit pulse
Function')
```

Output

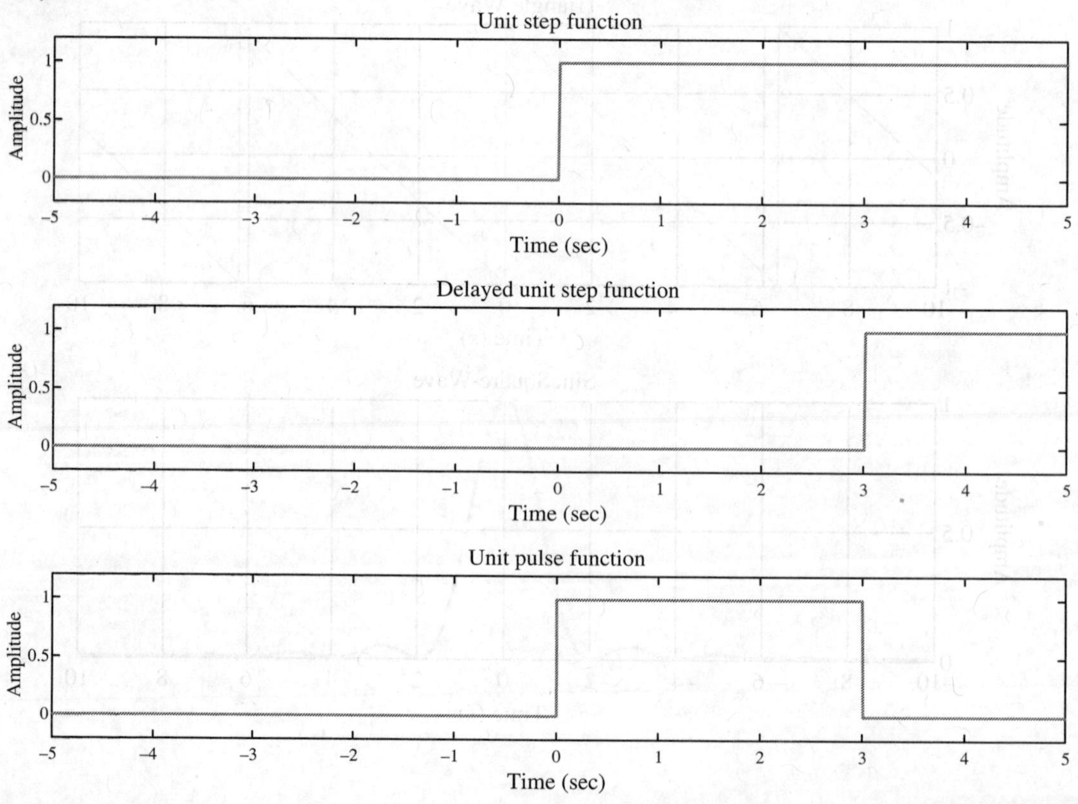

Example A5

Perform time shifting and time scaling operations using sinc function.
$x_1 = \text{sinc}(t)$, $x_2 = \text{sinc}(4t)$, $x_3 = \text{sinc}(2(t-5))$ and $x_4 = \text{sinc}(0.5t)$.

Program

```
% Time shifting and time scaling operations using sinc function
t = linspace(-10,10,1000);
x1 = sinc(t);
```

```
subplot(2,2,1), plot(t,x1);
xlabel('Time (sec)');ylabel('Amplitude'); title('Sinc Function')

x2 = sinc(4*t);
subplot(2,2,2), plot(t,x2);
xlabel('Time (sec)');ylabel('Amplitude'); title('Sinc Function')

x3 = sinc(2*(t-5));
subplot(2,2,3), plot(t,x3);
xlabel('Time (sec)');ylabel('Amplitude'); title('Sinc Function')

x4 = sinc(0.5*t);
subplot(2,2,4), plot(t,x4);
xlabel('Time (sec)');ylabel('Amplitude'); title('Sinc Function')
```

Output

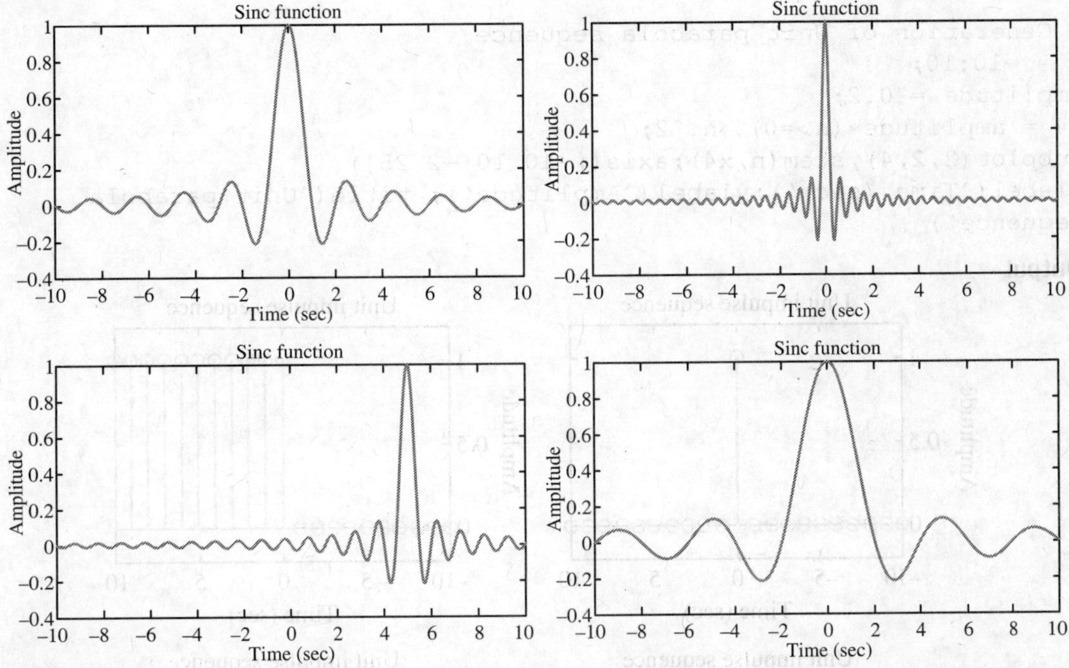

Example A6

Generate (a) unit impulse sequence; (b) unit step sequence; (c) unit ramp sequence and (d) unit parabola sequence.

Program

```
% Generation of elementary discrete time sequences.
% Generation of Unit Impulse sequence
n = -10:10;
amplitude = 1.0;
x1 = amplitude*(n==0);
```

```
subplot(2,2,1),stem(n,x1);axis([-10 10 -0.2 1.2])
xlabel('Time (sec)');ylabel('Amplitude'); title('Unit Impulse
Sequence')

% Generation of Unit Step sequence

x2 = amplitude*(n>=0);
subplot(2,2,2),stem(n,x2);axis([-10 10 -0.2 1.2])
xlabel('Time (sec)');ylabel('Amplitude'); title('Unit Step Sequence')

% Generation of Unit Ramp sequence

x3 = amplitude*(n>=0).*n;
subplot(2,2,3),stem(n,x3);axis([-10 10 -1 11])
xlabel('Time (sec)');ylabel('Amplitude'); title('Unit Ramp Sequence')

% Generation of Unit parabola sequence
n = -10:10;
amplitude = 0.2;
x4 = amplitude*(n>=0).*n.^2;
subplot(2,2,4),stem(n,x4);axis([-10 10 -2 25])
xlabel('Time (sec)');ylabel('Amplitude'); title('Unit parabola
Sequence')
```

Output

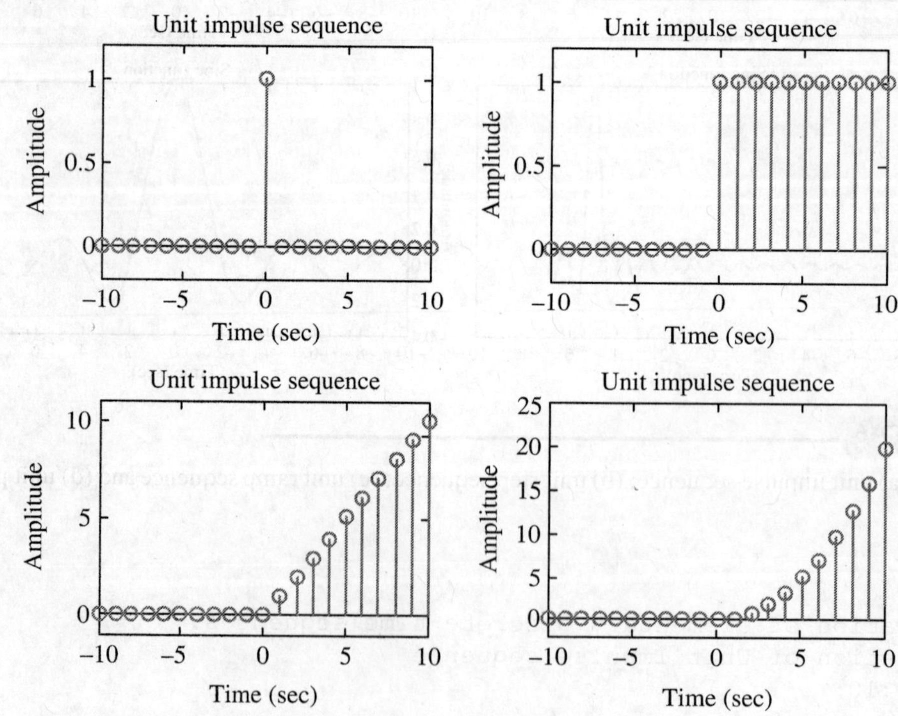

Example A7

Generate (a) a continuous time unit pulse signal; (b) an even function for the pulse signal; (c) odd function for the pulse signal.

Program

```
% Even and odd components of continuous time signal
t = linspace(-10,10,1000);

% Generation of unit step function using signum function
x1 = 0.5*sign(t)+0.5;

% Generation of delayed unit step function using signum function
x2 = 0.5*sign(t-3)+0.5;

% Generation of pulse function using above two functions
x3 = x1 - x2;
x4 = fliplr(x3);
xe = 0.5*(x3+x4);% Generation of even function
xo = 0.5*(x3-x4);% Generation of odd function

subplot(3,1,1),plot(t,x3);axis([-10 10 -1.2 1.2])
xlabel('Time (sec)');ylabel('Amplitude'); title('Unit pulse
Function')

subplot(3,1,2),plot(t,xe);axis([-10 10 -1.2 1.2])
xlabel('Time (sec)');ylabel('Amplitude'); title('Even Function')

subplot(3,1,3),plot(t,xo);axis([-10 10 -1.2 1.2])
xlabel('Time (sec)');ylabel('Amplitude'); title('Odd Function')
```

Output

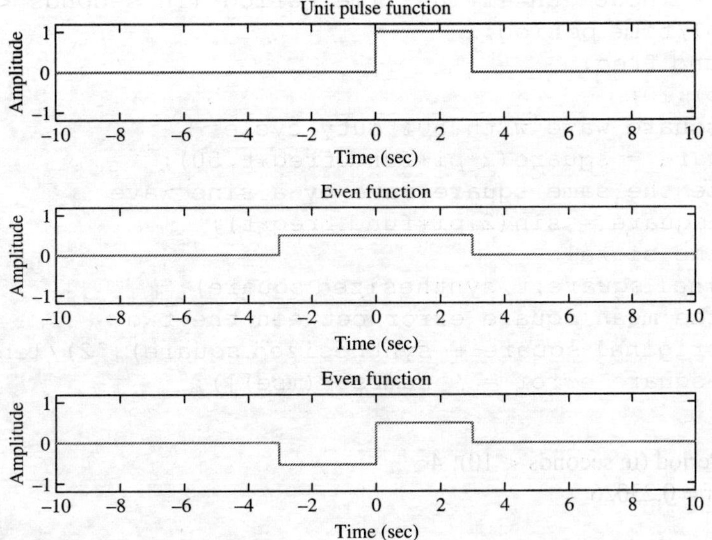

Example A8

Test the orthogonality between the signals $x_1(t) = \sin(2\pi t)$ and $x_21(t) = \cos(8\pi t)$ in the interval $0 < t < 2$. Condition for orthogonality is

$$\int_{t_1}^{t_2} x_1(t)x_2(t)dt = 0 \quad \text{or} \quad \int_0^2 \sin(2\pi t)\cos(8\pi t)dt = 0.$$

Program

```
syms t; % Define symbolic variable t;
x1 = sin(2*pi*t);
x2 = cos(2*pi*4*t);
t1 = 0;
t2 = 2;
% Perform integration of the product signal in the specified interval
y1 = int(x1*x2,t1,t2);
if y1==0
disp('Orthogonal');
else
disp('Not orthogonal');
end
```

Answer:

```
ORTHOGONAL
```

Example A9

Evaluation of mean square error between two square and sine signals.

Program

```
time_period = input('Enter The Time Period (in seconds <10): ');
fund_freq = 1/time_period;
Fsamp=100*fund_freq;
t=0:(1/Fsamp):10;
% Generate square wave with 50% duty cycle
original_square = square(2*pi*fund_freq*t,50);
% Approximate the same square wave by a sine wave
synthesized_square = sin(2*pi*fund_freq*t);
% Displays the signals
plot(t,original_square,t,synthesized_square);
% Now find the mean square error between the two
mse = sum((original_square - synthesized_square).^2)/length(t);
disp(['Mean square error = ' num2str(mse)]);
```

Output

Enter The Time Period (in seconds < 10): 4
Mean square error = 0.23026

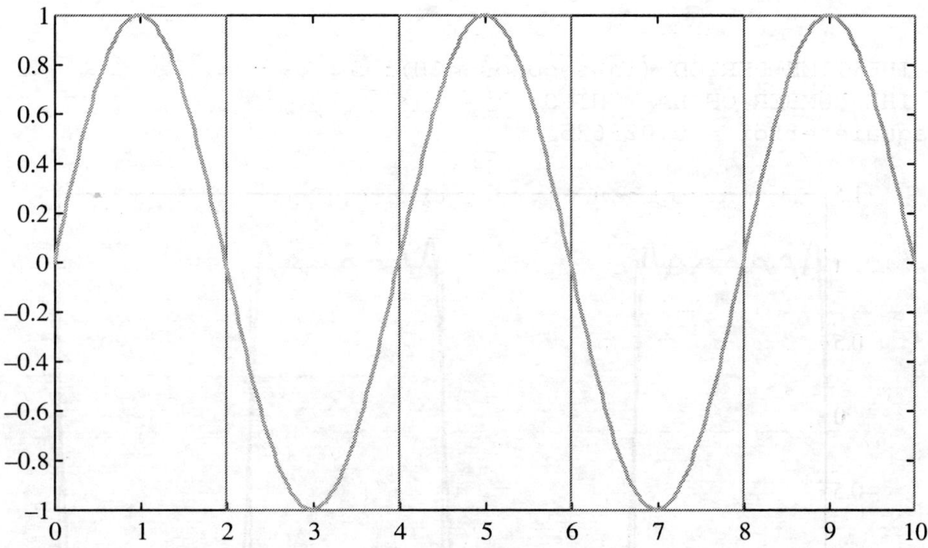

Example A10

Evaluation of the mean square with increasing harmonics for a square wave of period T

Program

```
time_period = input('ENTER THE TIME PERIOD (in seconds): ');
nharmonics=input('ENTER THE NUMBER OF HARMONICS: ');
fund_freq = 1/time_period;
Fsamp=100*fund_freq*nharmonics;
t=0:(1/Fsamp):1;
% Generate square wave with 50% duty cycle
original_square = square(2*pi*fund_freq*t,50);

% Synthesize same square wave by adding weighted odd hormonics of
sine waves
synthesized_square=zeros(size(t));
coeff = 1.27;
for hormonic=1:2:(nharmonics*2)
synthesized_square=synthesized_square+coeff*
sin(2*pi*hormonic*fund_freq*t)/hormonic;
end
% Displays the signal
plot(t,synthesized_square),hold on
plot(t,original_square),hold off
% Now find the mean square error between the two
mse = sum((original_square - synthesized_square).^2)/length(t);
disp(['Mean square error = ' num2str(mse)])
```

Output

```
ENTER THE TIME PERIOD (in seconds <10): 5
ENTER THE NUMBER OF HARMONICS: 7
Mean square error = 0.029675
```

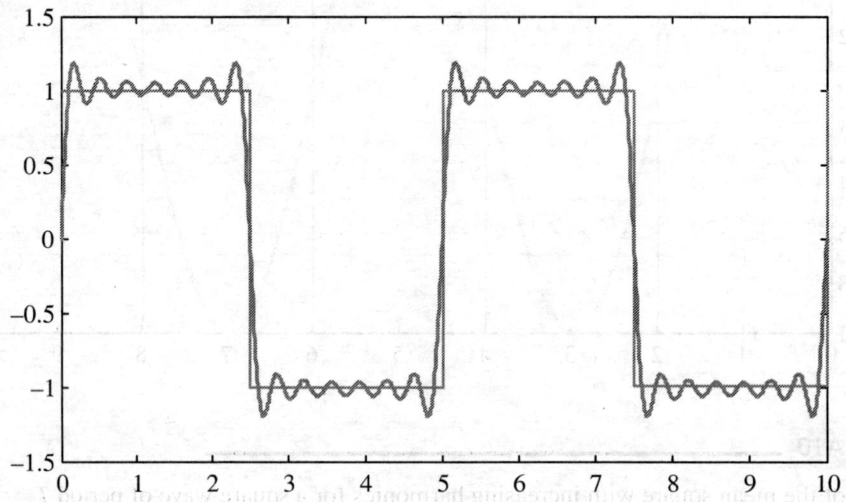


```
ENTER THE TIME PERIOD (in seconds <10): 3
ENTER THE NUMBER OF HARMONICS: 15
Mean square error = 0.013311
```

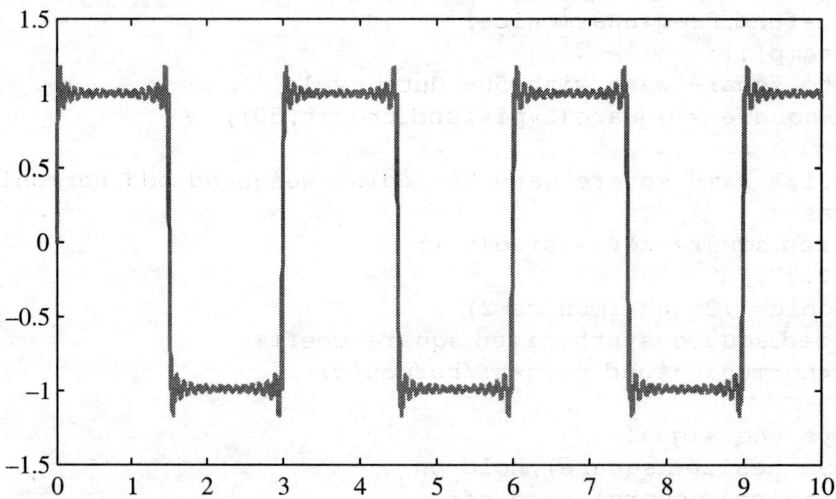

Example A11

Evaluate mean square error in the approximation of a sine wave with a unit pulse function in the interval 0 to 2 π.

Program

```
syms t
x1 = sin(t);
x2 = heaviside(t);
t1 = 0;
t2 = 2*pi;
mse = double(int((x1-x2)^2/(t2-t1),t1,t2));
disp(['Mean square error = ', num2str(mse)])
```

Answer:

Mean square error = 1.5

Example A12

Generate exponential Fourier series of periodic full-wave rectified signal.

Program

```
t = -3*pi:0.001:3*pi;
A = 20;
w = 1;
original = abs(A*sin(w*t));
plot(t,original),title('Original signal')
% Number of hormonics in the series
N = 25;
c0 = 2*A/pi;
coeff_idx = 1:N;
% Expression for the exponential Fourier series coefficients of full
wave
% rectifier waveform
% Negative indexed coefficients
cn = 2*A./(pi*(1-4*coeff_idx.^2));
% Full set of coefficients
coeff = [fliplr(cn),c0,cn(1:N-1)];
n=-N:N-1;
figure,stem(n, coeff),title('Discrete Fourier spectrum')
% Recostruction of original signal
y = repmat(coeff,length(coeff),1)*exp(1i*2*w*n'*t);
rec_sig = abs(sum(y))/(2*N);
figure,plot(t,rec_sig),title('Signal reconstructed from
coefficients')
```

Output

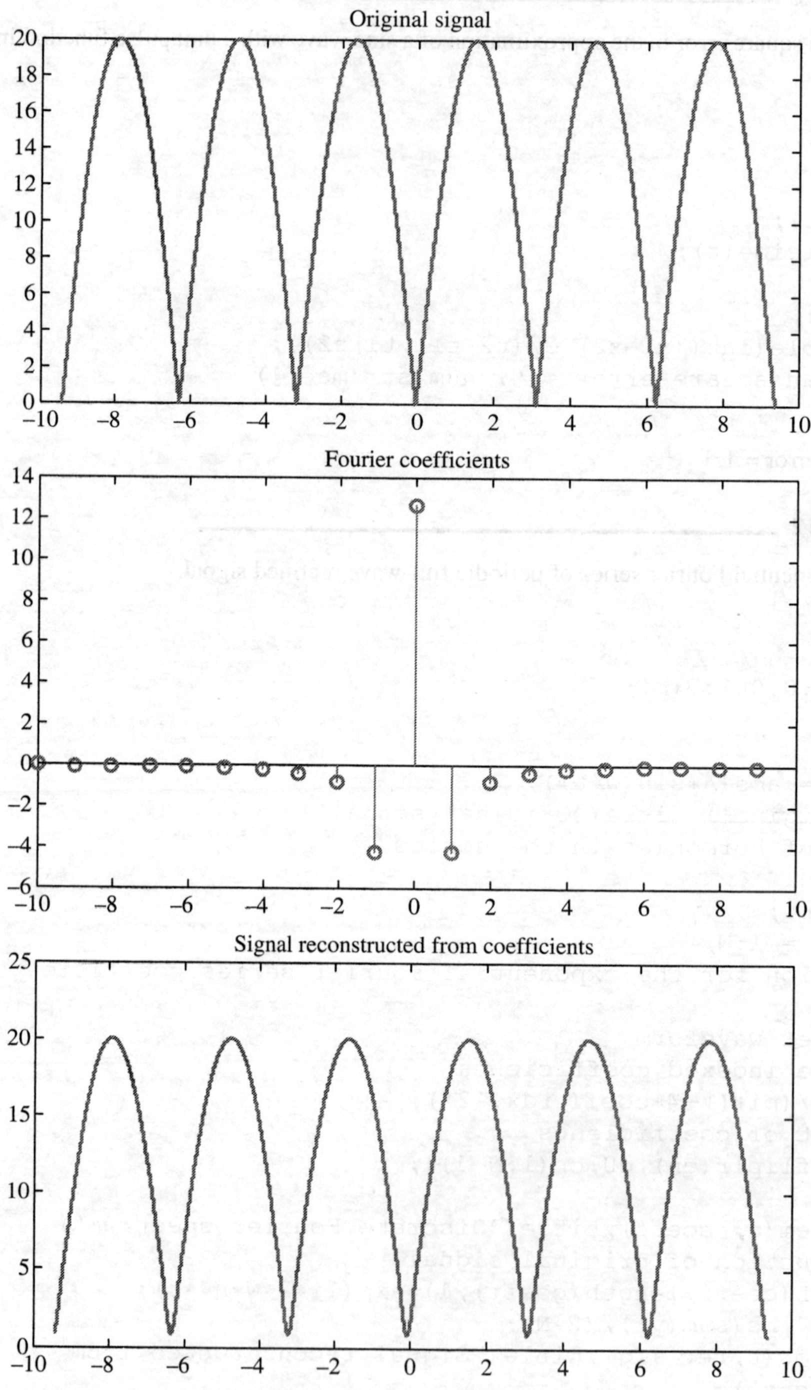

Example A13

Generate trigonometric Fourier series of periodic rectangular signal.

Program

```
t = -3*pi:0.001:3*pi;
A = 2;
f = 1/(2*pi);
y = A*square(2*pi*f*t);
plot(t,y),title('Original signal')
N = 25;
n = 0:N-1;
n1 = 2*n+1;
y1 = 4*A./(pi*n1);
y2 = repmat(y1,length(y1),1)*(sin(2*pi*f*n1'*t));
figure, stem(n1,y1),title('Discrete Fourier spectrum')
y3(1,:) = sum(y2)/N;
figure, plot(t,y3),title('Reconstructed signal')
```

Output

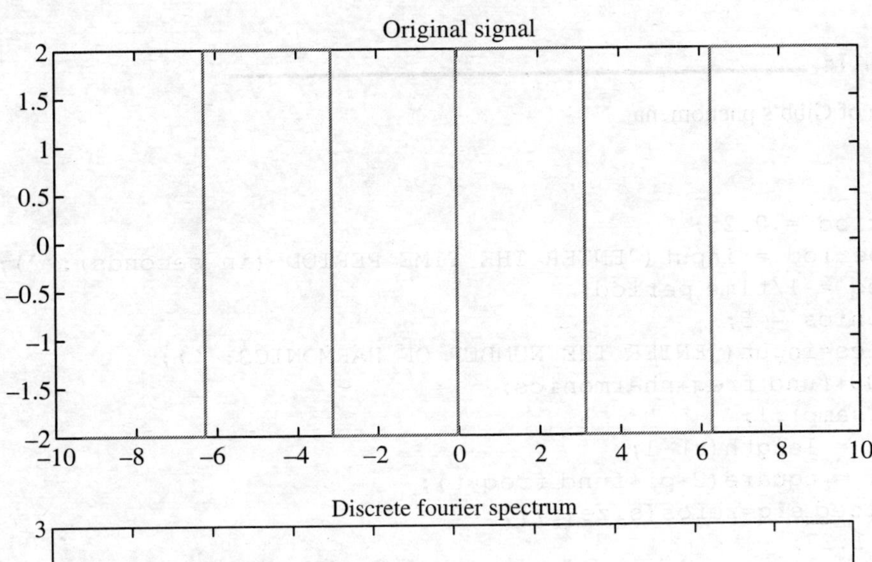

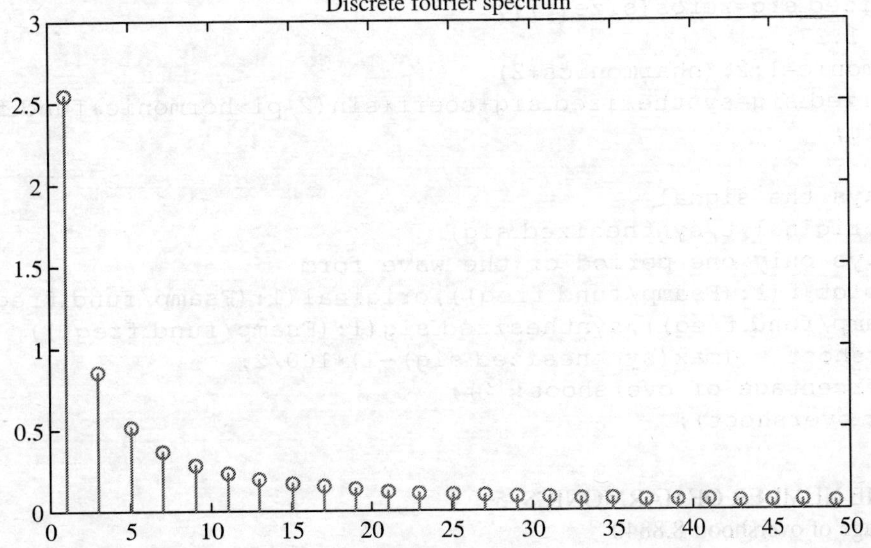

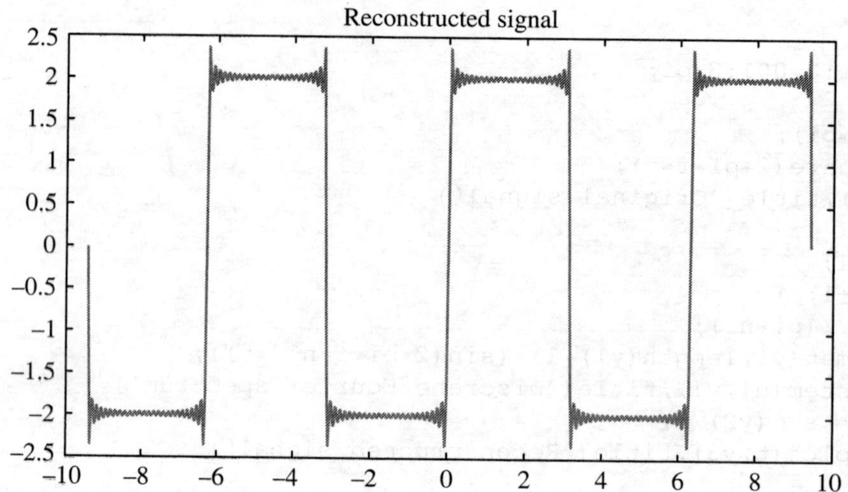

Example A14

Verification of Gibb's phenomena

Program

```
time_period = 0.25;
% time_period = input('ENTER THE TIME PERIOD (in seconds): ');
fund_freq = 1/time_period;
% nharmonics = 5;
nharmonics=input('ENTER THE NUMBER OF HARMONICS: ');
Fsamp=100*fund_freq*nharmonics;
t=0:(1/Fsamp):1;
duration = length(t)-1;
original = square(2*pi*fund_freq*t);
synthesized_sig=zeros(size(t));
coeff = 1.27;
for harmonic=1:2:(nharmonics*2)
synthesized_sig=synthesized_sig+coeff*sin(2*pi*hormonic*fund_freq*t)
/hormonic;
end
% Displays the signal
plot(t,original,t,synthesized_sig)
% Displays only one period of the wave form
figure,plot(t(1:(Fsamp/fund_freq)),original(1:(Fsamp/fund_freq)),
t(1:(Fsamp/fund_freq)),synthesized_sig(1:(Fsamp/fund_freq)))
Per_overshoot = (max(synthesized_sig)-1)*100/2;
disp('percentage of overshoot: ');
disp(Per_overshoot);
```

Output

ENTER THE NUMBER OF HARMONICS: 5

percentage of overshoot: 8.8840

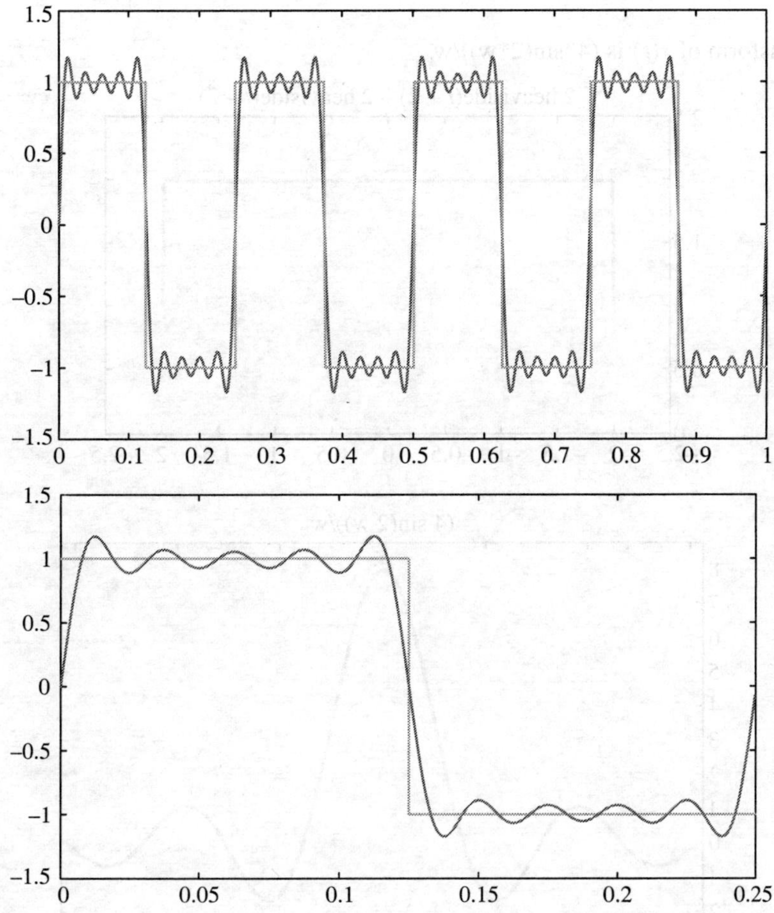

Example A15

Evaluate the Fourier transform of a rectangular pulse $x(t) = 2(u(t+2) - u(t-2))$ and plot the waveforms.

Program

```
syms t w      % Symbolic variables
x = 2*(heaviside(t+2)-heaviside(t-2));
subplot(3,1,1),ezplot(x,[-2,2]);
axis([-2.5 2.5 0 2.5]);
X = int(x*exp(-1i*w*t), t, -5,5);
X = simplify(X);
disp('The Fourier transofrm of x(t) is ')
disp(X)
subplot(3,1,2),ezplot(X)
Xm = sqrt((real(X).^2)+(imag(X).^2));
subplot(3,1,3),ezplot(Xm)
```

Note: The function heaviside $(t) = u(t)$

Output

The Fourier transform of $x(t)$ is (4*sin(2*w))/w.

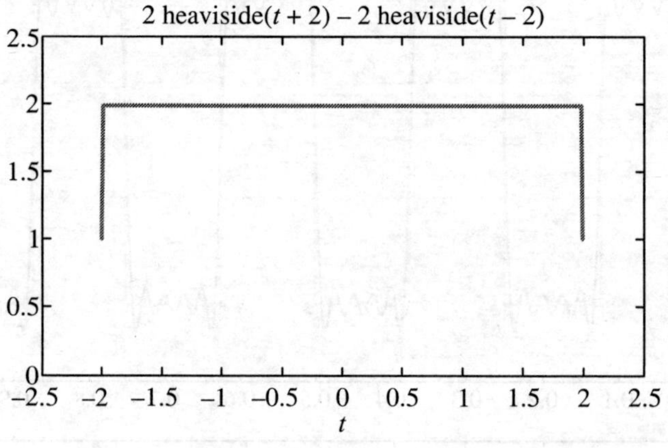

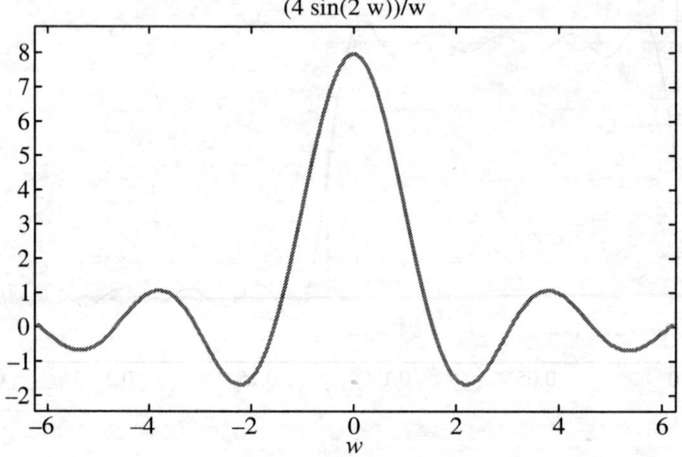

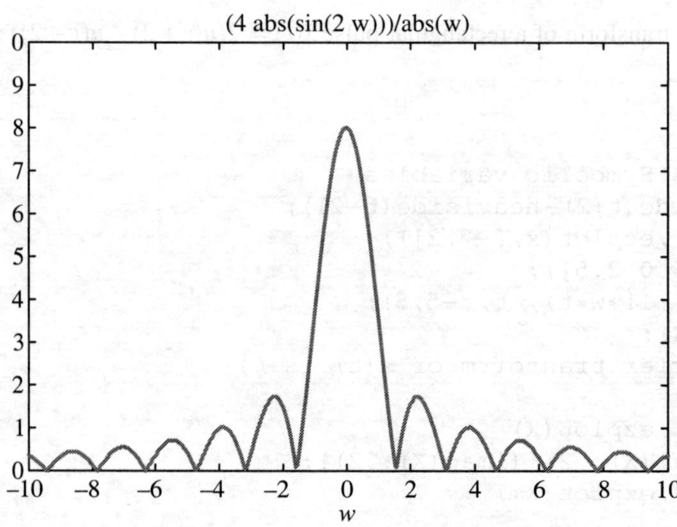

Example A16

Verify time scaling property of Fourier transform for sinc signal and plot the waveforms.

Program

```
%% time scaling property of Fourier transform for sinc signal
clear all
close all
clc
syms t w
a = input('Input scaling factor : ')
x = sinc(2*pi*0.1*t);
x1 = sinc(2*pi*a*0.1*t);
X = fourier(x); X = simplify(X);
X1 = fourier(x1); X1 = simplify(X1);
subplot(2,1,1),ezplot(x,[-20,20]);
title('without scaling'),axis tight
subplot(2,1,2),ezplot(X,[-20,20]);
title('Fourier transform without scaling'),axis tight
figure,subplot(2,1,1),ezplot(x1,[-20,20]);
title('with scaling'),axis tight
subplot(2,1,2),ezplot(X1,[-20,20]);
title('Fourier transform with scaling'),axis tight
```

Output

Input scaling factor : a = 2
The Fourier transform of x(t) without scaling is :
-(5*(heaviside(- w - pi^2/5) - heaviside(1/5*pi^2 - w)))/pi
The Fourier transofrm of x(t) with scaling is :
-(5*(heaviside(- w - (2*pi^2)/5) - heaviside(2/5*pi^2 - w)))/(2*pi)

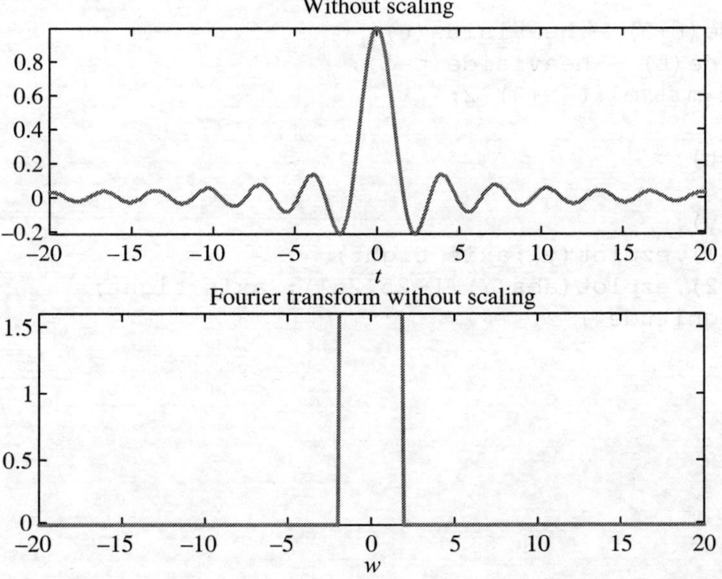

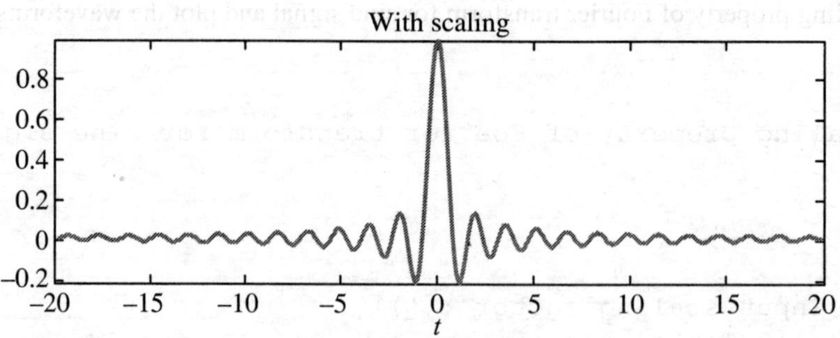

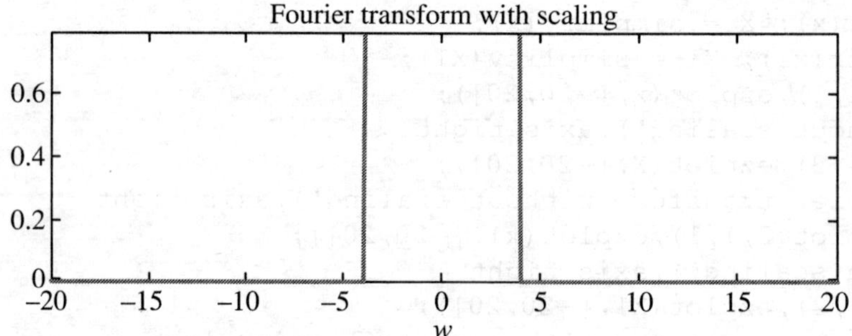

Example A17

Verify the modulation property of Fourier transform for sinc signal. A triangular wave is modulated by cosine function with 10 Hz frequency. Plot the waveforms.

Program

```
syms t w
m = heaviside(t+5) − heaviside(t);
m1 = heaviside(t) − heaviside(t−5);
x1 = (t+5)^2*m + m1*(−t+5)^2;
% x1 = x1/5;
x2 = cos(10*t);
y = x2.*x1;
Y = fourier(y);
subplot(2,1,1),ezplot(y);axis tight;
subplot(2,1,2),ezplot(abs(Y),[−25,25]); axis tight;
title('FT magnitude')
```

Output

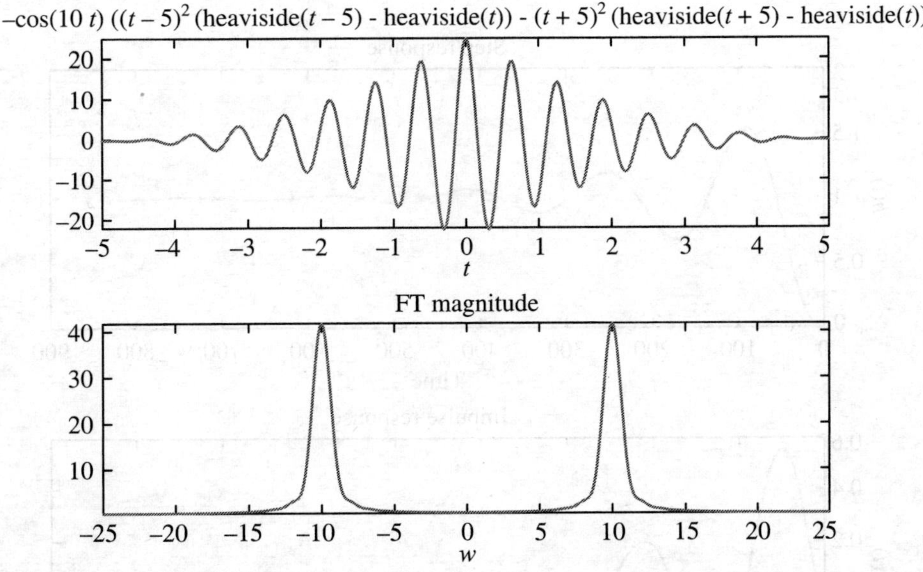

Example A18

(a) Generate a step response and; (b) impulse response of continuous time system with transfer function $H(s) = \dfrac{1}{s^3 + 2s^2 + s + 1}$.

Program

```
% Step response of continuous time system with transfer
function H(s)
num = [1];
den = [1, 2, 1, 1];
h1 =step(num,den);
subplot(2,1,1),plot(h1);title('Step response')
xlabel('Time');ylabel('h(t)')
h2 =impulse(num,den);
subplot(2,1,2),plot(h2);title('Impulse response')
xlabel('Time');ylabel('h(t)')
```

Output

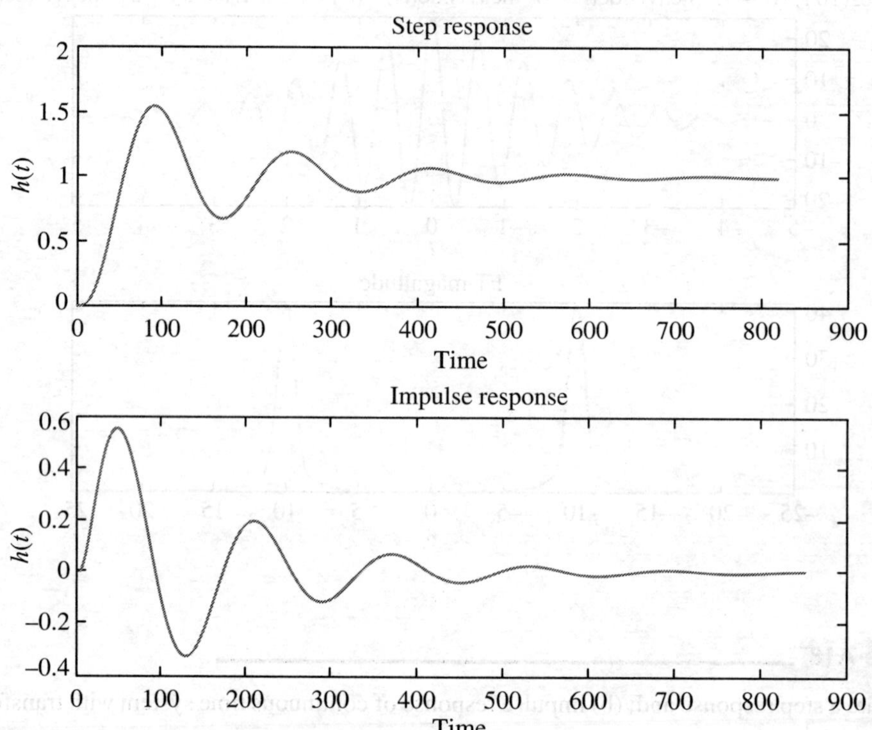

Example A19

Generate frequency and phase response of the system given by $H(s)=\dfrac{s^2-s+3}{s^2+\frac{s}{3}+\frac{1}{6}}$.

Program

```
% Frequency response of the given system
num = [1, −1, 3];
den = [1, 1/3, 1/6];

[H,omega] = freqz(num,den);
subplot(2,1,1), plot(omega/pi, 20*log10(abs(H)));
xlabel('Normalized freq.'),ylabel('Gain in dB'),title('Magnitude
response')
subplot(2,1,2), plot(omega/pi, angle(H));
xlabel('Normalized freq.'),ylabel('Phase in radians'),title('Phase
response')
```

Output

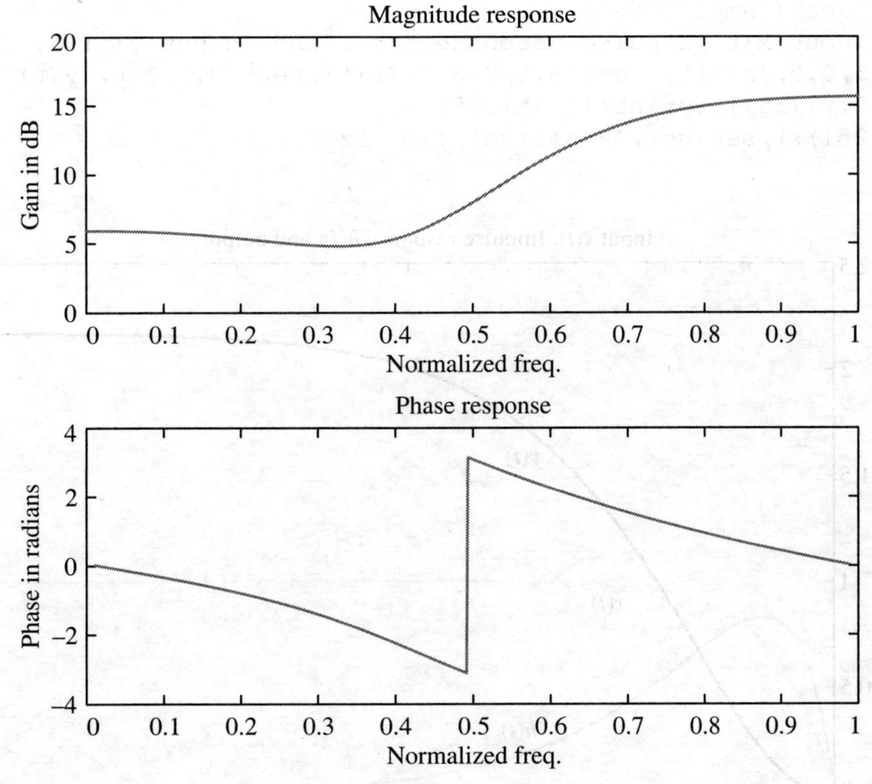

Example A20

Find the convolution between the functions $x(t) = 1 - e^{\frac{3}{2}t}$ and $h(t) = 2te^{-t} + e^{-2t} - e^{-3t}$.

Program

```
fig_size = [232 84 774 624];
t = 0:0.01:10; % time vector
nt = length(t);
dt = t(2);
h = 2*t.*exp(-t) + exp(-2*t) - exp(-3*t); % impulse response
x = 1 - exp(-1.5*t); % input signal
[numh,denh] = residue([0;2;1;-1;],[-1;-1;-2;-3],0); % calculation of
transfer function
[numx,denx] = residue([1; -1],[0; -1.5],0); % calculation of input
transform
[numy,deny] = series(numx,denx,numh,denh); % calculation of output
transform
[resy,poly,ky] = residue(numy,deny); % partial fraction expansion
for
output
y1 = lsim(numh,denh,x,t); % calculation of output using "lsim"
```

```
figure,clf,plot(t,y1,t,x,t,h,t(251),y1(251),'r*'),xlabel('Time
(s)'),ylabel('Amplitude'),...
title('Input x(t),Impulse Response h(t), and Output y(t)'),...
text(2.5,0.9,'x(t)'),text(3.5,0.3,'h(t)'),text(3.5,1.6,'y(t)'),...
text(2.7,y1(251),sprintf('y(t=2.5) =
%g',y1(251))),set(gcf,'Position',fig_size)
```

Output

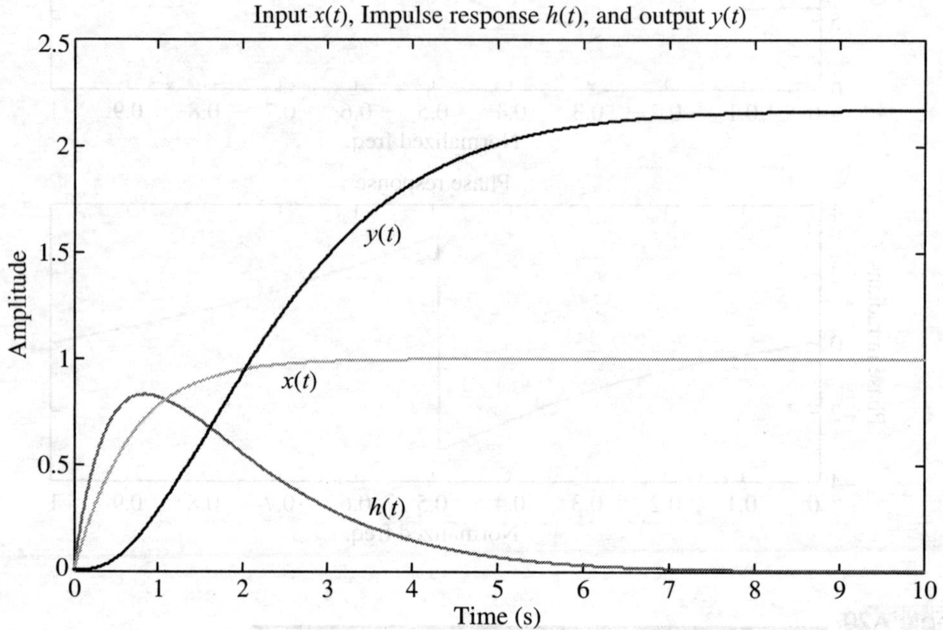

Example A21

Generate the frequency response of (a) a lowpass filter with a cutoff frequency of 900 Hz and; (b) a high-pass filter with a cutoff frequency 400 Hz.

Program

```
% Low Pass Filter response
f=100:20:8000;
fh=900;
k=length(f);
for i=1:k;
m(i)=1/sqrt(1+(f(i)/fh)^2);
mag(i)=20*log10(m(i));
end;
figure;
semilogx(f,mag);
title('magnitude response of analog of low pass filter')
```

```
xlabel('frequency------>');
ylabel('magnitude in db');

% High Pass Filter response
f=100:20:8000;
fl=400;
k=length(f);
for i=1:k;
m(i)=1/sqrt(1+(fl/f(i))^2);
mag(i)=20*log10(m(i));
end;
figure;
semilogx(f,mag);
title('magnitude response of analog of high pass filter');
xlabel('frequency------>');
ylabel('magnitude in db');
grid on;
```

Output

(a) Output

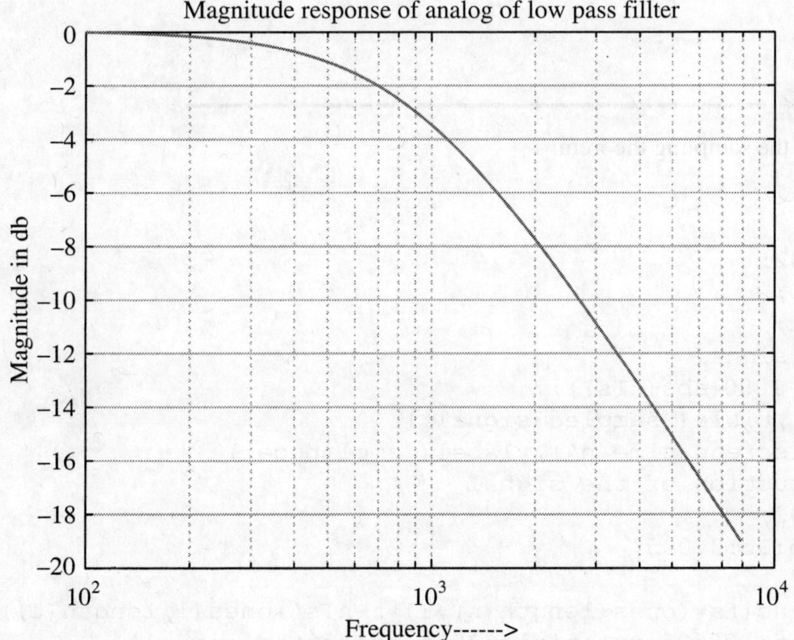

(b) Output

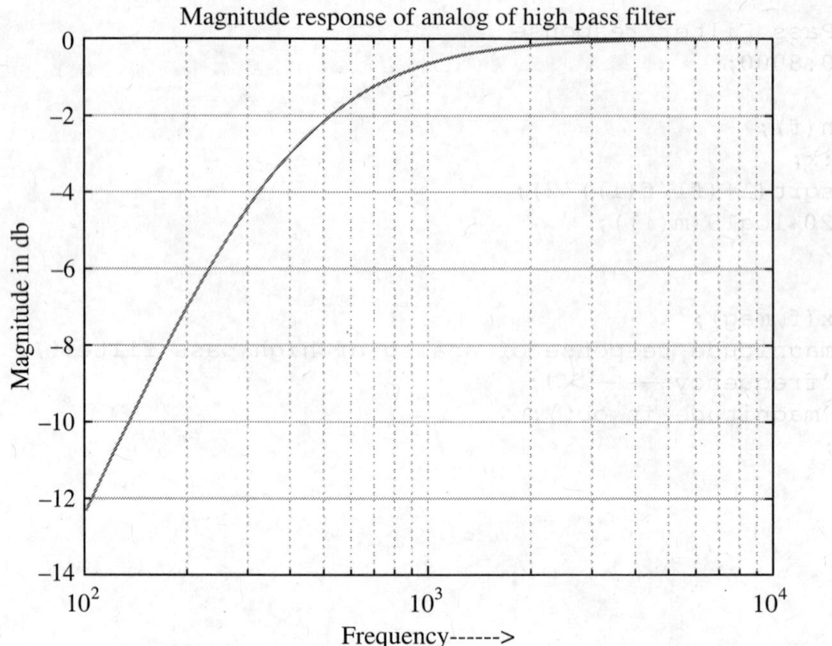

Example A22

Verification of the sampling theorem.

Program

```
Ts = 0.0002;
n = −25:25;
nTs = n*Ts;
% Sampled signal
xn = exp(−1000*abs(nTs));
stem(n,xn),title('sampled signal')
xlabel('discrete time n'),ylabel('amplitude')
% reconstruction of the signal
Dt = 0.0002;
t = −0.005:Dt:0.005;
fs = 5000;
xr = xn*sinc(fs*(ones(length(n),1)*t−nTs'*ones(1,length(t))));
figure,plot(−25:25,xr),title('Reconstructed signal')
xlabel('Cotinuous time t'),ylabel('amplitude')
```

Output

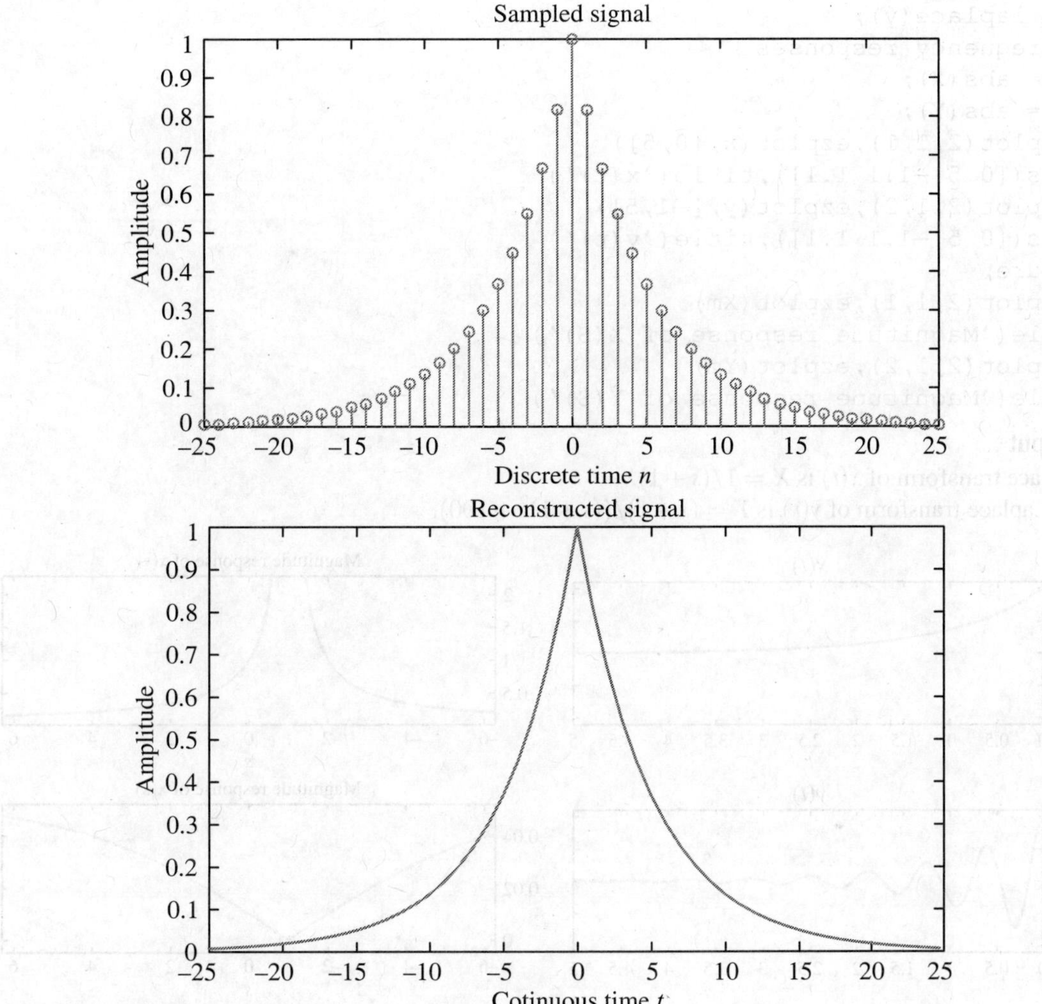

Example A23

Finding Laplace transform of signals $x(t) = \exp(-t); y(t) = x(t) * \cos(10 * t)$.

Program

```
% Finding Laplace transform of signals
syms t;
% input signals
x = exp(-t);
y = x*cos(10*t);
% displaying the laplace transform
disp('Laplace transform of x(t) is ')
X = laplace(x);
```

```
disp('Laplace transform of y(t) is ')
Y = laplace(y);
% Frequency responses
Xm = abs(X);
Ym = abs(Y);
subplot(2,1,1),ezplot(x,[0,5])
axis([0 5 -1.1 1.1]),title('x(t)')
subplot(2,1,2),ezplot(y,[-1,5])
axis([0 5 -1.1 1.1]),title('y(t)')
figure,
subplot(2,1,1),ezplot(Xm)
title('Magnitude response of X(S)')
subplot(2,1,2),ezplot(Ym)
title('Magnitude response of Y(S)')
```

Output

Laplace transform of $x(t)$ is $X = 1/(s+1)$.

Laplace transform of $y(t)$ is $Y = (s+1)/((s+1)^2+100)$.

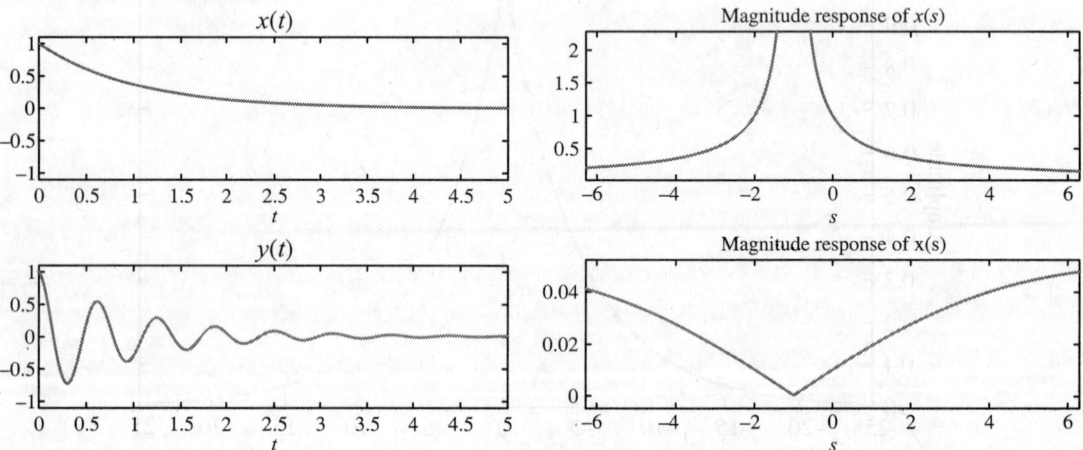

Example A24

Generate partial fraction expansion in Laplace transform for $X(s) = \dfrac{s+1}{s^3+4s^2+6s+4}$.

Program

```
% num = input('Enter the numerator coefficients : ');
num = [1 1]
% den = input('Enter the denominator coefficients : ');
den = [1 4 6 4]
[r,p,k] = residue(num,den);
disp('Residues in the partial fraction are : ')
disp(r)
```

```
disp('Poles of the system are : ')
disp(p )
```

Output

```
Residues in the partial fraction are :
-0.5000
0.2500 - 0.2500i
0.2500 + 0.2500i
Poles of the system are :
-2.0000
-1.0000 + 1.0000i
-1.0000 - 1.0000i
```

Example A25

Obtain convolution of $x_1(t) = e^{-2t}u(t)$ and $x_2(t) = u(t)$ using Laplace transform and draw the output.

Program

```
syms t
x1 = exp(-2*t);
x2 = heaviside(t);
disp('Laplace transform of x1(t) : ')
X1 = laplace(x1);
X1 =simplify(X1)
disp('Laplace transform of x2(t) : ')
X2 = laplace(x2);
Commented [T V 4]: AQ where is the output/answer?
X2 =simplify(X2)
% Convolution using Laplace transform
Y = X1.*X2;
disp('Convolution of two signals : ')
y = ilaplace(Y)
% Plotting the convolution
ezplot(y)
```

Output:

```
Laplace transform of x1(t) :X1 =1/(s + 2)

Laplace transform of x2(t) :X2 =1/s

Convolution of two signals :y =1/2 - exp(-2*t)/2
```

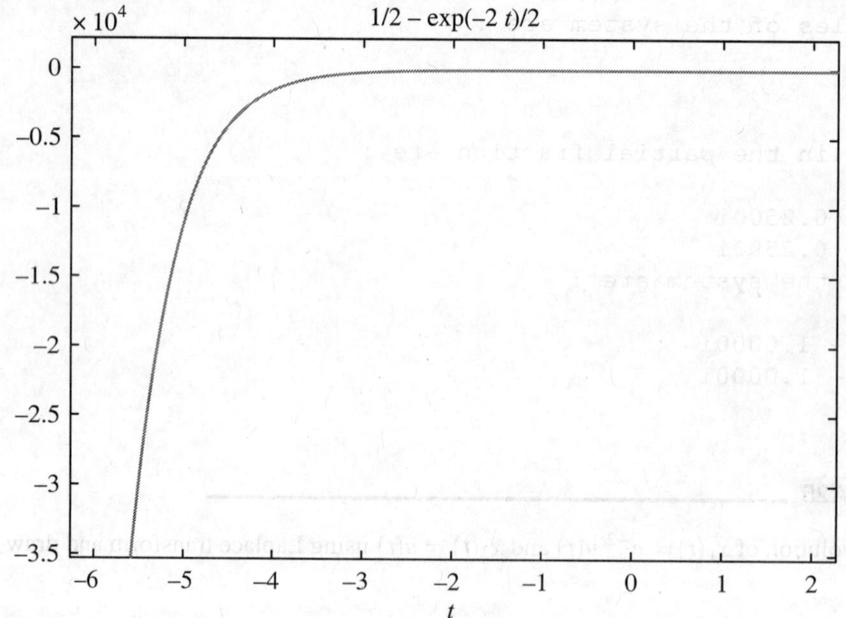

Example A26

Perform inverse Laplace transform of the given $X(s) = \dfrac{2s+3}{s^2+2s+4}$.

Program

```
syms s t w
% num = input('Enter the numerator coefficients : ')
num = [0 2 3];
% den = input('Enter the denominator coefficients : ')
den = [1 2 4];
zplane(num,den);
xlabel('Real part'),ylabel('Imaginary part')
title('Pole-Zero plot')
disp('Inverse laplace transform : ')
x = ilaplace((2*s+3)/(s^2+2*s+4))
figure, ezplot(x,[0,12])
xlabel('Time t'),ylabel('x(t)')
title('Time domain signal x(t)')
axis([0 12 -0.5 2.5])
```

Output

The inverse Fourier transform of $x(t)$ is $(2*(\cos(3^{1/2}*t)+(3^{1/2}*\sin(3^{1/2}*t))/6))/\exp(t)$

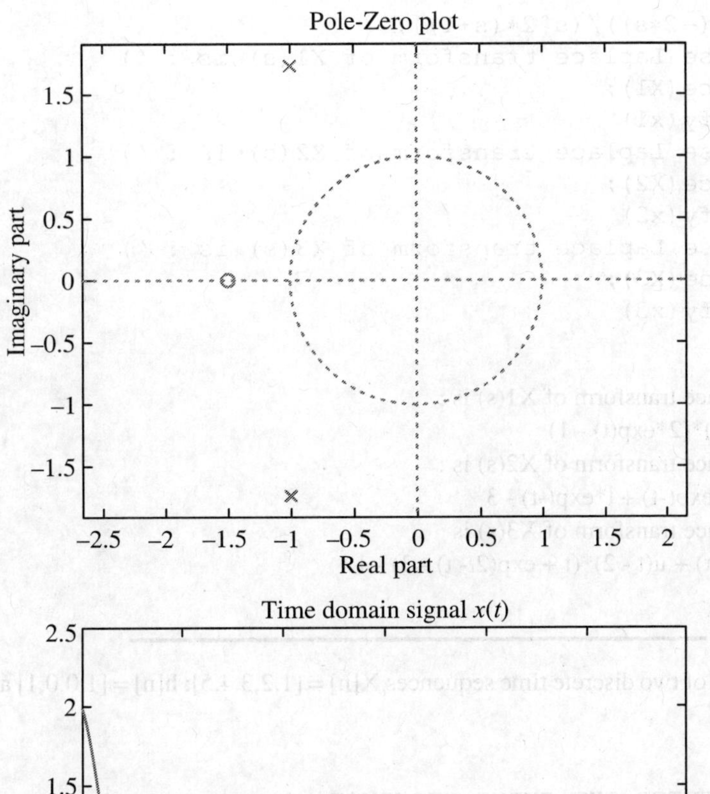

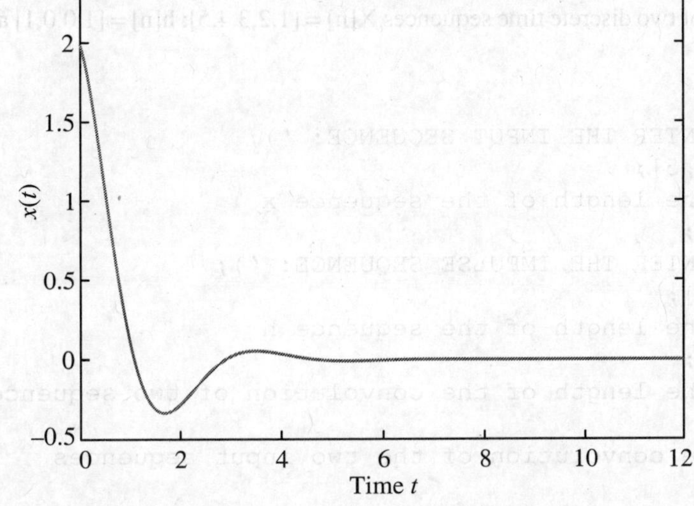

Example A27

Find inverse Laplace transform of the given functions:

$$X_1(S) = \frac{s+4}{s^2+5s+6};\ X_1(S) = \frac{s+2}{s^2(s+1)^2};\ X_3(S) = \frac{1+e^{-2s}}{s^2(s+1)}.$$

Program

```
symsscomplex
X1 = (s+4)/(s^2+5*s+6);
X2 = (s+2)/(s^2*(s+1)^2);
```

```
X3 = (1+exp(-2*s))/(s^2*(s+1));
disp('Inverse Laplace transform of X1(s) is : ')
x1 = ilaplace(X1);
x1 = simplify(x1)
disp('Inverse Laplace transform of X2(s) is : ')
x2 = ilaplace(X2);
x2 = simplify(x2)
disp('Inverse Laplace transform of X3(s) is : ')
x3 = ilaplace(X3);
x3 = simplify(x3)
```

Output

Inverse Laplace transform of X1(s) is :
x1 =exp(-3*t)*(2*exp(t) - 1)
Inverse Laplace transform of X2(s) is :
x2 =2*t + 3*exp(-t) + t*exp(-t) - 3
Inverse Laplace transform of X3(s) is :
x3 =t + exp(-t) + u(t - 2)*(t + exp(2 - t) - 3) - 1

Example A28

Find convolution of two discrete time sequences X[n] = [1,2,3,4,5]; h[n] = [1,0,0,1] and draw the output.

Program

```
% x=input('ENTER THE INPUT SEQUENCE: ');
x = [1,2,3,4,5];
%calculate the length of the sequence x
N1=length(x);
% h=input('ENTER THE IMPULSE SEQUENCE: ');
h = [1,0,0,1];
%calculate the length of the sequence h
N2=length(h);
%calculate the length of the convolution of two sequences
N=N1+N2-1;
% compute the convolution of the two input sequences
y=conv(x,h);
% kx, kh and ky are the durations of the x, h and y sequences
respectively
kx=0:1:N1-1;
kh=0:1:N2-1;
ky=0:1:N-1;
subplot(3,1,1),stem(kx,x);
title('INPUT SEQUENCE');
xlabel('TIME');ylabel('AMPLITUDE');
subplot(3,1,2),stem(kh,h);
title('IMPULSE RESPONSE SEQUENCE');
xlabel('TIME');ylabel('AMPLITUDE');
subplot(3,1,3),stem(ky,y);
```

```
title('RESULT OF CONVOLUTION');
xlabel('TIME');ylabel('AMPLITUDE');
disp(['CONVOLUTION OF x AND h IS : [' num2str(y) ']'])
```

Output

```
CONVOLUTION OF x[n] and h[n] IS : [1  2  3  5  7  3  4  5]
```

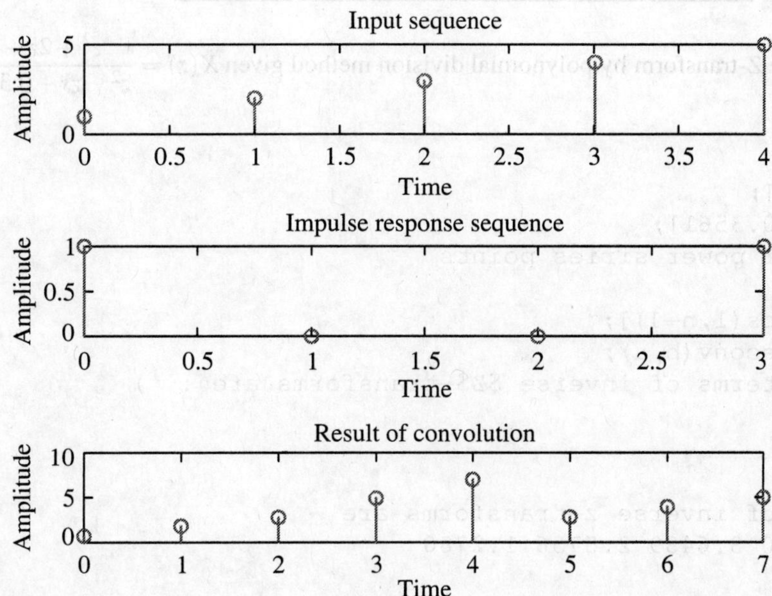

Example A29

Perform *Z*-transform and inverse *Z*-transform of given sequences:

$$x[n] = (n+1)u[n] \text{ and } x[n] = \cos[n\omega_0]u[n].$$

Program

```
syms n w0
x1 = n+1;
X1 = ztrans(x1);
x11 = iztrans(X1);
x2 = cos(w0*n);
X2 = ztrans(x2);
x22 = iztrans(X2);
disp('The input equations : ')
disp(x1)
disp(x2)
disp('The Z-transforms : ')
disp(X1)
disp(X2)
disp('The inverse Z-transforms : ')
disp(x11)
disp(x22)
```

Output

The input equations : $n+1, cos(n*w0)$

The Z-transforms : $z/(z-1)+z/(z-1)^2, (z*(z-cos(w0)))/\,(z^2-2*cos(w0)*z+1)$

The inverse Z-transforms : $n+1, cos(n*w0)$

Example A30

Perform inverse Z-transform by polynomial division method given $X(z)=\dfrac{z^2+2z+1}{z^2-zs+0.3561}$.

Program

```
b = [1 2 1];
a = [1 -1 0.3561];
% Numerator power siries points
n = 5;
b = [b zeros(1,n-1)];
[h,om] = deconv(b,a);
disp('The terms of inverse $Z$-transforms are : ')
disp(h)
```

Output

```
The terms of inverse z-transforms are
0000 3.0000 3.6439 2.5756 1.2780
```

Example A31

Finding inverse Z-transform using residue method.

Program

```
% Denominator coefficients
p1 = poly(0.5);p2 = poly(-0.5);
a = conv(p1,p2);
% Numerator coefficients
z1 = poly(0);z2 = poly(-1);
b = conv(z1,z2);
z = roots(b);
% Partial fraction expansion
[r, p, k] = residuez(b,a);
zplane(b,a)
title('Pole-Zero plot')
d = [1 zeros(1,99)];
% Generate x[n] from filter
x1 = filter(b,a,d);
figure, subplot(2,1,1),stem(x1)
axis([0 20 0 1.5])
title('Generate x[n] from filter')
% Generate x[n] from residue
```

```
n = 0:99;
x2 = r(1)*p(1).^n+r(2)*p(2).^n;
subplot(2,1,2),stem(x2)
axis([0 20 0 1.5])
title('Generate x[n] from residues')
```

Output

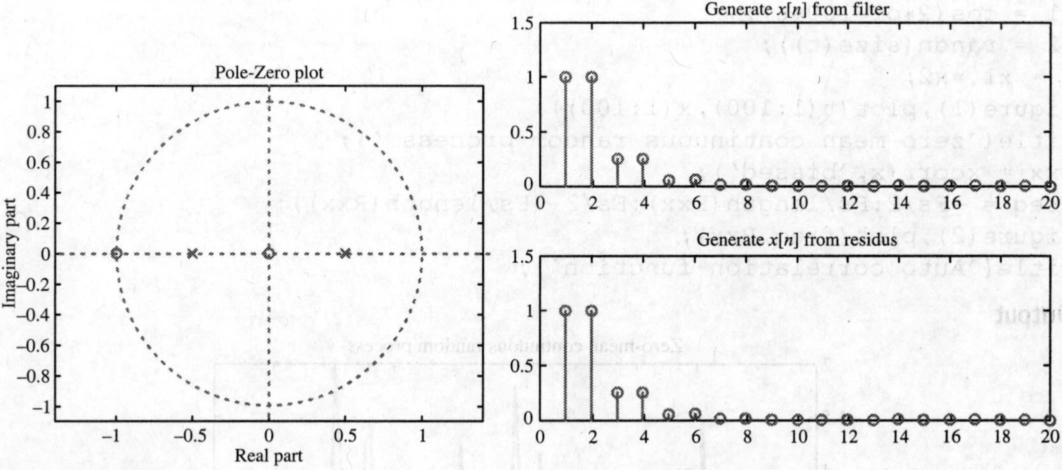

Example A32

Obtain the convolution of $x_1[n] = (2)^n u[n]$ and $x_2[n] = \left(\frac{1}{3}\right)^n u[n]$ using the convolution property of Z-transform.

Program

```
syms n z
disp('The first sequence x1[n] is : ')
x1n = 2.^n
disp('The second sequence x2[n] is : ')
x2n = (1/3).^n
X1Z = ztrans(x1n);
X2Z = ztrans(x2n);
YZ = X1Z*X2Z;
yn = iztrans(YZ);
disp('Convolution of two sequences is : ')
simplify(yn)
```

Output

The first sequence x1[n] is : $x1n = 2^n$.

The second sequence x2[n] is : $x2n = (1/3)^n$

Convolution of two sequences is : ans $=(6*2^n)/5 - 1/(5*3^n)$

Example A33

Generate a random process $Y(t) = X(t)\cos(2\pi 100t)$, where $X(t)$ is a zero-mean continuous and plot auto correlation function of $Y(t)$.

Program

```
clear all
close all
clc
Fs = 100;
t = 0:0.1/Fs:1-1/Fs;
x1 = cos(2*pi*100*t);
x2 = randn(size(t));
x = x1.*x2;
figure(1),plot(t(1:100),x(1:100));
title('zero-mean continuous random process ');
Rxx = xcorr(x,'biased');
freq = -Fs/2:Fs/length(Rxx):Fs/2-(Fs/length(Rxx));
figure(2),plot(freq,Rxx);
title('Auto correlation function');
```

Output

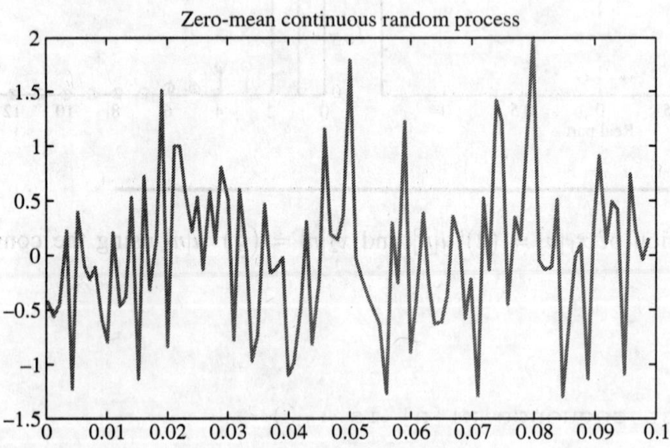

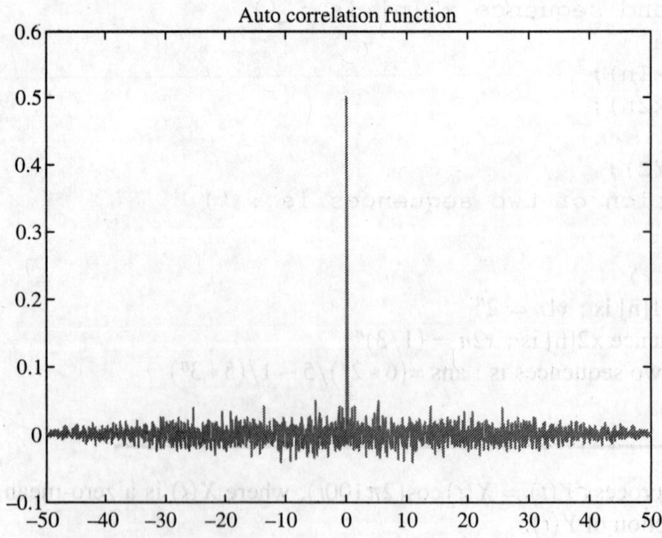

Example A34

Generate a pdf of the normal (Gaussian) distribution with mean 50 and standard deviation 100. Plot the autocorrelation and spectrum.

Program

```
clear all
close all
clc
Fs =100;
t= -400:400;
% pdf of the normal(Gaussian) distribution with mean 50 and standard
deviation 100
x = pdf('Norm',t, 50, 100);
figure(1); plot(t,x);
title('pdf of the normal(Gaussian) distribution');
Rxx =xcorr(x);
freq = -Fs/2:Fs/length(Rxx):Fs/2-(Fs/length(Rxx));
figure(2),plot(freq,Rxx);
title('Spectrum of Auto correlation function');
Rxxdft = abs(fftshift(fft(Rxx)));
figure(3),plot(freq,Rxxdft);
title('Spectrum of Auto correlation function');
```

Output

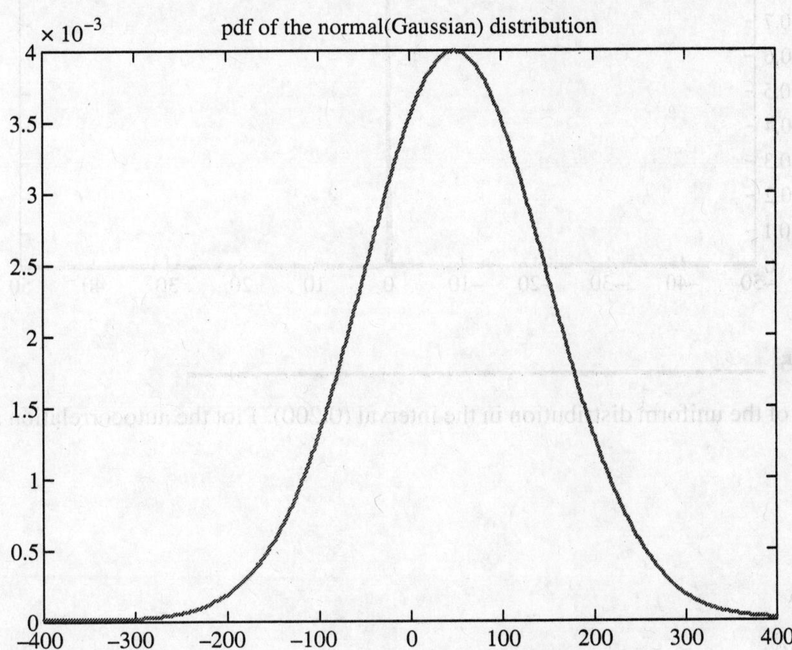

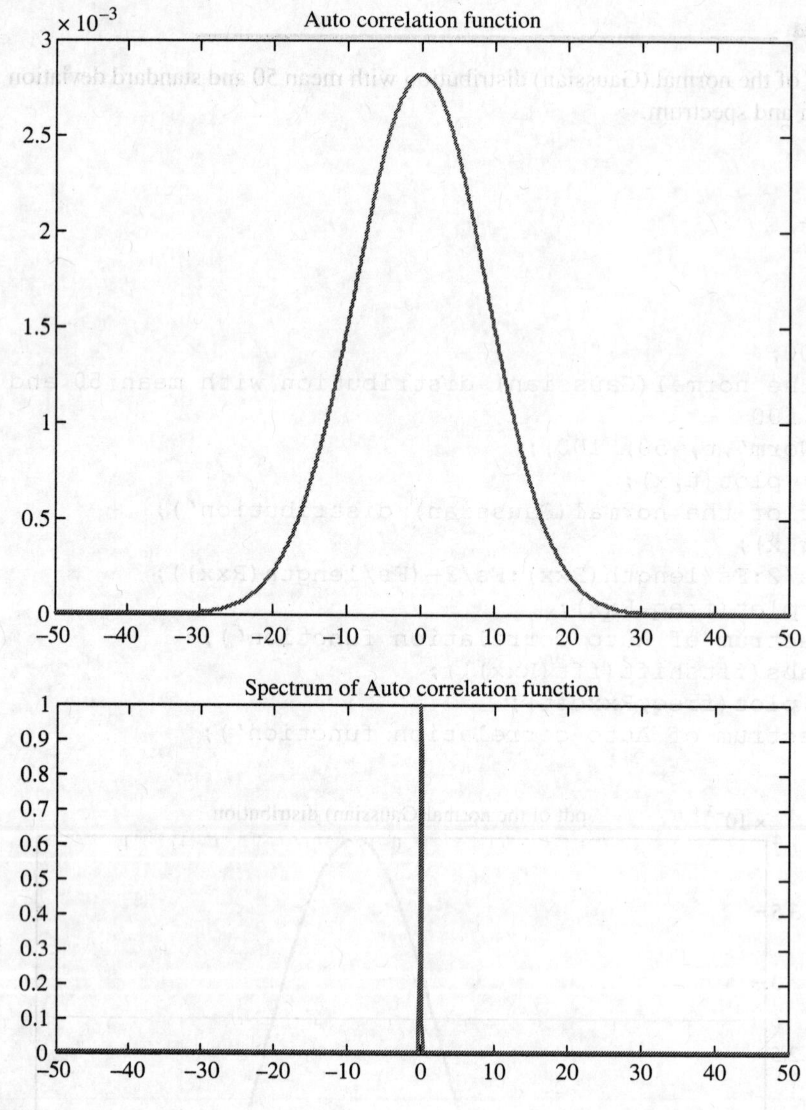

Example A35

Generate a pdf of the uniform distribution in the interval (0,200). Plot the autocorrelation and spectrum.

Program

```
clear all
close all
clc
Fs =100;
t= −400:400;
% pdf of the Uniform distribution in the interval (0.100);
x = pdf('Unif',t, 0, 200);
figure(1); plot(t,x);
```

```
title('pdf of the Uniform distribution');
Rxx =xcorr(x);
freq = -Fs/2:Fs/length(Rxx):Fs/2-(Fs/length(Rxx));
figure(2),plot(freq,Rxx);
title('Auto correlation function');
Rxxdft = abs(fftshift(fft(Rxx)));
figure(3),plot(freq,Rxxdft);
title('Spectrum of Auto correlation function');
```

Output

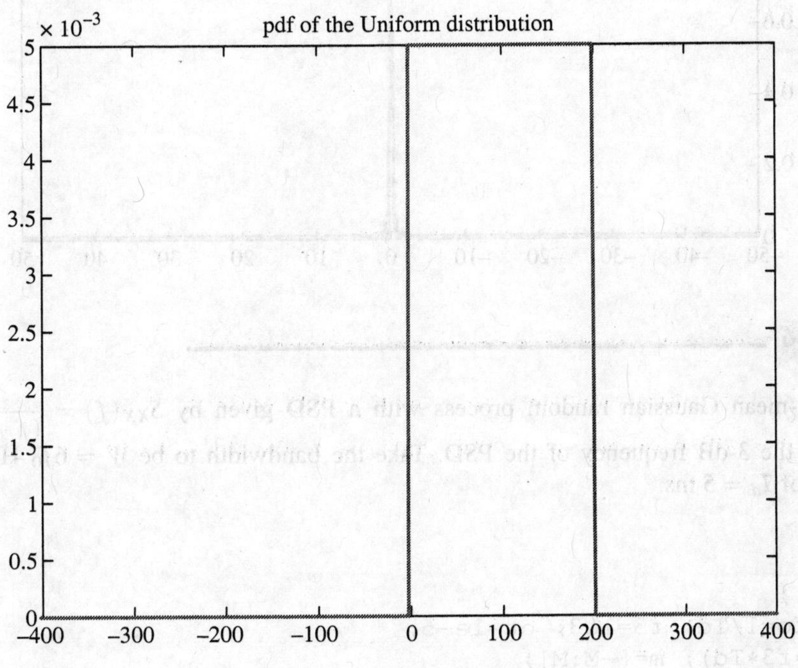

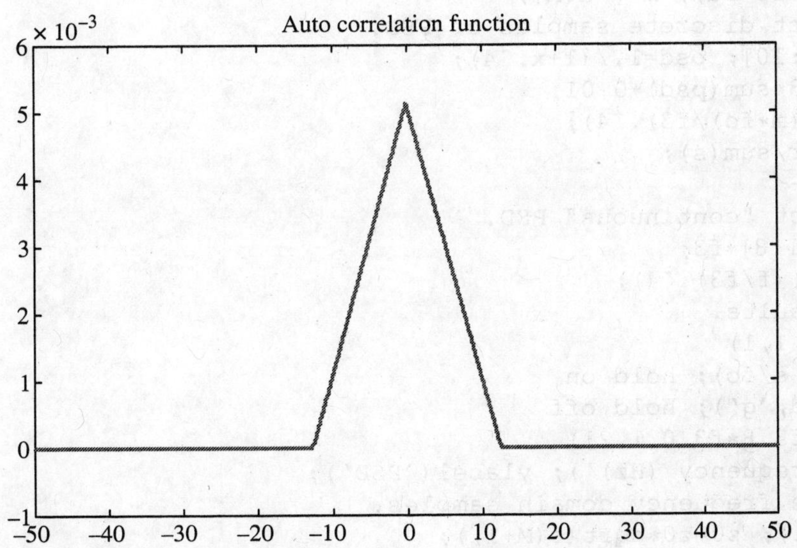

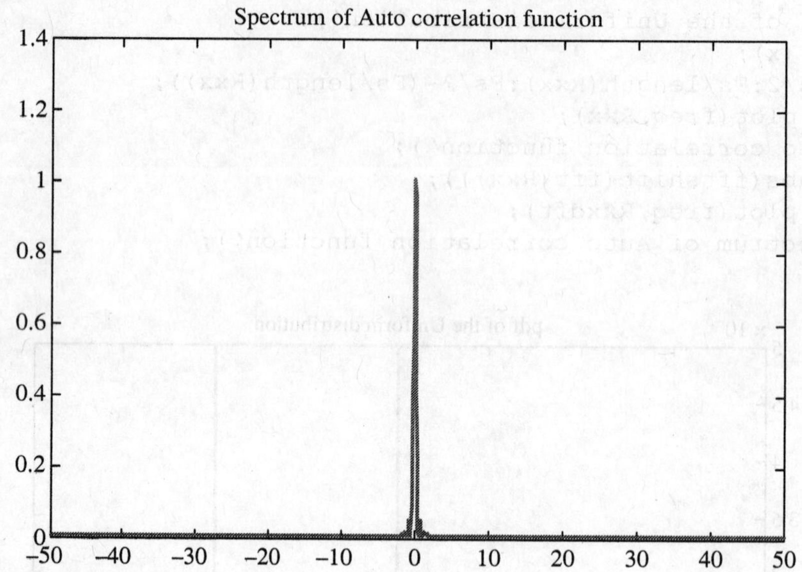

Example A36

Generate zero-mean Gaussian random process with a PSD given by $S_{XX}(f) = \dfrac{1}{(f+f_0)^4}$, where $f_0 = 1$ kHz is the 3-dB frequency of the PSD. Take the bandwidth to be $W = 6f_0$ Hz. Simulate a time duration of $T_d = 5$ ms.

Program

```
I=sqrt(-1);
Td=5e-3; fo=1/Td; f3=1e3; dt=1e-5;
M=floor(6*f3*Td); m=[-M:M];
% Construct discrete samples of PSD.
x=[0:0.01:10]; psd=1./(1+x.^4);
power=2*f3*sum(psd)*0.01;
s=1./(1+((m*fo)/f3).^4);
beta=power/sum(s);
s=beta*s;
% Construct "continuous" PSD.
f=[-8:0.01:8]*f3;
psd=1./(1+(f/f3).^4);
% Plot results.
subplot(2,1,1)
stem(m*fo,s/fo); hold on
plot(f,psd,'g'); hold off
axis([-8*f3 8*f3 0 1.2])
xlabel('frequency (Hz)'); ylabel('PSD');
% Generate frequency domain samples.
z0=randn(1); z0=z0*sqrt(s(M+1));
zplus=sqrt(s(M+2:2*M+1)/2).*(randn(1,M)+I*randn(1,M));
```

```
zminus=conj(fliplr(zplus));
z=[zminus z0 zplus];
% Create time domain process.
t=[0:dt:Td];
rp=zeros(1,length(t));
for m=-M:M
rp=rp+z(m+M+1)*exp(I*2*pi*m*fo*t);
end;
% Plot results.
subplot(2,1,2)
plot(t*1000,real(rp))
xlabel('t (msec)'); ylabel('X(t)')
```

Output

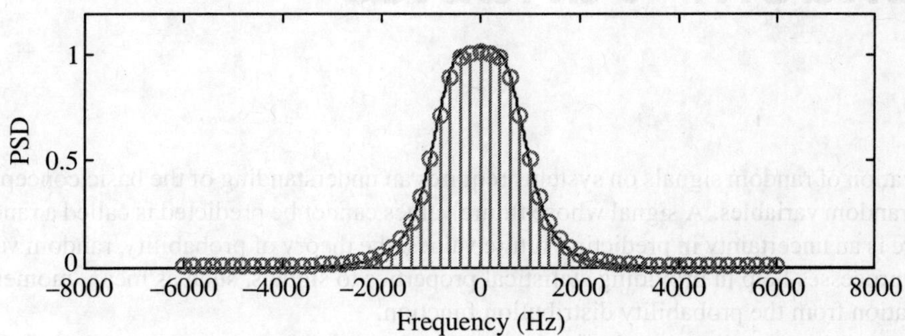

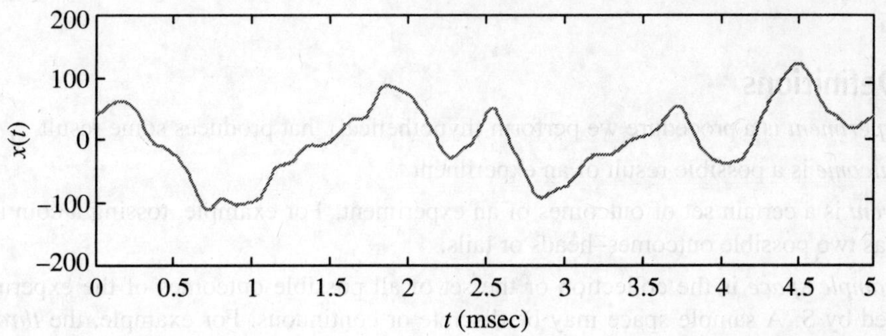

Appendix **B**

Review of Probability and Random Variables

The application of random signals on systems requires an understanding of the basic concepts of probability and random variables. A signal whose future values cannot be predicted is called a random signal. When there is an uncertainty in predicting future values, the theory of probability, random variables and stochastic processes help in providing statistical properties to signals, such as mean, moment, variance and correlation from the probability distribution function.

In this appendix we will review the fundamental tools needed for processing random signals through systems.

Simple Definitions

- An *experiment* is a procedure we perform (hypothetical) that produces some result.
- An *outcome* is a possible result of an experiment.
- An *event* is a certain set of outcomes of an experiment. For example, tossing a coin is an event that has two possible outcomes–heads or tails.
- The *sample space* is the collection or the set of all possible outcomes of the experiment. It is denoted by S. A sample space may be discrete or continuous. For example, the *throwing of a single die* has six outcomes. The discrete sample space is 1, 2, 3, 4, 5, 6.
- If any two events in an experiment have no common outcomes, then the events are said to be mutually exclusive.
- If in a given experiment one of the events in the sample space does not depend on another event, then the two events are called equally likely events.

Probability

Suppose a random experiment is repeated n times. If event A occurs $n(A)$ times, then the probability of event A is defined as the relative frequency of event A when the number of trials n tends to infinity. Mathematically,

$$P(A) = \lim_{n \to \infty} \frac{n(A)}{n}$$

where $n(A)/n$ is called the relative frequency of event A.

In the axiomatic definition, the probability $P(A)$ of an event is a real number which satisfies the following axioms.

Axiom 1:

$$P(A) \geq 0$$

The probability of event A is always a non-negative real number.

Axiom 2:

$$P(S) = 1$$

The event S is an universal set and it is a sure event. Therefore, its probability is always one.

$$P(\phi) = 0$$

The probability of a null event or an impossible event is always zero.

Axiom 3:
If a sample A_n has N events, $n = 1, 2, \ldots\ldots, N$, where N may be infinite, and if all the events are mutually exclusive, i.e., $A_m, \cap A_n = \{\phi\}$ for all $m \neq n$, then

$$P\left(\cup_{n=1}^{N} A_n\right) = \sum_{n=1}^{N} P(A_n)$$

$$\text{or } P(A_1 \cup A_2 \cup \cdots\cdots \cup A_N) = P(A_1) + P(A_2) + \cdots\cdots + P(A_N).$$

Properties of Probability

1. $0 \leq P(A) \leq 1$
2. $P(\overline{A}) = 1 - P(A)$
3. $P(A) \leq P(B)$, if $A \subseteq B$
4. $P(A \cup B) = P(A) + P(B) - P(A \cap B)$
5. $P(\overline{A} \cap B) = P(B) - P(A \cap B)$
6. If $S = A_1 + A_2 \ldots\ldots A_N$ are mutually exclusive events in S, then $\sum_{n=1}^{N} P(A_n) = 1$.
7. For null set $\{\phi\}, P\{\phi\} = 0$ and for universal set $S, P(S)$
8. For equally likely of n events in S, the probability of each event is

$$P_i = \frac{1}{n}, \quad i = 1, 2, \ldots, n.$$

Joint Probability and Conditional Probability

If an experiment repeats n times and the joint occurrence of events A and B is $n(AB)$ times, then the joint probability of events A and B is

$$P(A \cap B) = \lim_{n \to \infty} \frac{n(AB)}{n}.$$

The conditional probability of event A given event B is equal to the relative frequency of the joint occurrence $n(AB)$ with respect to $n(B)$ as n tends to infinity. It is expressed as

$$P(A|B) = \frac{\lim_{n\to\infty} \frac{n(AB)}{n}}{\lim_{n\to\infty} \frac{n(B)}{n}} = \frac{P(A\cap B)}{P(B)},\ P(B) \neq 0,$$

$$P(B|A) = \frac{P(A\cap B)}{P(A)} = \frac{P(AB)}{P(A)}\ P(A) \neq 0.$$

$$P(A\cap B) = P(A|B)P(B) = P(B|A)P(A).$$

Properties of Conditional Probability

1. For any two events A and B in a sample space S, if $B \subset A$, then $P(AB) = 1$.
2. If $B \subset A$ then $P(B|A) = \dfrac{P(B)}{P(A)}$.
3. If two events A and B are mutually exclusive, then the joint events, $A\cap C$ and $B\cap C$, are also mutually exclusive and hence $P((A\cup B)|C) = P(A|C) + P(B|C)$.
4. $P(A|B) \geq 0$, i.e., the conditional probability is always non-negative.
5. If two events A and B are in sample S, then

   $$P(S|A) = F(S|B) = 1,\ P(A|S) = P(A)\ \text{and}\ P(B|S) = P(B).$$

   The total probability of event A is expressed as

   $$P(A) = \sum_{n=1}^{N} P(A|B_n)P(B_n).$$

Bayes' Theorem

$$P(B_n|A) = \frac{P(A|B_n)P(B_n)}{P(A|B_1)P(B_1) + P(A|B_2)P(B_2) + \cdots\cdots + P(A|B_N)P(B_N)}$$

Two events are statistically independent if and only if

$$P(A\cap B) = P(A)P(B)\ \text{for}\ P(A) \neq 0\ \text{and}\ P(B) \neq 0.$$

$$P(A|B) = P(A)$$

and $P(B|A) = P(B)$.

Multiplication Theorem of Probability

It states that if there are N events in a given sample space, then the joint probability of all the events can be expressed as

$$P(A_1\cap A_2\cap A_3\cap \ldots\ldots \cap A_N) = P(A_1)P(A_2|A_1)P(A_3|A_1\cap A_2)\ldots\ldots P(A_N|A_1\cap A_2\cap \ldots A_{N-1})$$

and if all the events are statistically independent, then

$$P(A_1\cap A_2\cap A_3 \ldots\ldots \cap A_N) = P(A_1)P(A_2)P(A_3)\ldots\ldots P(A_N).$$

Random Variable

A random variable (RV) is a real function of the events of a given sample space S. Thus for a given experiment defined by a sample space S with events s, the random variable is a function of s. It is denoted with a capital letter such as $X(s)$.

For example, consider the experiment consisting of tossing two coins. Let the random variable be function X chosen as the number of heads shown. The random variable is $X = \{0, 1, 2\}$.

A random variable X is said to be *continuous* if it takes all the possible values between certain limits.

A random variable whose set of possible values is either finite or countably infinite is called *discrete* random variable.

Probability Distribution Function (PDF)

The probability distribution function (PDF) describes the probabilistic behaviour of a random variable. It is also called the cumulative probability distribution function of the random variable X and is given by

$$F_x(x) = P\{X \leq x\}.$$

For continuous random variable:

$$F_x(x) = \int_{\infty}^{x} f_x(x)dx.$$

For discrete random variable

$$F_x(x) = \sum_{i=1}^{N} P(x_i)u(x - x_i).$$

Probability Density Function

The probability density function of the random variable X is defined as the values of probabilities at a given value of x.

$$f_x(x) = \frac{d\,F_x(x)}{dx}$$

$$f_x(x) = \sum_{i=1}^{N} P(x_i)\delta(x - x_i)$$

Properties of Probability Distribution Function

If $F_x(x)$ is a probability distribution function of a random variable X, then

1. $F_x(-\infty) = 0$
2. $F_x(\infty) = 1$
3. $0 \leq F_x(x) \leq 1$
4. $F_x(x_1) \leq F_x(x_2)$ if $x_1 < x_2$
5. $P\{x_1 < X \leq x_2\} = F_x(x_2) - F_x(x_1)$
6. $F_x(x^+) = F_x(x)$

Properties of Probability Density Function

If $f_x(x)$ is a probability density function of a random variable X, then

1. $0 \leq f_x(x)$ for all x.
2. $\int_{-\infty}^{\infty} f_x(x)dx = 1$.
3. $F_x(x) = \int_{-\infty}^{x} f_X(x)dx$.
4. $P\{x_1 < X \leq x_2\} = \int_{x_1}^{x_2} f_X(x)dx$.

For Discrete Random Variables

The probability of $X, P(x_i), i = 1, 2, \ldots, \infty$ satisfies the following conditions:

1. $P(x_1) \geq 0$ for all i
2. $\sum_{i=1}^{\infty} P(x_i) = 1$

Conditional Distribution and Density Functions

The conditional distribution function of X, when the event B is known is denoted as

$$F_x(x|B) = P\{X \leq x|B\}.$$

The conditional density function of the random variable X is defined as

$$f_X(x|B) = \frac{d F_X(x|B)}{dx}.$$

Properties of Conditional Distribution Function

The properties of a conditional distribution function $F_X(x|B)$ are given below:

1. $F_X(-\infty|B) = 0$
2. $F_X(\infty|B) = 1$
3. $0 \leq F_X(x|B) \leq 1$
4. $F_X(x_1|B) \leq F_X(x_2|B)$ if $x_1 < x_2$
5. $P\{x_1 < X \leq x_2|B\} = F_X(x_2|B) - F_X(x_1|B)$
6. $F_X\{X^*|B\} = F_X(x|B)$

Properties of the Conditional Density Function

1. $f_X(S) = 1$
2. $\int_{-\infty}^{\infty} f_X(x|B) = 1$

3. $F_X(x|B) = \int_{-\infty}^{\infty} f_X(x|B)dx$

4. $P\{x_1 < X \leq x_2|B\} = \int_{x_1}^{x_2} f_X(x|B)dx$

Expected Value

If X is a continuous random variable with a valid probability density function $f_X(x)$, then the expected value of X or the mean value of X is defined as

$$E[X] = \overline{X} = \int_{\infty}^{\infty} x f_x(x)dx.$$

$$E[X] = \sum_{i=1}^{N} x_i p(x_i).$$

Properties of Expectation

1. If a random variable X is a constant, i.e., $X = a$, then $E[a] = a$, where a is a constant.
2. If a is any constant, then

   $$E[aX] = aE[X].$$

3. If a and b are any two constants, then

   $$E[aX + b] = aE[X] + b.$$

4. If $X \geq 0$, then $E[X] \geq 0$.
5. If X is any random variable, then the inequality

   $|E[X]| \leq E|X|$ exists.

6. If $g_1(x)$ and $g_2(x)$ are two functions of a random variable X, then

   $$E[g_1(x) + g_2(x)] = E[g_1(x)] + E[g_2(X)].$$

The nth Moment of X

The nth moment is defined as

$$m_n = E[X^n] = \int_{-\infty}^{\infty} x^n f_X(x)dx.$$

The first moment is equal to its mean value, $m_1 = E[X] = \overline{X}$.

$$m_1 = \int_{-\infty}^{\infty} x f_X(x)dx = E[X]$$

The second moment of X or the mean square value of X is

$$m_2 \int_{-\infty}^{\infty} x^2 f_X(x)dx = E\left[X^2\right].$$

The nth central moment of X is given as

$$\mu_n = E\left[(X - \overline{X})^n\right] = \int_{-\infty}^{\infty} (x - \overline{X})^n f_X(x)dx.$$

Variance of X

The variance of the density function is the second central moment of X.

$$\mu_2 = \sigma_X^2 = E\left[\left(X-\overline{X}\right)^2\right] = \int_{-\infty}^{\infty} \left(x-\overline{X}\right)^2 f_X(x)dx.$$

The standard deviation of a random variable is defined as

$$\sigma_X = \sqrt{E\left[X-\overline{X}\right]^2}.$$

Properties of Variance

1. The variance of a constant is zero, i.e., if k is a constant, then var $(k) = 0$.
2. If k is a constant, then for a random variable X,

   $$\text{var}(kX) = k^2 \ \text{var}(X).$$
3. For a given random variable X, the relationship between the variance and the moments is given by

   $$\sigma_X^2 = m_2 - m_1^2.$$
4. If X is a random variable and a, b are real constants, then

   $$\text{var}(aX+b) = a^2 \ \text{var}(X).$$
5. If two random variables X_1 and X_2 are independent, then

   $$\text{var}(X_1+X_2) = \text{var}(X_1)+\text{var}(X_2)$$

   and $\text{var}(X_1-X_2) = \text{var}(X_1)+\text{var}(X_2)$.

   Skew of the density function is given by

   $$\mu_3 = \overline{X^3} - 3\overline{X}\,\sigma_X^2 - \overline{X}^3 \quad \text{and}$$

   $$\text{Skewness} = \frac{\mu_2}{\sigma_X^3} = \frac{E\left[\left(X-\overline{X}\right)^3\right]}{E\left[\left(X-\overline{X}\right)^2\right]^{3/2}}.$$

Characteristic Function

Consider a random variable X with a probability density function $f_X(x)$. Then the expected value of the function $e^{j\omega X}$ is called the characteristic function. It is expressed as

$$\phi_X(\omega) = E\left[e^{j\omega X}\right].$$

It is a function of the real variable, $-\infty < \omega < \infty$, where j is an imaginary operator.

$$\therefore \quad \phi_X(\omega) = \int_{-\infty}^{\infty} e^{j\omega X} f_X(x)dx.$$

If $\phi_X(\omega)$ is a characteristic function of a random variable X, then the nth moment of X is given by

$$m_n = (-j)^n \left.\frac{d^n \phi_X(\omega)}{d\omega^n}\right|_{\omega=0}$$

Moment Generating Function (MGF)

Consider a random variable X with a probability density function $f_X(x)$. Then the moment generating function of X is defined as the expected value of the function e^{vX}. It can be expressed as

$$M_X(v) = E\left[e^{vX}\right],$$

where v is a real variable $-\infty < v < \infty$.

If $M_X(v)$ is a moment generating function of a random variable X, then the nth moment of X is given by

$$m_n = \left.\frac{d^n M_X(v)}{dv^n}\right|_{v=0}.$$

Distribution and Density Functions

Important density functions are

1. Gaussian density function or normal density function
2. Uniform density function
3. Exponential density function
4. Rayleigh density function
5. Binomial density function and
6. Poisson density function

The pdf, mgf, mean and variance of the density functions are given in Table B1.

Table B1

Sl. No.	Distribution	PDF ($P(X = x)$)	MGF	Mean	Variance
1.	Uniform	$f(x) = \begin{cases} \dfrac{1}{b-a}, & a < x < b \\ 0, & \text{otherwise} \end{cases}$	$\dfrac{e^{bt} - e^{et}}{(b-a)t}$	$\dfrac{a+b}{2}$	$\dfrac{(b-a)^2}{12}$
2.	Exponential	$f(x) = \begin{cases} \lambda e^{-\lambda x}, & x < 0, \lambda > 0 \\ 0, & \text{otherwise} \end{cases}$	$\dfrac{\lambda}{\lambda - t}$	$\dfrac{1}{\lambda}$	$\dfrac{1}{\lambda^2}$
3.	Rayleigh	$\dfrac{2}{b}(x-b)e^{-(x-a)^2/b} \quad x \geq a$		$a + \sqrt{b/2}$	λ
4.	Normal	$f(x) = \dfrac{1}{\sigma\sqrt{2\pi}} e^{\frac{-1}{2}\left(\frac{x-\mu}{\sigma}\right)^2}$	$e^{\mu t + \frac{t^2\sigma^2}{2}}$	μ	σ^2
5.	Binomial	$nc_x p^x q^{n-x}$	$(q + pe^t)^n$	np	npq
6.	Poisson	$\dfrac{e^{-\lambda}\lambda^x}{x!}$	$e^{\lambda(e^t - 1)}$	λ	λ

Joint Probability Distribution and Density Functions

The joint probability distribution function for the joint event $\{X < x, Y \leq y\}$ is defined as

$$F_{X,Y}(x,y) = P\{X \leq x,\ Y \leq y\} = P(A \cap B).$$

The joint probability density function of two random variables X and Y is defined as the second derivative of the joint distribution function. It can be expressed as

$$f_{X,Y}(x,y) = \frac{\partial^2 F_{X,Y}(x,y)}{\partial x \partial y}.$$

Properties of Joint Probability Density Function

The properties of a joint density function for random variables X and Y are given as

1. $f_{X,Y}(x,y) \geq 0$. A joint probability density function is always non-negative.
2. $\int_{-\infty}^{\infty}\int_{-\infty}^{\infty} f_{X,Y}(x,y)dxdy = 1$
   i.e., the area under the density function curve is always equal to one.
3. The joint distribution function is equal to
   $$F_{X,Y}(x,y) = \int_{-\infty}^{x}\int_{-\infty}^{y} f_{X,Y}(x,y)dxdy.$$
4. The probability of the joint event $\{x_1 < X \leq x_2, y_1 < T \leq y_2\}$ is given as
   $$P\{x_1 < X \leq x_2, y_1 < Y \leq y_2\} = \int_{x_1}^{x_2}\int_{y_1}^{y_2} f_{X,Y}(x,y)dxdy.$$
5. The marginal distribution function of X and Y are
   $$F_X(x) = \int_{-\infty}^{x}\int_{-\infty}^{\infty} F_{X,Y}(x,y)dxdy$$
   and $$F_Y(y) = \int_{-\infty}^{y}\int_{-\infty}^{\infty} F_{X,Y}(x,y)dxdy.$$
6. The marginal density functions of X and Y are
   $$f_X(x) = \int_{-\infty}^{\infty} f_{X,Y}(x,y)dy = \frac{\partial F(x,\infty)}{\partial x}$$
   and $$f_Y(y) = \int_{-\infty}^{\infty} f_{X,Y}(x,y)dy = \frac{\partial F(\infty,y)}{\partial y}.$$

Conditional Distribution and Density Functions

Conditional density function of X given Y, $f(x|y) = \dfrac{f(x,y)}{f(y)}$.

Conditional density function of Y given X, $f(x|y) = \dfrac{f(x,y)}{f(y)}$.

Conditional probability function of X given $Y, P\{X = x_i | Y = y_i\} = \dfrac{P(x,y)}{P(y)}$.

Conditional probability function of Y given $X, P\{Y = y_i | X = x_i\} = \dfrac{P(x,y)}{P(x)}$.

$$P\{X < a | Y < b\} = \frac{P(X < a, T < b)}{P(Y < b)}.$$

If X and Y are independent random variables, then

$$f(x,y) = f(x) \cdot f(y) \qquad \text{(for continuous random variable)}$$

$$P(X = x, Y = y) = P(X = x) \cdot P(Y = y) \qquad \text{(for discrete random variable)}.$$

The conditional density functions for independent random variables are

$$f_X(x|y) = f_X(x)$$

and $\quad f_Y(y|x) = f_Y(y)$.

The density function of the sum of N statistically independent random variables is equal to the convolution of their individual density functions.

$$Y = X_1 + Y_2 + Y_3 + \ldots + X_N$$

$$f_Y(y) = f_{x_1}(x_1) * f_{X_2}(x_2) * f_{X_3}(x_3) * \ldots\ldots * f_{X_N}(x_n)$$

Central Limit Theorem

Let N random variables $X_n, n =, 1, 2, 3, \ldots, N$ have probability density functions with mean $\overline{X}_n$ and variance $\sigma_{X_n}^2$. The central limit theorem states that the sum of the random variables $Y_N = X_1 + X_2 + X_3 + \ldots + X_N$ with mean $\overline{Y}_N = \overline{X}_1 + \overline{X}_2 + \overline{X}_3 + \ldots + \overline{X}_N$ and variance $\sigma_Y^2 = \sigma_{X_1}^2 + \sigma_{X_2}^2 + \ldots + \sigma_{X_N}^2$ have a probability distribution function asymptotically approaching a Gaussian distribution as N tends to infinity.

The Gaussian random variable is

$$W = \frac{1}{\sqrt{N}\sigma_X} \sum_{n=1}^{N} (X_n - \overline{X}).$$

Operations on Multiple Random Variables

For N random variables $X_1, X_2, X_3, \ldots, X_N$, with joint density function $f_{X_1 X_2 \ldots X_N}(x_1, x_2, \ldots, x_N)$, the expected value of the function $g(X_1, X_2, \ldots, X_N)$ is given as,

$$\overline{g} = E[g(X_1, X_2, X_3, \ldots, X_N)]$$

$$\overline{g} = \int_{-\infty}^{\infty} \int_{-\infty}^{\infty} \ldots \int_{-\infty}^{\infty} g(x_1, x_2, \ldots, x_N) f_{X_1 X_2 \ldots X_N}(x_1, x_2, \ldots, x_N)\, dx_1 dx_2 \ldots dx_N.$$

Joint Moments

For N random variables, the joint moments about the origin is defined as

$$m_{n_1,n_2,n_3....n_N} = E[X_1^{n_1} X_2^{n_2} .....X_N^{n_N}]$$

where $n_1, n_2, ..., n_N$ are all positive integers.

The first order moments are

$$m_{10} = E[X] = \overline{X} = \int_{-\infty}^{\infty}\int_{-\infty}^{\infty} x f_{X,Y}(x,y) dx dy.$$

$$m_{01} = E[Y] = \overline{Y} = \int_{-\infty}^{\infty}\int_{-\infty}^{\infty} y f_{X,Y}(x,y) dx dy.$$

The second order moments are $m_{20} = E\left[X^2\right]$

$$m_{02} = E\left[Y^2\right]$$

and $m_{11} = E[XY]$.

The point coordinate $(\overline{X}, \overline{Y})$ is the centre of gravity of the function $f_{X,Y}(x,y)$.

Correlation

Consider the two random variables X and Y, the second order joint moment m_{11} is called the correlation of X and Y. It is denoted as R_{XY}.

$$R_{XY} = m_{11} = E[XY] = \int_{-\infty}^{\infty} xy f_{X,Y}(x,y) dx dy.$$

For discrete random variables,

$$R_{XY} = \sum_m \sum_n x_m y_n P_{XY}(x_m, y_n).$$

(1) If two random variables X and Y are statistically independent, then X and Y are said to be uncorrelated. That is,

$$R_{XY} = E[XY] = E[X]E[Y].$$

(2) If the random variables X and Y are orthogonal, then their correlation is zero.

$$R_{XY} = 0$$

Joint Central Moments

For N random variables $X_1, X_2, X_3, \ldots, X_N$, the joint central moments are defined as

$$\mu_{n_1,n_2,......n_N} = E[(X_1 - \overline{X})^{n_1} (X_2 - \overline{X}_2)^{n_2}, \ldots, (X_N - \overline{X}_N)^{n_N}]$$

$$= \int_{-\infty}^{\infty} \cdots \int_{-\infty}^{\infty} (x_1 - \overline{X}_1)^{n_1} (x_2 - \overline{X}_2)^{n_2} \ldots (x_N - \overline{X}_N)^{n_N} f_{X_2,X_2,....,X_N}(x_1, x_2, ..., x_n) \cdot dx_1 dx_2, \ldots dx_N.$$

The order of the joint central moments is $n_1 + n_2 + n_3 + \ldots n_N$.

The first order central moments are

$$\mu_{10} = E[(X - \overline{X})] = E[\overline{X}] - E[\overline{X}] = 0$$

and $\mu_{01} = E[(Y - \overline{X})] = E[\overline{Y}] - E[\overline{Y}] = 0.$

The second order central moments are

$$\mu_{20} = E[(X-\overline{X})^2] = \sigma_X^2$$

$$\mu_{02} = E[(Y-\overline{Y})^2] = \sigma_Y^2$$

$$\mu_{11} = E[(X-\overline{X})(Y-\overline{Y})] = \sigma_{XY}^2.$$

Covariance

Consider the random variables X and Y. The second order joint central moment μ_{11} is called the covariance of X and Y. It is expressed as

$$C_{XY} = \sigma_{XY} = \mu_{11} = E[(X-\overline{X})(Y-\overline{Y})]$$

$$= \int_{-\infty}^{\infty} (x-\overline{X})(y-\overline{Y}) f_{X,Y}(x,y)dx, dy.$$

For discrete random variables X and Y,

$$C_{XY} = \sum_n \sum_k (x_n - \overline{X}_n)(y_k - \overline{Y}_k)P(x_n, y_k).$$

For the random variables X and Y, the normalised second order central moment is called the correlation coefficient. It is denoted by ρ and given by

$$\rho = \frac{C_{XY}}{\sigma_X \sigma_Y} = \frac{E[(X-\overline{X})(Y-\overline{Y})]}{\sigma_X \sigma_Y}.$$

Properties of Covariance

1. If X and Y are two random variables, then the covariance is

   $$C_{XY} = R_{XY} - \overline{XY}.$$

2. If two random variables X and Y are independent, then the covariance is zero, $C_{XY} = 0$, but the converse is not true.
3. For two random variables X and Y,

   $$\text{var}(X+Y) = \text{var}(X) + \text{var}(Y) + 2C_{XY}$$

   and $\text{var}(X-Y) = \text{var}(X) + \text{var}(Y) - 2C_{XY}$.
4. For two random variables, X and Y, the inequality

   $|\sigma_{XY}| \leq \sigma_X \, \sigma_Y$ is ture.

5. If X and Y are two random variables, then the covariance of $X+a, Y+b$, where a and b are constants, is

   $$\text{cov}(X+a,\, Y+b) = \text{cov}(X,Y) = C_{XY}.$$

6. If X and Y are two random variables, then the covariance of aX, bY is

   $$\text{cov}(aX, bY) = ab\,\text{cov}(X,Y) = abC_{XY}.$$

7. If X, Y and Z are three random variables, then

$$\text{cov}(X+Y, Z) = \text{cov}(X,Z) + \text{cov}(Y,Z).$$

Joint Characteristic Function

The joint characteristic function of two random variables, X and Y, is defined as the expected value of the joint function $g(x,y) = e^{j\omega_1 X} e^{j\omega_2 Y}$. It can be expressed as

$$\phi_{XY}(\omega_1, \omega_2) = \int_{-\infty}^{\infty} e^{j\omega_1 x + j\omega_2 y} f_{X,Y}(x,y)\,dx\,dy.$$

The joint density function is

$$f_{X,Y}(x,y) = \int_{-\infty}^{\infty} \phi_{X,Y}(\omega_1, \omega_2)\, e^{-(j\omega_1 x + j\omega_2 x)}\, d\omega_1\, d\omega_2.$$

Joint Moment Generating Function

The joint moment generating function of two random variables, X and Y, is defined as the expected value of the joint function $g(x,y) = e^{v_1 X + v_2 Y}$. It can be expressed as

$$M_{X,Y}(v_1, v_2) = \int_{-\infty}^{\infty}\int_{-\infty}^{\infty} e^{v_1 X + v_2 Y} f_{X,Y}(x,y)\,dx\,dy.$$

Joint Gaussian Density Function

If two random variables, X and Y, are said to be jointly Gaussian, then the joint density function is given as

$$f_{X,Y}(x,y) = \frac{1}{2\pi\sigma_X\sigma_Y\sqrt{1-\rho^2}} \exp\left\{\frac{-1}{2(1-\rho^2)}\left[\frac{(x-\overline{X})^2}{\sigma_X^2} - \frac{2\rho(x-\overline{X})(y-\overline{Y})}{\sigma_X\sigma_Y} + \frac{(y-\overline{Y})^2}{\sigma_Y^2}\right]\right\}$$

Transformation of Random Variables

One dimensional random variable $f_Y(y) = f_X(x)\,|dx|$

Two dimensional random variable $f_{UV}(u,v) = f_{XY}(x,y) \begin{vmatrix} \frac{\partial x}{\partial u} & \frac{\partial x}{\partial v} \\ \frac{\partial y}{\partial u} & \frac{\partial y}{\partial v} \end{vmatrix}$

Appendix C

Useful Mathematical Expressions

Rational Algebraic Functions

$$\int \frac{dx}{a^2+b^2x^2} = \frac{1}{ab}\tan^{-1}\left(\frac{bx}{a}\right)$$

$$\int \frac{xdx}{a^2+x^2} = \frac{1}{2}\ln(a^2+x^2)$$

$$\int \frac{x^2dx}{a^2+x^2} = x - a\tan^{-1}\left(\frac{x}{a}\right)$$

$$\int \frac{dx}{(a^2+x^2)} = \frac{x}{2a^2-(a^2+x^2)} + \frac{1}{2a^3}\tan^{-1}\left(\frac{x}{a}\right)$$

$$\int \frac{xdx}{(a^2+x^2)^2} = \frac{-1}{2(a^2+x^2)}$$

$$\int \frac{x^2dx}{(a^2+x^2)^2} = \frac{-x}{2(a^2+x^2)} + \frac{1}{2a}\tan^{-1}\left(\frac{x}{a}\right)$$

$$\int \frac{dx}{\sqrt{a^2-x^2}} = \sin^{-1}(x/a)$$

$$\int \frac{dx}{\sqrt{x^2-a^2}} = \log\left(x+\sqrt{x^2-a^2}\right)$$

$$\int \frac{dx}{\sqrt{x^2+a^2}} = \log\left(x+\sqrt{x^2+a^2}\right)$$

$$\int \frac{dx}{x^2-a^2} = \frac{1}{2a}\log\left|\frac{x-a}{x+a}\right|$$

Exponential Functions

$$\int e^{ax}dx = \frac{e^{ax}}{a}$$

$$\int xe^{ax}dx = e^{ax}\left[\frac{x}{a} - \frac{1}{a^2}\right]$$

$$\int x^2e^{ax}dx = e^{ax}\left[\frac{X^2}{a} - \frac{2x}{a^2} + \frac{2}{a^3}\right]$$

Definite Integrals

$$\int_{-\infty}^{\infty} e^{-a^2x^2+bx}dx = \frac{\sqrt{\pi}}{a}e^{b^2/(4a^2)} \quad a > 0$$

$$\int_{-\infty}^{\infty} x^2e^{-x^2}dx = \sqrt{\pi}/4$$

$$\int_0^{\infty} \text{Sa}(x)dx = \int_0^{\infty} \frac{\sin(x)}{x}dx = \frac{\pi}{2} \text{ where } \text{Sa}(x)dx = \frac{\sin(x)}{x} \text{ is a sampling function.}$$

$$\int_0^{\infty} \text{Sa}^2(x)dx = \pi/2$$

$$\int_{-\infty}^{\infty} \frac{x^2dx}{(a^2+x^2)^2} = \frac{\pi}{2a}$$

$$\int_0^{\infty} \frac{1}{a^2+x^2}dx = \frac{\pi}{2a}$$

$$\int_{-\infty}^{\infty} \frac{1}{1+x^2}dx = \pi$$

$$\int_{-\infty}^{\infty} \frac{1}{a+x^4}dx = \frac{\pi}{2\sqrt{2}a^3}$$

$$\int_{-\infty}^{\infty} \frac{x^2}{a+x^4}dx = \frac{\pi}{2\sqrt{2}a}$$

$$\int_{-\infty}^{\infty} \frac{1}{(a^2+x^2)^4}dx = \frac{5\pi}{16a^7}$$

$$\int_0^{\infty} e^{-a^2}x^2dx = \frac{\sqrt{\pi}}{2a}, \quad a > 0$$

$$\int_0^{\infty} x^n\, e^{-ax}dx = \frac{n!}{a^{n+1}}, \quad a > 0, n \text{ is integer}$$

$$\int_0^{\infty} e^{-ax}\cos(bx)dx = \frac{a}{a^2+b^2}, \quad a > 0$$

$$\int_0^{\infty} e^{-ax}\sin(bx)dx = \frac{b}{a^2+b^2}, \quad a > 0$$

$$\int_0^{\infty} e^{-a^2x^2}\cos(bx)dx = \frac{\sqrt{\pi}}{2a}e^{-b^2/4a^2}, \quad a > 0$$

$$\int_0^{\pi} \sin^2(mx)dx = \int_0^{\pi} \cos^2(mx)dx = \frac{\pi}{2}, \quad n \text{ is an integer}$$

$$\int_0^{\pi} \sin(mx)\sin(nx)dx = \int_0^{\pi} \cos(mx)\cos(nx)dx = 0, \quad m \neq n, \; n \text{ and } m \text{ are integers}$$

$$\int_0^{\pi} \sin(mx)\cos(nx)dx = \begin{cases} \dfrac{2m}{m^2 - n^2} & m+n \text{ odd} \\ 0 & m+n \text{ even,} \end{cases} \qquad n \text{ and } m \text{ are integers}$$

Series Expansion

Finite Series:

$$\sum_{n=1}^{N} n = \frac{N(N+1)}{2}$$

$$\sum_{n-1}^{N} n^2 = \frac{N(N+1)(2N+1)}{6}$$

$$\sum_{n=1}^{N} n^3 = \frac{N^2(N+1)^2}{4}$$

$$\sum_{n=0}^{N} x^n = \frac{x^{N+1} - 1}{x - 1}$$

$$\sum_{n=0}^{N} \frac{N!}{n!(N-n)!} x^n y^{N-n} = (x+y)^N$$

Infinite Series:

$$\sum_{n=1}^{\infty} a^n = \frac{1}{1-a}, \quad a < 1$$

$$\frac{1}{1-x} = 1 + x + x^2 + x^3 + \ldots$$

$$\frac{1}{1+x} = 1 - x + x^2 - x^3 + \ldots$$

$$e^x = 1 + x + \frac{x^2}{2!} + \frac{x^2}{3!} + \ldots$$

$$\sin x = x - \frac{x^3}{3!} + \frac{x^5}{5!} - \ldots$$

$$\cos x = 1 - \frac{x^2}{2!} + \frac{x^4}{4!} - \ldots$$

$$\log(1+x) = x - \frac{x^2}{2} + \frac{x^3}{3} - \ldots, \quad |x| < 1$$

$$\text{Sa}(x) = 1 - \frac{x^2}{3!} + \frac{x^4}{5!} - \ldots$$

Bibliography

1. Balmer, L., *Signals and System: An Introduction*, Prentice-Hall, Englewood Cliffs, NJ, 1991.
2. Bracewell, R., *The Fourier Transform and Its Applications*, McGraw-Hill, New York, 1965.
3. Carlson, A.B., *Communication Systems: An Introduction to Signals and Noise in Electrical Communication*, McGraw-Hill, New York, 2002.
4. Chen, C.-T., *System and Signal Analysis*, 2nd Edition, Saunders, Fort Worth, TX, 1994.
5. Cooper. G.R. and C.D. McGillem, *Probabilistic Methods of Signal and System Analysis*, Holt, Rinehart and Winston, New York, 2d ed., 1986.
6. Davenport, W. B. Jr., *Probability and Random Processes, An Introduction for Applied Scientists and Engineers*, McGraw-Hill, New York, 1970.
7. Davenport, W.B. Jr. and W.L. Root, *An Introduction to the Theory of Random Signals and Noise*, McGraw-Hill, New York, 1958.
8. Gardner, W.A., *Introduction to Random Processes with Applications to Signals and Systems*, 2d ed., McGraw-Hill, New York, 1990.
9. Gray, R.M. and L.D. Davisson, *Random Processes: A Mathematical Approach for Engineers*, Prentice-Hall, Englewood Cliffs, New Jersey, 1986.
10. Hsu, Hwei P., *Signals and Systems*, Schaum's Outline Series, McGraw-Hill, New York, 1995.
11. Kamen, E.W. and B.S. Heck, *Fundamentals of Signals and Systems Using MATLAB*, Prentice Hall, Upper Saddle River, NJ, 1997.
12. Lathi, B.P., *Linear Systems and Signals*, Berkeley-Cambridge Press, Carmichael, CA, 1992.
13. Lathi, B.P., *Signal processing and Linear Systems*, Oxford University Press, 1998.
14. Lathi, B.P., *Signals, Systems and Communication*, BS Publications, 2001.
15. Mallikarjuna Reddy Y., *Probability Theory and Stochastic Processes*, Universities Press, 2014.
16. McGillem, C.D. and G.R. Cooper, *Continuous and Discrete Signal and System Analysis*, 3rd Edition, Holt, Rinehart, and Winston, New York, , 1991.

17. Melsa, J.I., and A.P. Sage, *An Introduction to Probability and Stochastic Processes*, Prentice-Hall, Englewood Cliffs, New Jersey, 1973.

18. Oppenheim, A.V. and A.S. Willsky, *Signals and Systems*, 2nd Edition, Prentice Hall, Upper Saddle River, NJ, 1997.

19. Papoulis, A., *The Fourier Integral and Its Applications*, McGraw-Hill, New York, 1962.

20. Papoulis, A., *Probability, Random Variables and Stochastic Processes*, 3d ed., McGraw-Hill, New York, 1991.

21. Parzen, E., *Stochastic Processes*, Holden-Day, San Francisco, California, 1962.

22. Peebles, P.Z. Jr., *Communication System Principles*, Addison-Wesley, Reading, Massachusetts, 1976.

23. Peebles, P.Z. Jr., *Digital Communication System*, Prentice-Hall, Englewood-Cliffs, New Jersey, 1987.

24. Phillips, C.L. and J.M. Parr, *Signals, Systems, and Transforms*, Prentice Hall, Englewood Cliffs, NJ, 1995.

25. Prabhu, N.U., *Stochastic Processes, Basic Theory and Its Applications*, Macmillan, New York, 1965.

26. Taylor, F.J., *Principles of Signals and Systems*, McGraw-Hill, New York, 1994.

27. Zeimer, R.E., W.H. Tranter and D.R. Fannin, *Signals and Systems: Continuous and Discrete*, 4th Edition, Macmillan, New York, 1998.

Index